Mites Injurious to Economic Plants

Mites Injurious to Economic Plants

Lee R. Jeppson
University of California, Riverside

Hartford H. Keifer
(retired)
California State Department of Agriculture

Edward W. Baker
Agricultural Research Service
United States Department of Agriculture

UNIVERSITY OF CALIFORNIA PRESS
Berkeley *Los Angeles* *London*
1975

University of California Press
Berkeley and Los Angeles, California

University of California Press, Ltd.
London, England

ISBN: 0-520-02381-1
Library of Congress Catalog Card Number: 72-93523

Printed in the United States of America

This book is dedicated to those interested in pests that affect the health of plants used for food, fiber, or aesthetic purposes, and particularly to those who, by research, encouragement, suggestion, and cooperation, are aiding basic and applied investigations of such pests.

Contents

Preface

We have attempted to present in this book a digest of available information on the mites known to produce injury to plants of economic importance, that is, those plants used for food, fiber, or aesthetic purposes. The need for such a publication has been brought to our attention by students, entomologists, pesticide advisors, and growers. The intent of the volume is to provide general information on mite distribution, injury, biology, and response to chemical applications, and to aid in their identification in the field and from prepared specimens.

The book was first conceived and outlined by Lee R. Jeppson, who compiled the information in chapters 2 through 11, except for the taxonomic descriptions and drawings, which were prepared by Edward W. Baker. Chapter 12 was written by Hartford H. Keifer. The taxonomic descriptions in chapter 13 and all drawings of the Eriophyoidea are by H. Keifer, who also assisted on the biologies of some eriophyoid species. All three authors contributed to the writing of chapter 1.

Chapters 1 through 7 and 12 are designed to provide the student of agricultural acarology and entomologists with sufficient information concerning mites and their biology to stimulate interest in these animals and to understand more adequately the information given concerning each specific pest. Chapters 8 through 11 and 13 provide pertinent information on important mite species of the major groups injurious to economic plants. Information on the biology of many species is lacking.

Where specific information is available, the presentation is usually given in the following order: (1) general distribution; (2) nature of the plant injury produced by the mite; (3) life cycle, biology, and descriptions of the life stages; (4) brief statements on chemical and biological control, where such seems important; and (5) distinguishing characters as may be observed in specimens prepared for microscope examination.

The chemicals used to control mites are rapidly changing owing to the appearance of new acaricides and to the development of mite populations resistant to established acaricides. The predators and other agents that regulate mite populations vary from one locality to another. For these reasons the information on chemical and biological control of mites has been largely confined to the general discussions on mite control included in chapters 3, 4, and 5.

Critical studies of the taxonomy of the Eriophyoidea have not advanced as rapidly as have studies of the other injurious plant-feeding mites because of their relatively small size. It has been recognized that some taxonomic revision is advisable to adjust the generic names to conform with the Nomenclatural Code. This revision together with keys to the genera is included in appendix 3. Genus names in the text have been changed to conform with this revision.

GENERAL ACKNOWLEDGMENTS

Plant hosts of mites are given in the text under the common name where a common name is given in Bailey's *Manual of Cultivated Plants* (revised edition, eleventh printing, 1969, The MacMillan Co., Toronto, Ontario, Canada). The scientific names given by Bailey for each of these common names are listed in appendix 1. The scientific names of pesticides mentioned in the text are indicated in appendix 2.

References to original descriptions of most of the injurious species in the families Tetranychidae, Tenuipalpidae, and Tarsonemidae discussed in the text have been assembled in published works. Species assigned to the Tetranychidae are in Pritchard and Baker (1955) and Tuttle and Baker (1968); references follow chapter 8. Species of Tenuipalpidae may be found in Pritchard and Baker (1958), reference in chapter 9; species in Tarsonemidae recorded by Beer (1954), reference in chapter 10. The species of Eriophyoidea have not previously been assembled; they are brought together at the end of chapter 13.

Special appreciation is expressed to Bryan Croft and Robert Hobza, graduate students at the University of California at Riverside, for time spent on literature search on tetranychids and eriophyoids, respectively and to Robert L. Smiley, *ARS, USDA*, for checking the names of the tarsonemids. Gratitude is expressed to M. M. Barnes, L. R. Brown, R. H. Gonzalez, R. N. Jefferson, and J. A. McMurtry of the University of California, Riverside, and to F. F. Smith of the United States Department of Agriculture (USDA) laboratory at Beltsville, Maryland, who have provided pictures of mites and their injury.

In an endeavor to avoid errors the authors have submitted chapters of the manuscript to authorities in their special fields. We desire to thank the following for invaluable suggestions on the following chapters of this book: chapter 1, I. M. Newell; chapter 2, C. A. Fleschner and B. R. Bartlett; chapter 5, J. A.

McMurtry; and chapter 6, G. N. Oldfield; all are at the University of California, Riverside.

Special appreciation is expressed to M. M. Barnes for reviewing the manuscript and for his many suggestions and corrections; also to F. M. Summers, University of California, Davis, who provided information on the anatomy of the tetranychid gnathosoma.

ACKNOWLEDGMENTS FOR THE ERIOPHYOID SECTION

Much of the material on eriophyoids (chapters 12, 13) stems from information and specimens contributed by many people. We are primarily indebted to Richard A. Newkirk, formerly with the USDA. Newkirk not only collaborated in composing the generic key in appendix 3, but has provided otherwise unavailable information on the early history of the mycological phase of eriophyoidology. Newkirk secured copies of early and rare publications and has corrected types of *Eriophyes* von Siebold, of *Phytoptus* Dujardin, and of *Tegonotus* Nalepa. These type revisions bring these genera into compliance with the *International Code of Zoological Nomenclature,* and somewhat change the taxonomic meaning of these names. C. W. Sabrosky, of the U.S. National Museum, and member of the *Code* Editorial Committee, has instructed us in applying *Code* provisions.

W. E. Styer, Department of Entomology, Ohio Agricultural Research and Development Center, has supplied us with specimens and information. He sent prepublication copies of the article entitled "Fine Structure of *Aceria tulipae* (Keifer)," by Whitmoyer, Nault, and Bradfute. Styer also sent us scanning electron microscope pictures of *Eriophyes tulipae* K., taken at the Acarology Laboratory, Ohio State University, under the direction of G. W. Wharton. We also thank Clyde D. Wilson, formerly with the Department of Entomology, Berkeley, who enabled us to obtain scanning electron microscope pictures of certain eriophyids.

Tokuwo Kono, Plant Industry Division of the California Department of Agriculture, has secured both specimens and literature that have been most helpful. Professor C. C. Hall, Jr., Department of Biology, University of Texas at Arlington, has supplied many specimens, including mites from alder, maple, and elm trees. Gordon R. Nielsen, Vermont Department of Entomology, University of Vermont, Burlington, has sent many mite collections, especially galls and erinea from maple, ash, and birch trees.

H. K. Wagnon, Plant Industry Division, California Department of Agriculture, made very valuable collections of eriophyoids while in Turkey in 1968, and has collected extensively in the western United States.

Magdalena L. Briones, formerly of Guinobatan Experimental Station, in Albany province of Lyzon, Philippines, has submitted instructive species from the Philippines, and from certain North American areas. Many Arizona specimens have been supplied by Donald M. Tuttle, of the University of Arizona. Harold A. Denmark, Chief, Department of Entomology, Florida Department of Agriculture, has submitted many previously unnamed eriophyoids.

Robert O. Schuster, University of Calilfornia, Davis, has secured stereoscan pictures used in this book. In particular these include photographs of differences between protogynes and deutogynes of maple and walnut mites.

Ernesto Doreste S., Shell Foundation, has sent economic eriophyids from Venezuela.

G. Proeseler, Institute for Phytopathology, Ascherslaben, Germany, has sent grass mites and reprints with electron microscope pictures of featherclaws. Jan Boczek, Warsaw, Poland, has contributed grass mites and examples of the genotype of *Phyllocoptes.*

We are particularly indebted to J. A. Stevenson, mycologist, Plant Industry Station, Beltsville, Maryland, for allowing examination of the unique *Cecidotheca Italica* by Trotter and Cecconi, which includes folders with mite galls.

To John A. Weidhaas, Jr., Extension Entomologist, Virginia Polytechnic Institute, Blacksburg, we are indebted for examples of black walnut petiole gall mites, *Eriophyes caulis* Cook, collected in June. This early collection date enabled us to find and characterize the males and protogynes of this species. Weidhaas also sent early specimens of the persimmon gall mite, showing that the difference between protogynes and deutogynes consists in the shape of the microtubercles.

There are many others who have helped develop information that we present here, by sending in specimens of eriophyoids. Some are particularly thanked by having their names on the species they collected; others are cited in the references. Our appreciation goes to all of these people who have helped in the investigation of these tiny creatures. They have not only made our task pleasurable but have also contributed to the protection of agriculture.

Chapter 1

Introduction to the Acari

SIGNIFICANCE OF MITES TO MAN

The subclass Acari, which includes mites and ticks, forms an important part of the arthropodan class Arachnida, to which also belong the scorpions, spiders, and harvestmen. Mites have a worldwide distribution; they rival insects in the extent of their habitation. They live in salt and fresh water, in organic debris of all kinds, and on plants and animals. They are among the dominant animals in pastures and in arable soils. In forests they greatly outnumber all other arthropods. Some species even live in caves and some in thermal springs. Their associations with other animals include commensalism, predation, and true parasitism. Therefore, they may cause serious damage to livestock, agricultural crops, ornamental plants, and stored products; they may even bring sickness and death to man; or they may be parasites, predators, or saprophytes destroying animals, plants, or their products and adversely affecting man or his possessions. Thus their importance to man covers all phases of his life. This book, however, is limited to the plant-feeding mites; thus discussion is restricted to the Tetranychoidea, Eriophyoidea and to a few economically important species in the families Tarsonemidae, Tenuipalpidae, Tydeidae, Pyemotidae, Penthaleidae and in the Astigmata.

The tarsonemids feed on fungi, and on decaying plant and animal materials as well as on plants. Tarsonemids are commonly found on plants, but many appear to have no economic significance. Their importance in the fields of public health and medical acarology is a matter of record, yet to date irrefutable evidence incriminating tarsonemids as a direct cause of human suffering is lacking.

Mites belonging to the family Tenuipalpidae, the false spider mites, are primarily plant feeders. They are not readily detected by the casual observer because of their small size and sluggish activity; yet they may be found on most perennial plants. Their life cycle is relatively long, so populations increase slowly; therefore, it is conceivable that in many situations predators may prevent the development of injurious populations.

The Tydeidae largely feed on fungi and decaying materials, although some have been seen feeding on insects and mites, and others have caused damage to plants.

Two species in the family Penthaleidae, that is *Penthaleus major* (Dugès) and *Halotydeus destructor* (Tucker) feed on the young plants of a wide variety of economic plant species. They often cause destruction of newly planted crops, mature grasses, and grains.

People do not usually become aware of the existence of mites until they them-

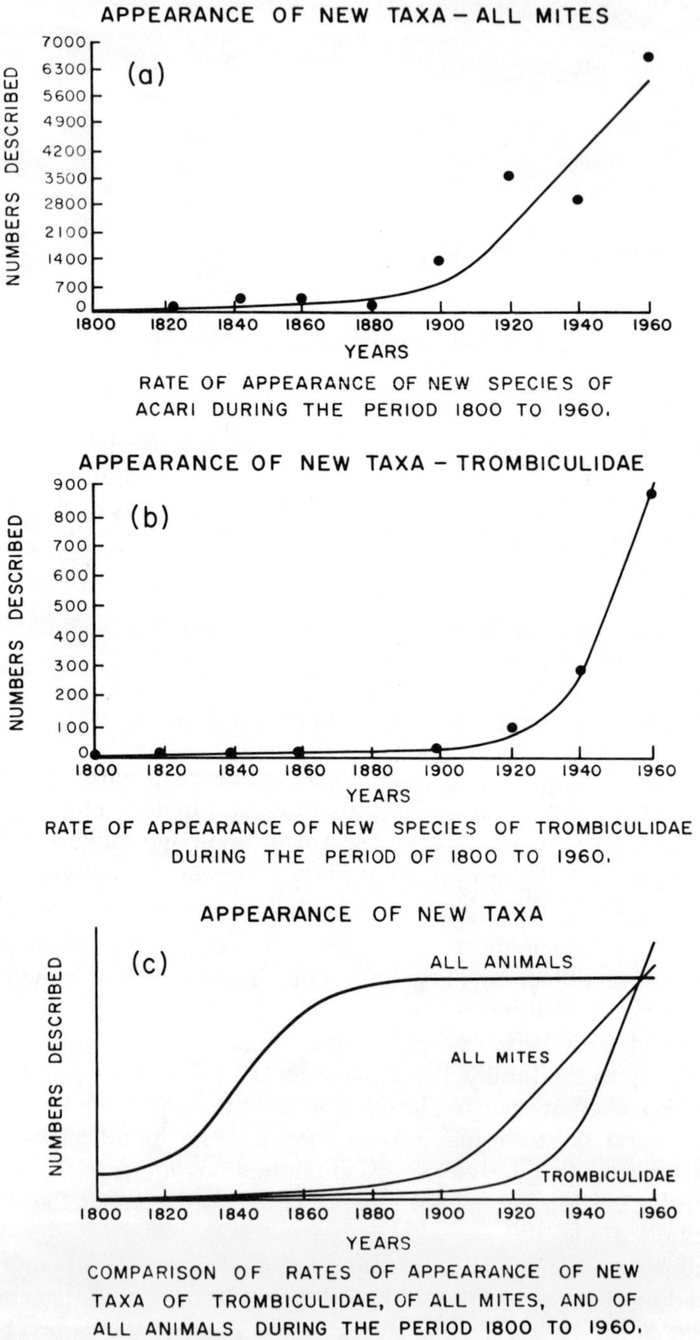

Fig. 1. Appearance of new taxa.

selves or their belongings are adversely affected. Even biologists had not systematically collected mites or recognized differences in their morphology and habits until about the middle of the nineteenth century. In an address to the First International Congress of Acarology, Wharton (1964) illustrated graphically the relative rate new taxa (genera and species) of mites are appearing in the literature in relation to the number of taxa of all animals (fig. 1). Such an increase in descriptions indicates the probability that many animals as yet undescribed are mites.

Although only a relatively small fraction of the total acarine fauna is known, acarology has contributed significantly to our understanding of biology. The demonstration that the cattle tick is the vector of Texas cattle fever was the first demonstration of an arthropod-borne pathogen. A knowledge of the Acari is now contributing significantly to a rapidly developing science of soil zoology. Their small size and terrestrial habitats have made mites and ticks excellent subjects for studies on water balance. The Acari have already made excellent experimental animals for the study of population dynamics.

Although many mite species are injurious, some are beneficial to man and his possessions: those that live in soil and water assist in the breakdown and decay of organic materials; many, no doubt, feed on fungi and other lower plants or on the animals living in soil and water. Beneficial or not, the activities of the vast majority of mite species are unknown.

SOME UNSOLVED BASIC PROBLEMS

The known distribution of the plant-feeding mites often coincides with the distribution of interested and trained acarologists rather than with the actual distribution of the species. The distribution, taxonomy, and biology of eriophyoid mites is particularly lacking in many parts of the world. These wormlike animals (pl. 1), the largest of which are 250 μ long, may differ in appearance between summer and winter sufficiently that mites of the same species have been placed in different genera depending on the season collected (see chap. 12). The study of the complicated life cycles of some of these mites should encourage other biological studies. The recent development of new instruments, such as the scanning electron microscope by which structures of these small organisms can be seen in depth and detail, should lead to increasing numbers of investigations on systematics and the relation of structure to the behavior of these small animals.

The ability of many of the tetranychid mites to adapt to new environments and to toxicants presents challenging problems in genetics; especially because the mites have few chromosomes and arrhenotokous reproduction. Physiological studies of mites, particularly the eriophyoids, have barely begun. The nature of the toxins or growth substances injected into the plants by mites needs investigation. It is hoped that the study of this compilation will aid in bringing the student to an awareness of what is known, and lead to a better appreciation of the many facets that need investigation.

PHYLOGENY OF THE ACARI

The small size and delicate structure of mites probably accounts for the lack of fossil remains. In a review on the phylogeny of mites, Woolley (1961) reports that mites are not found commonly as fossils, but they have been reported from Devonian, Carboniferous, and Tertiary formations. All the principal families are represented in the Tertiary period; and the specimens, most of them found in amber, show morphological features that demonstrate kinship with modern genera. The most ancient form, *Protacarus crani* Hirst, found in Devonian sandstone, appears to be too well specialized to be a common ancestor of the different groups of mites.

Snodgrass (1952) suggests that though mites and ticks do not resemble other arachnids in appearance, they do demonstrate fundamental arachnid features in both external and internal morphology, such as the chelicerae, palps, and four pairs of legs. Baker and Wharton (1952), however, consider the gnathosoma (capitulum), the absence of a prosomal-opisthosomal constriction, and the undivided prosomal region to be distinctive features of Acari.

Woolley (1961:282) concludes that "no known connections with other arachnids can be ascertained from the fossil record, but the chelicerae and other morphological features indicate that the Acari probably had an arachnid origin in opilionid-like ancestors, but developed with inherent peculiarities, principally the gnathosoma. They are at least diphyletic, probably polyphyletic, in origin, with two main groups, the older Anactinochitinosi (Notostigmata, Mesostigmata, Ixodides) and the younger Actinochitinosi (Trombidiformes, Sarcoptiformes). In general, the groups lost traces of segmentation, changed appendages, and compressed the neuromeres forward. Possibly the primitive Opilioacaridae are intermediate between the Opiliones and Mesostigmata or very close to the hypothetical 'proto-mesostigmatic.' However, Opilioacaridae and Holothyroidea, may be autonomous, independent groups."

The chelicerae indicate relationships within the Trombidiformes (Prostigmata) including primitive chelate forms and more recent specialized piercing types. The Eriophyoidae and Demodicidae represent specialized vermiform Trombidiformes of two separate derivations. See chapter 12 for additional discussion of the eriophyoid evolutionary development.

Since there are parasitic genera in families that have mostly free-living forms it seems evident that the parasitic habit of mites has originated independently a number of times. Also, mites show a different method of parasitism on different hosts. When free-living mites that ordinarily obtain food from fungi or decaying material find themselves on plants, in birds' nests, or on animals, it is not strange that some may be able to feed sufficiently to survive. Eventually, selection may produce offspring adapted to the new habitat.

HISTORY OF KNOWLEDGE OF THE ACARI

General Groups

Mites have always attracted considerable interest because of their small size and especially because of the remarkable habits of some species. Early observers examined them in a cursory way, and consequently much of the early literature is not reliable. For years the only work treating the mites as a whole that was accessible to American naturalists was Murray's *Economic Entomology; Aptera* (1877). In this book nearly 300 pages were devoted to the Acari. While helpful in bringing together specific information, much of Murray's treatise was a compilation without adequate analysis.

At first it was believed that most mites were parasitic, because of a lack of knowledge of their biology and feeding habits. It is now known that many of the species found on insects, animals, and plants are not truly parasitic. Recent studies have shown that soil and water provide numerous niches for mites.

Baker and Wharton (1952) and Krantz (1970) have noted that men were aware of Acari in Egypt as early as 1550 B.C., and that the early Greek writers recognized mites and ticks. At the time of Linnaeus about 90 species had been discussed in the literature. The tenth edition of *Systema Naturae* included fewer than 30 species, all of which were grouped in the genus *Acarus*.

In the 100 years following Linnaeus, major contributions to systematics of the Acari, particularly of higher categories, were made by Latreille, Leach, Dugès, DeGeer, and C. L. Koch. Michael (1884) summarizes these and related pioneer works.

The rise of acarology as a modern science came in late nineteenth- and early twentieth-century Europe with the historic researches of Kramer, Mégnin, Canestrini, Michael, Berlese, Reuter, Vitzthum, and Oudemans (Evans, Sheals and Macfarlane, 1961).

General books or reviews of mites in the English language include *An Introduction to Acarology* by Baker and Wharton (1952). This book gives the classification to the family category including keys and drawings of typical species in the family and a list of the known genera in each family. The manual called *Guide to the Families of Mites* by Baker et al. (1958) provides keys and descriptions of mite groups, including the families. The authors used such categories as supercohort, cohort, and superfamily. The drawings illustrating characters useful in identifying these categories are more detailed than in Baker and Wharton (1952).

More general valuable contributions written in English language include *The Terrestrial Acari of the British Isles*, I, by Evans et al. (1961) and *Mites or the Acari* by Hughes (1959). *The Terrestrial Acari of the British Isles* provides an overview of the mites including general morphology, techniques of collecting and mounting, some biology of free-living acari, mites associated with plants, and those associated with animals. *Mites or the Acari* and *A Manual of Acarology* (Krantz, 1970), are recent treatments of the subject, and are particularly detailed on mite structure, both external and internal. See these two texts for more detailed knowledge of the variations in structure and function among various groups of mites.

WATER MITES

Water mites have been known since mites were first discovered. Although of no economic importance they have been studied for many years, because of their habitat, large size, and coloration. Much comprehensive work has been done in recent years by Viets of Germany, Lundblad of Sweden, and Vitzthum of Germany. A few selected references are: Viets (1936), Lundblad (1941), and Vitzthum (1940–1942). Baker and Wharton summarized these works in their 1952 publication.

The Plant-Feeding Mites

Mites belonging to the Tetranychoidea and Eriophyoidea comprise most of the species feeding on higher plants. The taxonomy of tetranychid mites was reviewed by Ewing (1909), McGregor (1950) and Pritchard and Baker (1955). These works provide keys to the genera and species, information on synonymy and distribution, as well as references to the original taxonomic descriptions. A more recent, but more limited, work by Tuttle and Baker (1968) is of general interest because it brings more nearly up to date the generic names, and introduces subgeneric names into the classification.

There is no general review of the eriophyoid mites. References to the taxonomic papers by Nalepa and Keifer, who have done most of the authoritative work in this group of mites, are given at the end of chapter 13. The key to the genera of eriophyoid mites given in appendix 3 is the first key to give a careful treatment of genotypes; it is useful in defining the genera.

GENERAL CLASSIFICATION

One of the defining features of the class Arachnida is chelicerate mouthparts, which are basically forceplike feeding organs. Another anatomical feature is the complete lack of antennae.

Mites, as a subclass of Arachnida, differ in part from other arachnids by lacking abdominal segmentation. (Spiders also share this abdominal segmentation suppression.) The principal defining mite structure, however, is the distinctive anterior gnathosoma, or mouth. Most mites have forceplike chelicerae in this gnathosoma, often variously modified or reduced. Most of the plant-feeding mites in this text are examples of this needlelike cheliceral modification.

Krantz (1970) offers a list of orders, suborders, and lower categories in the subclass Acari. His arrangement of orders and suborders is shown below. Only superfamilies and families having economically important genera are included in this list. The genera and species of economic importance are treated later in this book.

Order Opilioacariformes
- Suborder Notostigmata (a small group with four dorsal pairs of abdominal spiracles).

Order Parasitiformes
- Suborder Tetrastigmata (a small group of larger mites ranging up to 7 mm long).
- Suborder Mesostigmata (an extensive series of mite families with diverse habits, including the predacious Phytoseiidae).
- Suborder Metastigmata (the ticks).

Order Acariformes
- Suborder Prostigmata
  - Tetranychoidea
    - (Tetranychidae: red spider mites; mites with palpal thumb-claw and long needlelike chelicerae).

(Tenuipalpidae: includes *Brevipalpus*; flat or red mites, lacking thumb-claw, but otherwise similar to tetranychids).
(Tuckerellidae: similar to tetranychids, posterior end of the mite with a series of long flagelliform setae).
Tarsonemoidea
(Tarsonemidae: light brown elliptical mites, females lacking claws on fourth pair of legs; short needlelike chelicerae).
Cheyletoidea
(Cheyletidae: largely free living predators).
Eupodoidea
(Penthaleidae: soft bodied, with tubercle on anterior portion of propodosoma bearing pair of setae; movable digits styletlike).
(Pyemotidae: similar to tarsonemids but females have claws on fourth pair of legs; itch mites).
Tydeoidea
(Tydeidae: soft bodied yellowish mites; short needlelike chelicerae; "honey dew mites").
Eriophyoidea (4-legged mites with wormlike shape and known as blister, rust, gall, and bud mites).
Suborder Astigmata
Acaroidea (Bulb and root mites; soft bodied semitransparent species with forceplike chelicerae).
Suborder Cryptostigmata (dark bodied "beetle mites" with forceplike chelicerae; oribatids).

MITE MORPHOLOGY IN RELATION TO MITE ACTIVITY

General Morphology

As the exoskeletal covering of mites relates to water loss and penetration of acaricides, it is important to visualize its general structure.[1] It begins its develop-

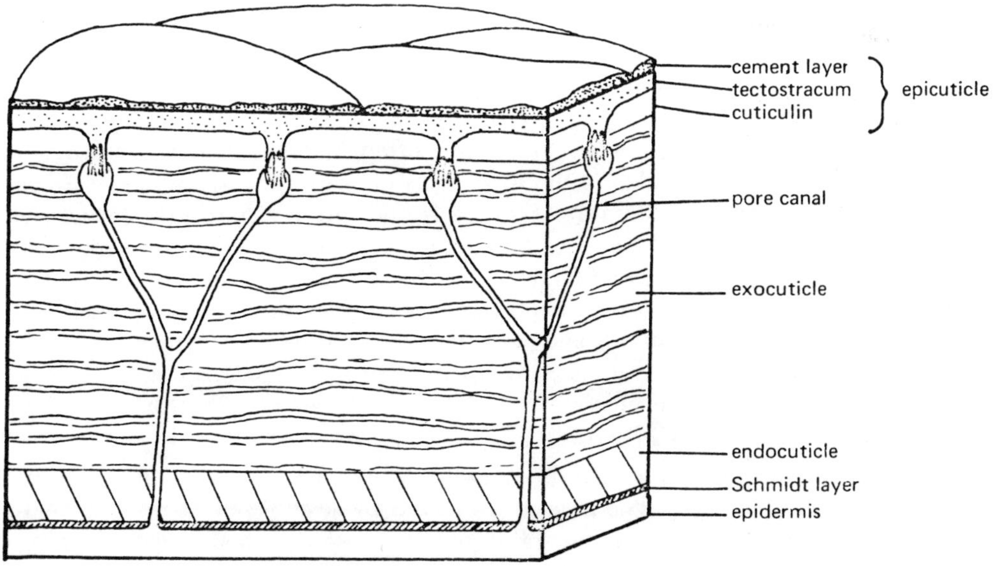

Fig. 2. Diagrammatic cross-section of cuticle of Acari (after Krantz, 1970).

ment as undifferentiated tissue covered by a thin layer of *cuticulin* (a mixture of waxes forming the cuticle) and separated from the underlying epidermis by an extremely thin, poorly defined granular Schmidt layer (fig. 2). As development proceeds, surface portions of the undifferentiated layer often becomes sclerotized to varying degrees through orthoquinone tanning. These portions, the *epicuticle* and *exocuticle*, are discrete shields or plates. The cuticulin surface layer may contain a profusion of micropores which are connected to *pore canals*. These canals appear to arise from the epidermal cells underlying the granular layer, and pass through both the endo- and exocuticular layers. A possible function of the pore canals may be to transport an epidermal secretion to the cuticulin surface layer, where the secretion forms a protective waxy coating as well as a thin overlying cement layer. The protective waxy coating and cement layers offer protection against excessive water loss from the body surface. Other pores occur in smaller numbers both on the body and appendages. These structures may also have a secretory function, since they too are connected internally to canals. Round, elliptical, or lyriform pores often occur dorsally, while only lyriform openings are found on the venter or on appendages.

Much of the beauty and form of mites is to be found in the ornamentations of the exoskeleton. These are the setae, pores, ridges or folds, or the pigments in the various layers of the exoskeleton.

The body of a typical mite is distinctly separated into an anterior gnathosoma and a posterior idiosoma. The idiosoma includes the propodosoma and the hysterosoma (fig. 3).

The gnathosoma resembles the head of the generalized insect only in that the mouthparts are appended to it. The brain is in the idiosoma and is behind the gnathosoma rather than in it: the eyes are dorsal or dorsolateral on the propodosoma, so the gnathosoma, internally, is little more than a tube through which food is carried into the esophagus. Above the mouth cavity are paired chelicerae, which generally are three-segmented. The chelicerae, along with the pedipalps on the gnathosoma, comprise the organs of food acquisition. The palps may be simple sensory structures equipped with chemosensory or thigmotropic hairs that aid in locating food. They are, however, often modified into a grasping or piercing raptorial organ similar to the mandibles of many predatory insects.

The chelicerae vary considerably in structure, but are never primarily sensory. Generally the terminal third segment of the chelicera is modified into a movable digit which opposes the fixed distal portion of the second segment. These opposed digits, or chelae, are edentate or toothed for grasping or grinding. The chelicera may be attenuate or elongate in some parasitic groups, thereby serving as piercing organs. Reduction of the fixed digit is not uncommon in predacious and phytophagous species. Accompanying this reduction is the development of the movable digit into a styliform piercing structure.

The mouth opens internally into the pharynx, which acts as a suction pump for the ingested food materials. The pharynx is served by several sets of muscles that, along with those muscles that control the movement of the chelicerae and palps, virtually fill the gnathosomal cavity. Salivary glands may be present, opening through paired ducts into the mouth cavity or through styli somewhat anterior to

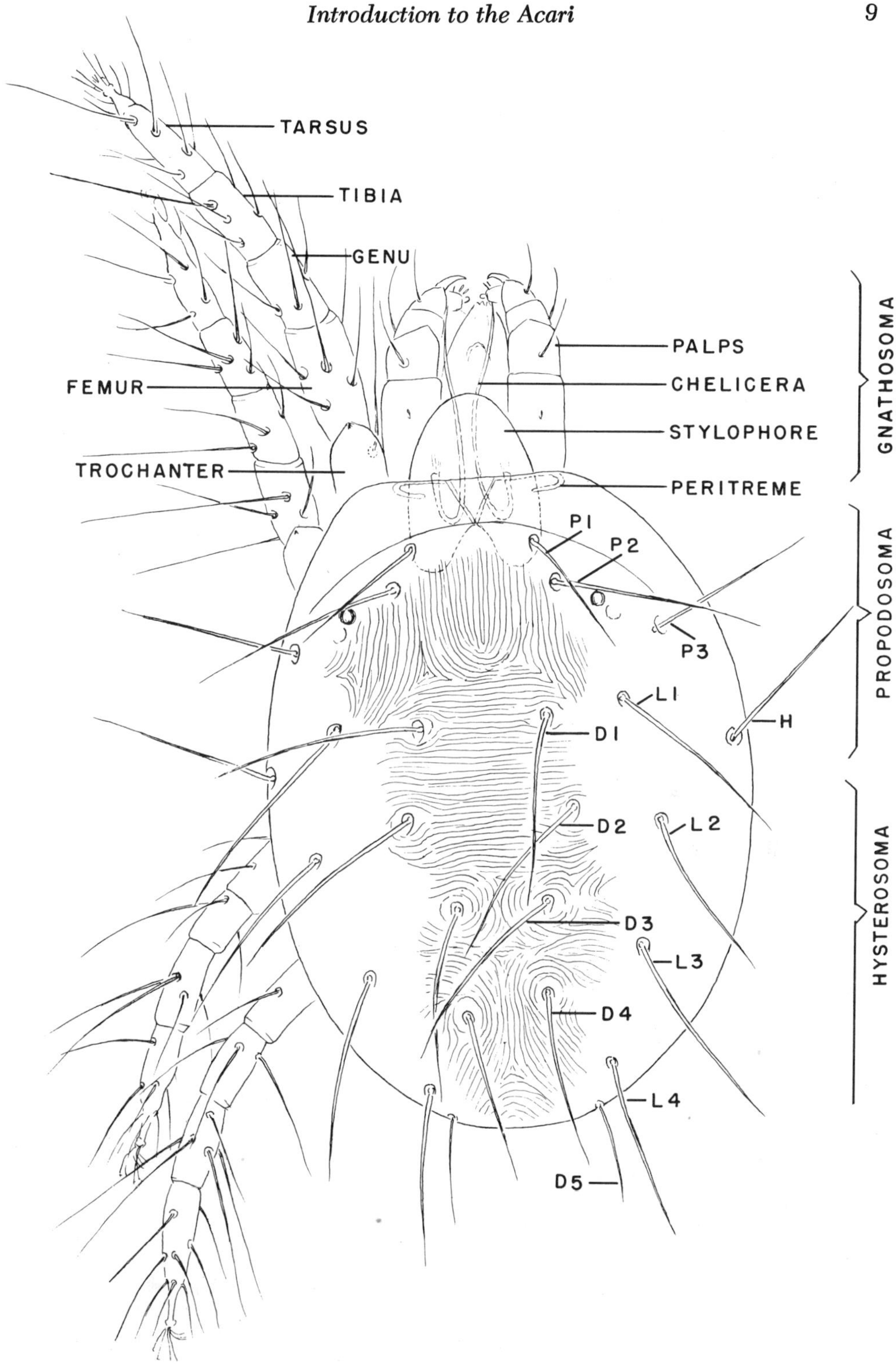

Fig. 3. *Tetranychus T. urticae* (Koch). Overwintering female showing arrangement of setae. P_1, P_2, and P_3 are dorsal setae; L_1 and L_4 are lateral setae.

the oral opening. These glands supply various enzymes that allow preoral digestion of food materials.

The function of the idiosoma parallels that of the abdomen, the thorax, and portions of the head of insects. The idiosoma may be covered with heavily sclerotized shields or it may be soft and virtually without sclerotization. Although the idiosoma is considered to be undivided, various grooves and furrows may occur in those groups where extensive idiosomal sclerotization has not occurred. The idiosoma includes an anterior propodosoma and a posterior hysterosoma, which may or may not be separated from each other by a furrow. The anterior two pairs of legs are inserted ventrally in the propodosomal regions, while the posterior two pairs of legs are located on the hysterosoma.

Shields or platelets commonly cover portions of the idiosoma. An anterior shield may cover the prodorsum or the entire propodosoma. A posterior shield or series of shields also may be present, while in some groups a single dorsal shield covers virtually the entire idiosoma. Ventrally, the idiosoma may be divided by furrows, and may or may not be provided with shields. The genital and anal orifices usually are set within a sclerotized shield or protected by paired valves. The genital and anal shields may, in many instances, be expanded so as to cover all or nearly all of the genitoanal region. Anteriorly, a sclerotized sternal area may be found which often is incorporated into an overall ventral shield complex. The primary external organs found on the idiosoma are locomotory, respiratory, copulatory, and sensory in function.

The general external structure has been described in detail in the taxonomic literature. The general morphology and anatomy of the Tetranychidae is given in chapter 7, and the morphology of the Eriophyoidea in chapter 12. These represent the major groups of plant-feeding mites.

The internal structure of mites also varies with the several mite groups. The general structure of mites belonging to the Tetranychidae is discussed in chapter 7. Little is known concerning the internal anatomy of the Eriophyoidea. The variations of internal anatomy of other mite groups is beyond the scope of this book.

Activity

Respiratory

The exchange of oxygen and carbon dioxide in mites is accomplished in several diverse ways. The presence or absence of stigmata, spiracular openings, and their relative position, provides a major diagnostic feature for the separation of orders within the Acari. Where stigmata occur, they open internally into a tracheal system that ramifies throughout the body to the various organ systems. In those mites that have no apparent stigmata or tracheal system, exchange of oxygen and carbon dioxide occurs through the integument.

Sensory

The idiosoma has many sensory receptors, most of which are setiform. Those on the idiosoma proper are primarily tactile in function. Tactile setae may be simple, plumose, or leaflike, but all lack protoplasmic extensions into the body of the seta itself. Specialized setae having protoplasmic nerve cell extension occur in some mite groups. Most groups have one or two pairs of simple eyes located laterally on the propodosoma. Several major mite groups have paired sensory organs located ventrally between coxae I and II. These variously shaped organs are thought to be humidity sensors. Nymphal and adult mites in the same groups have paired genital discs that probably serve the same function.

Locomotion

Adult and nymphal mites usually (the major exception being the Eriophyoidea) have four pairs of segmented legs, while the larvae have three pairs. The legs are divided into seven primary segments: the coxa, trochanter, femur, patella, tibia, and tarsus. The tarsus is usually terminated by a clawlike or suckerlike appendage, consisting of paired claws and a median padlike or clawlike empodium. The empodium may persist in the absence of true claws as a clawlike or suckerlike extension.

GENERAL BIOLOGY

Feeding

Mites feed on a wide variety of substances. Some are parasitic and feed on the blood or tissue fluids of vertebrates or invertebrates; some are fungivorous; some feed on decaying leaves of higher plants; and some feed on living plant tissue. Many species may be general feeders; the diet of a single species may include living and dead tissue of plants or animals.

The majority of mites feeding on higher plants belong to the suborder Prostigmata. The adaptations encountered in purely phytophagous forms are associated mainly with feeding organs, although certain changes in the life cycle are also clearly adaptive. The most highly specialized plant feeders are the Tetranychoidea and Eriophyoidea. In the former the bases of the chelicerae are fused to form an eversible stylophore and the movable digits are drawn out into flagelliform stylets, which are used for piercing the epidermis of the host. The mouthparts of the Eriophyoidea are discussed in chapter 12.

Reproduction and Development

Reproduction in the Acari generally follows the usual pattern of fertilization and subsequent production of male and female progeny but facultative parthenogenesis is known to occur throughout the Acari, and arrhenotoky (the production of males from unfertilized eggs) occurs in both the Mesostigmata and Prostigmata. Thelytoky (the production of females from unfertilized eggs) is common among certain Prostigmata and some other groups.

Mating or Sperm Transfer

Among the Acari mating is achieved in two ways. The spermatozoa are introduced into the female receptaculum by means of a sclerotized aedeagus; or a spermatophore or sperm sac is produced which is either picked up by the female or transferred to the female by the male chelicerae. Often, specialized structures in both male and female have been developed to achieve mating.

In mite groups where a male organ or aedeagus is present, transfer of sperm may be made directly into the female genital opening or into a special female copulatory structure called the bursa copulatrix. The bursa may be an extrusible tube on the idiosoma, which is connected internally to the female reproductive system, or it may be a posterodorsal opening.

In mite groups where the female picks up the sperm packet, the male places the spermatophore on a stalk that is secreted as a fluid by the male, and which hardens on contact with air. The spermatophore is picked up by the female upon contacting it with her genitalia (Krantz, 1970). Further discussion of this sperm transfer method in Eriophyoidea is discussed in chapter 12.

Krantz (1970) has reviewed the scanty information on the embryonic development in the Acari. It appears the total cleavage of the primordial cytoplasmic mass does not usually occur. The nucleus divides within the cytoplasm and migrates to the surface prior to cleavage. The nuclei continue to divide until they form an envelope called the blastoderm, within which is the yolk. A few blastoderm nuclei enter the yolk area where, as vitellophage cells, they liquefy the yolk and make it readily available to the developing blastoderm. An anterior polar cap appears, providing the site for subsequent development of the central nervous system. A germinal band then differentiates ventrally, and a ventricular (heart) area develops dorsal to the band. The germinal area eventually gives rise to both the head and body appendages, the latter usually consisting of only three pairs of limb buds. Four pairs have been observed in early developmental stages in some species, but the fourth pair is resorbed at the time of pedipalpal differentiation.

The mature embryo normally hatches by rupturing the eggshell. In many cases, however, a second membrane is formed inside the shell. This deutovarial membrane has been seen in Metastigmata, Cryptostigmata, and in some of the higher Prostigmata. Where the membrane occurs it expands after rupture of the eggshell, resulting in a prelarval or deutovum "stage." The developing larva thus has additional room for growth and expansion before actually hatching.

Oviposition and Life Stages

Typical oviposition occurs for the majority of Acari. The eggs are passed through the genital valves and dropped either singly or in clusters. Eggs may be smooth or sculptured and usually are opaque white, although eggs of many Prostigmata may be brightly colored in shades of red, orange, or green. A waterproof waxy coating may be applied to each egg before deposition. Phytophagous and graminivorous mites tend to oviposit haphazardly on their food susbstrate while predacious, soil-inhabiting mites generally deposit their eggs where loss from predators is at a minimum. Parasitic mites often are larviparous but those that oviposit generally choose a particular host tissue as an oviposition site. Others place their eggs in protected situations where access of hatching larvae to the host will be virtually assured. Females of the Cryptostigmata, and of some of the higher Prostigmata, possess an extrusible ovipositor through which the egg passes. By means of terminal "fingers" on the ovipositor, the female holds the egg while probing the substrate for a likely spot to place it. Aquatic mites of certain types utilize an ovipositor to insert eggs in aquatic plants (Krantz, 1970).

The developmental cycle of Acari includes one or more active immature stages between the egg and adult. The stage hatching from the egg is hexapod, except in the Eriophyoidea, which have only two pairs of legs in all active stages. Subsequent to the larval instar, there may be as many as three active octopod nymphal stages before the mite becomes sexually mature, but in many groups, particularly the Prostigmata, two of the active stages may be replaced by inactive resting stages.

Free-living mesostigmatid mites have four active stages in their life cycle, namely, a larva, protonymph, deutonymph, and adult. Several parasitic forms have some specialization of the life cycle.

On the basis of the number of active developmental stages, Evans et al. (1961) suggest that four distinct types of life cycles occur in the Prostigmata. In some tarsonemids only two stages follow the egg, namely, the larva and adult. The nymphal stages are probably passed within the larval skin. The Parasitengona, which includes the Trombidiidae, Erythraeidae, and Hydrachnellae in the suborder Prostigmata (none of which feed on plants), have a larva and one active nymph; the resting stage preceding and succeeding the active nymph are, respectively, the protonymph and tritonymph; the active nymph is equivalent to the deutonymph of a normal three-nymph life cycle. The Tetranychoidea pass through a larval and two nymphal stages before becoming adult. Finally, in many families such as the Bdellidae and Tydeidae and in the Endeostigmata (placed in the Prostigmata by Krantz; the latter has well-developed pseudostigmatic organs on the propodosoma) there are three nymphal stages, the proto- deuto-, and tritonymphal stages besides the larva.

The characteristic feature of the life cycles of many astigmatid mites is the presence of an optional hypopial stage between the proto- and tritonymphal stages. This stage is nonfeeding and may be active or inactive. It is often produced as a resistant stage during periods of adverse environmental conditions, and may also act as a dispersal stage.

Host Relationships

Relationships of plant-feeding mites to their hosts and host adaptations which involved bodily structure, such as the size of the mouthparts and adaptations to maintain water balance, are discussed under each family or species.

Water balance in the Acari, as in other terrestrial animals, can be divided into three basic processes: controlled intake of liquid water, controlled elimination of liquid water in the urine and feces, and the control of evaporative water loss. Those mites that are able to maintain a nearly constant ratio of water to solid matter in the face of changing moisture conditions and weight loss can be said to have an efficient control over water balance. Such plant-feeding species are able to develop populations under a wide variety of weather variations. Those with less efficient water balance may be restricted to areas having suitable weather conditions or be relegated to protected microhabitats.

Most mites can become trapped in the surface tension of droplets, therefore the intake of liquid water is a special problem; mites, however, have been observed to follow a behavior pattern that suggests they are drinking water. Many species may obtain sufficient water from their food, but this does not prevent desiccation during the inactive stages, that is, egg and molting stages.

Little is known regarding the control of water intake by mites, but one might assume that it is controlled, as it is in most animals, by some osmo-receptor system. It is possible that many mites can absorb water from air or liquid through the cuticle. Some mite groups whose food is not a liquid may reduce water loss by the reabsorption of water from the fecal mass in the rectum. Some mites, notably the tetranychids, may control water loss by regulating the time the spiracles are open, thereby reducing the interval that evaporation can take place from the tracheae. The prevention or regulation of water loss through the cuticle is important in the Acari, but the mechanism is not adequately understood (Winston, 1964).

Mites with restricted control over water balance may adjust their water loss by moving from less to more favorable situations of evaporation. Some eriophyoid mites (bud mites) inhabit areas beneath bud bracts, others (gall mites) inject a growth-promoting substance causing its host to produce an excess of hairs (erineum) or to form galls in which the mites live. Some tarsonemid mites work their way down to damp soil during periods of dry weather or during the day and come up at night or during periods of high humidity. Some tetranychid species feed on underleaf surfaces causing the leaves to "cup" or "curl under," providing protection from winds and thus utilizing the leaf transpiration to provide a sufficiently high humidity in their environment. Thus the habitat and activity of a plant-feeding mite species is controlled by adaptations to regulate water loss.

The mechanism possessed by a mite species for puncturing plant tissue often may limit its host range, the growth cycle of the host upon which it can feed, and its location on the host. Most species that feed on tender plant foliage also live in a habitat of high humidity. Such habitats may be favorable because either the plant tissue is tender, or the humidity in such locations is high; but a combination of both is most likely to produce a favorable habitat.

The location of the mites on the host frequently influences the kind and severity of the injury produced. Rust mites (free-living eriophyids) that feed on fruit (pl. 2) produce discoloration, resulting in "downgrading" of the fruit; but rust mites feeding on leaves often cause less economic loss. Bud mites (eriophyids) most frequently retard growth and may adversely influence the quality of flowers, seed, or fruit production. Mites that feed on young plants, though they may be relatively few in numbers, may destroy a newly planted crop.

Habits and Habitats

There is a great diversity in the morphology of Acari which is paralleled in many cases to behaviorial characteristics; that is, the specialization in habitat and habits often parallels specialization in structure. It is therefore essential that both habitats and habits be known for identification and classification.

Mites may be categorized as free-living or parasitic forms. The free-living forms are further categorized by Krantz (1970) as predacious, phytophagous, fungivorous, coprophagous or saprophagous, and phoretic mites. Phoretic species refer to those that utilize other arthropods as a means of dispersal.

Predacious mites are common in soil, moss, humus, and animal waste products where they feed on small arthropods or their eggs, on nematodes, and occasionally on other mite species. Predacious ground forms are commonly long-legged, fast-moving mites and are equipped with chelate-dentate or styletlike chelicerae. Where the chelicerae are styliform, the palps may be raptorial. Many have well-developed shields, and eyes often are present on the propodosoma. Predacious aerial species are usually long-legged and fast, and prey on plant-feeding mites or their eggs. While large dorsal and ventral shields occur commonly, they are weakly sclerotized and often difficult to see. A common but not consistent feature of the highly diverse aquatic mites is the presence of long "swimming hairs" on the legs. Many aquatic species are brightly colored: red, orange, green, or blue. Eyes usually are present and the palpi often are modified for grasping. Sclerotized shields may or may not be distinct. Adults and nymphs feed on other mites, and on small crustaceans, isopods, and insects. Larval forms commonly are parasitic on insects, mollusks, or fishes.

The phytophagous mites include the aerial species that feed on plants both above ground and in the soil. The economic forms are the subject of this book. Stored grains and other stored products often are infested by various kinds of mites, some of which feed on the product itself. This subject has been covered by Hughes (1961). Storage mites are white or brownish-white, and commonly slow-moving and saclike. The chelicerae are blunt and toothed, and are useful for scraping and gouging the food material. Graminivorous mites feed on the germ tissue of the grain and may move into the surrounding endosperm as well. Mites feed on dried fruit, stored tubers, and bulbs, injuring the stored products either by feeding activities or by excrement.

Aside from a tendency to be slow-moving or sedentary, the fungivorous mites do not lend themselves to categorization. They feed on fungi in habitats such as tree buds, stored grain, in soil, on woody plants, and in habitats of wood-boring insects. Some species may be pests of commercial mushroom houses.

Dung offers an attractive habitat for many mites, among which are mites that generally feed on animals that in turn feed on dung, but some mite species actually feed on dung. Some mites feed on dead or decaying plant and animal tissues.

Some free-living mites may utilize insects or other arthropods as a means of dispersal. Mites are very important as parasites of man and domesticated animals. They are involved in the transmission of pathogenic organisms and as carriers of internal parasites, such as tapeworms and filarial worms. Mite feeding on man or animals may cause irritation or provide sites for invasion of secondary disease organisms. The parasitic mites of medical importance have been reviewed by Baker et al. (1956).

SELECTED BIBLIOGRAPHY

Baker, E. W., T. M. Evans, D. J. Gould, W. B. Hull, and H. L. Keegan. 1956. A manual of parasitic mites of medical or economic importance. Natl. Pest Control Assoc. Tech. Publ. 170 pp.

Baker, E. W., J. H. Camin, F. Cunliffe, T. A. Wooley, and C. E. Yunker. 1958. Guide to the families of mites. Inst. Acarology Contrib. 3:1–242.

BAKER, E. W., and G. W. WHARTON. 1952. An introduction to Acarology. The Macmillan Co., New York. 465 p.

BANKS, N. 1915. The Acarina or mites. U. S. Dept. Agr. Rept. 108:1–153.

EVANS, G. O., J. G. SHEALS, and D. MACFARLAND. 1961. The terrestrial Acari of the British Isles. Vol. I. Introduction and biology. British Museum, London. 219 pp.

EWING, H. E. 1909. A systematic and biological study of the Acarina of Illinois. Bull. Univ. Illinois 7(14):434–436, 453–472, also Univ. Ill. Studies 3(8):1–120.

———. 1934. The suborders and superfamilies of Acarina. Proc. Helminthological Soc. Wash. 1(2):64–66.

HUGHES, A. M. 1961. The mites of stored food. Ministry Agric., Fish and Food Tech. Bull. 9, 287 p. + vi.

HUGHES, T. E. 1959. Mites or the Acari. Univ. of London, Athlone Press. 225 pp.

JOHNSTON, D. E. 1968. An atlas of Acari I. The families of Parasitiformes and Opiliocariformes. Acarology Lab., Ohio State Univ. 172:1–110.

KRANTZ, G. W. 1970. A manual of Acarology. O.S.U. Book Stores Inc., Corvallis, Oregon. 335 pp.

LUNDBLAD, O. 1941. Die Hydracarinenfauna sudbrasiliens und Paraguays. Kungl. Svenska VetensKapsaKademinens Handlinger, Ser. III, 19(7):1–183.

MCGREGOR, E. A. 1950. Mites of the family Tetranychidae. Am. Midland Naturalist 44 (2):257–420.

MICHAEL, A. D. 1884. British Oribatidae 1. Ray Society, London.

MURRAY, A. 1877. Economic entomology, Aptera. South Kensington Mus. Sci. Handbooks, London. Pp. 99–374.

OUDEMANS, A. C. 1937. Kritisch historisch Overzicht der Acarologie. E. Jedeelt Brill Co., Leiden, Holland. 8 v.

PRITCHARD, A. E., and E. W. BAKER. 1955. A revision of the spider mite family Tetranychidae. Pac. Coast Entomol. Soc. Mem. Ser. 2. San Francisco. 472 pp.

SNODGRASS, R. E. 1952. A textbook of Arthropod anatomy. Comstock Pub. Assoc., Ithaca, N. Y. 363 pp.

TUTTLE, D. M., and E. W. BAKER. 1968. Spider mites of southwestern United States and a revision of the family Tetranychidae. Univ. Arizona Press, Tucson, Arizona.

VIETS, D. 1936. Wassermilben oder Hydracarina (Hydrachnellae and Halacaridae). Die Tierwelt Deutschlands 31:1–288; 32:289–574.

VITZTHUM, H. G. 1940–1942. Acarina. Bronns' Klassen und Ordungen des Tierreiches. 5, Sect. 4, Book 5: 1–1011.

WHARTON, G. W. 1964. First International Congress of Acarology. Keynote Address. Proc. Extrait de Acarologia 6:37–43.

WINSTON, P. W. 1964. The physiology of water balance in Acarina. First International Congress of Acarology. Proc. Extrait de Acarologia 6:307–314.

WOOLLEY, T. A. 1961. A review of the phylogeny of mites. Ann. Rev. Entomol. 6:263–284.

Chapter 2

Population Ecology

The understanding of mite populations, their cycles, and outbreaks requires a knowledge of many factors. These factors include the biotic potential of the species, the influence of meteorological factors, the availability and relative susceptibility of hosts, competition between mite species, structural and chemical adaptations of each kind of mite, and pathogens and predators of mites.

Climate, host plants, and the economy of plant cultivation are likely to remain largely unchanged. Therefore, the management of predators and pathogens of mites offers the most promise for preventing populations of these tiny creatures from reaching injurious proportions and for minimizing the use of acaricides. For these reasons, chapter 5 is devoted to predators upon, and diseases of mites.

METHODS OF ESTIMATING AND EVALUATING MITE POPULATIONS

Many techniques have been devised for the sampling of field populations of mites, each of which has its advantages and disadvantages. Some workers devise methods to suit their particular preferences. The variations in mite size, however, the habits and habitats of mites, as well as the objectives for which the various evaluations are useful, encourage the development of an assortment of ways in which to estimate population density. Some of these methods have been reviewed by Morgan et al. (1955) and Pielou (1960).

One common method, usually presumed to be the most precise, is to collect samples of mite-infested leaves, and make actual counts of all stages present on these leaves. A dissecting microscope is used for this purpose. This is a time-consuming procedure, and may be less accurate than one might assume, especially when counting active mites. Handling the leaf may stimulate active mites to crawl around; and when the area under examination is larger than the microscope field of vision, moving mites may be counted more than once. Subsamples of these leaves, measured to microscope field size, may reduce this particular disadvantage.

There has been considerable use of the *imprint count method.* To make an imprint of mites present on a leaf, place the leaf between two papers of proper absorptiveness, and crush the mites against the papers. This method has the advantage of providing a semipermanent record of the mite infestation. It inactivates the mites at the most opportune time and counting is much easier. It allows a more accurate evaluation of all stages of the particular species present on the collected leaves. The imprint count method depends on distinctive colors to allow the worker to distinguish between two different mite species. If there are two different tetranychid species of the same color present on the leaves, it may not be possible to distinguish between them after crushing.

Field counts of mites as they occur in their habitats are possible with the unaided eye, if the mites are large enough. These field counts are rapid and do not require removal of favorite mite breeding sites from the host plant. In case of high mite populations, counts by tens are feasible. But these unaided eye counts primarily recognize adults, so more frequent sampling may be necessary when using this procedure. Another approach is to count mite-free leaves on a plant, after previous examinations have set up a reference standard.

A *brushing machine*, first built by Henderson and McBurnie (1943) (pl. 3), has had wide use in sampling populations of tetranychid mites on leaves. To use this machine, infested leaves are passed between rotating brushes; and, as mites are brushed off, they fall onto a revolving disc. The disc bears a sticky coating and has aliquot sections. This brush method has most of the advantages of imprints on paper; and it allows a better identification of mite species and stages that are removed from the leaves. The host plant sampled in this way must be large enough to permit removal of some leaves without materially disturbing the remaining mite population.

Beating branches is still another approach to population estimation, using a suitable surface for recovering the mites that fall from leaves. Beating is also suitable for counting mite species that live on both leaves and twigs, providing the mites do little webbing (Summers and Baker, 1952).

Sampling of brevipalpid mites (flat mites) and eriophyid rust mites is possible by examinations of one-half inch square grid areas on leaves by means of a linen counter.

To count bud mites it is necessary to dissect the bud scales and bracts, recording the presence or absence of mites in each bud. Bud mite populations may be estimated by macerating the buds, then separating the mites from bud tissue centrifugally, after which the mites may be counted (Sternlicht, 1966).

As mentioned above, the selection of an evaluation method depends both on the type of mite under study and on the objectives of the appraisal. To determine the adequacy of the size of the sample, take a large number of equivalent samples and calculate the coefficient of variation.

FACTORS INFLUENCING THE BIOTIC POTENTIAL OF MITE SPECIES

The rate of growth of a mite population depends in large part upon the adaptability of the species to changing conditions, upon food preferences, and upon egg-laying capacity. The rate at which the population of a mite can become injurious often determines its status as an agricultural pest.

The time required for the completion of the life cycle, from oviposition until the succeeding generation starts egg laying, and the potential egg-laying capacity, has been determined for most of the tetranychid mite pests, and for some others. In the treatment of individual species (see chaps. 8–11, 13), there is information on these topics as well as on the duration of life stages under constant temperature and under conditions in the field, insofar as data are available.

Climate and Population Cycles

The small size and soft bodies of most mites provide minimum protection against the annual weather cycles that occur in temperate continental areas. Even in the tropics there are wide humidity fluctuations. The various mite species have met these challenges in various distinctive ways. Adaptations to cyclic weather changes include diapause (for aestivation, or for hibernation, or for both), migration to protected niches, and stimulation of plant growth by injecting growth regulators either to keep plant parts unnaturally succulent or to cause gall formation.

Mite species are well adapted to cyclic weather changes through behavioral and anatomical modifications that enable them to survive. The species that find hot, dry summer undesirable retire to moist ground crevices, retreat under bud scales, or develop a generation with exterior anatomy more resistant to temperature and humidity extremes. Females of some species, with the onset of unfavorable conditions, enter a stage of arrested development; or they deposit weather-resistant eggs that survive during the unfavorable conditions. Species adapted to thrive in cool climates are frequently confined in their distribution to habitats within these specific temperatures ranges. In warmer climates, another group of species may be found to predominate, and under intermediate conditions there may be an interchange of species depending upon their adaptations to the climatic cycles.

Winter Diapause in Tetranychidae

Diapause has been defined (Dickson, 1949) as a physiological state of arrested development that enables an organism to survive more easily a period of unfavorable conditions. Once an organism has entered diapause it usually has to remain in that state for a certain period, regardless of the conditions of the environment.

Species Overwintering as Adult Females. Overwintering adult females of tetranychids and eriophyoids are often readily distinguishable in color or structure or both from actively breeding summer forms. This is especially, if not exclusively, true of mites that live on trees and shrubs. In the Eriophyoidea the term *deuterogyny* has been used to indicate these color and structural changes: *protogyne* for the summer and *deutogyne* for the winter form (See chapter 12).

Changes in Appearance and External Morphology. In species that live on deciduous hosts, tetranychid females destined for hibernation often possess a characteristic reddish or orange coloration, and are more deeply pigmented than summer females. Diapausing eriophyoid females on deciduous host plants are often distinctly darker brown than the nonresting females.

Reiff (1949) attributed the color changes in the tetranychid mites to alterations in the leaf metabolism. As the two-spotted spider mite, *Tetranychus* (*T.*) *urticae* Koch changes from summer to winter-type, the rapidity with which the color change occurs after the final molt apparently depends on the temperature at which the teneral (recently molted) females are kept. Linke (1953) observed that on hops in the autumn, fertilized winter-type females took about 3 to 8 days to change color, and Parr and Hussey (1966) found that when teneral females reared under diapause conditions at 13 C (55.4 F) were transferred to 25 C, the change took place in 3 to 5 days. During the period between maturity and color change many of the females fed, but no eggs were laid. Once the color changed, all feeding ceased, the contents of the hindgut were voided, and the mites migrated from the plants in search of winter quarters.

Pritchard and Baker (1952) demonstrated an important morphological difference between the summer and diapause forms in the structure of the integumentary striae on the dorsal body surface. They concluded that all mites belonging to the genera *Tetranychus* and *Eotetranychus* found in northern climates exhibited this difference. In the summer form these cuticular ridges possessed semicircular and triangular or even elongate lobes; but in the diapause form the ridges were devoid of lobes (pls. 4, 5) (Dosse, 1964).

Various gradations of these forms, however, developed when single and certain combinations of instars were exposed to "winter" conditions of 13 C and eight hours of daylight; development ordinarily occurs at 25 C and 16 hours of daylight: (1) Orange-red nonfeeding forms without lobes on the striae developed in a few cases where diapause was induced by the exposure of only a single instar to "winter" conditions. (2) Green, egg-laying females in which diapause had not been induced were either with or without lobes. Those on which either the combined protonymph and deutonymph or the entire nymphal development had been exposed to "winter" conditions always had striae without lobes; but those exposed only for a single instar, and others exposed as combined larval and protonymph instars, either had or were void of lobes indicating that individual mites vary in the amount of exposure required to change from summer to winter or diapause forms. (3) Pinkish, "intermediate" females, in which diapause appeared to have been only partly initiated, had lobes as did the summer forms (Parr and Hussey, 1966). The reason for this change in integumentary striation with the onset of diapause is thought to be related to the need of the mite to conserve water content during hibernation (Boudreaux, 1958).

Factors Influencing Onset and Termination of Diapause. Diapause in tetranychid mites appears to be facultative and controlled by three environmental factors —photoperiod, temperature, and nutrition. As noted above, exposure of immature stages to diapause-initiating influences produced the development of diapause in *T.* (*T.*) *urticae*. This undoubtedly applies as well to other phytophagous mite species that overwinter in the adult stage, including eriophyoids.

Parr and Hussey (1966) concluded that the tendency towards hibernation initiated by diapause-inducing conditions during nymphal development may be enhanced by returning the newly emerging females to these conditions. Alternatively, development may be diverted to the nondiapause condition by transferring the teneral females to long hours of daylight and high temperature. This reversal is sometimes only partial, and results in "intermediates." The effect of exposure of successive instars to conditions favoring diapause appears to be cumulative and may in some circumstances be reversed; this effect suggests that irrevocable diapause may be induced by a critical level of endocrine activity. Older instars possibly are more important in the evocation of diapause merely because when suitably stimulated they liberate greater quantities of hormone.

Studies by Lees (1953, *a*: 1953, *b*, 1955) indicate that the influence of light on diapause in the European red mite, *Panonychus ulmi* (Koch), (and probably other mites) is affected only by the duration of the photoperiod, not by the light intensity or total light energy, provided the intensity exceeds a threshold. Subthreshold illumination is presumably equivalent to darkness, which can markedly reduce the incidence of diapause. Lees concluded that the photoperiod reacts directly on the mites and not through the medium of the host plant. According to Bondarenko (1958*a*) the intensity threshold in *T.* (*T.*) *urticae* is about 3.5 lux (0.3 ft-c).

As in most arthropods, the dark phase of the illumination cycle is as important as the light phase in the determination of diapause. In the European red mite, *P. ulmi*, the light phase tends to prevent diapause and the dark phase to induce it, provided the cycle of illumination is short in relation to the sensitive stadia of the mite. The balance is determined by the absolute length of the phases and not by the ratio of light to darkness (Lees, 1953*a*, 1953*b*).

It has been found impossible to induce the development of diapause-type females in the generation of mites that hatch from eggs laid by females activated directly from diapause. In the two-spotted spider mite, *T.* (*T.*) *urticae*, generations that grow from eggs laid by nondiapausing females are progressively more responsive to diapause-promoting conditions the further the generations are from the original reactivated adults. This indicates that diapause-type females do not impart the ability to go into diapause to the generation they produce. Such experiments also suggest that annual rhythms have a basic part to play in the production of the orange-red resting generation. For the hawthorne spider mite, *Tetranychus* (*A.*) *viennensis* Zacher (= *crataegi* Hirst), the tendency to diapause increases until reaching a maximum in the eighth generation, after which it decreases to the fourteenth generation, following which the tendency again rises. Therefore, the performance of *T.* (*A.*) *viennensis* is based on mechanisms of an oscillator type (Nuber, 1961; Razumova, 1967).

Normally, diapause is terminated only by a period of chilling before the resting condition can be terminated by warming trends. At Leningrad, Bondarenko (1958*a*) found that the two-spotted spider mite, *T.* (*T.*) *urticae*, required 55 days at 3 to 6 C (38 to 42 F) to break diapause, and suggested that mites living in milder climates would need a longer period of chilling to break diapause than would those inhabiting colder regions. In England, Parr and Hussey (1966) found that 55 percent of newly diapausing females resumed feeding and egg-laying within 7 days without prechilling when placed at 25 C (77 F) and 16 hours of daylight. When similar females were chilled for two weeks at 2, 7, or 13 C (36, 45, or 55 F)

75, 73, and 90 percent respectively became active in 4 to 10 days at 21 C (70 F) and 16 hours of daylight. Results of these and other studies indicate that some females in diapause, when supplied with adequate food and warmth, readily revert to active feeding forms without chilling. Others do not break diapause until they have experienced a period of cold rest at temperatures below 10 C (50 F).

Observers in different localities have reported termination of diapause as early as January and as late as the end of April. Parr and Hussey (1966) obtained mites from 10 areas, each with different mean temperatures, and exposed them at 21 C (70 F) for 6, 10, and 15 weeks. They found no differences in the pattern of emergence between these strains, and concluded that differences in emergence probably result from differences in local environmental factors rather than strain characteristics.

Characteristics and Behavior of Diapausing Forms. There are numerous reports of the ability of diapausing females to withstand very low temperatures. In outdoor sheds where the temperature was often –27 C (–16 F), Bondarenko (1958*a*) found that only 10 to 15 percent of the diapausing mites died, but all succumbed at –32 C (–23.8 F). Parr and Hussey (1966) studied the effects of temperature and humidity on the survival of the two-spotted spider mite, *T.* (*T.*) *urticae*, and found that mites require low temperature and high humidity to survive winter diapause successfully. At 6 C (42 F), which is near the temperature threshold of development, many mites kept at relative humidities of 75 and 93 percent survived for more than 8 months, which was nearly twice as long as those kept at 40 percent RH. The higher temperatures did not favor survival, although here again mites showed a markedly favorable response to a moist atmosphere. Winter humidities are more likely to be higher outside than inside greenhouses; diapausing mites may be able to survive lower temperatures outside than inside greenhouses.

To survive, diapausing mites also seem to need protection from stagnant water (Helle, 1962). This may account for the selection of loose bark, and cracks in woody host plants for egg deposition or hibernation.

Tibilova (1932) found that females of the diapausing two-spotted spider mite remained alive after 100 hours submersion in water, whereas summer nondiapausing females perished after 10 hours in water. Poor "wetting" is probably related to differences in the cuticular structure of the two forms, and may account for reports that hibernating mites are more difficult to kill with chemical sprays than the summer forms.

Diapausing females of the two-spotted spider mite are positively geotactic and negatively phototropic. They leave their hosts by crawling or by dropping down on silk threads, and they wander in search of crevices in pieces of wood, under loose bark, or on fence posts. They may also seek shelter under ground cover and earth clods.

After winter chilling, when warm weather returns in the spring, the mites become active again and start to feed within a few hours. Females of the two-spotted spider mite often feed first upon low herbaceous plants in early spring. They spin a web on their first feeding site, and start egg deposition within three days. With the intake of food the orange-red overwintering color begins to fade to greenish, and the characteristic subdorsal black spots reappear. But these revived females do not regain the strial lobes that are on summer-type females.

Evidence indicates that for the two-spotted spider mite diapause termination depends primarily on a rise in temperature. Light intensity, photoperiod, and quality of food have a minimum influence at this point.

Species overwintering as diapause eggs. Some tetranychid mites, and some phytophagous species in other families lay two kinds of eggs. One kind hatches shortly after laying, and is characteristic of times of the year during which active mite breeding can take place. But when weather and changing conditions on the host plant make it necessary for these mites to adjust to survival stages, females of these species lay diapausing eggs, often in places different from those where they laid nondiapausing eggs. Diapausing eggs, then, carry the species through the unfavorable period, be it hot or cold, dry or wet. When the weather returns to conditions in which the mite can resume active life, diapausing eggs are stimulated to hatch.

The European red mite, *Panonychus ulmi* (Koch), and most species of the genera *Oligonychus* and *Bryobia* living in northern climates overwinter in the egg stage. The structure and formation of the summer and winter eggs of the European red mite were studied in considerable detail by Beament (1951). He indicated that the outer coverings of winter and summer eggs are identical and the mode of oviposition is similar. The summer egg, however, is vulnerable to desiccation up to six hours after deposition, but the winter egg is able to survive desiccating conditions as soon as it is deposited.

During the egg-laying process the egg receives its shell layer in the saclike ovary, then passes into a glandular ovipositing pouch that is evaginated through the genital aperture at oviposition. As the shell layer makes contact with the substrate, the pouch secretes cement over the rest of the shell layer; thus the egg adheres to the substrate by a ring of cement around the base. The outer wax is then secreted over the cement. The egg surface (shell layer) in contact with the substrate is without cement, but bordered by cement. In summer eggs the area of contact is not waterproofed, but the developing organism waterproofs the egg presumably by secreting a wax layer inside the shell. The shell of the winter egg is similarly composed and waterproofed, but winter eggs are held in the female until a stage of embryonic development at which time they are already waterproofed; then they are deposited on the bark. Summer eggs laid on leaves apparently are prevented from becoming desiccated by being firmly glued to the transpiring leaf; but the bark surface, where winter eggs are deposited, does not provide such protection.

The development of summer or winter eggs by *P. ulmi* is facultative and the same factors appear to engender females of the European red mite, *P. ulmi*, to lay overwintering eggs as those evoking diapausing females in the two-spotted spider mite, *T.* (*T.*) *urticae*, that is photoperiod, temperature, and nutrition. As with the two-spotted spider mite the lack of suitable food is, by itself, a factor engendering the laying of diapause eggs only when temperature and photoperiod approach conditions that favor initiation of diapause.

Lees (1953*a*, 1953*b*) found that at medium temperatures and daily photoperiods, of 6 to 13 hours, only winter females of the European red mite, developed; only summer females developed at photoperiods of 15 to 16 hours or continuous illumination. Also, the response is independent of light intensity provided the illumination exceeds a threshold of 1 to 2 foot-candles. Lees found that radiation in the near ultraviolet, blue, and blue-green regions of the spectrum is photoperiodically active with maximum sensitivity in the blue region. Wavelengths above 550 $m\mu$, that is, in the orange, red and infrared regions, are totally inactive. His studies indicated that developing *P. ulmi* are indifferent to diapause-inducing factors until the deutonymphal instar, but egg-laying females, if exposed to diapause-engendering conditions, can still be caused to "switch over" to the alternative egg type. Eggs intermediate in character may be laid during the period of "reversal." He further found that winter eggs in diapause never hatch at 18 to 25 C, but diapause can be broken by chilling (1 to 9 C) the eggs for 150 to 200 days.

Studies by Hueck (1951) convinced him that the winter egg of the European red mite enters diapause at a very early stage of development and hatching occurs three to four weeks after the breaking of diapause. He concluded that light seems to provide a stimulus to the fully developed embryo to break the eggshell: 86 percent of the eggs hatched during daylight.

Lees (1953*a*) found that diapause in *P. ulmi*, the European red mite, ended after an exposure of temperatures at 1 to 9 C (34 to 48 F) for 100 days. Exposure at 18 C (65 F), even for very long periods does not bring diapause to an end. Freezing at –5 C (23 F) also fails to break diapause and results in heavy mortality after about 100 days. Thus the upper limits of reaction lie between 10 and 18 C (50 and 65 F), and the lower limits between 1 and –5 C (34 and 23 F).

Summer Diapause

Some mites live near the ground or on low plants in areas where the winters are relatively mild, but the summers are particularly hot and dry. Such mites deposit heat-resistant (aestivating) eggs on clods or on sticks. The female of the red-legged earth mite, *Halotydeus destructor* (Tucker) dies at the onset of hot weather; the eggs survive, however, and remain inside her body for protection. Other mite species live in trees, and with habitats well above ground may find protection from summer heat among leaves. They too lay heat-resistant eggs that remain unhatched through the summer. Certain inhabitants of deciduous trees lay diapausing eggs on twigs, or on the bark of their host during the summer. These eggs must then go through summer heat and winter cold before they hatch the next spring.

Evidence indicates that females do not lay both nondiapausing and diapausing eggs at the same time. Females deposit the nondiapause-type eggs when conditions are favorable for such activity, but they switch to diapausing eggs exclusively if so stimulated by the onset of deteriorating weather.

The brown mite, or brown fruit tree mite, *Bryobia rubrioculus* (Scheuten), prefers trees. Mites of this group hatch from diapausing eggs in spring, go through 2 or more generations and then lay diapausing eggs on twigs in early summer. These eggs do not hatch upon temperature drop in the fall, but go through winter chilling before hatching.

The clover mite, *Bryobia praetiosa* Koch, which prefers to live on low plants, lays aestivating eggs in late spring. This mite has a relatively narrow temperature-humidity tolerance, so retires when conditions go above or below predetermined limits. All stages of nymphs and adults also frequently become inactive, usually after migrating off their hosts. They become active as soon as weather conditions become favorable, so do not enter diapause.

Certain species of the tetranychid genus *Petrobia* oviposit both nondiapausing eggs, and / or diapausing eggs, depending on climatic influences. The brown wheat mite, *Petrobia* (P.) *latens* Müller, lives close to the ground, and confines its attacks to low-growing agricultural crops. This mite does best in dry areas. Rising summer heat, with accompanying low relative humidities, cause females to lay glistening white heat-resistant eggs on ground litter, such as clods, sticks, and the like. When these females lay nondiapausing eggs, they quickly raise their abdomens up after depositing the egg, and that draws the wax out into the characteristic perpendicular stipe which projects up from the egg. For diapausing eggs they hold the genital opening in close contact with the egg for a time, thus topping the egg with a thick wax cap. Because excessive moisture and flooding will kill these eggs, brown wheat mite, thrives best in sandy soil areas (Brooking, 1957; Glancey, 1958; Lees, 1961).

In the family Penthaleidae there are two species of mites that can serve as further examples of survival through certain parts of the year by means of resistant eggs. One of these is the winter grain mite, *Penthaleus major* (Dugès). This mite prefers sandy soil for some of the same reasons as does the brown wheat mite, *P. latens.* The winter grain mite attacks low-growing plants. The winter eggs of the winter grain mite, *P. major*, hatch in from 25 to 35 days, and the aestivating eggs require 110 to 140 days as their incubation periods. The actvity of the winter grain mite is during the winter, as its name implies. Because of a peculiar *deutovum* stage, when the larva of *P. major* remains for a time only partially hatched, it is vulnerable to excess moisture at this critical stage. Sandy soils tend to drain away extra water.

A second penthaleid mite, the red-legged earth mite, *Halotydeus destructor* (Tucker), which lives in Australia and South Africa, is also rather limited to sandy soils that drain away excess water. This red-legged mite lays winter eggs in masses, mainly on the underside of leaves lying on the ground. These winter eggs are bright yellow or orange, and hatch in from 2.5 to 8.5 days. But the red-legged earth mite females do not deposit resistant summer eggs. Instead, these oversummering eggs, which are not as resistant to hot, dry conditions as are eggs of many small arthropods in the same areas, remain in the female body. When the female red-legged earth mite dies at the onset of hot weather, her carcass helps to protect the eggs (Wright, 1961).

Eriophyoids and Diapause

At present, it is only possible to comment upon diapause among these tiny eriophyoid mites as it is manifest in species that live on deciduous hosts in the northern hemisphere (see chapter 12). Diapause tactics by eriophyoids on nondeciduous hosts, and in the tropics, remain obscure at this time. As will be further explained in the section below on Regulation of Water Loss, ring microtubercles on some of these eriophyoid mites may have a function in water exchange through the skin, just as they do on tetranychids; alterations appear in these ring granules preparatory to diapause.

Eriophyoids are in general much more intimately attached to their hosts than are the common tetranychids. Eriophyoids lay no resistant eggs so far as is known. All immature stages go through their developments as soon as possible and adult females are usually the members of any species to travel, aestivate, or hibernate.

Deutogynes of these tiny quadrupeds are directly comparable to diapausing tetranychid females, and eriophyoid protogynes are comparable to the summer type, actively breeding tetranychid females.

Information as to how changes in photoperiod, or thermoperiod affect the appearance of diapausing females among eriophyoids is not now available. Experimental evidence indicates that once deutogynes—that is, the diapausing females—have developed in an eriophyoid colony, these deutogynes then cannot lay eggs the year they appear, but must go through the winter chilling before they are able to reproduce. Appropriate refrigeration of new peach silver mite deutogynes, *Aculus cornutus* (Banks), followed by exposure to warm laboratory conditions, has broken the diapause of these deutogynes ahead of the normal time at which they would revive (communication from G. N. Oldfield).

The presently accepted explanation for the appearance of new deutogynes in a colony of rust mites is that the weather acts upon the host, and the resulting seasonal cycles of the host plant induce mite deutogynes to develop. The California buckeye rust mite, *Tegonotus aesculifoliae* (Keifer), for example, develops deutogynes just before leaf drop in late spring or early summer and the pear leaf rust mite, *Epitrimerus pyri* (Nalepa) is sensitive to leaf maturity and its deutogynes appear in early summer. Further details on the deuterogyny of these mites is reported in chapters 12 and 13.

Rust mites (eriophyids) live in the open, and aside from crumpling leaves or causing deformities that do not altogther enclose them, they can exercise little control on their host plant. Many of them cause leaf discoloration only. Others, however, such as gall mites (eriophyids), not only live in microenvironments of their own engendering, which remove them from direct contact with sunlight and atmospheric humidity, but they also can to a considerable extent control the length of life and succulence of the tissues surrounding them in these galls, including erinea. This control is the result of continued injections of salivary growth directors; thus gall mites remain for much of the warm weather inside moist, partially darkened recesses or chambers.

Eventually leaves on deciduous hosts begin to deteriorate, and such deuterogynous eriophyoids as make leaf galls or erinea, must prepare for hibernation. They do this during the time the galls are in good condition by gradually filling them with deutogynes. For example, the California black walnut leaf pouch gall mite, *Eriophyes brachytarsus* Keifer, starts the production of deutogynes in early summer, but does not move out of galls until late summer or early fall when leaves begin to become old and dry (See chapter 12).

There would seem to be more of a cyclic urge to grow deutogynes in eriophyoid species making galls and erinea than is exhibited by rust mites and leaf vagrants. These latter types are more influenced by rapidly changing external conditions.

Genetic Basis for Diapause

The inherited character for nondiapause is usually dominant. Yet there is a wide variability in the diapause-controlling genes in spider mite populations. Inbreeding produces considerable diversity in photoperiodic responses in resulting generations, and this indicates that genetic deviations do exist in mite populations. This variability in tetranychid inheritance, as regards diapause, provides the mites with the means of adapting to wide climatic changes. It also enables them to live in unusual habitats, such as hothouses in temperate zones (Helle, 1968; Geyspitz, 1968).

The maintenance of genetic diversification as regards diapause response is favored because these characteristics are sex linked in arrhenotokous species such as tetranychid mites. Selections against unfavorable mutations act only in diploid females, and not in haploid males. This sex-linked limitation decreases the possibility of the elimination of deleterious genes and allows a load of unfavorable variants to persist even under male haploidy (Helle, 1968).

According to Helle (1965), "the adaptiveness of spider mites arises from interpopulation selection, rather than from intrapopulation variability. The data, however, indicate that intrapopulational variability is greater than expected." To explain the intrapopulational variability he proposes two hypotheses: (1) storage of sex-linked variants; and (2) high spontaneous mutation rate. Helle (1965) states that "adaptational dynamics in arrhenotokous species is set forth in a peculiar framework that differs in some principal aspects from that of species which have diploidy in both sexes. Theoretically, these dynamics can be deduced from male haploidy by: (1) Strong tendency to homozygosity (2) High mutational yield."

He explains that cross-fertilizing populations of species that have both sexes with diploid chromosomes can be loaded with recessive variants, even when these variants are deleterious. The possibility for storage of variants is realized by the fact that at low frequencies, recessive mutants occur mainly in the heterozygous condition, and are out of control of natural selection. The assumption is that this store of concealed genetic variety is often important for the adaptive potentialities of a population.

Conversely, in arrhenotokous species, the likelihood of mutational genetic overloads seems more restricted. In such speices, deleterious variants, whether recessive or dominant, remain even at low frequencies, under the direct influence of natural selection. The presence of hemizygous males gives harmful characters no opportunity to hide. Therefore, the strong tendency toward fixation of favorable alleles, will exist, with the consequence that positive limitations in genetic variety are to be expected.

Helle (1965) further states "that favorable mutations (and also chromosomal alterations) in arrhenotokous species are less liable to dispersive processes (like genetic drift), as compared to animals with both sexes diploid. Interaction between natural selection, and mutation, operate from the beginning in the haploid males. Consequently the chance of extinction of unfavorable mutants is much slighter, which results in a greater mutational yield, and subsequent divergence."

Mite Response to Weather Variations

Temperature Changes

Temperature has been the most extensively studied of all weather factors that bear on mite populations and it appears to have the most overall influence. Low temperatures can cause reductions in winter populations. High mortality also occurs when unseasonably low temperatures follow warm weather in early spring. Since the first spring generation of tetranychids cannot develop diapause-type females, and since most of the mites are immature in early spring, sudden lower temperatures under such circumstances will kill many mites.

The percentage of overwintering eggs that hatch is influenced by spring temperatures. A high percentage of mites successfully hatch when temperature permits an early and short hatching period (MacPhee, 1961).

The number of nonaestivating generations in species or strains of these mites is closely correlated with the temperatures to which they are subjected. Temperature preferences and tolerances vary with the species, which accounts for dissimilarities in distribution of these species, and also accounts for seasonal population cycles (Mori, 1961).

Mori (1961) compared the temperature zone of activity and the preferences of four tetranychid species:

Species Name	*Temperature C*	*Range C*	*Preference C*
Panonychus ulmi	5.0 – 41.0	36	25 – 28
Tetranychus viennensis	14.8 – 40.8	26	25 – 30
Tetranychus urticae	8.8 – 43.8	35	13 – 35
Bryobia rubrioculus	10.8 – 40.2	29	21 – 24

The potential progeny output from a mite increases exponentially as temperature rises. In a month at 15.5 C (60 F) a female tetranychid can produce 20 new individuals; she can produce 12,000 progeny at 21 C (70 F) in the same length of time; 13,000,000 at 26.5 C (80 F).

Atmospheric Humidity Levels

Infestations of most spider mites are favored by hot, dry weather. Continuous high humidities tend to depress population increase. High humidity tends to kill tetranychids during molting. In highly humid air these mites feed less vigorously, females slow their egg laying, and most mites experience shorter life spans (Harrison and Smith, 1961).

Optimum atmospheric humidity for the desert spider mite, *Tetranychus* (*T.*) *desertorum* Banks, is probably above 85 percent regardless of the degree of heat or cold (Nickel, 1960). In greenhouses, where temperatures and humidities are regulatable, populations of two-spotted mite, *T.* (*T.*) *urticae*, which are somewhat resistant to acaricides, have been more readily controlled by maintaining these enclosures at lower temperatures and relatively high humidities (Fritzsche, 1960). Extremely arid conditions, which extend over several days, may reduce the numbers of some tetranychid and eriophyid species, with greatest mortalities then occurring during hatching and molting (Jeppson, Complin, and Jesser, 1962; Hobza, 1970).

Rain

Most tetranychids behave as if rain does not adversely affect them, unless rain continues over prolonged periods, or is especially heavy. Heavy rains sometimes wash mites off their hosts, but most mites move during rains to leaf undersides or to other sheltered places. Natural pubescence on some plants helps mites to hold on during storms. But powerful rainfalls, such as Asiatic monsoons, do cause major reductions in some mite populations. The tea red spider mite, *Oligonychus* (*O.*) *coffeae* (Nietner), suffers high mortality during heavy rain, and a combination of wind and rain stops development of a tea red spider mite *Tetranychus* (*T.*) *kanzawai* Kishida in Japan (Das, 1959; Osakabe, 1965).

Experiments aimed at learning about rainwater as a depressing influence in mite populations have shown that European red mite eggs, *P. ulmi*, are not killed by immersion in water for 48 hours, but that some mortality takes place among active stages of these mites when they are given this same treatment. Immersion under water, however, does cause metabolism to stop; eggs do not hatch; and the mites stop feeding, molting, or ovipositing (Herne, 1968). The conclusion is that prolonged rainy periods are likely to lessen breeding in mite populations.

Response to Various Stimuli

Tetranychid mites generally show positive phototaxis during favorable times of the year, but this reaction is seasonal. Summer adult females show more sensitive and positive response to a beam of white light than do winter females. Males are variable in their reactions to white light. Winter females of the hawthorne spider mite, *T.* (*A.*) *viennensis*, respond more readily to the light than the males do. Summer females of the European red mite, *P. ulmi*, are more apt to move toward the beam of light than are the males (Suski and Naegele, 1968).

At moderate temperatures, that is 15 to 30 C (59 to 86 F), tetranychids show quicker positive reactions to stimulus caused by light than by temperature. But photokinesis lessens as the temperature rises above 38 C (100 F) (Mori, 1961, 1962*b*).

Mori (1962*a*, *b*) studied the photostatic response and thermal reaction of several species of spider mites. His results showed a positive phototaxis irrespective of sex, conditions of feeding, or the season at which mites were collected. The results of experiments on the effect of minimum illumination on phototactic response in three species of spider mites showed that the summer adult responds more sensitively to a light beam than does the winter adult. There is no difference in response to light between the sexes in the two-spotted spider mite, *T.* (*T.*) *urticae*. The winter male of the hawthorn spider mite, *T.* (*A.*) *viennensis*, however, is more susceptible to light than the female. In contrast, the European red mite, *P. ulmi*, summer female is more sensitive than the male.

A positive light response in the two-spotted spider mite, which varies with physiological state and changes slightly with light intensities ranging from 3 to 200 foot candles, has been demonstrated. Mites collected on freshly infested plants differ significantly in their behavior from those collected on plants severely injured by mite feeding. The name "sedentary phase" of the population was applied to mites on uninjured plants and "dispersal phase" to mites on injured plants. A distinct directional response could be demonstrated in the dispersal phase, but not in the sedentary phase (Suski and Naegele, 1963).

Mite response is also a function of wavelength with a maximum response in the near ultraviolet region (375 mμ) followed by another, much smaller, peak in the yellow green region (525–550 mμ), and a negative or indifferent response to wavelengths longer than 600 mμ (Naegele, McEnroe, and Soans, 1965). On the basis of anatomical and behavioral studies, the anterior pair of eyes are considered scanning point detectors containing independent ultraviolet and green photoreceptors. It is also suggested that *T.*(*T.*) *urticae* responds to the plane of light polarization (Hussey and Paar, 1963; McEnroe and Dronka, 1971; Naegele et al., 1965).

It has been demonstrated that increased time of starvation facilitates the ability of the mite to respond to light. The sign and pattern of the response, however, is controlled by the humidity accomodation of the mite. In the initial period of starvation (between the first and the fifth hour), mites demonstrate either photoindifferent or photonegative response, according to humidity conditions. Low humidities tend to facilitate a negative or indifferent response and high humidities an indifferent or positive light response. Later, (between the fifth and the eleventh hour) mites demonstrate photosensitive, photonegative, or phototelotaxis (compass light) reaction in different humidity conditions. In more advanced stages of starvation (between the eleventh and the twentieth hour), the positive compass light reaction is demonstrated according to humidity conditions (Suski and Naegele, 1968).

The implications of the various stimuli on the behavioral patterns of mites have received considerable study. A dense population soon destroys food reserves upon which it feeds and consequently would be exterminated if it were unable to find a new host plant. But special behavioral patterns enable the animal to search successfully for a new source of food. There is, however, too little known for a synthesis of the full chain of behavioral events accompanying progressive starvation. Some links, however, may be constructed with considerable accuracy on the basis of what is now known.

It is well known that fresh leaves, because of transpiration, produce a microenvironment of relatively high humidity. But leaves seriously damaged by mite feeding decrease their transpiration rate and soon dry out; so mites on severely injured plants not only become starved, but also accomodated to a low humidity. Normally mites wander around within a limited area of the leaf or plant surface, but the initial shortage of food stimulates them to move in a more direct path, often upward, dictated by stimuli such as light and humidity gradients. With the progress of time of starvation, mites develop the positive light response that leads them to the periphery of the plant where there is a higher probability of finding young, freshly-developed foliage. As soon as mites accomodated to a low humidity encounter a rising humidity gradient, whether because of a random search, or possibly other inhibitory stimulus, they become photonegative. Consequently, mites move toward the shadowed area; if the gradient was because of transpiration of the leaves, the mites ultimately find a new food supply. If, however, the gradient was produced by some factor other than transpiration of fresh leaves, and the behavioral pattern does not result in finding a new food source, the mites accomodate to the new humidity and random search starts anew. If no gradient of rising humidity is encountered, the positive light response results in aggregation of mites on the outermost parts of the plants. This enables the animals to leave the damaged plant by means of air currents. If chance brings them to a new, undamaged plant, they become photonegative or photoindifferent owing to increased humidity (or possibly other inhibitory stimulus) and search for food (Suski and Naegele, 1968). Geotactic responses in tetranychid mites are also variable.

Eriophyoid reactions to various stimuli are not nearly so well known, but in part they differ to some extent from the response to these stimuli displayed by tetranychids. These differences are largely because eriophyoids cannot crawl far under their own power, so must remain near prospective feeding sites to be sure of survival.

Mite-Host Relationships

Plant Growth Habits that Resist Mite Attack

The sizes of buds on various lemon varieties differ to some extent and lemon trees with larger buds are more suitable hosts for the citrus bud mite, *Eriophyes sheldoni* Ewing. The larger buds provide easier access for the mites to enter and better protection from weather extremes than do small buds. The feeding action of this bud mite causes attacked bud parts to blacken and die, which loosens more bud tissue. The mite is therefore able to move gradually through the bud until the bud is badly damaged, exposing the mites to drying conditions and stimulating them to migrate to new buds. In contrast, trees with small buds do not present this mite with satisfactory places to found colonies, so most orange trees and lemon varieties with small buds are not as severely injured by this mite.

Peach is another host that illustrates how growth habit can deter a mite from causing damage for a time. The mite is the peach silver mite, *Aculus cornutus* (Banks). Peach varieties differ as to the presence or absence of basal leaf glands, which secrete a sticky protective syrup over emerging embryonic leaves. On peach strains that lack these glands the silver mite is able to attack young leaves and to cause both yellow-spotting and longitudinal upward leaf curl. Glanded peach varieties do not show these injuries. Silvering of peach leaves as a result of mite action occurs, however, on both types of peaches, usually during the summer and early fall after the leaf glands have ceased to function.

Population Competition Within and Between Species

Under crowded conditions adult tetranychids develop strong tendencies to move. They leave damaged leaves, and because of their phototropic and geotropic responses (previously discussed) move to parts of the plant, or the substrate they are on, looking for means of transportation to new areas. Thus they relieve overcrowding and intraspecific competition.

An illustration of interspecific competition occurs in some circumstances between the brown fruit tree mite, *Bryobia rubrioculus* (Scheuten), and the European red mite, *Panonychus ulmi* (Koch). The brown fruit tree mite usually hatches before the European red mite in the spring and this advantage enables the brown fruit tree mite to depress development of the European red mite. But the brown fruit tree mite goes into aestivation in late spring to avoid summer heat, whereas the European red mite continues to feed and breed. Later in the season the two-spotted spider mite, *Tetranychus* (*T.*) *urticae*, may move to the leaves occupied by the European red mite and the webs made by the two-spotted spider mite may then further hamper development of the European red mite (Lienk and Chapman, 1958).

Populations of these mite species obviously become self-limiting in relation to food supply, if their increase in numbers is not controlled by some form of environmental or artificial resistance (Davis, 1952; Fleschner, 1952; Watson, 1964). As mite numbers swell, extensive feeding damages their hosts, especially the leaves, which become yellowish, and many begin to dry out. Various tetranychid species show divergent responses to overcrowding, and to changing environmental conditions (Boudreaux, 1963). In the face of failing food supply the European red mite is stimulated to premature deposition of diapause eggs. The two-spotted mite and some other species in the genus *Tetranychus* abandon their hosts when the feeding area deteriorates, and wander in search of new plants on which to feed. In the early fall failing food supply may cause premature diapause (McEnroe, 1969).

When leaves are injured to the point of becoming unfit for food, the six-spotted mite, *Eotetranychus sexmaculatus* (Riley), and the citrus red mite, *Panonychus citri* (McGregor), float away on silk strands (Ebeling, 1934). When confined to depleted leaves the citrus red mite and *Oligonychus* (O.) *punicae* (Hirst) lay few eggs, and there is considerable mortality among active members of the mite population (McMurtry and Johnson, 1966).

Several intraspecific changes can occur during a population cycle when the respective species develop to such numbers that they cause deterioration of their feeding sites. With banana squash as the host, certain *Tetranychus* species exhibit these changes: (1) a reduction in the number of eggs laid per female, even while total population increases; (2) an increasing percentage of nonviable eggs, following a population peak; (3) increase in mortality among immature stages.

Plant Nutrients and Mite Populations

Nutrients coming up from roots to leaves obviously vary with soil conditions and cultivation practices. Because of complexities concerning nutrient availabilities, our present knowledge of the influence of plant physiology and nutrition on these mites is relatively meager. Increase in osmotic pressure in plant sap by two or three times above normal, because of high fertilizer rates and excess soluble elements such as silicon, magnesium, and calcium in plants, tends to favor development of two-spotted mite populations (Huffaker, van de Vrie, and McMurtry, 1969; Leroux, 1954; Morris, 1961; Lord and Stewart, 1961, Post, 1962; Rodriquez, 1951, 1958, 1960; Rodriquez and Rodriquez, 1952; Storms, 1969).

Both field and laboratory studies of European red mite on apple indicate a positive correlation between nitrogen content in leaves and numbers of mites present on the leaves. Some workers have reported a shorter developmental time and increased egg production and mite longevity at higher leaf nitrogen levels (Breukel, and Post, 1959). It is possible that the insoluble nitrogen compounds such as glutamine and glutamic acid, in leaves stimulate mite reproduction.

The level of soil nitrogen, however, may not be correlated with the level of leaf nitrogen in plants growing on a particular site. Differences in reactions to nutrients between tetranychid species are illustrated by responses of the European red mite and the two-spotted spider mite. Studies in Japan have revealed that the European red mite is more prolific when nitrogen levels are raised, but the two-spotted mite produced more offspring on plants growing in nitrogen poor areas. In some experimental cases, altering the nitrogen levels seemed to have no influence on mite breeding (Hukusima, 1958).

Besides nitrogen, other elements including trace elements have a place in mite nutrition. Some of these are: potassium, phosphorus, calcium, magnesium, zinc, and cobalt.

Host Preference

Present taxonomical studies utilizing comparative anatomy indicate that many phytophagous mites are host specific, or will breed only on host plant species that are closely related to each other. This is particularly true for the great majority of eriophyoids that inhabit broad-leaved plants.

The intimacies of gall formation appear to limit eriophyoids [these four-legged species] to one or a few species of hosts. Some of the big-beaked rhyncaphytoptid mites, however, seem to have wider host ranges on broad-leaved plants than do eriophyoids, which have small rostra. An example is the big-beaked plum mite, *Diptacus gigantorhynchus* (Nalepa). This mite is common on cultivated plum and has a very wide geographical range. But mites of this *gigantorhynchus* type occur on a variety of rosaceous hosts, including *Prunus* spp., and on blackberry. In certain instances when infested blackberry is intertwined with grape, this big-beaked mite will occur sparsely on the grape leaves, indicating, for example, that non-preferred hosts will temporarily support certain arthropods if the secondary host is close enough to a heavily infested preferred host. In the eastern United States, mites of the *gigantorhynchus* type also seem to live on leaves of *Nyssa*, a non-rosaceous host. Such an occurrence remains unexplained. Another big-beaked mite, *Apodiptacus cordiformis* Keifer, has as primary hosts various juglandaceous trees in eastern North America. It is also recorded on sweet gum, peach, mulberry, and oak.

While some of the tetranychid mites are now known to infest only one host plant, the more indiscriminate species have no intimate relations with any of their hosts. They damage the plants they feed upon, and then crawl, or balloon away in search of fresh feeding sites.

Tetranychid mites are in part characterized by their preferences in breeding on the leaves or fruit of their preferred host plants. For example, the six-spotted mite, *E. sexmaculatus*, feeds on older citrus leaves located near the ground. Citrus red mite, *P. citri*, will not remain when placed on very young citrus leaves, but will proceed to search for older leaves that are somewhat yellow, though still not fully mature (Fleschner, 1952). The citrus flat mite, *Brevipalpus lewisi* McGregor, lives principally on citrus fruit.

The food plant can have much influence on the biology of a mite species. and mites like the two-spotted spider mite, *T.* (*T.*) *urticae*, show different rates of egg laying, depending on the host. On beans, the two-spotted spider mite completes a generation in 13 to 21 days at 22 C (71.6 F). At the same temperature *T.* (*T.*) *urticae* takes 16 to 26 days to complete a life cycle on tomato, and 22 to 29 days on cyclamen. During its adult lifetime the average two-spotted mite female lays 78.9 eggs on bean plants at 20 to 21 C (68 to 70 F), 111.8 eggs on calla, and 128.1 eggs on strawberry. The lowest number of eggs deposited by the average two-spotted female under these conditions occurred on beet, tomato, and cyclamen.

Under greenhouse conditions Bravenboer (1959), found grape to be the least suitable host for the two-spotted spider mite. Although the leaves may be damaged badly, very few mites will be found in most cases. Beans seem to be the most suitable host, whereas cucumber, peach, and plum are in between. Dabrowski, Rodriquez, and Chaplin (1971) found that foliage of strawberries was generally attractive to this species in the spring and early summer, unattractive during July and August, again attractive in September, but reverted to a relatively unattractive condition in October and November. The cause of these differences is not known, but the physiological condition of the plant seems to be an important factor.

COMPARISONS OF MITE MOUTHPARTS AND PLANT DAMAGE

The Tetranychidae (red spider mites) and the Tenuipalpidae (flat mites) are the two principal families in the Tetranychoidea. Feeding mechanisms in all tetranychoids consist of a stylophore at the upper gnathosome base, from the underside of which hang a pair of cheliceral or mouth stylets. The stylophore is protrusible, moving back and forth during feeding. The stylophore pushes the stylets down into the plant tissue, and upon retraction probably allows an opening through which plant fluids may be forced out of the cells by the hydrostatic pressure exerted by the cell (turgor pressure). Just behind the stylets the fused pedipalp tips and sensory pegs press on the leaf surface, and the pharyngeal pump, located close to the tip sucks up the plant juice. Stylets of the two-spotted spider mite, *T.* (*T.*) *urticae*, probably penetrate about 100 μ.

Although eriophyoid mouth stylets probably remain fairly stationary except when probing for sap, the tetranychid stylets, once they are inserted into plant tissue, continually punch and withdraw, causing mechanical damage. F. M. Summers, of the University of California at Davis, states that he has seen plant cells in the tetranychid gut. Indeed, the life habits of common tetranychids show that they do not depend on keeping the plant tissue alive. Instead, once they have depleted and yellowed the leaves they are on, they move to fresh sites.

Tenuipalpid mites are a little smaller than tetranychid mites, on the average, and more sluggish. While most of them behave similarly to tetranychids, the damage they do often takes longer to become apparent. A few tenuipalpids have become gall forming, and this means that they depend on keeping plant tissue alive for a time. Species of *Larvacarus* (Pritchard and Baker, 1958) make galls on Asiatic *Zizyphus*. But unlike eriophyoids, these tenuipalpids approach the gall-making habit from the eight-legged mite form, and show only the early stages of adaptation to life in such microenvironments. These adaptations are a slightly elongate body, and the loss of the rear leg pair. Eriophyoid mites, however, are all basically similar in regard to the specialized and elongate body, which lacks the 2 rear leg pairs. It may be theorized that mites in this group that now live in the open may represent a secondary development, having come out of microenvironments.

Mouth stylets in the eriophyoids are quite unlike those of the tetranychids. These tiny gall and rust mites have 5 mouth stylets. The front stylet pair are presumably the chelicerae, and they bore into plant tissue by pressure and by alternate forward impulses. Eriophyoid mouthparts are significantly smaller than those in tetranychid gnathosomes, and the average eriophyoid chelicerae can only penetrate plant tissue 15 μ to 35 μ. Some rust mites with larger rostra can penetrate 50 μ to 60 μ, but they are in the minority and do relatively little damage.

Minute eriophyoid mouthparts do very little mechanical damage; the damaging action on plant tissue is done by salivary enzymes. Some bud mites, as noted above, do kill buds and cause tissue shrinkage, but many eriophyoids not only do not kill plant tissue, but rather they depend on either not harming it or on diverting its growth and keeping the tissue alive and succulent. Salivary growth regulators cause bud clustering and succulent brooming. The plant galls and erinea these mites make often remain fresh and soft throughout the summer and provide microenvironments that are moist and suitable for mite food.

Eriophyoid rust mites that cause leaf damage have small mouthparts. These mites discolor fruit on occasion. While rust mites do not have the same anatomy as spider mites, they have to some extent invaded the spider mite ecosystem.

Mouthparts in eriophyoids seem to be more efficient than those in the tetranychids. The stylets of the eriophyoid mites are smaller and penetrate less than those of the tetranychids, therefore cause less mechanical injury. The auxiliary mouth stylets, just behind the chelicerae if they are indeed the salivary ducts, may deliver digestive enzymes into plant tissue. The fifth or oral stylet then moves down within the plant and sucks up the juice. Such action would not dislodge plant cells. Whether or not eriophyoids actually suck out entire plant cell contents remains to be discovered.

REGULATION OF WATER LOSS BY PHYTOPHAGOUS MITES

Water Loss from Postembryonic and Adult Stages

Mites, to survive, must be able to withstand considerable fluctuations of temperature, atmospheric humidity, and various other potential hinderances in their milieu which affect loss of water from the mite. Tetranychid mites regulate their water loss by virtue of a relatively impermeable cuticle and control of diffusion of water from the tracheal system by movements of the stylophore.

At the leaf surface—the usual environment of tetranychid mites—the relative humidity is usually high, and webbing produced by the mites has the further effect of maintaining the humidity immediately adjacent to the leaf surface. Boudreaux (1958) has shown a direct relationship of feeding rate and egg production to the relative humidity surrounding the food plant. From his work it can be assumed that the rate of water excretion governs the feeding rate, upon which egg production depends, and that the high water content of the food presents a problem in excretion rather than conservation.

McEnroe (1961*a*) indicated that the fluid intake of mites amounts to as much as 20 to 25 percent of the total weight per hour. With high humidity, therefore, large amounts of liquid excretory products must be eliminated when the mites are actively feeding. From his studies on water balance in the two-spotted spider mite, *Tetranychus (T.) urticae* Koch, McEnroe (1963) concluded that the female mite is able to pass the equivalent of 25 percent of its weight of water through the system at 30-minute intervals. McEnroe (1961*a*) showed by means of dyes that liquid could bypass the midgut and pass directly to the hindgut. This demands that the end of the esophagus be in contact with the entrance of the hindgut. Blauvelt (1945) showed that this is morphologically possible. The esophagus ends in a raised, funnel-shaped esophageal valve within the midgut which is larger in diameter than the opening to the hindgut. This esophageal valve could serve to make the necessary connection for a direct passage to the hindgut. Contraction of the dorsolongitudianal muscles would serve to lower the entrance to the hindgut down upon the esophageal valve. By this midgut shunt, particulate matter could enter the midgut and fluid could pass to the hindgut, thus relieving the digestive area of the burden of handling a large volume of water. Blauvelt (1945) stated that the food in a midgut appears as food balls plus recently acquired food material of chloroplasts and other material in a flocculent state, indicating ingestion of particulate food in the midgut. McEnroe (1963) pointed out the following advantages of the midgut shunt for mites that feed on cell sap: (1) Small molecules and ions in solution are not passed through the digestive area, but those required by the mite are directly adsorbed in the hindgut. (2) Digestion of a concentrated solid material can be conducted in the midgut for a longer time than would be possible if the shunt were not available. (3) Large volumes of water can be passed rapidly through the digestive system and excreted without requiring energy for selective adsorption and transport to the tracheal system for excretion. (4) The excretion of excess water as a fluid rather than as a vapor allows the mite to feed at high relative humidities.

As tetranychid mites do not live in a truly aquatic environment, the nonfeeding (diapausing) and molting forms must conserve water. These small animals have limited body capacities, and a relatively large external area in comparison to body volume. Consequently, these mites have had to adapt their mechanisms and chemistry to enable them to survive through growth changes and environmental fluctuations (Boudreaux, 1958; Lees, 1961; Winston and Nelson, 1965).

Blauvelt (1945) clearly showed the structural basis for the control of the openings of the peritremes by the positional adjustment of the stylophore, which in turn controls the area available for diffusion from the tracheal system. Oscillating movement of the stylophore, owing to its structural relationships with the peritremes (fig. 4, pl. 18), could also result in active ventilation of the tracheal system. This control of diffusion from the tracheal system is probably the most important variable factor in the mites' water balance at low relative humidities (McEnroe, 1961*a*).

McEnroe (1961*a*) suggests that the response of the stylophore to changes in humidity indicates the mites are sensitive to a vapor pressure deficit and are able to exercise a degree of control over the rate of diffusion from the trachea. This control of water vapor loss from the tracheal system is limited by the metabolic rate of the mite. Under severe water stress the mite can reduce its oxygen consumption and seal off its tracheal system for long periods of time.

According to McEnroe (1961*a*) mites treated with cyanide, chloroform, and silica gel showed differences in weight loss. There was a gradual and consistent weight loss by mites killed with cyanide which reflects diffusion from the tracheal system. The chloroform-treated mites showed more of a weight loss than did the cyanide-killed mites treated wth silica gel, although they survived, were unable to arrest their water loss, and under low humidity conditions their weight decreased rapidly. The chloroform and silica gel possibly changed the permeability of the lipid components in the cuticle, the presence of which was demonstrated by Gibbs and Morrison (1959). The cuticle normally serves as an effective evaporation barrier and to some extent as an agent in controlling water loss, unless temperatures exceed the critical range of 45 to 50 C (113 to 122 F).

A second type of external skin structure is the strial lobe on tetranychids (pls. 4, 5A, 5B), and the microtubercle on eriophyoid rings, also called ring granules. Skin striae are the fine, delicate tracery on red spider mites; on actively feeding individuals these striae are thickly studded with fine lobes in various shapes according to the species. These minute lobes may serve as evaporative structures providing water loss through the cuticle. Perhaps water can be absorbed from the air or excreted depending on the circumstances. Nonfeeding, diapausing tetranychid females lack these lobes over much or all of the body surface, and the absence of lobes indicates that since they are not taking in water, this absence of strial lobes may help them conserve body water. Such a change in skin anatomy may thus help resting females retain their water. Diapausing females may obtain water during hibernation by metabolism of accumulated body stores (Boudreaux, 1963).

Certain nitrogenous concretions have been found in the hindgut of *T.* (*T.*) *urticae*. McEnroe (1961*b*) determined the presence of guanine in mite excretory products and suggested that this excretory product may serve the nonfeeding diapause form and molting forms that are subject to water stress. Eriophyoid mites, which do not have well developed excretory systems, have what appears to be excretory products visible in their bodies when the gut tissues are stained and magnified (see chapter 12).

Behavioral patterns also contribute to survival of diapausing tetranychids, and to water conservation. These mites can, under certain circumstances, change their resting locations if harmful conditions develop after they have chosen a site. Many diapausing mites, however, after they have assumed the full torpor, are unable to revive until after a predetermined period and will perish if harmful changes take place.

Various specific examples show how active stages of these tetranychids differ with regard to suitable atmospheric humidity requirements, and under what conditions they can conserve water. The desert spider mite, *Tetranychus (T.) desertorum* Banks, succeeds better at relatively higher humidities than are necessary for the well-being of the carmine spider mite, *T. (T.) cinnabarinus* (Boisduval). In

California the strial lobes on the desert spider mite apparently allow water to escape so rapidly that mortality results when there is a considerable drop in humidity. The tumid spider mite, *T. (T.) tumidus* Banks, behaves as if its strial lobes permit it to live only within a relatively narrow humidity range (Iglinsky and Rainwater, 1954; Nickle, 1960).

The tumid spider mite has relatively large strial lobes that cover more of the ventral body surface than do the lobes on the desert spider mite. These larger lobes, distributed over a greater ventral surface area on the tumid mite, allow easier water escape, and this logically accounts in part for the narrow humidity tolerance shown by this mite.

Satisfactory water retention by diapausing female tetranychids, which do not feed, is attained in part by complete absence of strial lobes. The two-spotted spider mite, *T. (T.) urticae*, entirely lacks strial lobes when diapausing; this species is an illustration of water-impervious skin anatomy (Boudreaux, 1958).

Some mites that have thin interscutal membranes but no skin striae, such as the tarsonemids, move down into inner recesses on their host plant to avoid adverse effects of dry periods.

Certain nontetranychid mites protect immature stages from external influences by retaining them within the female body until they are adult or nearly so. The red-legged earth mite, *Halotydaeus destructor* (Tucker), of the Penthaleidae, is one of these: the female is able to move immediately on emergence to protected locations (Wright, 1961).

Certain pyemotid mites have females that attach themselves to their food source, in some cases an insect, after which the abdomen balloons into a globular brood chamber. Within this brood chamber the young are protected from outside influences and do not emerge until mature. They may even mate within this chamber. One such pyemotid is the cereal mite, *Siteroptes cerealium* (Kirchner).

Eriophyoid mites that live on deciduous plants in the northern hemisphere also have elaborate body adjustments that help them survive unfavorable periods. Their ability to retain water is comparable to that of the tetranychids. While eriophyoids lay no resistant eggs after the manner of some tetranychid species, they have a greater variety of epidermal and body adjustments to aestivation and hibernation than do the larger mites. The tiny eriophyoids are from one third to one fifth the size of the larger mites; thus, they can find their way into niches and crevices that are too small for the tetranychids. Therefore these smaller mites can sequester themselves close to places on plants where they can resume feeding upon return of favorable conditions. Tetranychids that hibernate near feeding sites remain largely exposed.

Water Loss from Mite Eggs

Owing to the small size of mite eggs, the problem of satisfying oxygen requirements without losing too much water is at an extreme. Because of these demands the tetranychid egg has evolved a peculiar respiratory system. According to Dittrich (1971), the maturing egg has beneath the transparent shell an air-filled duct system with two branches. The ducts are kept open with the aid of a multitude of "mircopillars" that expand between the shell and the underlying intermediate lamella providing an air space (0.1 to 0. 45 μ) between the shell and lamella. The micropillars arise on the lamella in the course of development of the embryo. The intermediate lamella is composed of laminae that frequently show a substructure of arched fibrils. Because of morphological adaptations, the micropillars, and the perforation organs, the intermediate lamella is a most important mediator of embryonic gas exchange.

Two perforation organs penetrate the eggshell, causing air to enter the system. They probably originate within the embryo and rise to the embryo's surface, where they fuse with the intermediate lamella, possibly initiating the differentiation of its superficial zones of micropillars. After perforation, the shell admits air and the perforation organs become part of the air duct system. Perforation is effected by a slender cone approximately 5 μ long and 1.8 μ wide at its base. The perforatory organ has an internal cavity that is closed toward the embryo but continuous with fine ducts leading to the tip of the cone, making possible the conduct of respiratory gases from the orifices of the shell through their chamber into the connecting canal and the air duct system.

These specialized adaptations in the mites and their eggs illustrate the importance of mechanisms for regulating body moisture in these small animals. Without such special adaptations or habits, mites would be severely limited in their distribution.

AGRICULTURAL PRACTICES THAT INFLUENCE MITE ECOLOGY

The influence of some tetranychid mite populations in agriculture has greatly increased in importance in the past three decades (1940–1970). Species particularly include the European red and citrus red mites, *Panonychus ulmi* and *P. citri*, and several species in the genus *Tetranychus*, such as *urticae*, *cinnabarinus*, *turkestani*, *pacificus*, *tumidus*, *desertorum*, *mcdanieli*, and *viennensis*. The influence of populations of most of the mite pests belonging to the other genera of tetranychid and other mite families have not similarly increased during this period. The many changes in agricultural practices may contribute to or cause the increase or decrease in the mite population levels in a local or regional area. The agriculturist should be aware that the relationships of organisms to each other are constantly changing. In a near constant environment these oscillations tend to vary around a mean, like a balance scale without friction loss but, except for drastic changes like those caused by fire or by man, environmental changes beyond normal oscillations are gradual and therefore not usually evident within two or three decades.

Man, however, has drastically changed the plant environment in agricultural and in some forest areas. Many changes are obvious, but many are not: the equivalent changes taking place among the arthropod complex, that is, the parasites, predators, and competitors. It is possible at this time to suggest some of the environmental changes that influence mite populations; but no attempt to evaluate their importance is intended. Such an evaluation, it appears, is possible only on a specific basis and then only after much more information has been acquired.

New Varieties and Bud Selections

Gradual changes in plantings of cultivars may alter the intensity of mite infestations. For example, new citrus varieties and bud selections have to some extent replaced old plant stocks; but little information is available concerning their relative favorability as mite hosts. In California, Troyer seedlings have recently been used as rootstocks for oranges. Troyer seedlings in citrus nurseries usually become infested with citrus red mite before other citrus varieties in the nursery. Populations of citrus red mite seem to be more prevalent and difficult to manage in young orange orchards planted with the Troyer rootstock than in orchards planted with the same variety of buds on other rootstocks. The infestation; however, may be the result of more vigorous growth produced the first few years by trees budded on Troyer roots.

Absence of Predators

Mites that accompany a crop grown in a new area may become a major pest because of a lack of predators. Some of the enemies and competitors are left behind, and those transferred with their host may not be able to survive the new habitat. Recognition of this has been the basis of extensive explorations by entomologists —and more recently by acarologists—into the native areas of the host plants, or to areas where the pest is effectively regulated by natural means.

Plant Vigor

Crop culture designed to produce increased plant vigor and production has changed during the past three decades. The more effective reduction of weeds by mechanical cultivation or by chemicals has not only eliminated plant competition and resulted in vigorous plant growth, but it has often drastically reduced the habitat for the general predators. The extensive new growth has often provided a very suitable food supply, particularly for mites that increase most rapidly during growth cycles, such as the European and citrus red mites. Extensive studies have indicated that the populations of European red mite on apples are favored by cultivation, fertilization, and pruning (Post, 1962).

Monoculture

Planting an area with a single crop provides extensive food supplies for a mite pest and limits the reservoirs of mite enemies and competitors. Monoculture predisposes the rapid development of pest populations over extensive areas before the regulating agencies can effectively operate. Large acreage plantings, therefore, increase the difficulties of employing pest management by biological agents and at the same time provide more convenient situations for the application of chemicals.

Pesticides

Chemicals applied to crops for pest and disease control, for minor element deficiencies, for fruit setting or thinning, or for defoliation may sometimes provide more favorable conditions for the development of some phytophagous mites (Eichmeier and Guyer, 1960). Of course, broad spectrum insecticides are very destructive to the predators of phytophagous mites. These factors no doubt have been major contributors in the general world increase of certain tetranychid mites on fruit trees, cotton, and other plants where such pesticides have been regularly applied (Clancy and Pollard, 1952; Klostermeyer and Rasmussen, 1953; Lochar, 1958; Tew and Groves, 1957).

Applications of chemicals to the foliage may also produce physiological effects on or in the plants that favor the development of plant feeding mites. There is ample experimental evidence to indicate that DDT applications may result in population increases of some tetranychid mite species in the absence of natural enemies (Baleviski, 1960; Klostermeyer and Rasmussen, 1953; Post, 1962; Rodriguez, Chen, and Smith, 1957). The physiological changes brought about by such applications in the plant or in the mites have not been established. Some studies have indicated that increased fecundity is induced in the mites by DDT, whereas other studies suggest that DDT and other insecticides cause a change in the plants making them more favorable for mite development, such as changing the K: Ca ratio. The effects of chemical applications on mite populations thus may result from a combination of factors, or they may differ according to mite species and host plant.

Inert Dusts

It has frequently been observed that crops grown near dusty roads have been the first to become heavily infested with tetranychid mites. These "inert" dusts may be lethal to predaceous insects or mites either directly or indirectly. They may cause death through dessication directly by increasing the permeability of the cuticle and thus increasing evaporation from the body, and indirectly by deterring the parasite or predator from finding or reaching its prey. Dusts may aid the development of tetranychid mite populations on some smooth plant surfaces by providing bases for attaching webs. In the laboratory, dusts and kapok have thus aided in the rearing of mites on citrus (Bartlett, 1951; Fleschner, 1958).

Soil Organisms

It is known that changes in the soil flora and fauna produced by soil fumigation influences plant growth and vigor, yet little is known concerning the effect of such changes on the development of populations of plant-feeding mites. Several common soil organisms may, under proper cultural conditions, produce chemicals that adversely influence the reproductive capacity of tetranychid mites, notably *Streptomyces griesus* (Kyansky) Waksman and Henrici. Knowledge of the amounts of these toxins or antibiotics produced by such organisms and their influence on mite development may aid in understanding certain mite population increases (Harries, 1963; Jeppson, Jesser, and Complin, 1966).

SELECTED BIBLIOGRAPHY

ANDERSON, N. H., and C. V. G. MORGAN. 1958. Life histories and habits of the clover mite, *Bryobia practiosa* Koch, and the brown mite, *B. arborea* M. & A. in British Columbia. Can. Entomol. 90(1):23–42.

BAKER, E. W., and G. W. WHARTON. 1952. An introduction to Acarology. The Macmillan Co., New York. 465 pp.

BALEVISKI, A. 1960. Experiments to determine the effect of DDT on fecundity of the brown apple mite (*Bryobia redikorzevi* Rech.) Past. Zasht. 8 (6) : 59–64.

BAPTIST, B. A., and D. J. W. RANAWERRA. 1955. The scarlet mites of the genus *Brevipalpus* as pests of tea in Ceylon. Tea Quarterly 26(4):127–136.

BARTLETT, B. R. 1951. The action of certain "inert" dust materials on parasitic Hymenoptera. J. Econ. Entomol. 46(6):891–896.

BEAMENT, J. W. L. 1951. The structure and formation of the egg of the fruit tree spider mite, *Metatetranychus ulmi* Koch. Ann. Appl. Biol. 38(1):1–24.

BLAUVELT, W. E. 1945. The internal morphology of the common red spider mite. *Tetranychus telarius* L. Cornell Univ. Agr. Expt. Sta. Mem. 270:1–11.

BONDARENKO, N. V. 1950. The influence of short days on the yearly cycle of development of the spider mite *Tetranychus urticae* Koch. Doklady Akad. Nauk. U.S.S.R., 70:1077–1080.

———. 1958*a*. Characteristics of the diapause in *Tetranychus urticae* Koch. [In Russian] Zool. Zhur. 37:1012–1023. (From Rev. Appl. Entomol. Ser. A. [1960] 48:395).

———. 1958*b*. Peculiarities in the origin of diapause in different geographical populations of *Tetranychus telarius*. Dokl. Akad. Nauk. USSR. 119(6):1247–1250. (From Rev. Appl. Entomol. Ser. A. [1959] 47:230.

BOUDREAUX, H. B. 1958. The effect of relative humidity on egg-laying, hatching, and survival in various spider mites. J. Insect Physiol. 2(1):65–72.

———. 1963. Biological aspects of some phytophagous mites. Ann. Rev. Entomol. 8:137–154.

BOULANGER, L. W. 1958. The effect of European red mite feeding injury on certain metabolic activities of red delicious apples. Maine Agr. Expt. Sta. Bull. 570:1–32.

BRAVENBOER, L. 1959. De Chemische en Biologische Bestrijding van de Spintmijt *Tetranychus urticae* Koch. Landbouwpublikaties (Pudoc) Landbouwdocumentatie Wageningen.

BREUKEL, L. M., and A. POST. 1959. The influence of the manurial treatment of orchards on the population density of *Metetranychus ulmi* (Koch). Entomol. Expt. and Appl. 2(1):38–47.

BROOKING, B. C. 1957. A study of the evocation and termination of diapause in *Petrobia apicalis* (Banks). Master's thesis, La. State Univ., Baton Rouge.

CLANCY, D. W., and H. N. POLLARD. 1952. The effect of DDT on mite and predator populations in apple orchards. J. Econ. Entomol. 45(1):108–114.

DABROWSKI, Z. T., J. G. RODRIGUEZ, and C. E. CHAPLIN. 1971. Studies in the resistance of strawberries to mites. IV. Effect on season on preference or nonpreference of strawberries to *Tetranychus urticae*. J. Econ. Entomol. 64(4):806–809.

DAS, G. M. 1959. Bionomics of the tea red spider, *Oligonychus coffeae* (Nietner). Bull. Entomol. Res. 50(2):265–274.

DAVIS, D. W. 1952. Influence of population density on *Tetranychus multisetis*. J. Econ. Entomol. 45(4):652–654.

DICKSON, R. C. 1949. Factors governing the induction of diapause in the Oriental fruit moth. Ann. Entomol. Soc. Am. 42(4):511–537.

DILLON, L. S. 1958. Reproductive isolation among certain spider mites of the *Tetranychus* complex, with preliminary notes. Ann. Entomol Soc. Am. 51(5):441–448.

DITTRICH, V. 1971. Electron-microscopic studies of the respiratory mechanism of spider mite eggs. Ann. Entomol. Soc. Am. 64(5):1134–1143.

DOSSE, G. 1964. Boebachtungen über Entwicklungstandenzen in *T. urticae-cinnabarinus* complex. Pfanzenschutzberichte, 31:113–128.

EBELING, W. 1934. Observations on a method of dissemination employed by mites. Pan-Pacific Entomol. 10(2):89.

EICHMEIER, J., and G. GUYER. 1960. An elevation of the rate of reproduction of the two-spotted mite reared on gibberellin-treated bean plants. J. Econ. Entomol. 53(4):661–664.

FLESCHNER, C. A. 1952. Host plant resistance, as a factor influencing population density of citrus red mites on orchard trees. J. Econ. Entomol. 45(4):687–695.

———. 1958. The effect of orchard dust on the biological control of avocado pests. Calif. Avocado Soc. Ybk. 42:94–98. Los Angeles, Calif. 4438 Everett Ave.

FOOTT, W. W. 1964. Geotactic response of the two-spotted spider mite, *Tetranychus urticae* Koch. Proc. Entomol. Soc. Ont. 95:106 108.

FRITZSCHE, R. 1960. Morphologische, Biologische, und Physiologische Variabilität und ihre Bedeuting für de Epidemiologie und Bekämpfing von *Tetranychus urticae* Koch. Biol. Zentralbl. 79(5):521–576.

GEYSPITZ, K. F. 1968. Genetic aspects of variation in photoperiodic adaptations. [In Russian]. *In* A. S. DANILEVISKA, chap. 2, pp. 52–79. Leningrad State Univ., Leningrad. 269 pp.

GIBBS, K. E., and F. O. MORRISON. 1959. The cuticle of the two-spotted mite, *Tetranychus telarius*. Can. J. Zool. 37(5):633–637.

GLANCEY,B. M. 1958. Studies of embryonic diapause in the legume mite, *Petrobia apicalis* (Banks). Master's thesis, La. State Univ., Baton Rouge.

HARRIES, F. H. 1963. Effects of some antibiotics and other compounds on fertility and mortality of orchard mites. J. Econ. Entomol. 56(4):438–441.

HARRISON, R. A., and A. G. SMITH. 1961. The influence of temperature and relative humidity on the development of eggs and on the effectiveness of ovicides against *Tetranychus telarius* (L.). New Zealand J. Sci. 4:540–549.

HARVEY, W. R. 1962. Metabolic aspects of insect diapause. Ann. Rev. Entomol. 7:57–80.

HELLE, W. 1962. Genetics of resistance to organophosphorus compounds and its relation to diapause in *Tetranychus urticae* Koch. Tydschr. Pl. Ziekt. 68:155–195.

———. 1965. Population genetics of arrhenotokous mites. Boll. Zool. Agric. Bachicoltura. Ser. II, 7:219–225.

———. 1968. Genetic variability of photoperiodic response in arrhenotokous mite *(Tetranychus urticae)*. Entomol. Expt. and Appl. 11(1):101–113.

HENDERSON, C. F., and H. V. MCBURNIE. 1943. Sampling techniques for determining populations of the citrus red mite and its predators. U. S. Dept. Agr. Circ. 671.

HERNE, D. H. C. 1968. Some responses of European red mite, *Panonychus ulmi* (Koch) to immersion in water. Can. Entomol. 100(5):540–541.

HOBZA, R. F. 1970. Interrelationships of temperature and humidity on development of citrus rust mite, *Phyllocoptruta oleivora* (Ashm.) Master's thesis, University of California, Riverside.

HUECK, H. J. 1951. Influence of light upon the hatching of winter eggs of the fruit tree red spider. Nature 167:933–944.

HUFFACKER, C. B., M. VAN DE VRIE, and J. S. MCMURTRY. 1969. The ecology of tetranychid mites. Ann. Rev. Entomol. 14:125–173.

HUKUSIMA, S. 1958. The effect of varying nitrogen levels upon the arthropod fauna in young apple trees, with special reference to the increase of the mite and aphid populations. Bull. Fac. Agr. Hirosaki Univ. Japan 4:72–79.

HUSSEY, N. W., and W. J. PARR. 1958. A genetic study of the color forms found in populations of the greenhouse red spider mite, *Tetranychus urticae* Koch. Ann. Appl. Biol. 46(2):216–220.

———. 1963. The effect of glasshouse red spider mite (*Tetranychus urticae* Koch) on the yield of cucumbers. J. Hort. Sci. 38:255–263.

IGLINSKY, W., Jr., and C. F. RAINWATER. 1954. Life history and habits of *Tetranychus desertorum* and *bimaculatus* on cotton. J. Econ. Entomol. 47(6):1084–1086.

JEPPSON, L. R., J. O. COMPLIN, and M. J. JESSER. 1962. Effects of application programs on citrus red mite control and development of resistance to acaricides. J. Econ. Entomol. 55(1):17–22.

———. 1954. Seasonal weather influences on efficiency of Systox applications for control of mites on lemons in southern California. J. Econ. Entomol. 47(3):520–525.

———. 1966. Cycloheximide derivatives and mite control, with special reference to mites on citrus. J. Econ. Entomol. 59(1):15–19.

KLOSTERMEYER, E. C., and W. B. RASMUSSEN. 1953. The effect of soil insecticide treatment on mite population and damage. J. Econ. Entomol. 46(5):910–912.

LEES, A. D. 1950. Diapause and photoperiodism in the fruit tree red spider mite, *Metatetranychus ulmi* (Koch). Nature 166:874–875.

———. 1953*a*. Environmental factors controling the evocation and termination of diapause in the fruit tree red spider mite, *Metatetranychus ulmi* (Koch). Ann. Appl. Biol. 40(3): 449–486.

———. 1953*b*. The significance of the light and dark phases in the photoperiodic control of diapause in *Metatetranychus ulmi* (Koch). Ann. Appl. Biol. 40 (3) :487–497.

———. 1955. The physiology of diapause in arthropods. Cambridge Univ. Press, London.

———. 1961. On the structure of the eggshell in the mite *Petrobia latens* Müller. J. Insect Physiol. 6(2):146–151.

LEROUX, E. J. 1954. Effects of various levels of nitrogen, phosphorus, and potassium in nutrient solution on the fecundity of the two-spotted mite reared on cucumber. Can. J. Agr. Sci. 34(2):145–151.

LIENK, S. E., and P. J. CHAPMAN. 1958. Influence of presence or absence of the European red mite on two-spotted mite abundance. J. Econ. Entomol. 44(5):623.

LINKE, W. 1953. Investigation of the biology and epidemiology of the common spider mite, *Tetraychus altheae* von Hanst with particular consideration to hop as the host. Höfchenbriefe 4(4):181–232.

LOCHAR, F. J. 1958. Der enfluss Von dichlordiphenyl trichlormethyl methan (DDT) auf einigen Tetranychiden. Z. Angew. Zool. 45(2):201–248.

LORD, F. T., and D. K. R. STEWART. 1961. Effects of increasing the nitrogen level of apple leaves on mite and predator populations. Can. Entomol. 93 (10) : 924–927.

MCENROE, W. D. 1961*a*. The control of water loss by the two-spotted spider mite *Tetranychus telarius* (L.) Ann. Entomol. Soc. Am. 54(6):883–887.

———. 1961*b*. Guanine excretion by the two-spotted spide mite, *Tetranychus telarius*. Ann. Entomol. Soc. Am. 54(6):925–926.

———. 1963. The role of the digestive system in the water balance of the two-spotted spider mite. Advances in Acarology 1:225–231. Cornell Univ. Press, Ithaca, N.Y.

———. 1969. Spreading and inbreeding in the spider mite. J. Heredity 60 (6) :343–345.

MCENROE, W. D., and K. DRONKA. 1971. Photobehavioral classes of the spider mite, *Tetranychus urticae*. Entomol. Expt. and Appl. 14(4)419–424.

MCMURTRY, J. A., and H. G. JOHNSON. 1966. An ecological study of the spider mite, *Oligonychus punicae* (Hirst) and its natural enemies. Hilgardia 37(11):363–402.

MACPHEE, A. W. 1961. Mortality of winter eggs of the European red mite *Panonychus ulmi* (Koch) at low temperature, and its ecological significance. Can. J. Zool. 39(3): 299–343.

MORGAN, C. V. G., D. A. CHANT, N. H. ANDERSON, and G. L. AYRE. 1955. Methods of estimating orchard mite populations, especially with the mite brushing machine. Can. Entomol. 87(5):189–200.

MORI, H. 1961. Comparative studies on thermal reaction in four species of spider mites. J. Fac. Agr. Hokkaido Univ. 51 :574–591. Sapporo, Japan.

———. 1962*a*. Seasonal difference of phototactic response in three species of spider mites.

J. Fac. Agr. Hokkaido Univ. 52 :1–9. Sapporo, Japan.

———. 1962*b*. The effects of photo-stimulus on the thermal reaction of four species of spider mites. J. Fac. Agr. Hokkaido Univ. 52 :10–19. Sapporo, Japan.

Morris, O. N. 1961. The development of the clover mite *Bryobia praetiosa* Koch in relation to the nitrogen phosphorus and potassium nutrition of its host plant. Ann. Entomol. Soc. Amer. 54(4):551–575.

Naegele, J. A., W. D. McEnroe, and A. B. Soans 1965. Spectral sensitivity and orientation in the two-spotted spider mite, *Tetranychus urticae* K. from 350 mμ to 700 mμ. J. Insect Physiol. 12(9):1187–1195.

Nickel, J. L. 1960. Temperature and humidity relationships of *Tetranychus desertorum* Banks with special reference to distribution. Hilgardia 30(2):41–100.

Nuber, K. 1961. Zur Frage der Uberiwinterung der Gemeinen Spinnmilbe *Tetranychus urticae* Koch in Hopfenbau. Höfchenbriefe 14(11):6–15.

Osakabe, M. 1965. Seasonal fluctuations of population density of the tea red spider mite, *Tetranychus kanzawai* Kishida, in the tea plantation. Jap. J. Appl. Entomol. Zool. 9(3):206–210.

Parr, W. J., and N. W. Hussey. 1960. Further studies on the reproductive isolation of geographical strains in the *Tetranychus telarius* complex. Entomol. Expt. and Appl. 3(2):137–141.

———. 1966. Diapause in the glasshouse red spider mite *Tetranychus urticae* Koch. A synthesis of present knowledge. Hort. Res. 6(1):1–21.

Pielou, D. P. 1960. Contagious distribution in the European red mite, *Panonychus ulmi* (Koch), and a method of grading population densities from a count of mite-free leaves. Can. J. Zool. 38(3):645–653.

Post, A. 1962. Effect of cultural measures on the population density of the fruit tree red spider mite, *Metatetranychus ulmi* Koch. Tydschr. Pl. Zeikt. 68(1):1–110.

Pritchard, A. E., and E. W. Baker. 1951. The false spider mites of California. Univ. Calif. Pub. Entomol. 9(1):1–93.

———. 1952. A guide to the spider mites of deciduous fruit trees. Hilgardia 21(9):253–287.

———. 1958. The false spider mites. Univ. Calif. Pub. Entomol. 14(3):175–274.

Razumova, A. P. 1967. Variability of the photoperiodic reaction in a number of successive generations of spider mites. Entomol. Rev. (Entomologicheskoe Obozrenie) 46(2): 159–169.

Reiff, M. 1949. Physiologische Merkmale bei Spinnmilben nach Veränderung des Blattstoff wechsels Verh. Schweiz Naturf. Ges. 165–166.

Rodriguez, J. G. 1951. Mineral nutrition of the two-spotted spider mite, *Tetranychus bimaculatus* Harvey. Ann. Entomol. Soc. Am. 44(4):511–526.

———. 1958. The comparative NPK nutrition of *Panonychus ulmi* (Koch) and *Tetranychus telarius* (L.) on apple trees. J. Econ. Entomol. 51(3):369–373.

———. 1960. Nutrition of the host and reaction to pests. Biological and chemical control of plant and animal pests. Am. Assoc. Adv. Sci. Washington, D. C.: 149–163.

Rodriguez, J. G., H. H. Chen, and W. T. Smith, Jr. 1957. Effects of soil insecticides on beans, soybeans, and cotton, and resulting effect on mite populations. J. Econ. Entomol. 50(5):587–593.

Rodriguez, J. G., and L. D. Rodriguez. 1952. The relation between minerals, B-complex vitamins, and mite populations in tomato foliage. Ann. Entomol. Soc. Am. 45(2):331–338.

Shinakji, N. 1959. Experimental studies of the temperature and relative humidity influencing the development of the citrus red mite *Panonychus citri* (McGregor): II. On the influence of temperature and relative humidity upon the postembryonic development, adult life span, and egg-laying potency of the citrus red mite. [In Japanese with English summary]. Bull. Hort. Div. Tokai-Kinki Agr. Expt. Sta. 5:129–142.

Smith, F. F. 1957. A resume of spider mite control in greenhouses. Florist Exchange

128(18):10, 36–40.

Solomon, M. E. 1937. Experiments on the effects of temperature and humidity on the survival of *Halotydeus destructor*. Aust. J. Expt. Biol. Med. Sci. 15(1):1–16.

Sternlicht, M. 1966. Trials in the control of the citrus bud mite, *Aceria sheldoni* Ewing. Israel J. Agr. Res. 16(3):115–124.

Storms, J. J. H. 1969. Observations on the relationship between mineral nutrition of apple rootstocks in gravel culture and the reproduction rate of *Tetranychus urticae*. Entomol. Exp. and Appl. 12(3):297–311.

Summers, F. M., and G. A. Baker. 1952. A procedure for determining relative densities of brown almond mite populations on almond trees. Hilgardia 21(13):369–382.

Suski, Z. W., and J. A. Naegele. 1963. Light responses in the two-spotted mite. Advances in Acarology 1:435–444, 445–453.

———. 1968. Environmental determinants of white light response in the two-spotted spider mite, *Tetranychus urticae* K. Mass. Expt. Sta. Bull. 571:1–43.

Tew, R. P., and J. R. Groves. 1957. Some observations on the effects of formulations on the persistence of DDT and its toxicity to certain orchard mites and insects. 44th Rept. East Malling Res. Sta. 1955–1956:152–160.

Tibilova, A. A. 1932. Biology of the red spider mite in the Tashkent District. Byult mauchnoisslet Inst. Khlopkov, 1. Tashkent.

Vasser, R. E. 1938. L'influence de la témperature et de l'humidite de l'air sur le development du tetranyche due coton. Plant Protect. USSR. 17:39–51.

Watson, T. F. 1964. Influence of host plant condition on population increase of *Tetranychus telarius* (L.). Hilgardia 35(11):273–322.

Whitcomb, W. D. 1932. Influence of temperature on development of red spider. Bull. Mass. Agr. Expt. Sta. Ann. Rept. 293:34.

Winston, P. W., and V. E. Nelson. 1965. Regulation of transpiration in the clover mite, *Bryobia praetiosa* Koch. J. Expt. Biol. 43(2):257–269.

Wright, W. E. 1961. Red-legged earth mites. Agr. Gaz. N. S. Wales 72(4):213–215.

Chapter 3

History of Chemical Control and Mite Resistance to Acaricides

Chemical control of mites has had a dynamic history, especially during the past two decades. These changes have been so closely correlated wth acaricide resistance that the two are discussed together.

ACARICIDES USED FOR CONTROL OF AGRICULTURAL MITES

Inorganic Acaricides

Sulfur

Sulfur was essentially the only acaricide used to control mites until about 1920. Sulfur is still extensively employed to control many mite species, as certain mite groups seem to be more susceptible to sulfur than others. Most of the mites belonging to the families Eriophyoidae and Tenuipalpidae (false spider mites Brevipalpus or flat mites) are susceptible to sulfur, lime-sulfur sprays, or sulfur dusts. Certain tenuipalpid species are more readily controlled by some of the newer acaricides than by sulfur. The majority of the tetranychid mites (red spider mites) belonging to the genera *Oligonychus, Eotetranychus,* and *Eutetranychus,* and tarsonemid mites of the genus *Polyphagotarsonemus* are usually most economically controlled by sulfur applications.

Most of the spider mite species belonging to the genera *Tetranychus* and *Panonychus* are not readily controlled by sulfur applications. Two notable exceptions are the Atltantic spider mite *Tetranychus* (*T.*) *turkestani* Ugarov and Nikolski (= *T. atlanticus* McGregor), and the desert spider mite *T.* (*T.*) *desertorum* Banks. Some tarsonemid mites as well as mite pests in the families Pyemotidae and Penthaleidae do not respond to sulfur. Because this element was used at one time for control of the *Tetranychus* and *Panonychus* mites, important questions have remained unanswered, that is, did these mites develop resistance to sulfur, was an inferior degree of control tolerated, or was the control obtained adequate because biological and cultural factors regulated populations more effectively at that time. These questions will probably never be answered: mite strains not previously exposed to sulfur are not available, demands for better quality and quantity of produce have changed agricultural practices, including pest control.

Sulfur has many advantages for use as an acaricide. It is not toxic to the applicator or to those who consume crops on which it is applied. After application it releases vapors that are toxic to many mite species. Bud mites (Eriophyidae), whose habitat is confined to surfaces between bud bracts or other places protected from direct spray or dust applications, may succumb to the vapor. Sulfur is also relatively inexpensive, is available in most areas, and may be formulated with many other pesticides. It is relatively specific and therefore does not interfere with the development of most predators and parasites of plant feeding pests, that is, it has a minimum influence on the biological complex.

Sulfur is relatively nontoxic to man, although it produces eye irritation, and is uncomfortable to apply. It is phytotoxic to many plant species when applications are made during hot weather, yet ineffective when applied during cool weather. Plant injury occurs when treatments of sulfur precede or follow applications of petroleum oil. Sulfur may also taint such crops as tea; when treated produce is canned, the inside of tin-coated cans may become tarnished.

Petroleum Oils

Tar oil dormant spray applications came into use as a means of killing aphid eggs, the eggs of European red mite, *Panonychus ulmi* (Koch), and *Bryobia* spp., in about 1920. As the eggs of these pests are deposited on the twigs, injurious spring and summer infestations were mitigated by a reduction in overwintering egg populations. The tar oils were followed by the dormant oils; then in the 1930s, dinitro compounds were incorporated in the oils for more effective egg mortalities. More recently other acaricides, particularly those toxic to mite eggs, have been added to dormant oils.

Additional refinements in the distillation of petroleum oils and the discovery that the phytotoxic properties of the oils could be negated by treatment with sulfuric acid or other processes permitted the use of oils on plant foliáge and initiated the use of the well-known summer oils. These oils, with improvements and modifications, are now effectively and extensively used for control of mites on many trees and shrubs. The incorporation of the proper balance of emulsifying and wetting agents to provide wetting of foliage and pests, without the deposition of unnecessary and injurious quantities of oil, has extended the use of these oils to other crops. These supplements have also permitted the use of oils during seasons or periods where oil applications had previously caused injury.

Recent studies have led to the development of more effective and even less phytotoxic spray oils than those used in the 1930s. The high molecular weight oils were found to produce certain adverse tree effects and the low molecular weight oils were noneffective as ovicides. Improvement in oil distillation techniques has provided oil fractions with a narrow range of molecular weight. Also, effective oil fractions that cause minimum adverse tree response may be selected for a particular mite species, host plant, or even season of the year. For example, a petroleum oil having a distillation midpoint of 415 C (A.S.T.M. method D-1160) at 10 mm pressure and a 10–90 degree range of not over 60 C, appears to be just as effective as the wide distillation range or high molecular weight oils, yet produces little adverse effect on orange trees in the form of decreased fruit set the spring following the application of the oil.

Summer oils have been applied extensively to deciduous and citrus fruit trees as well as to ornamental plants for control of *Panonychus* spp., *Bryobia* spp., *Tetranychus* (*T.*) *urticae*, *T.* (*A.*) *viennensis* (tetranychids) and some bud mites (eriophyoids). Oils have not been particularly effective in the control of bud mites deep in bud tissue, or tetranychid species that feed in colonies on the undersides of leaves, such as the six-spotted mite on citrus. These mites can be effectively controlled on leaves of fruit trees with petroleum oil sprays only by spray coverage sufficient to contact both the mites and their eggs. This necessitates the use of spray applications that distribute the oil to all plant surfaces inhabited by the mites. Such applications are usually expensive in time, manpower, or equipment. The effectiveness of oils against mites as well as important insect pests, such as the scale insects, however, and the inability of mite populations to develop resistance to the oils, has resulted in the extensive usage of oils over a long period of time in spite of the rapid development of new acaricides.

SYNTHETIC ORGANIC ACARICIDES[1]

The Dinitrophenyl Acaricides

The development of the dinitrophenol (DN) type acaricides, which occurred during the 1930s, resulted in the first of the organic acaricides (Boyce et al., 1939). As indicated, above, dinitro compounds were incorporated in dormant oil sprays, but related analogues became extensively used as summer sprays for a wide variety of important mite species. DNOCHP, known by growers as DN-dry mix No. 1, was used in Florida for more than 10 years for control of the citrus red mite. Citrus in California and some deciduous fruits were injured by DNOCHP, but the dicyclohexalamine salt, known in the trade as DN-111, was extensively used for red spider mite control on citrus and deciduous fruit trees and ornamental plants until the late 1940s.

The DN compounds did not replace sulfur for the control of most mite species that were susceptible to sulfur. These species include the bud and rust mites (Eriophyidae) and most mites belonging to the genera *Eotetranychus*, *Eutetranychus*, *Oligonychus*, or *Polyphagotarsonemus*. The DN acaricides were used for mite control on citrus fruits during periods of the year when oil applications produced adverse tree effects, for summer control of mite species infesting deciduous fruit trees, and for control of infested vegetable and field crops that were not controlled by sulfur.

The DN acaricides possessed several disadvantages. Like sulfur, they produced phytotoxic effects when applied during hot weather and were relatively ineffective during cool weather. Also, plant injury occurred when DN acaricides were applied near petroleum oil applications. Residue persistence was short, thus ineffective in killing mites that hatched from eggs missed by the application; therefore, repeated treatments were often required.

Appearance of Synthetic Acaricides

The surge of new acaricide development began in the late 1940s. Table 1 includes a list of chemicals evaluated as agricultural acaricides during the period from 1940 to 1970, together with the year of their first appearance for field evaluation.

TABLE 1

Introduction of Chemical Acaricides

Acaricide	*Approximate date of introduction as a potential acaricide*	*Acaricide*	*Approximate date of introduction as a potential acaricide*
Neotran®	1945	chlorbenside	1953
DMC	1946	Fenson®	1953
Schradan®	1946	dioxathion	1954
parathion	1947	tetradifon	1954
ovex	1949	carbophenothion	1955
EPN	1950	ethion	1955
Aramite®	1950	binapacryl	1960
dinocap	1950	Morestan®	1960
Sulphenone®	1952	Omite®	1960
demeton	1952	Pentac®	1960
chlorobenzilate	1952	Plictran®	1969
dicofol	1952	Galecron®	1969

Chlorinated Hydrocarbons and Related Acaricides

During the late 1940s, chlorinated hydrocarbon compounds appeared as potential acaricides. DCPM (Neotran®) was found to be more effective than DN-111 in the control of citrus red mite, *Panonychus citri* (McGregor), in California and safe to apply in hot weather or near oil applications. It was, however, least effective during periods of hot weather. Neotran was not extensively used on other mite pests or on citrus outside of California. In 1949 and 1950, several chlorinated hydrocarbon and other compounds were evaluated as acaricides. Among these were ovex or chlorofenson (Ovotran®), Aramite®, dinocap (Karathane®) and DMC or DCPC (Dimite®).

Ovex and Aramite have been used extensively throughout the world for control of tetranychid mites infesting vegetable and field crops; and fruit, ornamental and shade trees. DMC has had relatively limited use. Dinocap, a DN compound, has the same limitations as DN-111 except it has the combined advantage of being less phytotoxic and is an effective fungicide. Dinocap has been used extensively where fungus diseases and mite infestations occur concurrently, especially when repeated applications were necessary to control diseases or other pests (Armstrong, 1953; Ascher and Cwilich, 1960; 1962; Bouron and Mimaud, 1956; Harries, 1960; Hofmaster and Greenwood, 1953; Isely, 1941; Kirby and Tew, 1952; Madsen and Borden, 1955; Newcomer and Dean, 1953; Parent, 1963; Taylor, 1952; Taylor, Henneberry, and Smith, 1959).

The extensive use of ovex and Aramite was curtailed because populations that had received repeated treatments developed resistance to the acaricides (Smith, 1960); mites treated infrequently, however, or populations in areas where these acaricides have not been extensively used may still respond to treatment. There has been no evidence that citrus red mite or European red mite, *Panonychus ulmi* (Koch), populations have become resistant to Aramite. After the discovery that Aramite is potentially carcinogenic, it has not been used on food crops where residues were likely to be present at harvest. Neither Aramite nor ovex were effective in controlling eriophyoid, tarsonemid, or most brevipalpid mites. Aramite is most toxic to the active mite stages (Ebeling and Pence, 1954), but residues are sufficiently persistent to kill mites that develop from eggs present at the time of treatment. Mites exposed to a lethal dose of Aramite pass through several hours of slow paralysis and disorganized movement of the appendages and in the process often become dislodged from the plant. Residues of Aramite and ovex penetrate through leaves of many plants sufficiently that mites living on the side opposite to the deposit are killed by the translocated chemical (Cooke, 1964; Ebeling and Pence, 1954). This translaminar property may partly account for the effectiveness of Aramite and ovex in control of mites living on the undersides of leaves of low-growing plants when applications are made by aircraft. Aramite has a low order of toxicity to insects or mites other than the Tetranychidae; it also appears to have minimum influence on predators or parasites of mites or insects. Residues soon become ineffective when applications are made with lime or other highly basic materials.

Ovex. Ovex is primarily toxic to the egg and larval mite stages (Ebeling and Pence, 1954). The larvae are most susceptible to this acaricide just after they hatch from the egg. Residues remain toxic to eggs and larvae for several weeks after an application; therefore, offspring from surviving adult females are usually killed by a single treatment. Ovex is relatively nontoxic to warm-blooded animals, to phytoseiid and other mite predators, and to insect predators or parasites. This acaricide has no objectionable odor, is not readily influenced by variations in water composition, and is compatible with most pesticides. Residues are not readily washed from the plants by rain or decomposed by high temperatures.

The extensive use of ovex stimulated interest in evaluating related compounds for mite control. Chlorbenside, Fenson (CPBS or PCBS), and Sulphenone® were found to be effective and were used rather extensively (Kirby and Tew, 1952; March, 1958). Mite populations resistent to ovex, however, were found to be cross-resistant to these acaricides, or repeated treatments soon in developed resistant populations (Jeppson, 1960, 1964; Jeppson and Jesser, 1962).

Dicofol. Studies to evaluate the effectiveness of dicofol (Kelthane®) began in 1952. This acaricide has had extensive use for control of a wide variety of mite pests on agricultural or ornamental plants (Kirby and Tew, 1952; March, 1958). It has a relatively low order of toxicity to adult mites, but it is toxic to all mite stages. Mites exposed to a lethal dose of dicofol have their legs extended, but otherwise appear lifelike and in normal feeding position for a day or more. Its range of effectiveness includes not only the Tetranychidae (red spider mites), but also mites belonging to the family Tenuipalpidae (false spider mites) and the genera *Steneotarsonemus* and *Polyphagotarsonemus* of the Tarsonemidae.

Dicofol has most of the acaricidal advantages possessed by ovex, but it is adversely affected by combinations with basic compounds; dicofol is toxic to predatory mites, some insect parasites, and some predators. European red mite, *Panonychus ulmi* and citrus red mite, *P. citri* (McGregor), populations have been slower in developing resistance to dicofol than to the organophosphorus acaricides (OP). Resistance to dicofol by the McDaniel mite, *Tetranychus* (A.) *mcdanieli* McGregor, was first noted in 1958. Mite populations resistant to OP compounds or to other chlorinated hydrocarbon acaricides were not cross-resistant to dicofol (Hoyt and Kinney, 1964).

Chlorobenzilate. The evaluation of chlorobenzilate as an acaricide occurred in early 1952. This acaricide has been used rather extensively for the control of tetranychid mites, but its use has been overshadowed by other chlorinated hydrocarbon acaricides, partly because of price differential, but also because of the relatively ineffective control of the citrus red and European red mites. Chlorobenzilate is still effective in controlling certain eriophyid mites, such as the citrus bud mite, *Eriophyes sheldoni* Ewing, citrus rust mite, *Phyllocoptruta oleivora* (Ashmead), and other rust mites.

Tetradifon. Tetradifon (Tedion®) became available for field evaluations as a potential acaricide in 1954. This acaricide, like ovex, is primarily toxic to the egg and larval stages, but newly hatched larvae are the most susceptible. Spray residues are toxic to the eggs and larvae for several weeks; thus, a single application is often effective. The translaminar migration by tetradifon is not sufficient to kill mites on the underleaf surfaces; therefore, it has limited use for control of mites that inhabit the undersides of the leaves of dense or low-growing crops. Thus tetradifon is especially unsuitable to air applications to dense or low-growing crops.

Tetradifon has been used extensively for control of red spider mites on deciduous and citrus fruits and on ornamental plants. Phytotoxicity has rarely occurred even when applied to the most sensitve plants; mites, however, developed resistance after 3 or 4 applications. It has not been particularly toxic to brevipalpid or eriophyoid mites and is ineffective in the control of eupodid or tarsonemid mite pests.

Omite®. Omite is closely related to Aramite, but Omite is not known to be carcinogenic. Mites exposed to a lethal dosage have reactions similar to mites exposed to Aramite. Omite is as effective as Aramite in controlling many red spider mite species. The translaminar movement of Omite is less than that of Aramite, however, and for this reason it has not proven as useful as Aramite in controlling mites on cotton, beans, and other low-growing crops. Because populations of European red mite or citrus red mite have so far remained susceptible to Aramite, it is anticipated that the response to Omite will be similar. Resistance studies indicate that citrus red mite populations are slow in becoming resistant to Omite; citrus red mites in these studies remained susceptible even though exposed to 10 field applications (Jeppson, Jesser, and Complin, 1967).

Binapacryl. Binapacryl appeared as a potential acaricide in 1960, after red spider mite populations had developed resistance to the available chlorinated hydrocarbon and OP acaricides. This DN type compound proved less phytotoxic than the other acaricides of this chemical group. It became the acaricide of choice in the control of the McDaniel mite in the Pacific Northwest. Its use in the control of other mites has been largely owing to the shortage of other acaricides rather than to its outstanding properties.

Several representatives of new types of compounds have appeared during the 1960s as potential acaricides, namely *Morestan®*, *fentin hydroxide* (Plictran®) and *Galecron®*. An analogue of Morestan, namely thioquinox (Eradex®) has proven to be an effective acaricide, but some workers have developed dermititis while handling the formulated acaricide; therefore, it has been withdrawn in favor of Morestan. Morestan, like tetradifon, apparently is not readily adsorbed into plant tissue and consequently surface residues are available to mites for a relatively long period. This acaricide kills slowly, so some mite species exposed to its residues appear to be stimulated to migrate from treated plants. Such stimulation occurs at much lower dosages than will actually produce mortality of citrus red mite (Jeppson et al., 1967). Morestan is effective against most red spider mites, brevipalpid, and many eriophyoid mites. It is likewise toxic to phytoseiid and perhaps other predacious mites. Morestan sprays applied during high temperatures or shortly before or after petroleum oil applications may injure young plant tissue.

The potential acaricidal effectiveness of Plictran® has not been fully explored. Like Morestan, it is not readily adsorbed in plant tissue, therefore it appears to have more promise in the control of fruit tree mites than those that live on low-growing crops. Plictran's most serious restriction, however, appears to lie in its high phytotoxicity. Resistance studies indicate that at least some red spider mite populations are relatively slow in developing resistance to Plictran and Morestan.

The potential value of Galecron® as an acaricide has not been explored. It appears to be relatively nonphytotoxic, but spray residues are less persistent than some acaricides. It may be most valuable in controlling mites on crops where relatively frequent applications are economically feasible.

Organophosphorus and Carbamate Acaricides

Field evaluations of organophosphorus type (OP) acaricides started in 1946 (Dean and Newcomer, 1948; Hofmaster and Greenwood, 1953; Huckett, 1948; Kirby and Tew, 1952; March, 1958; Mukerjea, 1962; Smith, 1960). Tetraethyl pyrophosphate (TEPP) was used to provide immediate reduction of active mites under field conditions, but since repeated treatments were necessary, its field as an acaricide was limited. These compounds kill very quickly and cause mites to shrivel and dry rapidly.

Evaluations of parathion as a potential acaricide began about 1947. It was soon extensively used for control of mites that were not satisfactorily controlled by sulfur, particularly species belonging to the genus *Tetranychus* that inhabit deciduous fruit trees. Although extensively used on citrus for scale insect control, parathion was not sufficiently effective to be a reliable treatment for citrus red mite. After two or three years of extensive parathion usage, mite populations developed sufficient resistance that it no longer provided satisfactory mite control.

Other OP acaricides were rapidly developed. EPN proved to be an effective acaricide, but was soon overshadowed by the development of the systemic acaricides. Systemic type compounds were so named because they translocated through plants from the roots to leaves. Schraden® (OMPA) and demeton (Systox®) were the first to be evaluated. By the early 1950s, California citrus red mite populations had developed resistanec to ovex, and demeton was the only available acaricide. Mite populations in orchards that received four to five applications of demeton during the first two years of its commercial usage developed resistance to this acaricide (Jeppson, 1960). Erratic control resulted from soil applications, and applications to the tree trunk and limbs produced cracking of the bark.

Demeton came to be extensively used for control of the two-spotted mite and European red mites on deciduous fruit trees. It also effectively controlled the southern two-spotted spider mite, *Tetranychus* (*T.*) *cinnabarinus* (Boisduval), tetranychid mites on vegetable, field, and ornamental crops. Acting as a systemic, demeton translocated into pear and apple buds effectively killing pear and apple bud mites in their confined habitat.

Other acaricides became available by the time mite populations had developed resistance to demeton and parathion (Leigh, 1963; Schuster, 1959). Among the OP acaricides appearing in the mid 1950s were carbophenothion (Trithion®), dioxathion (Delnav®) and ethion. Mite populations resistant to parathion and demeton, however, had a degree of cross-resistance to these three acaricides (Jeppson, 1960; Jeppson and Jesser, 1962). Thus, they were effective only against nonresistant mite populations. OP acaricides have generally not proven effective for control of brevipalpid mites.

During the last decade (1960–1970) many OP and carbamate type compounds have been submitted and evaluated to ascertain their possibilities as acaricides. Most of these have been less toxic to OP-resistant than to susceptible mite strains. A few of these compounds that were equally toxic to both strains, when subjected to repeated treatments in the field, developed measurable resistance by the second to the fourth applications (Bravenhoer and Theume, 1961; Brown, 1963, 1964; Dittrich, 1960; Hansen, 1958; Herne, 1967; Huwald, 1965; Jeppson, 1960; Jeppson and Jesser, 1962; Jeppson, Jesser, and Complin, 1958; Newcomer and Dean, 1953; Nomura, Tomita, and Nakagashi, 1965; Smith, 1960; Tsugawa et al., 1964; Van Zon, Overmeer, and Helle, 1964; Vrie, 1959, 1963). Thus the prospects for the development of new acaricides from the organophosphorus and carbamate types of compounds are not very encouraging.

Control of Species in Other Mite Families

Acaricides most effective against tetranychid mites sometimes have no value for control of injurious mites belonging to the families Eriophyidae, Tenuipalpidae, and Tarsonemidae. As earlier indicated, chlorobenzilate has effectively controlled citrus bud and rust mites. Zineb, a fungicide, has been used extensively against the citrus rust mite, but it is ineffective for control of bud mites. Diazinon, endosulfan (Thiodan®) and carbaryl (Sevin®), which are primarily insecticides, have been most useful in keeping bud mites on deciduous fruits and berries and ornamental plants under control. The best chemicals for control of *Steneotarsonemus pallidus* (Banks) include endrin, endosulfan, dicofol, and Eradex®.

ACARICIDES USED FOR CONTROL OF GREENHOUSE MITES

Control of greenhouse mites has been reviewed by Henneberry et. al. (1961), and Smith (1960) from which the information of this section has been taken.

Because of the lack of adequate control of spider mites, before 1947 the cropping season of greenhouse flowers and vegetables was limited to the cooler parts of the year, that is, from fall to early spring. Growers tried to control mite populations by destroying all mite hosts before planting the fall crop both by means of a summer cleanup of plant material and by fumigation with burning sulfur. Before 1929, the only acaricides available for mite control on most crops were sulfur sprays and dusts; although table salt solutions were sometimes used on carnations. Syringing roses and carnations with water throughout the summer and winter was the general practice. Many new chemicals were tried between 1929 and 1947, but they were too phytotoxic or lacked effectiveness. Included among these were rotenone, thiocyanates, cyclohexylamine derivatives, dinitrophenol compounds, polysulfides, selenosulfides, and oil emulsions. Fumigants, such as naphthalene and methyl bromide, were also tried.

Sprays containing ammonium potassium selenosulfide (Selocide®) and rotenone marked the first advance in spider mite control in greenhouses. These sprays were highly successful at first, but soon became ineffective. Resistance to Selocide was first encountered in 1937. This resulted in its use being discontinued by 1947. Azobenzene vapors were found to provide effective control of red spider mites. Even though applications of azobenzene caused fading of flower color and loss of flowers for several days following the applications, it was used extensively as a vapor for about two years. Treatments were accomplished by painting the acaricide on steam pipes at one- to two-month intervals.

The introduction of organophosphorus acaricides in 1947 resulted in virtual elimination of spider mites in greenhouses, first by TEPP and later by parathion. Year-round flowering of roses and carnations then became the general practice. This resulted in a fivefold increase in summer production and a year-round crop increase of 20 to 40 percent. Resistance to parathion and TEPP became evident in 1949, and by 1950 resistant mites were present in a large percentage of the rose houses in the eastern United States.

New acaricides that became available from 1948 to 1958 were each in turn used as aerosols or sprays until resistance developed. The acaricides and their period of use in commercial greenhouses in the United States are:

TEPP	1946–1947	Aramite	1950–1954
parathion	1947	chlorobenzilate	1953–1954
sulfotepp	1948–1950	chlorbenside	1953–1954
Schradan	1949–1953	dicofol	1954–1956
ovex	1949–1953	tetradifon	1956–1957
demeton	1950–1953		

In Europe, resistance to OP compounds became evident in the Netherlands by 1950; in Demark, Norway, and France by 1951; in Germany by 1952; in Belgium and Italy by 1955, and in Switzerland by 1958.

Different stages of red spider mites are susceptible to different toxicants; therefore combinations of two or more acaricides came into use, especially in the early stages of resistance development. Combinations of Aramite and Schraden came into use in 1953 and an Aramite-tetradifon combination was used in 1958. These remained effective beyond 1960.

The extensive development since 1960 of the organophosphorus compounds and later the carbamate type insecticides have not produced many useful acaricides. Either these pesticides were nontoxic to mites, or the existent resistant populations had the capabiliy of developing resistance after being exposed to two or three applications of these acaricides.

Mite research turned to the possibilities of obtaining ways of killing mites by physical means. Micronized powders, such as the silica gels, proved effective under certain conditions by causing mite dehydration. Cellosize, which literally caused mites to burst by excessive moisture absorption, was used in some greenhouses.

In the early 1960s, Pentac® was found to be an effective acaricide for tetranychid mite control under greenhouse conditions. This acaricide kills slowly; the mites were often found dying with eggs partly protruding. Pentac proved to be effective against mite populations resistant to other acaricides. Although Pentac has been used for 5 to 10 years, treatments appear to be as effective as they were initially.

Other acaricides such as Morestan and binapacryl, which are effective against existing resistant strains, are generally too phytotoxic to be applied extensively under greenhouse conditions.

RESISTANCE TO ACARICIDES

The history of acaricide development has shown that spider mites belonging to the genera *Panonychus* and *Tetranychus* are capable of rapidly developing resistance to a wide variety of toxicants. Mite populations resistant to a toxicant are often cross-resistant to chemically related and to some nonrelated compounds; thus the development of effective alternate acaricides has not kept pace with the development of resistance by mites. As a result, resistance in spider mites has become more serious a problem of control than for most other agricultural pests.

Resistance has been defined as the development of an ability of a strain to tolerate doses of toxicants which are lethal to the majority of individuals in a normal population of the same species. The acaricides cause resistance by repeatedly screening out the most susceptible among the population leaving the more tolerant. Body structures that may be used to differentiate resistant from nonresistant strains of mites have not been found.

Environmental Tolerance and Resistance Reversion

Several studies indicate that the two-spotted spider mite populations that developed resistance to an acaricide by repeated treatments may have become more susceptible to adverse conditions in the environment. This may be reflected in lower egg production or viability, a longer life cycle, decreased ability to survive weather extremes for exposed mites than for mite populations not having been exposed to repeated treatment by the acaricide. This phenomenon is more evident in the larger animals and in plants. Cattle bred or selected for milk production, dogs selected for house pets, or plants for large flowers are often not as able to survive unless they are kept in protected environments. Since for mites, the original acaricide-susceptible population had the best characteristics to survive weather extremes, the tendency is for the population rapidly to revert to susceptibility to the selecting agent (acaricide). When the population or strain is subjected again to selection pressure it rapidly develops resistance to the acaricide, after which it no longer possesses decreased fitness to the environment (Schulten, 1968).

These studies indicate that reversion of resistance cannot be of great importance for the control of red spider mites. Even a population with a completed reversion can regain the original level of resistance in nearly half the number of selections (treatments) that were necessary for the initial development of resistance. This accelerated regaining of resistance does not appear to be accompanied by a decreased fitness, and then a reversion to susceptibility is probably not to be expected.

More information is needed on the relative susceptibility of resistant and susceptible strains to mite food shortages and weather extremes. Theoretically, some mite populations, having become resistant to an acaricide, may not be as tolerant to weather extremes as the original susceptible strain. Such populations might be expected to become gradually less resistant to the acaricide after treatment has been discontinued. It is not easily determined whether a decrease in the resistance of any field populations is because of survival ratio or dilution by susceptible mites that migrate or become transported into the population.

Development of Resistance

The number of treatments or selections required to produce resistance seems to vary with the acaricide, the species, or strain of spider mite (Brown, 1963, 1964; Dittrich, 1960; Herne, 1967; Huwald, 1965; Jeppson 1960, 1964; Jeppson and Jesser, 1962; Lemin, Goyack, and MacDonald, 1965; Nomura et al., 1965; Schulten, 1968; Smith, 1960; Vrie, 1959, 1963). Most mite populations appear to be capable of rapidly developing resistance to organophosphorus and carbamate acaricides, but no such regular pattern is evident among the other acaricides. Greenhouse and field populations of *Tetranychus* species soon became resistant to Aramite, but field populations of the European red mite, *Panonychus ulmi*, (Koch), have not developed resistance. There was no change in the susceptibility of a field population of the citrus red mite after 21 Aramite applications (Jeppson, Jesser, and Complin, 1962).

Cross-Resistance

Cross-resistance refers to cases in which a pest population that has developed resistance to a pesticide is also resistant to a toxicant to which the population has not been exposed; or more specifically, to cases in which one mechanism confers protection against various toxicants. Populations resistant to one organophosphorus acaricide often show various degrees of resistance to other OP and carbamate type compounds. Mite populations that have developed resistance to OP compounds remained unchanged in their susceptibility to the organochlorine acaricides, or to Aramite, Karathane, Pentac, Morestan, Plictran, and Phenoflurazol®. Populations resistant, or repeatedly exposed to the specific acaricides such as chlorobenzilate, Aramite, and dicofol, show some degree of cross-resistance to some of the OP acaricides but no cross-resistance to the other specific acaricides, except those very closely related chemically (Jeppson and Jesser, 1962; Smith, 1960).

Economic Importance of Resistance

The economic importance of the resistance phenomenon in mites has not only stimulated research on the genetics of red spider mites, but has served as a marker in such studies (Helle, 1965; Schulten, 1968). Resistance against organic phosphates appears to be based principally on a single mutated gene that carries an incompletely dominant character that can be transmitted by both sexes. Modifiers, transmitted by both sexes, can enhance the resistance level of OP compounds.

Genetics of Resistance

The genetics of red spider mites, however, is still in its infancy. Such studies have been made with only two species, that is, the common red spider mite, *Tetranychus* (*T.*) *urticae*, and the Pacific mite, *T.* (*A.*) *pacificus*. These have been made with organophosphorus compounds and tetradifon (Ballantyne, 1969; Overmeer, 1967). The finding of visible markers in several species has indicated that mites may serve as animals to study basic genetic principles (Helle, 1965; Helle and van Zon, 1966; Henneberry et al., 1961).

It has been known from rearing data that most species in the Tetranychidae are bisexual and their reproduction takes place by means of arrhenotokous parthenogenesis, but only recently has this been confirmed by cytological studies (Helle and van Zon, 1966). Fertilized females usually produce eggs with a diploid chromosome number, and they also produce eggs with a haploid chromosome set. Females (diploid) and males (haploid) develop from these eggs. Unfertilized females exclusively produce eggs with a haploid number resulting in male offspring.

All genera of red spider mites have relatively few chromosomes, 2 being the lowest and 7 the highest haploid number (Schulten, 1968). In 13 species studied, 12 showed the existence of haploid and diploid eggs with the following number of chromosomes: *Neotetranychus rubi* (Tragardh) with 7 and 14; *Eurytetranychus buxi* (Garman) with 5 and 10; *Bryobia sarothamni* Geijskes, *Eotetranychus tiliarium* (Hermann), and *E. carpini* (Oudemans) with 4 and 8; *Panonychus ulmi* (Koch), *Schizotetranychus schizopus* (Zacher), *Oligonychus* (*O.*) *ununguis* (Jacobi), *Tetranychus (T.) hydrangea* Pritchard and Baker, *T. (A.) pacificus* McGregor, *T.* (*T.*) *urticae* Koch and *T.* (*T.*) *cinnabarinus* (Boisduval), with 3 and 6. The chromosomes of all species are very small, ranging in length from 1.0 to 2.0 microns. Consequently, details regarding constrictions and finer particularities with respect to shape are beyond the resolving power of the light microscope (Helle and Bolland, 1967).

Spontaneous mutations forming suitable markers have been found in three tetranychid mites. These markers are aberrations in the pigmentation of the haemolymph and / or eyes. Investigations into the inheritance of the markers confirmed genetical haplo-diploidy, namely the females are biparental and males uniparental and impaternate, regardless of whether the male offspring originated from fertilized or unfertilized females; also, all behave as single recessives. Linkage studies between different pigment mutations and mutations for resistances indicate that more factors segregate independently from each other than can be expected from organisms with three pairs of chromosomes, therefore, the recombination index must be large. The occurrence of intragenic crossing-over of mutations located within a complex-locus for albinism in both *T. (A.) pacificus* and *T. (T.) urticae* have been found to be of sufficiently high order of frequency to indicate that allelic replacement of linked material during backcross procedures with recurrent parential types must occur. Such procedures are used in genetic work on resistance (Helle and van Zon, 1966).

SELECTED BIBLIOGRAPHY

Armstrong, T. 1953. Summary of studies on new acaricides in Canada, 1948–1953. 84th Rept. Entomol. Soc. Ont. pp 38–45.

Ascher, K. R. S., and R. Cwilich. 1960. Laboratory evaluation of acaricides against *Tetranychus telarius* L. on sugarbeet and on beans. Ktavim 10(3–4):159–163.

———. 1962. Field demonstrations on the control of red spider mites on cotton. Israel J. Res. 12(3):107–120.

Ballantyne, G. H. 1969. Genetic fine structure and complementation at albino locus in spider mites (*Septranychus* species). Genetica 40(3):289–323.

Bouron, H., and J. Mimaud. 1956. Essais de traitement sur les acariens des arbres fruitiers en 1955. Phytoma. 8(81):25–27.

Boyce, A. M., D. T. Pendergast, J. F. Kagy, and J. W. Hansen. 1939. Dinitro-o-cyclohexylphenol in the control of mites on citrus and Persian walnuts. J. Econ. Entomol. 32(3):445–467.

Bravenhoer, L., and D. Theume. 1961. Resistance of spider mites to acaricides. Meded. Landb. Hogesch. Gent. 26:1024–1032.

Brown, A. W. A. 1963, 1964. Insect resistance. Pt. I. Nature and prevalence of resistance. Farm Chemicals 126(10):21–28, 126(11):24–28; 127(1):58–69.

Cooke, V. A. 1964. The translaminar effect of Kelthane on three strains of glasshouse red spider mite, *Tetranychus urticae* Koch, on cucumber. Ann. Appl. Biol. 51(3):485–488.

Dean, F. P., and E. J. Newcomer. 1948. Comparative efficiency of materials for controlling orchard mites. Wash. State Hort. Assoc. Proc. 44:131–133.

DITTRICH, V. 1960. Populationsgenetische Undersuchungen an normalen und phosphorsäureester-resistenten Stämmen von *Tetranychus urticae* Koch. Ph.D. thesis, Institut für Pflanzenkrankheiten, Univ. Bonn.

EBELING, W., and R. J. PENCE. 1954. Susceptibility to acaricides of two-spotted spider mites in the egg, larval, and adult stages. J. Econ Entomol. 47(5):789.

HANSEN, C. O. 1958. Cross-resistance induction in the two-spotted mite *Tetranychus telarius* L. Ph.D. thesis, Cornell Univ.

HARRIES, F. H. 1960. Laboratory studies on orchard mites with emphasis on their development and control. Proc. Wash. State Hort. Assoc. 1959 meeting p. 165.

HELLE, W. 1962. Genetics of resistance to organophosphorus compounds and its relation to diapause in *Tetranychus urticae* Koch. Tydschr. Pl. Ziekt. 68:155–195.

———. 1965. Resistance in the Acarina: Mites. Advances in Acarology 2:71–93.

HELLE, W., and H. R. BOLLAND. 1967. Karyotypes and sex determination in spider mites. Genetica 38(1):43–53.

HELLE, W., and A. Q. VAN ZON. 1966. A search for linkage between genes for albinism and parathion resistance in *Tetranychus pacificus* McGregor. Genetica 37(2):181–185.

HENNEBERRY, T. J., E. A. TAYLOR, F. F. SMITH, A. L. BOSELL, and R. V. TRAVIS. 1961. Combination acaricide-insecticide-fungicide sprays on outdoor roses. J. Econ. Entomol. 54(3):420–422.

HERNE, D. H. C. 1967. Organophosphorus resistance in the two-spotted spider mite, *Tetraanychus urticae* Koch. Ph.D. thesis. Univ. Western Ont. London, Ont., Canada.

HOFMASTER, R. N., and D. E. GREENWOOD. 1953. Control of mites on strawberries in Virginia. J. Econ. Entomol. 46(2):224–233.

HOYT, S. C., and J. R. KINNEY. 1964. Field evaluation of acaricides for the control of the McDaniel spider mite. Wash. Agr. Expt. Sta. Circ. 439:1–13.

HUCKETT, H. C. 1948. Control of the two-spotted spider mite on lima beans on Long Island. J. Econ. Entomol. 41(2):202–206.

HUWALD, KARL VON. 1965. Resistenzentwicklung bei *Tetranychus urticae* Koch in Abhangigkeit von der Vorgeschichte der population und dem physiologischen Zustand der Wirtsplanze. Z. Ange. Entomol. 56:1–60.

ISELY, DWIGHT. 1941. Control of the common red spider on cotton. J. Econ. Entomol. 34(2):323–324.

JEPPSON, L. R. 1960. Resistance of mites attacking citrus. Misc. Pub. Entomol. Soc. Am. 2(1):13–16.

———. 1964. The resistance of citrus mites to control chemicals. Agric. and Veterinary Chem. and Agric. Engineering 5(4):95–96.

JEPPSON, L. R., and M. J. JESSER. 1962. Laboratory studies on resistance of the Pacific spider mite to acaricides. J. Econ. Entomol. 55(1):78–82.

JEPPSON, L. R., J. O. COMPLIN, and M. J. JESSER. 1958. Resistance of the citrus red mite to organic phosphates in California. J. Econ. Entomol. 51(2):232–233.

———. 1962. Effects of application programs on citrus red mite control and development of resistance to acaricides. J. Econ. Entomol. 55(1):17–22.

JEPPSON, L. R., M. J. JESSER, and J. O. COMPLIN. 1967. Response of citrus mites to selected quinoxaline cyclic di- or trithiocarbonates. J. Econ. Entomol. 60 (4) :994–999.

KIRBY, A. H. M., and R. P. TEW. 1952. Agricultural acaricides. Repts. Prog. Appl. Chem. 37:263–276. Soc. Chem. Ind. 46, Victoria St., London S. W.

KREMER, F. W. 1956. Studies on the biology, epidemiology, and control of *Bryobia praetiosa* Koch. Höfchenbriefe 9(4):189–252.

LEIGH, T. F. 1963. Control of certain insects and mites on cotton with three systematic organophosphorus compounds. J. Econ. Entomol. 56(3):326–333.

LEMIN, A. J., G. A. GOYACK, and R. M. MACDONALD. 1965. Carbamate insecticides, Alkyl and Halophenyl N methylcarbamates plus toxicity to insects and mites. J. Agr. and Food Chem. 13 (3) :214–215.

MADSEN, H. F., and A. D. BORDEN. 1955. Prebloom treatments to control European red mite eggs on pears in Northern California. J. Econ. Entomol. 48(1):103–105.

MARCH, R. B. 1958. The chemistry and action of acaricides. Ann. Rev. Entomol. 3:355–376.

MISTRIC, W. J., JR. 1957. Chemical control of *Tetranychus telarius* (*L.*) and *T. cinnabarinus* (Boisduval) on cotton J. Econ. Entomol. 50(6):803–805.

MUKERJEA, T. D. 1962. New acaricides for control of red spider mite, *Oligonychus coffeae* (Nietner) on tea. Bull. Entomol. Res. 53(1):59–74.

NEWCOMER, E. J., and F. P. DEAN. 1953. Control of orchard mites resistant to parathion. J. Econ. Entomol. 46(5):894–896.

NOMURA, K., J. TOMITA, and S. NAKAGOSHI. 1965. On insecticide resistance of carnation red spider mite, *Tetranychus telarius* L. to Methyldemeton and Kelthane. [In Japanese with English summary] Tech. Bull. Fac. Hort., Chiba Univ. 13 :1928.

OVERMEER, W. P. J. 1967. Genetics of resistance to tedion in *Tetranychus urticae* Koch. Arch. Needrlandaisas Zool. 17(3):295–350.

PARENT, B. 1963. Efficacité comparée de plusieurs pesticides contre le tetranyque rouge du pommier, *Panonychus ulmi* (Koch) dans le sud-ouest Quebeck. Phytoprotection 44(2):78–95.

SCHULTEN, G. G. M. 1968. Genetics of organophosphate resistance in the two-spotted spider mite. Comm. No. 57, pp. 1–57. Koninklijk Inst. Voor de Tropen, Amsterdam.

SCHUSTER, M. F. 1959. Chemical control of *Tetranychus marianae* McGregor on tomatoes in the lower Rio Grande valley. J. Econ. Entomol. 52(4):763–767.

SMITH, F. F. 1960. Resistance of greenhouse spider mites to acaricides. Misc. Pub. Entomol. Soc. Am. 2(1):5–11.

TAYLOR, E. A., T. J. HENNEBERRY, and F. F. SMITH. 1959. Control of resistance spider mites on greenhouse roses. J. Econ. Entomol. 52(5):1026–1027.

TAYLOR, G. G. 1952. Spray treatments for control of mites in apple orchards. New Zealand J. Sci., Tech. Ser. A 34(1):36–46.

TSUGAWA, E., M. YAMADA, S. SHIRASKI, and N. OYAMA. 1964. Studies on insecticide resistance in apple orchard insects, 1. On the influence of acaricide applications on *Panonychus ulmi* (Koch) and some other insects. Jap. J. Appl. Entomol. Zool. 8(3):191–202.

VAN ZON, A. Q., W. P. J. OVERMEER, and W. HELLE. 1964. Resistance to tedion in haploid and diploid offspring of *Tetranychus urticae* Koch. Entomol. Expt. and Appl. 7(3): 270–276.

VRIE, M. VAN DE. 1959. Waarnemingen over het optreden van resistentie tegen bestrijdingsmiddelen bijde fruitsspintmijt *Metatetranychus ulmi* Koch. Mededelingen van de Landbouwhogeschool en de Opzoekingstations van de staat to gent. 24(3):986–993.

———. 1963. De bestrijding van de fruitspintmijt *Betatetranychus ulmi* Koch, in verband met de ontwikkeling van resistentie tegen acariciden. [The control of the fruit tree red spider mite, *M. ulmi* Koch in connection with the development of resistance against acaricides.] Parasitica 19(2):41–55.

Chapter 4

Principles of Chemical Control of Plant Feeding Mites

The processes employed in the cultivation of plants tend to create a biological imbalance in nature. These processes include the removal of competitive plants, the use of strains obtained by selection, area planting to a single crop, fertilization, irrigation, pruning, and pest control. Man has been unable to counteract this imbalance by natural means and still obtain the desired quantity and quality of production. Therefore, the use of chemicals has been necessary to provide the protection against decreased production or crop destruction. Much of the information in this chapter has been summarized from a review on this subject by Jeppson (1965).

EVALUATING NEED FOR CHEMICAL CONTROL

Chemical applications to control pests are often a major factor in upsetting the natural balance; therefore, knowledge of the effects of pesticide treatments on subsequent mite infestations may aid in providing the most economical control of mite populations. Effects of pesticide usage on mite infestations are discussed in chapter 2.

The use of pesticides against pests of annual crops usually has minimum influence on mite infestations the following year. The crop is destroyed at the end of the season, thus reducing mite populations to low levels unless other plant hosts are available. In some cases, such as late infestations on cotton, mite injury may not have an adverse influence on the crop, and may even aid in defoliation, thereby facilitating the harvesting of the cotton. Under these conditions the decision to apply chemical treatments depends primarily on the economics at the time the infestation occurs.

The criteria for evaluating the desirability of applying chemical treatments to perennial crops should take into account not only the influence of the application on current mite infestations, but also the effects on the host and mite populations during succeeding years. Premature leaf drop of deciduous fruit trees, even though it occurs after harvest, may adversely affect tree growth and vigor and consequently the production obtained by subsequent crops. As mites and many of their predators overwinter on the trees either as adults or eggs, the effects of chemical applications on the next season's population need be taken into account in determining the desirability of treatment.

Mite infestations of subtropical or tropical plants require special consideration before the initiation of a chemical treatment. Because there are no severe winters or annual loss of leaves, population development may continue unless arrested by biological enemies, by adverse weather, or by defoliation. Thus, continuing heavy populations during one year may reduce tree vigor and production over a period of years. If the mite pest is largely regulated by predators or disease, any chemical control methods that eliminate the regulating factor may initiate a need for repeated chemical applications in an otherwise relatively constant situation. Suppose, for example, populations of a phytophagous mite species kept low by a combination of predators were treated by a chemical that eliminated the effectiveness of one of the predator species; then the regulating ability of the complex might become insufficient to maintain the population below economic levels and a cycle of repeated treatments would be required to prevent economic injury.

Continuous development of high populations of a mite pest on subtropical crops may be prevented or arrested by frequent or regular weather extremes. In California, for example, the citrus rust mite, *Phyllocoptruta oleivora* (Ashmead), is a pest in coastal citrus areas and only in between periods of unusually low humidity. Since autumn weather brings low humidity to coastal districts, damage by this mite normally occurs only during the summer period of high humidity.

In the absence of periods of low humidity, however, populations continue to increase during the winter and following summer and cause damage to crops during two seasons. In such situations weather rather than predators or parasites becomes the major regulating agent in reducing or preventing the development of injurious populations. Thus chemical treatments are not likely seriously to upset the regulating mechanisms.

Some mite species tend to regulate the density of their own population peaks. Populations of the citrus red mite, *Panonychus ciri* (McGregor), will decrease after a certain amount of leaf injury has occurred (Henderson and Holloway, 1942). When favorable conditions for plant growth exist, the amount of injury thus produced may be tolerated; that is, the effect on plant growth or production is not sufficient to justify a chemical treatment. During hot, dry weather, however, the same degree of mite feeding may result in defoliation and loss of fruit. Often twig dieback occurs which effects production for several years. Therefore, it may be advisable to take into account weather factors as well as population density before deciding to use acaricides.

Populations of bud mites (Eriophyidae) may restrict the amount of bud injury by limiting the availability of suitable habitats. These mites, as indicated above, are poorly protected from water loss and survive only under the high humidity conditions existing under the bud bracts of their host. For example, feeding by high populations of the citrus bud mite kills the bud tissue that dries and exposes the mites to fatal atmospheric conditions; most mites are killed before the bud is completely destroyed; thus the remaining bud tissue can develop. This unsymmetrical growth pattern results in twisted or deformed leaves or fruit (Boyce and Korsmeier, 1941). As there are no effective natural enemies, the decision to use chemical control must be based primarily on the cost of treatment versus the loss in the quantity and the quality of production.

These few examples illustrate that research, experience, and judgment are necessary to ascertain the need and overall effects of chemical control measures. Methods of evaluating mite population densities are discussed in chapter 2; but the economic threshold and overall effects of artificial control methods are difficult to evaluate. The choice of the methods of mite control, and the factors involved in the interpretation of the results, indicate the importance of special training and experience in evaluating the need, as well as the methods, of chemical control of plant-feeding mites.

FACTORS INFLUENCING EFFECTIVENESS OF ACARICIDES

The control obtained by an acaricide depends on the contact and residual toxicity to the mites directly. Control also depends on the chemical and physiological factors bound up in the protoplasm of the plant, the morphological structure and features of the plant, the physical factors in the environment during and after application, the formulation and application of the acaricide, the biological and physical factors influencing the rate of mite development, and repopulation after treatment.

To achieve the most efficient regulation of mite populations the grower should recognize (1) the acaricide's specificity, that is, the most susceptible species or mite group; (2) the properties of the acaricide as they relate to methods of formulation and application; (3) the relationship of the application to variations in mite distribution on the host plant; (4) the effects of seasonal plant growth and mite activity on the initial deposit and the duration of the effective residues.

Specificity of Acaricides

Closely related mite species differ in their physiological susceptibility to acaricide applications; but toxicity is usually measured by laboratory evaluations. Habitat, life history, activity, and feeding habits, however, may influence the effectiveness of field applications.

Correlation has been observed between physiological susceptibility and mite classification. Mites of the family Tetranychidae, for example, have been susceptible to at least initial applications of many organophosphorus type toxicants and to aromatic halogenated thioethers and their oxidation products (sulfoxide, sulfone), such as ovex, chlorbenside, tetradifon, and related sulfones. Mites belonging to the family Tenuipalpidae have not generally been susceptible to the organophosphorus acaricides and have been only moderately susceptible to the halogenated thioether group. Neither of these groups of acaricides has been particularly toxic to mites of the genus *Steneotarsonemus* (Allen, Nakakihara, and Schaefers, 1957), but both have been more effective against *Polyphagotarsonemus latus* (Banks). Morestan® and other related cyclodene carbonates have exhibited a wide range of toxicity to mites. These materials have been effective in controlling mites of the families Tetranychidae and Tenuipalpidae, as well as many species of the family Eriophyidae. Conversely, mites of the family Tarsonemidae are susceptible to endrin and endosulfan, those of the families Euopodidae and

Phytoseiidae are susceptible to DDT compounds that are completely ineffective in the control of tetranychid species.

Taxonomically closely related mite groups may exhibit wide differences in response to toxicants. Sulfur has been used effectively for the control of the majority of species in the family Eriophyidae, the genera *Eotetranychus* and *Oligonychus* in the family Tetranychidae, and the genus *Polyphagotarsonemus*, but not for the genus *Steneotarsonemus*, in the family Tarsonemidae. Other species of tetranychid mites are susceptible to sulfur, but the patterns of susceptibility to this element in other genera seem to be less evident. Differences in susceptibility may sometimes result from the development of a tolerance or resistance to sulfur.

Acaricide Properties, Formulation, and Application

The toxicity of an acaricide, as measured by ordinary bioassay techniques, often has little correlation with its economic effectiveness under field conditions; other properties than toxicity may influence the value of an acaricide under the varied field conditions encountered in actual use.

There is wide variation in the susceptibility of the egg, larva, nymphal, and adult stages to the different specific acaricides (Ebeling and Pence, 1954). An effective acaricide need be toxic to only one active tetranychid mite stage providing residues are toxic throughout the life cycle of the mite. The larva, nymph, and adult stages of most tetranychid mites move about sufficiently to contact toxic residues that are well distributed over the plant surface.

Acaricides toxic only to adult mites and whose residues degrade rapidly after application are usually ineffective in controlling mites under many field conditions. Such acaricides may be the only means of chemical control on food crops infested with mites near the time of harvest. Such acaricides, especially those toxic in the vapor phase, have been successfully used for mite control in greenhouses and under field situations where frequent applications are feasible. Usually single applications of acaricides having short residual life fail to provide adequate control of mite populations in the field. Many mites either are in a tolerant stage (often the egg) at the time of application, or are not contacted directly; thus a rapid reinfestation is possible. Acaricides with more persistent residues are essential for economical mite control when the cost prohibits frequent treatments. Persistent residues that remain toxic to mites tend to retard the development of mite populations on plants by either killing the mites that move out from locations protected from direct contact, or those that develop into a more susceptible stage after treatment.

The type of action of an acaricide often limits its use as an acaricide. As indicated earlier, a toxicant that kills mites only by direct contact during application is relatively ineffective for mite control on dense or low-growing crops. Coverage of all aerial parts of the plants normally is necessary on other crops. In contrast, acaricides that penetrate through the leaves, or those that are toxic in the vapor phase, require a minimum of spray coverage to achieve adequate mite control. This permits the use of more economical methods of application, such as low volume ground or aerial applications in the field or atomized mists in greenhouses.

The purpose of formulation and application is to distribute lethal quantities of the toxicant to the maximum number of individuals and with minimum cost—and ideally with the least effect on predators and parasites. The methods and techniques for applying acaricides are essentially the same as those for applying insecticides. The formulation and application techniques should be tailored to the mite, the crop, and the stage of each to which the acaricide is to be applied. Loss of efficiency and often effectiveness occurs when a single formulation is used for many mite species in different situations and by several methods of application, or when several pesticides are combined in a single application.

Some of the requirements for effectiveness and efficiency are diametrically opposed to one another. It has not, therefore, been possible to obtain a single acaricide, formulation, or type of application ideally suited to control two-spotted spider mite populations under the diversity of situations in which this polyphagous species lives, not to mention the situations that may occur with a variety of different mite species.

Distribution

Mites that inhabit the undersides of leaves of low-growing crops, such as strawberries, or fast-growing dense foliage crops like cotton, are not economically controlled except by acaricides that are either toxic in the vapor phase or translocated at least through the plant leaves. This may account for the lack of effectiveness of tetradifon in controlling the tetranychid mites on these crops. Acaricides that penetrate readily into the citrus plant may concentrate in the citrus oil where toxic ingredients are not available to the mites either by contact or ingestion. When the amount of residue as measured by chemical analysis is compared with results obtained by bioassay it becomes evident that loss in toxicity to mites by such acaricides is much more rapid than the decrease of toxicant on and in the citrus leaves or fruit (Jeppson and Gunther, 1971). Formulation solvents used in emulsifiable formulations and spray adjuvants appear to be of only minor importance in altering these characteristics of acaricides.

Chemicals toxic only to the egg stage are usually not sufficiently effective when low volume application methods are used. Slow and costly full coverage application methods are required to distribute the ovicide sufficiently for contact with enough mite eggs to reduce the mite population. Also, eggs are frequently deposited in places on the plant not ordinarily reached by spray or dust treatments. Without persistent residues toxic to the mobile stages, reinfestation soon occurs.

Acaricides may come in contact with mites by several routes: (1) as a surface poison by direct contact and penetration through the mite integument, (2) as a stomach poison by ingestion if the mite feeds on surface or subcuticular residues, and (3) as a fumigant. Earlier it was believed necessary for an acaricide to contact mites directly or to encompass them with vapor, but it is now known that certain acaricides penetrate through leaf surfaces (Blauvelt and Hathaway, 1950; Ebeling and Pence, 1954). Acaricides applied to the upper surface of leaves may produce mortality of mites and eggs confined to the lower surfaces. It appears that the mortality of adults and larvae on opposite sides of a leaf results from action of the acaricide as a stomach poison.

Seasonal Variability

Adequate information is not available concerning the physiological variations within the plant which alter the effectiveness of acaricide applications. The toxicity of acaricides to mites on their host plants varies from one time to another even when tested under controlled laboratory conditions. This variability has been mimimized in laboratory evaluations of larger insects by injection or by topical applications. In an attempt to eliminate the plant surface source of test variability, mites have been placed on their backs on sticky tape and then dipped in the acaricide (Voss, 1961). If the effects of temperature, humidity, and sunlight on the plant can be responsible for variation in toxicity under the most constant laboratory conditions, it is evident that the day-to-day or season-to-season variability in the field can account for the successes and failures obtained with comparable acaricide applications. Although the interrelationships between plant host and toxicity are important with larger insect pests, this importance is magnified many times with mites, which are so intimately associated with plants.

Interrelated with the internal physiological changes are the plant surface differences brought about by weather variations, seasonal changes, and plant development, which alter the deposition, retention, and penetration of the acaricide and thus the control obtained. Differences between the number and size of spray drops retained by young as compared with mature leaves are evident. Spray applications short of runoff are not as subject to deposition variations as full coverage sprays. This probably accounts for the degree of effective mite control obtained by low volume applications despite the lack of complete distribution.

Sources of Reinfestation and Treatment Schedules

There are two sources of reinfestation after an acaricide application: (1) reproduction of mites missed by the application, and (2) movement from untreated hosts within or outside the orchard or field. In tree crops the former appears to be the most important source, especially for such host-specific species as the European and citrus red mites. Under these conditions, the rate of reinfestation is directly related to the number of mites that survive the application. The number of survivors is influenced by the toxicity of the acaricide and persistence of its residues, the application, and the density of the population being treated. Mites in certain protected "niches" are not reached by the treatment, and when these areas are heavily populated the reinfestation rate is more rapid. Therefore, where regular treatments are necessary it is generally advisable to apply chemical control measures when populations are low and before plant injury by the mites occurs.

Treatment of all the host plants in the area may be advisable where reinfestation comes from outside the crop area or from neighboring fields, as often occurs each summer with annual crops. Any need for a subsequent application that season would likely result from populations that developed from the progeny of mites missed during the previous treatment.

Successive timed treatments are necessary when residues of an acaricide are of short duration and toxic only to the active stages of the mite. If the first treatment kills only the active stages, for effective control, the second treatment must be applied after all the eggs have hatched and before developing mites begin to deposit eggs. Properly timed applications are often difficult to achieve because of such interfering conditions as cultural operations, irrigation, cultivation, or weed control, and such adverse weather conditions as winds, rain, and adverse temperatures. These often prevent proper timing of the second application. Timed treatments may cause reduction of predacious and parasitic insects or mites, resulting in upsets of pest populations normally maintained below injurious levels by natural enemies.

The most successful use of a scheduled treatment program for control of mites has been in greenhouses. Although vapor sprays are not always toxic to mite eggs, they may be applied easily and economically without leaving unsightly residues. The destruction of natural enemies is not generally important under such artificial conditions. Because temperature conditions are kept nearly constant, the life cycle of the mite and the time interval required between such treatments is relatively consistent. Successive treatments may be made according to a schedule.

Pest Management and Acaricides

The judicial use of biological, cultural, and chemical control presents the ideal approach to the control of all pests. This requires the use of specific pesticides; that is, those pesticides that effect reduction of the target pest without eliminating predators or parasites of the pest or other potential pests in the treated area. Because specific pesticides are effective against fewer pests than broad-spectrum pesticides, their usage and therefore the quantity of the sales of specific pesticides is more limited.

Governments have required that toxicological data be obtained on the important biological organisms in the environment before the pesticide can be applied. This is enforced by requiring that each pesticide be registered with the appropriate government agency. The registrant is required to supply sufficient information to satisfy the registering agency that the application of the pesticide to the crop will not adversely effect individuals who apply the treatment or enter the treated area, or the people who handle or consume the treated crop, or the useful nontarget animals or plants in the surrounding area. Limits on the amount of deposit or residues remaining on food crops are frequently indicated by the registering agency as pesticide tolerances.

The cost of obtaining adequate toxicological data on all these factors in the ecosystem discourages companies from developing the required information for registration of the specific pesticides and favors the development of broad-spectrum pesticides. As these broad-spectrum toxicants reduce populations of a wide variety of parasites and predators as well as target pests, their application is not inimical to proper pest management practices. Consequently the rate of development of proper pest management practices will be considerably curtailed unless either considerable reductions are made in the cost of developing chemicals for market use on food crops, or the current chemical approach to the pest control problem is somehow drastically changed.

Meteorological Conditions

Meteorological conditions before and after acaricidal treatments frequently determine the success or failure of the applications. General climatic conditions influence the growth cycles (i.e., the foliage density, maturity of fruit, and other conditions), the stage of development, the activity of the mite pests or their predators, and the rate of acaricide weathering. All of these may alter the effectiveness of the applications (Boudreaux, 1958; Ebeling and Pence, 1954; Hamstead and Gould, 1957; Jeppson, Complin, and Jesser, 1961; Jeppson, Jesser, and Complin, 1953; Mistric and Martin, 1956; Post, 1962). Extremes of temperature and humidity at the time of and after the application often have critical influence on effectiveness (Becker, 1952; Boudreaux, 1958). Winds and low humidity especially affect the amount of deposit and rate of weathering of dust applications. Weather also has a direct effect on the control achieved by certain acaricides, such as sulfur.

Extreme weather conditions may sometimes be used to advantage in the control of mites by chemicals. The mobile and molting stages of the citrus red mite are sometimes killed by hot dry winds. Since the egg stage is not usually affected by such weather extremes, only the active stages are reduced, and when the eggs hatch the population is composed mostly of larvae and nymphs, the stages most susceptible to acaricides.

The total surface area to be treated, and also the prevalence of "niches" that protect the mites from contact with the application, frequently vary according to the growing season. Efficient mite control is often based on timing in relation to mite biology, growth of the plant, or presence of fruit. Recognition of these factors is important in the development of treatment schedules.

The dormant deciduous tree is less susceptible to injury and requires less spray for coverage than when in full foliage. The overwintering eggs, however, are not only more difficult to kill than active stages but are often protected beneath the bark, buds, and other parts. At the pink bud stage, the overwintering eggs have hatched and most of the population is active (Underhill, 1950). The trees are still not in full foliage thererfore require less spray than is necessary after leaves reach mature size. Thus, this period has been extensively used to apply acaricides for control of *Panonychus* and *Bryobia* mite species.

Bud mites usually migrate to new growth buds at a certain stage of twig development. The citrus bud mite, *Eriophyes sheldoni* Ewing, migrates from the old injured buds to new ones as soon as the new buds are sufficiently developed to provide adequate protection from exposure to adverse weather. For this reason control of this mite by acaricides is usually most effective when applied soon after new growth is formed. The mites are killed by the residue as they crawl over the twig surface to the new bud (Jeppson, Jesser, and Complin, 1958).

Populations of the majority of tetranychid mite species become most abundant in orchards during the growth cycles of trees. The exact dates on which these cycles occur in a particular orchard vary with climatic influences, as well as with irrigation, pruning, and other cultural practices. Applications of residual type acaricides made during the growth cycles have often been less effective than similar treatments before or after these cycles. Mites not contacted by the treatment may move to the expanding leaves and there find untreated surfaces to initiate the reinfestation.

Inert Dusts and Residues

Tetranychid mite infestations appear to be most serious along dusty roads; spray deposits of copper, zinc, and manganese sulfates, lime, soda ash, thiram, and sulfur tend to increase mite populations (Griffiths and Fisher, 1950; Holloway, Henderson, and McBurnie, 1942). Several factors may be responsible for these effects of so-called inert materials on mite populations. Such materials have an adverse influence on natural enemies, either by killing them directly by abrasion resulting in dehydration, or by retarding their activity and thus their ability to find prey.

Heavy dust deposits on foliage absorb some of the petroleum oil applied for insect and mite control, thus reducing the deposit available to kill the pests. Some workers have suggested that dust deposits may change the physiology and nutrition of the host plant sufficiently to influence mite populations indirectly. These claims, however, have not been adequately substantiated.

Pesticide Combinations

It is generally not advisable in agricultural mite control to combine in one application two acaricides to control one mite pest even though each is toxic to one mite stage only. Effective control of mites under field conditions requires that residues be toxic for a sufficient interval after deposition to reduce the active mite stages that develop after treatment. Because the acaricides in such combinations are usually at reduced dosages the residual life of each acaricide is usually low.

Under certain conditions value is received by combining acaricides, one an adulticide and the other primarily an ovicide or larvicide. The value of such combinations is optimal when actively feeding populations are producing economic plant injury. The adulticide serves to prevent continued plant injury and egg deposition until the egg hatch is complete and the larvae succumb to the larvicide. Under proper control practices, however, the mite treatment should be made before mites are sufficiently abundant to damage the crop. Then the use of an adulticide is unnecessary.

Two toxicants applied to control the same pest may have different ranges of activity against beneficial insects and mites. Any slight gain in immediate reduction of the mite population may be more than offset by (1) the rapid development of resistance by mites to both toxicants, and (2) the adverse effect on the biological factors that tend to delay or prevent reinfestation.

Combinations of an insecticide with an acaricide may frequently be advisable when insect and mite pests are not susceptible to the same toxicant. Knowledge of their compatibility should be available, as well as knowledge of the effect on wetting and depositing properties that may result from the combination. Combining toxicants in one application likely would have less influence on natural enemies than two applications closely spaced. The treatment should be timed to obtain most effective control of the major pest or the one having the most limited timing restrictions. Any change in the pesticide program brought about by the development of a new pesticide, or the populations becoming resistant to those in use, may necessitate changes in the timing or season for treatment, the combination of pesticdes to use, and the methods of application. When lead arsenate was used for control of the codling moth, *Laspeyresia pomonella* L., complete coverage and exact timing for this pest was more critical than for the newer pesticides. By the same token the increased importance of mites in deciduous fruit orchards may justify careful attention to proper scheduling for more effective control of the tetranychid mites. A similar change in situation has occurred with the control of the citrus red mite and the scale insect, *Aonidiella aurantii* (Maskell), on navel oranges in southern California. When petroleum oil was the only scalicide used, the application was best made in August and September. New treatments for scale, such as parathion and malathion, are effective when applied throughout the spring, summer, and fall months. This provides opportunity to apply combination treatments at a time more suitable for control of the citrus red mite, infestations of which normally peak during the spring and fall.

SELECTED BIBLIOGRAPHY

ALLEN, W. W., H. NAKAKIHARA, and G. A. SCHAEFERS. 1957. The effectiveness of various pesticides against the cyclmen mite on strawberries. J. Econ. Entomol. 50(5):648–652.

BECKER, H. 1952. Uber den Einfluss Konstanter Temperturen, relativer Luftfeuchtigskeiten und Licht auf die Frühjahrsentwicklung der Wintereier der Obstbaumspinnmilbe, *Paratetranychus pilosus* C. et F. Anz. Schädlik. 25:116–118.

BLAUVELT, W. E., and W. B. HATHAWAY. 1950. K-6451 aerosol for greenhouse mite control. Down to Earth 5(4):2–4. (Dow Chemical Co.)

BOUDREAUX, H. B. 1958. The effect of relative humidity on egg-laying, hatching, and survival in various spider mites. J. Insect Physiol. 2(1):65–72.

BOYCE, A. M., and R. B. KORSMEIER. 1941. The citrus bud mite, *Eriophyes sheldoni*. J. Econ. Entomol. 34(6):745–756.

EBELING, W., and R. J. PENCE. 1954. Susceptibility to acaricides of two-spotted spider mites in the egg, larval, and adult stages. J. Econ. Entomol. 47(5):789–795.

GARMAN, P., and B. H. KENNEDY. 1949. Effect of soil fertilization on the rate of reproduction of the two-spotted spider mite. J. Econ. Entomol. 42(1):157–158.

GRIFFITHS, J. T., and F. E. FISHER. 1950. Residues on citrus trees in Florida. J. Econ. Entomol. 43(3):298–305.

HAMSTEAD, E. O., and E. GOULD. 1957. Relation of mite populations to seasonal leaf nitrogen levels in apple orchards. J. Econ. Entomol. 50(1):109–110.

HENDERSON, C. F., and J. K. HOLLOWAY. 1942. Influence of leaf age and feeding injury on the citrus red mite. J. Econ. Entomol. 35(5):683–686.

HOLLOWAY, J. K., C. F. HENDERSON, and H. V. MCBURNIE. 1942. Population increase of citrus red mite associated with the use of sprays containing inert granular residues. J. Econ. Entomol. 35(3):348–350.

JEPPSON, L. R. 1965. Principles of chemical control of phytophagous mites. Advances in Acarology 2:31–51.

JEPPSON, L. R., J. O. COMPLIN, and M. J. JESSER. 1961. Factors influencing citrus red mite populations on navel oranges and scheduling of acaricides applications in southern California. J. Econ. Entomol. 54(1):55–60.

JEPPSON, L. R., and F. A. GUNTHER. 1971. Acaricide residues on citrus foliage and fruits and their biological signicance. Residue Rev. 33:101–136.

JEPPSON, L. R., M. J. JESSER, and J. O. COMPLIN. 1953. Timing of treatments for control of citrus red mite on orange trees in coastal districts of California. J. Econ. Entomol. 46(1):10–14.

———. 1958. Factors affecting populations of the citrus bud mite in southern California lemon orchards and acaricide treatments for control of this eriophyid. J. Econ. Entomol. 51(5):657–662.

MISTRIC, W. J., JR., and D. F. MARTIN. 1956. Effects of climatic conditions on the chemical control of certain sucking pests of cotton. J. Econ. Entomol. 49(6):760–763.

POST,A. 1962. Effect of cultural measures on the population density of the fruit tree red spider mite. *Metatranychus ulmi* Koch. Tydschr. Pl. Zeikt. 68 (1) :1–110.

UNDERHILL, G. W. 1950. Timing early sprays for summer control of European red mite. J. Econ. Entomol. 43(5):637–641.

VOSS, G. 1961. Ein neves akarizid-austestungsverfahren für spinnmelbe. Auz. f. Schädlingskunde. 24(5):76–77.

Chapter 5

Biological Enemies of Mites

The information in this chapter (except for virus diseases) is largely summarized from a comprehensive review by J. A. McMurtry, C. B. Huffacker, and M. van de Vrie (1970). Original sources of data may be found in the review.

The reproductive capacity of many plant-feeding mites is sufficient to destroy or seriously reduce plant growth or crop production in the absence of mortalities produced by adverse weather, climate, biological enemies, or man's intervention. Weather and climate may retard mite development and produce or delimit population cycles. Biological enemies are very important agents in reducing or regulating populations of injurious plant-feeding mites within these limits or cycles. More than 65 predators have been recorded for the European red mite, *Panonychus ulmi* (Koch), alone. Among the more important of these biological agents are predatory mites and insects, but others include spiders and disease-producing pathogens.

PATHOGENS

Pathogens have, at times, reduced populations of a few injurious plant-feeding mites; particularly hibernating populations of the two-spotted spider mite, *Tetranychus* (*T.*) *urticae* Koch, the European red mite, the citrus red mite, *Panonychus citri* (McGregor), the avocado red mite, *Oligonychus* (*O.*) *yothersi* (McGregor), the Texas citrus mite, *Eutetranychus banksi* (McGregor), the six-spotted mite, *Eotetranychus sexmaculatus* (Riley), and the citrus rust mite, *Phyllocoptruta oleivora* (Ashmead). The fungal mite diseases appear to be more infectant when humidity is high.

VIRUS DISEASES

A noninclusion virus disease of citrus red mite is effective in rapidly destroying not only laboratory-reared colonies but high field populations of this citrus mite. But this reduction usually occurs after much plant injury has been produced by the mites. Diseased mites have been found in more than 80 percent of the orchards sampled in California, but the virus remains in an occult state until conditions are suitable for infestation (Shaw, Tashiro, and Dietrick, 1968).

A rod-shaped noninclusion virus infects the European red mite in Ontario, Canada. Many of the symptoms of this virus are very similar to those in the citrus red mite in Californa, especially the crystalline, spheroidal nonviral inclusions and rigid extension of the legs, but the virus in citrus red mite has smaller, somewhat spheroidal or polyhedral viral particles. In the European red mite infection and death can occur in any postovarial stage. Most infected mites contain, in the midgut, more-or-less spheroidal inclusions with a radiating crystalline structure. Infected mites deposit inoculum on the leaves, probably in excreta or oral secretions at feeding sites, which is picked up orally by uninfected mites while feeding. The inoculum on the leaves is very unstable, seldom remains infective for more than a week, and is almost immediately inactivated after exposure to water. Suspensions of infected mites, triturated in water or in various solutions, were inefficient inocula. Introduction of virus into orchard populations of the mites induced epizootics that rapidly reduced the population density; like the virus in citrus red mite, however, natural epizootics are found only in dense populations (Putman, 1970).

PREDATORY MITES AND SPIDERS

Phytoseiidae

Mites of the family Phytoseiidae are probably the most effective and widespread predators of injurious plant-feeding mites. They are distributed throughout the world from the arctic to the tropics and one species or another has adapted to each situation. Thirty years ago this group of predators was classified as 1 or 2 species, but studies since then have revealed many more species. Most of the identification and classification of mites in this group has come in the past decade (1960–1970). As a result, some species and generic relationships have not been established. For example, one specialist recognizes many predacious species in the family Phytoseiidae as one genus, that is, *Typhlodromus*, while another specialist separates them into 43 genera.

The first published records relating to the possible value of phytoseiids as predators of plant-feeding mites were those of Parrott, Hodgkiss, and Schoene in 1906. From 1930 to 1950, entomoligists became aware that these predators may have a significant influence on populations of certain important phytophagous mite pests, such as the European and citrus red mites and the cyclamen mite, *Steneotarsonemus pallidus* (Banks). Table 2 is a list of some of the more important predacious species in this family.

Statements in the literature concerning the importance of some of these predators have been based on field observations, some on routine counts, some on laboratory studies, and others on experiments involving the addition or removal of predators from a population of plant feeding mites.

Phytoseiid mites pass through the same developmental stages as do the tetranychids; namely, larva with 6 legs, the protonymph, and deutonymph, each having 8 legs. Most species require food to develop through the larval stage. The larvae are less motile than the protonymphs, and are not as able as the protonymphs to find or capture prey; thus species with large larvae may have the advantage.

TABLE 2
Some phytoseiid mites considered to be important predators of tetranychid mites on various crops

Phytoseiid	*Tetranychid*	*Crop*	*Location*
Amblyseius aberrans (Oudemans)			
	Panonychus ulmi	Grape	Switzerland
	Eotetranychus carpini	Grape	France
A. aurescens Athias-Henriot			
	Steneotarsonemus pallidus	Strawberry	California
A. cucumerus (Oudemans)			
	S. pallidus	Strawberry	California
	Tetranychus (T.) urticae	Alfalfa	Washington
	T. (T.) cinnabarinus	Cotton	Egypt
A. fallacis (Garman)			
	P. ulmi	Apple	Wisconsin
		Apple	West Virginia
	T. (T.) urticae	Apple	Wisconsin
A. finlandicus (Oudemans)			
	P. ulmi	Apple	England
A. hibisci (Chant)			
	Oligonychus punicae	Avocado	California
	P. citri	Citrus	California
	Eotetranychus sexmaculatus	Avocado	California
A. largoensis Muma			
	P. citri	Citrus	Japan
A. libanesi Dosse			
	T. (T.) cinnabarinus	Citrus	Lebanon
		Castor bean	
A. limonicus Garman and McGregor			
	E. sexmaculatus	Avocado	California
	O. (O.) punicae	Avocado	California
A. longispinosus (Evans)			
	T. (T.) urticae	Clover	Japan
A. potentillae (Garman)			
	P. ulmi	Apple	Netherlands
A. rademacheri Dosse			
	T. (T.) urticae	Soybean	Japan
A. rubini Swirski and Amitai			
	Eriophyids	Citrus	Palestine
A. similis (Koch)			
	P. ulmi	Plum	Netherlands
A. tsugawai Ehara			
	T. (T.) urticae	Soybean	Japan
Phytoseiulus persimilis Athias-Henriot			
	T. (T.) urticae	Bean	Canada (indoors)
			Germany (indoors)
		Cucumber	Netherlands
			USSR (indoors)
		Cucumber	England (indoors)

Phytoseiid	Tetranychid	Crop	Location
		Peach	Netherlands (indoors)
		Roses	United States (indoors)
		Strawberry	California
	T. (T.) cinnabarinus	Vegetables	Lebanon
P. macropilis (Banks)			
	P. ulmi	Plum	Poland
	T. (T.) urticae	Strawberry	California
Typhlodromus caudiglans Schuster			
	P. ulmi	Peach	Ontario, Canada
		Apple	Wisconsin
	T. (T.) urticae	Apple	Wisconsin
T. athiasae Porath and Swirski			
	Eriophyids	Citrus	Palestine
T. floridauns Muma			
	E. sexmaculatus	Citrus	Florida
T. longipilus Nesbitt			
	T. (T.) urticae and	Cucumber	Netherlands (indoors)
	other tetranychids	Orchards	Washington
T. occidentalis Nesbitt			
	E. willametti	Grape	California
	Tetranychus spp.	Cotton	California
	T. (A.) mcdanieli	Apple	Washington, Utah
	T. (A.) pacificus	Grape	California
	T. (T.) urticae	Strawberry	California
T. pomi (Parrott)			
	P. ulmi	Apple	West Virginia
T. pyri Scheuten (= tiliae)			
	P. ulmi	Apple	England
			Netherlands
			Germany
			Nova Scotia, Canada
		Apple	New Zealand
		and other	Canada
		deciduous	Switzerland
		fruits	Nova Scotia, Canada
			New Zealand
T. rhenanus (Oudemans)			
	B. rubrioculus	Apple	Nova Scotia, Canada
	T. (T.) urticae		Quebec, Canada
			Illinois
T. richeri Chant			
	T. (A.) viennensis	Fruit trees	Tambow, USSR
	Tetranychids, eriophyids	Citrus	California
T. soleiger Ribaga			
	T. (A.) viennensis	Fruit trees	Tambow, USSR
T. subsolidus Beglyarov			
	P. ulmi	Fruit trees	Latvia, USSR
Species not specified			
	P. ulmi	Apple	West Virginia
			Ohio
			Connecticut

Life Cycle

The life cycle of phytoseiids may be shorter than that of tetranychids under comparable conditions, but relatively few direct comparisons have been made. Although the minimum developmental time for most species appears to be about six to seven days, species belonging to the genus *Phytoseiulus* appear to develop more rapidly. It is obvious that development is prolonged at low temperatures, and that extreme high temperatures are detrimental. The quantity and quality of food also influence developmental rate. Some predators appear to develop best on tetranychid mites alone, some on a combination of tetranychids and eriophyids, others on mites and pollen, a few on pollen alone.

The preoviposition period in some phytoseiids is relatively short (24 to 30 hours), but under optimum conditions most species require 3 to 5 days. This period, however, may be longer for field-collected overwintering females of a few species. *Typhlodromus pyri* (Scheuten), for example, required 15 days at 25 to 26 C (77 to 79 F). There are reported instances where the preoviposition period decreases as winter progresses. This may be a form of diapause.

Phytoseiid predators appear to produce fewer eggs per time unit than their prey. An average of about 2 eggs per female per day seems to be the maximum productivity for most species, but productivity is dependent on temperature and food supply. The total number of eggs laid per female likewise is dependent on food and climatic factors. The more effective species appear to be capable of producing 30 to 60 eggs during their lifetime, which is generally equivalent to the egg-laying ability of their prey. The number of prey the predators can find appears to be a very critical factor in determining egg production; therefore these predators are capable of rapidly increasing their numbers when plant-feeding mite populations are high.

Only the mated females overwinter in temperate climates. Arboreal species may be found in deep crevices, canker wounds, and beneath scale or bark coverings, that is, in the same protective situations as their prey. As with plant-feeding mites, the winter mortality may be high. Winter mortality is influenced by temperature extremes and the availability of protected places. Early frost sometimes kill the phytoseiids before they reach protected winter quarters, although some species may survive temperatures of –29 to –31.5 C (–20 to –24 F). Most conditions unfavorable for the predators, however, are likely to be unfavorable for the plant-feeding mites. These predacious mites are active throughout the year on evergreen trees and shrubs in warmer climates. Under these conditions the limiting factor is probably the availability of the food supply.

Feeding Habits

Phytoseiid mites show a great diversity in feeding habits; some are strict carnivores, others prefer plant or plant-derived foods such as pollen and nectar. Species like *Phytoseiulus persimilis* Athias-Henriot have reached such a degree of specialization that they are dependent on tetranychid mites for food (pl. 6). Some of the mite predators are rather specific as to the mite species upon which they feed, such as *Amblyseius fallacis* (Garman), which readily feeds on the two-spotted spider mite, *T.* (*T.*) *urticae*, and the McDaniel mite, *T.* (*A.*) *mcdanieli*, but not on the European red mite, *Panonychus ulmi* (Koch), or the brown mite, *Bryobia rubrioculus* (Scheuten). Other species such as *Typhlodromus occidentalis* Nesbitt (pl. 7) appear to prefer strongly webbing plant-feeding mites. *T. floridanus* Muma and *P. persimilis* prefer species that live in colonies rather than those distributed over the leaf surfaces. *T. caudiglans* Schuster and *Amblyseius hibisci* (Chant) prefer to feed on species that are more generally distributed. These last two species appear to be hindered by the webbing produced by the two-spotted spider mite. They serve as better predators of the European red and citrus red mites that produce little webbing. The smaller predators, as well as the younger stages of the larger species, have difficulty in capturing prey comprising larger or more active plant-feeding mites.

Food Requirements and Sources

The number of mites a predator requires influences its effectiveness in regulating plant-feeding mite populations. The average number of spider mite prey consumed per predator during development seems to average less than 20, even though some workers have reported as many as 114 or 119. A high percentage of predators complete development when fed only 2 larvae per day, consuming an average of 22 larvae during a developmental period of 12 days.

Eriophyoid mites, especially the rust mites, are commonly utilized as prey by phytoseiid mites. Some predators, such as *Amblyseius hibisci* (Chant), *A. limonicus* Garman and McGregor (pl. 8), and *A. rubini* Swirski and Amitai, feed on eriophyoids, but are unable to reproduce on them in the absence of other food. Some, including *Typhlodromus caudiglans* Schuster, *A. fallacis* (Garman), *T. longipilis* Nesbitt, and *T. rickeri* Chant, reproduce equally well on eriophyoids and on tetranychids; and finally some mite predators, such as *T. pyri* Scheuten, *T. rhenanus* (Oudemans), seem to prefer eriophyoids.

Other plant-feeding mites are used as food by the phytoseiid predators. *A. cucumeris* (Oudemans) and *A. aurescens* Athias-Henriot and *A. chilenensis* (Dosse) will accept *Brevipalpus* species as prey.

As indicated above, a major decline in fecundity and general development of populations of predacious mites results from lack of food supply. An insufficient number of plant-feeding mites to supply food for normal predator development may result from adverse weather conditions, from off-season levels of suitable plant food, or from intervention of acaricides. Predators that can survive on food sources other than tetranychids may have a better chance of surviving such adverse conditions.

Some phytoseiid mites can use eggs or immature stages of certain insects as food. Some mite predators can survive on scale insect crawlers, eggs of certain species of moths, white flies, or thrips. Some species can extract plant juices from the host plant, notably, *Typhlodromus pyri* Scheuten, *T. rhenanus* (Oudemans), and *A. finlandicus* (Oudemans). Other phytoseiid predators may use, or even prefer, pollen as a source of food. *T. caudiglans* Schuster and *T. rickeri* Chant develop and reproduce on avocado pollen, but at a slower rate than on mites. The development time and oviposition rate of *A. limonicus* Garman and McGregor is the same when fed pollen or mites. Pollen actually induced a higher rate of reproduction than spider mites when each were fed to four phytoseiid speices: namely, *A. rubini* Swirski and Amitai, *A. swirski* Athias-Henriot, *T. athiasae* Porath and Swirski and *A. hibisci* (Chant). In the absence of mites, avocado pollen is the source of food for spring increases of *A. hibisci*.

Various kinds of fungi may be a source of food for some predacious mites. *Amblyseius aberrans* (Oudemans) and *A. unbraticus* (Chant) and others are able to complete development on apple powdery mildew, *Podosphaera leucotricha* (Ellis and Everhart) Salmon. Populations of *A. finlandicus* (Oudemans) can develop when fed entirely on fungus, but the individual's life span is shorter than when mites are supplied as food. Evaluations so far indicate that fungi may be satisfactory hosts for just a few predacious mites, but further study is needed in this area.

Plant nectar and honeydew from aphids, soft scales, white flies, and mealybugs are consumed by some predacious mites when their preferred mite species is scarce. With a few species such as *A. hibisci* (Chant), the combination of honeydew and mites results in less mortality during the developmental and preoviposition period than a diet of mites alone. *A. limonicus* may survive 60 days when nectar from orange blossoms is the only food supplied; but they die in a few days without food. The more general feeders, such as *A. hibisci* and *A. limonicus* arc able to dcvelop and reproduce on artificial foods containing yeast and carbohydrate; more specialized predators of mites, however, such as *T. occidentalis* Nesbitt, *T. rickeri* Chant, and *P. persimilis* Athias-Henriot benefit little from such substances.

It is evident that not all the phytoseiid predators of plant-feeding mites are entirely dependent on mites for their food, and some require other food for maximum reproduction. The presence of various alternate or supplemental foods available in the field may have an important impact on the predator-prey interaction. Therefore, it may be unrealistic to depend on laboratory studies, where only mites are supplied as the food source, to determine the developmental rate of predatory mite species when field populations are likely able to utilize several sources of food, especially where these food supplements increase productivity and survival. This suggests the possibility of supplementing the food supply of these predators during the low period in the prey population cycle; such supplement would provide an adequate supply of predators at the time conditions become favorable for the injuriuos prey species. In this way injurious phytophagous mite populations may be delayed or entirely prevented from developing.

Food Availability and Searching Capacity

For predators to keep their prey at a low population level, they must (1) be able to search out and capture their prey, and (2) be adapted to the type of habitat where their preferred prey lives. Phytoseiids generally feed and deposit their eggs close to junctions of leaf veins on the undersurfaces of leaves, a pattern of behavior that probably is related to low photokinesis or high thigmokinesis. Thus the tendency of the mites is to remain on the lower side of horizontal surfaces. This may be a disadvantage in contacting certain phytophagous species, such as the European red mites that roam over the entire surface and especially the upper leaf surfaces. Even these tetranychid species are susceptible to predation because eggs are deposited, and developmental stages live near veins. It is evident that each predatory species has its own special habitat and habits. For example, *A. cucumeris* (Oudemans) lives on low-growing plants where it is a good predator of clover mite eggs, *Bryobia praetiosa* Koch, but unimportant as a predator of the brown mite, *B. rubrioculus* (Scheuten), which inhabits trees. *Typhlodromus occidentalis* Nesbitt exhibits a wide scope of activity; thus it is adapted to be a more general predator. But the biology of only a few species has been adequately studied.

The external morphology of the plant may have an influence on predatory mite activity and its resulting effectiveness. For example, *T. caudiglans* Schuster is more numerous on Spartan and McIntosh apples than on Delicious apples. Leaves of the former two are pubescent and have pronounced and rough fruit spurs; these qualities provide a greater number of sheltered areas for phytoseiids throughout the year than do Delicious apples. Sometimes the habits, distribution and activities of predatory mites do not coincide with the most common plant-feeding mite prey. *A. hibisci* (Chant) deposits the majority of its eggs on the undersides of downward curled leaves. This provides maximum protection from adverse weather, but, because such leaves often contain no prey, the newly hatched progeny are not ideally situated to obtain food unless there is a high density of prey.

Some phytoseiid mites adapt their habits and activities to coincide with the availability of their prey. In California and the Netherlands, *P. persimilis* Athias-Henriot will not colonize trees or grapes. As this mite is successful on trees in Lebanon and France, it appears that behavorial strains occur. Species that prefer to oviposit in certain habitats may leave such preferred areas and move over considerable areas of the plant in search of food.

Other Acarine Predators

Predatory mites other than the phytoseiids include certain species in the families *Bdellidae, Anystidae, Stigmaeidae,* and *Cheyletidae*. The taxonomy, biology, and effectiveness of most of these have not been adequately studied.

Bdellidae

Bdella depressa Ewing is an important enemy of the clover mite in the grassy areas of western United States. Another *Bdella* species preys on the same host on cover crops in British Columbia orchards.

Anystidae

Species of *Anystus* and *Balaustium* are predators of the European red mite in Canada, but having only 2 generations per year, their increase is relatively slow.

Stigmaeidae

There are several known mite predators among the *Stigmaeidae. Zetzellia mali* (Ewing) is a predator of the two-spotted spider mite, the European red mite, the brown mite, and other mites on fruit trees in North America, Europe, and Israel. This species by itself is usually not able to keep the tetranychid mites in check, but it is occasionally able to maintain populations of apple rust mite, *Aculus schlechtendali* (Nalepa), at low densities in the northwestern United States. Because there are but 2 generations a year in temperate climates, *Z. mali* can do no more than assist other predators in regulating populations of important mite pests of agricultural crops.

The genus *Agistemus* in the family Stigmaeidae includes a number of known predators of mites. These and their known prey include the following:

Predator	*Prey*
Agistemus fleschneri Summers	*Panonychus ulmi, Tetranychus (T.) cinnabarinus, T. (T.) kanzawai*
A. floridanus Gonzales	*Eotetranychus sexmaculatus*
A. exsertus Gonzales	*P. citri*
A. faneri Dosse	Tetranychids in Lebanon
A. longisetus Gonzales	*P. ulmi* and *Bryobia rubrioculus*

Agistemus spp. are also predators of tarsonemid, brevipalpid, and tydeid mites, but their biology and importance as predators need to be more fully ascertained.

Spiders

Spiders are almost ubiquitous and have long been known to be predators of insects. More than 30 species of spiders are known to feed on phytophagous mites in apple orchards in Canada; also as many species of spiders are known in Japan to be mite predators. The small or young spiders feed on mites, but evidence of real importance is lacking.

INSECTS AS MITE PREDATORS

Coleoptera

Two families in the Coleoptera contain important mite predators namely, the *Coccinellidae* and *Staphylinidae.*

Coccinellidae

Stethorus: Insects of the genus *Stethorus* (pl. 9, *a, b*) are effective predators only of mites, but these predators may utilize other food for survival. Many species are relatively small and remarkably well adapted to live and search for prey in the habitats of plant-feeding mites. Species have been reported in almost all areas where mite predators have been studied. Some of the *Stethorus* species, their prey, and plant host are listed below.

Predator	*Prey*	*Crop and Area*
Stethorus picipes Casey	Various tetranychids	Citrus, avocados, walnuts, melons, apples in western United States
S. punctum Leconte	tetranychids	deciduous fruit trees
S. punctillum Weise	tetranychids	apple, beets, in greenhouses in Europe, Asia

S. *utilus* Horn	*Eotetranychus sexmaculatus*	citrus in Florida
S. *atomus* Casey	*Eutetranychus banksi*	on citrus and other crops in Texas
	Oligonychus (R.) pratensis	
S. *japonicus* Kamiya	*Panonychus citri*	citrus in Japan
S. *gilvifrons* Mulsant	many	various crops in Lebanon
S. *bifidus* Kapur	tetranychids	England

Biological studies have been conducted on a few species of *Stethorus*. Under the most favorable temperature conditions, development may be completed in 2 weeks, a slightly longer time than is required for the development of most plant-feeding mites. There is also a longer oviposition period, yet when food is abundant, their rate of oviposition is higher than their mite prey. Repeated matings are required for continued production of fertile eggs, but the sex ratio is about equal. *Stethorus* species are capable of consuming large numbers of mites, in excess of 40 adult or large immature spider mites per day or a total during the larval development of more than 200 per insect. These beetles require 15 adults and 50 to 100 eggs or young mites to have sufficient food for egg production. Most *Stethorus* species are general mite feeders; S. *punctillum* Weise, however, does not feed on *Bryobia*, and S. *gilvifrons* Mulsant ceases oviposition when *Bryobia* is the only source of food. In the absence of mites, *Stethorus* will feed on raisins, nectar, or honeydew; or they may even become cannibalistic. Host preferences of this genus need to be more adequately studied.

Climate and weather effect the seasonal history of both *Stethorus* and their prey similarly. In temperate climates, for example, species such as S. *punctillum* Weise hibernate as adults and have two or three generations per year. Others, including S. *gilvifrons* Mulsant, S. *picipes* Case, and S. *bifidus* Kapur, hibernate in temperate climates; but continuous development occurs throughout the year in semitropical climates.

Stethorus has no apparent ability to find its prey except by contact. S. *picipes* searches in a random pattern for prey and on finding and consuming a mite it subsequently searches more intensively in the immediate area. *Stethorus* species exhibit positive phototropism similar to their mite prey and tend to deposit their eggs among mite colonies.

Some plants with hooked trichomes are favorable for mite development and even for the activities of adult *Stethorus*, but unfavorable for *Stethorus* larvae. For example, bean varieties with hooked trichomes impede movement and feeding activity or even may kill *Stethorus* larvae.

Stethorus species are specialized predators of spider mites; but they require high prey densities before a rapid increase in numbers occurs. For this reason these beetles rarely exert a suppressive effect on mite populations before economic population levels are reached. *Stethorus* will not remain on the plants when mites are extremely scarce; therefore, unless reservoirs of prey are readily available, the chances of large numbers of immigrants reaching the prey populations are remote. But there are situations in which *Stethorus* species effectively control mite populations. Avocado trees will tolerate high densities of the avocado brown mite, *Oligonychus* (*O.*) *punicae* (Hirst), in southern California and *Stethorus picipes* is generally able to increase in adequate numbers to suppress mite populations before serious injury to the plant is produced.

Although the larger species of Coccinellids, that is, *Hyppodamia*, or lady beetles are not primarily mite predators, they sometimes exert a suppressive effect especially when gravid females move in from other plants and lay eggs among heavy mite populations. In this case there may be sufficient offspring to reduce mite populations adequately. But these coccinellids are usually only of minor importance in helping to regulate plant-feeding mite populations.

Staphylinidae

Staphylinid beetles belonging to the genus *Oligota* are the only mite predators in this family. Observations or studies have been made on but a few species that seem to be specialized mite feeders. Some of these and their prey are:

Predator	*Prey*	*Predator Distribution*
O. flavicornis	Mites on deciduous fruits	Europe
O. pussilima (Gravenhorst)	Tetranychids	Switzerland
O. oviformus (Casey)	*P. citri*	Japan
O. flavicornis (Boisduval)	*P. citri*	Japan
Oligota spp.	*P. ulmi*	Japan
O. oviformus (Casey)	Subtropical fruit mites	California
O. pygmaea Solier	Apple mites	Chile

The time required for the larvae of *O. flavicornis* to develop under field conditions in England is 8 to 15 days. Pupation occurs just below the ground surface. Females may live five weeks and lay 40 to 50 eggs that have an incubation period of 4 to 7 days. Thus the complete life cycle from egg to adult occurs in about 28 days. Both adults and larvae prefer active stages of spider mites rather than eggs. There is only 1 and a partial second generation per year after which the beetles overwinter as adults. The incubation period of *O. oviformis* in the laboratory at 27 C (80 F) is 4 days, and the larvae seek pupation sites in the soil after a minimum of 4 days feeding. The duration of the larval plus prepupal stage is 8 to 13 days, the pupal stage 9 to 13 days, and the oviposition period 30 days. A female is capable of producing as many as 300 eggs. Each larva may consume 20 mites per day or 200 to 300 during development and each adult about 10 mites per day. Neither of these species are considered effective mite predators, but they aid other predacious insects and mites in regulating injurious mite species.

NEUROPTERA

Chrysopidae

Chrysopids are mainly aphid predators; but, being general feeders, their prey include mites. *Chrysopa carnea* Stephens is a general predator of mites in both Europe and North America where it is considered an important member of a complex of predators that influence European red mite populations in some districts. It feeds on citrus and avocado mites in California especially when attracted to these trees by other prey. Several species studied were able to develop to adults on mites alone, but *C. carnea* is reportedly unable to complete development when given only European red mites. Chrysopids are voracious feeders. For example, last instar larvae of *C. carnea* may consume an average of 1000 to 1500 citrus red mites daily, and complete development in 13 to 19 days at 26.7 C (80 F). *C. vulgaris* Schneider larvae may consume 30 to 50 European red mite larvae per hour, This species requires 18 to 31 days to complete development at 35.6 and 25 C (96 and 76 F), respectively.

Chrysopids have a better searching capacity than *Stethorus*. They can find their prey at extremely low mite densities so are usually the first predators to appear following applications of DDT.

Hemerobidae

The Hemerobidae are general predators that feed on mites. But observations indicate that they are of no significant importance.

Coniopterygidae

The Coniopterygids also have general feeding habits and thus may utilize many mite species as food. Three species are common on citrus in California, namely *Parasemidalis flaviceps* Banks, *Conwentizia nigrans* Carpenter, and *Coniopteryx angustus* Banks. They are occasionally important as predators of citrus red mite *Coniopteryx vicina* Hagen feeds on citrus rust mite and six-spotted mite on citrus in Florida. *C. vicina* Hagen, *Conwentizia hageni* Banks, and *C. psociformes* (Curtis) feed on the European red mite on peaches on Ontario, Canada. *Conwentiza pineticola* Enderlein and *C. psociformes* are predators associated mainly with outbreaks of European red mite in England and Finland. *Coniopteryx tineiformis* Curtis and *Semidalis aleyrodiformis* (Stephens) are associated with European red mite in England, and *Semidalis albata* Enderlein is a predator of the European and citrus red mites in Japan. Feeding studies indicate that some coniopterygids generally require sources of food other than mites to complete their development. Other species in this family fed only mites develop normally. For example, *Conwentzia pineticola* Enderlein completes its life cycle on the European red mite in 16 days and females lay about 5 to 7 eggs per day; the average number laid per female is about 107. *C. pineticola* may consume 30 to 40 European red mite third instar larvae per day, and complete two and a partial third generation per year. Some species fed entirely on a mite diet developed to maturity, but not without adverse influences on their development. *Coniopteryx vicina* is able to complete its life cycle on the six-spotted mite, *Eotetranychus sexmaculatus* (Riley) and the citrus rust mite, *Phyllocoptruta oleivora* (Ashmead). When fed entirely on the six-spotted mite, the total developmental time averaged 36.4 days, but mortality was high and larvae were trapped in the mite webbing. When fed on the citrus rust mite the developmental time was 43.5 days at 26 C (80 F), but again mortality was high. Females produce from 2 to 5 eggs per day over a period of 16 to 26 days and deposit a maximum number of 266 eggs per female.

Hemiptera

Many predacious Hemiptera exist on some crops; and many are suspected or known to feed on red spider mites. The majority belong to two families, the *Anthocridae* and *Miridae.* Few, if any, appear to be specialized predators of mites, but where mites become abundant in sprayed orchards they provide a large part of the diet of these bugs.

Anthocoridae

Anthocorus musculus Say actively feeds on the European red mite *Panonychus ulmi* (Koch) in Nova Scotia, and *A. nemorum* L. feeds on the two-spotted mite *Tetranychus (T.) urticae* Koch living on beans in England. *Orius minutus* L. is listed as a predator of the European red mite in many places in Europe including England. *O. insidiosus* (Say) and *O. tristicolor* (White) are well known predators of the two-spotted spider mite, the European red mite, the citrus red mite, *P. citri,* and of mites infesting cotton in North America.

Miridae

Blepharidopters angulatus (Fallen). Among the Miridae, this species appears to be one of the most important insect predators of mites in orchards in England. During a period of high population of European red mite the predator thrives and often reduces mite populations late in the season. It is partially phytophagus so may be able to survive when mite populations are low and thus is able to reduce the European red mite populations to very low levels. This may result in low overwintering egg populations and consequently low initial infestations the following spring. *B. angulatus* has but one generation per year, which tends to cause cycles of European red mite populations that is high and low infestations in alternate years.

Eggs of *B. angulatus* are laid from July to October in the wood of trees where they remain embedded until the following spring. The nymphal stages develop in 35 to 39 days. Adult females deposit an average of 43 eggs during their life span of about 51 days. Females consume as many as 4,000 adult mites during their lifetime or about 50 per day during the adult stage. This species feeds on plant tissue, but it apparently causes no plant injury.

Campylomma verbasci (Meyer). This species occurs in orchards in Europe and North America. It feeds on the European red mite and *Bryobia* larvae, but it may also feed on the apple fruit causing some damage to the crop.

Hyaloides harti Knight. This species is common in apple orchards in North America. It has a single generation per year so tends to produce cyclic fluctuations in mite populations.

Nabidae and Lygaeidae

Bugs in the families Nabidae and Lygaeidae are general predators that consume mites during their feeding, but their value has not been ascertained.

Thysanoptera

At times predacious thrips may reduce mite populations rapidly. The six-spotted thrips, *Scolothrips sexmaculatus* Pergande, is a specialized predator of mites in North America. It preys on several tetranychid mite species living on several important crops, often causing rapid reduction in populations of these injurious mites, but its rate of population increase is slower than that of spider mites. Thus predator population increase is too slow to prevent development of injurious mite populations. The time required for development of each stage of the six-spotted thrips in the laboratory is: egg, 10 days; larva, 5 days; prepupa, 1 day; pupa, 5 days; and the adult lives 2 to 3 weeks.

Other species of thrips predacious on mites include *Scolothrips longicornis* Priesner, which feeds on the two-spotted spider mite and other mites in East Germany and Austria; *Crypothrips nigripes* Reuter, a predator of hibernating mites of the same species in East Germany; and *Haplothrips faurei* Hood, now considered one of the most important predators of the European red mite and the brown mite, *Bryobia rubrioculus* (Scheuten) in Ontario, Canada. Thrips do not use their legs to grasp prey; therefore, they are limited to nonmotile forms of mites. The larvae of *H. faurei* consume an average of 143 eggs of the European red mite during a developmental period of eight to ten days, and the adult female consumes an average of 43.6 eggs per day. About 33 days are required to complete their life cycle. In Nova Scotia there are two to three generations a year.

Diptera

The larvae of a *Cecidomyed* fly, *Arthrocnodax occidentalis* Felt, feeds particularly on the six-spotted mite in California. It appears to be adapted to mites that live in colonies. A larva may consume 380 mites in 17 days. Other species of *Arthrocnodax* as well as the larvae of *Syrphidae*, *Dolichopodidae*, and *Empididae* frequently feed on tetranychid mites, but their effectiveness has not been evaluated.

REQUIREMENTS FOR EFFECTIVE PREDATION

Among other characteristics, a predator needs some of the following to be effective:

1. Short developmental period—shorter than that of the prey.
2. High reproductive potential.
3. Capability of consuming many prey or ability to survive on very few.
4. Plant host preference the same as the prey.
5. Effective searching capacity at low densities.
6. Micro-habitat preference the same as the prey.
7. Seasonal cycle corresponding with that of the prey.
8. Ability to tolerate weather extremes as well as the prey.
9. Ability to tolerate pesticides as well as the prey.

None of the predators meet all of these qualifications, but each possess some. Many of the phytoseiids have a shorter life cycle than their prey, equivalent reproductive potential, good searching capacity, and ability to survive on relatively few prey. This is especially true of those capable of feeding on other food sources, but most phytoseiids are limited in the amount of prey they can consume.

Stethorus species are unable to survive on low populations, which limits their ability to keep the prey at low density. Staphylinids have a long developmental period and lack food-searching ability.

The life cycle of the larger insect predators is too long to match the reproductive potential of the plant-feeding mites. They consume a large number of prey and thus are capable of reducing high populations, but are generally unable to prevent the development of injurious mite populations. *Chrysopa*, with their excellent searching ability and wide host range, may be able to prevent the development of high mite populations. But this only occurs when there are sufficient other prey to maintain the *Chrysopa* population when mite populations are low.

EFFECTS OF CHEMICAL TREATMENTS ON PREDATORS

Agricultural sprays and dusts—fungicides, insecticides, acaricides, and nutritional sprays—may have drastic effects on natural enemies of mites. Many of the commonly used pesticides have relatively broad spectrum toxicity to insects and mites, yet in certain situations, some are relatively innocuous to predators. Predator reduction by pesticides may arise through direct mortality, reduced natality, or decreased amount of the prey or other food sources. The predators may die or be unable to complete their seasonal cycle depending on the amount of prey or other food available.

Deposits of some inert dust or spray residues are toxic to, or impair movement of, or retard reproduction of certain mite predators. Some fungicides, such as sulfur, are toxic to or adversely affect development of a few predators. Other fungicides, such as glyodin, zineb, maneb, ziram, ferbam, and selbar are relatively noninjurious to the development of most predator populations. Most of the chlorinated aryl hydrocarbons and DDT relatives are toxic to predators; some species, however, are tolerant to each of these, even DDT. Tolerance to DDT has been noted in *Chrysopa* spp., *Anthocoris musculus* Say, the mite species *Amblyseius fallacis* (Garman), *A. cucumeris* (Oudemans), *Typhlodromus caudiglans* Schuster, and *T. pyri* Scheuten, as well as predacious stigmaeids. Dieldrin is only moderately toxic, endosulfan is relatively nontoxic, and some specific acaricides as dicofol, ovex, Aramite, tetradifon, Omite®, and chlorobenzilate have low toxicity to most of the phytoseiid mites.

The organophosphorus compounds, such as parathion, malathion, azinphosmethyl, and TEPP, are highly toxic to insect predators and predaceous mites; *Typhlodromus occidentalis* Nesbitt, however, is relatively immune to these compounds. Like the organophosphorus compounds, the carbamates generally reduce populations of insect and mite predators. Isolan, however, has relatively little effect on phytoseiid mites.

The Nitrophenols and their derivatives are generally detrimental to predacious mite populations. Dinitrocresols are particularly toxic to *Phytoseiulus macropilis* (Banks), *Typhlodromus tiliarum* Oudemans, *A. finlandicus* (Oudemans), and *T. pyri.* Binapacryl causes reductions of *T. pyri* in the field and Dinocap is toxic to *P. persimilis* (= *riegeli*), *T. pyri* Scheuten, and *T. caudiglans* Schuster. Morestan® reduces most predacious mites and *Stethorus* populations.

Predatory mite and insect eggs and their active stages are killed when they are contacted by a lethal amount of petroleum oil during the spray operation, but oil residues have little effect on predators. Winter eggs of some species are susceptible to oil applications, but many eggs are located out of reach of the sprays.

Systemic compounds, applied to the plant parts not inhabited by predators, can be harmful indirectly as the predator feeds on its prey. Plant root drenches containing dimethoate or thionazin are toxic to *Phytoseiulus persimilis* Athias-Henriot that feed on resistant two-spotted spider mites infesting the treated plants. The destruction of competitive predators may result in an increase of a predatory species not affected by the application.

SELECTED BIBLIOGRAPHY

McMurtry, J. S., C. B. Huffaker, and M. van de Vrie. 1970. Ecology of tetranychid mites and their natural enemies. Pt. I. Tetranychid enemies: their biological characters and the impact of spray practices. Hilgardia 40(11):331–390. (396 references)

Parrott, P. J., H. E. Hodgkiss, and W. J. Schoene. 1906. The Eriophyidae, Pt. I. The apple and pear mites. N. Y. Agr. Expt. Sta. Bull. 283:302–303.

Putman, W. L. 1970. Occurrence and transmission of a virus disease of the European red mite, *Panonychus ulmi* (Koch). Can. Entomol. 102(3):305–321.

Shaw, J. G., H. Tashiro, and E. J. Dietrick. 1968. Infections of the citrus red mite with virus in central and southern California. J. Econ. Entomol. 61(6):1492–1495.

Chapter 6

Mites and Plant Diseases

The term plant disease in its broadest meaning includes all injuries or abnormalities generated from sources outside the plant regardless of the cause. In this chapter the meaning is restricted to abnormalities caused by biological organisms including viruses. The feeding by some mites appears to have little more effect on the plant than to puncture epidermal tissue and remove the plant cell contents. Feeding by other mites appears to inject toxins or growth regulators into the plant. Feeding by still others transmits virus diseases, or sometimes spreads fungus diseases. Specific mite-host relationships are discussed in chapters 8, 9, 10, 11, and 13.

LOCAL INJURIES

Removal of Cell Contents

Mite species that produce injury primarily by removing cell contents cause economic damage only when sufficient plant material has been removed or lost as a result of the feeding over a period of several days by high mite populations. The population and feeding interval required to produce visible or measurable injury by these species is influenced by the vigor of the plant, the food and moisture supplied from the roots, and the transpiration rate, all of which are affected by weather conditions.

It is often difficult to determine whether the unhealthy plant symptoms or plant abnormality caused by mites result entirely from physical means or whether some local toxin is secreted into the plant during feeding. Such mites as the avocado mite, *Oligonychus* (*O.*) *punicae* (Hirst), *Bryobia* species, some eriophyoids, and other plant-feeding species appear to affect the plant only by such physical means.

Modification of Developing Tissue

Eriophyoid mites that feed on buds or growing plant tissue may injure but rarely cause complete destruction of stem, leaf, or fruit primordia. Auxiliary primordia may develop and as each newly developed primordium is injured, the resulting growth pattern becomes irregular and sometimes even bizarre (pl. 10, *a*, *b*). Such symptoms are often similar in appearance to those caused by virus diseases, by systemic toxins injected by mites, or by feeding by other plant pests. Typical symptoms include irregular deformities of growth pattern, rosette type growth, irregular leaf or fruit growth, and even the total destruction of the growing tips; but this total effect is rare. Leaves occasionally may become chlorotic but no necrotic tissue occurs except within the buds or on young tissue. Eriophyoid mites that produce plant injury, such as the citrus bud mite, *Eriophyes sheldoni* Ewing, the bud strains of the pear leaf blister mite, *Phytoptus pyri* Pagenstecher, the grape erineum or blister mite, *Colomerus vitis* (Pagenstecher), as well as the cyclamen mite, *Steneotarsonemus pallidus* (Banks), the broad mite, *Polyphagotarsonemus latus* (Banks), of the family Tarsonemidae, and many other mites cause this type of injury. Sometimes these and other mite species feed on partially developed leaves causing the destruction of some tissues. Uninjured tissue continues to grow, causing the development of irregular leaf patterns.

Mites that feed as colonies on the undersides of leaves, such as the six-spotted mite, *Eotetranychus sexmaculatus* (Riley), on citrus, retard local tissue growth on the underleaf surface. Growth continues on the upper surface, which results in inverted saucer-shaped areas with the upper surface raised and usually chlorotic. The broad mite feeds on the underleaf surface near the periphery; such feeding causes the edges of the leaves to roll under.

It is difficult to account for the relatively severe injury produced by low populations of some plant-feeding mites except by assuming that they inject toxins or growth regulators, into the plant during the feeding process. The serious effect on pear trees produced by relatively low populations of the Pacific mite, *Tetranychus* (*A.*) *pacificus* McGregor, indicates an involvement of a toxin with at least localized effects. Plant galls and some localized deformities produced as a result of the feeding by a few mites appear to be caused by an injected chemical; the symptoms of the latter have been sometimes reported as plant diseases.

Gall-Forming Species

Feeding of the gall-forming eriophyoid mite species produces local deformities of the plant which are so regular for each species and so different between species that students of galls have concluded that each species must inject a specific growth regulator into the plant to cause such regular or characteristic growth responses. The galls produced by each mite species are often so uniform in shape and size that the appearance of the gall served originally as a means of identification (see chapter 12). The galls produced by the various gall-producing mites vary from those that merely produce hairiness on localized areas of the leaf to those that cause wartlike protrusions (pl. 11).

INJECTION OF SYSTEMIC OR PERSISTENT TOXINS

A few mites appear to inject a systemic and persistent toxin or growth regulator into their host which causes a disruption or increase in localized growth at some distance from the feeding area. Such growth regulators seem to stimulate continuous growth, perhaps similar to the way cancer causes growth in animals. Such injury is produced by *Brevipalpus* species (Vergani, 1945; see also chap. 9) and

Acalitus phloecoptes (Nalepa) (see chap. 13). *Calacarus citrifolii* Keifer appears to inject a chronic toxin into citrus trees which appears to remain more localized. *Eriophyes tulipae* Keifer, besides transmitting viruses and other diseases, may inject a systemic toxin into corn plants and some grasses. See chapter 13 for additional discussions of these eriophyid mites and an index to species of injurious eriophyoids.

Kernel Red Streak of Corn

Kernel red streak of corn appears to be caused by the injection of a salivary toxin by *Eriophyes tulipae* Keifer. The symptoms on corn are reported from north central North America and southern Canada where *E. tulipae* is abundant on wheat. Similar symptoms on corn occur in France, Bulgaria, Rumania, and Yugoslavia.

Almond Gall Mite Injury

The growth regulator injected by feeding of the almond gall mite, *Acalitus phloecoptes* (Nalepa), on almond trees appears to cause permanent irregular galls around the buds, and even deformed fruit spurs and irregular woody tissue. The entire physiology of the trees seems to be upset, as indicated by the failure of fruit buds to form, the loss of tree vigor, and the early death of the tree. Damage appears to be progressive and irreversible, all of which suggests the mite injects a growth regulator that produces a more-or-less systemic effect in almond trees. Only local symptoms occur, however, when the almond gall mite feeds on plum trees.

Grey Mite Injury

The grey mite, *Calcarus citrifolii* Keifer, produces toxicogenic symptoms referred to by the descriptive name *Concentric Ring Blotch* (pl. 12, *a*, *b*) (Dippenaar, 1956). This mite initiates symptoms primarily on actively growing tissue. The injected toxin penetrates through the leaf and produces necrotic tissue on the opposite leaf surface. The toxin also appears to spread laterally enough that the small spots of necrotic tissue unite to form a blotch over much of the leaf. Mite feeding on the midrib of young leaves results in the development of a concentrically marked blotch of an oakleaf pattern, which usually covers more area of the leaf than other discrete blotches on the same leaf, suggesting greater spread of the toxin along and out from the veins. Lesions tend to develop into oval rather than circular blotches on twigs and stems. The toxin appears to penetrate into the tissue of the twig, causing the lesions exposed to direct sunlight to appear as darkened resinous rings. These symptoms are followed by longitudinal splitting of the twigs and copious gum exudation. Eventually scaly bark occurs and sometimes shoots and branches are girdled. Many other symptoms may occur, but this series of symptoms provides evidence that as the mites feed a relatively chronic toxin is injected which spreads over localized areas of the citrus tree. Studies have strongly indicated that a virus is not involved.

Brevipalpus Mite Injuries

The disorders produced by several species of *Brevipalpus* mites on citrus and other plants indicate that these mites introduce a persistent systemic toxin into the plant during the feeding process. A disorder of sweet orange called Leprosis (pl. 13; Vergani, 1945) has symptoms resembling those of concentric ring blotch. The disease symptoms first appear as small irregular platelets of dry gummy material on leaves, fruit, and twigs. As injury increases and the injured twigs grow into large limbs, the lesions increase proportionately in size until they resemble the scrofulous shelling of the bark, known as "shell bark." Feeding by these mites on sour orange seedlings produces proliferation of lesions along main stems. An abscission of the initial leaves occurs and adventitious buds sprout, but are successively killed, producing hypertrophies at the bud loci. These woody protuberances or galls may be up to 5 mm in diameter. They appear as axils that have proliferated to a point where bud-studded cushions develop instead of leaves.

VIRUS TRANSMISSION

Mites belonging to the Eriophyoidea have been known since 1933 to transmit plant viruses (Amos et al., 1927), but symptoms produced by mite feeding are often similar to those resulting from virus infection. In addition, the very small size of these mites and their inclination and ability to work their way into buds and other protected places make it difficult to ascertain whether the symptoms result from an injected toxin or a virus transmitted by the mites. Thus special techniques must be used to establish proof that the observed abnormalities result from a virus transmitted by the mites rather than from effects of a toxin injected as the eriophyoid mites feed on the plant.

It has been suggested, therefore, that three steps be completed before assuming that an eriophyoid mite is a vector of a plant virus: (1) The presence of the mite should be correlated with the appearance of the disease in nature; (2) The development of disease symptoms must not depend on the continued presence of the mites. Evidence that the causal virus can also be transmitted by artificial means without mites is preferable. (3) The mites must not be able to induce the disease symptoms on healthy plants until after they have fed on diseased plants or have acquired virus in another way. Mite-transmitted viruses that are not sap transmissible present a particular problem in fulfilling the second requirement, as it is very difficult to certify that all the plant tissue has been freed from mites (Oldfield, 1970; Proeseler, 1967, 1968, 1971; Slykhius, 1960, 1963, 1965; Smith, 1957).

Wheat Streak Mosaic

Wheat streak mosaic was found in 1929 to be caused by a virus that could readily be transmitted in the laboratory by sap inoculation. In 1952, transmission in the field by the wheat curl mite, *Eriophyes tulipae* Keifer, was demonstrated. Plants inoculated with wheat streak mosaic virus may show visible symptoms in 6 to 8 days. When mites reared on diseased plants are individually transferred to healthy plants, about ⅓ of the plants become infected with the virus (as confirmed by sap transmission). All stages of the mites, except the eggs, can transmit the disease, although the older adults lose their ability as vectors. Viruliferous mites remain infective for 6 days on a host immune to the virus and the ability to transmit the virus is not lost during molting. Nymphs can become viruliferous during a 10- to 30-minute feeding period on diseased wheat. Apparently adults may acquire the virus, but are unable to transmit it unless the mites had access to the virus before reaching the adult stage.

Symptoms are not a reliable criteria for identification of wheat streak mosaic. Factors affecting symptom expression include the time of infection, temperature, soil moisture, soil fertility, the strain of the virus, and the variety of wheat. Symptoms of the disease are seldom recognizable through the fall and winter, but brief periods of warm weather in late March and April cause the symptoms to appear; and injury increases in severity as long as rapid plant growth continues.

The first symptom seen under proper light conditions is a faint green mottling of the expanding leaves, followed by light-colored streaks along the veins, and a general stunting of the plants. As the disease progresses, chlorosis becomes more pronounced, appearing as yellowish-green to markedly yellow-mottled striping; also, infected plants tend to become less upright in growth than healthy plants. Moderately infected plants may produce poorly filled heads containing shriveled kernels, but heads are not formed on severely infected plants. Although these symptoms may be caused by factors other than the virus, the presence of the virus may be determined by rubbing juice from freshly diseased plants to healthy plants in the 2- to 3-leaf stage. After 6 to 8 days at 20 to 25 C (68 to 77 F), chlorotic dashes and streaks become evidence of infection.

Wheat streak mosaic virus may be transmitted by *E. tulipae* to barley, oats, corn, rye, certain wild annual grasses, as well as to wheat. The mite can colonize on some common perennial plants, including Wheeler blue grass, Western wheat grass and Foxtail barley, which are apparently immune to this virus. *E. tulipae* may also colonize on *Agropyron trichophorum* (Link) Richt, Canada wild rye, Virginia wild rye, Indian rice grass and Canada blue grass, all of which may become infected with the virus. Some of these may be natural reservoirs of the virus, but as yet no perennial plant that harbors the virus has been shown to be a source of infection for wheat in the field.

Although wheat is the main plant to consider in the epidemiology of wheat mosaic virus, barley and rye have been known to be an occasional source from which *E. tulipae* may transfer the virus to spring wheat fields (Slykhius, 1960, 1963, 1965).

Currant Reversion Disease

Early in the twentieth century, normal black currants in Britain and Holland changed completely in character and became unproductive. The bushes were thought to have reverted to the wild ancestral type, so the condition became known as "reversion." It was later found that unproductive vines resulted from a virus disease transmitted from one bush to another by an eriophyid mite now known as *Cecidophyopsis ribis* (Westwood and Nalepa).

The most reliable diagnostic symptoms of reversion are the reduced numbers of submain veins and the coarsely toothed margins of the leaves. Other symptoms include the development of a crowded woody growth or "nettlehead" from lateral buds, abnormal flower color, flower drop, and the shriveling and dropping of immature fruit, but any of these symptoms may result from other causes. Although *C. ribis* is always associated with reversion the mite alone may cause the gall-like swellings called "big bud." Reversion virus is apparently not sap transmissible, but is readily transmitted by grafting diseased shoots, or even wood without buds, onto healthy bushes. About 18 months are required, however, for the symptoms to appear.

There is symbiotic relationship between the virus disease and the mites. One effect of the virus is to decrease the hairiness of the leaves and young stems. Mites are impeded in progress by the hairs during spring migration. The hairs on the stipules at the base of the petioles are especially effective in protecting the most susceptible buds from mites. As a result of the natural protection provided by these hairs, only "reverted" bushes with their decreased hairiness develop many galls and the healthy bushes appear to have resistance, that is, infection with the virus is in proportion to the hairiness of the flowers and vegetative parts. Therefore, currant varieties having the fewest number of hairs are the most susceptible to mite infestation and thus to virus infection. This striking adaptation between mites and the virus is of considerable mutual advantage. The effects of the virus are particularly subtle in that susceptibility of the plants to the mites is increased but vegetative vigor or the number of buds available for colonization by mites is not decreased. The favorable conditions for mite population development provided by the virus increases the possibilities for mite dispersion and therefore the spread of the virus.

Wheat Spot Mosaic

Wheat spot mosaic virus is reported only in Alberta, Canada. During studies on wheat streak mosaic it was discovered that transfers of indiviidual *Eriophyes tulipae* Keifer from naturally infested wheat to test plants caused some plants to develop chlorotic spots, severe chlorosis, stunting, and necrosis instead of the expected streak symptoms. No sap transmissible virus could be detected, and symptoms continued to develop even after the plants were freed from mites. The possibility of a toxin or feeding injury was eliminated when it was found that the progeny from mites that produced the symptoms did not induce symptoms unless they were first colonized on diseased plants.

Different isolates of the wheat spot mosaic virus differ in severity; some cause extreme chlorosis and kill the plants quickly. Mites can carry both wheat streak mosaic and wheat spot mosaic viruses, but plants infected with both viruses become more severely diseased than plants infected with either virus alone. The epidemiology and control of wheat spot mosaic are the same as for wheat streak mosaic. All stages of *E. tulipae* except the eggs carry the virus. Mites reared on diseased wheat remain infective for 13 days after transfer to a plant immune to the virus. The hosts of wheat spot mosaic include barley, corn, rye, and several wild grasses (Oldfield, 1970; Slykhuis, 1953).

Fig Mosaic

Fig mosaic, described in California in 1933, is the first tree virus demonstrated to be mite transmitted (Condit and Horne, 1933). In 1951 the virus was transmitted experimentally from diseased to healthy plants using the fig mite, *Eriophyes ficus* Cotte. The mites and the effects of the disease are known also in Australia, the British Isles, Bulgaria, New Zealand, Italy, and India.

The mosaic symptoms first appear on leaves and fruit as uniform spots of different sizes. The light yellow-colored spots on leaves contrast strongly with the natural green color of the foliage, yet the borders of the spots are indefinite. The combined spots appear as irregular light green patches diffused widely throughout the leaf blade with no regular relationship to the veins. The leaves may be formed into an infinite variety of shapes and sizes. The appearance of mosaic spots on fruit is similar to that on leaves. The effects on the fruit and foliage may result in premature dropping of figs, but seldom does the disease cause a necrosis of the leaf tissue.

Different fig varieties vary in their susceptibility to the virus. All varieties except an entire leaf form of *Ficus palmata* are susceptible, but Kadota and Calimyrna are only slightly affected, and White Adriatic and Brown Turkey are relatively resistant under good cultural conditions, but Mission is severely affected. *Cudranea tricuspidata*, a member of the family *Artocarpaceae*, is the first susceptible host found outside the genus *Ficus*.

The virus is not transmitted through seed or by sap inoculation, but *E. ficus* is a very efficient vector. Introduction of one viruliferous mite per test plant resulted in seven of ten plants developing mosaic. The minimal feeding time for infection is 5 minutes and the virus may be retained in the vector for more than 20 days at 5 C (41 F). The acquisition of the virus by the vector is possible 4 days before the appearance of virus symptoms. Nonviruliferous mites fed on symptomless leaf sections of diseased plants are able to transmit the virus. Virus transmission was more successful when nonviruliferous mites acquired the virus from the under than the upper side of a leaf with symptoms. Most infections are caused by the mites on the terminal buds (Proeseler, 1971). So efficient is the vector that all field-grown trees in California are infected. The disease is not transmitted through the egg.

Rye Grass Mosaic

Rye grass mosaic virus disease of grasses in northern Europe and North America is known to be transmitted by the eriophyid mite, *Abacarus hystrix* (Nalepa). The disease is common on perennial rye grass and Italian rye grass, but perennial rye grass appears to provide the main widespread, permanent reservoir of both virus and vector. The virus may be transmitted from Natel grass or infected wheat to winter wheat (Mulligan, 1960). The symptoms include green to yellow mottling and streaking of the leaves. Some isolates cause severe brownish necrosis of leaves and reduce growth and vigor of some strains of rye grass. There are several other virus diseases that have similar symptoms on rye grass, such as cocksfoot streak virus transmitted by aphis, a mosaic in Germany transmitted by nematodes, and a virus exhibiting similar symptoms, but with an unknown vector in British Columbia.

Rye grass mosaic virus is sap transmissible. The mite is a relatively inefficient vector of this disease. Nonviruliferous mites may acquire the virus in 2 to 12 hours, but lose their infectivity within 24 hours when placed in immune hosts. All stages of the mite except the eggs carry the virus (Oldfield, 1970).

Peach Mosaic

Peach mosaic was described as a new virus from Texas in 1932. It was experimentally proven in 1955 to be transmitted by the mite species subsequently named *Phytoptus insidiosus* (Keifer and Wilson). Peach mosaic is also known to occur in Colorado, California, Utah, Mexico, Arizona, Oklahoma, and Arkansas in the United States. It was so highly contagious and destructive that commercial peach plants became unproductive in three to six years. The symptoms of the disease include shortening of the internodes, profuse growth on leaf axil buds, and mosaic patterns on leaves, which are often small, narrow crinkled, and irregular in outline. The symptoms appear, under greenhouse conditions, in 14 to 100 days after inoculation. Field grown trees inoculated when breaking dormancy in the spring may develop mosaic symptoms in 3 to 6 weeks, but if trees are in foliage when inoculated the symptoms do not develop until the next spring. In the absence of a virus source infective mites retain their ability to transmit the virus for at least 2 days.

The host range of this virus includes all peach and plum varieties as well as apricots, almonds, and nectarines and some noncommercial *Prunus*; including *P. angustifolia* Marsh in Texas, *P. munsoniana* Wight and Hedrick in New Mexico, and some other species native to areas east of the Rocky Mountains. Most of the *Prunus* species native to western North America, however, appear to be immune, as are cultivated cherries.

The spread of peach mosaic virus has been prevented by controlling the mite vector with a yearly application of diazinon at petal fall time (Oldfield, 1970).

In most commercial peach varieties the vector of peach mosaic is usually limited to retarded adventitious buds found near the base of large scaffold branches where they cause considerable cell hypertrophy. The mites may occasionally be found unprotected on petioles and green stem tissue near leaf axils and buds of wild plums and flowering peaches.

Latent Virus of Plum

A latent virus of different varieties of plum trees and *Prunus* species is transmissible by the gall mite *Aculus fockeui* Nalepa and Trouesart. A single mite may transmit the virus, but more mites increase the infection rate and the numbers of local lesions per leaf. Infection may occur after mites feed a few minutes, but an hour provides time for optimal infection and at 5 C (41 F) the virus may persist in the mite for more than 20 days. The virus is transmitted as efficiently by immature as by adult stages and is transmissible by mechanical inoculation tests. Mites from diseased plants induce local lesions on many *Chenopodium* species but never a systemic infection (Proeseler, 1971). A species of *Chenopodium* is the best test plant; mites, however, usually perish 24 hours after transfer to *Chenopodium*. *A. fockeui* is the first species of eriophyid mite producing a deutogyne that has been reported to transmit a virus (Gilmar and McEwen, 1958).

Cherry Mottle Leaf

Cherry mottle leaf is a virus disease of sweet cherries that occurs in several western states in the United States and in British Columbia in Canada. When transfers of *Phytoptus inaequalis* (Wilson and Oldfield) were made from infected wild bitter cherry to peach seedlings or bing cherries, virus transmission was effected (Oldfield, 1970).

Viruses Transmitted by Tetranychid Mites

Potato Virus Y. There is considerable evidence that a virus disease known as Potato virus Y is transmitted by the two-spotted spider mite. The disease is found in potato, tomato, tobacco, and other Solanaceae. The disease occurs on all continents, but its severity as well as symptoms differ in the various potato-growing areas. The virus is readily transmitted by *Myzus persicae* (Sulzer) and some other aphids. Its transmission by the two-spotted spider mite has been demonstrated under greenhouse conditions, but more study is necessary to evaluate the role of this mite as a vector.

The symptoms appear about 19 days after infection, first as blotchy mottling on the topmost leaves; the mottled appearance then apparently spreads outward from the veins. A fine necrosis appears along the veins on the undersides of the leaves; this is followed by necrosis of the upper leaf surface. Necrosis develops down the petiole to the stem where brown longitudinal lesions appear. During the year of infection the leaves may become completely necrotic—withered, yet they remain attached as by a thread. In subsequent years there may be little necrosis or leaf drop, but affected plants are stunted, their leaves and stems become brittle, internodes are short, leaves become mottled, twisted, and bunched together. The whole plant is dwarfed and rosetted, but the tubers remain normal (Oldfield, 1970; Thomas, 1969).

Other Viruses

Tobacco ring spot virus, *tobacco mosaic* virus, *southern bean mosaic* virus and *cotton curliness* are apparently transmitted by the two-spotted spider mite. When macerates of mites previously fed either on plants infected with southern bean mosaic virus or tobacco mosaic virus were rubbed on healthy leaves, the leaves became infected. Presence of the viruses was verified by electron microscopy. When healthy mites were placed on leaves previously sprayed with diseased plant sap, the plants developed virus symptoms; but similarly treated plants bearing no mites remained symptomless.

Apparently the two-spotted spider mite can carry and excrete some viruses, but it does not inject the causative agent during feeding. Nevertheless, sufficient injury is provided by the mites while feeding for the agent to enter the leaf; apparently the mite stylets carry the virus present on the leaf surface into the plant (Fritsche, Schmeler, and Schmidt, 1967; Thomas, 1969).

Suspected Cases of Virus Transmission by Mites

Several plant pathogenic viruses have been suspected of being transmitted by mites:

1. *Vine Panaschure* is a mosaic disease of grape vines, cucumber, and tobacco. It has been reported, but not confirmed, that the disease can be transmitted from vine to vine by *Colomerus vitis* (Pagenstecher) (Oldfield, 1970).

2. *Agropyron mosaic* virus is a disease of *Natel* grass which occurs in widespread areas of North America, including Virginia, Nebraska, South Dakota, Saskatchewan, Ontario, and eastern provinces of Canada. Infection developed in healthy seedlings, adequately screened against entrance by thrips, when placed near diseased wheat. *Abacarus hystrix* (Nalepa) has been incriminated as a vector of this virus. The disease causes a light-green yellow mosaic on the leaves, frequently in the form of dashes or streaks. Ordinarily chlorosis is mild and sometimes the plants become stunted (Slykhius, 1969; Oldfield, 1970).

3. Cadang-Cadang disease of the coconuts is a destructive degenerative disease of coconut palm that is slowly spreading in the Philippines. It is suspected that the symptoms result from infection by a virus disease. *Notostrix attenuata* Keifer and other eriophyid mites are closely associated with diseased palms, but proof of transmission has not been obtained (Slykhius, 1960).

4. Pigeon pea sterility is a disease characterized by a mosaic pattern on leaves, and partial sterility of the flowers of pigeon pea, *Cajanus cajan* (L.) Millspaugh. Some evidence suggests that a mite may transmit a virus that causes pigeon pea sterility; the evidence, however, is currently inconclusive (Oldfield, 1970).

5. Mosaic symptoms on wheat, corn, and rye and attributed to the mite, *Tetranychus sinhai* Baker. The symptoms appear as a darkening, followed by yellowing and withering in patches along the midrib. There are no experimental data as yet to show that the symptoms are produced by a virus. The symptoms might result from the injection of a toxin by the mites (Slykhuis, 1963, 1965).

6. Rose rosette is reported to be transmitted by *Phyllocoptes fructiphilus* Keifer. A common manifestation of this disease on *Rosa multiflora* Thumberg is the breaking of most or all axillary buds on an otherwise normal stem. A new shoot arising from a basal axillary bud is the first symptom to appear. Such shoots develop consecutively toward the stem apex, frequently grow at an accelerated rate, are thicker than normal, and ususally have shortened internodes, particularly toward the shoot apex. The leaves emerging on these shoots may be normal in size and shape basally, but apically they are small and misshapen. Short secondary shoots bearing very small leaves usually emerge from the auxiliary buds of affected shoots. This shoot proliferation with the crowding of nodes and leaves produces a "witches-broom" or rosette appearance in infected plants. A dwarfed plant results when young plants are infected. A striking symptom is the bright red color of the leaves, which may occur over the entire leaf or on part of the leaflets. Leaves without this red pigment usually exhibit a degree of interveinal chlorosis. Most of these symptoms are typical of those produced by bud mite feeding on plants (Allington, Staples, and Viehmeyer, 1968; Oldfield, 1970).

7. Cherry mottle leaf was first reported in 1920 and later (1935) found to be caused by a virus. It occurs in northwestern United States, often in close association with wild bitter cherry. The disease symptoms have been produced by transferring *Phytoptus inaequalis* (Wilson and Oldfield) from a peach (a symptomless carrier of the disease) to bing cherry and wild bitter cherry. Verification of these

results by graft transmission to other healthy cherries is in progress (Oldfield, 1970).

MITES AS CARRIERS OF OTHER DISEASES

Mites, like other organisms that move about, can be agents that spread fungus spores or other plant diseases. Because mites are limited in their activities as compared to flying insects, their involvement in the spread of disease is usually of minor importance. At times, however, mites can be an important factor in increasing the rate diseases are spread.

Stewart's Bud Rot

Stewart's bud rot or central bud rot of carnations is caused by *Fusarium poae* (Peck) Wollenweber and spread by *Siteroptes cerealium* Kirchner. Young carnation buds may outwardly appear normal, but when opened, show a moist, brownish, decayed mass of the inner floral organs. Stamens, stylets, and petal bases may be completely rotted by the fungus that is generally visible as a heavy, or sparse, white cottony growth (Cooper, 1940).

Bulb Diseases

Eriophyes tulipae Keifer is implicated in the spread of the fungus-causing rot of garlic bulbs in the field and in storage. *Rhizoglyphus* mites are also closely associated with the spread of various bulb diseases caused by *Fusarium*, *Stromatinia*, and *Pseudomonas* fungi (pl. 14). It is reported that *Rhyzoglyphus* species, for example, feed on the causal agent of *Stromotinia* rot of gladiolus, and spread the scab organism, *Pseudomonas* (Forsberg, 1965; Jefferson, Bald and Morishita, 1956) (pl. 15). Mites of this genus congregate on the roots of onions, which causes the onions to tip over and fail to enlarge. The stems of Easter lilies are tunneled at ground level causing the lilies to break at that point. Lily plants that fail to develop properly may also be infested in the lower stem region with *Rhizoglyphus* mites. Infested plants remain small or stunted and have few or no roots (Latta, 1939); such plants also seem frequently to become infected by mosaic. *Rhizoglyphus* mites also attack bulbs of Tuberose, especially when they are incompletely cured or dried (Weigel and Nelson, 1936). Mites enter the neck of the bulb and there are associated with the decaying tissue, which causes the bulbs to rot.

Diseases of other bulbous plants are known to be spread by soil-inhabiting mites; the mite taxonomy and the exact relationship of the mites to the diseases, however, has not been adequately studied.

SELECTED BIBLIOGRAPHY

ALLINGTON, W. B., R. STAPLES, and G. VIEHMEYER. 1968. Transmission of Rose Rosette Virus by the eriophyid mite, *Phyllocoptes fructiphilus*. J. Econ. Entomol. 61(6):1137–1140.

AMOS, J. M., R. G. HATTON, R. C. KNIGHT, and A. M. MASSEE. 1925. Experiments in the transmission of reversion disease in black currants. Ann. Rept. East Malling Res. Sta., Kent, pp. 126–150 (1927).

CONDIT, I. J., and W. T. HORNE. 1933. A mosaic of the fig in California. Phytopathology 23(11):887–896.

COOPER, K. W. 1940. Relations of *Pediculopsis graminum* and *Fusarium poae* to central bud rot of carnations. Phytopathology 30(10):853–859.

DIPPENAAR, B. J. 1956. Concentric ring blotch of citrus–its cause and control. So. African J. Agr. Sci. 1(1):83–106.

FORSBERG, J. L. 1965. The relationship of *Pseudomonas marginata, Stromatina gladiola,* bulb mites, and chemical control treatments to the occurrence of scab and Stronatinia rot of gladiolus. Phytopathology 55(10):1058. (Annual abstract)

FRITSCHE, R., K. SCHMELER, and H. B. SCHMIDT. 1967. Prüfung der Eignung von *Tetranychus urticae* Koch als Vektor Planzenpathogener Viren. Arch. Pflschutz 3(2):89–100.

GILMER, R. M., and F. L. MCEWEN. 1958. Chlorotic fleck, an eriophyid mite injury to myrobalan plum. J. Econ. Entomol. 51(3):335–337.

JEFFERSON, R. L., J. G. BALD, and F. S. MORISHITA. 1956. Effect of vapam on *Rhizoglyphus* mites and gladiolus diseases. J. Econ. Entomol. 49(5):584–589.

LATTA, R. 1939. Observations on the nature of bulb mite attack on Easter lilies. J. Econ. Entomol. 32(1):125.

MARAMOROSCH, K. 1963. Arthropod transmission of plant viruses Ann. Rev. Entomol. 8:401–402.

MULLIGAN, T. E. 1960. The transmission by mites, host range, and properties of rye grass mosaic virus. Ann. Appl. Biol. 48(3):575–579.

OLDFIELD, G. N. 1970. Mite transmission of plant viruses. Ann. Rev. Entomol. 15:343–380. (168 references)

PROESELER, G. 1967. Ubertragung phytopathogener Viren durch Gallmilben. Arch. Pflschutz. 3(3):163–175.

———. 1968. Ubertragungsversuche mit dem latenten Prunus Virus und der Gallmilbe *Vasates fockeui* Nal. Phytopath. Z. 63(1):1–9.

———. 1971. Gallmilben (Eriophyoidea) als Virusüberträger unter besonderer Berüchsichtigung ihrer Morphologie, Okologie und Bekämpfung. Nova Acta Leopoldina, Suppl. 4, Bd. 36. Johann Ambrosius Barth, Leipzig. 123 pp.

SCHULTZ, J. T. 1963. *Tetranychus telarius* (L.), a new vector of virus Y. Plant Disease Rept. 47:594–596.

SLYKHUIS, J. T. 1953. The relation of *Aceria tulipae* K. to streak mosaic and other chronic symptoms of wheat. Phytopathology 43 (9) :484–485.

———. 1960. Current status of mite-transmitted plant viruses. Proc. Entomol. Soc. Ont. 90:22–30.

———. 1963. Mite transmission of plant viruses. Advances in Acarology 1:326–340.

———. 1965. Mite transmission of plant viruses. SMITH, K. M., Advances in Virus Res. 11: 97–137. Lauffer, Academic Press, N. Y. and London.

———. 1969. Transmission of Agrophyon mosaic virus by the eriophyid mite, *Abacarus hystrix*. Phytopathology 59(1):29–32.

SMITH, K. M. 1957. Textbook of Plant Viruses. 2d Ed. Churchill, London. 652 pp. Little, Brown, and Co., Boston.

THOMAS, C. E. 1969. Transmission of tobacco ringspot virus by *Tetranychus* sp. Phytopathology 59(5):633–636.

VERGANI, A. R. 1945. Transmission y naturaleza de la "Lepra explosiva" del Naranjo Argentina Inst. Sanidad Veg. 1, Series A (3):10 p.

WEIGEL, C. A., and R. H. NELSON. 1936. Heat treatment for control of bulb mite on Tuberose. J. Econ. Entomol. 29(4):744–749.

Chapter 7

The Tetranychidae Donnadieu

SYSTEMATICS

Concepts of the family Tetranychidae have undergone considerable evolution since the group was given the suprageneric name Tétranycidés by Donnadieu and raised to family status by Murray in 1877. According to Pritchard and Baker (1955), only two genera, *Tetranychus* Dufour and *Bryobia* Koch were recognized before 1887; at that time certain species of predacious mites properly belonging to several families of the Prostigmata were assigned to the genus *Tetranychus.* Even Donnadieu, in proposing the family name Tétranycidés in 1875, considered eriophyids to be immature stages of tetranychids, and some contemporary workers of that time believed chiggers to represent their young.

Predacious mites properly belonging to several genera of the Raphignathidae were once included in tetranychid genera, and the raphignathid genus *Neophyllobius* Berlese, 1886, was only recently removed from the Tetranychidae (McGregor, 1950).

During recent years the tetranychoid mites have been segregated into separate units of family or subfamily status. The genera *Tenuipalpus* Donnadieu, *Brevipalpus* Donnadieu, *Phytoptipalpus* Tragardh, *Pseudoleptus* Bruyant, *Raoliella* Hirst, *Tuckerella* Womersley, and *Tegopalpus* Womersley, all previously referred to the Tetranychidae, are now placed in the Tenuipalpidae or other families.

Ewing (1913) recognized the importance of the shape of the aedeagus in the male for tetranychid species identification. Subsequently, McGregor utilized this character for his many descriptions of North American spider mites. This work culminated in his paper (McGregor, 1950) on mites of the family Tetranychidae. Pritchard and Baker (1955) used the aedeagus and the leg and body chaetotaxy for species recognition, as well as the higher categories. This was carried further by Wainstein (1960) in his revision of the family. Setal patterns and striation patterns were used to raise the mites to subgeneric and generic levels. More recently, Tuttle and Baker (1968) reviewed the family, including genera and species in Arizona, using the system developed by Wainstein.

Comprehensive systematic surveys are now being conducted in several countries; included are surveys by Carlos Fletchtman, Brazil; Roberto Gonzalez, Chile; M. Zaher and A. A. Attiah, Egypt; Magdalena Meyer, South Africa; Jean Gutierrez, Madagascar; A. T. Bagdasarian and B. A. Wainstein, USSR; David Manson, New Zealand; J. J. Davis, Australia; and Schôzô Ehara, Japan.

TAXONOMIC CHARACTERS

Tuttle and Baker (1968) defined the tetranychids as follows:

> The Tetranychidae Donnadieu possess long recurved whiplike movable chelae set in the stylophore or fused basal segments of the chelicerae [pl. 16, *a*]; the fourth palpal segment bears a strong claw; the tarsi I and II, and sometimes the tibiae, usually bear specialized duplex setae; the claws possess tenent hairs, and the empodium may or may not have tenent hairs [pl. 16, *b*]; the female genitalia is characteristic of the family as well as of the species. Normally there are three pairs of propodosomal, four pairs of marginal, five pairs of dorsal, and one pair of humeral setae. Setae may shift, drop out, or extra pairs may be added.

Characters used for identifying the tetranychids are (fig. 3): (1) the type of tarsal claws and empodia—either padlike or clawlike; (2) the peritremes, which may end in a simple bulb, a distal hook, or an anastomosing pattern; (3) the dorsal setae pattern and type of setae, which may be simple or broadly clavate and serrate; (4) the type of striation pattern in the dorsum of the female hysterosoma; (5) the number and position of the leg setae; (6) the shape of the male aedeagus; (7) and the presence and types of lobes on the female striae. Colors of adult females vary from species to species and genus to genus; hibernating forms are differently colored from actively feeding mites. In northern species the females are usually greenish while in the southern forms they are reddish (in the genus *Tetranychus*).

Tetranychid mites have tactile and chemosensory setae. The tactile setae are pubescent, slender, finely pointed, and have thick walls. The chemosensory setae have thin walls in which transverse striations may be evident. Setae on the tarsal appendages having a knob or hook are referred to as tenent hairs. They are always found on the claw. The dorsum of the tarsus of leg I of the adult bears 2 pairs and the dorsum of tarsus II 1 pair of intimately associated setae which are called duplex setae. The distal and longest member of each duplex setae is sensory and the proximal member is tactile. Placement of these setae are sometimes valuable for recognition of species groups or higher categories.

The legs have sensory setae other than the duplex setae, on all tarsi, the anterior tibiae, and sometimes on the tibia of legs II to IV. Other leg setae on all segments are tactile, but their number is often variable. The tarsal appendages consist of a pair of true claws laterally, and a central empodium. The claw is primitively clawlike or else padlike with lateroventral tenent hairs. Claw differences are used in defining generic and higher category classification.

The hysterosoma always bears 5 pairs of setae mediodorsally called the dorsocentrals or D setae. Laterad of the first pair of dorsocentrals are 2 setae on each side; the outermost is called the humeral (H) and the inner is the first dorsolateral or L_{-1}. Laterad of each of the second, third, and fourth dorsocentrals in certain of the more generalized tetranychids are the other dorsolateral setae. The generalized setal pattern of the hysterosoma is 5 pairs of D setae, 4 pairs of L setae, and a single pair of humeral setae (fig. 3).

Chaetotaxy of the venter is constant within the family, except for the opisthosoma. Females of the higher Tetranychids usually have 2 pairs rather than 3 pairs of anal setae, and normally males have 4 pairs rather than 5 pairs of genitoanal setae. Certain genera of the tribe Tetranychini have the caudal set of the 2 pairs of paraanal setae displaced to appear as a terminal pair of dorsal setae (the postanals), but these are lacking in some genera.

The dorsal texture of the integument of the body of a tetranychid may be smooth, except for large folds in some of the more generalized forms. There may be a development of mediodorsal areas bearing areolate or punctate impressions on the propodosoma and opisthosoma. The higher tetranychids, however, bear integumentary striations, irregular and widely spaced in some species but similar to that af a fingerprint pattern in most. The striae may or may not possess lobes (pl. 5; Gasser, 1951; Grandjean, 1948; Pritchard and Baker, 1955; Tuttle and Baker, 1968).

Host plants may be helpful for species identification, because many species are host specific, and only a few mite species are found on any given host in an area. The distribution and hosts of many species are not yet completely known, however, and some care should be taken in identifying mites in an "unknown" area or on a "new" host. It should again be stressed here that whenever possible males should be included in the collection since most specific identifications depend upon the study of the male aedeagus.

The Tetranychidae are divided into 2 subfamilies, the Bryobiinae and the Tetranychinae. These are further subdivided into tribes and genera (Pritchard and Baker, 1955).

Key to the Tetranychid Genera

This key to the tetranychid genera includes all genera now known, including species not considered to be of econominc importance. The genera dealt with in this book are marked with an asterisk.

Key to the Genera of Spider Mites Based on the Females

1. Empodium with tenent hairs (Bryobiinae) 2
   Empodium absent, or if present without tenent hairs (Tetranychinae) 24
2. True claws uncinate; empodium padlike . . . . (Bryobiini) . . 3
   True claws padlike; empodium padlike or uncinate 6
3. With 4 pairs of propodosomal setae 4
   With 3 pairs of propodosomal setae 5
4. With prominent projections over rostrum; D_4 setae marginal; coxal setal formula 2–1–1–1 . . . . Bryobia Koch*
   Without prominent projections over rostrum; D_4 setae in normal dorsal position; coxal setal formula 2–2–1–1 . . . . *Pseudobryobia* McGregor
5. Tarsus I with normal 2 sets of duplex setae; Para-anal setae ventral . . . . *Parabryobia* McGregor
   Tarsus I without duplex setae; para-anal setae dorsal . . *Bryobiella* Tuttle and Baker
6. Claws and empodia padlike (Hystrichonychini) 7
   Claws padlike and empodia uncinate 20
7. With 3 pairs of propodosomal setae 8

With 4 pairs of propodosomal setae .... *Tetranycopsis* Canestrini
8. With 5 pairs of dorsocentral hysterosomal setae 9
With 4 pairs of dorsocentral hysterosomal setae .... *Porcupinychus* Anwarüllah
9. Fourth pair of dorsocentral setae marginal or nearly so 10
Fourth pair of dorsocentral setae in normal dorsal position 13
10. Without propodosomal projections over rostrum 11
With propodosomal projections over rostrum 12
11. Body with normal striae; setae set on tubercles, long, strong; dorsocentral setae 1, 2, 3, and 5 contiguous .... *Beerella* Wainstein
Body covered with a tuberculate pattern; without projections over rostrum; without dorsal shields .... *Reckiella* Wainstein
12. With 2 anterior projections over rostrum; some or all 3 pairs of posterior body setae on tubercles .... *Mesobryobia* Wainstein
With 3 anterior projections over rostrum; posterior setae not set on strong tubercles .... *Monoceronychus* McGregor
13. With 10 pairs of hysterosomal setae 14
With 12 pairs of hysterosomal setae .... *Hystrichonychus* McGregor
14. With the normal number of ventral and coxal setae 15
With many ventral and coxal setae .... *Taurioba* Livshitz and Mitrofanov
15. Female with normal 2 sets of duplex setae on tarsus I 16
Female with 3 sets of duplex setae on tarsus I .... *Parapetrobia* Meyer and Ryke
16. Some or all dorsal body setae set on strong tubercles 17
Dorsal body setae well separated and not set on strong tubercles 18
17. Dorsocentral setae set on tubercles and well separated except for the fourth pair .... *Aplonobia* Womersley*
Dorsocentral setae on strong tubercles, the second, third, and fourth pairs contiguous .... *Georgiobia* Wainstein
18. Fourth pair of hysterosomal dorsocentrals not in normal position 19
Fourth pair of dorsocentrals in normal position; peritreme simple .... *Paraplanobia* Wainstein*
19. Fourth pair of dorsocentrals closer together than first 3 pairs; leg setae may be strongly plumose .... *Anaplonobia* Wainstein
Fourth pair of dorsocentrals further apart than first 3 pairs; leg setae finely serrated .... *Neopetrobia* Wainstein
20. With the normal 3 pairs of ventral setae (Petrobiini) 21
With many ventral body setae (Neotrichobiini) .... *Neotrichobia* Tuttle and Baker
21. With 2 sets of duplex setae on tarsus I 22
With a single set of duplex setae on tarsus I; duplex setae present on tibia I of male .... *Schizonobiella* Tuttle and Baker
22. Without projections over rostrum 23
With 3 setal bearing projections over rostrum .... *Mezranobia* Athias-Henriot
23. Empodium with a row of tenent hairs .... Petrobia Murray*
Empodium with a single pair of tenent hairs .... *Schizonobia* Womersley*
24. Tarsus I without closely associated duplex setae, or duplex setae absent; empodium clawlike when present (Eurytetranychini) 25
Tarsus I with 2 pairs of duplex setae; empodium clawlike or split distally 28
25. Empodial claw present 26
Empodial claw absent 27
26. Empodial claw small .... *Eurytetranychus* Oudemans*
Empodial claw obviously enlarged .... *Synonychus* Miller
27. With 1 pair of anal setae in female; dorsocentral setae 4 marginal .... *Aponychus* Rimando
With 2 pairs of anal setae in female; dorsocentral setae 4 in normal position *Eutetranychus* Oudemans*

28. Hysterosoma with dorsocentral setae 4 marginal (Tenuipalpoidini) 29
Hysterosoma with dorsocentral setae 4 in normal dorsal position (Tetranychini) . . 30
29. With 2 pairs of dorsal hysterosomal setae; tarsus II with distal member of duplex setae a short solenidion . . . . *Tenuipalpoides* Reck and Bagdasarian
With 9 pairs of hysterosomal setae; the duplex setae are of normal length. . . . *Eonychus* Gutierriez
30. With 2 pairs of para-anal setae 31
With 1 pair of para-anal setae 40
31. Empodium clawlike 32
Empodium ending in a tuft of hairs 37
32. Empodium with proximoventral hairs 33
Empodium without proximoventral hairs 34
33. Empodial claw shorter than proximoventral hairs, which are at less than right angles to claw . . . . *Allonychus* Pritchard and Baker*
Empodial claw as long as or longer than dorsoventral hairs which are at right angles to the claw; dorsal body setae on strong tubercles. . *Panonychus* Yokoyama*
34. Empodium a simple hook 35
Empodium split into 2 parts . . . . *Schizotetranychus* Trägårdh*
35. Dorsum of body covered with striae 36
Dorsum of body covered with spinules . . . . *Tylonychus* Miller
36. Dorsum of body with lobed striae . . . . *Anatetranychus* Womersley
Dorsum of hysterosoma with striae forming a basket weave pattern . . . . *Mixonychus* Ryke and Meyer
37. Hysterosomal striae transverse dorsomedially 38
Hysterosomal striae longitudinal between third pair of dorsocentral setae. . . . *Mononychellus* Wainstein*
38. Striae normal; dorsal setae not set on tubercles; empodium split near middle. . . . 39
Striae forming a basket weave pattern in female; setae set on strong tubercles; empodium split distally . . . . *Neotetranychus* Trägårdh
39. Dorsal setae very short, not as long as intervals between bases . . . . *Platytetranychus* Oudemans*
Dorsal setae as long as or longer than intervals between bases . . . . *Eotetranychus* Oudemans*
40. Empodium clawlike with proximoventral hairs; duplex setae of tarsus I distal and approximate 41
Empodium split distally, usually into three pairs of hairs; duplex setae of tarsus I well separated . . . . *Tetranychus* Dufour*
41. With a single pair of anal setae . . . . *Atrichoproctus* Flechtmann
With two pairs of anal setae . . . . *Oligonychus* Berlese*

MORPHOLOGY AND ANATOMY

Mouthparts and Feeding

The mouthparts of tetranychids and the muscles associated with them (fig. 4; p. 16, a; 17, *a*, *b*) consist of a pair of recurved stylets, the stylophore (mandibular plate of Blauvelt) and the rostrum. The stylophore is composed of 2 segments. The thicker basal segment to which a group of muscles are attached is embedded in the stylophore proper. The narrower distal segment at its origin is directed caudally, but then it curves downward and forward and approaches the apex of the rostrum. The two stylets fit into a V-shaped groove on the dorsal surface of the rostrum. This groove runs the length of the rostrum and serves to help keep the stylets aligned during feeding. The stylets are subcircular in cross-section and contain no grooves or tubes. They have an average length, from the curved portion to the tip, of 130 μ.

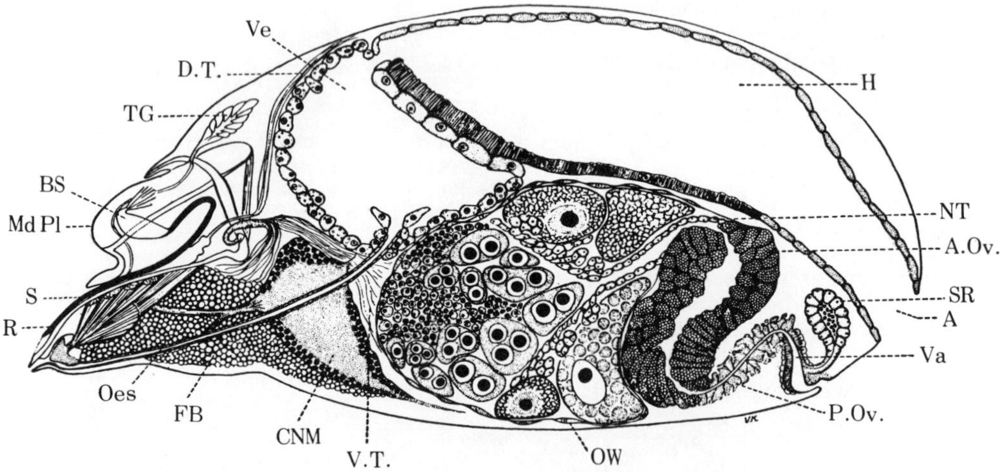

Fig. 4. Median sagittal section of a tetranychid mite: Ve = Ventriculus; D.T. = Dorsal trachea; TG = Tracheal salivary gland; BS = Basal segment of stylet; Md Pl = Mandibular plate; S = Stylet; R = Rostrum; Oes M Esophagus; FB = Fat Body; CNM = Central nerve mass; V.T. = Ventral tracheae; OW = Ovary wall; P.Ov. = Posterior oviduct; Va = Vagina; A = Anus; SR = Seminal receptacle; A.Ov. = Anterior oviduct; NT = Nutritive tissue; H = Hindgut and excretory organ (after Blauvelt, 1945).

The stylophore (fig. 3; pl. 18) is considered to be the fused basal segments of the chelicerae. It is a large bulbous structure that is movable along the dorsal surface of the rostrum. Motion of the stylophore provides the principal method by which protraction and retraction of the stylets occur (Baker and Connell, 1963; Blauvelt, 1945). It also serves to control the exposure of the peritreme groove (pl. 18), which aids in regulating water loss.

The conically shaped, ventral portion of the mouthparts is usually termed the rostrum (pl. 17, *a*, *b*). This structure represents the fused maxillae. The rostrum houses the pharynx and its dilators and bears at its tip the mouth opening. The tip of the rostrum is capable of a slight movement during feeding.

When feeding, the mite is tipped up in such a position that the third and fourth pairs of legs are off the leaf surface and the mite is supported by the first and second pairs of legs as well as by the rostrum. This position appears to give the mite the necessary leverage for most efficient penetration of the stylets. Tetranychids often pivot back and forth around the rostrum, the tip of which is positioned against the leaf tissue. The stylets penetrate various plant cells which exude their contents owing to turgor pressure. The "flaps" at the tip of the rostrum apparently overlap against the plant surface providing a plungerlike cup that permits the vacuum produced by the pharyngeal pump to draw up the exuded cell contents (fig. 4, pl. 17, *b*). When the mites are placed in the right light position under the low power of the microscope the movement of the fluid from the plant cell into the mite appears as a discontinuous green stream of protoplasm moving up through the leaf tissue to the pharynx.

The Digestive System

The digestive system of tetranychids is composed of a pharynx, esophagus, midgut or ventriculus, and an organ functioning both as a hindgut and Malpighianlike excretory organ (fig. 4). The information on the digestive system is largely summarized from detailed studies by Blauvelt (1945) and Ehara (1960).

Pharynx

In *Tetranychus* the walls of the pharynx have thickened intima; the dorsal posterior wall is especially thickened and elastic forming a cup-shaped chamber. From the anterior part of the pharynx chamber a tube runs forward and opens into the dorsal median rostral groove near the tip. In dorsal view the tube appears like an oval chamber. In lateral view it is a narrow passage, greatly constricted just behind the point where it enters the dorsal median rostral groove. The esophagus extends back from the ventral posterior part of the pharynx. The pharyngeal plunger is actuated by muscles inserted by tendons arranged in a longitudinal row on each side of the dorsal posterior surface of the plunger. When relaxed, the plunger fills the pharyngeal chamber; a nipplelike projection of the plunger blocks the anterior opening. When the muscles contract, the central part of the plunger is drawn back and upward, inverting the plunger on itself and reducing the pressure on the pharyngeal chamber so that the liquid food enters. During the sucking action, the return of the fluids from the esophagus or stomach is prevented by an esophageal valve located at the junction of the esophagus and stomach. When the pharyngeal muscles relax, the plunger returns to the relaxed position because of its own elasticity and thus forces food into the stomach. The nipplelike projection of the plunger reaches and fills the anterior opening before the rest of the plunger is completely relaxed, preventing the food from being forced forward again. In *Tetranychus* and *Panonychus* the pharyngeal muscles are inserted on the plunger at a single level, while in *Bryobia* they are not concentrated at a single level but are widely attached.

Esophagus

The esophagus is a long slender tube leading from the ventral posterior of the pharynx to the ventral surface of the central part of the ventriculus. It passes obliquely backward and upward through the central nerve mass, accompanied by tracheae. At the entrance of the ventriculus there is a valve. The wall of the ventriculus is invaginated where the esophagus enters.

Ventriculus

The midgut or ventriculus is a large saclike organ occupying most of the body cavity except that part occupied by the combined hindgut and excretory organ. The outer wall is indented at intervals by groups of dorsoventral muscles that divide it into broad caeca. The compression of these muscles against the walls of the ventriculus brings about a movement of food material in the stomach.

The epithelium of the ventriculus, except that part bordering the hindgut, is composed of a layer of glandular cells of variable size and shape. Some are oval or cuboidal, others are pear-shaped or round. The living cytoplasm contains many vacuoles caused by the presence of droplets and larger globules. In sections, free cells are often seen in the lumen, usually near the margin of the epithelial layer. As the mode of secretion is holocrine, that is, passage of an enzyme into the lumen of a gut through disintegration of parent cells, the walls of the free cells undoubtedly break, releasing the contents. That the cells of the epithelium are evidently renewed is indicated by the differences in size among them. The epithelium bordering the hindgut and excretory organ is strikingly different from the other midgut epithelium. The cells are much larger, with light-staining protoplasm without vacuoles, each with a large nucleus and dark-staining nucleolus.

Until recently it was believed that the gut of trombidiform mites had no anal opening and that their hindgut, separated from the gut, functions exclusively as an excretory organ with a caudal uropore. Evidence from studies of three genera of Tetranychidae indicates that communication exists between the hindgut and ventriculus in this family. Thus the hindgut that has a true anus serves for the removal of food residues as well as for the elimination of excretory matter.

There appear to be some differences in the ventriculus among tetranychid genera. The epithelial cells of the ventriculus of *Bryobia* are similar to those of *Panonychus*, but somewhat different from those of *Tetranychus*: they are much elongated to form a thin epithelium in *Bryobia* and *Panonychus*, but round or oval in *Tetranychus*. Food balls are often present only in the anterior 1 or 2 pairs of the lateral caeca in *Tetranychus*, but in *Bryobia* they are found in all parts of the ventriculus.

Hindgut and Excretory Organ

The hindgut and excretory organ serves the dual purpose of food evacuation from the stomach, and as a specialized nephritic excretory organ for the elimination of urates. It is a V-shaped tube lying horizontally along the median line of the midgut dividing the gut for most of its length so that only the posterior extremity of it is free from the midgut. This part runs back and downward between the tips of the posterior caeca of the ventriculus to the anus.

The epithelium of the lateral walls that are in contact with the epithelium of the midgut consist of a layer of thick glandular cells with ill-defined cross walls and only occasional nuclei. The dorsal wall consists of a layer of slender cells; the cytoplasm of which is evenly stained in microscopic preparations. The posterior part consists of a thin layer of cells much like the dorsal wall, and may be the only part that acts as a true hindgut; the anterior portion may be formed from several Malpighian tubes stretched out between the midgut and hindgut.

Respiratory System

The respiratory system consists of a pair of tubular peritremes with slitlike stigmata; a pair of tracheal trunks with a chitinous supporting structure—the sigmoid piece; a pair of small double accessory trunks and many small trachea running to the various organs and appendages. Although the tracheal trunks rise to the surface through a cleft in the stylophore they are continuous with tubular processes, the peritremes. These peritremes lie partially embedded in the fold of skin known as the epistome (a dorsal anterior projection that extends over the mouthparts or ganathosoma) (pl. 18). The shape of the peritremes differs considerably in different genera of Tetranychidae; therefore the peritremes have been used as a taxonomic character, being designated "collar tracheae." The peritreme in surface view appears to be divided into 6 or 7 chambers by heavy cross walls, which are really heavy riblike braces in the wall. In addition there are fine annular thickenings of the intima similar in appearance to those in the trachea.

In *T.*(*T.*) *urticae* Koch there is a continuous slitlike opening in the outer wall of the peritreme, extending from its junction with the tracheal trunk to the tip where it widens. The peritremes, with their continuous slitlike stigmata, allow communication with the outer air at all normal positions of the stylophore. When the stylophore is fully retracted the peritremes are completely inverted and enclosed between the epistome and stylophore and shut off from contact with outer air. This may account for the resistance of this mite to toxic gasses as well as its ability to conserve moisture during dry periods.

Reproductive System

Female

The female reproductive system is composed of an ovary, an oviduct, a vagina, and a seminal receptacle or spermatheca (fig. 4); these almost fill the ventral half of the interior of the hysterosoma. The ovary, as represented by *Bryobia*, is a large sac pressing on all neighboring organs. The anterior tip of the ovary is located above the posterior part of the central nervous mass; its dorsolateral and lateral parts adjoin the ventriculus, while the dorsomedian portion is immediately ventrad of the hindgut.

The ovary contains germ cells arranged in successive stages of development, with younger germ cells in the anterior part; the oogonia occur in the anterior tip and the anterior central portion, surrounded by larger oocytes. In the ovary there are usually 2 or 3 ova in advanced stages of maturity. The most mature ovum ready to be deposited occupies the posterior part of the ovary, the second ovum is located laterad and cephalad of the former, and the third lies laterad and cephalad of the second ovum on the opposite side.

The oviduct is a large, fleshy organ, leading the posterior portion of the ovary into the anterior portion of the vagina. The epithelium of the oviduct is formed of large, glandular columnar cells. These cells stain well with haematoxylin in the proximal portions where the oval nuclei are situated; the distal parts of the cells are weakly stained and greatly vacuolated, the vacuoles frequently occupying the greater part of the cells. Since the inner surface of the epithelium is usually pressed interiorly, the central lumen appears to be almost nonexistent; when a mature ovum to be next deposited is located in the oviduct, the lumen is strikingly expanded.

The vagina is a short, terminal part of the genital canal, leading the posterior portion of the oviduct into the caudoventral genital opening (pl. 19). There is a wide lumen in the organ. The vaginal epithelium is composed of a layer of columnar cells which contain weak-eosinophilic cytoplasm and proximal elongate nuclei.

The spermatheca is a very small, clavate blind organ entering the posterior part of the vagina via a short curved duct. The epithelium of the spermatheca consists of a single layer of columnar cells and surrounds the lumen.

Male

The male reproductive system consists of a pair of testes, a pair of vasa deferentia, a seminal vesicle, and an ejaculatory duct entering the aedeagus. In each side of the caudal ventral portion of body there is a testis that varies in shape among different specimens, being ovoid, spherical, or subconical. The germ cells are arranged in successive stages of development, the younger ones occurring caudally.

The vasa deferentia are large, thick ducts, located cephalad to the testes. The vas deferens runs almost straight with little undulation to enter the seminal vesicle via a short and narrow distal portion that is not definitely marked off from the rest. The lumen of the vas deferens is very large, containing spermatozoa and secretion that stain dark during slide preparation. The epithelium of the vas deferens consists of a single layer of glandular cells that have proximal nuclei that also stain dark.

The seminal vesicle is a single, median, more-or-less spherical organ lying slightly behind the center of the body, occurring caudad of the central nervous mass and ventrad of the hindgut. This vesicle is larger in diameter than each testis, and receives one vas deferens in each of the post-lateral portions. Issuing postmedially is an ejaculatory duct that enters the aedeagus.

The walls of the seminal vesicle bear a thin, areolate outer layer of which the cells are arranged in one stratum and are free from one another in the distal portion. Interior to this layer there is a thick, stout muscular layer that stains deeply with eosin and shows a horseshoe arrangement in longitudinal and horizontal sections. The contraction and expansion of the muscular layer probably play an important role in the ejaculation of spermatozoa during coitus. The innermost layer is very thin, its cells contain elongate nuclei. The central lumen harbors dark-staining spermatozoa.

The Egg

Beament (1951) studied the structure and formation of the egg of the European red mite. He could find no reliable factor by which the summer and the overwintering (diapausing) eggs may be distinguished on inspection as both types have a common basic shell structure. The outer surface is exceptionally hydrofuge. It consists of an outer thick wax layer with a very high melting point, and a cement layer of oil and protein that attaches the wax to the underlying "shell" layer enclosing the living material.

The shell layer is extremely resistant to penetration and to attack by chemicals or solvents; it appears to be composed of a material similar to keratin. This layer is formed in the ovary, towards the end of yolk accumulation. The egg is apparently fertilized precociously, and there are no associated nurse or follicle cells. Smith and Boudreaux (1972) studied fertilization in spider mites and hypothesized that the spermatozoa leave the spermatheca, and travel through the haemolymph to the ovary where eggs are fertilized.

According to Beament (1951), the egg receives its shell layer in the simple saclike ovary, and then passes into a glandular ovipositing pouch that is evaginated through the genital aperture at oviposition (pl. 19). The shell layer makes contact with the substrate, and the pouch secretes the cement over the rest of the shell layer so that the egg adheres to the substrate by a ring of cement around the base. The outer wax, is then secreted over the cement, but no wax is present at the base of the egg that is in contact with the substrate.

The female leaves the egg by rotating into an almost vertical position. In this process the ovipositing pouch is drawn off the egg, and withdrawn into the female; consequently, the wax, which is plastic when first secreted, is drawn up into the characteristic spike or stipe that surmounts the egg. The spike is of no further physiological significance to the egg. Owing to the absence of wax at the base of the egg, the total shell structure is not waterproof. The developing organism waterproofs the summer egg by secreting a wax layer into the inside of the shell about 6 hours before laying, before which the egg will only survive in humidities greater than 85 percent RH. After waterproofing, the egg readily develops, in humidities of 30 percent RH. The shell of the winter egg is similarly composed and waterproofed, but winter eggs are held up in the female until a later stage of embryonic development, so that they are waterproofed at the base of the egg when deposited on the bark.

Embryological Development

The initial phase of the development of the spider mite *Tetranychus* (*T.*) *urticae* Koch, according to Dittrich (1965), is characterized by two divisions of the total, equal type. After each of the first two cleavages a complete or partial partitioning can be observed within the egg. Before the third cleavage the cytoplasm of the blastomeres assumes a peripheral position. Thereafter cleavages up to the completed blastoderm are superficial. At room temperature all divisions occur at 30-minute intervals and at the 1024 cell stage the blastoderm is completed.

In the early germ band the factors of the extremities appear ventrally, while nuclei can still be perceived at very shallow light incidence. Between the apical part of the primordial extremities a median furrow can be observed. While further developing, the extremities stretch until the distal parts are in a parallel position. By the time the median furrow has disappeared the germ band extends in all directions: front and tail end grow along the periphery of the spheric egg until they almost touch. The partitioned head lobe develops at the front end as the embryo loses its wormlike shape through central growth and simultaneous contraction. The embryo assumes a short and compact appearance.

In this developmental stage a system of brightly contrasted lines appear, consisting of two lateral branches connected by a frontal branch. While this system develops, two embryonic stigmata connected with the frontal part through semicircular extensions grow through the eggshell achieving direct connection with the atmosphere.

The embryo now grows around the yolk, which is subdivided in spherules. The egg loses its spheric shape, it becomes flat at the front end and appears oval when viewed from the top. Dorsolaterally from the stigmata the eye-spots become visible. The extremities are completely developed by this time.

When the embryo has completed its internal organization, air enters between the shell and the space around the extremities rendering this portion nontransparent. Soon after the larva hatches.

The Mite Cuticle

Gibbs and Morrison (1959) studied the cuticle of a typical tetranychid mite, *T.* (*T.*) *urticae*. They found the cuticle to be a thin layer with a pattern of external ridges 1.25 μ thick in the troughs and twice that thickness measured at the ridges. Ridges are 1.0 to 1.6 μ apart. There is no tectocuticle but an outside lipoid layer and a dark-staining nonchitinous epicuticular layer from 0.1 to 0.2 μ thick. A double-layered inner procuticle remains unstained in contrast to the dark-staining inner layer. The inside surface of cuticle bears elevations or ridges opposing the external troughs. The epicuticle only is shed at molting. Either one or both layers of the procuticle contain chitin.

Electron micrographs of transverse and sagital sections of the opisthosoma of the two-spotted spider mite taken by Henneberry, Adams, and Cantwell (1965) showed a layered cuticle and a hypodermal layer with a basement membrane. The entire cuticle measured 0.70 to 0.91 μ (av. 0.81 μ), thick, the epicuticle being 0.01 μ, exocuticle 0.09 to 0.11 μ (av. 0.1 μ), and endocuticle to 0.61 to 0.78 μ (av. 70 μ). Cuticular extension, lobed at the apices, were 1.37 to 1.74 μ high and about 0.09 μ thick at the narrowest point. The hypodermal layer was about 1.19 to 1.71 μ thick and was delineated by the basement membrane.

The Setae

Some setae of Tetranychidae, as other Acari, are presumably tactile receptors. McEnroe (1969) photographed the base of a tactile seta of *Tetranychus* (*T.*) *urticae* from which he indicated that the seta has only one direction of movement as it is rigidly hinged at the base and the flexible membrane surrounds the base for only 270°. When the seta is bent back on its hinge, it rests in a socket. A single large cell, presumably a sensory neuron, lies directly below the seta. As the seta is solid, the neuron must terminate on the base of the seta. The neuron is encapsulated by a cellular mass that isolates it. The single neuron can only respond to movements of the seta. Because of the hinge arrangements of the seta, however, a directional component is added to this receptor. The hinges of these setae lie at different angles so that movements in various directions can be resolved by the mite.

The Eyes

Tetranychid mites have 2 pairs of nonfaceted eyes on the propodosoma, 1 pair close together on each side. Externally they appear as striated lenses, the anterior lens being more convex than the lens of the posterior eye with transverse striations. The striations on the posterior lens are a continuation of the body striations, but on the eye are smaller and closer together than on the body (pl. 20).

Silk Glands

The silk glands in tetranychids are located in the palpus. They are in the form of large sacs beginning posterior to the base of the palpus and traversing the entire palpus, ending in a nipplelike projection referred to as the "terminal euphathid" (pl. 17, *a*, *b*). This structure, which serves as the spinneret, has one or more small openings (6 to 8 M in diameter) in its tip through which the silk is discharged.

COLLECTION AND PREPARATION

Collecting

Tetranychid mites are found on most plants, usually in small numbers, but occasionally in such large populations that leaf defoliation results. Many of the species appear to be host specific, while others, such as the economic species *Tetranychus* (*T.*) *urticae* Koch and *T.* (*T.*) *cinnabarinus* (Boisduval), infest a wide range of plants. The mites feed on either leaf surface, on the bark, in grass sheaths, and the like. Some may produce heavy webbing, others form characteristic colonies, and others are general surface feeders. Some species overwinter in the egg stage on the host, others overwinter as living mites on the host or in the soil about the host.

Both sexes should be collected. Tetranychid males often appear similar to the nymphal stages, being small, slender, and greenish. All sizes and shapes should be included in the collection. The collection should include a large series of specimens, since more than one species may infest the same plant. Females are usually dominant during most of the season, and these usually may be determined only to subgenus. Alcohol (50 to 80 percent) may be used as preservative. Data should be taken on host, locality, position of mite on the leaf, webbing, color of the female, and damage.

Usually the mites are removed from the leaves and placed in small vials containing alcohol. Leaves with heavy infestations may be placed in vials without removing the mites. Leaf and twig materials need be brought into the laboratory for examination. Plastic or paper bags are used for storing the material until it is examined. Refrigerated bags may be stored several days before examination. In the field the material should be kept from the sun; wet paper towels or burlap over the bags will help protect the collected material in hot dry weather.

Preparation for Examination

The collected material may be examined directly with a dissecting microscope or the mites may be separated from the leaves by placing the collected material in a modified Berlese funnel or beaten into a white photographic tray. Both leaves and branches should be examined. Small mites, such as the Tydeidae, do not go through the funnel but become lost within the debris. Direct observation is needed to collect these and other tiny plant mites.

Plant mites may be also collected by beating leaves, twigs, and branches through a USDA seed sieve, number 20 mesh. A plastic funnel fits into the sieve sleeves, and a dry vial fits around the neck of the funnel. After mites have been collected in the vial, alcohol or AGA fluid (see below) is added and a paper or cardboard slip with collection data is placed in the vial, which is then corked and taken to the laboratory for future examination. A leaf or twig and flower sample is kept for identification of the plan. Large numbers of samples can be collected using this method, and the sorting and slide preparation can be accomplished at leisure. Color and habits of the living mite cannot be determined, however, as can be done under the dissecting microscope. The AGA mixture prevents the mites from hardening, and they can be properly oriented on the slide after preservation in this liquid for a long time. The AGA solution as used for mites by D. M. Tuttle of the University of Arizona is as follows:[1]

8 parts 70% isopropyl or ethyl alcohol
1 part glacial acetic acid
1 part glycerine
300 grams of sorbitol may be added to 1 gallon of the above

Mounting Media and Slide Preparation

Plant mites are best mounted in a modified Hoyer's solution, consisting of:

40 grams distilled water
30 grams gum arabic
200 grams chloral hydrate
20 grams glycerin

Female tetranychids should be oriented dorsoventrally with the legs spread. This makes it possible to study the dorsal and ventral striation pattern of the body, and the leg setae. Since the tarsal claws and empodia are imporant generically and specifically, profile mounts should also be made to study these organs. Male spider mites must be mounted in profile since the shape of the aedeagus is of specific importance. Some difficulty may be encountered because of the tendency of the mite to roll upon heating the slide. After the cover slips have been placed over the Hoyer's fluid, the slides are gently heated until the solution begins to show bubbles—this heating expands and clears the specimens. The specimens should be checked before the slides are placed in the oven; if specimens are not properly oriented, they can be rolled into position by gently moving the cover slip. Then slides should be placed in a warming oven at 45 to 50 C (122 F) for 24 hours, and kept flat until the Hoyer's solution is dry. They should not be mailed to a specialist for identification if there is any possibility that the solution is still soft.

There has been some difficulty with the permanency of slide preparations made with Hoyer's solution. Ringing the cover slip with various preparations has been of some help; ringing with excess Hoyer's solution has proved to be even better; also, we have lessened the amount of water in the solution by 20 percent, which has prevented most of the evaporation problems. Experiments are continuing with polyvinyl alcohol and Hoyer's solutions, and it is hoped that satisfactory mounting fluids can be developed.

Mites other than Tetranychidae—the Tenuipalpidae, Tuckerellidae, Tydeidae, and others—can also be treated in this manner.

BIOLOGY OF TETRANYCHID MITES

A general knowledge of how tetranychid mites live on their hosts is necessary in order to evaluate their economic significance. The general nature of their injury to plants and many of the factors influencing populations have been discussed in chapter 2. The specific biologies of the injurious species are reviewed in chapter 8. Hence, this account is necessarily superficial, but it may help those interested to discern some of the general patterns displayed by tetranychid mites.

Life History

Tetranychid mites develop through egg, larva, protonymph, deutonymph, and adult stages. The nymphal and adult stages are initiated during intervening periods of inactivity called protochrysalis, deutochrysalis, and teliochrysalis. During these periods the mite anchors itself to a leaf or to its webbing. The legs are bent upon themselves and a new cuticle is prepared before the exuviae is cast off. It is not known whether there is an increase in somatic cell number during these "chrysalis" stages. No mitotic figures have been found in any stage past that of the larva; therefore, it is suggested that all the somatic cells are formed in the larval instar, and that any increase in size of the mites is the result of cell size increase.

Developmental Rate and Life Span

The length of the incubation period of each species and the time required to develop to adult is primarily correlated with temperature, except where extreme temperatures and / or humidity cause the mites to enter a diapause or aestivation phase. There is, however, considerable variation in developmental rate between genera and to a less degree between species within a genus; however, comparative data are not available. The time required for tetranychid species to complete the life cycle appears to be longest in *Bryobia*, *Petrobia*, and *Panonychus* with *Oli-*

gonychus, *Tetranychus*, and *Eotetranychus* requiring less time to complete a generation. The life span of adults appears to be related to the rate of development, but again comparative data are not available.

Fertilization and Sex Determination

In most species of Tetranychidae two sexes occur. Several species of the Bryobiinae have no, or only a sporadic appearance of males; reproduction on these species appears to be based on thelytokous parthenogenesis. Rearing data, largely with species of *Panonychus* and *Tetranychus*, have shown that reproduction is based on arrhenotokous parthenogenesis, that is, unfertilized females give only male offspring and fertilized females produce both females and males. Genetic evidence of the correctness of this was provided by interspecific crosses (Boudreaux, 1963) and by genetic studies of resistance (Taylor and Smith, 1956) and of visible marker genes (Helle and van Zon, 1966).

The males have an aedeagus, and a spermatheca in the female has been described as containing nondescript inclusions assumed to be spermatozoa, but the presence of sperm in this structure has not been confirmed.

The mating process is usually accomplished immediately after the last molt of the female. The males detect the teliochrysalis by contact and then remain waiting until the exuviae are cast, or may aid in removing it. The male then crawls head first under the posterior end of the teneral female and arches the end of the abdomen upward to accomplish coupling. The female is held by the 2 pairs of fore limbs of the male in the process (Boudreaux, 1963; Evans, Sheals, and MacFarlane, 1961; Gasser, 1951).

The chromosome number in males is haploid. The sex ratio produced by any female seems to depend on the amount of spermatozoa introduced during the mating act. The amount of spermatozoa introduced in turn depends both on the length of time spent in copulation and on the sperm supply of the male. After the first matings females always produce a preponderance of female offspring, while subsequent matings result in a preponderance of males. This suggests that the mating act results in the establishment of some unknown barrier that prevents further effective insemination (Boudreaux, 1963).

Cytological confirmation of haploid-diploidy in spider mites was first obtained from studies with *T.* (*T.*) *urticae*. Schrader (1923) observed that 2 classes of eggs occur in this species, that is, with 3 and with 6 chromosomes. The number of 6 was the result of fusion of egg nucleus and sperm nucleus, while 3 chromosomes were found in eggs in which no spermium was present. The 2 classes with 3 and 6 chromosomes could be demonstrated during cleavage and later embryonic development. Schrader was also able to show the existence of the haploid number in the spermatogonia, as well as the diploid number in the oogonia.

Helle and Bollard (1967) determined the number of chromosomes in 13 species of spider mites and found that 12 of these had haploid and diploid eggs and determined the haploid and diploid number in each (see chapter 2). They found that *Tetranycopis horridus* was not bisexual: only females could be collected from the field; adult females in laboratory culutres produced eggs that developed into females only, and the eggs always showed four chromosomes. Helle and Bollard

(1967) concluded that this species has thelytokous parthenogenic reproduction and found it likely that 4 chromosomes is the diploid number.

Helle (1967) studied the stage of oocyte development at which fertilization occurred in the two-spotted spider mite. He concluded that virgin females produced haploid eggs first, after mating then diploid eggs. From this and other facts, it was concluded that fertilization of the oocytes occurs in a very early stage of egg development. Fertilization is assumed to take place in the ovary and not in the oviduct. By using genetic markers Helle confirmed the observations of Boudreaux (1963) that most first matings are effective, that later matings usually are ineffective, and that the sperm supply in the first mating determines the likelihood of later matings.

Mite-Host Plant Relationships

Tetranychid plant hosts have a wide range of toleration to mite populations, ranging from those that show very little injury response to high mite populations, to those that are severely damaged by a very few mites. These differences may partially result from the nature of the substances injected into the plant during the feeding process. For example, the feeding by a few *Tetranychus* (*A.*) *pacificus* mites on pear leaves causes the leaves to turn dark and dry as though they had been scorched by fire. But low population levels of this species on apple, grape, and bean leaves causes the typical stippled type of injury produced by feeding of many tetranychid mite species. These observations suggest that plant species respond differently to an injected toxin. It seems unlikely that the substances secreted by a mite species would differ when the mite feeding on different host species. High populations of other species, such as the avocado brown mite, *Oligonychus* (*O.*) *punicae* (Hirst) may feed on the leaves of its host over a considerable time period without producing serious plant damage.

Host Preferences

Although tetranychid mites are less host specific than the eriophyoids, most species have a relatively narrow host range. A few of the best-known species, however, such as the two-spotted spider mite, *Tetranychus* (*T.*) *urticae* Koch, the Atlantic or strawberry spider mite, *T.* (*T.*) *turkestani* (Ugarov and Nikolski), and the vegetable mite, *T.* (*T.*) *neocalidonicus* Andre, may live and cause injury to plants in a wide variety of plant genera.

Some mite species and genera inhabit a restricted group of plants. Known mites belonging to the genus *Platytetranychus* are found only on conifers. The honey locust mite, *Eotetranychus multidigituli* (Ewing), has been found only on honey locust. Most injurious species in the genus *Schizotetranychus* live on monocotyledonous plants; *S. baltazari* Rimando, however, damages citrus in Taiwan.

Location on the Host

Many tetranychid species have a preferred location on their host; but they will move out to other parts of the plant when populations become high. The six-spotted mite, *Eotetranychus sexmaculatus* (Riley), for example, feeds in colonies on the undersides of citrus leaves, while *E. lewisi* McGregor populations usually increase most rapidly on citrus fruit that has turned yellow (mature fruit). In contrast, *E. yumensis* (McGregor), occurs about equally on fruit, leaves, and occasionally on young twigs. The strawberry spider mite feeds in colonies on the underleaf surfaces of its hosts, the two-spotted spider mite prefers the upper leaf surfaces of the plants, whereas the Pacific mite lives and deposits eggs on both leaf surfaces.

A mite species may have different habitat preferences on different host plants. The two-spotted spider mite inhabits both leaf surfaces of many plant hosts, but on citrus it lives almost entirely in colonies on the underside of the leaves.

Most tetranychid mites tend to deposit more eggs near the midribs of the leaves of their host irrespective of the leaf surface they prefer. Species that overwinter in the egg stage deposit the overwintering eggs on young twigs. A few species inhabit the fruit, but even these tend to cause a minimum of direct damage to this part of the plant.

Plant Injury

Removal of Cell Contents

Tetranychid mites feed by penetrating the plant tissue with sharp stylets and removal of the cell contents. The chloroplasts disappear and the small amount of remaining cellular material coagulates to form an amber mass. In the palisade layers, only the penetrated cells are damaged; adjacent cells show no evidence of injury. There appears to be no damage to the conducting elements of leaf veins; much damage, however, is caused to parenchyma cells in the well-differentiated vascular bundle sheaths of plum varieties. In all apple varieties, cells immediately adjacent to the bundles are damaged.

High citrus red mite populations on citrus were shown by Wedding, Riehl, and Jeppson (1958) to cause substantial changes in photosynthesis and transpiration rates. They indicated that transpiration increases during heavy feeding, but decreases below normal after mites have been removed. The amount of chlorophyll in the leaves may be decreased as much as 60 percent.

Liesering (1960) estimated that *T.* (*T.*) *urticae* could exhaust about 18 to 22 cells per minute while feeding and suggested that certain substances are secreted into the plant tissue during feeding. Puncturing of new cells proceeds from one spot to another in the form of a circle which results in the formation of the small rounded chlorotic spots. Continued feeding leads to irregular spots formed by the integration of primary suction spots; finally the typical picture of tetranychid injury appears (pl. 21). It has been demonstrated by use of a potentiometer that the water balance of mite-attacked leaves is greatly disturbed. Transpiration is highly accelerated, which finally leads to the drying out and dropping of leaves (pl. 22). That mite feeding also causes inhibition of photosynthesis is shown by Warburg experiments. Strongly injured leaves may exhibit no photosynthesis at all. Liesering (1960) further suggests that feeding by *T.* (*T.*) *urticae* decreases the amount and changes the composition of leaf pigments, indicating that the black excretion products of the mites are mostly leaf pigments and their digestion products.

The extent of speckling seen from the upper leaf surface is generally believed to reflect the total feeding damage to the leaf. This is not necessarily correct because such speckling results only from upper surface feeding. The relative intensities of feeding from the two surfaces may differ under different environmental conditions and among different species and varieties. In the daytime there are normally more mites on the lower surface of the leaves, suggesting that feeding may be predominantly from this surface. Larger speckles occur when mites feed during the period of leaf expansion. This causes cells adjacent to those directly damaged to develop abnormally, resulting in local distortion.

Histological studies reveal that mite feeding on the lower surface causes damage to spongy mesophyll cells (Baker and Connell, 1963). In some cases the cells of the lowest palisade layer, but not those of the uppermost palisade layer, collapse. Mites on the upper surface feed on all the palisade layers and sometimes damage a few of the adjacent spongy mesophyll cells. In all the host varieties, stylet penetration from the upper surface is to a greater depth than from the lower surface. Leaves of the apple varieties studied show that more cells are damaged in the second palisade layer than in the first layer, but the reverse was true of plums.

It was believed that the mite stylets penetrated through the epidermal cells causing some collapse of these cells. The cells of the epidermis, especially in apple leaves, often become twisted during sectioning owing to lack of support from the underlying injured spongy mesophyll. If such is the case, either the stylets must pass through the epidermal cells without causing noticeable injury or they are directed between these cells.

Bronzing, which develops later, is associated with damage to mesophyll cells further from the veinlets. Differences in the intensity of bronzing caused by mites on different varieties may result from, in some cases at least, damage of different tissues. Thus leaves of a variety in which little damage is done to the uppermost palisade cells may appear quite green even though underlying cells are damaged.

Chemical Injection

There is a great deal of observational evidence that toxins or growth regulators are injected into plant tissue while certain species of mites are feeding on plant tissue. Little, however, is known concerning the nature of these chemicals or the precise mechanism by which they are introduced into plant cells.

Tetranychid mites have semicircular stylets capable of penetrating into the palisade cells of leaves. Scanning electron microscope photographs indicate that the stylets are spearlike, suggesting that their only function is to pierce the cell walls of host plants. The apex of the rostrum appears externally to have flexible "flaps" capable of overlapping around the stylets to form a plungerlike formation, such that plant sap released from the cells by stylet penetration can be taken up by the mite by means of the suction pump. There seem to be no special facilities to inject a toxin into the plant. It is conceivable, however, that the mite may release a toxin onto the plant surface during the feeding process. The toxin would be confined within the area covered by the plungerlike flaps, then carried into the plant tissue as the stylets puncture the cells. Evidence that tetranychid mite stylets are capable of carrying materials into plant cells is indicated by the ability of these mites to transmit certain viruses artificially placed on leaf surfaces (see chapter 6).

The structure of the mouthparts provides no clue to explain why different plants respond so differently to feeding by the same species. For example, why are pear leaves so severely "scorched" by feeding of a relatively few Pacific mites, whereas leaves of apple and other host plants tolerate high populations with only the typical stippling injury? One may only assume that plant fluids react differently to the chemical reaching the cells. More information is needed concerning the precise feeding methods and the fluids involved during feeding by different species of phytophagous mites.

Economic Effects of Injury to Plants

The damage resulting from mite populations on plants is sometimes obvious, but at other times the precise degree of injury is rather obscure. Thus precise measurements of crop losses caused by various tetranychid mite species are not generally available. As earlier indicated, however, the effects of mite feeding on plants is so interdependent on weather and plant vigor or vitality that measurements of the amount of plant injury, such as "stippling," do not always indicate the degree of crop damage (pl. 21). For example, orange trees can tolerate considerable mite feeding during periods of high humidity; but even low mite populations may increase water stress during periods of maximum transpiration sufficiently to cause serious tree injury in the form of leaf and fruit drop and twig die back (pl. 22). Apparently healthy trees may be unable to supply sufficient moisture to leaves and fruit during these periods owing to poor root systems, dry soil, plugged xylem vessels, or other cultural conditions related to tree vigor.

Feeding by the European red mite, *Panonychus ulmi* (Koch) (Tetranychidae), not only causes defoliation, but the production of unripe, sour, undersized, and poorly colored fruit prone to early dropping. There is a reduction in the number of fruit buds set for the following season. Stored food becomes depleted, which increases susceptibility to frost and winter injury. Heavy infestations on young trees may result in smaller tree trunk girth. Bearing trees may have decreased bloom and lower yields for a season or two following such infestations. Thus, mite injury may delay tree development as well as adversely influence crop production (Boulanger, 1958; Blair, 1961; Lienk, 1953).

European red mite densities on plum of one to two mites per square centimeter may cause decreased rate of shoot extension; but the growth rate of the root system decreases before that of the shoots. A decrease in photosynthesis occurs only when leaves are severely damaged or bronzed. The initial injury appears to result from an imbalance in the growth-controlling substances caused by mite feeding. Radioactivity from C^{14}-labeled mites may be detected in growing regions remote from the infested mature leaves. This suggests that retarded growth may result from the effects of substances injected into the plant during feeding (Avery, 1962; Avery and Briggs, 1968, *a*, 1968, *b*).

In a well-replicated experiment, Barnes and Moffitt (1962) demonstrated that heavy infestations during summer of the European red mite produced no effect on size of crop or quality of English walnuts until the third year when a 30 percent reduction in crop was demonstrated.

Mite-feeding injury on truck and field crops often results in complete destruction of crops in parts of the field or even over whole fields where infestations have been developing over an extended period of time. Therefore, on these crops the effects are readily visible. Feeding on fruit may result in discoloration, so that a direct influence on crop quality is immediately evident resulting in downgrading of the fruit.

The effects of feeding by mites on forest and shade trees ar often difficult to evaluate. Mites may retard growth and adversely influence the quality of flowers, seed, or fruit production; therefore the degree of economic loss, especially from moderate infestations, is difficult to ascertain.

Seasonal Cycles

The seasonal cycles of tetranychid mites differ with the species, climate, and host plants. Some species, especially those living in semitropical climates, remain on their host plants throughout the year, stopping egg production and development only as temperatures fall below their developmental threshold. Development continues when winter temperatures rise above their threshold. All stages may be found on the host plants throughout the year.

Tetranychid species inhabiting temperate climates usually enter a diapause stage during the winter months. Diapause may take place as either fertilized females or as overwintering eggs. A number of economically important species in the genera *Eotetranychus* and *Tetranychus* overwinter as fertilized females. These are variously called hibernating forms, deutogynes, or simply overwintering females. These are generally uniformly colored, commonly yellow or straw-colored in the genus *Eotetranychus* or pellucid orange-red or orange in the genus *Tetranychus*. As indicated in chapter 2, these hibernating mites do not feed and are normally devoid of dorsal lobes. Once these overwintering forms have developed, they will not change to the summer form until they have undergone a period of chilling—even though environmental conditions become favorable for development. Low temperatures and decreased length of daylight are conditions favorable for development of overwintering forms.

Some species living in temperate climates overwinter in the egg stage, notably the European red mite, *Panonychus ulmi*, the brown mite, *Bryobia rubrioculus*, and species of *Oligonychus*. As cold weather approaches, and day length lessens, females of the European red mite, for example, develop winter eggs that are retained by the female until a later stage of embryonic development than those of the summer form in order that waterproof waxes may be secreted before the eggs are deposited. The development of winter eggs is facultative so mites kept under continuous summer conditions of photoperiod and temperature fail to develop diapause eggs. The overwintering eggs are deposited on the bark rather than on leaves, and they also must undergo a period of chilling before rising temperatures in the spring cause resumption of activity.

Some tetranychid mite species living in semitropical climates, such as *Petrobia latens* and *P. apicalis*, are active during the cooler period of the year and hibernate or aestivate in the egg stage during the hot dry summers. Such diapause eggs are better protected from weather extremes than are summer eggs. More detailed biologies are reported under each species in chapter 8.

Dispersal

Injurious plant-feeding mites have become widely distributed in the world as man has transported their hosts, especially those most valuable for shade, ornamental, and agricultural purposes. The small size of the mites and their habit of living or depositing their eggs in secluded places has protected them against detection during transportation.

Agreement has not been reached on the importance of tetranychid mite dispersal to new hosts by crawling along the ground. Experience has shown that barriers against crawling are necessary to keep uninfested plants in greenhouses from becoming infested with the two-spotted spider mites being reared in the same or nearby greenhouses. Species of *Bryobia praetiosa* and *Petrobia latens* frequently crawl from the host plant to deposit eggs on various surfaces, and do not return to the same hosts for feeding. *P. apicalis* leaves the host plant to deposit diapause eggs on tree trunks, fence posts, and the like, and returns to feed on its host plant.

Several mite species, of the family Tetranychidae, including the European red mite, the citrus red mite, the six-spotted mite, the avocado brown mite, and the legume mite, will lower themselves from the host plant on silk threads. A slight breeze releases the mites from the host and the threads serve as balloons or parachutes to carry them considerable distances. Unfavorable host conditions and favorable weather, that is, almost calm, seem to be the stimuli for mass "ballooning." The authors are not aware of any "ballooning" habits of mites belonging to the genus *Tetranychus*, although when populations are high and food is in short supply, these mites may form in mass at the top of their host from which they are no doubt dispersed by winds. It has been reported that *T.* (*T.*) *cinnabarinus* (Boyle, 1957) will perish with the plant rather than crawl off a potted host plant after depleting available food. A moderate breeze is, however, effective in causing mites of this species to drift to nearby plants. It is likely that insects and birds may act as incidental means of mite dispersal.

SELECTED BIBLIOGRAPHY

Avery, D. J. 1962. The vegetative growth of young plants of Brompton plum infested with fruit tree red spider mites. E. Malling Res. Sta. Ann. Rept. 1961, pp. 77–80.

Avery, D. J., and J. B. Briggs. 1968*a*. Damage to leaves caused by fruit tree red spider mite, *Panonychus ulmi* (Koch). J. Hort. Sci. 43(4):463–473.

———. 1968*b*. The aetiology and development of damage in young fruit trees infested with fruit tree red spider mite, *Panonychus ulmi* (Koch). Ann. Appl. Biol. 61(2):227–228.

Baker, J. E., and W. A. Connell. 1963. The morphology of the mouthparts of *Tetranychus atlanticus* and observations on feeding by this mite on soybeans. Ann. Entomol. Soc. Am. 56(6):733–736.

BARNES, M. M., and H. R. MOFFITT. 1962. Influence of walnut aphid and European red mite on walnut quality and production. Diamond Walnut News 45(3):6–7.

BEAMENT, J. W. L. 1951. The structure and formation of the egg of the fruit tree spider mite, *Metatetranychus ulmi* Koch. Ann. Appl. Biol. 38:1–24.

BLAIR, C. A. 1961. Damage to apple leaves by the fruit tree red spider mites. E. Malling Res. Sta. Ann. Rept., pp. 152–154.

BLAUVELT, W. E. 1945. The internal morphology of the common red spider mite (*Tetranychus telarius* L.). Cornell Univ. Agr. Expt. Sta. Mem. 270:1–11.

BOUDREAUX, H. B. 1963. Biological aspects of some phytophagous mites. Ann. Rev. Entomol. 8:137–154.

BOULANGER, L. W. 1958. The effect of European red mite feeding injury on certain metabolic activities of red delicious apples. Maine Agr. Expt. Sta. Bull. 570.

BOYLE,W. W. 1957. On the mode of dissemination of the two-spotted spider mite, *Tetranychus telarius* (L.) Proc. Hawaiian Entomol. Soc. 16:261–268.

DITTRICH, V. 1965. Embryonic development of tetranychids (In 5th European symposium of Acarology at Milan, 23 to 25 Sept. 1965 proceedings) Boll Zool. Agric. Bachic Ser. (2) 7 :101–104.

EHARA, SHÔZÔ. 1960 Comparative studies on the internal anatomy of three Japanese Trombidiform Acarinids. Fac. Sci. Hokkaido Univ. Series VI, Zool. 14(3):410–434.

EVANS, G. O., J. G. SHEALS, and D. MACFARLANE. 1961. The terrestrial Acari of the British Isles, an introduction to their morphology, biology, and classifications. 219 pp. British Museum Press, London.

EWING, H. E. 1913. The taxonomic value of characters of the male genital armature in the genus *Tetranychus* Dufour. Ann. Entomol. Soc. Amer. 6:453–460.

GASSER, R. 1951. Zur Kenntnis der Gemelinen Spinnmilbe, *Tetranychus urticae* Koch. I. Mitteilung: Morphologie Anatomie, Biologie und Oekologie. Mitt. Schweiz. Entomol. Ges. 24:217–262.

GIBBS, K. E., and F. O. MORRISON. 1959. The cuticle of the two-spotted spider mite, *Tetranychus telarius* (Linnaeus) Can. J. Zool. 31:633–637.

GRANDJEAN, F. 1948. Quelques caractères des Tétranyques. Bull. Museum d'histore naturelle, Paris. Serie 2, 20:517–524.

HELLE, W. 1967. Fertilization in the two-spotted spider mite, Entomol. Expt. and Appl. 10(1):103–110.

HELLE, W., and H. R. BOLLARD. 1967. Karyotypes and sex determination in spider mites. Genitica 38:43–53.

HELLE, W., and VAN ZON. 1966. Pigment mutations in *Tetranychus pacificus*. Entomol. Expt. and Appl. 9:402–403.

HENNEBERY, T. J., J. R. ADAMS, and G. E. CANTWELL. 1965. Fine structure of the integument of the two-spotted spider mite, *Tetranychus telarius*. Ann. Entomol. Soc. 58(4):532–535.

LIENK,W. 1953. Investigation of the biology and epidemiology of the common spider mite, *Tetranychus altheae* von Hanstein with particular consideration to hop as the host. Höfchenbriefe. 4:181–232.

LIESERING, V. R. 1960. Beitrag zum Phytopathologischen Wirkungmechanisms of *Tetranychus urticae* Koch. Z. Pflkrankh. 67:524–542.

MCENROE, W. D. 1969. A tactile seta of the two-spotted spider mite, *Tetranychus urticae*. Acarologia 11(1):29–31.

MCGREGOR, E. A. 1950. Mites of the family Tetranychidae. Amer. Mid. Nat. 44(2): 257–420.

PRITCHARD, A. E., and E. W. BAKER. 1955. A revision of thc spider mite family Tetranychidae. Pac. Coast Entomol. Soc. Mem. 2:1–172.

SCHRADER, F. 1923. Haploidie bei einer Spinnmilbe. Arch. Mickrosk Anat. 97:610–622.

SMITH, J. W., and H. B. BOUDREAUX. 1972. An autoadiographic search for the site of fertilization in spider mites. Ann. Entomol. Soc. Am. 65(1):69–74.

Taylor, E. A., and F. F. Smith. 1956. Transmission of resistance between strains of two-spotted spider mites. J. Econ. Entomol. 49 (6) :858–859.

Tuttle, D. M., and E. W. Baker. 1968. Spider mites of southwestern United States and a revision of the family Tetranychidae. 143 pp. Univ. Ariz. Press, Tucson, Ariz.

Wainstein, B. A. 1960. Tetranychid mites of Kazakhsten (with revision of the family). [In Russian.] Kazakh, Akad. Sel'sk. Nauk. Mauch.-Issled. Inst. Zash. Rast. Trudy. 5: 1–276.

Wedding, R. T., L. A. Riehl, and L. R. Jeppson. 1958. Red mite on citrus, experiments designed to measure damage give bases for further studies. Calif. Agr. 12(8):9, 10, 12.

Chapter 8

Injurious Tetranychid Mites

BRYOBIINAE BERLESE

The empodium has tenent hairs; the female has 3 pairs of anal setae, and the male has 5 pairs of genitoanal setae. Since it is difficult to find the genitoanal setal pattern, it is recommended that the empodium be used to separate the *Bryobiinae* from the *Tetranychinae*. These are generalized members of the family none of which is known to produce silken strands.

The Bryobiinae is divided into 4 tribes—the Bryobiini, Hystrichonychini, Petrobiini, and Neotrichobiini. Of these, the most important economic species are to be found in the Bryobiini and Petrobiini.

BRYOBIINI RECK

The true tarsal claws are uncinate or hooked, and the empodia are padlike. This tribe contains the important genus *Bryobia* of which several species cause considerable damage to cultivated plants.

Bryobia Koch

The true claws are uncinate and have tenent hairs; the empodium is padlike and also has tenent hairs; there are 4 pairs of propodosomal setae and 2 pairs of prominent projections over the rostrum; the coxal setae pattern is 2-1-1-1 (2-1-1-1 means 2 setae on coxae I, 1 setae on coxae II, 1 seta on coxa III, and 1 seta on coxa IV).

Bryobia eharai Pritchard and Keifer. This mite (fig. 5) is an occasional pest of chrysanthemum in Japan and Lyallpur, Pakistan. Feeding by this mite causes the leaves to turn yellow, then brown, then finally retarding the growth of young plants (Gahi, 1968).

The dorsal body setae of *B. eharai* are broadly clavate and lightly serrate, the bases or stems of the setae being more elongate than in the other species. The body is broadly rounded and there are a few rounded wrinkles rather than striae dorsally. Empodium I has 2 pairs of tenent hairs and the other empodia each possess a double row of ventrally directed tenent hairs. Legs I and the body are more-or-less similar in length. The duplex setae of tarsus III are aprroximate and of about equal length; the duplex setae of tarsus IV are well separated, the solenidion being short.

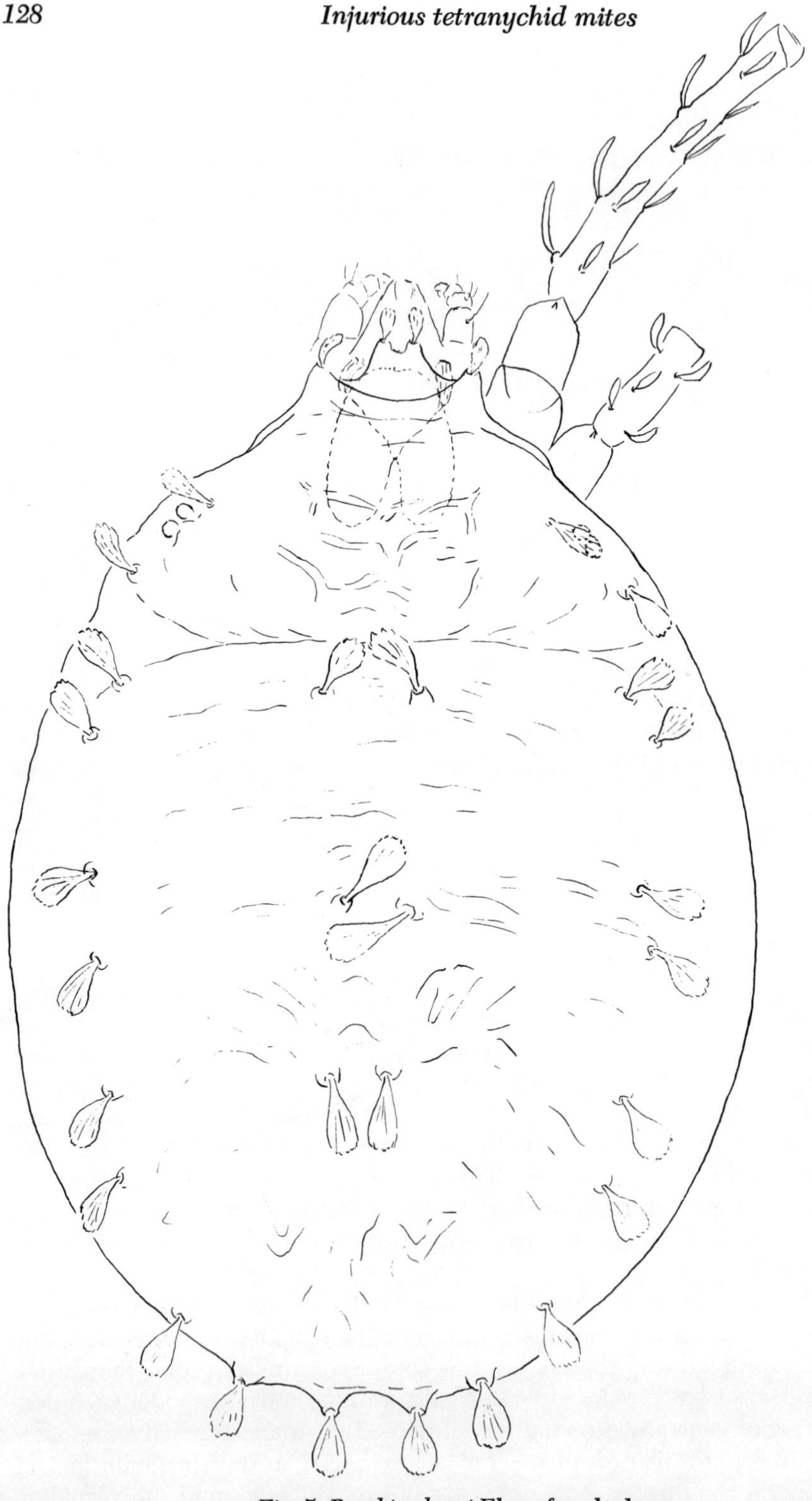

Fig. 5. *Bryobia eharai* Ehara female dorsum.

Bryobia cristata (Dugès). *B. cristata* occurs in Japan, France, Switzerland, North Africa, Australia, and New Zealand on grasses and other herbaceous plants and occasionally on fruit trees. All stages hibernate in cracks and crevices of walls and tree trunks. Throughout its life cycle this species develops best when relative humidities are below 60 percent. There may be 5 to 7 summer generations per year. During the period of maximum activity (May and June) females often penetrate the interior of buildings to deposit their eggs. Some of these develop into diapause or winter eggs, but a proportion develop normally (Evans, Sheals, and MacFarlane, 1961; van Eyndhoven, 1957).

B. cristata is a distinct species in that the empodium of tarsus I of the female consists of a short stub bearing 2 rows of ventrally directed tenent hairs, and the dorsal setae of the body appear to be more slender than those of *B. praetiosa* Koch.

Clover mite, *Bryobia praetiosa* Koch. This mite (fig. 6) has been regarded as a complex of closely related races that are similar in morphological character, but vary slightly in life history, host plant specificity, and habits. At least 4 biotypes have been recognized. Those that feed only on fruit trees are multivoltine and winter in the egg stage are now known as *B. rubrioculus* (Scheuten). Those that feed on a wide range of herbaceous plants, are either univoltine or multivoltine, and winter in all stages, are now called *B. praetiosa.* Those that are specific to ivy, are multivoltine, and winter in all stages, are designated *B. kissophila* van Eyndhoven. Finally, those that infest gooseberry, are univoltine, and winter in the egg stage, are called *B. ribis* Thomas. Because these groups differ biologically they will be considered separately. Reports made before the late 1950s concerning injury, host plants, distribution, and biology refer to the *praetiosa-rubrioculus* complex, so data summarized here, especially on *B. praetiosa,* are largely from later studies (Mathys, 1954).

Bryobia praetiosa, now recognized under the common name of clover mite, is widely distributed in North and South America, Europe, Asia, Africa, and Australia. It feeds on a wide variety of herbaceous plants from which it may move to dwellings, but seldom is it found on aerial portions of trees. Injury to plants by this mite is first observed as a winding, etiolated trail, similar to the mines of certain leaf miners. Extensive damage to clovers, lawn grasses, ornamental flowers, alfalfa, wheat, rye, barley, and other grains may result from feeding of high populations of this species. Foliage and floral injury may also occur on flowering plants. The foliage usually turns yellow or brown and then wilts.

In warmer climates the clover mite overwinters in both the egg and active stages; consequently spring development follows two courses. The winter eggs hatch and give rise to a spring generation, or the overwintered adults deposit spring eggs, the progeny of which also become a portion of the spring generation. The overwintering larvae and nymphs mature in the spring, their progeny forming a slightly delayed portion of the spring generation. Embryonic development in winter eggs appears to be a continuous process as long as weather conditions are suitable.

The mites oversummer in the egg stage. With a drop in temperature, often in

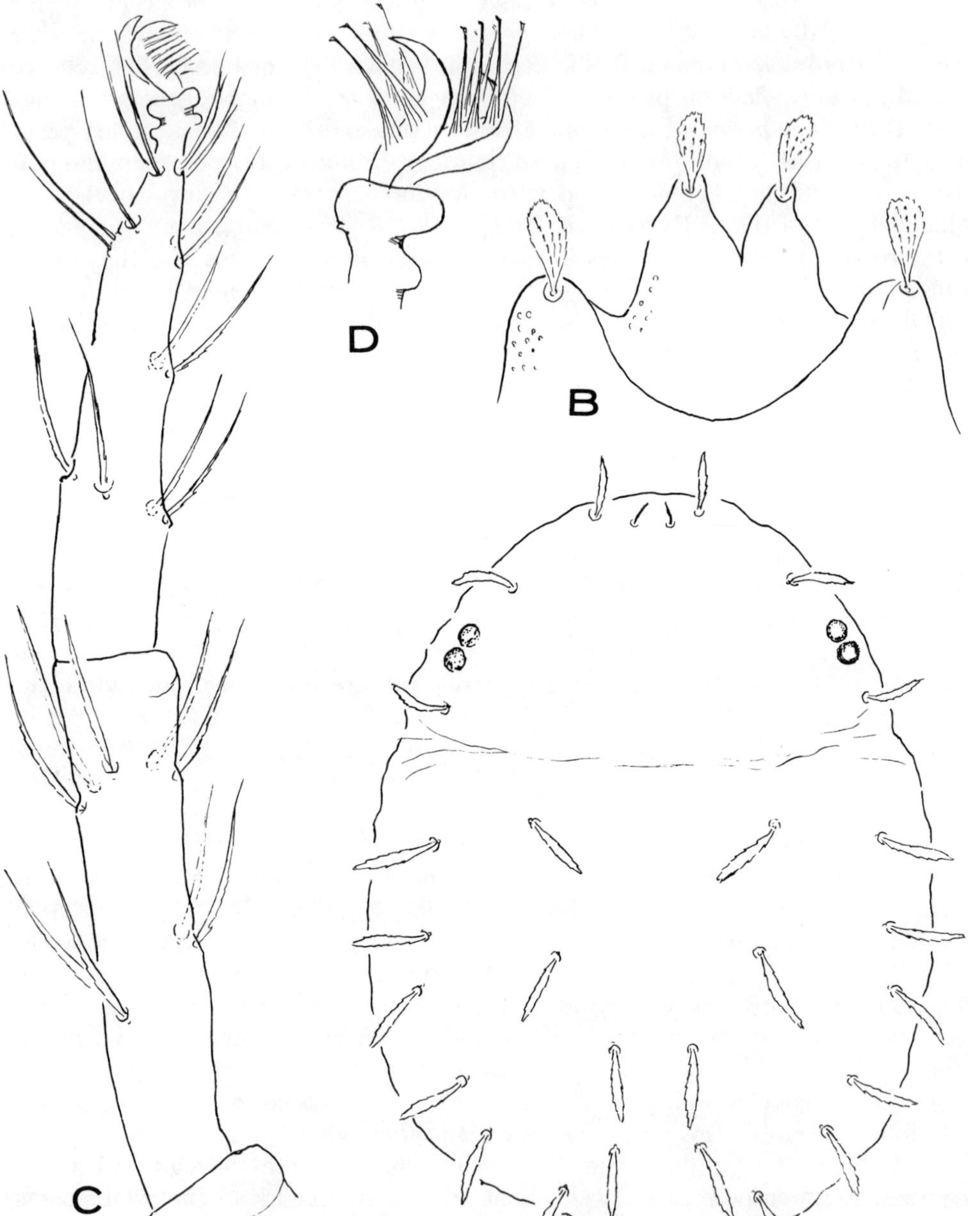

Fig. 6. *Bryobia praetiosa* Koch.: *a*, detail of larval seta arrangement; *b*, female protosoma, enlarged; *c*, leg IV; *d*, appendages of tarsus IV.

late September, the eggs start hatching. The mites feed and eggs continue to hatch until stopped by cold weather; thus all stages overwinter. Hatching commonly occurs on sunny days during the winter on the south side of dwellings and trees. The mites migrate to their host plants as mite activity increases in the spring. Feeding continues until early June, when the active stages die and eggs become dormant (English and Snetsinger, 1957*a*).

In northern areas the winter eggs hatch soon after the snow disappears in early spring, and before other species of phytophagous mites become active (fig. 7). Eggs may hatch after a week with temperatures at –2 to 8 C (28 to 46 F). Overwinter eggs exposed to temperatures of 18 to 24 C (65 to 75 F) will hatch in 12 18 hours. The winter adults also have a very low threshold of activity, so become active and begin depositing eggs about the same time as winter eggs begin to hatch. Overwintering adults continue to oviposit until mid-April. The hatch may extend to the middle of June, when most of these adults die.

The bright red, disc-shaped larvae are positively geotropic so they either migrate or drop from the egg site to herbaceous plants where they begin to feed, soon becoming dark green and almost spherical. All stages migrate to sheltered

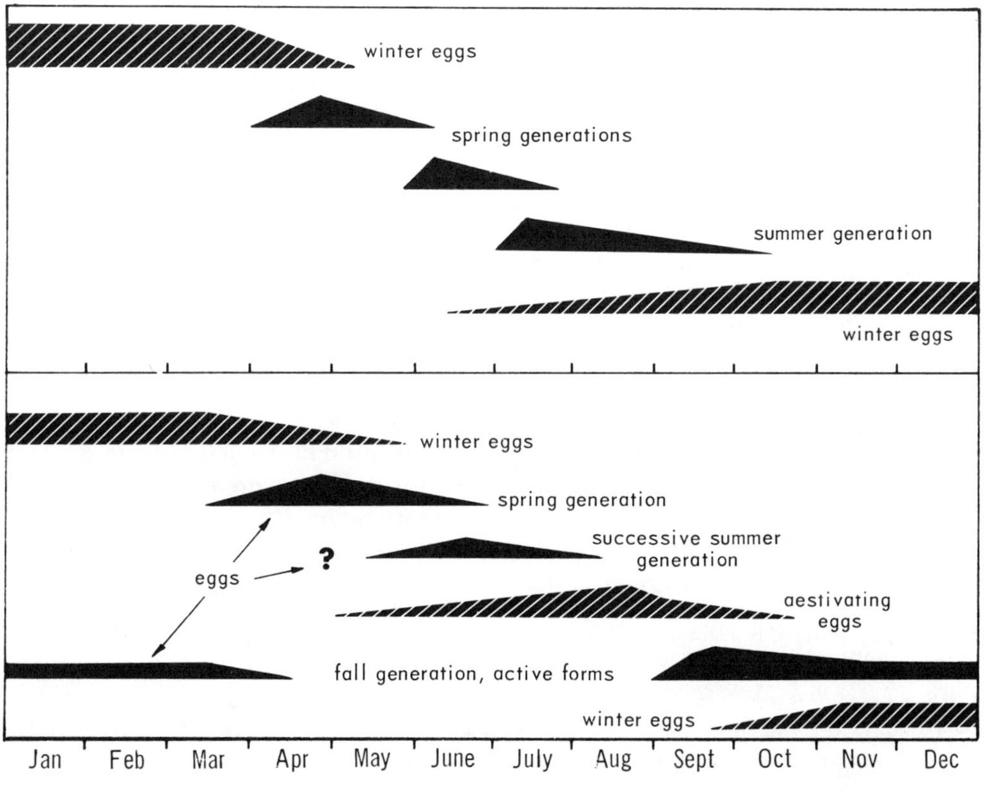

Fig. 7. Life histories of *Bryobia rubrioculus* (Scheuten) (*B. arborea* Morgan and Anderson) (top) and *B. praetiosa* Koch (bottom), based principally on data obtained in 1954, Summerland and Oliver (redrawn after Anderson, N.H. and C. V. G. Morgan. 1958. Can. Entomol. 90(1):23–42.)

locations to molt or oviposit; thus very few quiescent individuals are found on herbaceous plants. The mites respond readily to disturbance, whereupon they drop from the plants, curl their legs, and remain quiet.

A few of the eggs deposited by the spring adults do not aestivate, but hatch in early summer giving rise to continuous generations. Aestivating eggs deposited in May and June may hatch in late summer or early fall; the peak emergence may follow the first period of cool, fall weather, that is, maximum temperature below 21 C (70 F) and continue even after the first frosts (Anderson and Morgan, 1958; Gäbele, 1959).

The clover mite is sometimes a pest in dwellings. As indicated, the hibernating mites and the mites that develop from overwintering eggs migrate to protected places to molt or lay eggs. Where dwellings are surrounded by or near clovers, grass, weeds, or other hosts the mites may enter the dwellings through cracks in the floor, walls, or masonry as they seek protected places to molt or oviposit. They do not injure people or pets nor do they feed on anything in the buildings; but they leave stains when crushed. In this way they cause concern to the inhabitants, especially when tremendous numbers infest the walls, windows, and floors.

Clover mites move into the buildings from the lawns, therefore any means of discouraging their reaching the building will aid in preventing the nuisance. This may be accomplished by creating a grass-free band 18 to 24 inches wide around the dwellings. The lawns may also be treated with a nontoxic acaricide, such as malathion, dicofol, or chlorobenzilate, to obtain the quickest response (English and Snetsinger, 1957).

B. praetiosa is distinctive in the larval form in that the dorsal body setae are long, slender, and serrate. The adult is distinctive in having typical propodosomal protuberances, broadly rounded serrate dorsal body setae, and in that legs I are longer than the other legs and about as long as the body. The duplex setae of tarsi III and IV are similar, arising from a common base, the proximal seta being about ⅔ as long as the distal rodlike seta.

Bryobia kissophila van Eyndhoven. This species lives commonly on ivy in Europe. It does not survive on fruit trees, gooseberry, or on herbaceous plants. Six to 8 successive generations occur throughout the year, but development ceases at 0 C (32 F), so no eggs are laid during cold weather; there is, however, no true diapause. This species is at present distinguishable from *B. rubrioculus* and *B. ribis* only in its biology (Evans et al., 1961).

Bryobia ribis Thomas is found in Europe on gooseberry, but in cultures it is able to colonize on fruit trees. It can also be maintained on herbaceous plants, such as clover and various grasses; but not on ivy.

These mites overwinter as eggs, which start hatching in early March. Populations peak in mid-May and in June, postembryonic stages disappear as diapause eggs are laid; thus this species is univoltine. It is distinguishable from *B. rubrioculus* and *B. kissophila* only by differences in biology (Evans et al., 1961).

Bryobia graminum Schrank. *Graminum* lives on grasses throughout most of the year, but in Germany it occasionally becomes a pest of apple and pear trees during late summer.

This mite overwinters primarily in the egg stage, although all stages may be found throughout the winter on the lower parts of trunks of apple and pear trees, especially among lichens, or on walls or fences near these trees. The overwintering eggs hatch in late March, about one month earlier than those of *B. rubrioculus*. The spring generations feed on grasses, but generations developing in mid-July to mid-September ascend the trees and feed on the leaves. Summer generations develop in 27 to 57 days and survive for 18 to 103 days. Females deposit an average of 30 to 35 eggs during their life span. There are about 7 generations per year.

Another biological race of this species lives entirely on grasses and weeds near the bases of fruit trees. The eggs hatch the last part of January, but development is slow because of cold weather. Populations decline as grasses become mature in mid-June (Gäbele, 1959).

Bryobia repensi Manson. This mite was first recorded during 1967 in New South Wales, Australia, on herbage, apple, and strawberries. It has since been reported to be causing injury to perennial rye grass, vetches, clovers, and lucerne. The mites feed on leaves causing bleaching of the blade which, together with dry conditions, may result in severe injury to these crops (Manson, 1967). The mites are most active from midday until late afternoon, mainly feeding on the upper leaf surfaces; this gives the foliage a mottled appearance. Development of heavy infestations appears to be favored by extremely dry weather (Hamilton, 1972).

B. repensi has the propodosomal lobes well developed and empodium I of the female with a single pair of tenent hairs. Specifically, the female is distinctive in that the outer propodosomal lobes are large and teatlike, and in that femur I usually bears 24 to 26 setae, rather than 16 to 21 as in the other species. The larva is distinctive in that the dorsocentral hysterosomal setae and the anterior member of the posterior dorsolateral setae are distinctly smaller than most of the other dorsal body setae.

Brown mite, *Bryobia rubrioculus* (Scheuten) (pl. 23; fig. 8). This fruit tree species has long been confused with the clover mite, *B. praetiosa* Koch, which feeds on grasses, clover, and low-growing plants. It is indistinguishable from *B. redikorzevi* Reck, under which name its biology has been reported in the USSR. Because of some differences in the biology reported, the information from that area will be discussed separately. *B. rubrioculus* is distributed through North and South America, Europe, South Africa, Australia, and Asia. It is a pest of apples, pears, peaches, and other deciduous fruits.

The larvae emerging from overwintering eggs migrate to buds where their sucking activities cause whitish-grey spots on the upper surface of young or spur leaves (pl. 24), chiefly close to the base of the veins. As the infestation and season continues, the injury spreads, causing the whole tree to become lighter in color. Early infested leaves fail to grow to normal size. Bronzing of leaves, typical of the European red mite, *Panonychus ulmi* (Koch), is not produced by this spe-

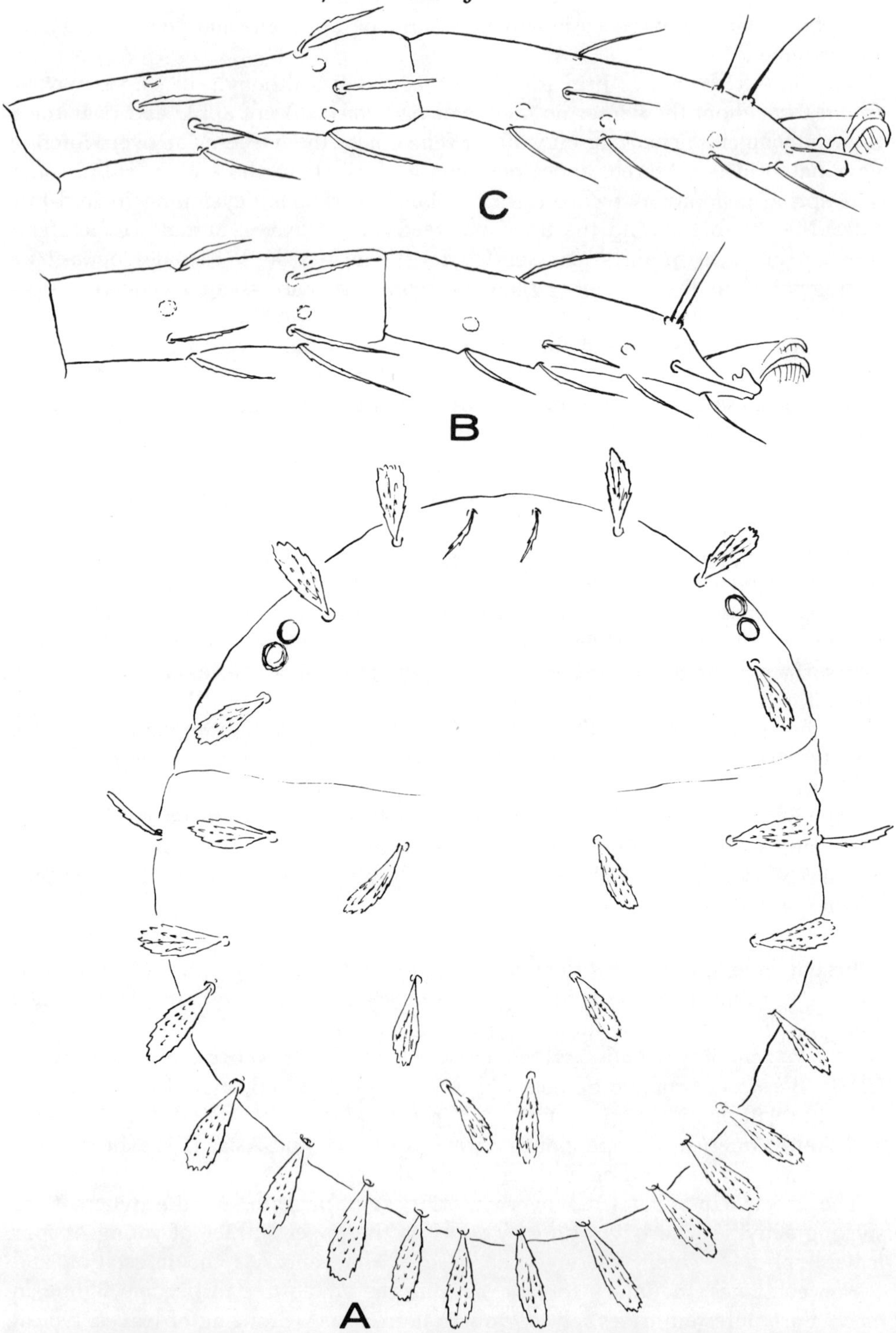

Fig. 8. *Bryobia rubrioculus* (Scheuten): *a*, female, detail of seta; *b*, leg III; *c*, leg IV.

cies. Cast skins are not prevalent over the apple leaf surfaces, but occur near the midrib or on the branches where these mites migrate to molt, oviposit, or rest. Dessert peaches may be made unsightly by cast skins, egg shells, and excreta that accumulate in the hollow of the stem end; this aspect, however, is not important on apples and pears. This mite rarely if ever causes the leaf drop typical of many other mite species infesting fruit trees.

Spring deposition of brown mite eggs occurs on the undersides of apple leaves either near the prominent veins or toward the base of the leaf. Other favorite sites are the leaf petioles, twig spurs, and the wood of the younger shoots. The substrate upon which the eggs are laid is either covered with fine hairs or very roughened. Later in the season more eggs are laid along the midrib of the upper leaf suface, as well as in the calyx end of the apples. Winter eggs are deposited on the spurs, along the undersides of twigs and forks of branches, and on the side of the trunk and branches; eggs are usually packed close together in folds and crevices.

Eggs of the brown mite (fig. 7) begin hatching 1 to 2 weeks earlier in the spring than eggs of the European red mite; the hatch is completed before the fruit trees bloom. The greatest percent of eggs hatch when temperatures range from 19 to 27 C (66 to 80 F). Both low and high temperatures cause a decrease in percent of eggs that hatch, also, the percentage hatch decreases sharply as humidities are increased about 80 percent.

The rate of hatching on almonds accelerates rapidly as the trees begin to bloom. Maximum rate occurs between full bloom and the completion of petal drop. Larvae hatching from diapausing eggs in the spring are positively phototactic, which stimulates upward migration to the first buds that open at the tips of the shoots. The first generation eggs hatch in 6 to 18 days into bright orange red larvae that, after feeding on the undersides of leaves, become dark green. Molting usually takes place in close proximity to the midrib and prominent veins of the leaf. The larvae and nymphs feed mainly on the underleaf surfaces, but they move freely over the whole leaf and onto the twigs. The adults feed irregularly rather indiscriminately over the entire leaf surface (Anderson and Morgan, 1958; Dosse, 1963; Georgala, 1958).

The brown mite has 2 generations per year, but in some apple orchards in Nova Scotia, there is a partial third. The first generation females in the bivoltine and trivoltine populations lay summer eggs on the leaves and twigs, and diapause eggs are laid only on twigs. Bivoltine populations lay both summer and diapause eggs, but eggs laid by females of the third generations are the first to hatch in the spring (Herbert, 1965).

There are 3 to 4 generations of the brown mite per year in Lebanon and Austria and as many as 6 in South Africa. Deposition of winter eggs may begin in July; eggs are always laid on the wood instead of the leaves.

Temperatures up to 36 C (97 F) in Nova Scotia apparently have no harmful effect on adults of the brown mite, except to cause them to move to shaded areas; in Lebanon, however, hot weather in July and August destroys many of the active stages, thus reducing the total mite population. Extended periods of rain adversely influence the development of populations by killing the mites remaining in water drops for even a few hours, by mechanically washing mites from the leaves, and by retarding development of the immature stages. High temperatures early in the season provide favorable conditions for population increase by accelerating mite development and by promoting sufficient foliage growth to supply food. Later in the year the availability and quality of food are the major factors influencing population development.

The brown mite does not freeze at temperatures above –27 C (–16 F). Each time the period of exposure is doubled at this temperature, there is an increase of 8 to 10 percent in mortality (MacPhee, 1963).

Chemical control of the brown mite can be directed against the winter eggs, the postembryonic stages of the first generation, and against the summer adults and eggs. Winter sprays do not always prevent the development of injurious summer infestations because it is practically impossible, owing to concealed oviposition sites, for the spray to reach an adequate proportion of the eggs. Winter or dormant sprays are valuable in preventing damage by the first mite generation, especially when heavy oviposition has occurred. Prebloom applications are particularly effective as the mites at this time are primarily in active stages, and active mites are more readily killed by acaricides than are eggs. Also, the sparse amount of foliage provides more suitable conditions for efficient and effective application than can be achieved later in the summer. Control measures applied at prebloom prevent the early generations from retarding the generative development of the fruit tree or adversely affecting the number of fruit set and their ultimate size. Single summer applications frequently fail to affect economic control because stages are present and the eggs are frequently laid in niches difficult to contact by spray applications. Residual type acaricides or repeated applications of those having shorter residue life are therefore necessary to kill the mites that emerge from these eggs. Even then, effective summer sprays may be too late to prevent injury to the trees or the crop.

The presence of both the brown mite and European red mite in commercial apple orchards currently require chemical control applications to prevent economic damage to the trees or crop. Dormant spray oils applied at 3 to 4 % in the spray, or decreased amount in combination with an effective ovicide, have been useful in reducing winter egg populations. Summer type oils, or one of the available acaricides, may be successfully used against the spring populations (Kremer, 1956).

B. rubrioculus larva is distinctive in having large, broadly clavate, serrate dorsal body setae. The adult is similar to that of *Bryobia praetiosa*, but is distinctive in the arrangement of the duplex setae in tarsi III and IV. The setae on tarsi III are similar to those of *B. praetiosa* in arising from a common base, but those on tarsus IV are well separated, each on its own base.

Brown fruit mite, *Bryobia rubrioculus redikorzevi* Reck. This species cannot be differentiated morphologically from *B. rubrioculus* (Scheuten). It is a serious pest in many areas in Asia including Azerbaidzhan, Moldavia, the Ukraine, Kazakhstan, Bulgaria, Serbia, Hungary, and Yugoslavia where it damages apples, apricot, cherry, peach, plum, almond, walnut, and sloe trees. In Yugoslavia, this mite attacks particularly apples, plum, and cherry; and to somewhat less degree, pear, almond, and peach. Active stages are rarely found on walnut, apricot, and quince even when the susceptible fruits in the same vicinity are severely injured. The most serious infestations occur when the average relative humidity is 61 to 69 percent, and populations do not become serious in areas with a mean relative humidity above 70 percent. Consequently the brown fruit mite is more frequently found in the steppe zone of the Ukraine and in the Crimea, Soviet Moldavia, the Caucasus, Soviet central Asia, and southern Kazakhstan, areas where relatively dry conditions are most favorable for population development.

These mites prefer the crown of the trees. In contrast to *B. rubrioculus*, about 90 percent of the first generation live on the twigs or branches and 10 percent on the leaves; during the night or early morning, however, the mites move to the foliage where they feed largely on the upper surfaces. They molt and lay their eggs on the branches and undersides of leaves.

The mites of this subspecies hibernate as winter eggs deposited in protected places in cracks of the bark, at axils of buds, in recesses of fruit-bearing offshoots, and at forks of branches. The winter eggs begin to hatch in early spring (March and April) after the mean temperature reaches 10 to 15 C (50 to 59 F). Hatching extends over a period of about 16 to 49 days depending on the climate. Embryonic development of both spring and summer generations requires 9 days at 24.5 C (76 F), 16 days at 19.5 C (67 F), and 27 days at 15 C (59 F). Development from larvae to mature female requires 14 to 17 days and the preoviposition period 2 to 3 days. The first generation produces the most eggs—averaging 20 per female. The number laid decreases until the last generation, which averages only 7 eggs per female. The sum of effective temperatures above 10 C (50 F) required to complete development of the overwintering generation is 221 to 258 C (397 to 464 F). There are 6 to 7 generations per year in southern and 4 in northern USSR. Females of the generations appearing in August, or later, oviposit overwintering eggs that remain unhatched until the following spring.

Treatment in early spring with 8% oil emulsion is effective. Selected acaricides applied in the pink-bud stage protects the apple leaves from damage for 70 to 90 days (Ramakaev, 1966; Tomašević, 1965).

Hystrichonychini Pritchard and Baker

This tribe belongs to the subfamily Bryobiinae. The empodia and true claws are padlike, however, with tenent hairs.

Tetranycopsis Canestrini

The claws and empodia are padlike and bear tenent hairs. There are 4 pairs of propodosomal setae; all dorsal body setae are long and strong and set on strong tubercles. There are 12 pairs of hysterosomal setae, including the humeral setae.

Tetranycopsis horridus (Canestrini and Fanzago). *Horridus* (fig. 9) is the only species in this genus which appears to be injurious to economic plants. This mite species feeds on the upper surface of hazelnut trees in southeastern Europe,

England, and Russia. A persistent infestation occurs near Niles, California, on hazel trees imported from Europe.

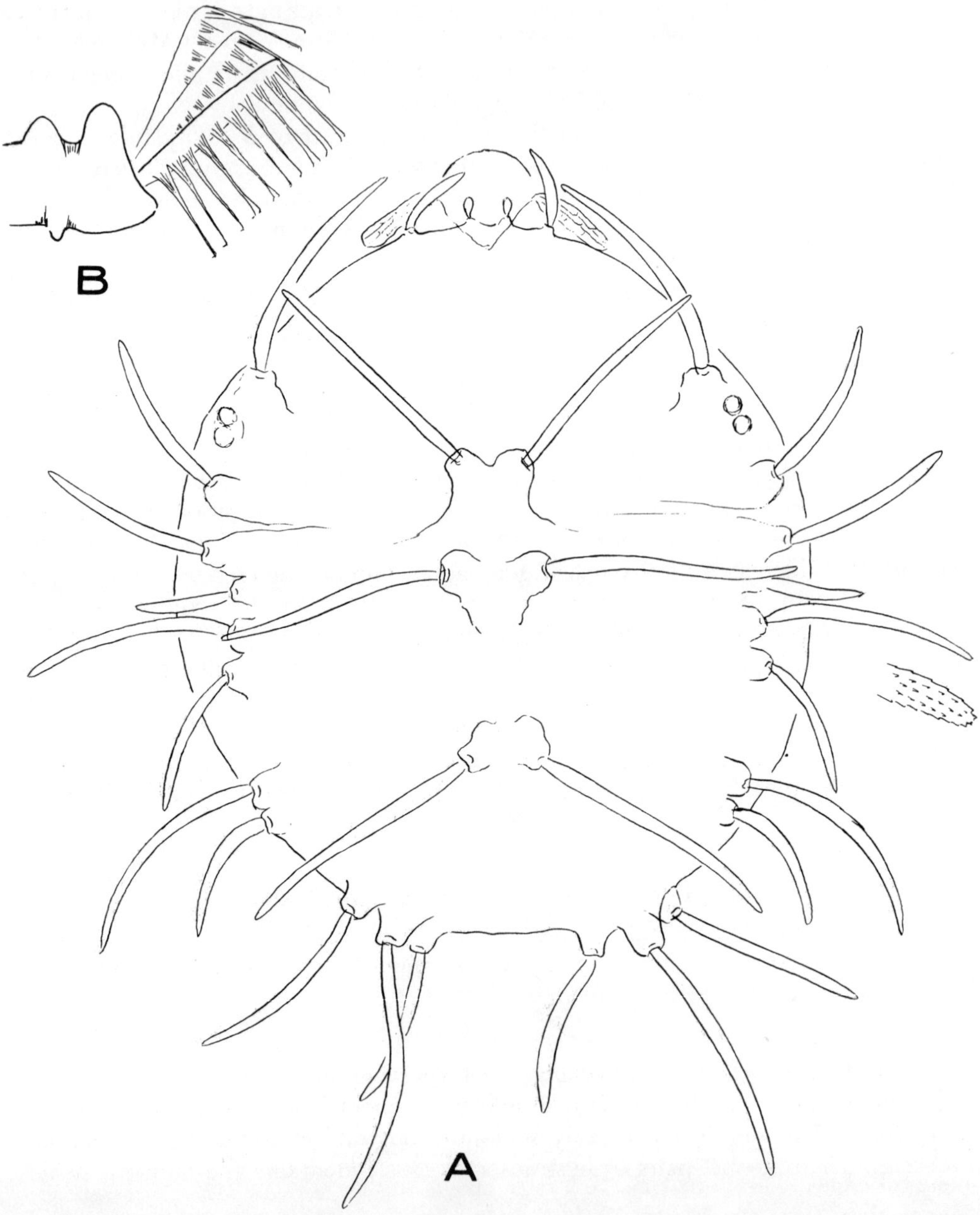

Fig. 9. *Tetranycopsis horridus* (Canestrini and Fanzago): *a*, female dorsum showing setae; *b*, claw.

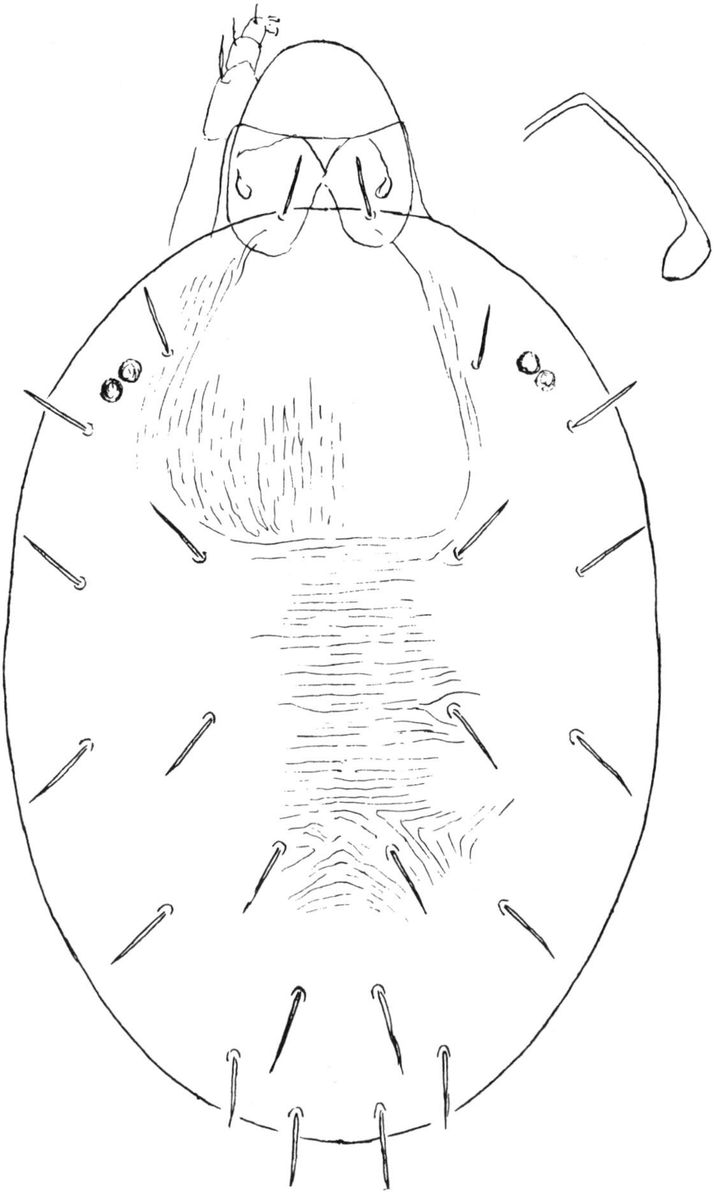

Fig. 10. *Paraplonobia (P.) myops* (Pritchard and Baker). Female dorsum showing setae and striations, inset, collar trachae.

This species has distinctively long and strong dorsal body setae set on prominent tubercles. The peritremes are free, anastomosing distally into units that are longer than they are broad. The anterior pair of propodosomal setae are very small; the dorsal setae of the hysterosoma are much longer than the others.

Paraplonobia Wainstein

The dorsal body setae are not set on tubercles and are well separated; the peritreme in *Paraplonobia* (s. str.) ends in a simple bulb; the tarsal claws and em-

Fig. 11. *Aplonobia histricina* (Berlese). Dorsal view showing setae, striations, tarsal appendages, and collar traechae.

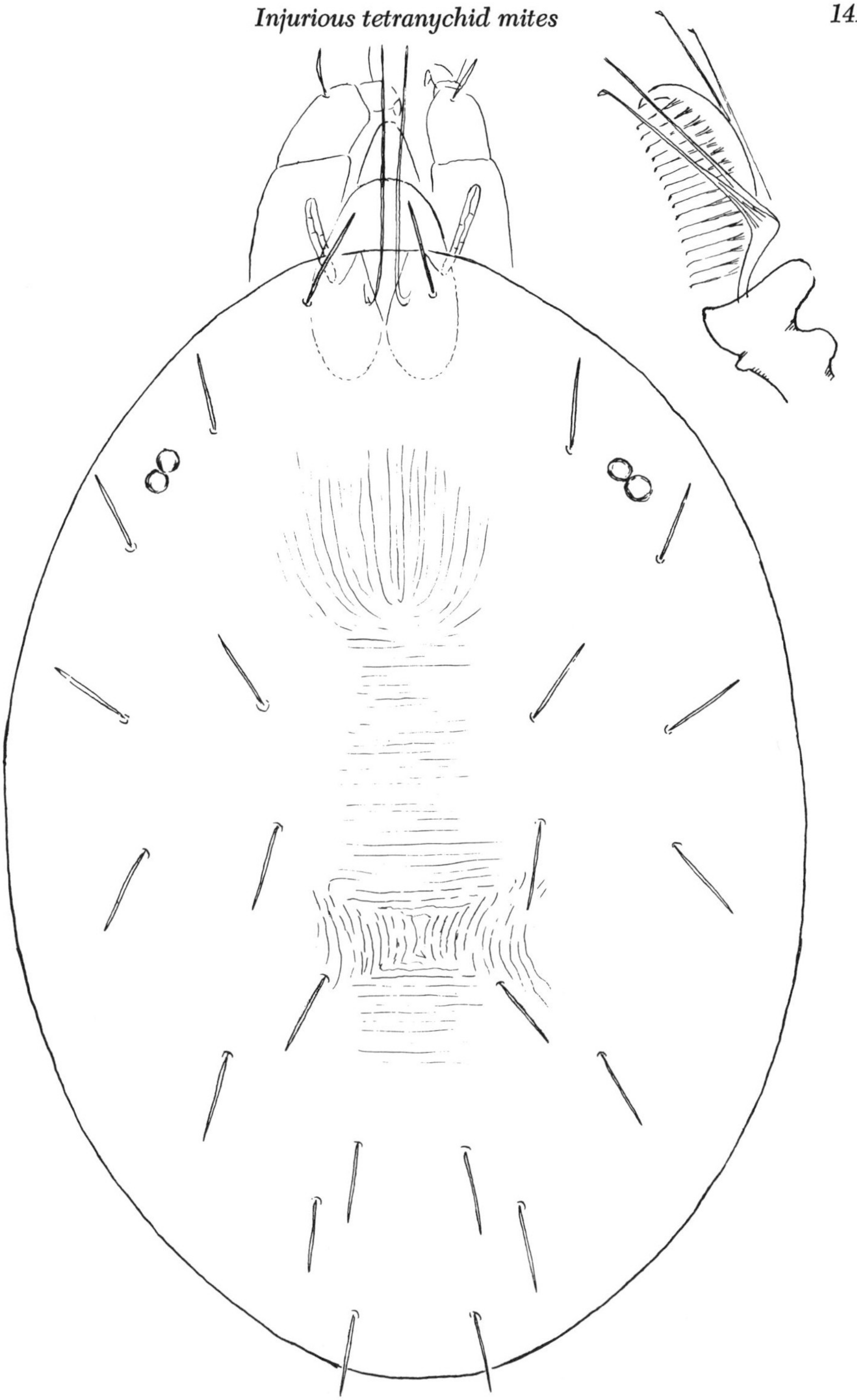

Fig. 12. *Perobia (Petrobia) latens* (Müller) female dorsum showing setae and striations, claw inset.

podium are padlike, with tenent hairs; the empodium is much stronger than the claws. The dorsal setae count is similar to that of *Aplonobia.*

Paraplonobia (*Paraplonobia*) *myops* (Pritchard and Baker). This mite (fig. 10) has caused considerable damage to commercial asparagus fields in the San Joaquin and Imperial valleys of California. These are relatively active mites, so on being disturbed dislodge themselves and drop from the plant. The mites tend to feed along the stem, rather than at its tip; feeding produces a chlorotic spotting and silvering effect typical of tetranychid mite injury.

Along with the characters of the genus, the stylophore of *P.* (*P.*) *myops* is rounded anteriorly; the peritreme ends in a simple bulb; the dorsal body setae are of equal length, much shorter than the distance between their bases, and serrate; the propodosomal shield is only slightly differentiated from the rest of the propodosoma.

Aplonobia Womersley

This genus is distinctive in having 3 pairs of propodosomal setae, 10 pairs of hysterosomal setae all set in strong tubercles, the peritremes anastomosing distally, and the claws and empodia padlike, with tenent hairs. The empodium is much stronger than the claws. The dorsal body setae are well separated.

Aplonobia histricina (Berlese). Injurious infestations have occurred on fruit in South Australia, New South Wales, and South Africa. The dorsal body setae of this species (fig. 11) are strong, serrate, and subequal in length, being about as long as the distance between their bases; all setae, except the first pair of propodosomals, are set on prominent tubercles. The peritremes, as figured, anastomose distally (Pritchard and Baker, 1952).

Petrobiini Reck

These mites are in the subfamily Bryobiinae; the empodium and claws possess tenent hairs, and 3 pairs of anal setae in the female. The Petrobiini are distinctive in that the true claws are padlike and the empodium is clawlike, with tenent hairs.

Petrobia Murray

There is a row of tenent hairs on the empodium; there are no projections over the rostrum; there are 2 sets of duplex setae on tarsus I, and the peritremes anastomose distally. The dorsal body setae are not set on tubercles.

Brown wheat mite, *Petrobia (Petrobia) latens* (Müller). *P. (P.) latens* (fig. 12) was first described from specimens collected from under stones in Denmark. It is now recognized as a pest of small grains in the United States, Europe, Turkey, China, Africa, India, South Rhodesia, Japan, England, and Australia. Injury has been recorded on fruit trees in England; on sorghum and onions as well as grains in India; and on grass, strawberry, and stone leek in Japan. It has caused considerable damage in the United States, to onions, carrots, cotton, endive, lettuce, iris, gladiolus, sorghum, alfalfa, bur clover, and wild onion; the mite is also a nuisance in houses.

This mite is strictly a dry weather pest and its damage appears to be similar to that caused by drought. The plants suffer from the loss of plant sap, which is withdrawn by the mites as food; therefore heavily infested fields appear to be dried out, even when there appears to be sufficient soil moisture. Feeding by this species causes a very fine mottling of the leaves and, at a distance, a bronzing or yellowing effect may be seen. Injury to wheat may be evident as early as May 15 in western United States.

Injury to carrot fields usually starts in small localized areas and then spreads rapidly through the fields. The foliage turns yellow in the early stages of infestation, and soon becomes dry and brittle. The mites produce no webbing and, when disturbed, may run rapidly over the leaf surface or drop to the ground.

The eggs are not deposited on vegetation but either on soil particles or any solid object in the area of the plants. Both nondiapause and diapause eggs are laid which differ in general appearance. The cherry red nondiapause eggs develop in about seven days at 22 C (72 F). They are subspherical, lightly ribbed, and possess an inconspicuous central stalk. At hatching, the shell spilts near the equator. The glistening white diapause egg consists of a basal rounded portion and an expanded cap. The top of the cap bears 20 to 30 distinct ridges expanded radially to form a small elevated "boss" in the center.

Summer eggs require an incubation period of 8.5 to 11.2 days under field conditions. The time from egg hatch until the adult stage is reached varies from 8.0 to 11.0 days. Adults have a preoviposition period of 1 to 2 days, and an average adult life of 2 to 3 weeks. Summer females may lay 70 to 90 summer eggs within a 3-week period, but females that produce winter eggs lay about 30 in the same time interval. In no case does any female lay both kinds of eggs. This species is no doubt parthenogenetic since no males have been found.

Diapause eggs of this mite remain unhatched for indefinite periods during hot, dry weather. Soil moisture will generally initiate hatching within a few days during warm conditions. This indicates that moisture and increased humidities rather than decreased temperatures cause the diapause eggs to hatch, provided the eggs have been in diapause for sufficient period. Although hatching requires moisture, subsequent development of the population is dependent on a paucity of heavy rainfall. Heavy rains may destroy most of the hatching mites unless they are laid in well-drained soil or soil with gravel and stones (Cox and Lieberman, 1960; Fenton, 1951).

Sulfur dusts or sprays have provided control of this mite when applications were made under warm weather conditions. The organophosphorus compounds such as demeton, parathion, and Metacide®, have resulted in the most consistent control. See *Bryobia praetiosa* for methods of preventing infestations from entering buildings. Late sowing, deep ploughing, and crop rotation aid in avoiding development of injurious populations (Chung, Wei, and Tieng, 1963; Reynolds and Swift, 1951).

P. (*P.*) *latens* dorsal body setae are not on tubercles, are serrate, and are shorter than the distance between their bases; the anterior propodosomal pair are the longest. Leg I is longer than the body. The distal anastomosing portion of the peritremes is longer than it is broad.

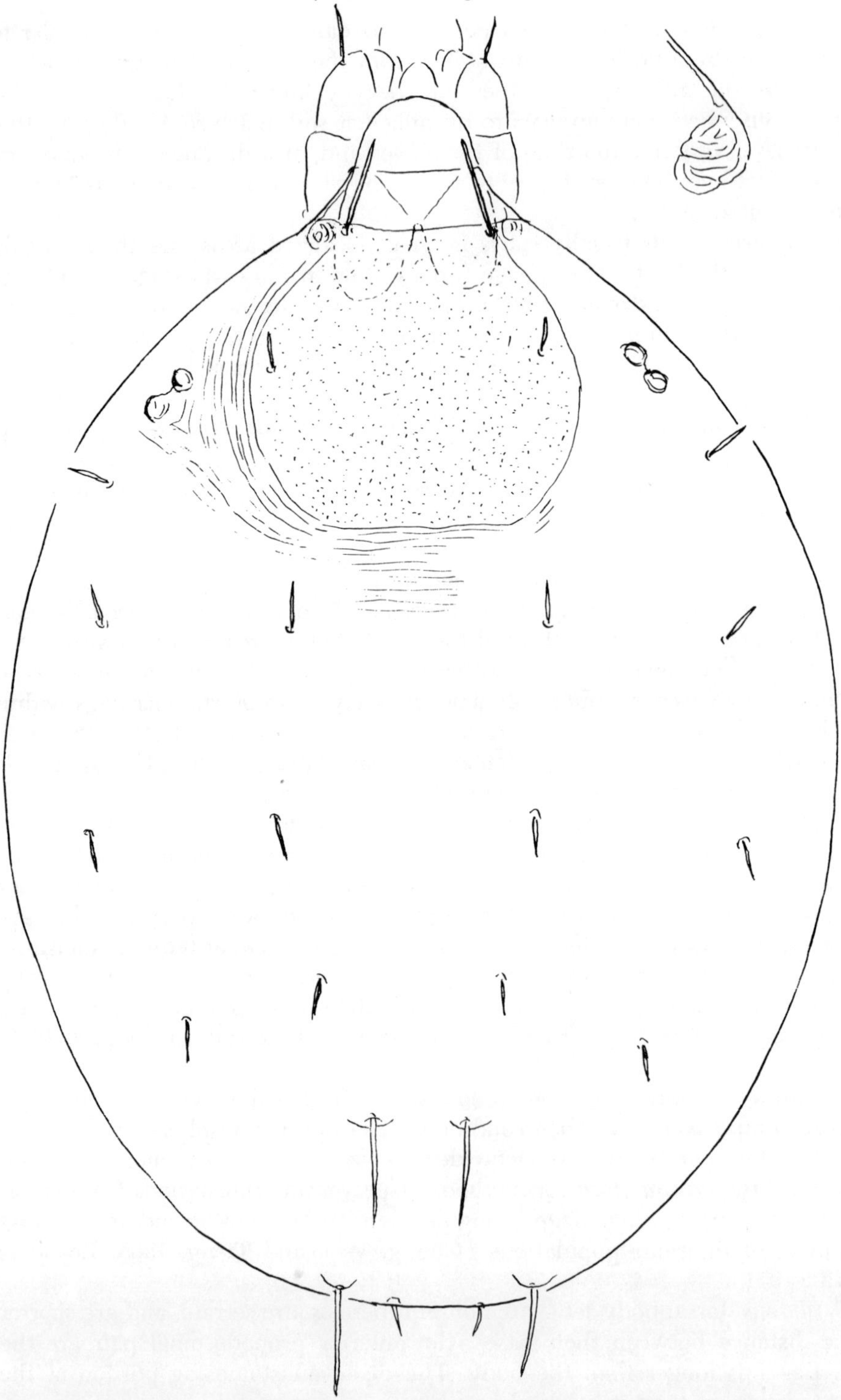

Fig. 13. *Petrobia* (*Tetranychina*) *apicalis* (Banks) dorsum. Inset: collar tracheae.

Petrobia (*Tetranychina*) Banks

Some or all dorsal body setae are set on prominent tubercles. Otherwise this subgenus is similar to *Petrobia* s. str.

Legume mite, *Petrobia (Tetranychina) apicalis* (Banks). This mite (fig. 13) is a serious pest of clovers, vetches, and winter peas in Louisiana. Favorite hosts in order of preference are winter pea, white clover, hop clover, common vetch, crimson clover, Persian clover, onions, and hairy vetch. It is probably distributed throughout the Gulf States as far north as Missouri, and has recently been reported in Portugal.

These mites feed mainly on the upper surface of the leaves. Severe damage causes the leaves to become greyish and no longer appetizing as forage. The plants produce little or no seed and often die; consequently, stands of these self-seeding plants may be eliminated in two or more successive years of infestation. This mite may also be a nuisance during the period of summer egg deposition because great numbers often enter buildings, residences, offices, factories, and the like, causing concern to the occupants (Smith and Webber, 1954).

The legume mite is active from November to May and passes the summer as diapuse eggs. These summer eggs often are laid in large numbers on rough surfaces such as in cracks or loose bark, or on trees, bushes, and fences. Each egg is composed of a white, rigid, cup-shaped receptacle full of cherry red embryonic material securely covered with a lid. The receptacle has 20 to 30 superficial longitudinal sections, each of which corresponds to a radial section of the lid.

Females that produce diapause eggs develop mainly in response to long photoperiods during most of their immature development, although crowding and starvation may evoke diapause earlier in the year. Diapause is apparently facultative, as the development of such eggs may be induced artifically by controlling environmental conditions. Development is arrested at an early embryonic stage and hatching does not occur until the eggs have been in diapause for at least ten weeks. Cool to moderate or fluctuating temperatures and moisture are necessary to break diapause. Hatching of diapause eggs does not occur above a temperature of about 21 C (71 F). The hatching rate rises sharply as the temperature drops until a temperature of 18 C (65 F) is reached; then it levels off and declines at lower temperatures.

The initiation of egg laying by diapausing females appears to be independent of such environmental factors as temperature, photoperiod, and food availability; but egg laying appears to be related to the age of the diapause eggs at the time of their hatching. The age of such eggs also influences the initiation of diapause in the females of the progeny of these eggs, that is, the older diapause eggs, the fewer number of active generations, and the sooner diapause is initiated in their progeny. This suggests that the evocation of diapause in this species is influenced by the inherent physiological conditions of the female at the time of egg laying. If such is the case, eggs that hatch last in the spring are laid by fall mite populations that develop under warm, dry weather, rather than fall populations that develop during cool, wet conditions. Consequently, there are likely to be fewer

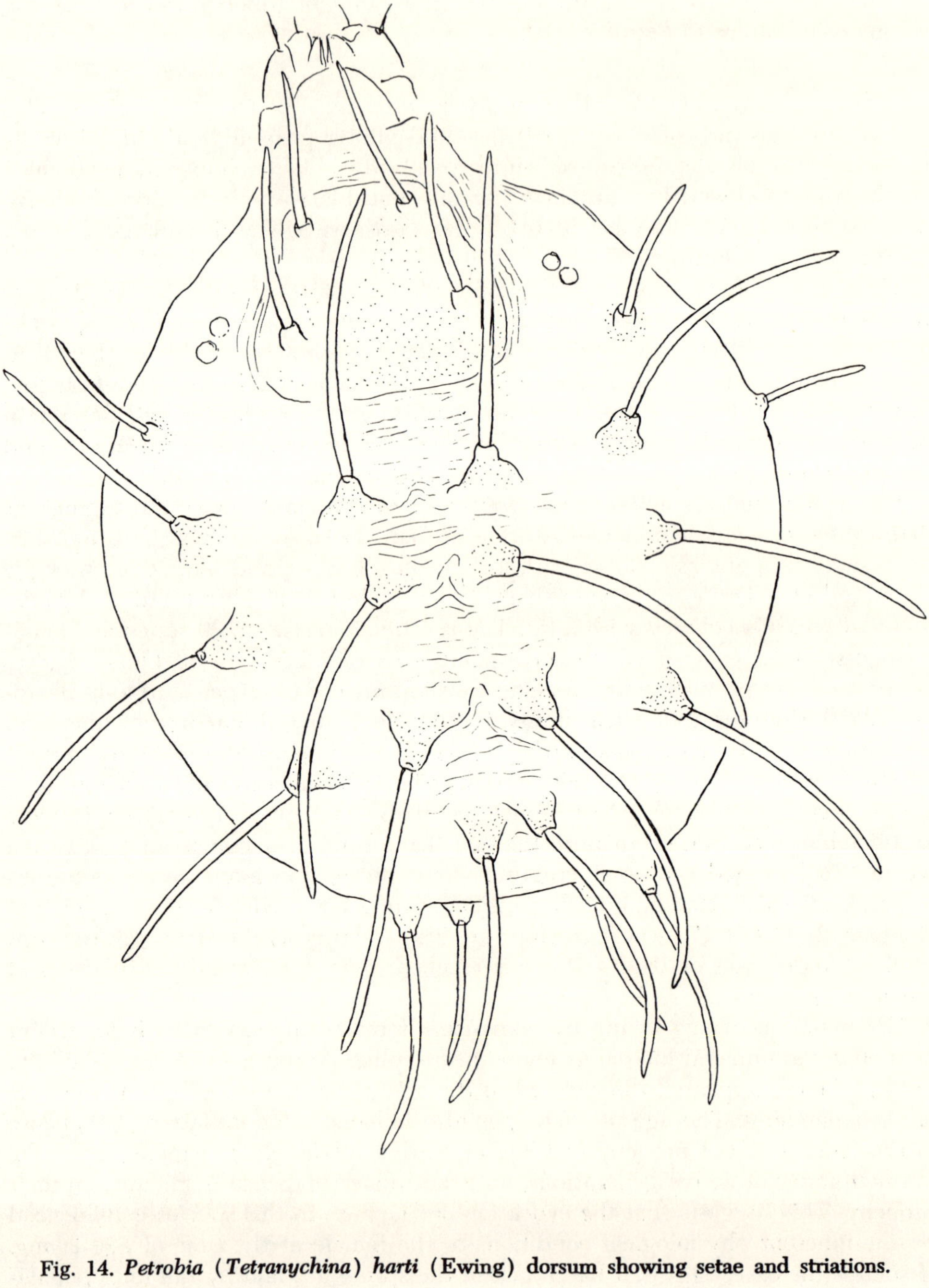

Fig. 14. ***Petrobia*** **(*Tetranychina*)** ***harti*** **(Ewing) dorsum showing setae and striations.**

generations per year after a warm and dry fall than after a cool and wet, one.

The nondiapause egg has a pearly, transparent, and minutely striated shell containing cherry red embryonic material that gives the egg its color. It is almost hemispherical with a straight, finely wrinkled, and sticky base. The larva starts feeding immediately after hatching; its scarlet color soon darkens, eventually becoming dark greenish brown. The feeding punctures are arranged in straight, bent, or criss-cross lines. This pattern is helpful in distinguishing field infestations of this species from other mite species found on the same hosts (Brooking, 1957; Glancey, 1958).

The 8-legged protonymph is ovoid with a dark greenish brown body. The deutonymph is similar in color, but a third pair of preanal setae have developed. The adult female body is ovoid or frequently oblong and dark reddish brown. The nondiapause eggs are laid singly or in masses on debris or on the host plant. Each female is capable of producing 70 or more eggs, and 9 or more generations may be produced per year.

Dissemination of this mite is through crawling, floating on surface water, and being carried in air currents. The most effective dispersion is probably accomplished by the larval stage capable of spinning a thread or web. Aided by this thread, larvae may "balloon" some distances depending on their altitude and the velocity of the air current. Diapause eggs have been observed deposited on trees at heights exceeding 12 feet from the ground. Upon hatching, larvae will often climb even higher thereby increasing their potential dispersion range when carried away by air currents (Zein-Elden, 1956).

The females of *P.* (*T.*) *apicalis* have very short dorsal body setae except for the first pair of propodosomal setae. The fourth pair of dorsal setae and the fourth pair of marginal setae are 4 to 5 times as long as the other body setae. The propodosomal shield is punctate; the rest of the body has the typical transverse striation patterns. The peritremes anastomose distally.

The male is similar to the female except for the much longer legs in relations to the body; also, the difference between the setal lengths is not so obvious.

Petrobia (*Tetranychina*) *harti* (Ewing). This mite (fig. 14; pl. 25) appears to be distributed throughout the world for it has been collected in North and South America, Africa, Asia, Australia, and Japan. *Oxalis* is the major host, but populations occur on *Crotalaria* in India, clover in the United States, sugarcane in Mauritius, and citrus in Mozambique. Injury appears as fine stippling on leaves. Although the leaves turn yellow, no serious damage has been observed. (Moutia, 1958).

The tarsal appendages of *P.* (*T.*) *harti* are typical for the genus. The dorsal body setae are set on prominent tubercles, and are much longer than the intervals between them; the fifth pair of dorsal setae are much shorter than the others. Legs I are longer than the body. The peritreme ends in a simple bulb. The setae of the male are shorter than those of the female, and the legs are much longer in relation to the body.

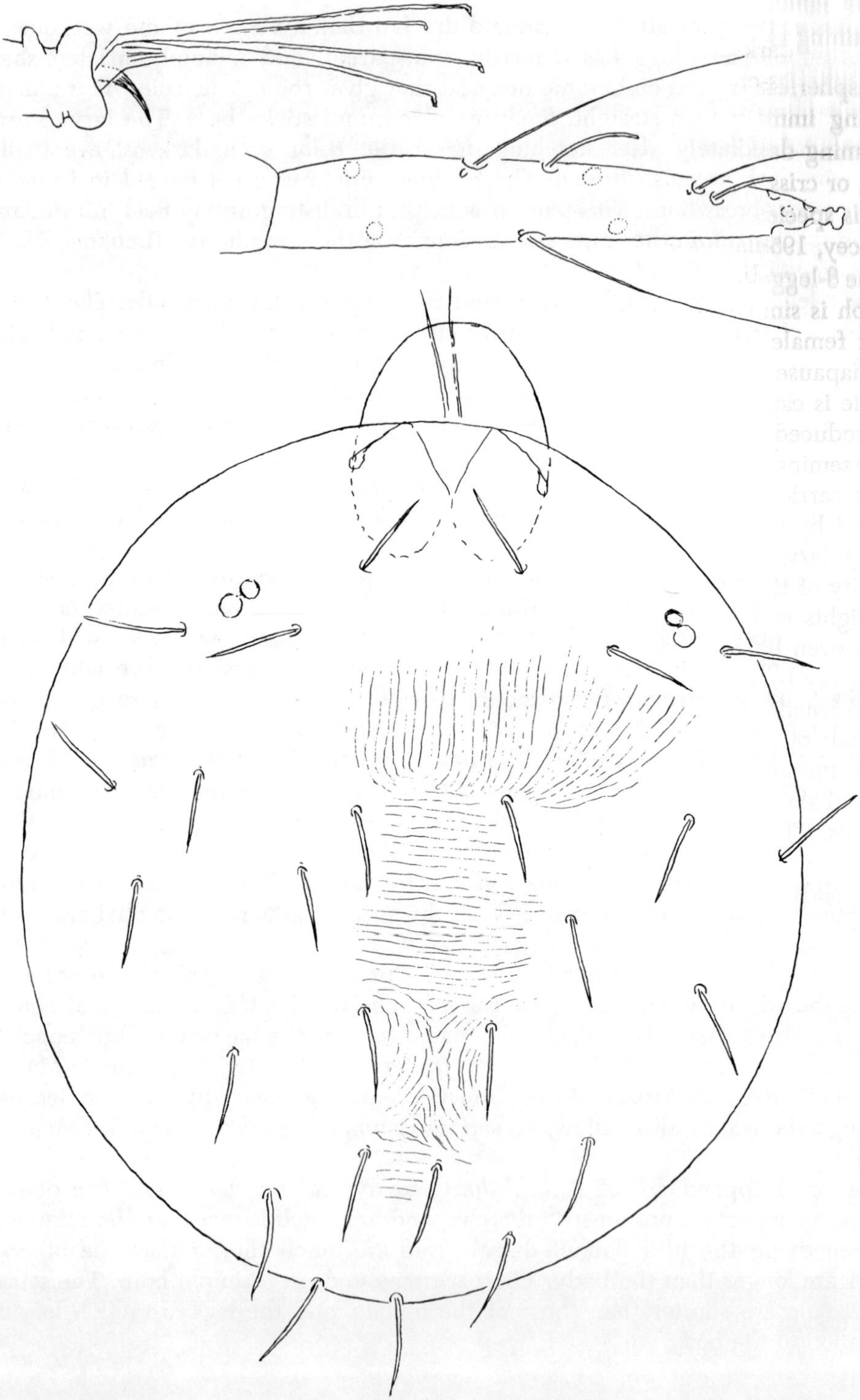

Fig. 15. *Eurytetranychus buxi* (Garman), dorsum and tarsus and tarsal appendages.

Schizonobia Womersley

The true claws are padlike, the empodium is clawlike, each bears a pair of tenent hairs; the peritremes are broadly expanded distally.

Schizonobia sycophanta Womersley. This species causes injury to grasses in Australia and Tasmania. Typical damage consists of a drastic shortening of the internodes and reduced leaf blades and sheaths. The finer grasses seem to be the most readily attacked and considerable areas are often killed by the mites.

Mites live in large numbers at the bases of the sheaths where eggs are laid and the 6-legged larvae feed. All developmental stages overwinter in small colonies. The incubation period of the eggs may be as short as 7 days under ideal temperatures (Goldsmid, 1962).

Lime sulfur applied at 16 oz per 100 gal of water per acre has resulted in effective control on grasses.

TETRANYCHINAE BERLESE

The empodia do not possess tenent hairs; there are tenent hairs on the true claws. Usually there are 2 pairs of anal setae in the female and 4 pairs of genito-anal setae in the male, but in some genera of noneconomic importance, setae may be lacking.

Eurytetranychini

Tarsus I may have a single set of duplex setae; or the duplex setae may not be closely associated, or may be absent. Empodia may be present or absent.

Eurytetranychus Oudemans

The empodium is very small, and clawlike; the duplex setae are not associated. The body is large, globular, and without propodosomal shields.

Boxwood mite, *Eurytetranychus buxi* (Garman). *E. buxi* (fig. 15) is primarily a pest of boxwood, *Buxus sempervirens* L. It occurs in many states in the United States, including New York, Connecticut, Michigan, Ohio, Virginia, Georgia, North Carolina, California, and Oregon; it also occurs in England and in Italy.

Mite feeding on boxwood first causes a mottling, followed by yellowing and browning near the midvein of underleaf surfaces. Small comma-shaped spots are evident on the upper surface of the leaves. Severely injured leaves may become entirely bronzed and finally drop, leaving a scraggly looking plant. During heavy infestations all stages of this mite may be found on both leaf surfaces (pl. 26).

This mite overwinters in the egg stage. Egg hatching begins in early April to May, and the life cycle is completed in 18 to 21 days. The 3 developmental stages are completed in 2 to 4 days, 2 to 3 days, and 3 to 7 days respectively. A female may lay 3 to 4 eggs in a day and 25 to 35 during her lifetime. The eggs are lemon yellow, lozenge-shaped, and covered with shallow, yet distinct sculpturing that

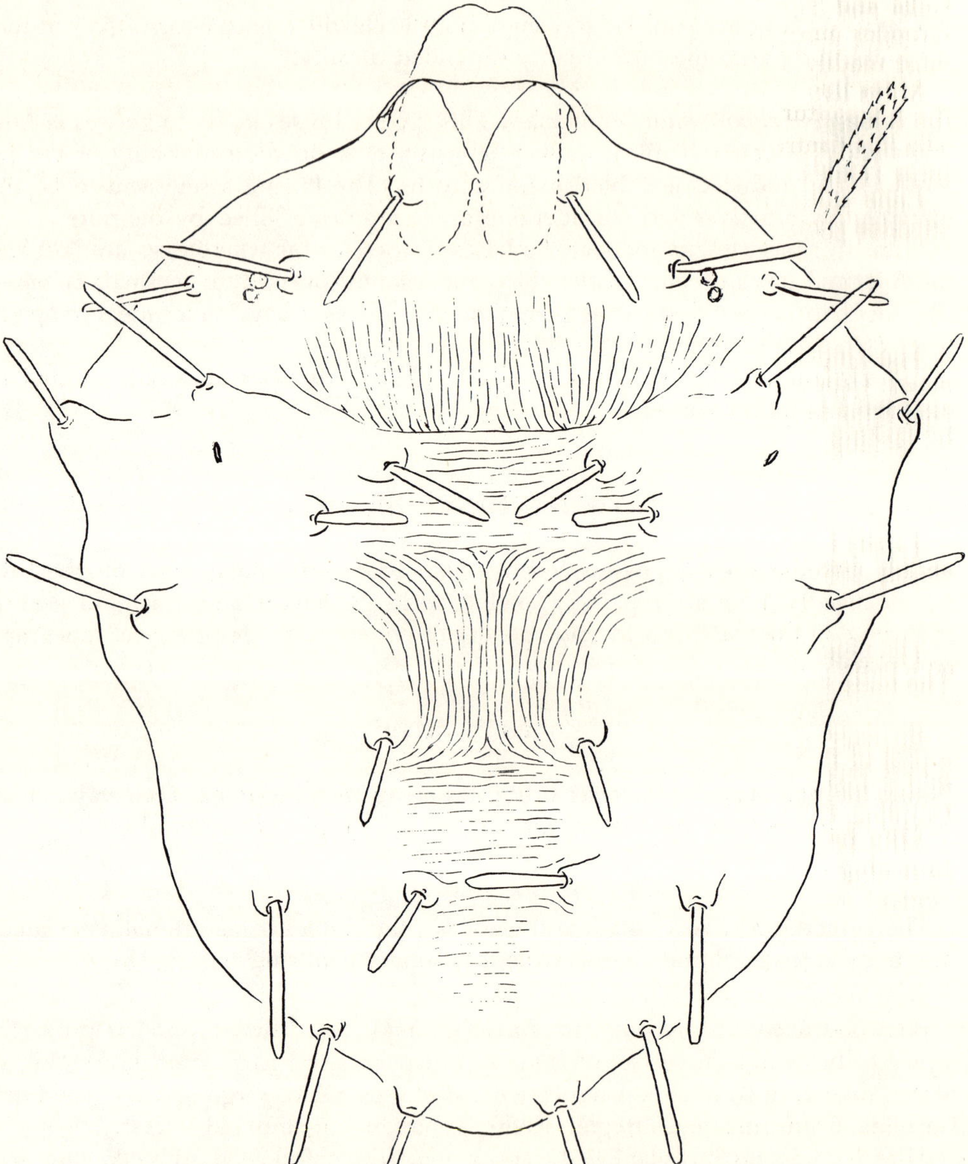

Fig. 16. *Eutetranychus africanus* (Tucker), dorsum showing striations and setae.

extends in ridges from the base to the center top. The eggs hatch in 6 to 10 days into yellowish green larvae. The nymphal instars are deep green and green to yellowish brown, respectively. There are at least 8 generations per year. The overwintering eggs are laid during the latter part of September and early October.

High temperatures and low humidities provide the most satisfactory conditions for mite development. These mites prefer to feed on the tender shoots and the upper surface of the young leaves. The life span of the adults varies from 2 to 5 weeks under constant weather conditions (Reeves, 1963; Ries, 1935).

The empodia of *E. buxi* are present and clawlike; the solenidia are absent on tibia III and IV; the dorsal body setae are almost as long as the distance between their bases; the peritreme ends in a simple bulb. Tarsi I of both sexes have a single set of duplex setae that are somewhat separated.

Eutetranychus Banks

The characteristic duplex setae are not present; the tarsi do not possess empodia; the claws are short, padlike, and have tenent hairs.

Eutetranychus africanus (Tucker). *Africanus* (fig. 16) is commonly found on citrus, frangipani, peach, and loquat in South Africa, India, and Natal; and on a cherry-leaf tree (*Eriobotrya japonica*) in Mauritius. Heavy infestations produce many fine stipplings on the leaves causing them to drop prematurely without turning brown. Heavy rain is a limiting factor in the distribution of this species (Baker and Pritchard, 1960; Moutia, 1958).

The dorsal body setae of *E. africanus* are short, strong, serrate, and set on prominent tubercles; the dorsal striae of the hysterosoma are normal and longitudinal between dorsal setae D_2 and D_3; setae D_1 and D_4 are shorter than the other dorsal body setae.

Texas citrus mite, *Eutetranychus banksi* (McGregor). This mite (fig. 17) occurs in North, Central, and South America on citrus, almond, croton, fig, castor bean, and others. Injury caused by this mite on citrus is similar to that produced by the citrus red mite, *Panonychus citri* (McGregor).

The Texas citrus mite deposits its eggs along the midrib and near the lateral margins of the leaves. The eggs are flat and disc-like, with a fine rolled rim or edge; they vary in color from light yellow when first laid, through tan and green to reddish brown just prior to hatching. Newly hatched larvae are light yellow to tan with pale legs. Adult females and nymphs are similar in color, varying from tan to brownish green with dark brown to greenish spots and bars near the lateral margins. The legs are pale, becoming tan to brown on the basal segments. The females are broad, robust, and flattened with moderately strong legs. The males are triangular, and similar in color to the nymphs and adults.

Eggs of this species are found throughout the year in Texas; lowest numbers occur from February to April, however, and the largest numbers during May and June. Low relative humidity, 8 to 10 hours per day of temperatures above 27 C (80 F), and the relative absence of rain, are conditions most favorable for the development of this species. Unless populations are high the mites favor the south

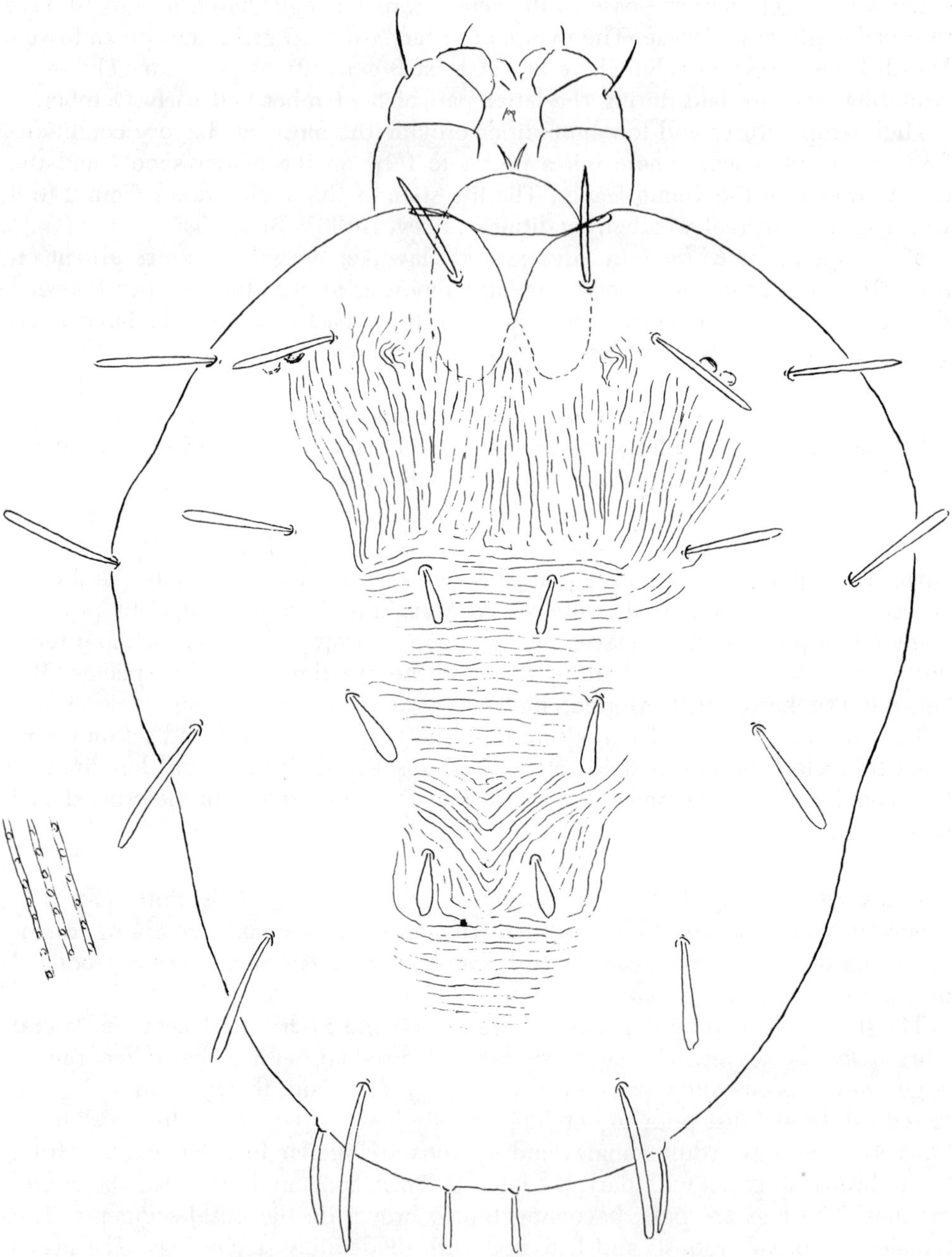

Fig. 17. *Eutetranychus banksi* (McGregor), dorsum showing striations (from Central America).

quadrants of the tree (Dean, 1959).

The dorsal body setae of *E. banksi* are variable in length, usually broader distally, and not as long as distance between their bases. Hysterosomal setae D_3 are closer together than the D_2 setae. The striae on the propodosoma are longitudinal and broken, and those on the hysterosoma are transverse except for the V-pattern between setae D_2 and D_3.

Eutetranychus enodes Baker and Pritchard. This species occurs on fig, peach, *Raffia*, *Vignia* and *Maihot* in Zaire (formerly the Congo). The dorsal integument of the body has smooth setae, and the dorsal setae are very short and borne on short tubercles.

Oriental red mite, *Eutetranychus orientalis* (Klein). This mite (fig. 18) is primarily a pest of citrus. It causes injury to this host in Israel, Turkey, Jordan, Iran, Egypt, Cyprus, Sudan, Afghanistan, India, South Africa, Formosa, East Transvaal, Thailand, Pakistan, Philippines, and Taiwan. Other hosts include cotton, squash, frangipani, pear, grapevines, quince, walnut, and *Euphorbia*.

Feeding by this species on the upper leaf surface produces a multitude of gray spots, which gives leaves a chlorotic appearance. Infested leaves weaken and finally drop; twigs dry, which results in bare trees in the nursery or young neglected orchards. Injury is more severe in the fall, especially if the trees lack moisture. The combined effect of insufficient water and a few mites causes as much leaf drop and twig dieback as does a heavy mite population.

Female mites of this species lay up to eight eggs per day. Summer eggs are deposited mainly along the midrib on the upper side of the leaf; but eggs occur on both leaf surfaces when infestations are high or during the winter. The developmental threshold for this species is 11 C (52 F) and the thermal constant, above which temperature development becomes slower, is 26 C (79 F). The preoviposition period lasts 1 to 2 days at 23 C (73 F) or above; 2.5 to 3 days at 20 to 22 C (68 to 72 F); and 4 to 8 days at 14 to 15 C (57 to 59 F). The longevity of the adults is about 12 days in summer, 14 to 18 days in the spring and autumn, and up to 21 days during the winter.

The eggs are very sensitive to low humidities. Such humidities cause a decrease in development rate and often death of the embryo. All stages show increased mortality with decreased temperatures; thus, high temperature and medium air humidity are optimum for the development of all stages. The optimum conditions are 21 C (70 to 71 F) and 59 to 70 percent relative humidity (RH); development, however, proceeds between 18 to 30 C (65 to 86 F) and 35 to 72 percent RH. Conditions beyond these limits are unfavorable for the development of this species (Bodenheimer, 1951).

The time of year populations peak and most injury occurs is largely determined by prevailing temperatures and humidity. In areas where summer humidities are low, populations do not increase from April to September, but may increase later sufficiently to cause injury. Where summer humidities are medium, as in coastal districts, populations increase during the summer. This species seems to prefer sour lemon stock to sweet lemon, mandarin, and orange. The structure and physiology of these varieties suggest a positive correlation between rate of population development on leaves and both the number of oil glands and the amount of tannic acid in the leaves, and a negative correlation with the thickness of the leaf cuticles (Mohamed, 1965).

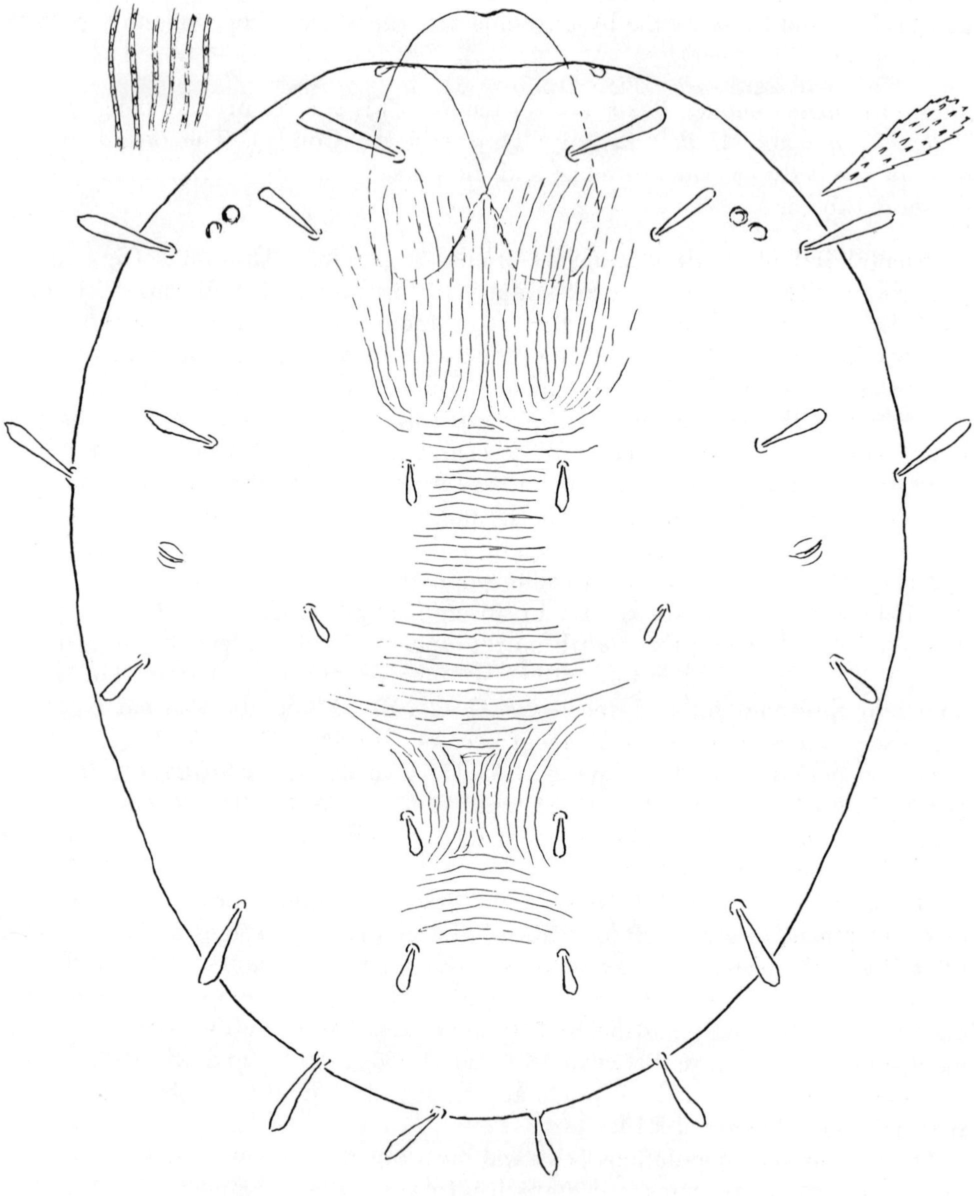

Fig. 18. *Eutetranychus orientalis* (Klein), dorsum showing setae and striations.

The Oriental red mite is susceptible to sulfur dusts and sprays. It also may be reduced by applications of the petroleum oils refined for use on citrus to control scale insects and other mite pests. The specific acaricides effective in controlling other tetranychid mite species are generally effective against the Oriental red mite.

The dorsal body setae of *E. orientalis* are short and broad distally, and set on small tubercles; setae D_3 and D_4 of the hysterosoma form a rectangle; the marginal body setae are longer than the dorsal median setae. The striae are longitudinal between the hysterosomal setae D_3.

Eutetranychus sudanicus Elbadry. It has recently been reported that this mite is a pest of citrus and a number of oramental plants throughout the Sudan. Other economic host plants include papaya, *Cassia* spp., watermelon, hibiscus, castor beans, morning-glory, jasmine, and banana.

Plant injury is characterized by severe premature bronzing of the leaves and stunting of plant growth. Eventual death of young plants may occur when populations are high.

The length in days of the winter and summer stages of this mite averaged: egg, 5.7 and 4.3 days; larva 3.7 and 2.9 days; protonymph, 2.4 and 1.7 days; deutonymph female, 2.7 and 2.2 days; and deutonymph male, 2.4 and 2.0 days. The preoviposition, oviposition, and postoviposition periods averaged 2.4 and 1.0, 12.2 and 10.4, and 3 and 2.4 days in winter and summer, respectively. Fertilized females laid an average of 22 and 32 eggs. The average life of the females was 15.2 and 12.8 days and that of the males, 11.9 and 10.3 days in winter and summer, respectively. The generation period lasted 14.5 days in winter and 11.2 in summer. There were as many as 27 generations per year. Relative humidity did not seem to influence the biology of this species significantly (Siddig and Elbadry, 1971).

TETRANYCHINI RECK

Panonychus Yokohama

There are two species of the genus *Panonychus* which cause serious injury to agricultural crops, namely the European red mite, *P. ulmi* (Koch), and the citrus red mite, *P. citri* (McGregor).

The genus may be characterized by its clawlike empodium with three pairs of ventrally directed hairs; there are two pairs of anal and two pairs of para-anal setae; the dorsal body setae are very strong and set on strong tubercles.

The European red mite can be distinguished from the citrus red mite in the field by the presence of light-colored tubercles at the base of the dorsal body setae; whereas the citrus red mite is uniformly reddish.

Raspberry red mite, *Panonychus caglei* (Mellott). This species (fig. 19, *a*) is a pest of raspberry, dewberry, and blackberry and populations live on snap bean, lima bean, soybean, common mallow, Kudsu-vine, *Potentilla*, and *Rosa*. It occurs in Delaware and is widespread in Virginia. The mites feed on both surfaces of raspberry leaves; immature mites feed mostly on the lower surfaces and the adults feed on the upper surfaces. Feeding injury appears as lines or short dashes on the upper leaf surface.

In early spring most of the eggs are deposited on the lower surfaces of the leaves, but 75 to 80 percent of the third- to ninth-generation eggs are on the upper leaf surfaces. Toward fall a higher percentage of eggs are again oviposited on the lower leaf surfaces and some are laid on the canes. Some of the eggs laid on

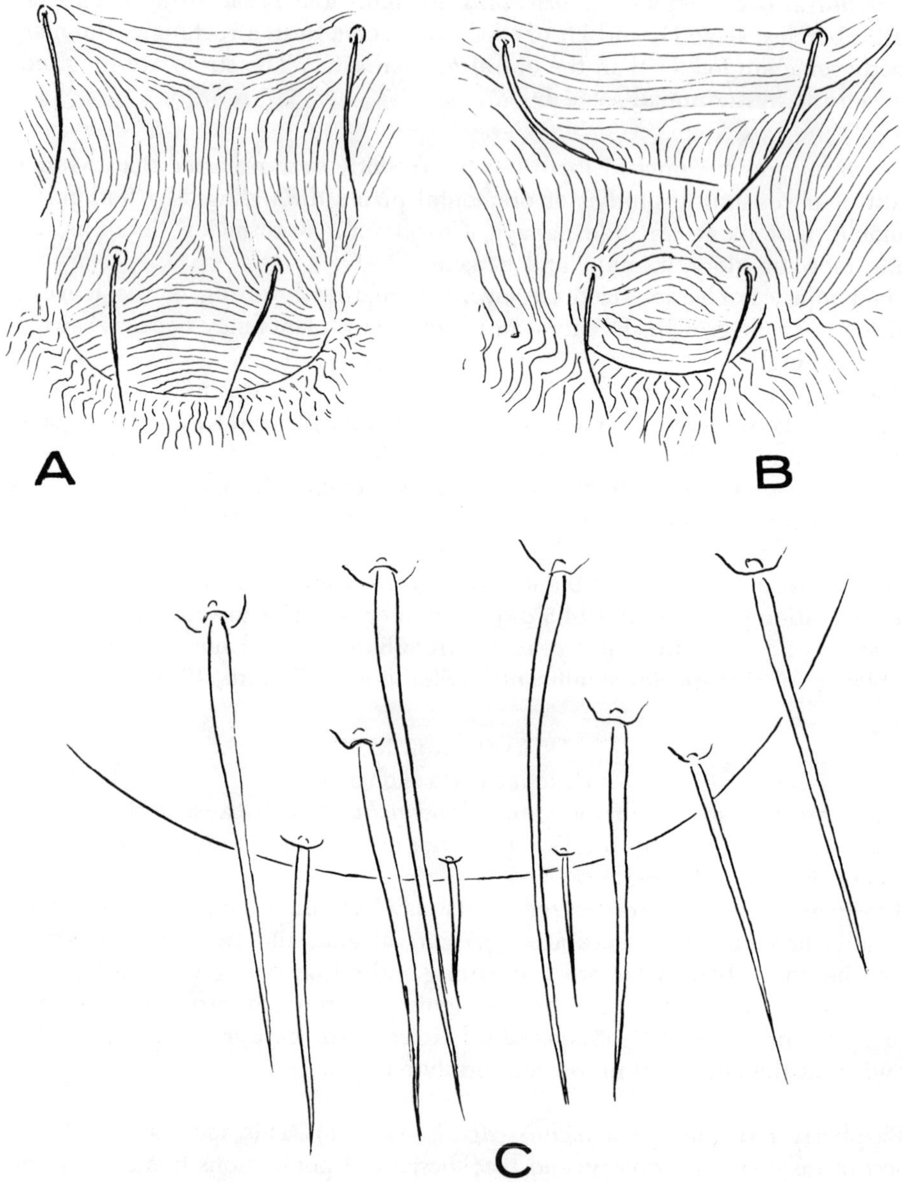

Fig. 19. *a, P. cageli* McIlott genital plate; *b, P. citri* (McGregor) and *P. ulmi* (Koch), genital plates; *c, P. ulmi* (Koch) dorsal-posterior setae.

canes in September may hatch in the usual time of summer eggs. During mild winters hatching may occur throughout the winter, but under colder winter conditions hatching does not start until March 25. A warm early winter followed by a cold period in late winter causes high mortalities of the active mites, resulting in low spring infestations. There may be 9 or more overlapping generations, although the length of the egg and developmental stages are dependent on the prevailing temperatures.

The eggs of the raspberry red mite are round, slightly flattened, and stalked. Striations are indistinct on unhatched eggs, but are distinct on the empty shells of hatched eggs. There are no regular guy fibrils; but silk strands from the stalk to the leaves may sometimes occur. Newly hatched larvae are yellowish to reddish orange, but become pale yellowish, to dark green when feeding. Dark areas, caused by food in the digestive tract, later appear along the sides and across the front, often causing the mites to appear black. The nymphal stages vary in color depending on their food; the abdomen may be green, amber, brownish green to dark reddish brown, or black. The body of the adult female is oval, arched above with setose body setae located on tubercles. The color of newly molted females is variable, being grayish or greenish brown, dark brown, or brownish red. Later the color changes to bright red, often appearing very dark brownish red, especially along the sides (Cagle, 1962; Mellott, 1968).

The body of the adult female of *P. caglei* is deep red; the dorsal tubercles are light red to pink rather than white as in *P. ulmi*; the striae of the female genital region are distinctive in that the striae are longitudinal both on the genital plate and in the area anterior to the plate; setae D_5 and L_4 of the hysterosoma are of equal length (as in *P. citri*) and shorter than D_4; there are 2 setae on genu IV.

Citrus red mite, *Panonychus citri* (McGregor). *P. citri* (fig. 19, *b*, 20; pl. 27) is the most serious citrus pest of California, South Africa, and Japan. It also occurs in Florida, China (Canton), South America, USSR, and India. Hosts other than citrus include rose, almond, pear, castor bean, and several broadleaf evergreen ornamentals.

Injury to citrus leaves produced by this mite includes stippling, light colored spots, and a greyish or silvery appearance, very similar to the injury produced by the European red mite on apple trees (pl. 28). Injury is most severe when there exists a combination of high mite infestation and either a high transpiration rate owing to low humidities and wind or a deficiency in leaf moisture owing to drought, poor root system, or other influences that reduce the plant's ability to supply adequate moisture to the leaves. Under these conditions the combined influence of mites and weather may result in heavy leaf and fruit drop, twig dieback and even death (pl. 29) of large limbs. The direct effect on the quality of the fruit is usually of minor importance.

Populations of this mite on orange trees increase most rapidly on the new growth during plant growth cycles that occur during the spring and fall, unless weather conditions during these periods are unfavorable. In addition, the mites develop on orange fruit. They seem to prefer the green to the yellow fruit, but during the winter months in California the mature navel orange fruit rather than the leaves harbor most of the mites. Populations of this species may increase rapidly on lemon trees any time weather conditions are sufficiently favorable to produce new growth and green fruit.

Extremely hot days, at 40.5 C (105 F), or several days of hot, dry weather (5% RH and 32 C [90 F]) accompanied by wind, usually cause high mite mortality.

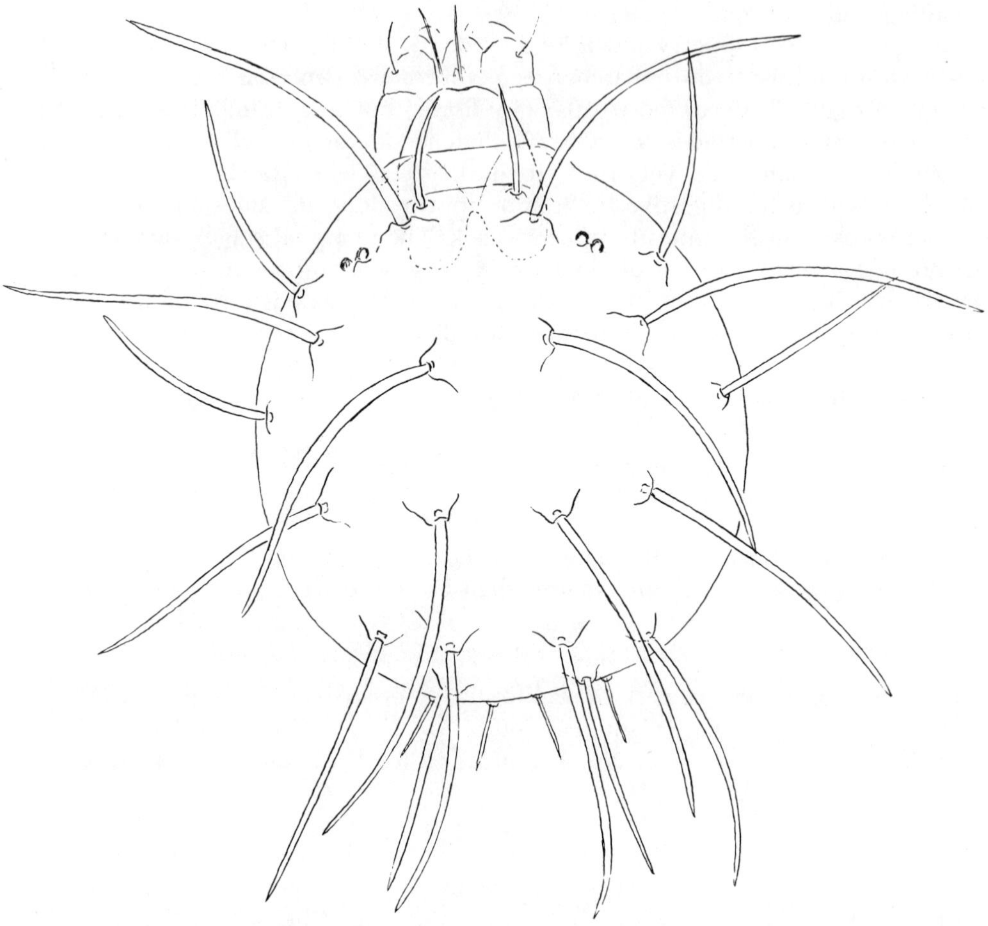

Fig. 20. *Panonychus citri* (McGregor) dorsal view.

Such conditions may occur during the summer, in southern California but they occur more frequently during the fall months. Prolonged periods of high humidity are also unfavorable for the development of populations of *P. citri*. The susceptibility to such extremes of both temperature and humidity limit the distribution of this species and alter the seasonal population trends.

The biology and time required for the development and duration of the life stages at different temperatures are shown graphically in figures 21, *a*, *b*. At a constant temperature of 26 C (78 F), the life cycle is completed in 14 days, but at 10 C (50 F) development is more than five times as long. The life span at 10 C (50 F), however, is 9 times longer than at 27 C (80 F); consequently the rate of development in hot weather is offset by the longer life in cool weather (English and Turnipseed, 1941; Shinkaji, 1954).

The eggs of the citrus red mite are red, nearly spherical, somewhat flattened, and stalked with guy fibrils radiating from the tip of the stalk to the substrate. The developing stages and the adults are deep red to purplish in color. The dorsal spines arise from tubercles. The tips of the tubercles are red in contrast to the whitish tubercles of the European red mite. As this difference can be seen by a 10X hand lens, it provides an excellent means of distinguishing these species in the field.

Citrus red mite populations like those of the European red mite have been difficult to keep from being injurious to their host crops. Sulfur was first used to control this mite, but its use was replaced by the summer grade petroleum oils. The time of year and the frequency that these oils may be applied without producing injurious effects on the crop, however, limit their use on such varieties as navel oranges and tangerines. Therefore, other acaricides are also required to prevent injury to these trees; but this mite species has proven capable of rapidly developing resistance to such acaricides.

The setae of *P. citri* are strong and are borne on prominent tubercles of the same color as the rest of the integument; the striae on the genital plate are transverse and those anterior to the plate are longitudinal; the hysterosomal setae D_4 are not as long as the other dorsal setae, but are about three times as long as the D_5 and L_4 setae, which are subequal in length. There are 3 setae on genu IV.

Panonychus elongatus Manson. This mite has been recorded from Burma and Queensland, Australia, on citrus and other hosts. It is believed to be widespread in tropical and subtropical areas, having been previously confused with the citrus red mite, *P. citri* McGregor (Manson, 1967).

P. elongatus is smaller than *P. citri* and has shorter setae. In the male of this species the distal portion of the aedeagus is elongate, slender, tapering directed posteriorly, similar to that of *P. citri*, but distinctly longer.

European red mite, *Panonychus ulmi* (Koch). This species (fig. 19, *c*; pl. 30) is found in most deciduous fruit orchards in the United States, southern Canada, Europe, Georgia in the USSR, China, Bermuda, Argentina, South Africa, India, Tasmania, New Zealand, and Japan. This mite is a major pest of apple, pear, plum, quince; and may cause injury to peach, walnut, cherry, almond, grape, raspberry, hawthorn, mountain ash, elm, rose, and chestnut. As this species is the most important mite pest on deciduous fruit trees, many publications have dealt with its biology and control (Cagle, 1946; Cutright, 1963; Doreste, 1964; Parent and Bealieu, 1957; Tsugawa and Shirasaki, 1961).

The mites feed by withdrawing juices and chlorophyll from the foliage, causing the leaves to become pale and assume a bronze color that impairs photosynthesis and respiration and temporarily increases transpiration. The immature mites feed primarily on the undersurfaces of the leaves, but the adults may feed on both leaf surfaces, especially when populations are heavy. Continual feeding by heavy populations will cause the leaves to turn brown and fall; therefore the earlier in the season mite injury occurs, the greater the damage to fruit trees. Trees injured in June will not form the normal number of flower buds for the following year's

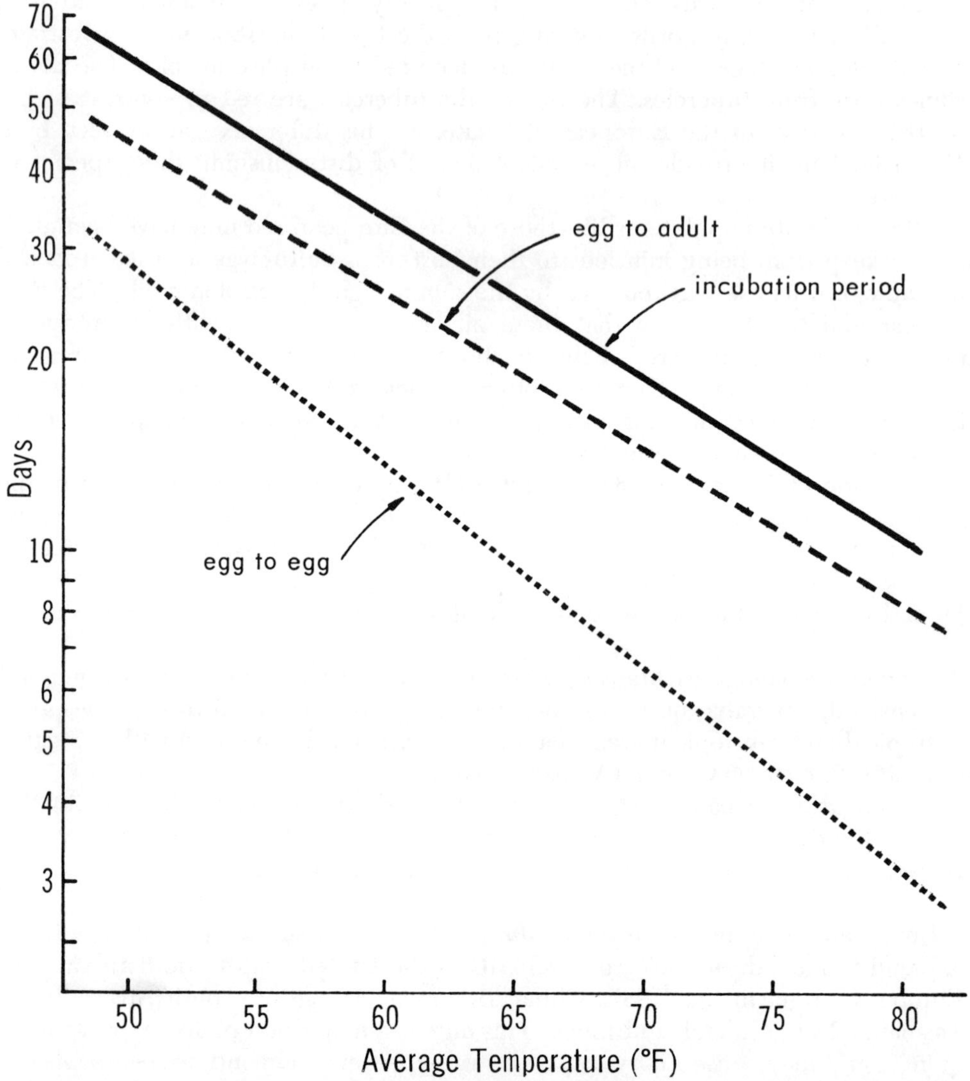

Fig. 21. *a,* Relationship between average temperature and development of the citrus red mite (1) incubation period, (2) egg to adult, and (3) egg to egg (redrawn after English and Turnipseed); *b* (p. 161), relationship between average temperature and the adult life span of citrus red mite.

crop, and severe attacks may leave the tree barren the next season. Injury in July and August causes the fruit to be small and of poor quality.

The European red mite passes the winter in the egg stage (fig. 22). The eggs are placed in groups on roughened bark areas at the base of buds, spurs, in wounds, and points that mark the beginning of new growth. Overwintering egg deposition starts in August and extends to October or November, depending on the climate. Development of these diapause eggs does not begin until temperatures will average at least 7 C (45 F). As the average daily temperatures increase

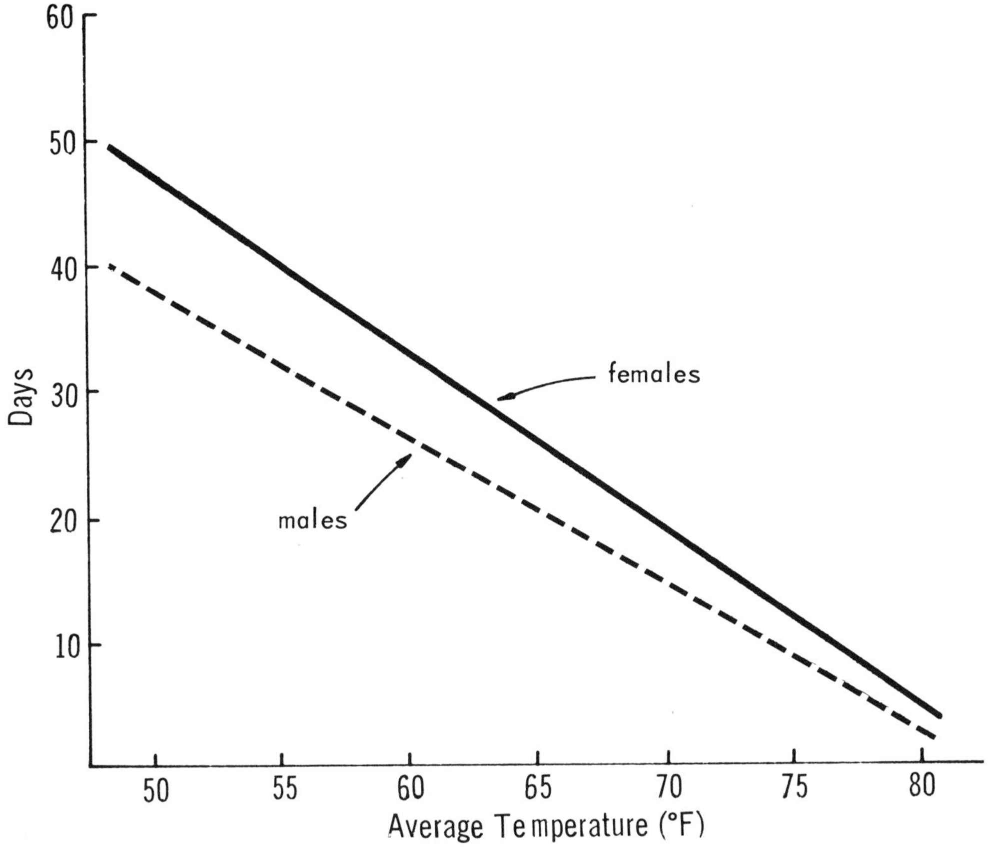

above this point, development within the eggs is accelerated, and hatching takes place when a total of 195 C degrees have accumulated. For further information on the diapause eggs refer to chapter 2.

The percentage hatch of diapause eggs is not greatly influenced by their exposure to winter weather conditions above –31 to –37 C (–24 to –35 F), but high humidities (80 to 100% RH) at the time of hatching greatly reduce the percentage of eggs hatched. The first hatch usually occurs when the earliest blossom buds show pink. There is a short period in which fewer eggs are present following the hatching of the last overwintering eggs, and before the eggs of the first summer generation are laid. This provides an excellent opportunity for control applications; the time of hatching in a given orchard, however, may vary as much as 30 days.

Larvae from overwintering eggs seek the young as yet unfolded leaf clusters where they find protection. The life cycle requires 3 to 4 weeks depending on prevailing temperatures. There may be 5 to 6 generations per year in Canada and as many as 9 to 10 in Virginia (fig. 22). Nondiapause or summer eggs are laid on the lower leaf surfaces from the early part of May until September or October,

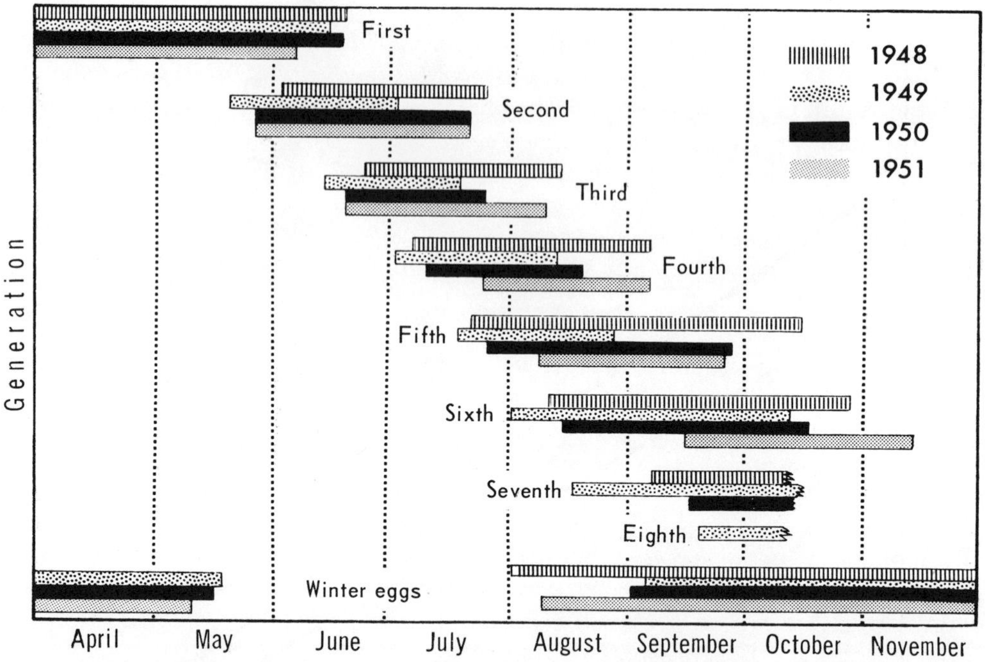

Fig. 22. Numbers and durations of generations of European red mite, *Panonychus ulmi* (Koch), in Virginia (broken lines indicate generations without adults) (after Parent and Beaulieu).

depending on the climate of the area. The incubation period under constant temperatures is about 5 days at 23.5 C (75 F) and 20 days at 13 C (56 F). The time required for development from egg to adult is 4 days at the higher and 19 to 22 days at the lower temperature. The average life span of the adult female is about 19 days during which time the female may deposit 10 to 90 eggs—the average is about 45; figures given for maximum and average number per female, however, vary greatly between research studies. Mites reared from eggs deposited by females previously exposed to males are about 63 percent females and 37 percent males, but this ratio also varies.

The egg of the European red mite is globular and somewhat flattened. Under magnification many grooves are visible which run from the base toward the top center where a slender, tapering stipe arises, which is about 100 μ long. Summer eggs apparently vary in color depending on the area in which they live and their stage of development, ranging from pale green, dark green, reddish brown, dark orange and bright red. The winter eggs are all deep red.

The larva is pale orange when newly hatched, but it becomes reddish brown with a pale front margin. The protonymph also varies in color, but shows the first evidence of pale spots at the base of the setae—a coloration most distinct in the deutonymph. The newly molted adult female is dark velvety green, velvety brown or brownish green. The color later changes to brownish red. The white spots at the bases of the dorsal setae are conspicuous—a characteristic that distinguishes them from the uniformly brownish citrus red mite (Cagle, 1946; Cutright, 1963;

Parent and Bealieu, 1957).

The European red mite has no doubt become widely spread throughout the deciduous fruit-growing areas by transporting the overwintering eggs to new plantings on nursery stock. Local distribution may occur by various means, but this mite and its close relative the citrus red mite are particularly adapted to dispersal by wind. The mites spin a silken thread, one end of which is attached to a leaf or twig. The mite supports itself on this thread, drops 2 to 5 inches, and remains suspended, waiting for a breeze to break the thread—a characteristic called "ballooning." Ballooning usually occurs as a result of overpopulation or unfavorable food supply.

Factors that influence populations of the European red mite and others are discussed in chapter 2. These factors include weather, cultural practices, predators, insecticide applications, and host plants. Insect and mite predators may have an influence on the development and decline of mite populations. The predacious species and their effect differ in the various areas where this mite is found. This subject is considered in chapter 5.

Summer weather conditions influence the food supply and rate of population development, which determines the number of overwintering eggs deposited and the initial population the following spring. Trees that have been severely attacked in midsummer carry only a few eggs because the populations become reduced before winter egg deposition begins, either because of poor food or abundance of predators; conversely, trees with foliage in good condition in August will carry far greater numbers of diapause eggs. Young mites are very susceptible to low temperatures following the spring hatch. Night temperatures slightly below freezing at this time may significantly reduce mite populations. Studies have shown that diapause eggs taken from one area are killed by higher subfreezing temperatures than those from other areas; within the lethal range of each, however, the mortalities increase by about 10 percent for each doubling of the exposure period. Resistance to freezing temperatures can be increased by laboratory selection (Doreste, 1964; Tsugawa and Shirasaki, 1961).

Evidence indicates that agricultural practices, such as fertilization, pruning, and cultural operations (which induce greater amount of tree growth and production), are favorable for development of mite populations. Insecticide applications for other pests may reduce mite predators; and some insecticides, in the absence of predators, seem to cause a more rapid development of mite populations. Certain hosts also appear to be less favorable for mite development than others. For example, in the same locality mortalities of the protonymph and deutonymphal stages are much higher on walnut than on pear or plum; the comparative figures are 20 percent on walnut, 5.7 percent on pear, and 3.8 percent on plum. The life cycle is shorter on walnut, but fewer eggs are laid, and the life span of the adult female is shorter than on pear or plum.

Applications of acaricides seem to be necessary in most commercial deciduous fruit orchards to prevent injury to the trees or the crop by the European red mite. Timing of the applications is often as important as the choice of an acaricide in achieving mite control. As the overwintering eggs approach hatching, their rate of development increases and they become more susceptible to spray applications

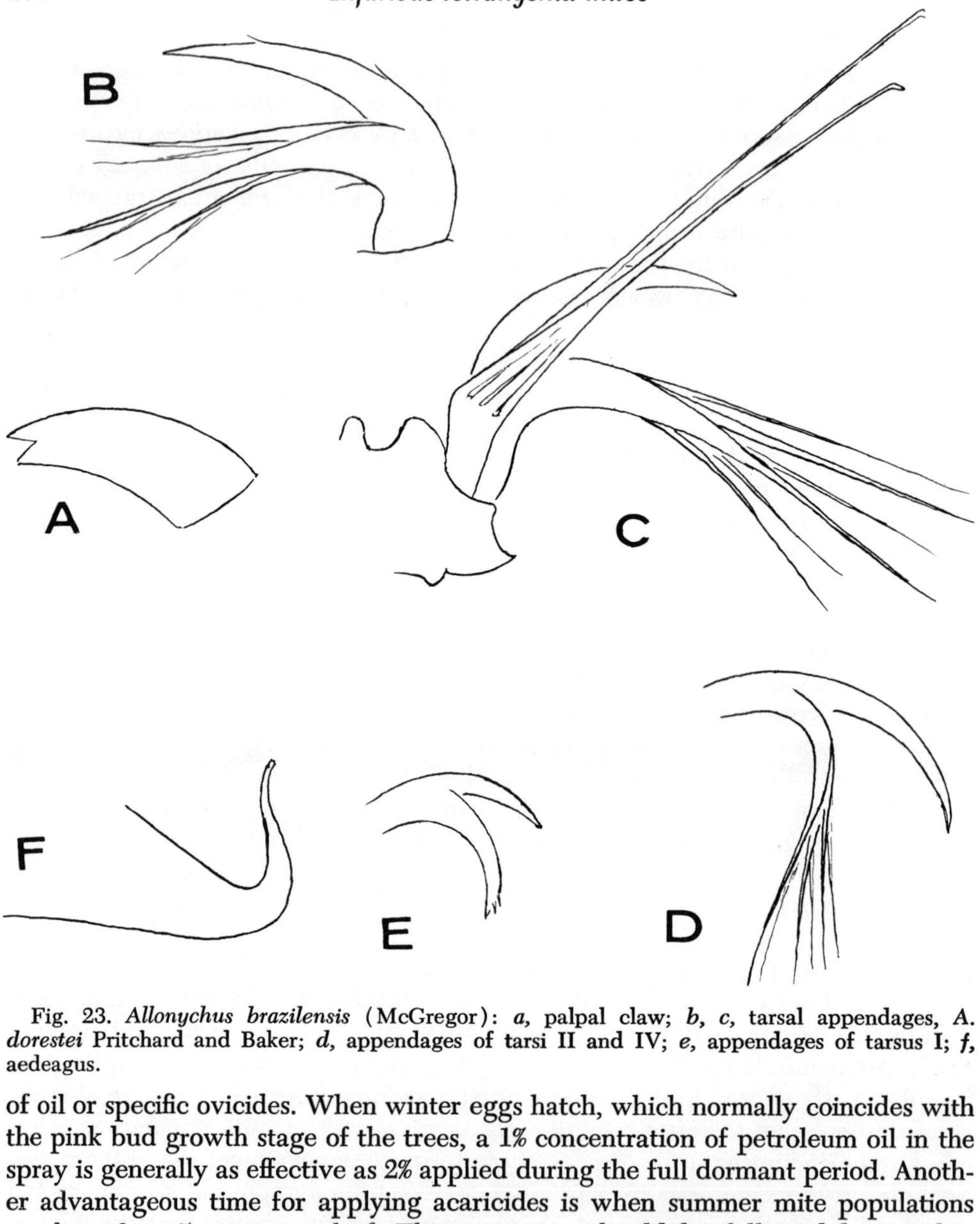

Fig. 23. *Allonychus brazilensis* (McGregor): *a*, palpal claw; *b*, *c*, tarsal appendages, *A. dorestei* Pritchard and Baker; *d*, appendages of tarsi II and IV; *e*, appendages of tarsus I; *f*, aedeagus.

of oil or specific ovicides. When winter eggs hatch, which normally coincides with the pink bud growth stage of the trees, a 1% concentration of petroleum oil in the spray is generally as effective as 2% applied during the full dormant period. Another advantageous time for applying acaricides is when summer mite populations are low, 2 to 5 mites per leaf. This treatment should be followed by another about 8 to 10 days later. It is important to have 2 summer spray applications spaced relatively close together.

The adult females of *P. ulmi* are reddish; the strong dorsal body setae are borne on prominent tubercles much lighter in color than the rest of the body; the striae of the genital region of the female are similar to those of *P. citri*, being transverse on the genital plate and longitudinal in the region anterior to the plate; setae D_5 of the hysterosoma are shorter than the L^4 setae, which are shorter than the D_4 setae. There are 3 setae on genu IV.

Allonychus Pritchard and Baker

On *Allonychus*, Pritchard and Baker (fig. 23), there are 2 pairs of para-anal setae. The palpal claw is bifurcate distally. The lobes of the dorsal striae of the female, which are transverse on the hysterosoma, are sharp and obviously taller than they are broad. The empodium bears a medioventral spur with 3 pairs of proximoventral hairs dissimilar in size and not at right angles to the spur. The species are very similar and males are needed for specific identification.

Three species of the genus *Allonychus* appear to be potential pests of agricultural crops. *Allonychus braziliensis* (McGregor) (fig. 23, *a-b*) has been found in Nicaragua on banana, in Brazil on quince and pear, and in Australia on Buffalo grass. The female is typical for the genus. The aedeagus of the male, which is similar to that of *A. dorestei*, is upturned, slender, sigmoid, and "swan necked" in shape. The empodia of legs I to IV are all similar: all have the split ventral hairs.

A. dorestei Baker and Pritchard occurs in Venezuela on bananas; and *A. littoralis* (McGregor) occurs in Ecuador, Guatemala and Honduras on cotton and avocado. The female of *dorestei* (fig. 23, *d-f*) is typical for the genus. The aedeagus of the male is upturned, slender, sigmoid, and "swan necked" in shape. The ventral hairs of empodium I of the male are coalesced and the other empodia are free distally. The aedeagus of *A. littoralis* is much less angulate than that of *A. dorestei* (Baker and Pritchard, 1955; Estebanes and Baker, 1968).

Eotetranychus Oudemans

The duplex setae of tarsus I on *Eotetranychus* Oudemans (fig. 24) are distal and approximate; the empodium is split into 3 pairs of ventrally directed hairs; there are 2 pairs of para-anal setae. The striae are longitudinal on the propodosoma and transverse on the hysterosoma. The striae of the genital flap and the area anterior to this is of taxonomic importance in the female. The genus can be divided into 4 groups: (1) with transverse striae on the flap and in the area anterior to the flap; (2) with transverse striae on the flap but with longitudinal striae on the anterior area; (3) with transverse striae on the genital flap and with irregular striae on the anterior area; and (4) with longitudinal striae on the anterior portion of the flap as well as on the anterior area. The dorsal body setae are long and slender in both sexes.

Eotetranychus species normally are tiny, slender straw-colored or somewhat greenish mites with several spots of dark pigment along each side of the body. They inhabit trees, shrubs, sometimes berries, and typically lay pearly, globular eggs that possess a tiny stipe. Species found in temperate regions hibernate as bright yellow females in protected places on the bark or in limb crotches of the host. Most species feed on the undersurfaces of leaves, primarily along the veins, forming small colonies where they generally produce webbing. *E. yumensis* (McGregor) and *E. lewisi* (McGregor) are exceptions as the former feeds on all surfaces of citrus leaves and fruit and the latter feeds mostly on the fruit; but both web profusely.

Members of this genus are primarily considered economic pests of ornamental and shade trees, although a few species are pests of citrus, deciduous trees, and

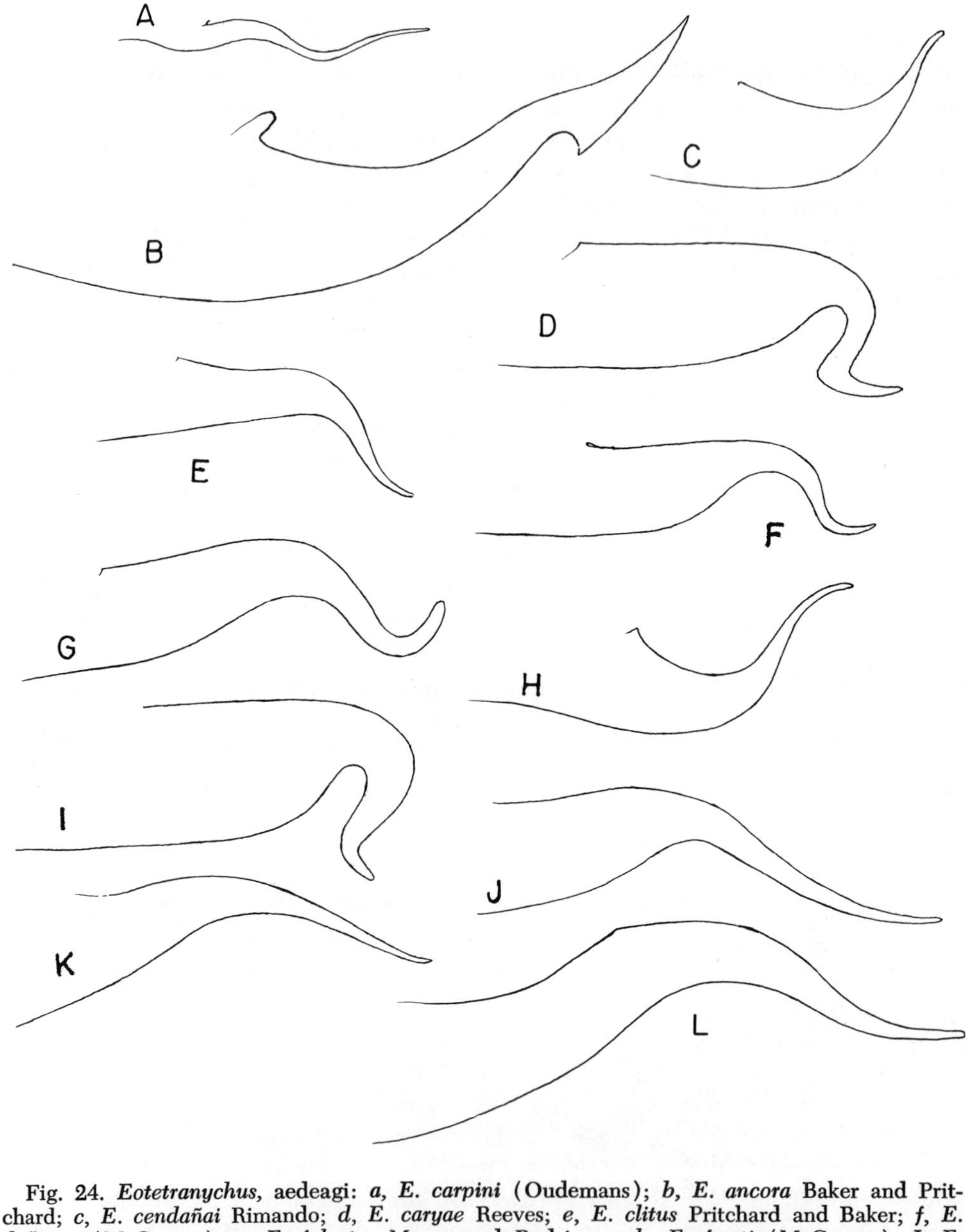

Fig. 24. *Eotetranychus,* aedeagi: *a, E. carpini* (Oudemans); *b, E. ancora* Baker and Pritchard; *c, E. cendañai* Rimando; *d, E. caryae* Reeves; *e, E. clitus* Pritchard and Baker; *f, E. deflexus* (McGregor); *g, E. falcatus* Meyer and Rodrigues; *h, E. frosti* (McGregor); *I, E. hicoriae* (McGregor); *j, E. hirsti* Pritchard and Baker; *k, E. kankitus* Ehara; *l, E. lewisi* (Mc Gregor).

berries. These mites are generally not evaluated as major pests, probably because of their habits of feeding in colonies where they become readily available to predacious mites. In addition, most of the species are susceptible to spray and dust applications of sulfur.

Eotetranychus ancora Baker and Pritchard. This mite (fig. 24, *b*) has occasionally caused some injury to apple leaves in Mauritius where it lives in small colonies near the veins on the undersurface of the leaves. Feeding in colonies produces crinkling of the leaves (Moutia, 1958). Predacious thrips, *Scolothrips*, and mites, *Typhlodromus*, appear to be effective predators; they reduce infestations enough that this species is assumed to be of little importance (Moutia, 1958).

Tibia I of the female of *E. ancora* has 9 tactile setae; tibia II has 8 tactile setae; there are 5 tactile setae proximal to the duplex setae on tarsus I; the striation pattern of the female genital flap is transverse, as is the area anterior to the flap. Tibia I of the male has 5 tactile setae; tibia II varies from 5 to 8 tactile setae, there is 1 tactile seta proximal to the duplex setae on tarsus I. The aedeagus is distinctive in that the shaft is strongly curved dorsad and tapering; there is a short, ventrally directed terminal angulation that is slender and acute distally.

Eotetranychus carpini carpini (Oudemans). This species (fig. 24, *a*) lives on apple, hornbean, hazelnut, oaks, willow, maple, alder and *Carpinus* in Germany, England, Mexico, and New York. In England it is only found on unsprayed apple trees.

The mites live in small, well defined colonies on the underside of the leaves, forming light patches visible only from below. They spin heavy webbing over the colony area beneath which all stages of the mite live and feed. Eggs are laid either in this webbing or beneath it on the leaf surface. When females mature they often leave the colony and form a new colony. This species is found most abundantly on apple varieties that have a hairy underleaf surface.

The eggs are spherical, about 100 μ in diameter, and are considerably smaller than those of the two-spotted spider mite, *Tetranychus* (*T.*) *urticae* Koch. Eggs are very pale green and do not change color before hatching. The immature stages are all pale green or greenish yellow.

This species overwinters as adult females on the tree beneath bark and in crevices. The females become active in April and migrate to young leaves to lay eggs. The incubation period varies from 14 days in August to 26 days in September and October. Each female deposits an average of more than 1 egg per day. The developmental period from hatching to adult requires 12 to 18 days. In England, 6 summer generations occur before overwintering females appear in October.

Serious infestations of this species are usually prevented by predacious mites (*Typhlodromus*), cecidomyiid larvae, and other predators. Chemical control measures used for other tetranychid mites prevent injury by this species (Collyer and Groves, 1955).

E. carpini has nine tactile setae on tibia I of the female; tibia II has 8 tactile setae; there are 5 tactile setae proximal to the duplex setae on tarsus I. The striation pattern on the genital flap and the anterior area is transverse; the proximal members of the duplex setae are very short, as in the male. The distal end of the peritreme ends in a simple bulb in both sexes. The aedeagus of the male is long, slender, acutely tapering distally, and strongly undulate near the middle.

Yellow vine mite, *Eotetranychus carpini vitis* (Oudemans). *Carpini vitis* differs from *E. carpini carpini* in its biology and preferred host. It is a serious pest of grapes in Italy and France where it produces crinkling, spotting, and discoloration of the leaves; this causes the vines to assume a dusty appearance.

Fertilized females overwinter in colonies under the bark of vines at or above ground level or sometimes in the upper soil. In March or early April they feed first on the opening buds and later along the veins on the lower surfaces of the leaves. They deposit their eggs at the junction of the midrib and veins. Oviposition begins in April and females lay an average of 30 to 40 eggs during their life span. The eggs hatch in about 7 days depending on the climate. The summer generations start in April or May and 4 to 6 generations may develop before the hibernation forms are produced in October.

This species appears to be sensitive to heat, alternate heavy rains, and hot sun in the summer, any of which may result in high mortalities, especially of the eggs.

Predacious mites, particularly *Amblyseius aberrans* (Oudemans), and predacious insects feed on this species. Lime-sulfur applications in the early spring delay development of populations. Summer sprays of acaricides, however, may be necessary to prevent injury from summer populations (Ambrosi and Lenarduzzi, 1959, Rambier, 1958).

Yellow spider mite, *Eotetranychus carpini borealis* (Ewing). This species is a pest of apple and pear in the Pacific northwestern area of North America from central California into British Columbia. Other hosts include cherry, raspberry, blueberry, spirea, alder, and willow. It is easily confused with *T. willamettei* (McGregor), as its appearance and the injury produced are similar. Feeding injury under the leaves somewhat resembles mildew spots.

The yellow spider mite is small and slender, flesh color to pale yellowish or greenish, but later in life 2 or 3 pairs of small dusky spots usually appear on the body The eggs are spherical and clear with a dorsal stipe.

This subspecies overwinters as bright yellow females, principally under the bark or in crevices on the smaller limbs of the trees. In late winter or early spring mites begin moving into expanding fruit buds that serve as the initial ovipositional sites. An ovarian diapause in the overwintering females is also indicated because mites brought under laboratory conditions before February 15 failed to lay eggs, and yellow mites collected after that date laid but 6 eggs per female over a 3-week period; yet overwintering two-spotted spider mite females, *T.* (*T.*) *urticae*, averaged 38 eggs per female under the same conditions.

Eggs laid by overwintering yellow mites under southern Oregon conditions hatch in about 2 weeks. Fecundity of summer females is higher than the winter forms, but much lower than summer females of the two-spotted spider mite. At 27 C (80 F), yellow mite females laid an average of 36 eggs per female over a maximum life span of 27 days; whereas two-spotted spider mite females deposited 100 eggs over a span of 30 days.

Yellow mites in pear orchards in Oregon are rarely encountered in the lowest three feet of the tree, but the two-spotted spider mite is common on the lower branches. Active yellow mite stages feed principally along the midrib and veins underneath the leaf. Initially they live in small colonies, but become more generally distributed as populations increase. Very little webbing is produced. In the Pacific Northwest, the yellow mite is normally active later in the fall than other tetranychid mites occurring on apple and pear trees (Westigard and Berry, 1970).

Eotetranychus caryae Reeves. *E. caryae* (fig. 24, *d*) infests pecan, hickory, horse chestnut, blackjack oak, and Asiatic and European chestnut in New York. The injury produced is similar to that caused by *E. hicoriae* (McGregor) (Reeves, 1963).

E. caryae female tibia I has 9 tactile setae; tibia II has 8 tactile setae; there are 5 tactile setae proximal to the duplex setae on tarsus I. The striation pattern of the female genital flap is transverse, and that of the anterior area is longitudinal. The peritremes of both sexes are hooked distally. Tibia I of the male has 9 tactile setae; tibia II has 8 tactile setae; there are 4 tactile setae proximal to the duplex setae on tarsus I. The aedeagus of the male is sharply downcurved and sigmoid, somewhat similar to that of *E. pallidus* Garman, but the axis of the knob of the aedeagus is parallel to that of the shaft.

Eotetranychus cendanai Rimando. This mite (fig. 24, *c*) appears to be a potential pest of citrus in the Philippines where it has produced injury to young trees previously treated with DDT. It also occurs in Cambodia and Thailand. This species is pale yellowish green with silvery setae. The mites feed on the dorsal surface of leaves and produce little if any webbing (Rimando, 1962).

E. cendanai female tibia I has 8 tactile setae; tibia II possess 5 tactile setae. There are 4 tactile setae proximal to the duplex setae on tarsus I. The striae of the genital flap and the area anterior to the flap are transverse. The peritremes of both sexes end in simple bulbs. The aedeagus of the male is similar to that of *E. mandensis* Manson but is strongly sigmoid.

Eotetranychus clitus Pritchard and Baker. This mite (fig. 24, *e*) is known from the southeastern United States where it is found on low-growing hosts, such as blackberry, azalea, and passion flower. This mite has 9 tactile setae on tibia I of the female; tibia II has 6 tactile setae; there are 5 tactile setae proximal to the duplex setae on tarsus I. The striation pattern is transverse on the genital flap and on the area anterior to the flap. Tibia I of the male has 9 tactile setae; tarsus I has 4 proximal tactile setae. The aedeagus of the male tapers distally and ends in a gentle sigmoid curve.

Eotetranychus deflexus (McGregor). This species (fig. 24, *f*) has been collected from snowberry in Oregon, Arizona, and Texas, from *Cercocarpus* in California, and from oak in Mexico. It has more recently been reported to be a pest of cotton in Uganda, where it appears to be well distributed (Tuttle and Baker, 1968).

E. deflexus striae on the genital flap and on the area anterior to the flap are transverse in the female. The aedeagus of the male is distinctive in that it is strongly bent ventral near the middle, with the bent portion tapering and sigmoid; the terminal sensillum of the male palpus is rudimentary.

Eotetranychus falcatus Meyer and Rodrigues. *Falcatus* (fig. 24, *g*) is a potential pest of cotton in South Africa. It has also been collected from peanut, tapioca plant, Indigo, hibiscus, and *Millettia*. *E. falcatus* feeds in colonies on young foliage. It is very light green with dark irregularly shaped inclusions in the body (Meyer and Rodrigues, 1966; Rodrigues, 1967).

E. falcatus bears 9 tactile setae on tiba I of the female; tibia II bears 8 tactile setae; there are 5 tactile setae proximal to the duplex setae on tarsus I. The striae on the genital flap and the area anterior to the flap are transverse. Tibiae I and II of the male are similar to those of the female; tarsus I of the male bears 4 tactile setae proximal to the duplex setae. The aedeagus bends strongly ventrad, narrowing distally and is strongly sigmoid, the distal end being upturned, scythelike.

Eotetranychus frosti (McGregor). *E. frosti* (fig. 24, *h*) has been collected from rose, blackberry, and raspberry in Arizona, California, Louisiana, Missouri, North Dakota, Ohio, and New York.

E. frosti has 8 tactile setae on tibia II and 9 tactile setae on tibia I; tarsus I bears 5 tactile setae proximal to the duplex setae. The striation pattern is transverse both on the genital flap and on the area anterior to the flap. The aedeagus of the male is sharply bent dorsad to form a slender sigmoid curve.

Eotetranychus hicoriae (McGregor). This species (figs. 24, *i*, 26, *b*) is a pest of pecan and chestnut trees throughout the eastern United States. It also infests hickory and oak trees. Injury is evident by the presence of scorched areas that first appear as dark brown or liver-colored blotches on the leaflets. Old scorch injury appears as areas of irregular size and pattern on the leaves. Heavy infestations cause leaves to fall prematurely starting at the lower branches. The foliage that remains on the trees may be russeted or scorched, especially during dry, hot weather (Pierce, 1953; Reeves, 1963).

E. hicoriae may be found on both surfaces of pecan leaflets, but it seems to feed and reproduce principally on the undersurface. The mites begin attacking foliage soon after the trees initiate growth in the spring. Leaf scorching may be observed as early as April 10, but this injury is confined to leaves on shoots arising from small branches on water sprouts attached to the main branches and trunks of the trees. Injury does not appear on mature leaves until about June. By this time populations have reached 10 to 60 mites per leaflet. Heavy losses of foliage in August and early September are usually responsible for failure of the trees to bloom properly the following spring.

Injurious infestations of this mite have been effectively reduced by spray applications of 5 lbs of wettable sulfur plus 0.63 lb of 40% dinitro-o-cyclohexyl phenol per 100 gallons of spray. Parathion and other acaricides are also effective. One or 2 applications should be applied during the latter part of May (Reeves, 1963; Pierce, 1953).

Tibia I of the female of *E. hicoriae* bears 9 tactile setae; tiba II bears 8 such setae; there are 5 tactile setae proximal to the duplex setae of tarsus I; and the peritremes are hooked distally. The striation pattern of the genital flap is transverse, and that of the area anterior to the flap is longitudinal. The aedeagus of the male is bent sharply ventrad, the bent portion is strongly sigmoid and at an acute angle to the shaft.

Fig spider mite, *Eotetranychus hirsti* Pritchard and Baker. This mite (fig. 24, *j*) is a serious pest of fig in India and Pakistan. Injurious infestations have been reported from several areas in India and at Lyallpur in Pakistan. Initial feeding injury results in transparent patches on the leaves when viewed in cross light. These patches turn yellowish green then brown and become rough and dry. The leaves, as well as the fruit, may drop prematurely, sometimes resulting in almost complete defoliation.

The mites feed mostly on the densely haired undersides of the leaves, but occur on the upper sides and on the fruit when trees are heavily infested. Injured fruit may fall prematurely. Mites are active from May to October at Lyallpur. When leaves dry in November the gravid females migrate to the branches and lodge themselves in terminal buds where they hibernate until February or early March, at which time activity is resumed.

Sulfur sprays as well as other acaricides, such as phosphamidon and chlorobenzilate have effectively controlled this mite on figs (Binda and Varma, 1966; Kanta Rai, and Rattan, 1963).

E. hirsti is similar to *E. frosti* in the setal pattern of the legs and in having transverse striae on the female genital flap and on the area anterior to the flap. The empodium I of the male is a simple claw; the aedeagus bends ventrad, with the downturned portion slender, tapering, and slightly sigmoid.

Eotetranychus kankitus Ehara. This species (fig. 24, *k*) has caused injury to citrus on the island Osaki-Shimojima in the Inland Sea, Japan. The injury produced on citrus, as well as the general appearance of this mite, is apparently similar to that of the six-spotted spider mite, *E. sexmaculatus* (Riley) (Ehara, 1964).

The female tibia I of *E. kankitus* bears 9 tactile setae; tibia II bears 8 tactile setae; there are 5 tactile setae proximal to the duplex setae on tarsus I and 3 proximal tactile setae on tarsus II. The peritremes are slightly curved distally but are not hooked. The striation pattern on the female genital flap is transverse on the posterior portion and longitudinal on the anterior central portion; that of the area anterior to the genital flap is longitudinal. The empodium I of the male is split distally into 3 pairs of ventrally directed hairs; the aedeagus is slightly downcurved, tapering distally, and slightly sigmoid.

Lewis spider mite, *Eotetranychus lewisi* (McGregor). *E. lewisi* (fig. 24, *l*) occurs in southwestern United States, Central America, Washington, Michigan, and Massachusetts. It is an occasional pest of citrus in southern California exclusive of the desert areas, of greenhouse poinsettias in California and Washington state, and on papaya in Mexico and Central America. It also has been collect-

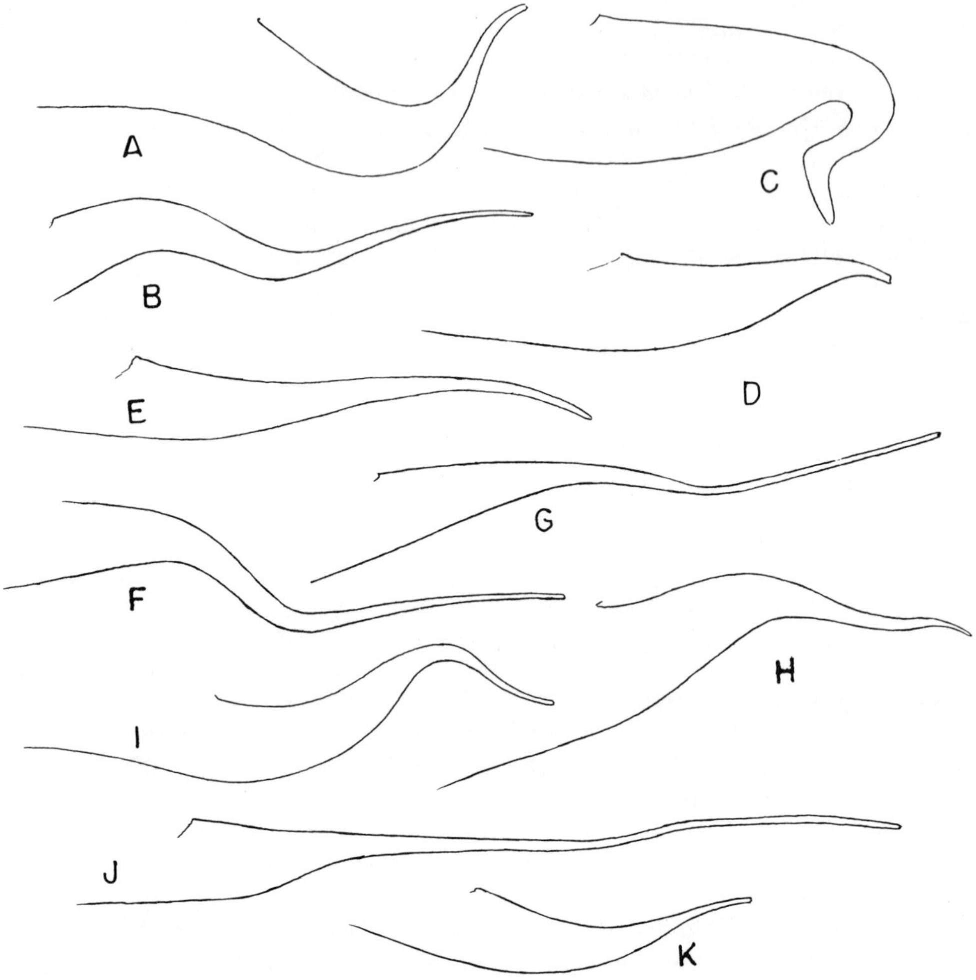

Fig. 25. *Eotetranychus*, *aedeagi*: *a*, *E. mandensis* Manson; *b*, *E. matthyssei* Reeves; *c*, *E. pallidus* (Garman); *d*, *E. pamelae* Manson; *e*, *E. populi* (Koch); *f*, *E. pruni* (Oudemans) (after Reeves); *g*, *E. querci* Reeves (after Reeves); h, *E. sexmaculatus* (Riley); *i*, *E. smithi* Pritchard and Baker; *j*, *E. tiliarium* (Herman); *k*, *E. yumensis* (McGregor).

ed on castor bean, olive, clover, *Ceanothus*, and *Euphorbia marginata* Pursh.

Feeding on citrus by this mite results in a stippling of the citrus rind. Heavy infestations produce silvering on lemons and either a silvering or russeting of oranges. The profuse webbing collects dust that makes detection of heavy infestations easy. There is no injury to the leaves of citrus.

The mites feed on the underside of poinsettia leaves producing a speckled or peppered effect on the foliage. Continual feeding of a large number of mites results in profuse webbing; leaves become chlorotic owing to loss of chlorophyll; and finally, extensive leaf drop may occur (Doucette, 1962).

The eggs are laid in the depressions of citrus fruit surface. They have a short stalk, but no threads extend from the top of the stalk to the host. The life cycle from egg to adult averages 12 days for the males and 14.5 days for the female. The female oviposits 2 to 3 eggs per day (D. Stiller, Whittier College, Pomona, Calif., personal communication).

Harvesting of citrus fruit often removes most of the infestation, and injurious populations may be readily controlled by treatments used for other tetranychid mites.

E. lewisi has 9 tactile setae on tiba I of the female and 8 tactile setae on tibia II; tiba I of the male has 9 tactile setae; tarsus I bears 5 tactile setae proximal to the duplex setea. The peritremes are hooked distally; the striation pattern is transverse both on the genital flap and on the area anterior to the flap. The aedeagus of the male gradually tapers distally and forms a broad sigmoid ventral bend.

Eotetranychus mandensis Manson. *Mandensis* (fig. 25, *a*) occurs on citrus in India. Tibia I of the female possesses 9 tactile setae; tibia II bears 6 tactile setae; tarsus I has 5 tactile setae promixal to the duplex setae. The striation pattern of the female genital flap and the area anterior to the flap is transverse. Tibia I of the male bears 9 tactile setae; tibia II bears 6 tactile setae. The terminal sensillum of the palpus is not present. The aedeagus abruptly turns upward at almost a right angle to the shaft, and is slightly sigmoid and tapers distally.

Eotetranychus matthyssei Reeves. This species (fig. 25, *b*) is a pest of elms, but it also occurs on hackberry in South Dakota, Arizona, and Central America, and on black locust in New York State. It causes severe browning and cupping of the undersides of American elm leaves. Injury appears first on the leaves close to the trunk and near the ground. These mites produce only a small amount of webbing primarily near the leaf veins. They overwinter as adult females on the bark of the trees only, and not in the debris around the bases of the trees. These mites are similar in general appearance to *E. pruni* (Oudemans), the females being greenish yellow and relatively oblong (Reeves, 1963).

Tibia I of the female of *E. matthyssei* bears 9 tactile setae; tibia II bears 8 tactile setae; there are 5 tactile setae proximal to the duplex setae of tarsus I. The peritreme ends in a simple bulb, the distal portion being slightly bent but not hooked. The striation pattern of the female genital flap and the area anterior to the flap is transverse. Tibia I of the male also bears 9 tactile setae, and tiba II 8 similar setae; there are 4 tactile setae proximal to the duplex setae on tarsus I. The aedeagus is long, slender, and tapers distally, and it is strongly undulate near the center.

Eotetranychus pallidus (Garman). This mite (figs. 25, *c*; 26, *c*) occurs on beech and alder in Connecticut, New York, and probably other eastern states. The mites feed close to the midvein on the undersides of the leaves, causing injured areas to turn brown (Reeves, 1963).

Tibia I of the female of *E. pallidus* has 9 tactile setae; there are 8 tactile setae on tibia II; tarsus I has 5 tactile setae proximal to the duplex setae. The striae of the

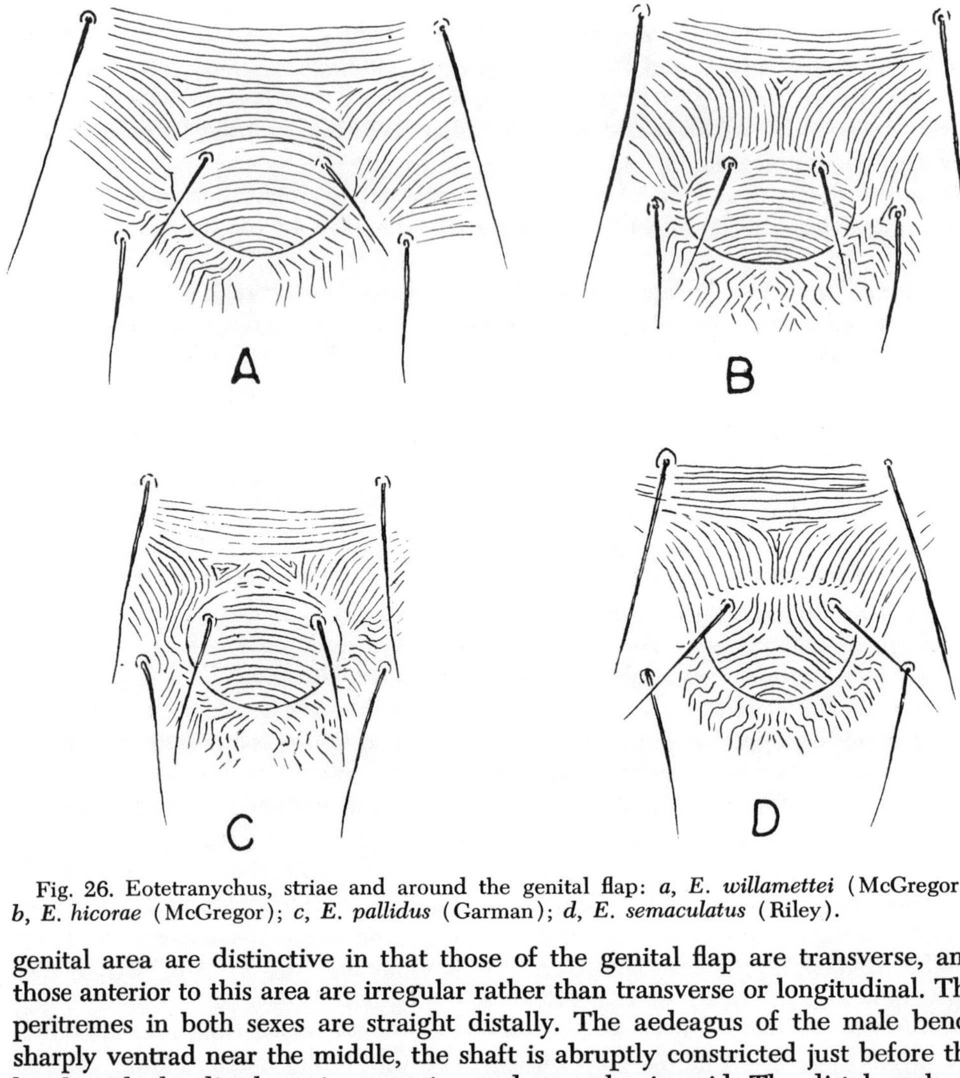

Fig. 26. Eotetranychus, striae and around the genital flap: *a*, *E. willamettei* (McGregor); *b*, *E. hicorae* (McGregor); *c*, *E. pallidus* (Garman); *d*, *E. semaculatus* (Riley).

genital area are distinctive in that those of the genital flap are transverse, and those anterior to this area are irregular rather than transverse or longitudinal. The peritremes in both sexes are straight distally. The aedeagus of the male bends sharply ventrad near the middle, the shaft is abruptly constricted just before the bend, with the distal portion tapering and strongly sigmoid. The distal angle of the aedeagus is acute.

Eotetranychus pamelae Manson. This species (fig. 25, *d*) occurs on citrus in Assam, India. Tibia I of the female bears 9 tactile setae and tibia II has 8 tactile setae; there are 5 tactile setae proximal to the duplex setae on tarsus I. The peritremes are bent distally. The striae of the genital flap are transverse posteriorly and longitudinal on the central anterior portion; that on the area anterior to the flap is longitudinal. The aedeagus of the male is short, strong, directed dorsally, and slightly sigmoid posteriorly.

Eotetranychus populi (Koch). *Populi* (fig. 25, *e*) is a pest of aspen, poplar, and willow in Russia, Serbia, Germany, and England, as well as the eastern United States as far west as South Dakota. The mites first attack the sucker growth where the newly developing leaves are heavily pubescent. They infest the lower surface of these leaves, spreading a layer of silk over the pubescence that provides a protective covering above them. Cast skins and dirt particles adhering to the webbing increase mite protection. Damage is easily discernible because the leaves become chlorotic or brown and are frequently curled.

The females of this mite overwinter beneath old bark on the trunk or in the forks of branches. During mild winters they may be found among dry leaves and in the crown of the tree. They leave hibernation quarters between the end of March and mid-April and settle among the young leaves where eggs may be found by the last of April. Population development continues on leaves until cold weather occurs. In the laboratory the duration of the egg stage is 7.1, 4.3, 2.5, and 2.2 days at 16, 24, 29, and 34 C (61, 76, 85, and 94 F) respectively, and the postembryonic development requires 7.9, 4.9, and 4.4 days at 24, 26, and 35 C (75, 78, and 95 F). At an average daily field temperature of 22 C (72 F) the egg stage lasted 5 to 7 days, the postembryonic periods 3 to 5 days, and the preoviposition period 2 to 3 days (Reeves, 1963; Tomašević, 1964).

This mite may be controlled by summer applications of dicofol or by some organophosphorus acaricides.

E. populi female tibia I bears 9 tactile setae; tibia II bears 8 tactile setae; there are 5 tactile setae proximal to the duplex setae on tarsus I. The peritremes form an irregular anastomosing pattern distally. The dorsal body setae are longer than the distance between their bases. The striation pattern of the genital area is transverse on both the genital flap and the area anterior to the flap. The aedeagus of the male is strong and only slightly undulate.

Eotetranychus pruni (Oudemans). *E. pruni* (fig. 25, *f*) is found in England and the United States. It is a pest of apple, cherry, prune, grape, and some species of shade and forest trees throughout Europe and the USSR. Some well-known synonyms include *E. pomi* Sepasgosarian, *E. coryli* (Reck), *E. viticola* (Reck), *E. aceri* (Reck) and *E. aesculi* (Reck). Feeding injury results in premature leaf drop and reduction in fruit size.

E. pruni overwinters as fertilized females primarily in the axils of branches. Activity is resumed in April or early May. The eggs are laid among the webbing mainly on the lower leaf surfaces where the mites generally congregate. The average developmental cycle in the field requires 27.4 days in the spring, 25.3 days in summer, and 31.1 days in the autumn. There are 3 to 4 generations per year. Female mites complete their life cycle under greenhouse conditions in 30.5, 20.5, and 30 days in the spring, summer, and autumn, respectively; there are 6 generations per year. The optimum temperature for development is 27.5 C (81.5 F). The egg stage requires more than 40 percent of the total development period. Females lay from 25 to 61 eggs per individual under field conditions and 27 to 98 eggs under greenhouse conditions—the averages are 37 and 48.1. The ratio of females to males is about 7:1 in the spring, 1.5:1 in the summer, and 0.7:1 in the autumn.

Spring temperatures often determine the ultimate summer and fall populations. Damp cool weather in May and June hinders reproduction sufficiently to prevent injurious populations from occurring in August and September.

E. pruni is susceptible to the acaricides applied during the summer to control tetranychid mites in apple orchards. Winter spray applications fail to reach sufficient hibernating females to provide adequate reduction of populations. *Typhlodromus* mites destroy up to 40 percent of summer mites while other mites, anthrocorids, and coccinellids prey upon this species (Sepasgosarian, 1956).

E. pruni has 9 tactile setae on tibia I of the female; tibia II bears 8 tactile setae; tarsus I bears 5 tactile setae proximal to the duplex setae. The peritremes are simply bent distally. The striation pattern is transverse both on the genital flap and on the area anterior to the flap. Tibia I and II of the male are similar to those of the female; there are 4 tactile setae proximal to the duplex setae of tarsus I. The aedeagus is long, slender, tapering, and strongly undulate.

Eotetranychus querci Reeves. *E. querci* (fig. 25, *g*) is a pest of oaks and white birch. This mite causes injury similar to that produced by frost, that is, leaves become distorted, discolored, and finally drop, resulting in defoliation of the trees. Initially these mites attack the leaves of pin oak close to the base of the midvein. If populations occur early in the season and become sufficiently high to produce severe injury in these areas, the leaves turn brown and fail to develop normally. Leaves less severely injured often have normal distal portions, but are distorted around the bases of the midveins. Mite populations on northern red oak are usually less damaging; damage appears as yellowed areas on the upper surfaces and as depressed areas viewed from below, but with little leaf distortion (Reeves, 1963).

E. querci has 9 tactile setae on tibia I of the female; there are 8 tactile setae on tibia II; tarsus I possesses 5 tactile setae proximal to the duplex setae. The peritremes are bent distally but are not U-shaped. The striation pattern of the genital flap and the area anterior to the flap is transverse. Tibiae I and II of the male are similar to those of the female; there are 4 tactile setae proximal to the duplex setae on tarsus I. The aedeagus is long, slender, tapers distally, and is undulate at the middle.

Six-spotted spider mite, *Eotetranychus sexmaculatus* (Riley). This mite (figs. 25, *h*; 26, *d*; pl. 31) is a pest of citrus, avocado, and several trees and shrubs, such as maple, pyracantha, azalea, camphor, and *Elaeagnus*. It is a periodic pest of citrus and avocado or both in California, Florida, and Formosa, and of grape vines in Arizona and New Zealand.

This mite feeds in colonies on the underside of the leaves near the midrib or larger veins of citrus. Feeding injury by the colonies results in yellow depressions on the undersides of the leaves which become covered with webbing. The upper surface of the leaves, opposite the mite colonies, becomes raised and yellow or yellowish white (pl. 32). As the infestation increases the yellowish areas converge, the leaves become entirely yellow, distorted, or misshapen, and drop prematurely. The mites normally do not inhabit the fruit of citrus except during high infestations.

The globular, colorless or transparent, or pale greenish yellow eggs are loosely attached to the web that covers the colony, or they are attached to the surface of the leaf underneath the web. The eggs bear a stalk, but have no radiating guy fibrils. Twenty-five to 40 eggs are deposited by each female during a period of 10 to 20 days. The incubation period during the summer is 5 to 8 days. The developmental stages require 2 to 3 days each, and the adult begins laying eggs 2 to 3 days after emergence. The adults are lemon yellow, usually with blackish spots grouped in 3 blotches along each side of the body; these spots, however, may be inconspicuous or absent.

Six-spotted mite populations are adversely influenced by dry weather conditions. Injurious populations of this species only occur in the more humid coastal districts in California. Leaves near the ground are infested first, and only when infestations on the tree have persisted for some time may injury be observed on leaves more than 4 to 5 feet from the ground. Dry winds, which commonly occur in the fall in southern California, usually reduce populations of this mite.

The webbing covering the colonies in leaf depressions probably protect the mites from insect predators; this protection and the concentration of mites, however, provides a suitable habitat and a ready source of food for predatory mites. Typhlodromid mites together with unfavorable weather have reduced this destructive mite pest to one of occasional importance on citrus in California and Florida.

The six-spotted mite is susceptible to sulfur sprays and dusts as well as to acaricides used for control of the citrus red mite. Sprays must reach the undersurfaces of the lower leaves of the trees. Normal applications of petroleum oils only temporarily reduce populations of this species (Ebeling, 1959).

E. sexmaculatus female has 9 tactile setae on tibia I; tibia II has 8 tactile setae; there are 5 tactile setae proximal to the duplex setae on tarsus I. The peritreme is hooked distally. The striae are longitudinal on the anterior central portion of the genital flap and longitudinal on the area anterior to the flap. Tibia I and II of the male are similar to those of the female. The aedeagus is slightly curved dorsad near the middle of the shaft; the distal portion is directed caudoventrally and the tip is characteristically deflexed.

Eotetranychus smithi Pritchard and Baker. *E. smithi* (fig. 25, *i*) was first collected in 1951 on rose and various berries (*Rubus*) in the northeastern United States and later on the same hosts in Japan. It has become a serious pest of grape in northern Kyushu, Japan, and on cotton in Tennessee in the United States.

The mites feed in colonies near the midrib of cotton leaves producing injury similar to that caused by the carmine spider mite, *Tetranychus (T.) cinnabarinus* (Boisduval). The mites produce very little webbing. Both males and females feeding on cotton are rosy red and the eggs uniformly amber, but on roses they are described as straw-colored or greenish with several spots.

Acaricides used for control of tetranychid mites on cotton or roses usually control this species (Caldwell, 1967; Ehara, 1960).

E. smithi female has 9 tactile setae on tibia I; tibia II has 8 tactile setae; there are 4 tactile setae proximal to the duplex setae on tarsus I. The peritremes are strongly hooked distally. The striae on the genital flap are transverse; those an-

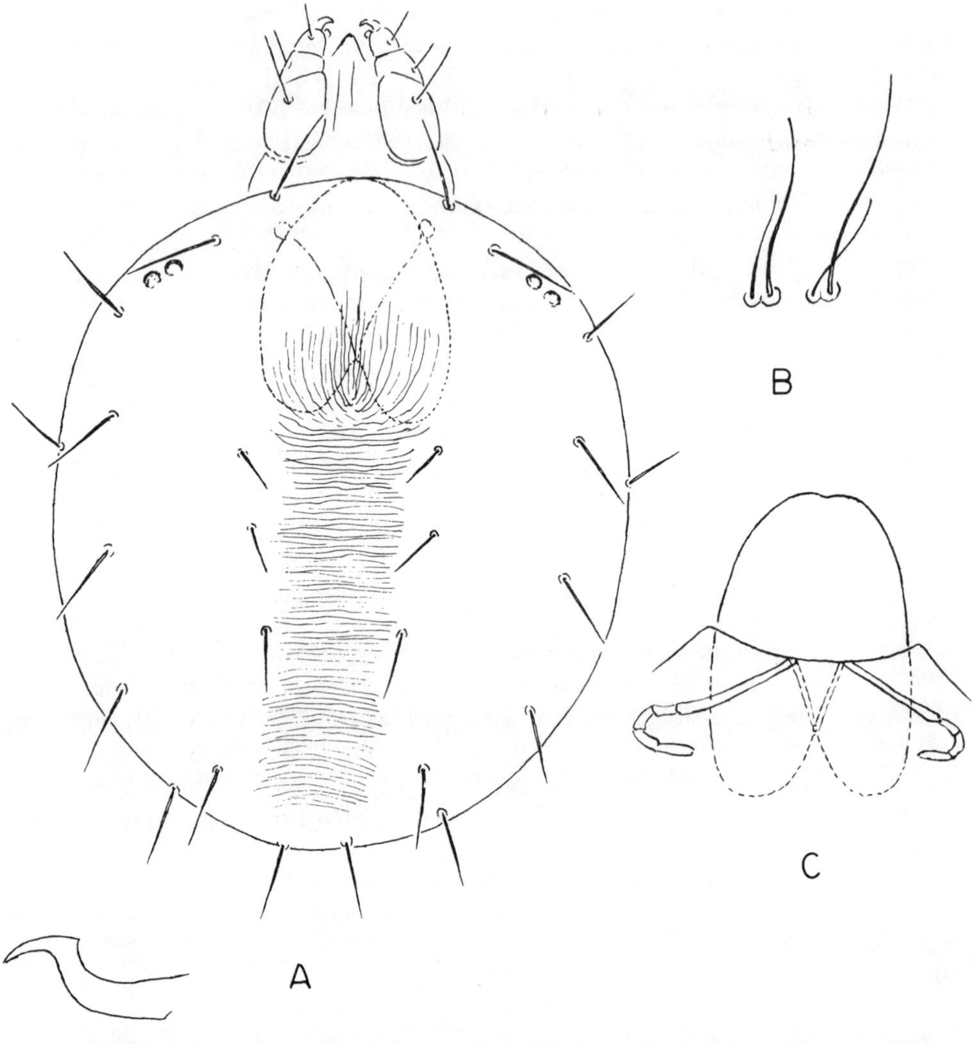

Fig. 27. *a, Platytetranychus multidigituli* (Ewing), female dorsal view and aedeagus (not a typical platytetranychid); *b, Eotetranychus willamettei* (McGregor), duplex setae, female leg I; *c, Eotetranychus uncatus* Garman, hooked peritremes.

terior to the flap are longitudinal. Tibiae I and II and tarsus I of the male have the same tactile setal pattern as the female. The proximal half of the aedeagus has the ventral margin directed somewhat dorsad, the distal portion abruptly narrowing to form a fine, caudally directed stylet.

Eotetranychus tiliarium (Herman). This mite (fig. 25, *j*) has been reported from Europe and the Atlantic coast of the United States. Its hosts include linden and lime (*Tilia*), sycamore, horse chestnut, hawthorne, hazel and willow. In England mites of this species have been observed congregating in large masses on the trunks of lime trees. The mites feed in large numbers on the undersides of leaves between the veins of lime and linden trees, where they produce severe discoloration of the leaves (Reeves, 1963).

E. tiliarium female has 9 tactile setae on tibia I, and 8 tactile setae on tibia II; there are 5 tactile setae proximal to the duplex setae on tarsus I. The striae on the genital flap and on the area anterior to the flap are transverse. The aedeagus of the male is long, slender, and only slightly undulate; the tip of the aedeagus is rounded.

Eotetranychus uncatus Garman. This species (fig. 27, *c*) is a serious pest of apple and stone fruits in eastern United States. It has been recorded from Utah and southern California. Populations have also been reported on birch, hornbeam, and linden. Feeding by this species causes the characteristic stippling, yellowing, and eventual bronzing of foliage. Because these mites feed primarily on the underleaf surfaces, they frequently cause an uneven growth of the two surfaces, resulting in cupping or crinkling of the leaves. Feeding by this species on the undersurface of alder leaves results in patches of yellow on the upper surfaces. Spring damage to newly developing apple leaves appears similar to frost injury. Adult females and deutonymphs spin considerable webbing, which is worked into a regular pattern over the leaf. Severely attacked trees may become completely defoliated and devitalized, indirectly affecting the size of the fruit (Reeves, 1963).

This species overwinters as the adult female in clusters under loose bark or other protected niches on the tree; mites swarm out of their hibernating quarters as the days warm in the spring. This activity starts during the time known as the delayed dormant stage of apple bud development. The mites make their way over the branches to the spur leaves where they feed for several days before oviposition begins. The time required for the development of each of the stages and the life span of adult females under two constant temperature conditions are reported in days as follows:

	Days at 21–22 C (69–72 F)	*Days* at 27–28 C (80–82 F)
Average incubation period	9	2
Average for each developmental stage	6	2
Maximum from hatching to adult	20	5
Average preoviposition	6	2
Average adult life span	14	8

The eggs of *E. uncatus* are spherical, clear, and watery when first laid, becoming opaque and pearly white. As incubation progresses they become straw-colored with the reddish eyespots of the embryo clearly visible just prior to hatching. The eggs of this mite can be distinguished from those of the two-spotted spider mite by the presence of a short stipe at the dorsal center. The larva of *E. uncatus* first are colorless, except for the red eyes. As feeding occurs they become pale green, but the propodosoma and legs remain colorless and transparent. Four small black spots appear on each side of the hysterosoma. The proto- and deutonymphs are similar in color except the spots become more pronounced. The adult female is more elongate than the European red mite, is flesh color or pale lemon yellow, sometimes with a tinge of green, and has small black spots posterior to the eyes (Reeves, 1963; Ubertalli, 1955).

E. uncatus female has 9 tactile setae on tibia I and 8 tactile on tibia II; tarsus I has 5 tactile setae proximal to the duplex setae. The peritremes are strongly hooked distally which distinguished this species from *E. carpini*. The striae of the genital flap and the area anterior to the flap are transverse. The aedeagus of the male is similar to that of *E. carpini*; it is long, slender, and tapers distally, and is strongly undulate.

Willamette mite, *Eotetranychus willametti* (McGregor). This mite (fig. 26, *a*; 27, *b*) is a serious pest of grapes throughout the grape-growing areas of central and northern California. It also occurs on elm, white oak, apple, pear, box elder, serviceberry, and antelope brush in Washington and Oregon.

The mites remain in colonies feeding on the upper surface of grape leaves. This localized feeding produces injured areas that turn straw-colored and then become scarious, an effect visible on the upper leaf surface as chlorotic green or straw bronze discolorations. The initial feeding is confined to areas enclosed by the larger net veins, but a narrow strip of leaf tissue close to the larger veins remains green, giving the leaf a mottled appearance. Damage from this mite on buds or leaves as they unfold from the growing tip of the shoot appears as sharply defined injured areas owing to rapid growth after injury. The characteristics of these areas can be used to distinguish the work of the Willamette mite from the damage caused by the Pacific mite, *Tetranychus* (*A.*) *pacificus* McGregor, since the latter species does not produce injury in the early spring, but only on mature leaves in midsummer. The yellow spot on the leaf gradually expands as each Willamette mite increases in numbers and size until the yellow center shades imperceptibly into the green uninjured tissue. Feeding effects on tender developing shoots may cause dwarfing and deformation of leaves, angularity at nodes, shortened internodes, and spindle stems. The tips of the slow-growing axial shoot may be killed by summer or fall. Canes on severely injured plants may be short, fruitfulness is impaired, clusters and berries are few, berries are small, lacking uniformity, of poor quality, and somewhat delayed in maturity.

Willamette mites hibernate only as adult females. They become amber to yellow in contrast to the very much paler yellowish white or ivory color of summer generations. Hibernating females can be found singly or in groups under the bark of 3-year-old or older vines, mainly on the undersides of the vines, although a few may reach the main trunk. These hibernating individuals become active when disturbed, but probably do not feed. The mites slowly move toward the spurs as the buds swell in the spring but remain under the bark on the 2-year-old wood until the leaves begin to expand. At this time the entire population migrates to the swelling buds or leaves simultaneously even though buds do not break at the same time. As these hibernating females start to feed they spin a sparse web and become much paler than before.

The eggs of the Willamette mite are tiny, round, transparent and shiny, and are oviposited anywhere on the undersurface of the leaf, glued to the epidermis, leaf hair, or vein. The eggs hatch in 7 days or less and the total nymphal period lasts about 10 days. About 15 days after migration to the leaves some of the overwintering mites abandon the basal 2 or 3 leaves and migrate up toward the younger leaves. All stages are present on grape foliage until about the first of November when migration to winter quarters begins.

Banding the bases of the grape spurs with a sticky material and summer applications of acaricides are means of controlling this mite on grapes (Frazier and Smith, 1946).

The *E. willamettei* female has 9 tactile setae; tibia II bears 8 tactile setae; there are 5 tactile seta proximal to the duplex setae on tarsus I. The peritreme is straight distally, ending in a simple bulb. Both sexes of this species are distinctive in that the paired duplex setae are much more equal in length than in the other members of the genus. The terminal sensillum of the palpus of the female is about 4 times as long as it is broad. The aedeagus of the male is similar to those of *E. carpini* and *E. uncatus*: it is long, sinuous, and tapers distally.

Yuma spider mite, *Eotetranychus yumensis* (McGregor). This species (fig. 25, *k*) is confined to a limited desert area of California, Nevada, and Mexico. It occurs on all citrus varieties as well as castor bean, grain sorghum, grapes, primrose, puncture vine, and *Atriplex lentiformus* (Torrey) Watson; the last appears to be a native host. When these mites feed on the leaves, fruit, and green twigs of citrus trees they produce a silvering of mature fruit and quantities of webbing to which dust adheres, making it easy to detect dense mite populations. Such populations accentuate the development of dieback, which is caused by low humidities, wind, and other factors that either increase leaf transpiration or otherwise cause the leaves to have a moisture deficiency. The leaves dry, fruit drops, and young twigs and even whole limbs become bare of leaves and may be killed.

Yuma spider mite populations increase in October and November, remain high throughout the winter, then decrease in spring or early summer. The mites aestivate during the summer under bark of the trunk and limbs or in cracks and crevices of more shaded areas. A few eggs may occur during the summer, but no young have been observed.

Laboratory studies show that relatively high temperatures are required for the development of populations of this mite. No eggs are laid at constant temperatures of 10 C (50 F) and at 15.5 C (60 F) deposited eggs fail to hatch if the relative humidity is below 50 percent. Complete development occurs at temperatures between 21 C (70 F) and 38 C (100 F), but at 43.5 C (110 F) eggs fail to hatch. These unusual temperature parameters probably confine this species to the hot, desert areas where average daytime temperatures of 21 C (70 F) or below are of relatively short duration.

The Yuma spider mite is similar to the six-spotted spider mite and the Lewis spider mite in general appearance. The adult females vary in color, depending on their food, from strawcolor to dark pink. The habits and distribution of these species on citrus in California are considerably different. The six-spotted spider mite occurs in the cooler, more humid coastal areas, feeds in colonies on the un-

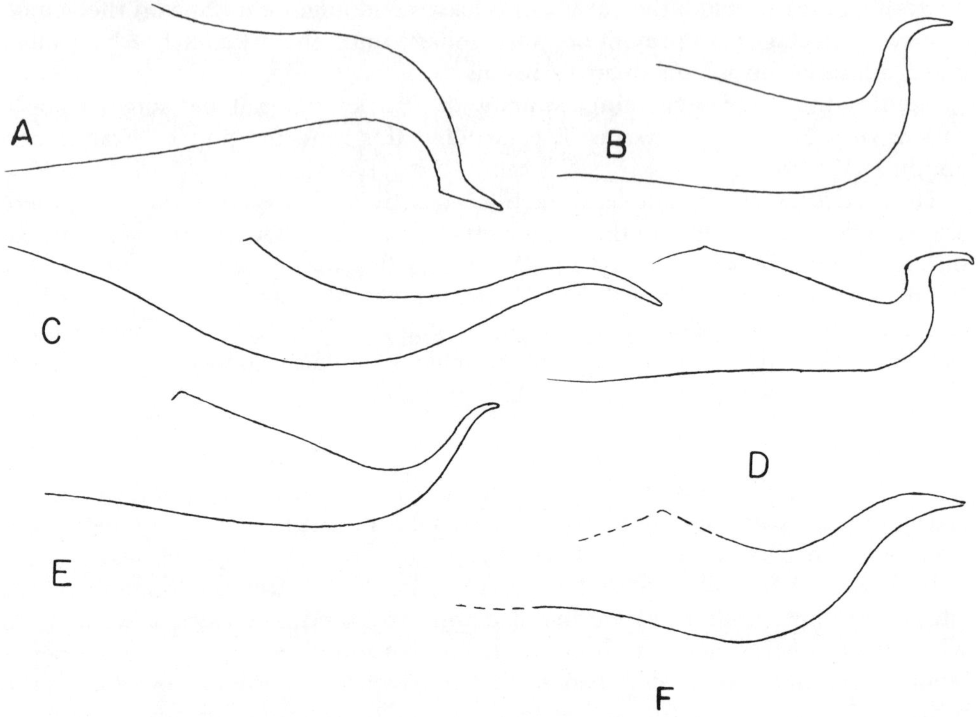

Fig. 28. *Schizotetranychus,* aedeagi: *a, S. asparagi* (Oudemans); *b, S. baltazari* Rimando; *c, S. celarius* (Banks); *d, S. hindustanicus* (Hirst); *e, S. oryzae* Rossi de Simons; *f, S. andropogoni* (Hirst).

dersides of leaves, and is not commonly found on fruit. Lewis spider mite populations concentrate on mature or yellow fruit and do not survive desert conditions. The Yuma spider mite feeding and activity is not confined, but occurs on leaves, fruit, and twigs.

Populations of the Yuma spider mite on citrus can be effectively reduced to below injurious levels by applying sulfur as a dust or spray during the fall, winter, or early spring. Sulfur applications should be made during cool weather periods in order to avoid plant injury by the sulfur (Elmer, 1965, *b*).

E. yumensis female has 9 tactile setae on tibia I; tibia II has 8 tactile setae; there are 4 tactile setae proximal to the duplex setae on tarsus I. The peritremes are strongly hooked distally. The striae are transverse on the genital flap and longitudinal in the area anterior to the flap. The setal pattern of the legs of the male is similar to that of the female. The aedeagus is distinctive in forming a shallow, nearly sigmoid dorsal curve; the distal end gradually tapers to a caudally directed point.

Schizotetranychus Trägårdh

There are 2 pairs of para-anal and anal setae. The duplex setae are distal and on tarsus I are approximate to each other; the empodia are strong, split, and claw-like; the 3 pairs of empodial hairs may or may not be present. The striae on the propodosoma may be longitudinal, or may be transverse on part or in whole; that of the hysterosoma is usually transverse, but may be longitudinal on the anterior portion.

Web-spinning mite, *Schizotetranychus andropogoni* (Hirst). This mite (fig. 28, *f*) occurs in India and Mexico. It has become a sufficiently serious pest of sugar cane in India to warrant the use of acaricides. The alternate host upon which populations are maintained during the rainy season in India appears to be *Dicanthium annulatum* Stapf. Spray applications of dicofol were not effective in controlling this mite, but several of the organophosphorus compounds resulted in adequate mite control (Agarival, 1957; Rattan, Rai, and Kanta, 1962).

The dorsal setae of the female of S. *andropogoni* are about as long as the distance between their bases; the setae of the male are relatively shorter, the setae are broadened proximally and are finely tapering distally. Tibia I of the female has 6 tactile setae; tarsus I has only 1 setae proximal to the duplex setae, tibiae II and III have 5 tactile setae each. The male setal count is similar. The aedeagus of the male forms a dorsally directed sigmoid curve that is not acutely angled. The striae of the propodosoma are longitudinal; those of the hysterosoma are transverse.

Pineapple mite, *Schizotetranychus asparagi* (Oudemans). This species (figs. 28, *a*; 29) occurs on asparagus ferns grown under greenhouse or lathhouse conditions in the United States, Germany, Portugal, Holland, Puerto Rico, and Hawaii. It frequently does severe damage to newly set pineapple plants. Infested plants remain small and produce little or no fruit; serious infestations kill the plants. Field injury by this mite usually may be prevented by planting mite-free nursery stock.

S. *asparagi* female body setae are ½ as long as the distance between their bases; the setae are slender, strongly pubescent, and subequal in length. The striae are longitudinal on the propodosoma and transverse on the hysterosoma. Tibia I of the female has 9 tactile setae; tibia II has 7 tactile setae; there are 5 tactile setae proximal to the duplex setae on tarsus I and 2 tactile setae proximal to the duplex setae on tarsus II. The aedeagus of the male turns abruptly ventrad, the distal end is nearly sigmoid and has an anterior angulation.

The citrus green mite, *Schizotetranychus baltazari* Rimando. *Baltazari* (fig. 28,

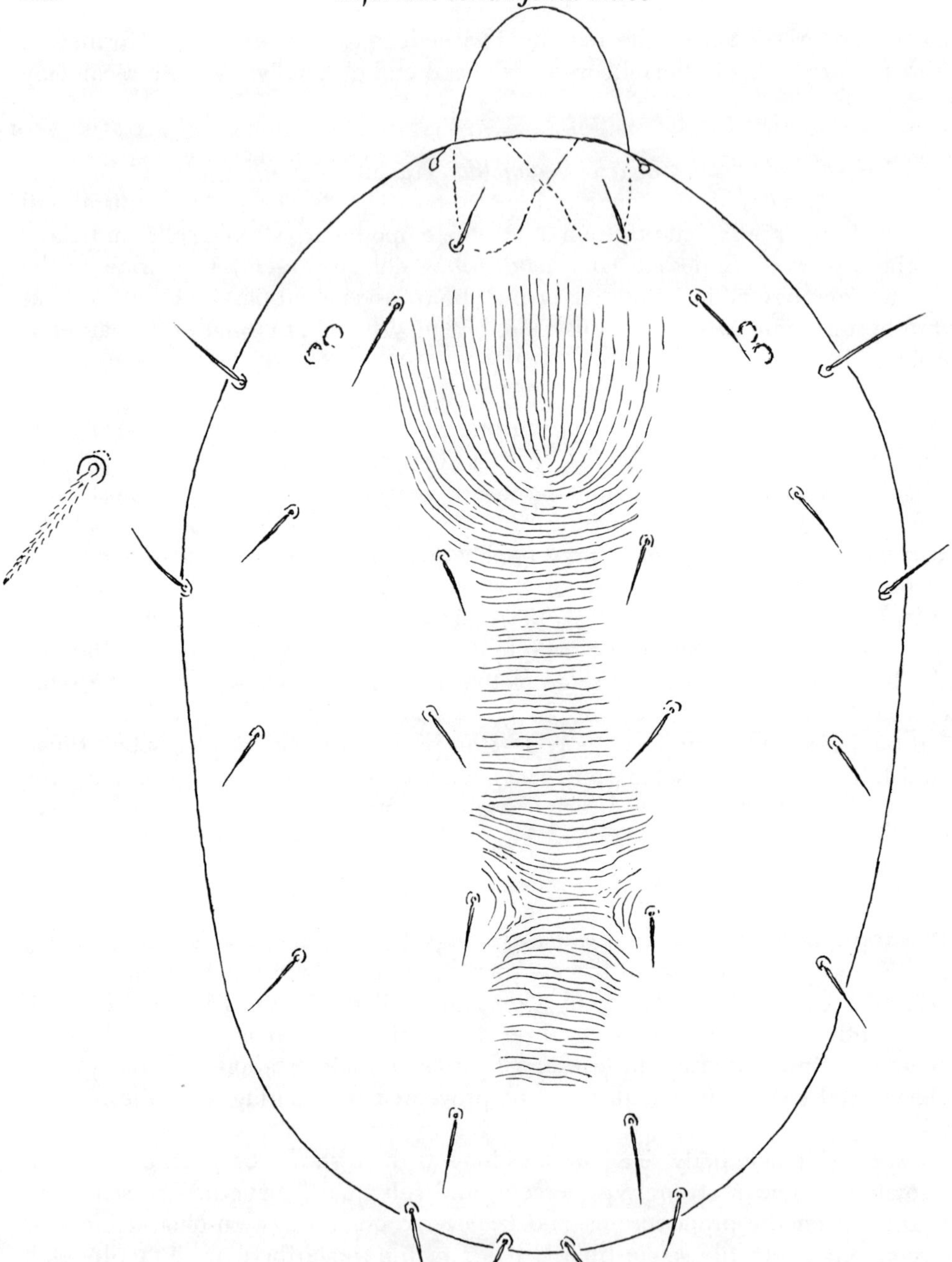

Fig. 29. *Schizotetranychus asparagi* (Oudemans), dorsal view, setae, and striations.

b) is a pest of citrus in Taiwan. This species feeds on leaves and fruit causing large, discolored, or grey spots on both sides of the leaves and on the fruit. All stages are present throughout the year, but the quiescent period is prolonged in the cooler season.

Females deposit eggs on or near the medial vein on both the upper and lower surfaces of the leaves. The eggs are milky white, pearllike, circular and about 120 microns in diameter when viewed from above, but elliptical in lateral profile. The larvae are nearly oval, greyish yellow when first hatched, but become yellow or yellow green. The nymphal and adult stages are yellow or yellow green with dark spots along each side. The adults have short legs and a somewhat flattened body. They actively move about over leaves and fruit and the females spin webbing over the eggs they deposit. The mean duration of the life stages in days are: egg 3.4, larvae 6.6, protonymph 5.4, deutonymph 5.2, and the preoviposition period 3.2. Thus total life cycle requires about 23.8 days. The oviposition period lasts about 20 days during which the females oviposit an average of 31.6 eggs per female (Lo and Hsia, 1968).

S. *baltazari* female propodosomal striae are longitudinal and continue posteriorly to the D_1 setae of the hysterosoma; the setae are about ½ as long as the distance between their bases, and are broad proximally and narrow distally; the D_4 setae are well separated. Tibia I has 6 tactile setae; tarsus I has 2 tactile setae approximate to the duplex setae. The peritremes end in simple bulbs. Tibia I of the male has 8 tactile setae; tarsus I has 3 proximal tactile setae; tibia II possesses

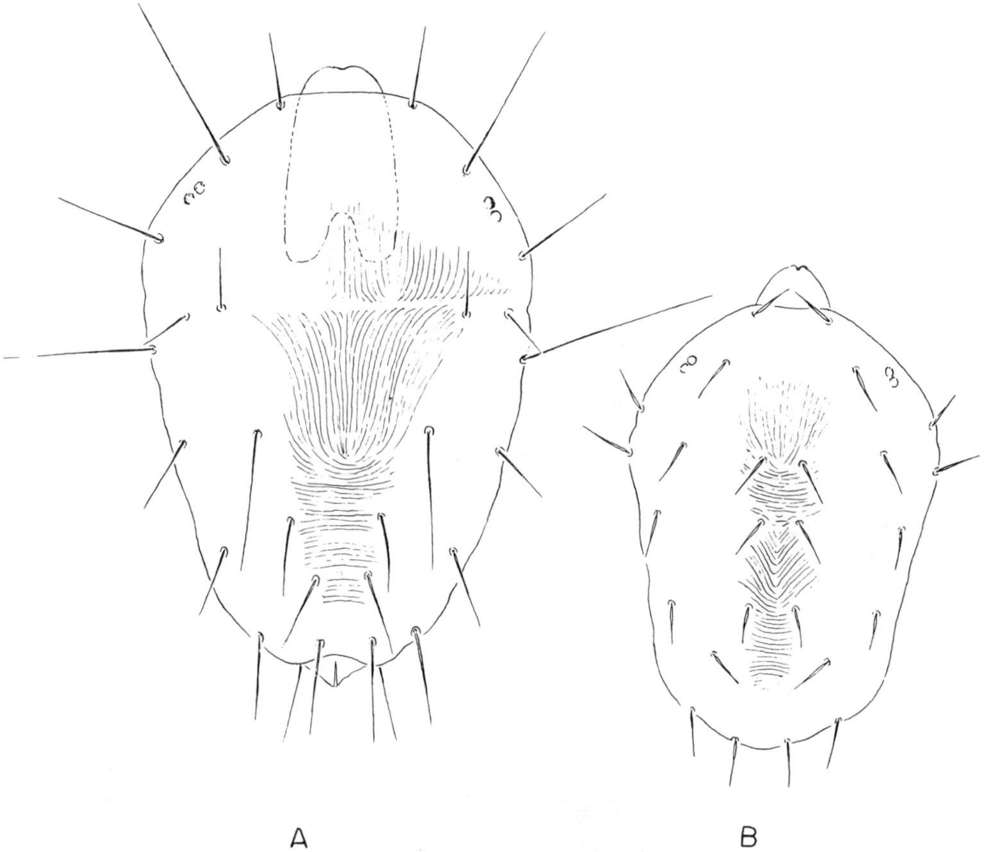

Fig. 30. *Schizotetranychus,* dorsal views showing setae and striations: *a,* S. *celarius* (Banks); *b,* S. *spiculus* Baker and Pritchard.

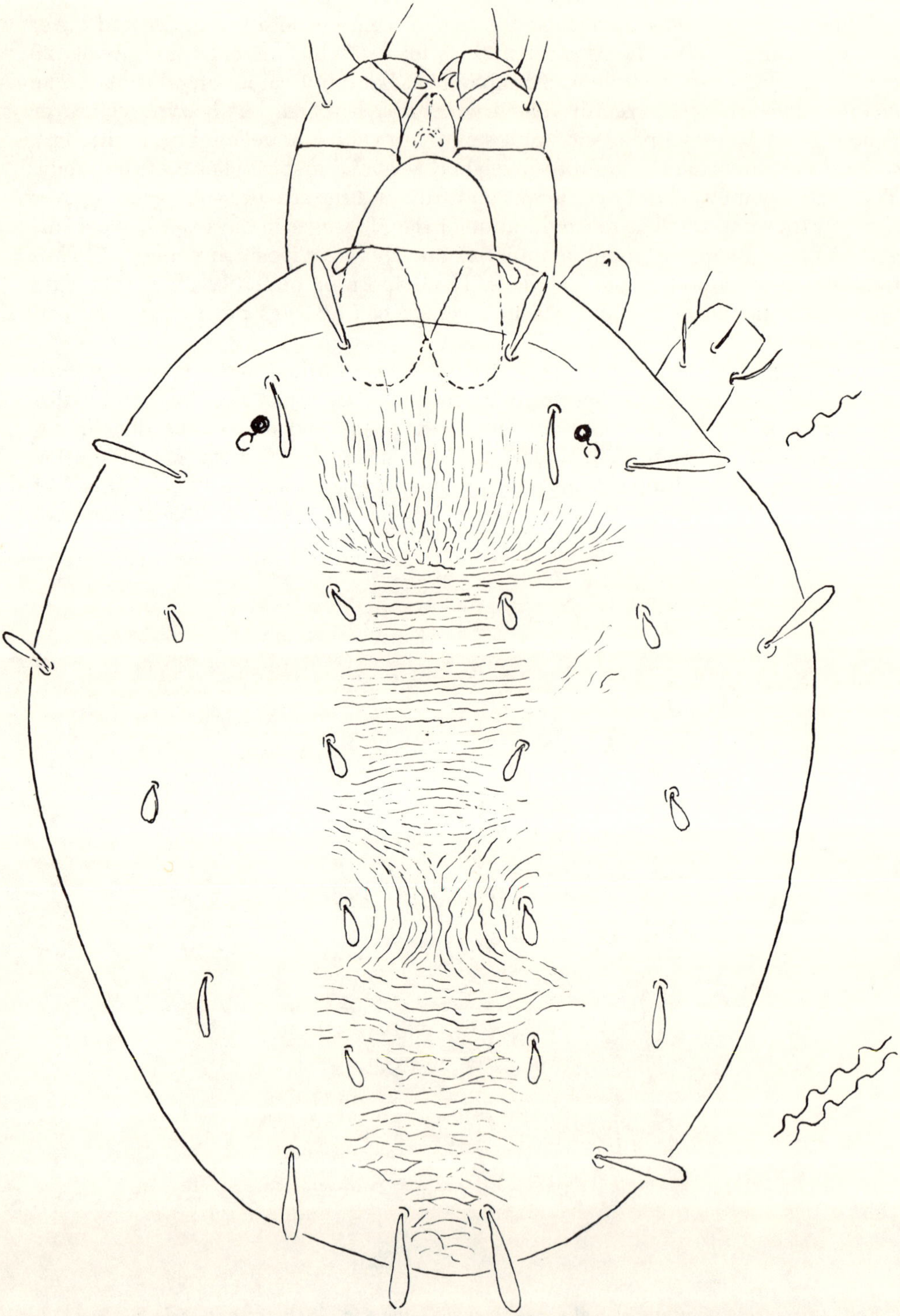

Fig. 31. *Monoychellus caribbeannae* (McGregor), dorsal view showing setae and striations.

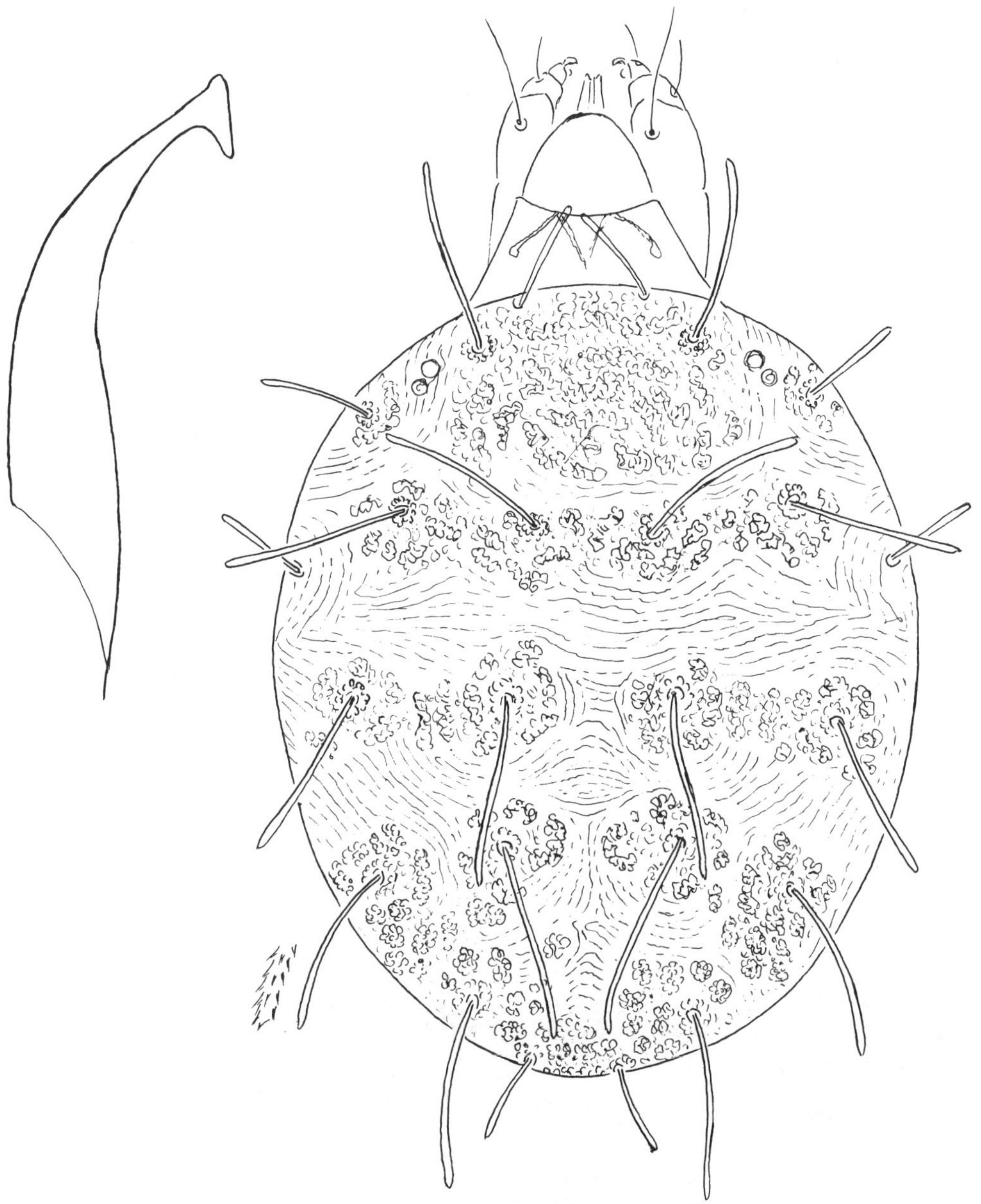

Fig. 32. *Mononychellus planki* (McGregor), dorsal view of female and aedeagus of male.

5 tactile setae; and tarsus II has 1 proximal tactile seta. The terminal sensillum of the palpus is absent. The aedeagus is bent dorsad at right angles with the main shaft, and has a short distal sigmoid curve.

Schizotetranychus celarius (Banks). This species (figs. 28, *c*; 30, *a*) is known from Hokkaido, Honshu, Shikoki, and Kyushu in Japan and Florida, Georgia, and California in the United States. It is a common pest of bamboo in Japan and an

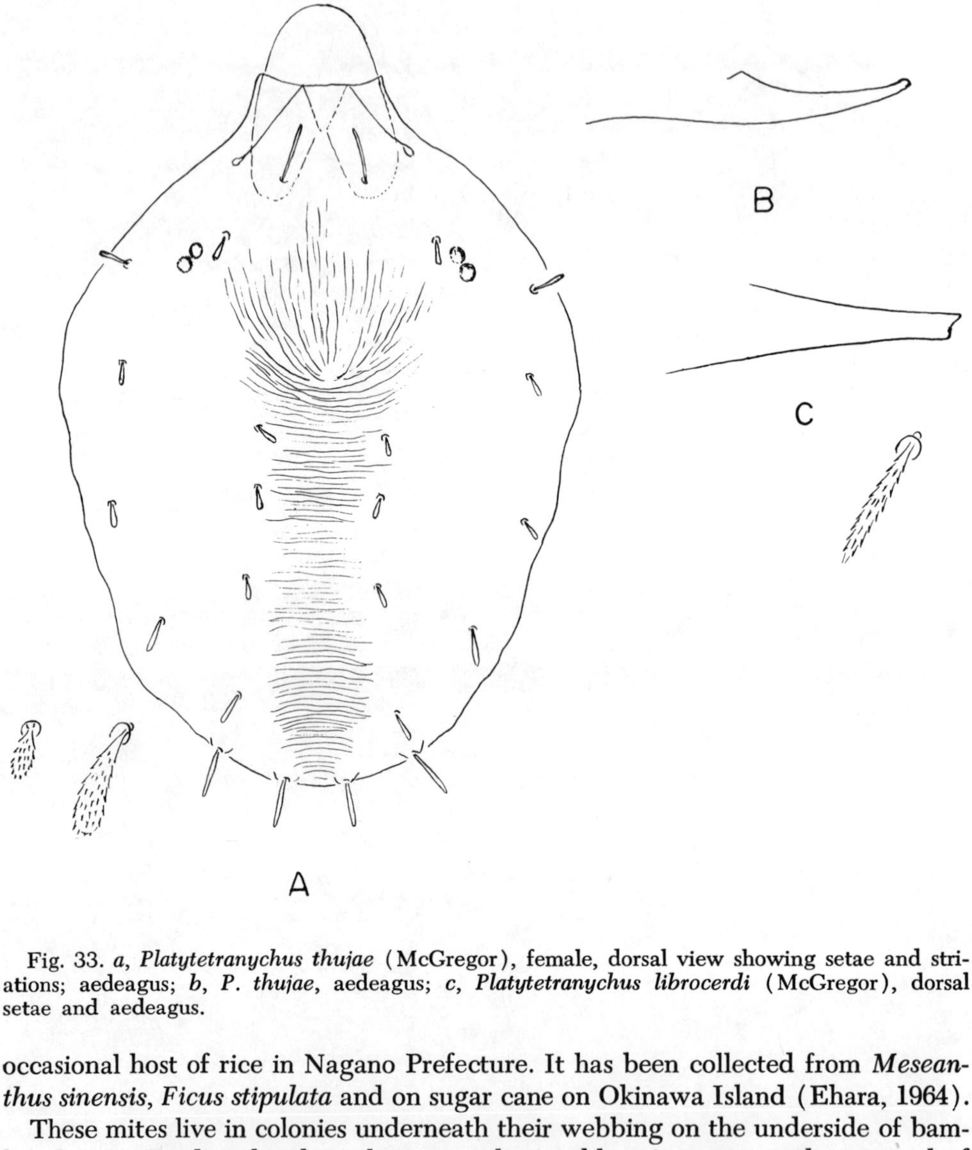

Fig. 33. *a*, *Platytetranychus thujae* (McGregor), female, dorsal view showing setae and striations; aedeagus; *b*, *P. thujae*, aedeagus; *c*, *Platytetranychus librocerdi* (McGregor), dorsal setae and aedeagus.

occasional host of rice in Nagano Prefecture. It has been collected from *Meseanthus sinensis*, *Ficus stipulata* and on sugar cane on Okinawa Island (Ehara, 1964).

These mites live in colonies underneath their webbing on the underside of bamboo leaves. Feeding by the colonies results in chlorotic spots on the upper leaf surface opposite the mite colony.

The first pair of dorsocentral hysterosomal setae (D_1) of the female of S. *celarius* are about ½ as long as the second pair and equal in length to the first pair of lateral setae (L_1). The striae are longitudinal on the propodosoma and on the anterior portion of the hysterosoma, reaching to the D_2 setae. The aedeagus of the male is broadly curved dorsad and slightly sigmoid.

Other potential *Schizotetranychus* pests of economic plants include S. *hindustanicus* (Hirst) (fig. 28, *d*), S. *oryzae* Rossi de Simons (fig. 28, *e*) (Rossi de Simons, 1966), and S. *spiculus* Baker and Pritchard (fig. 30, *b*).

S. *hindustanicus* occurs on citrus in south India. It is known only from the male. The dorsal setae are about ⅔ as long as the distance between their bases; the D_4 setae are much further apart than the other D setae. The aedeagus is distinctive; the distal portion turns dorsad to form a sigmoid curve; it is slightly hooked at the tip, and the axis of the knob or barb is almost parallel with the shaft.

S. *oryzae* occurs on rice in Argentina, Brazil, Columbia, and Texas. The female peritremes are slightly bent distally; all dorsal body setae of both sexes are short, subequal in length, and about ½ as long as the distance between their bases. The striae are longitudinal only on the propodosoma. The male does not have a palpal terminal sensillum; the aedeagus is bent dorsad, tapers gently to the tip, and is sigmoid in shape.

S. *spiculus* inhabits citrus in Kenya. It is known only from the female. The dorsal setae are short, a little more than ½ as long as the distance between their bases, subequal in length, broad at their bases, and narrow distally. The longitudinal propodosomal striae extend to the D_1 setae of the hysterosoma; the striae between the hysterosomal setae D_2 and D_3 form a V-pattern. Setae D_4 are much further apart than the other D setae.

Mononychellus Wainstein

There are 2 pairs of para-anal setae; the duplex setae of tarsus I are distal and adjacent. The empodium is split into 3 pairs of ventrally directed hairs. The striae are longitudinal between the third pair of dorsocentral hysterosomal setae; the lobes of the striae may be prominent; the striae may form a netlike pattern. And the setae may be borne on weak tubercles.

Mononychellus caribbeanae (McGregor). This mite (fig. 31) occurs on cassava and *Platymiscium* in Mexico, Costa Rica, Florida, and Puerto Rico (Estebanes and Baker, 1968). This species is distinctive in having a reticulate striation pattern around the dorsal body setae; these setae are much shorter than the distance between their bases.

Mononychellus planki (McGregor). *M. planki* (fig. 32) has been reported from Brazil, Puerto Rico, Colombia, and Trinidad. It occurs on soybean, peanut, bean, and other plants, and is a pest of cotton in Brazil (Flechtman and Baker, 1970).

M. planki is distinctive in having a reticulate pattern on the dorsum of the body, especially surrounding the setal bases. The dorsocentral and marginal setae of the hysterosoma are as long as the intervals between their bases and are subequal in length. Tibia I of the female bears 9 tactile and one sensory setae.

Platytetranychus Oudemans

The dorsal body setae are much shorter than the distance between their bases; the striae are longitudinal on the propodosoma and transverse on the hysterosoma. The peritreme ends in a single bulb. Tibia II has 5 tactile setae; and the duplex setae of tarsus I are distal and adjacent.

The genus *Platytetranychus* contains two species found on conifers. *P. thujae*

(McGregor) (figs. 33, *a*; 33, *b*) is common on junipers and cupressaceous conifers in eastern, southeastern, and midwestern United States as far north as New Hampshire and New York state (Tuttle and Baker, 1968).

The peritreme of *P. thujae* female is simple; the dorsocentral hysterosomal setae are very short; tibia I has 7 tactile setae; tibia III has 3 to 4 tactile setae; the dorsal body setae are short, broadening distally. The aedeagus of the male is a narrowing rod with a rounded, distal tip.

P. libocerdri (McGregor) (fig. 33, *c*) occurs in western and southwestern United States on junipers, arborvitae, cypress, tamarisk, and in Mexico on pine and fir (Tuttle and Baker, 1968).

The female of *P. libocerdri* is similar to *P. thujae* except the dorsal setae of the female are short and taper distally; the dorsolateral setae are also longer than the dorsocentral setae. Tibia I has 9 tactile setae. The aedeagus of the male is distinctive and consists of a rod narrowing distally and emarginate at the tip.

Honey locust spider mite, *Platytetranychus multidigituli* (Ewing). This mite (figs. 27, *a*) has been found only on honey locust. It is distributed throughout Connecticut, Washington, D.C., North Carolina, Louisiana, Ohio, Illinois, South Dakota, east Texas, and probably wherever its host is found. Feeding by this mite results in stippling and bleached spots on dead leaf tissue, mainly on the undersides of the leaflets. Heavy infestations produce severe yellowing and browning and finally dropping of the leaves.

The summer males and females are light to dark green in color. As cool weather approaches, the female mites move to the twigs and branches and gradually become orange red and almost inactive; thus the mites pass the winter under bud scales, in crevices of the rough bark, at the base of twigs, and even under shells of dead scale insects. Oviposition begins as early as April 11 and larvae may appear by May 1. Populations develop rapidly in May and June and remain high until late August or September.

Egg incubation requires 4 to 9 days and development from larvae to adults varies from 4 to 11 days, averaging 4.9 days in June and 8.5 in September. The total development time is 8 to 17 days, averaging 9.5 in June and 14.6 days in September. The preoviposition period and time for each molt is about 1 day each. A female may lay as many as 72 eggs over an 11-day period. About 55 percent of the eggs are laid on the undersides of the leaflets, 17 percent on the upper surfaces, and 28 percent on leaflet petioles.

The eggs of the honey locust spider mite are minutely striated above and without a central stipe; they are circular in dorsal view and slightly depressed or flattened in lateral view. The color is pearly with a slight yellowish tinge. At first the red eyes of the larvae are conspicuous, but as feeding occurs the larval color changes from a pale yellow to dark green. After each molt the nymphal stages are greenish yellow, but upon feeding they become dark green with vague spots along the sides.

Acaricides effective against other tetranychid mites appear to be effective in controlling the honey locust spider mite (Reeves, 1963; English and Snetzinger, 1957, *b*).

P. (P.) multidigituli is distinctive in that the peritremes form an anastomosing chamber distally. Tibia I of the female has 7 tactile setae; tibia II has 5 tactile setae; there are 3 tactile setae proximal to the duplex setae on tarsus I. The dorsal body setae are shorter than the intervals between their bases in both sexes. Tibia I, II, and tarsus I of the male are similar to those of the female. The aedeagus is distinctive in that it bends sharply ventrad and forms a distal knob, the distal angulation being much stronger than the anterior angulation. This species is not typical for the genus.

Oligonychus Berlese

The true claws are padlike with tenent hairs; the empodium is clawlike and has proximoventral hairs set at right angles to the empodium. The dorsal body setae, with few exceptions, are not set on tubercles. There are 2 pairs of anal setae and a single pair of para-anal setae.

Oligonychus (*Oligonychus*) s. str. (fig. 34)

The hysterosomal striae are transverse; the dorsal body setae are longer than the distance between their bases; the aedeagus turns down at an obtuse angle. These mites generally feed on the upper surfaces of the leaves of broadleaf plants; a few feed on conifers.

Oak mite, *Oligonychus (Oligonychus) bicolor* (Banks). This species (fig. 34, *b*) is a serious pest of oak and other ornamental trees throughout the eastern United States and Canada. It has also been found in Arizona, Kansas, and South Africa (western province), Iran, and Transvaal. This mite is a major pest of most oak species, but it may also be a serious pest of grape, chestunt, birch, beech, elm, spruce, maple, and hickory.

The mites feed on the upper leaf surface and occasionally on the bark of young twigs. The barrel-shaped rather squatty eggs are laid on the upper leaf surface along the midrib and lateral veins. Overwintering eggs are laid in crevices and around axils of small twigs where they remain dormant until the following May. Damage is manifested by leaf discoloration. These injured areas later become reddish, and when the population is high, the injury, webbing, and numerous cast skins produce a greyish rust appearance to the surface of the leaves.

O. bicolor resembles *O. newcomeri* (McGregor) and *O. coffeae* (Nietner) in that its basic color is reddish brown. On chestunt leaves the legs and anterior ⅓ of the body are light yellowish orange with a narrow yellowish band along the midposterior dorsum. The remaining portion of the dorsum is a dark reddish brown (Specht, 1963).

Tarsus I of *O.* (*O.*) *bicolor* has 3 tactile setae proximal to the duplex setae, of which the proximal member is very short; tibia I has 7 tactile setae and 1 sensory seta; the dorsal body setae are longer than the intervals between their bases; the hysterosomal setae D_4 and L_4 are subequal in length. The bent distal portion of the aedeagus forms an obtuse angle with the shaft, and abruptly narrows distally.

Tea red spider mite, *Oligonychus (Oligonychus) coffeae* (Nietner). This mite (fig. 34, *c*) has been known as a pest of tea in India from the early days of tea cultivation. It was first discovered in Assam in 1868 but is now known from

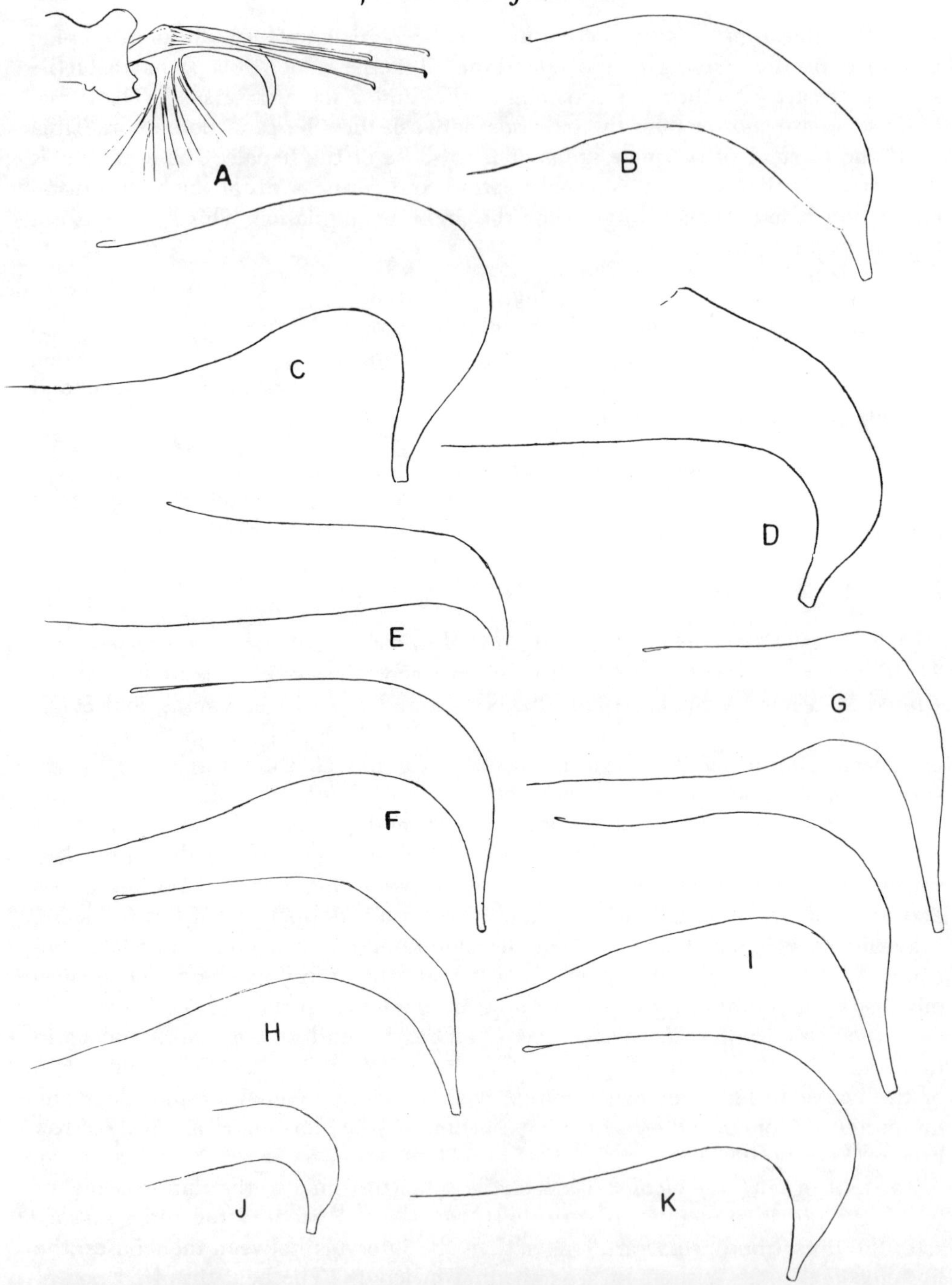

Fig. 34. *a*, *Oligonychus* (*O*) S. str., empodium and claws; *b*, *O.* (*O.*) *bicolor* (Banks), aedeagus; *c*, *O.* (*O.*) *coffeae* (Nietner), aedeagus; *d*, *O. ilicis* (McGregor), aedeagus; *e*, *O.* (*O.*) *coniferarum* (McGregor), aedeagus; *f*, *O.* (*O.*) *ununguis* (Jacobi); *g*, *O.* (*O.*) *yothersi* (McGregor); *h*, *O.* (*O.*) *newcomeri* (McGregor); *i*, *O.* (*O.*) *viridis* (Banks); *j*, *O.* (*O.*) *punicae* (Hirst); *k*, *O.* (*O.*) *mangiferus* (Rahman and Sapra).

India, Sri Lanka (Ceylon), south and east Africa, Indonesia, Australia, Zaire, Mauritius, Florida, the Middle East, and Formosa. It is considered the most serious of all tea pests. It is also a pest of jute in Bangladesh (East Pakistan) and of cotton in Egypt. It is known to attack coffee, rubber, indigo, grape, cashew nut, citrus, mango, camellia, camphor, mulberry, oil palm, and many other tropical plants.

Injury on tea is first evidenced as a yellowish spotting along the midrib and veins of leaves and occasionally on the petioles. As feeding continues these spots turn brown and eventually coalesce, causing large areas and even the entire leaves to become deeply bronzed, necrotic, and often to fall from the plant; thus growth of tea bushes is retarded.

The mites live in colonies, preferably on the upper surface of older leaves, but will inhabit both sides during severe infestations and conditions of drought; mites will even move to younger leaves, which are less preferred owing to their higher turgidity. During drought, younger leaves become less turgid and are more susceptible to attack.

Mites can be found throughout the year on tea in northeast India; populations begin to increase in early March and reach their greatest density during late March and early April. Injury becomes most severe during May and June or until the monsoon rains wash off or kill all active forms on the leaves. Eggs hatch and infestations are renewed after the rainy season, but populations never become as injurious as during the premonsoon period. During the cool weather of December and January, populations become extremely low and damage is relatively uncommon.

Temperature and humidity may be limiting factors in the development of populations of this mite. Eggs fail to hatch under constant temperatures of 34 C (94 F) and 17 percent relative humidity (RH). At 33 C (92 F) none of the eggs hatch if humidities are below 72 percent, but some hatch when the humidity remains just above 72 percent. Maximum daytime temperatures in tea-growing areas of northeast India may reach 38 C (100 F), yet drop to 26 to 30 C (79 to 86 F) at night. Exposures at 37 C (98 F) and 72 to 77 percent RH, and 37 C (98 F) and 90 to 94 percent RH for six hours results in a substantial decrease in egg hatching. The optimal conditions for development are 20 to 30 C (68 to 81 F) and 49 to 94 percent RH. Under optimal conditions there may be 22 generations in a year, but developmental rates vary inversely with temperature.

The eggs of the tea red spider mite are scarlet, spherical, and possess a stipe. The propodosoma of the adult female is bright red and the hysterosoma wine colored. The duration of the life stages in days at constant temperatures of 22 C (71 to 72 F) are: larva 8.0, protonymph 40, deutonymph 2 to 3; thus 14 to 15 days are required for development from egg to adult. Females are capable of laying 40 to 50 eggs during their life span.

All stages of this mite are found during the winter on the remaining few old leaves and on small leaves at the base of the shoots (janams) of the tea bushes. This persisting population is primarily responsible for the attack in the spring. Pruning removes many of the old leaves and the janams and thus many of the mites. Pruned tea is less attacked than unpruned tea or skiffed tea (where a little

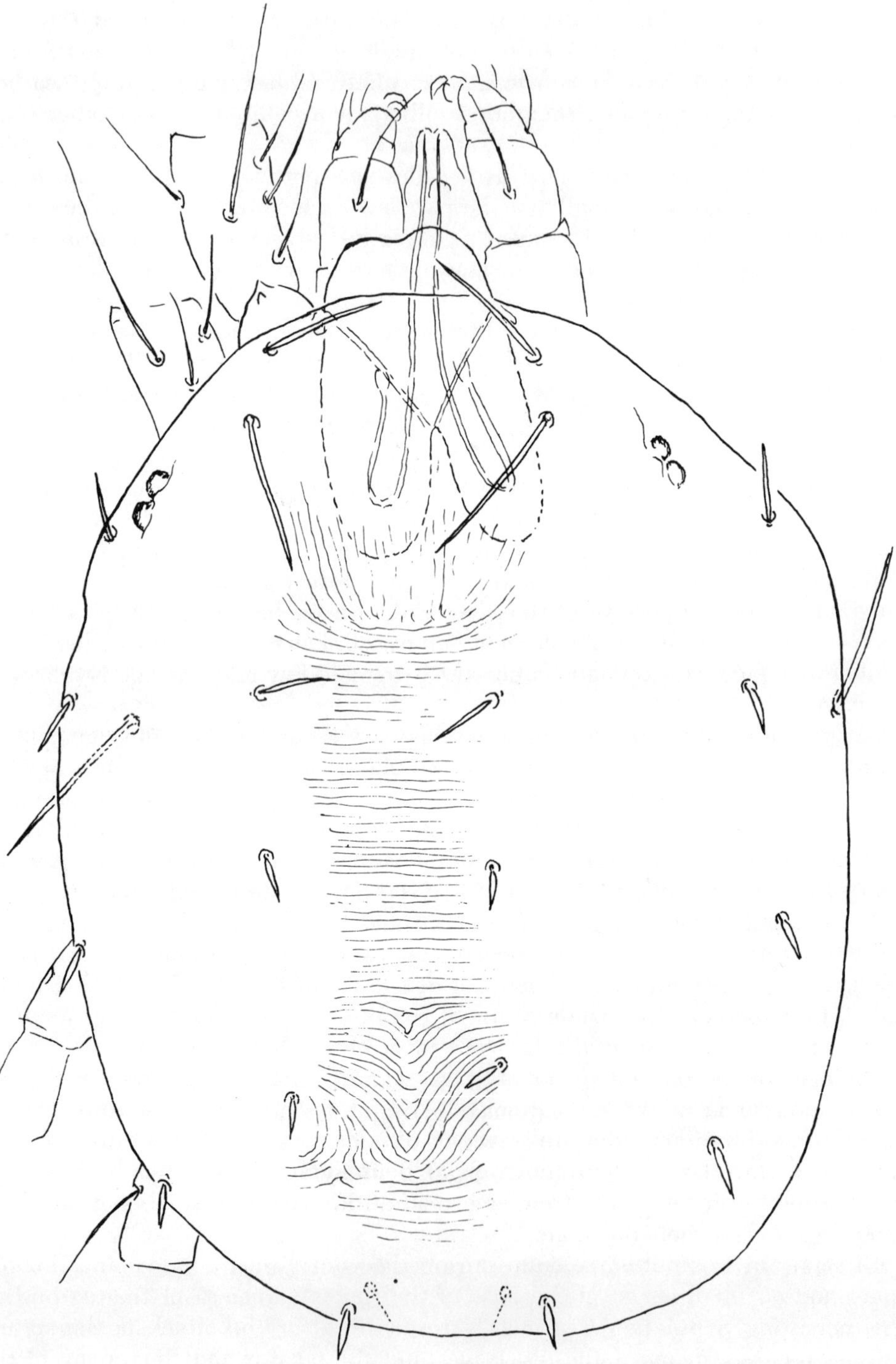

Fig. 35. *O.* (*O.*) *hondoensis* (Ehara), female dorsal view.

of the tops are removed). Early pruned tea is more subject to attack than late-pruned bushes. Thus the winter removal of the leaves tends to reduce the possibility of summer infestations. Bushes defoliated as a preventive measure against subsequent attacks usually remain free of mites. Abandoned tea is severely attacked the first summer after abandonment, but as trees grow larger, the degree of injury progressively decreases. This mite is virtually absent in tea seed gardens except on occasional border trees (Hu and Wang, 1965; Das, 1959, 1960; Das and Das, 1967).

Chemical control during the summer is not practical because the thick canopy of leaves during the growing season makes distribution of acaricides difficult. Also, toxicants having persistent residues cannot be used during the plucking season unless leaves are discarded for at least two plucking rounds after treatment, then there is danger of tainting the tea or endangering the consumers. Acaricides may be effectively used, however, to reduce the overwintering populations (Das, 1960).

O. (*O.*) *coffeae* has 7 tactile setae on tiba I and 3 proximal tactile setae on tarsus I; the proximal members of the duplex setae are very short. Hysterosomal setae D_4 and L_4 are equal in length. The aedeagus is bent ventrally at a right angle, is not sigmoid, and abruptly narrows distally.

Oligonychus (*Oligonychus*) *coniferarum* (McGregor). This mite (fig. 34, *e*) occurs on coniferous species in Arizona, Florida, and Texas. Tibia I has 7 tactile setae; tarsus I has 4 tactile setae proximal to the duplex setae; the proximal members of the duplex setae are very short. The dorsocentral hysterosomal setae are typcially longer than the distance between their bases. The male, as in the female, is similar to *O.* (*O.*) *ununguis*, differing only in that the aedeagus forms a short, truncate, caudolaterally direct bend.

Oligonychus (*Oligonychus*) *hondoensis* (Ehara). This species (fig. 35) occurs in Japan and in New York on Japanese cedar.

This species is one of the most destructive pests of this host in Japan. It is so important that it has been designated as a legal pest by the government, a dubious honor that no other forestry pest has been awarded (Ehara, 1964).

O. (*O.*) *hondoensis* described from New York as *Oligonychus weidhaasi* Reeves, is distinctive in its dorsal setal pattern. The first and second propodosomal setae and the humeral setae are subequal in length and much longer than the intervals between their bases; the other dorsal body setae are much shorter than the distance between their bases and are also subequal in length, although setae D_1 of the hysterosoma are somewhat longer than the setae D_3 and D_4. The stylophore is indented anteriorly.

Southern red mite, *Oligonychus* (*Oligonychus*) *ilicis* (McGregor). This mite (figs. 34, *d*; 36) was first described in 1917 from American holly in South Carolina. Later it was found to be a pest of plane trees and was called the "Plane tree spider mite." It is a pest of conifers in eastern and of azalea and camellia in southeastern United States, of cranberries in Massachusetts, of walnut and syca-

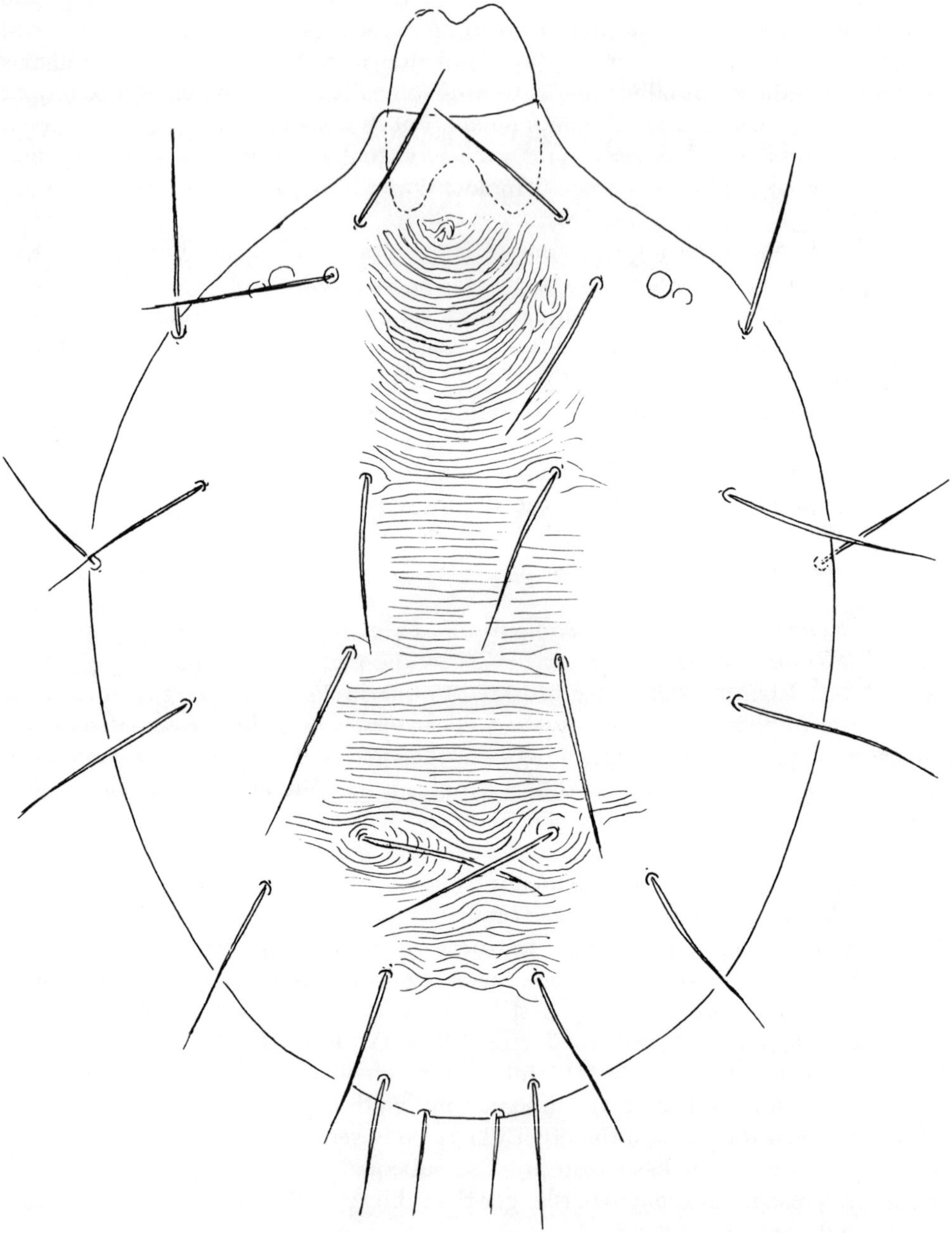

Fig. 36. *O.* (*O.*) *ilicis* (McGregor), female striae, setae.

more in California, of coffee in Brazil, and tea, rice, laurel, holly, and boxwood in Japan. It also attacks camphor, eucalyptus, oak, spruce, loquat, pear, and quince (Ehara, 1963).

The females are basically purplish or reddish, paler anteriorly with a pale spot medially. They live almost exclusively on the top side of the host leaves, causing a bronzing and browning along the midrib of the upper leaf surface of sycamore, loquat, azalea, camellia, and other hosts. A dirty, ashy grey appearance of the upper leaf surface results from injury by this mite to walnut. This white flecking is present where groups of cells have been injured owing to mite feeding on the epidermis and palisade layer of the upper leaf surface (pl. 33) (Smith, 1939).

Leaf infestations often are initially localized along the midrib, but other portions of the leaf may have few mites and be relatively free of injury. Lower parts of trees seem to be most severely injured, especially trees showing drought symptoms. Feeding by this mite apparently produces less toxic effects to its hosts than many economic mite pests because heavy infestations can be tolerated before leaf drop occurs (Smith, 1939).

The southern red mite is widely distributed on coffee in Brazil. Old and young plants are readily attacked resulting in webbing and bronzing around the leaf veins. Migrating mites on interleaf webbing are dispersed by wind currents to new host plants. Observers in Brazil and in Massachusetts have noted that dry climatic conditions are favorable for population development. During rainy seasons or in low damp areas these mites increase slowly and control measures are not necessary.

In a Brazilian study, females deposited 10 to 15 eggs and the mites completed development in two weeks at temperatures between 22 and 24 C (72 and 75 F). The oviposition period lasted 3 days and adult females lived for 15 days (Calsa and Sauer, 1952). Hatching of overwintering eggs occurs in late April and early May in Massachusetts; there populations peak in June at about the time migration to new growth occurs. There is often a population decline in late June and July followed by an increase in September resulting in the deposition of large numbers of overwintering eggs (Matthysee and Naegele, 1952).

On woody ornamentals a single spray of residual acaricide applied in mid-May usually prevents severe summer injury. Sulfur, as well as recommended acaricides, may be applied as sprays.

Tibia I of *O. (O.) ilicis* has 7 tactile setae; there are 3 tactile setae proximal to the duplex setae of tarsus I; the proximal member of the duplex setae is very short. The dorsal body setae are on small tubercles (not always seen in mounted specimens). The aedeagus is bent ventral, narrows distally, and is somewhat flattened dorsoposteriorly.

Oligonychus (Oligonychus) mangiferus (Rahman and Sapra). *Mangiferus* (fig. 34, *k*) is widely distributed throughout the tropics, and is recorded from India, Mauritius, Hawaii, Peru, and Egypt. It is a pest of cotton, mango, loquat, peach, quince, pear, and pomegranate. It also occurs on grapes, roses, and *Eugenia*. Damage is often severe on mango where feeding produces a drying effect and premature leaf drop.

In Egypt *O.* (*O.*) *mangiferus* is a pest of cotton and is considered the second most serious pest of pomegranate. Infestations occur on the upper leaf surfaces. Populations reach their maximum in July on quince, in August on pear, and in September on pomegranate (Mohamed, 1963; Moutia, 1958).

O. (*O.*) *mangiferus* has 7 tactile setae on tibia I; on tarsus I there are 4 tactile setae proximal to the duplex setae; the proximal members of the duplex setae are short. The aedeagus is bent ventrally, the distal bent portion forms an acute angle with the shaft and abruptly narrows at the tip.

Oligonychus (Oligonychus) newcomeri (McGregor). *Newcomeri* (fig. 34, *h*) is probably indigenous to the northeastern United States, but was originally described from Washington State. Its favorite hosts are serviceberry and hawthorn, but other hosts include apple, pear, *Prunus*, and a few other members of the family Rosaceae (Pritchard and Baker, 1952).

Damage caused by this mite on serviceberry is so severe that the upper surface of the leaves is often completely brown by fall; apple may also be seriously injured. Mites overwinter in the egg stage. Spring activity starts in early May, and by June adults may be found primarily on the upper leaf surfaces, although quiescent forms and males may commonly be seen on the lower leaf surfaces. There seems to be no sharp population increases during the summer, but rather a gradual buildup in numbers, depending on the amount of fresh foliage available. Overwintering eggs are laid in September and active mites disappear by October. This species spins very little webbing.

These mites may be recognized by their globular shape and their dorsal setae, not set on tubercles; they are purple color with a somewhat heart-shaped, pinkish area anteriorly. The anterior legs are pale like the anterior portion of the body, but the posterior legs are somewhat darker. The egg is strongly flattened, striate above, and bears a dorsal stipe. Summer eggs are pale until a day or two before hatching, but winter eggs are amber (Reeves, 1963).

Tiba I of *O.* (*O.*) *newcomeri* has 7 tactile setae; there are 3 tactile setae proximal to the duplex setae on tarsus I; the proximal member of the duplex setae are short; the hysterosomal setae D_4 and L_4 are similar in length; the setae are not set on tubercles. The aedeagus is distinctive, bent ventrally, gradually narrows to the tip, and is distinctly flattened dorsoposteriorly.

Avocado brown mite, *Oligonychus (Oligonychus) punicae* (Hirst). This mite (fig. 34, *j*; pl. 34) is a pest of avocados in southern California and also of grapes and pomegranate in tropical Asia, Central, and South America. Feeding is first confined to the upper surfaces of avocado leaves; first along the midrib, then along the smaller veins, and during heavy infestations, over the entire leaf surfaces. The areas along the veins become brownish and are covered with myriads of whitish hatched eggs and cast skins. The destruction of the chlorophyll reduces the value of the leaf to the tree, but if the leaves are not severely brown when the mites are removed, the green color returns. This species causes relatively minor damage compared to the severe injury produced by similar populations of the six-spotted spider mite.

The stalked eggs and the immature mites resemble those of *O.* (*O.*) *yothersi* (McGregor). The propodosoma of the adult female is pinkish, the lateral area of the hysterosoma and sometimes the median area is occupied by many blotches of purplish brown. The hysterosoma of older individuals may be solidly blackish brown. The forelegs and palpi are rusty pink, with the other legs pale. The dorsal body setae are strong, pale, but do not arise from tubercles. The color of the male is paler than that of the female.

The mites are active and populations may remain high all winter in California but characteristically, old avocado leaves drop during the blooming period taking with them many of the overwintering mites. The annual peak levels of abundance of this mite range from 8 to 84 adult females per leaf. The seasonal increase usually begins in early summer, peaking in late summer followed by an abrupt decline. The increase is sometimes delayed until late summer; then the peak occurs in fall or early winter.

The length of time required for development of each stage does not differ significantly from the avocado red mite *O. (O.) yothersi* under similar temperature conditions. In the summer there may be 2 complete generations within a month. Only 7 days are required to complete a generation at a constant temperature of 25 C (77 F). At a constant temperature of 32.5 C (91.4 F) all stages died, including the eggs (Ebeling, 1959).

Biotic factors in California usually keep the population of this mite below injurious numbers. Abrupt population declines are associated with predation, or intraspecific competition or both. The most abundant natural enemies are the lady bird beetle, *Stethorus picipes* Casey, and 2 predatory mites, *Amblyseius hibisci* (Chant) and *A. limonicus* (Garman and McGregor). Other predators are *Scolothrips sexmaculatus* (Pergande), a cecidomyid fly, *Arthrocnodax occidentalis* Felt, a staphylinid beetle, *Oligota oviformis* (Casey) and a green lacewing, *Chrysopa carnea* Stephens; but these do not keep the populations of this mite from increasing early in the season (McMurtry and Johnson, 1966).

The female of *O.* (*O.*) *punicae* is typical for the subgenus, with seven tactile setae on tibia I and four tactile setae on tarsus I proximal to the duplex setae; it is indistinguishable from females of *O.* (*O.*) *yothersi* and *O.* (*O.*) *mangiferus*, all of which are found on broadleaf tropical plants. The male is distinctive in that the ventrally directed aedeagal hook is rather broad and the distal end abruptly narrows to form a fingerlike projection.

Spruce spider mite, *Oligonychus (Oligonychus) ununguis* (Jacobi). This mite (fig. 34, *f*) is a serious pest of conifers throughout the world. Studies on the biology and injury were made in 1923 by Garman. Species of spruce, hemlock, fir, arborvitae, juniper, larch, redwood, yew, cypress, false cypress, incense cedar, *Cryptomeria*, and others are attacked.

Mite feeding causes the needles to turn brown. During severe infestation the trees appear brown and drop their needles until trees are bare. Areas of hemlock damaged by this mite appear whitish or pale. Damage is most severe in the lower parts of large trees. Seedlings and small trees are often killed, and in a few instances large acreages of mature trees have been destroyed by this mite pest. Three-year-old spruce trees appear to be the most preferred host. Moderate to severe damage by this mite has occurred in most of the conifer areas of the world.

This species overwinters as eggs deposited near the base of the needles and other protected areas, but never on the needles themselves. Eggs hatch in April and adults complete development in 11 to 23 days. Summer eggs are also laid in sheltered portions of the tree. The young immature stages prefer the needles and shoots in the lower part of the crown, but adults show no particular preference. Adult females may lay 45 eggs in their lifetime, 60 to 80 percent of which become females. Mites spread out from localized infestations on a tree very slowly. The threshold for activity is 6 to 7 C (43 to 45 F). Diapause will terminate after a resting period when eggs are exposed to temperatures of 20 C (68 F).

The eggs of this species have a stipe and are greyish brown on first being laid, but change to darker orange brown in a few days. Larvae are pinkish at first, then turn greenish. The adults are orange to blackish (Reeves, 1963; Serafimovski and Thalenhorst, 1962).

O. (*O.*) *ununguis* has 7 tactile setae on tibia I; there are 4 tactile setae proximal to the duplex setae on tarsus I; the proximal members of the duplex setae are short. The aedeagus is bent ventrad at right angles to the shaft, and tapers gradually to the acute tip.

Oligonychus (Oligonychus) viridis (Banks) (fig. 34, *i*) is a pest of the pecan in South Carolina, Texas, and Louisiana. It has also been collected in New York, Georgia, and Florida from hickory. This mite prefers the upper leaf surfaces, but during heavy infestations it occurs on both sides of pecan leaves. Such infestations cause the leaves to become greyish white owing to feeding by the mites and the presence of old cast skins. Severe browning of the upper leaf surfaces has been observed on shagbark hickory (Pierce, 1953; Reeves, 1963).

O. (*O.*) *viridis* has 7 tactile setae on tibia I; tarsus I has 3 tactile setae proximal to the duplex setae. The proximal members of the duplex setae are short; the hysterosomal setae D_4 and L_4 are similar in length. The aedeagus bends ventrally, the distal portion is slightly sigmoid, and tapers evenly to the tip.

Avocado red mite, *Oligonychus (Oligonychus) yothersi* (McGregor). This mite (fig. 34, g), probably of tropical Asian origin, was first observed in Florida in 1909 where it has become a pest of avocado. It now occurs in the South American countries of Brazil, Columbia, Ecuador, and Argentina, in Central America, and in New Jersey and Maryland in the United States. It has been collected from pomegranate, grape, apple, mango, litchi, boxwood, camphor, and eucalyptus. The avocado red mite feeds on the upper leaf surface, initially causing a white spotting; but with increased feeding, the leaves turn a reddish brown in the area of the midrib. Heavy infestations may cause the entire upper surface and a portion of the lower surface to become bronzed followed by partial defoliation.

The eggs of this species are globose, smoky amber with a stalk at the apex. Eggs are oviposited singly, at first along the midrib, but as populations increase eggs are scattered over the entire leaf surface. The adult female is oval, rusty red, and larger than the male. Females lay an average of 35 eggs during their life; the eggs complete development in 7 to 10 days; the average life cycle is 14.2 days. In Florida the mites remain active throughout the year, but in New Jersey they overwinter in the egg stage. There are 5 to 6 generations per year.

High populations of this mite are necessary before serious injury to avocado occurs; weather and natural enemies generally keep the populations from becoming serious. In Florida, a natural leaf drop in March and April often contributes to a reduction in mite populations, and well-timed spring and summer rains often keep populations at below economic levels. Seasonal increase usually begins in early summer with peak populations occurring in late summer, followed by an abrupt decline; the increase, however, is sometimes delayed by rains until late summer, then the populations peak in fall or early winter.

The active stages of the avocado red mite can be killed by thorough dusting with sulfur. The sulfur residue retained on the leaves by webbing also kills the mites that subsequently hatch from the eggs (Hamilton, 1926; McKenzie, 1935).

Tibia I of *O.* (*O.*) *yothersi* has 7 tactile setae; tarsus I has 4 tactile setae proximal to the duplex setae; the proximal members of the duplex setae are short. The aedeagus bends ventrally at a slight obtuse angle, the bent portion as long as the distal part of the shaft; the distal bent portion of the aedeagus is long and slender.

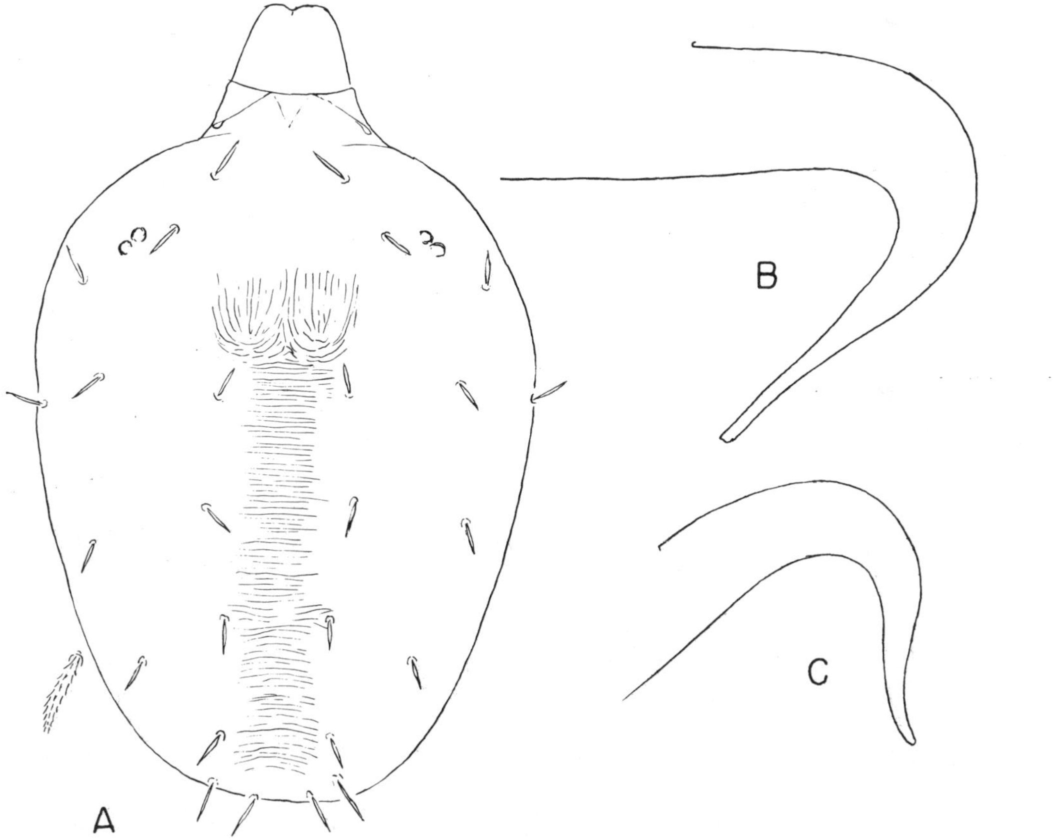

Fig. 37, *Oligonychus* (Wainsteiniella): *a*, *O.* (*W.*) *subnudus* (McGregor), dorsum of female; *b*, *O.* (*W.*) *milleri* (McGregor), aedeagus; *c*, *O.* (*W.*) *subnudus* (McGregor), aedeagus.

Oligonychus (Wainsteiniella) Tuttle and Baker

The dorsal body setae are much shorter than the intervals between their bases. The hysterosomal striae are transverse, and the aedeagus is downturned.

Oligonychus (Wainsteiniella) milleri (McGregor). *Milleri* (fig. 37, *b*) occurs on several conifers and is widely distributed in the United States, including California, Idaho, Utah, Wisconsin, Louisiana, Florida, North Carolina, Delaware, Arizona, Virgina, Alabama, and New York (Pierce, 1953). Recently it has been reported to have caused injury to nursery seedlings of Caribbean pine in Jamaica; damage included yellowing and occasionally bronzing of the needles on seedlings and young pines, and sometimes the death of the trees (Muma, 1970).

The female has the typcially short dorsal body setae, the hysterosomal setae D_1 to D_4 increase in length progressively. The aedeagus of the male is downturned at an acute angle; the distal portion is long and narrows evenly from a broad curvature. The setal leg counts of the female are: tibia I has 6 tactile setae and one solenidion; tarsus I has 1 proximal tactile seta and 1 solenidion; tibia II has 4 tactile setae; tarsus II has 1 proximal tactile seta.

Oligonychus (Wainsteiniella) subnudus (McGregor). This species (figs. 37, *a*; 37, *c*) infests pine, fir, milkweed, and pussy-toes, in Mexico, Arizona, California, and Washington. It has become a serious pest of Monterey, canary, and aleppo pines in California, especially on ornamental plantings. Seriously injured young pines are not suitable for the Christmas and landscape tree market (Garman, 1923).

Damage by *O.* (*W.*) *subnudus* begins as characteristic stippling of the foliage, but when infestations are heavy, the stippled areas coalesce resulting in permanent discoloration of the entire needle. The needles do not drop prematurely, so affected trees retain an unsightly appearance for several years. This mite, unlike the spruce spider mite, *O.* (*O.*) *ununguis* (Jacobi), does not produce webbing.

Weather conditions, plant growth cycles, and the season of the year all influence the population density and distribution of this species on pine trees. Mite densities during February and March are highest near the middle of the past season's growth, but by April and May, maximum densities occur at the tip of the past season's growth and the mites are just beginning to move onto the base of the current season's growth. Summer populations move up on the current seasons growth, and by September population density is maximum on branches near the middle of the current season's growth. September populations on young trees are most commonly on the tree leader and uppermost whorl.

This species is most abundant from May through October, varying with climatic conditions. The cooler the growing season the later the period of greatest abundance of mites (Garman, 1923; Matthysse and Naegele, 1952).

O. (*W.*) *subnudus* is similar to that *O.* (*W.*) *milleri* except the hysterosomal setae D of the former are short and more-or-less equal in length. The aedeagus of the male is distinctive in that the ventrally bent distal portion is not more than ¼ as long as the dorsal portion of the shaft; the distal portion is acuminate and slightly sigmoid.

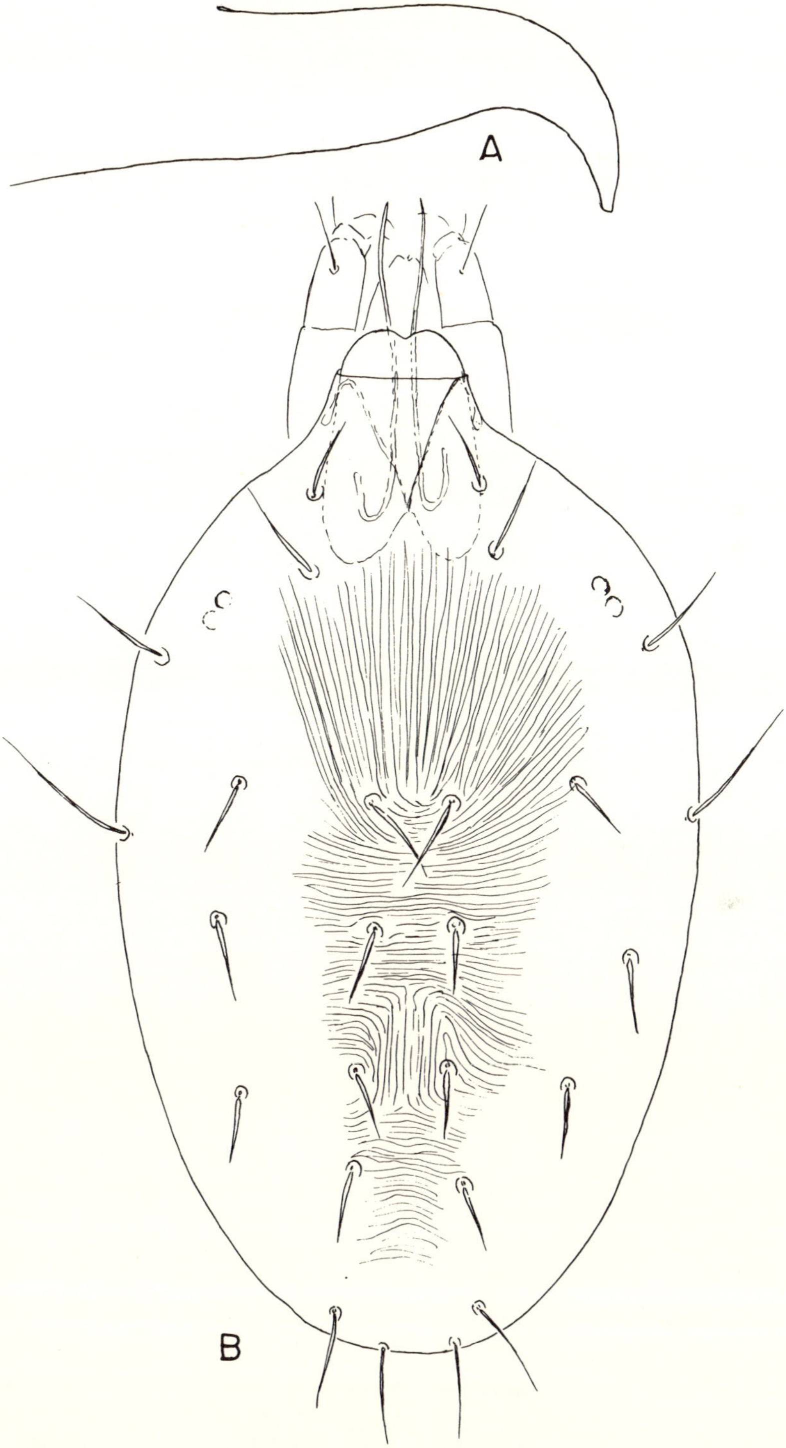

Fig. 38. *Oligonychus* (*Homonychus*) *peruvianus* (McGregor): *a*, aedeagus: *b*, female dorsal view.

Oligonychus (*Homonychus*) Wainstein

The hysterosomal striae are transverse except for the longitudinal pattern between the setae D_3. Tibia I setae vary in number. The aedeagus is downturned.

Oligonychus (*Homonychus*) *peruvianus* (McGregor). This mite (fig. 38, *a*, *b*) is reported to be a pest in California, Texas, Guatemala, Mexico, Peru, Trinidad, and Venezuela. It is known to infest cotton, grape, willow, and carob in Peru,

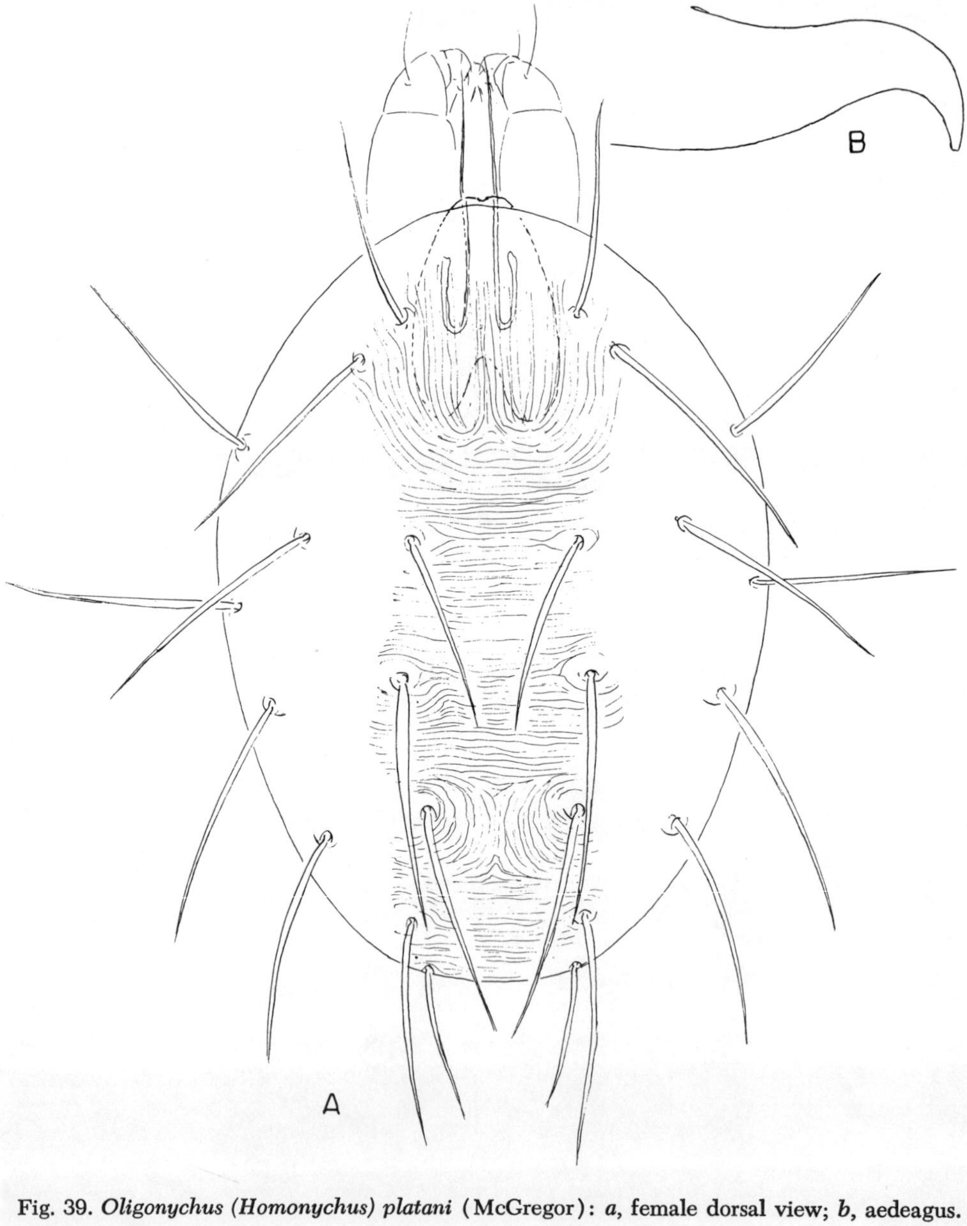

Fig. 39. *Oligonychus (Homonychus) platani* (McGregor): *a*, female dorsal view; *b*, aedeagus.

but it is not found on cotton in California and Trinidad. It has also been reported on carrot, avocado, and clover. These mites feed in restricted colonies on the undersides of the leaves of the host plants (Baker and Pritchard, 1953; Estebanes Baker, 1968).

Tibia I of *O.* (*H.*) *peruvianus* has 9 tactile setae; tarsus I has a single anterior ventral seta below the duplex setae and 4 tactile setae proximal to the duplex setae. Tibia II usually has 7, seldom 6 tactile setae. The dorsal body setae are short, nearly nude, and strongly lanceolate, broadest near the base. The aedeagus is downbent, the distal portion is short.

Oligonychus (Homonychus) platani (McGregor). *Platani* (fig. 39, *a, b*) is a serious pest in the hot interior valleys of California where it infests sycamore, London plane tree, *Toyon*, broad leaf evergreen, and oak. In Arizona it is a serious pest on pyracantha, and causes bronzing and unsightly heavy webbing on both leaf surfaces. It has also been collected in Texas and in Mexico on avocado and oak.

The eggs of this species are flattened at the base and domelike. They are deposited singly on the upper leaf surface along the midrib, but may be found over the entire leaf surface during heavy infestations. The eggs are pale when laid, changing to orange owing to the appearance of the larval eyes. The eggs are radially striate dorsally with a stipe. The color of the adult female body appears greenish or brownish with pronounced black spots, but the legs are semitransparent with brown tarsi.

The average time required for mites to develop from egg to adult in the field was 12.8, 12.1, and 9.8 days during June, July, and August, respectively. Under constant temperature conditions of 15, 24, and 34 C (59, 76, and 93 F), the life cycle required 32, 10.2, and 9.6 days respectively (Butler and Abid, 1965).

Tibia I of *O.* (*H.*) *platani* has 7 tactile setae; tarsus I has 3 tactile setae proximal to the duplex setae, and 1 ventral seta below the duplex setae. The proximal members of the duplex setae are nearly as long as the other members. The aedeagus is downbent, the distal portion is relatively short, and gradually tapers to the tip.

Oligonychus (*Metatetranychoides*) Wainstein

These mites are similar to *Oligonychus* s. str. but differ in having an irregular striation pattern between the hysterosomal setae D_3. The aedeagus is downturned. The dorsal body setae are set on tubercles. These mites feed on the lower, rather than the upper leaf surfaces.

Oligonychus (Metatetranychoides) aceris (Shimer). *Aceris* (fig. 40, *a, b*) is commonly found on maple throughout the eastern United States. It is also known to occur in Kansas and Washington. The mites occur primarily on the lower surfaces of the leaf, except during heavy infestations when they may move to the upper surfaces and cause the leaves to become yellow (Reeves, 1963).

Tibia I of *O.* (*M.*) *aceris* has 6 tactile setae; the solenidion or sensory seta of tibia I is more than ½ as long as the dorsal tactile seta. The body setae are long, slender, and set on tubercles. The striae of the hysterosoma are transverse except

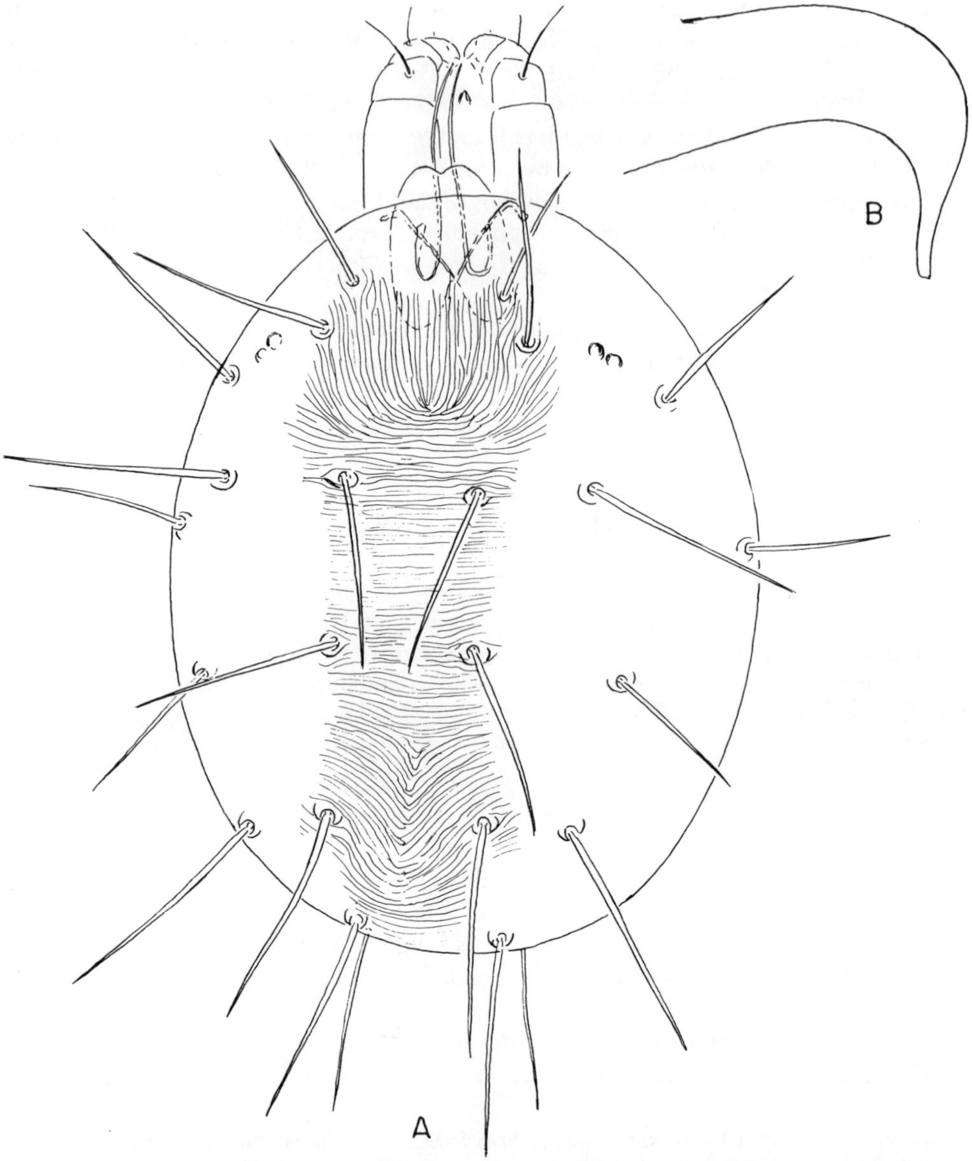

Fig. 40. *Oligonychus* (*Metatetranychoides*) *aceris* (Shimer): *a*, adult female showing striations and setae; *b*, aedeagus.

for an irregular V-like area between setae D_2 and D_3. The aedeagus turns vertically at right angles to the shaft and tapers to a longish slightly sigmoid tip.

Oligonychus (*Metatetranychoides*) *endytus* Pritchard and Baker. This mite (fig. 41, *a, b*) attacks oaks and is a pest of chestnuts in California. Tibia I on this species has 6 tactile setae; tarsus I has 3 tactile setae proximal to the duplex setae; the proximal members of the duplex setae are short; there is a single seta ventral

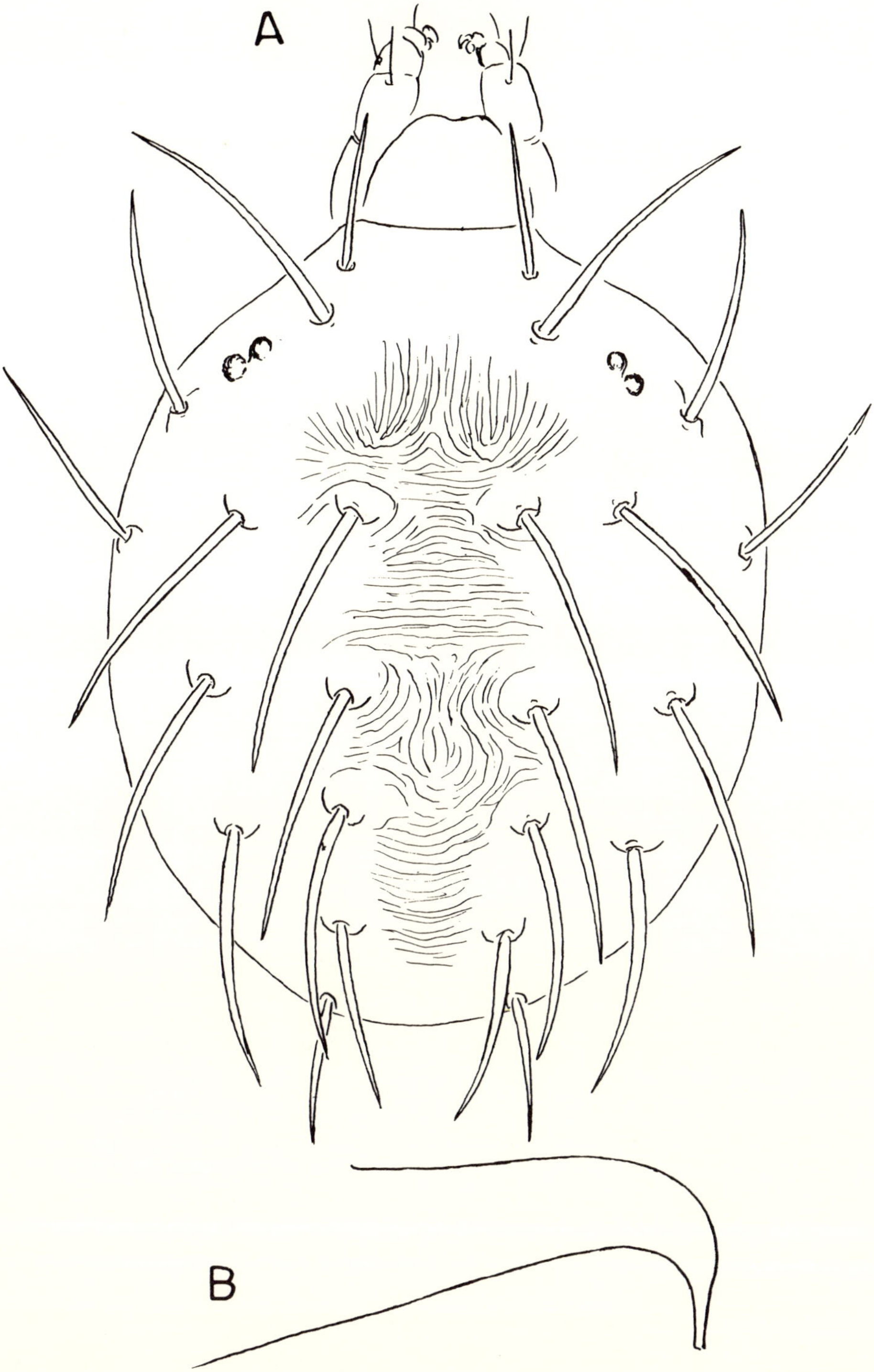

Fig. 41. *Oligonychus* (*Metatetranychoides*) *endytus* Pritchard and Baker; *a*, adult female showing striations and setae; *b*, aedeagus.

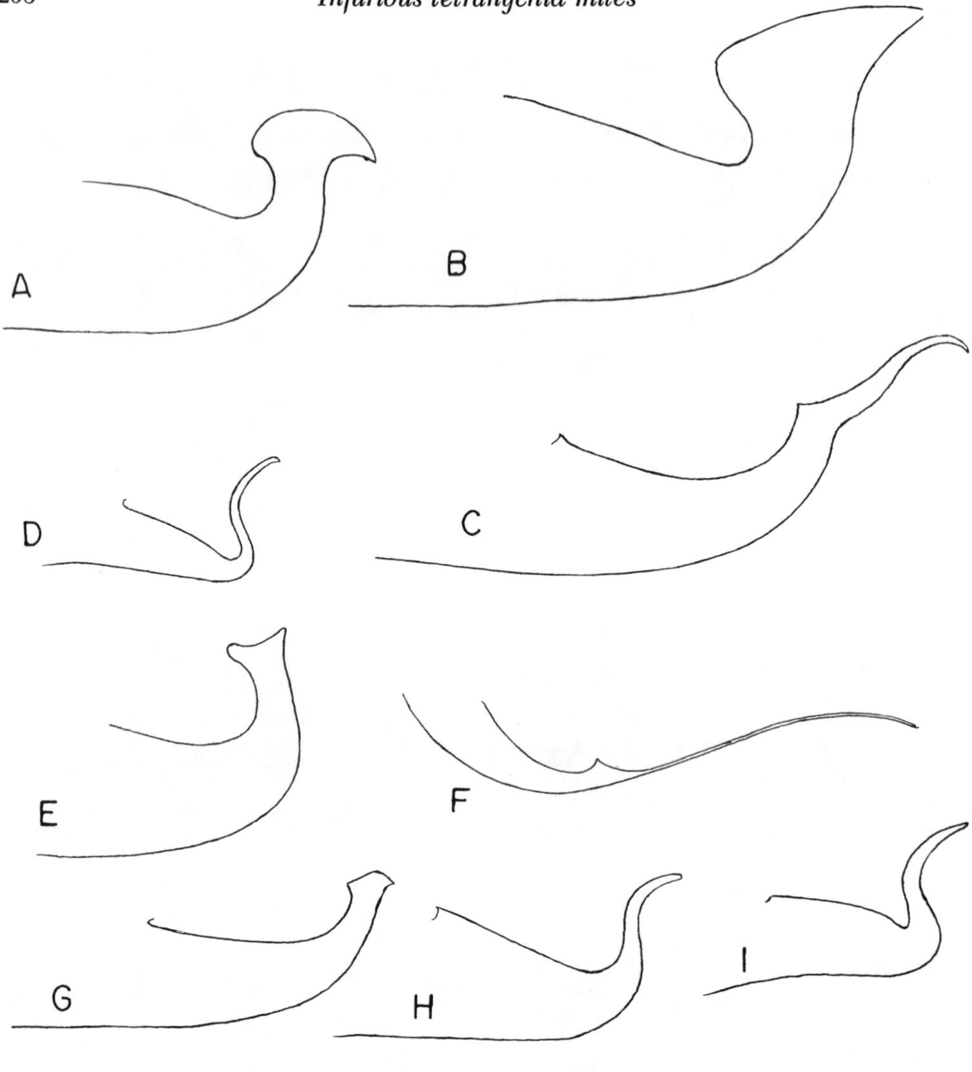

Fig. 42. *Oligonychus* (*Reckiella*); aedeagi: *a, O.* (*R.*) *afrasiaticus* (McGregor); *b, O.* (*R.*) *exsiccator* (Zehntner); *c, O.* (*R.*) *gossypii* (Zacher); *d, O.* (*R.*) *grypus* Baker and Pritchard; *e, O.* (*R.*) *indicus* (Hirst); *f, O.* (*R.*) *mcgregori* Baker and Pritchard; *g, O.* (*R.*) *modestus* (Banks); *h, O.* (*R.*) *orthius* Rimando *i, O.* (*R.*) *oryzae* (Hirst).

to the duplex setae. The hysterosomal striae are transverse except for a strongly irregular pattern, almost longitudinal, between setae D_2 and D_3. The dorsal body setae are long, strong, and set on prominent tubercles. The aedeagus is downturned, and the distal third narrows gradually to the tip.

Oligonychus (*Reckiella*) Tuttle and Baker

Tibia I has 9 tactile and tibia II 7 tactile setae. There are 4 tactile setae proximal to the duplex setae on tarsus I; there are 2 setae ventral to the duplex setae on tarsus I. The peritreme ends in a simple bulb. The striae on the hysterosoma of the female are transverse except for a longitudinal pattern between setae D_4. The aedeagus is upturned and usually has an anterior and posterior angulation.

The females are all similar and the aedeagus of the male is necessary for specific determination. Tarsus I of the male has 4 tactile and 3 sensory setae proximal to the duplex setae.

Mites of subgenus *Reckiella* feed on grasses and other monocotyledonous plants.

Oligonychus (Reckiella) afrasiaticus (McGregor). This species (fig. 42, *a*) is sometimes a serious pest of dates in North Africa, the Middle East, Algeria, Iraq, and Iran. Besides the injury produced by feeding, these mites form large amounts of webbing that covers the dates. Dust is frequently entangled in the webbing, and provides sufficient shade to prevent the fruit beneath from coloring. A fig wasp, *Polistes olivaceus* (DeGeer) appears to aid dispersal. Orchards near unkept areas are most severely injured. In Mesopotamia date-growing areas, infestations appear to be more serious near bodies of water, probably because of the increased humidity (Buxton, 1920–1921).

The female of *O.* (*R.*) *afrasiaticus* is typical for the subgenus in having longitudinal striae between the fourth pair of dorsocentral setae of the hysterosoma. The male is distinctive in that the terminal knob of the aedeagus is moderately large and about ⅕ as long as the dorsal portion of the shaft; the axis of the knob is parallel to that of the shaft, the anterior projection of the knob is broadly rounded, and the posterior angulation is deflexed at the tip and about as long as the anterior angulation.

Oligonychus (*Reckiella*) *araneum* Davis and *O.* (*R.*) *digitatus* Davis. These mites are recorded as pests of grasses in eastern Australia. The mites occur in very large populations, yellowing the grass in a ring-shaped area as the infestation spreads outward; they also spin conspicuous webbing. The two species occur together and females are indistinguishable from one another or from other species of the *Reckiella* group, but males are identified by the shape of the aedeagus (Davis, 1969).

Oligonychus (Reckiella) exsiccator (Zehntner). *Exsiccator* (fig. 42, *b*) is reported to cause injury to sugar cane in Java; little, however, is known concerning its biology. Mites causing injury to sugar cane in Hawaii reported as *O.* (*R.*) *exsiccator* appear to have been misidentified [see *O.* (*R.*) *pratensis* (Banks)].

The female of *O.* (*R.*) *exsiccator* is similar to the others in the subgenus; the mite is characterized by the longitudinal striae between the fourth pair of dorsocentral setae. The male is distinctive in that the aedeagus is abruptly upturned posteriorly and has a very large knob about half as long as the dorsal margin of the shaft; the neck is short and stout; the anterior projection is broadly rounded and the posterior projection is acute; the axis of the knob is at a slight angle to that of the shaft.

Oligonychus (Reckiella) gossypii (Zacher). This species (fig. 42, *c*) is known only from Central America, Brazil, and equatorial Africa where it is widespread and common. It was first recognized from Togo, West Africa, as a pest of cotton. It has also been found on cassava, beans, rose, citrus, and peach (Baker and Pritchard, 1960).

The female is typical for the subgenus. In the male the aedeagus is upturned, with an anterior angulation and an elongate sigmoid posterior projection. The size of the aedeagus may vary between African and Central American specimens.

Oligonychus (Reckiella) grypus Baker and Pritchard. *O. (R.) grypus* (fig. 42, *d*) is a pest of sugar cane and grass in Zaire and South America (Baker and Pritchard, 1960).

The female of this species is typical for the subgenus. The aedeagus of the male is distinctive in not having an anterior angulation; the shaft narrows distally, the dorsally bent portion is as long as the shaft and tapers to the tip; there is an obtuse angulation near the middle, and the distal end is directed dorsocaudad.

The sugar cane or cholam mite, *Oligonychus (Reckiella) indicus* (Hirst). This mite (fig. 42, *e*) is probably well distributed worldwide. In India it is a sporadic pest of bananas, sugar cane, sorghum, and *Panicum* spp. Damage to sugar cane and sorghum is first observed as red spots on leaves where the mites have fed, but as attacks increase the red spots spread in size and coalesce, forming large red patches. Finally the leaves become red, necrotic, and dry.

Eggs of the sugar cane mite are dull white when freshly laid, becoming brownish as development proceeds. Just before hatching the red eyes of the larvae become visible and transparent areas appear along the sides of the encased mite. Females lay from 1 to 7 eggs daily and an average of 53 eggs during their lifetime. Development from egg to adult is closely correlated with the season, ranging from 4 to 7 days during the warm season and 26 to 53 days during the colder months. Preoviposition lasts from 1 to 6 days. The life cycle may be completed during the summer in 9 to 12 days. Generations commonly overlap and all stages of the mite are present throughout the year. Under controlled conditions there may be more than 30 generations per year.

Populations of these mites develop on Johnson grass ("buru") through the noncrop season of January and February; they reach a peak in April, at which time they disperse to the developing sugar cane or sorghum. Populations increase on these cultivated plants until the onset of the monsoon rains in July, which kill all stages except the eggs. After the monsoons, populations increase on sugar cane and sorghum sufficiently to cause severe damage to these crops. Populations also increase on Johnson grass where they survive and serve as the source of infestation for the following year's crops.

The major means of mite dispersal is by cane transfers and in the field by air currents. Observers have reported movements of over 200 feet, but it is believed they are carried much longer distances. Infestations spread in sugar cane and sorghum fields by migrations from plant to plant, especially where leaves touch each other.

Sugar cane varieties with soft leaves appear to be more susceptible to attack than those with harder leaves. Nitrogen fertilizer applications have been shown to induce heavy infestations in Punjab, India.

Infestations of the sugar cane mite on sugar cane may be avoided by destroying the wild host plants, such as *Panicum* L. and *Paspalum distichum* L. from the edges of fields. Canes to be transported from infested areas should be soaked in water at 46 C (115 F) for ½ hour. Infested fields may be treated with sulfur dusts, sulfur or lime-sulfur sprays, or recommended specific acaricides (Cherian, 1933; Harbans and Sudhu, 1961; Rahman and Sapra, 1940).

The female of *O.* (*R.*) *indicus* is typical for the subgenus. The aedeagus is strongly upcurved, with an anterior and posterior angulation of about equal size; the knob is concave dorsally.

Oligonychus (*Reckiella*) *mcgregori* Baker and Pritchard. This mite (fig. 42, *f*) attacks cotton, *Ficus*, and other plants in Central America and Mexico. The female is typical for the subgenus. The aedeagus of the male is quite distinctive; the proximal portion of the shaft is curved dorsally and strongly narrows distally, and there is a small dorsad angulation before the midpoint; the entire shaft is slightly sigmoid.

Oligonychus (Reckiella) modestus (Banks). This species (fig. 42, *g*) occurs on corn and bamboo in Arizona and Washington, D.C. The female of this species is typical for the subgenus. The aedeagus of the male is distinctive in that the distal end is scarcely enlarged, the anterior angulation is barely evident, and the dorsal surface of the tiny knob is curved and at a slight angle to the axis of the shaft.

Oligonychus (Reckiella) orthius Rimando. *Orthius* (fig. 42, *h*) has been reported as a pest of sugar cane and *Imperata* in the Philippines and on Okinawa Island. This species is similar to *O. (R.) zeae* (McGregor). The female is typical for the subgenus. The male is distinctive in that the distal end of the aedeagus is obtusely curved rather than sharply curved caudally; the terminal shaft of the aedeagus is not as long as that of *O.* (*R.*) *zeae* (Rimando, 1962).

Oligonychus (Reckiella) oryzae (Hirst). *Oryzae* (fig. 42, *i*) may reduce rice yields in India up to 25 percent. These mites live in colonies under webbing on the undersides of leaves of both nursery and transplant crops. Infested seedlings become pale, stunted in growth, and invariably die soon after transplanting. First evidence of injury on transplant crops appears as white patches on the upper surface of the leaves. Infested plants develop poorly, but are seldom killed. High populations usually occur from July through January.

The active stages are susceptible to sulfur, the organophosphorus compounds, and the specific acaricides (Nagarajan, 1957).

The female of *O.* (*R.*) *oryzae* is typical for the subgenus. The male is distinctive from the other species by the shape of the aedeagus in which the slender, sigmoid distal tip is broadly angled dorsally.

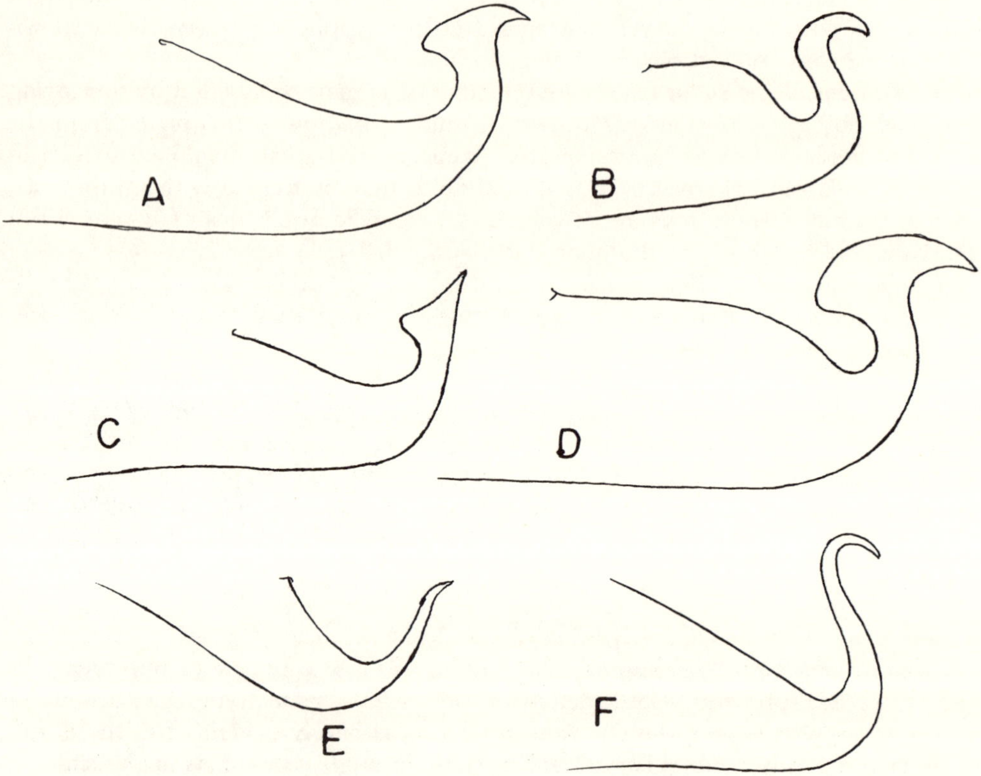

Fig. 43. *Oligonychus* (*Reckiella*); aedeagi: *a*, *O.* (*R.*) *pratensis* (Banks); *b*, *O.* (*R.*) *sacchari* (McGregor); *c*, *O.* (*R*). *simus* Baker and Pritchard; *d*, *O.* (*R.*) *stickneyi* (McGregor); *e*, *O.* (*R.*) *plegas* Baker and Pritchard; *f*, *O.* (*R.*) *zeae* (McGregor).

Oligonychus (Reckiella) plegas Baker and Pritchard. This species (fig. 43, *e*) is found on coconut leaves, on fataque or Guinea-Grass, on maize in the field, and on sugar cane grown in greenhouses in Mauritius. Small colonies are usually found on the lower surface of coconut leaves along the midrib. Severe feeding produces long yellowish patches in the area of each colony. Infestations are sporadic, probably because a cecidomyiid fly and a *Typhlodromus* mite aid in keeping populations in check (Moutia, 1958).

The female of *O.* (*R.*) *plegas* is typical for the subgenus. The male is distinctive in that the shaft of the aedeagus narrows distally, the dorsally directed portion is abruptly bent, tapers and is slightly sigmoid distally and about as long as the dorsal margin of the shaft.

Banks grass mite, *Oligonychus (Reckiella) pratensis* (Banks). This mite (figs. 43, *a*; 44), originally reported in Hawaii as *O. exsiccator* (Zehnter), has been called the timothy mite in the northwestern United States and the date mite in southern California. It is a serious pest of many grass species in the southern and western United States, Puerto Rico, Central America, Mexico, Hawaii, and Africa.

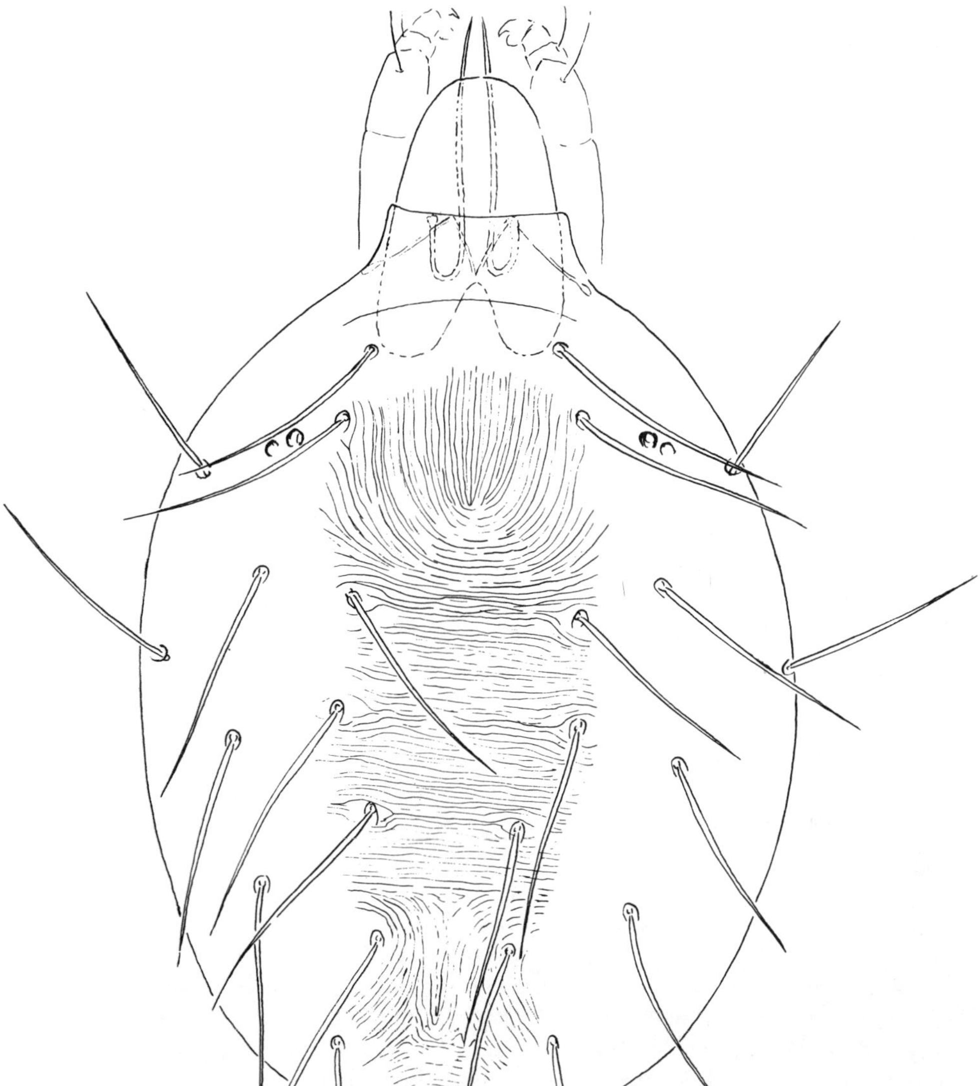

Fig. 44. *Oligonychus* (*Reckiella*) *pratensis* (Banks), female, dorsal view.

Economic hosts include wheat, corn, sugar cane, sorghum, maize, dates, bluegrass, and Bermuda grass. This species causes serious damage to wheat and corn in western Kansas, to bluegrass seed fields in dry areas in Washington State and Oregon; to maize in the Sacramento and San Joaquin valleys of California, and to dates in the southern desert areas of California.

In New Mexico, for example, mite colonies may be present when the fall wheat is in the 2- to 5-leaf stage. The leaves first turn yellow and, as infestations increase, the foliage where mites have fed becomes necrotic and brownish yellow. In the fall the mites move to the crown near ground level where they feed throughout the winter. Young leaves become covered with webbing, discolored, curled inward, and die. Occasionally heads become infested as they emerge from the boot. When populations continue to increase kernels may shrink or the plant may even fail to develop.

Feeding by the Banks grass mite on sugar cane produces tiny spots on the undersides of leaves. Numerous spots in proximity to each other appear as yellow streaks that redden with age. Eventually the edges and tips of the leaves become dry. Small red blisters are produced on the surface of young internodes still in the sheath canes, finally resulting in a reddish brown corroded appearance of the surface of the cane, especially above the eye. Often one or more internodes are so heavily infested as to appear rusty red. Infestations cause premature drying of foliage, increased stalk breakage, a reduction in the moisture content of the grain at harvest, kernel shrinkage, and a subsequent reduction in grain yields.

Banks grass mite injury to sorghum appears as discoloration, and premature death of the leaves. Mite infestations develop on the underside of the lower leaves and progress to the upper leaves and head. Heaviest populations occur along the midrib beneath the webbing. Symptomatology appears in the following sequence: (1) small, white stipuled spots occur on the leaf along the midrib; (2) the spots increase in size, especially on the basal half of the leaf, followed by red or brown discoloration of the spots; (3) leaves fold downward along the midrib nearest the base of the leaf; (4) symptoms proceed from leaf to leaf up the plant. Damage to heads consists of shriveled seeds and extensive webbing. There is some evidence that the mites weaken the plant to the point that disease organisms partially overcome the natural resistance of many plants (Ward et al., 1972).

Banks grass mite feeding on dates produces scar tissue on date skin, causing it to harden, crack, and shrivel with subsequent reduction in the grade of the fruit. A heavy deposit of fine webbing collects dust, making even moderate infestations easily visible. Populations on dates begin to increase in June and peak in July and August; all stages of development are found throughout the year, although activity and numbers generally decrease during the winter. Mites live during the cooler winter months on late-maturing or off-season fruit, on the date palm foliage, or on grasses, particularly Bermuda grass (Elmer, 1965, *a*).

In Mauritius this species feeds along the midrib on the lower surface of coconut leaves. It attacks jointly with *Raoiella indica* Hirst causing yellowish patches at the points of attack. During the dry months mites may be found on sugar cane growing in greenhouses and in the field on fataque and on maize.

Certain areas in the northwestern United States have warm periods during the winter. These warm days provide favorable conditions for mite activity, and mite feeding on grasses or fall grains often produces considerable injury. Feeding injury appears on the leaves as white stippled areas followed by chlorosis and eventual weakening of the plants. Severe infestations usually result in high percentage of plant mortality. The mites need not be present in large numbers to reduce seed production greatly. Serious damage to grasses by this mite normally occurs in areas of meager moisture or rainfall; thus susceptible lawn grasses sprinkled frequently are usually not attacked.

The egg of the Banks grass mite is relatively small, pearly white when first laid, but it soon becomes "glassy" and finally light straw colored. A short time before hatching the carmine eyespots of the embryonic larva are plainly visible. The empty chorion is clear and brilliantly iridescent. The larva is nearly whitish and transparent to light salmon at first, then becomes light and later dark green. The nymphal stages are pale green or bright green depending on their food. Upon feeding, the adults become deep green, except for the palpi and first pair of legs which remain light salmon. Reflections of light from striations on the mite often imparts on iridescent hue. Overwintering forms are bright orange salmon.

The average length of the life cycle varies from 8 to 25 days depending on the temperature and the life span of adults averages 23 days. There may be 6 or 7 generations per year in northern United States and more in southern areas.

Mites of this species spin copious amounts of webbing. During heavy infestations the lower portions of the grass become matted with these webs. This matting provides a protected place for eggs, which are laid either on the webbing or on leaf surfaces beneath it.

In Washington state the active stages of this mite overwinter near the base of host plants. The adult females change from their customary green to yellow and finally to the orange color typical of overwintering forms. No eggs are deposited by these overwintering females even though weather remains warm. Injury to plants may be observed early in March, indicating that activity starts early in the spring. Periods of drought are apparently passed in the immature stages in the dry foliage near the base of the plants.

The rapid growth of grasses in the spring reduces the amount of damage as well as the effectiveness of acaricides. Late summer high temperatures favor feeding and reproduction, resulting in serious damage. Therefore, control measures are most advantageous during this period (Malcolm, 1955).

Sulfur and parathion appear to be the most toxic of the present acaricides to the Banks grass mite. Combinations of sulfur and parathion dusts may produce effective control; adequate coverage, however, is difficult to achieve because of the copious webbing near the base of the plants. Sulfur dusts are effective in controlling this mite on dates, but several applications are usually necessary as mites near the center of the date bunches are difficult to contact by the applications (Depew, 1960).

The female of *O.* (*R.*) *pratensis* is similar to other members of the subgenus. The male is distinctive in that the distal knob of the aedeagus is about twice as wide as the stem of the knob; the axis of the knob is at a distinct angle to the axis of the shaft; the dorsal margin of the knob may be nearly straight with the tip slightly turned down, or curved, or angulate; the anterior projection of the knob is rounded and the posterior angulation acute.

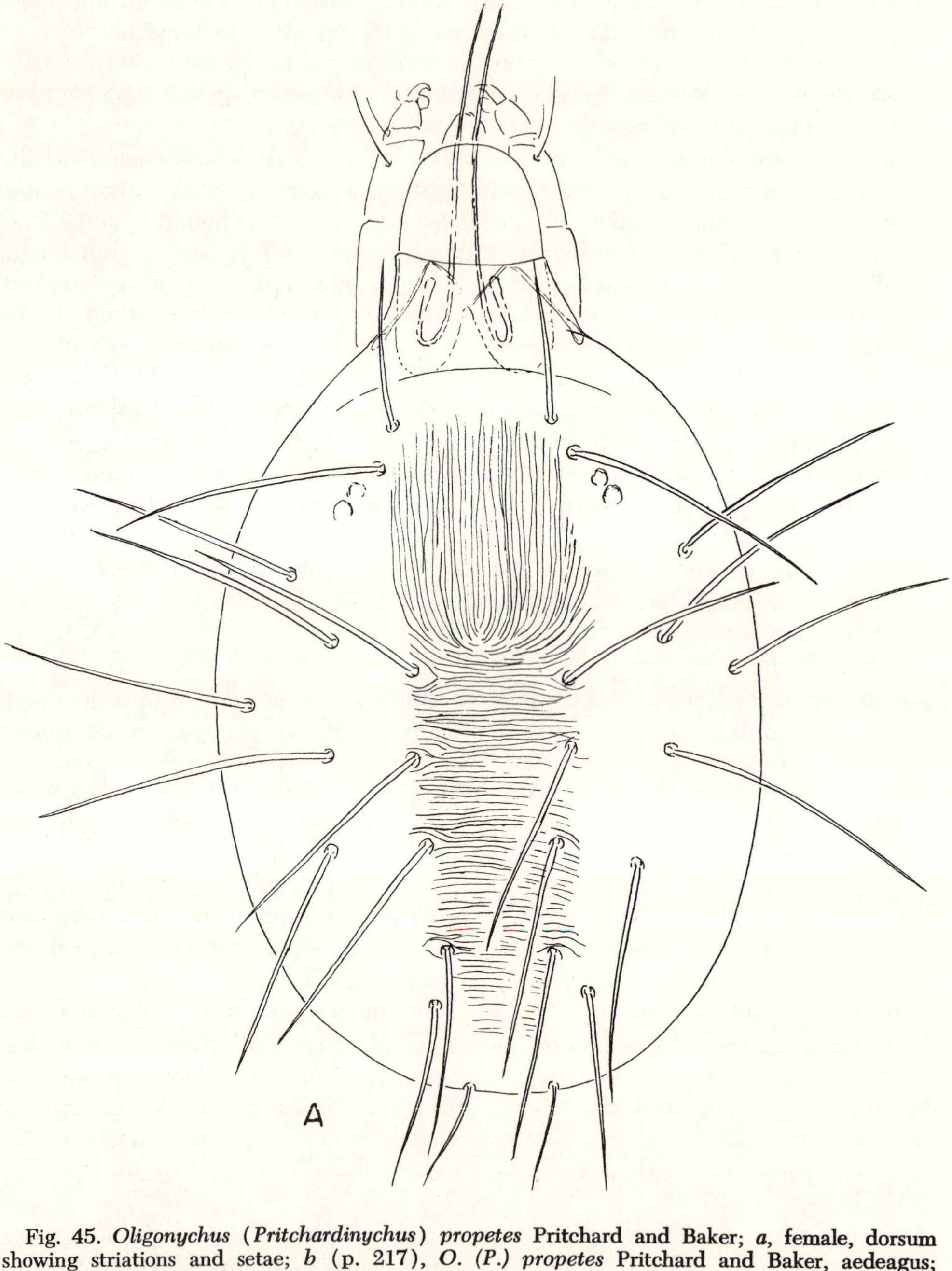

Fig. 45. *Oligonychus* (*Pritchardinychus*) *propetes* Pritchard and Baker; *a*, female, dorsum showing striations and setae; *b* (p. 217), *O. (P.) propetes* Pritchard and Baker, aedeagus; *c* (p. 217), *O. (P.) biharensis* (Hirst), aedeagus.

Oligonychus (Reckiella) sacchari (McGregor). This species (fig. 43, *b*) occurs on sugar cane in Puerto Rico and on dendrobium orchids and foxtail millet in the New Hebrides.

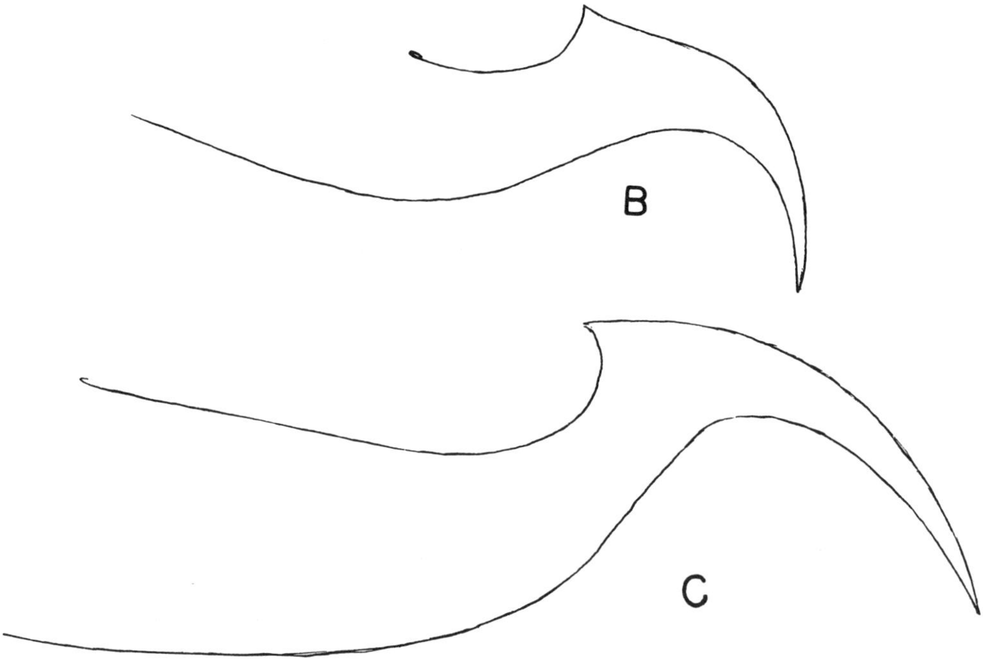

The female of this species is similar to the others in the subgenus. The male is distinctive in that the distal end of the aedeagus is bent dorsad, is tapering and strongly sigmoid; the distal tip is turned ventrally.

Oligonychus (*Reckiella*) *simus* Baker and Pritchard. This mite (fig. 43, *c*) occurs on sorghum in Nyasaland. The female of this species is similar to the others in the subgenus. The male is distinctive in that the shaft of the aedeagus is very broad at the base and evenly narrows to the dorsally directed bend; the distal end is slightly widened, and has an obtuse angulation anteriorly and a dorsally directed angulation posteriorly; the dorsal portion of the knob is slightly concave.

Oligonychus (Reckiella) stickneyi (McGregor). *Stickneyi* (fig. 43, *d*) causes injury to corn, rye, maize, and sorghum in California, Arizona, Mexico, and Florida. The female of this species is similar to the others in the subgenus. The male is distinctive in that the terminal enlargement or knob of the aedeagus is very large, about ⅓ as long as the dorsal portion of the shaft; the anterior angulation of the knob is broadly rounded and the posterior end is angulate; the axis of the knob forms about a 30° angle with the axis of the shaft.

Oligonychus (Reckiella) zeae (McGregor) (fig. 43, *f*) lives on corn, banana, and other hosts in Ecuador, Honduras, and Mexico (Estebanes and Baker, 1968).
The female is typical for the subgenus. The male is distinctive in that the aedeagus is slender distally, bending upward at slightly more than a right angle, and is recurved distally to form a "goose neck"; the base of the shaft is broad.

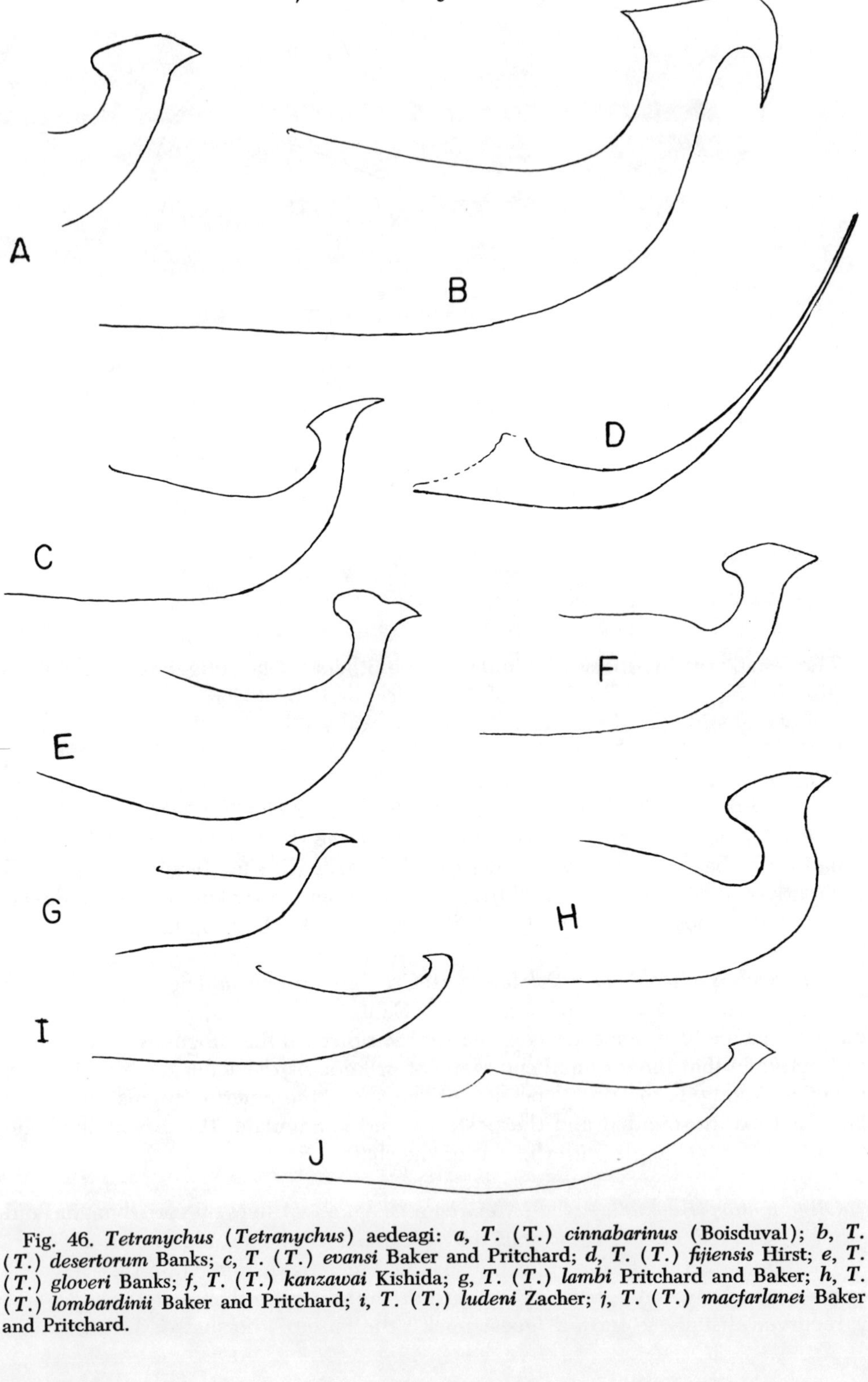

Fig. 46. *Tetranychus* (*Tetranychus*) aedeagi: *a*, *T.* (T.) *cinnabarinus* (Boisduval); *b*, *T.* (*T.*) *desertorum* Banks; *c*, *T.* (*T.*) *evansi* Baker and Pritchard; *d*, *T.* (*T.*) *fijiensis* Hirst; *e*, *T.* (*T.*) *gloveri* Banks; *f*, *T.* (*T.*) *kanzawai* Kishida; *g*, *T.* (*T.*) *lambi* Pritchard and Baker; *h*, *T.* (*T.*) *lombardinii* Baker and Pritchard; *i*, *T.* (*T.*) *ludeni* Zacher; *j*, *T.* (*T.*) *macfarlanei* Baker and Pritchard.

Oligonychus (*Pritchardinychus*) Wainstein

The dorsal striae of the hysterosoma are transverse; the aedeagus of the male is upturned and then ventrally directed, with both anterior and posterior angulations. There are 9 setae on tibia I. The members of this subgenus are green and feed on the ventral leaf surface, except for *O.* (*P.*) *biharensis.*

Oligonychus (Pritchardinychus) propetes Pritchard and Baker. *Propetes* (fig. 45, *a, b*) occurs on oak in Arizona and is a pest of deciduous oaks in the northeastern United States. It prefers the underside of the leaves where feeding injury to white oak appears on the dorsal leaf surface as yellow oblong or barlike spots along the veins. This species can easily be mistaken in the field for *Eotetranychus uncatus* Garman, since both infest the undersides of oak leaves and both are yellowish-green. This species, however, has several black spots dorsally and produces less webbing than *E. uncatus.*

The peritremes of *O.* (*P.*) *propetes* are straight distally. The proximal members of the duplex setae are very short. The aedeagus has a strong knob that is convex and has an acuminate tip directed ventrally.

Oligonychus (Pritchardinychus) biharensis (Hirst). This species (fig. 45, *c*) is injurious to mango in Mauritius. Its hosts also include rose, loquat, litchi, cotoneaster, camphor, and *Euphorbia longana* Steudel, but infestations are sporadic. This mite is also known in India, Hawaii, Thailand, Malaya, Philippines, Antigua, Brazil, and Mexico. Its feeding causes numerous white spots on upper leaf surfaces of mango leaves and a characteristic dark bronzing of loquat leaves. Natural enemies of this pest in Mauritius include *Stethorus vinsoni* Kapur and a cecidomyiid fly (Moutia, 1958).

O. (*P.*) *biharensis* keys out to the subgenus *Pritchardinychus*, but is not typical; it has distally hooked peritremes, is much larger than other mites in the subgenus, with longer legs, and different feeding habits. It is tentatively placed here. The peritreme is hooked distally in both sexes. The aedeagus of the male is long and slender; the axis of the distal enlargement is parallel to the shaft, but the dorsal margin is convex and the tip bends ventrally.

Tetranychus Dufour

These mites usually feed on the undersurface of leaves of broadleaf plants. Many produce profuse webbing. The northern species are yellowish to green and the more southern species are reddish. There is a single pair of para-anal setae. The empodia of the females are split distally into 3 pairs of ventrally directed hairs, except for *T. fijiensis* Hirst, which has only 2 pairs; the empodia of tarsi I are widely separated on the long, tapering tarsus. The dorsal body setae are long and slender. The aedeagus bends dorsad and is of specific importance. The subgenera are separated by the striation pattern on the dorsum of the hysterosoma.

Tetranychus s. str.

The dorsal hysterosomal striations form a diamond-shaped pattern between setae D_3 and D_4. Most of the species belong to this subgenus and the females are

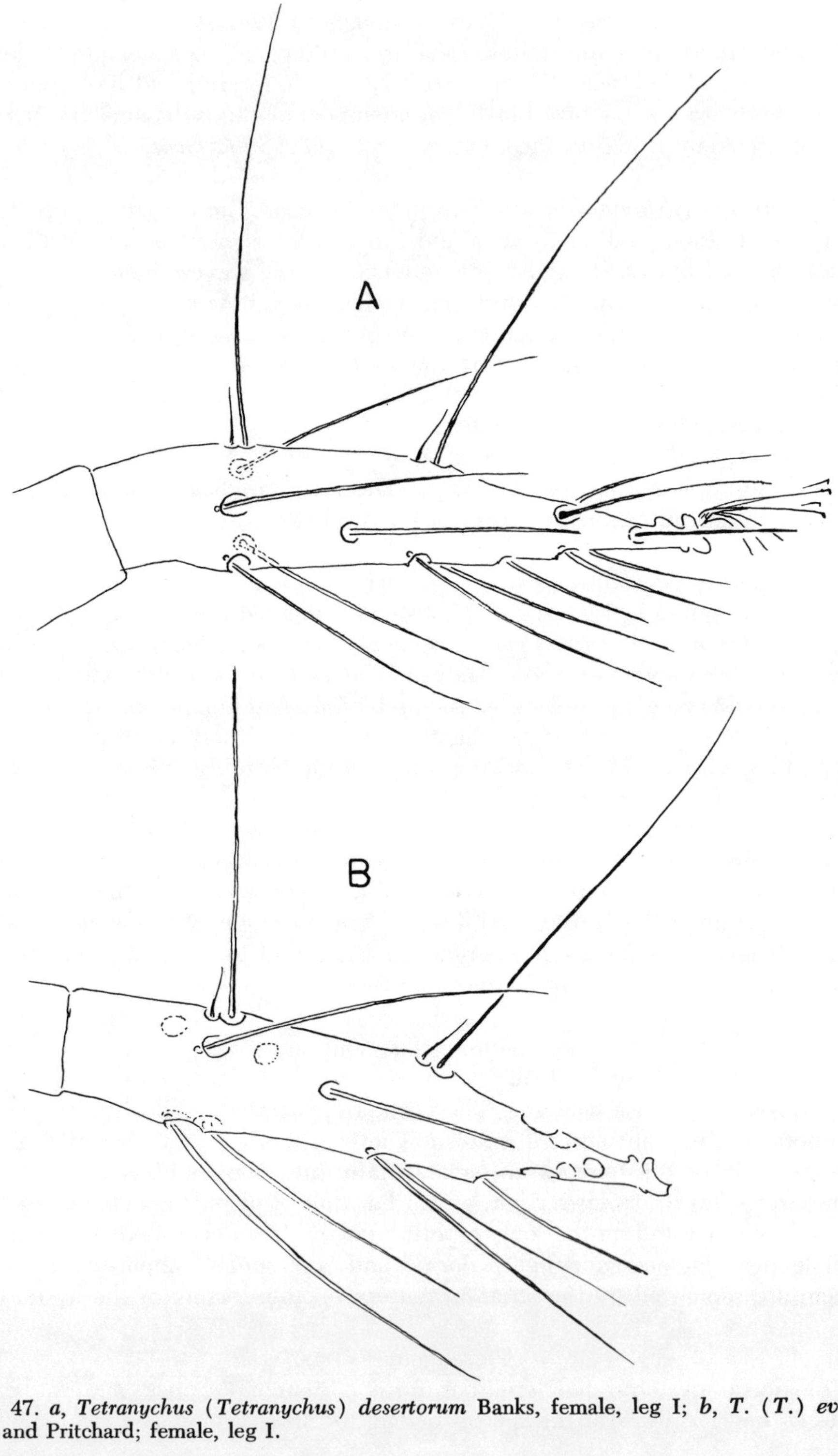

Fig. 47. *a*, *Tetranychus* (*Tetranychus*) *desertorum* Banks, female, leg I; *b*, *T.* (*T.*) *evansi* Baker and Pritchard; female, leg I.

difficult to separate. Consequently, males mounted in profile so as to study the shape of the aedeagus are needed.

Carmine spider mite, *Tetranychus (Tetranychus) cinnabarinus* (Boisduval). This mite (fig. 46, *a*) has been made a separate species from the two-spotted spider mite, *T.* (*T.*) *urticae* Koch, because of differences in morphology, habits, host preferences, geographic distribution, and crossbreeding results (Smith and Baker, 1968).

This mite is primarily a pest of low-growing plants in semitropical areas of the world. It is a major pest of cotton and seems to thrive in the climatic areas most suitable for cotton production. It does not appear to be a pest outside the greenhouse in Continental climates. This species remains on its host plants and does not enter into diapause during the winter. Throughout the winter, warm weather periods may stimulate activity, feeding, and egg production. Attempts to crossbreed mites of this species with the two-spotted spider mite result only in male offspring indicating that fertilization does not occur. Although the immature stages of these two species appear similar, summer populations of the carmine spider mite appear more brick red or ferruginous red than those of two-spotted spider mite, but color varies with the host.

T. cinnabarinus feeding on oranges in Lebanon first results in a blackish area around the navel end of the fruit, but as populations increase the whole fruit becomes a dirty grey. Populations on foliage commonly develop in colonies on the undersides of young citrus leaves. The leaves buckle at the colony site; the upper leaf surface becomes raised and yellow. When mites feed on fruit, they attain a bright red color, but those living on leaves have the characteristic black spots (Dosse, 1964).

The eggs of the carmine spider mite are deposited singly, directly on the undersurfaces of the leaves of cotton and other plants, or attached to the fibrils of webbing. The duration of the life stages and the fecundity of the females compare with the two-spotted spider mite. Because the carmine spider mite reproduces throughout the year, there may be nearly 20 broods per year in the field. The optimum temperature for development is about 32 C (89 F), but mites are able to reproduce at temperatures well above 35 C (95 F) if exposed for limited periods of time (Düzgünes, 1965; McGregor and McDonough, 1917).

Conditions of extremely high relative humidity cause all ages of this species to go into an extended period of quiescence, which may last up to 10 days beyond the normal period required for development; but any time after the normal quiescent period, mites exposed to dry environment molt within a few minutes. Acaricides used for control of the two-spotted spider mite are effective against this species.

The body of the female of *T.* (*T.*) *cinnabarinus* is carmine in color and contains dark lateral internal markings. The lobes of the dorsal striae of the female are taller than they are broad. The tactile setae on tarsus I of the female are well proximad to the posterior set of duplex setae. The aedeagus of the male has a small knob or head; the angulations are similar to one another or (usually) the anterior angulation is rounded and the posterior angulation sharp.

Desert spider mite, *Tetranychus (Tetranychus) desertorum* Banks. This mite (figs. 46, *b*; 47, *a*) is a serious pest of cotton in the southern United States. It is also a pest in Argentina, Brazil, Nicaragua, Paraguay, Peru, Australia, Japan, and Mexico. Its favorite hosts are clover, cotton, and evening primrose, but it may be a serious pest of alfalfa, barley, bean, brome grass, corn, dill, onion, melon, eggplant, celery, carrot, turnip, gladiolus, and papaya. The injury is similar to that produced by the carmine spider mite *T.* (*T.*) *cinnabarinus* (Boisduval) (Nickel, 1958).

The developmental time of this mite from egg to adult is relatively short, that is, 5.8 to 11.2 days, averaging 8.3 days. The average duration of the egg stage throughout the year is 3.3 days, being 2 days in summer and 4 to 5 days during the winter. Each developmental stage requires 1.0 to 1.5 days in summer and 1.6 to 3.0 days in February and March.

The mites occur during the winter months on native host plants such as evening primrose, sow thistle, horehound, and *Verbena bipinnatifida* Nuttall in Texas, and *Verbena lenuisecta* Briquet in Paraguay. Plants protected by a cover of grass are most attractive as winter host plants. By mid-March mites are abundant in localized patches on bur clover, and by the last of April they may be found most abundantly on bloodweed and seedling cotton. Maximum reproduction on cotton occurs in July. Summer infestations occur on horsemint, tievine, and Johnson grass; cocklebur becomes the principal host in October and from cocklebur the mites move to their winter hosts in November and December. Spring rains adversely affect populations of this mite on cotton, often causing infestations to disappear.

Mites are dispersed from one host to another by crawling and by winds. The mites congregate at the tops of host plants from which they may be carried by winds to new hosts. That they are airborne is indicated by tanglefoot trap studies.

Mites of this species are not able to survive at temperatures below 10 C (50 F). Studies on development of this mite have been made at temperatures of 17, 25, 30, and 33.5 C (62, 77, 86, and 92.5 F) at 25 to 30 percent and 85 to 90 percent relative humidities (RH). Optimum conditions for most rapid population increase were found to be 30 C (86 F) and 85 to 90 percent RH. At all temperatures, high humidity resulted in greater longevity and fecundity, more rapid development, and less mortality of immatures than lower humidities. The effects of temperature and humidity on this mite probably limit its distribution to warmer climates such as the southern part of the United States. The arid conditions of central and southern California and Arizona appear to be the factor preventing this species from being a major cotton pest in these important cotton-producing areas.

Where populations of this mite cannot be prevented from developing on cotton by controlling the winter hosts, acaricides recommended for control of tetranychid mites on cotton may be used to control this species (Hightower and Martin, 1956; Iglinsky and Rainwater, 1954).

The female of *T.* (*T.*) *desertorum* is greenish with darker shoulder and posterior spots; the overwintering females are bright orange. The tactile setae on tarsus I of the female are on a line with the posterior set of duplex setae. The body of the female is reddish in color; the lobes of the dorsal body striae are taller than they are broad. The aedeagus of the male is distinctive in that the dorsal margin is sigmoid; the anterior angulation is small and acute, and the posterior angulation is also acute and curved ventrad to a variable extent; the width of the knob is not more than one-fourth as long as the dorsal margin of the shaft.

Tetranychus (Tetranychus) evansi Baker and Pritchard. This species (figs. 46, *c*; 47, *b*) was described as a new species in 1960 from specimens collected on Mauritius island (pl. 35). It is now also known from Brazil, Texas, and California. Recorded hosts include tomato, eggplant, potato, peanut, rose, Plaine de Papayes, nightshade, *Solanum*, sweet potato, *Asystasia coromandeliana* Nees, red pepper, *Nicotiana glauca* Graham, *Phacelia* sp., lily-of-the-valley vine, and *Salpichora rhomboidea* Miers (Moutia, 1958).

Infested tomato plants turn yellow green and then brown. Plants generally show a bleached yellow orange or russeted appearance. Mites may kill their hosts very rapidly.

The first eggs laid by a female are deep orange, but later eggs become lighter orange to colorless and transparent. As the embryo develops, the egg becomes opaque and rust red prior to hatching. Newly emerged larvae are cream colored, but turn greenish yellow after feeding. Feeding nymphs are greenish yellow and adult females are reddish orange or carmine.

The incubation period at 23 C (73 F) and 49 to 50 percent RH last 60 to 72 hours and the duration of the larvae and the two nymphal stages averages 43 and 39 hours. About 21 to 25 hours of this time is spent in the resting stages. The total time from egg to adult females ranges from 8.2 to 9.3 days, averaging 9.0 days. At summer temperatures averaging 22.8 C (73 F) and winter temperatures of 19.4 C (66 F) the life cycle requires 6.5 days and 18.5 days respectively. Oviposition begins after the first day and females reach their maximum egg-laying capacity the fourth day after reaching maturity. At this time they may oviposit up to 30 eggs per day, after which the oviposition capacity decreases; yet 12 females averaged 6.1 eggs per day over a 26.7-day period.

Preferred oviposition sites are on the unexposed lower leaf surfaces at the junction of veins. More eggs are laid on normally pubescent nightshade leaves than on those leaves with light pubescence. High humidities result in increased oviposition. The longevity of fertilized females is 13 to 32 days, averaging 26.2 days; unfertilized females live longer, however, 27 to 39 days, averaging 34.5 days.

Reproduction is continuous throughout the year; thus there may be 24 to 30 generations per year. Populations increase and spread rapidly with the aid of profuse webbing or by wind (pls. 36, 37) (Oatman, Fleschner, and McMurtry, 1967; Quershi, Oatman, and Fleschner, 1969).

Although this mite is carmine in color with relatively long legs, it can be distinguished from the carmine spider mite, *T.* (*T.*) *cinnabarinus*, by the absence of green forms and the dark areas on the sides of the body.

Control has been achieved using high volume sprays of carbophenothion, chlorobenzilate, and diazinon. Ethion, carbophenethion, and dicofol dusts may continue to reduce mite infestations over a 7-day period (Schuster, 1959).

The female of *T.* (*T.*) *evansi* is reddish. The proximal tactile setae of tarsus I

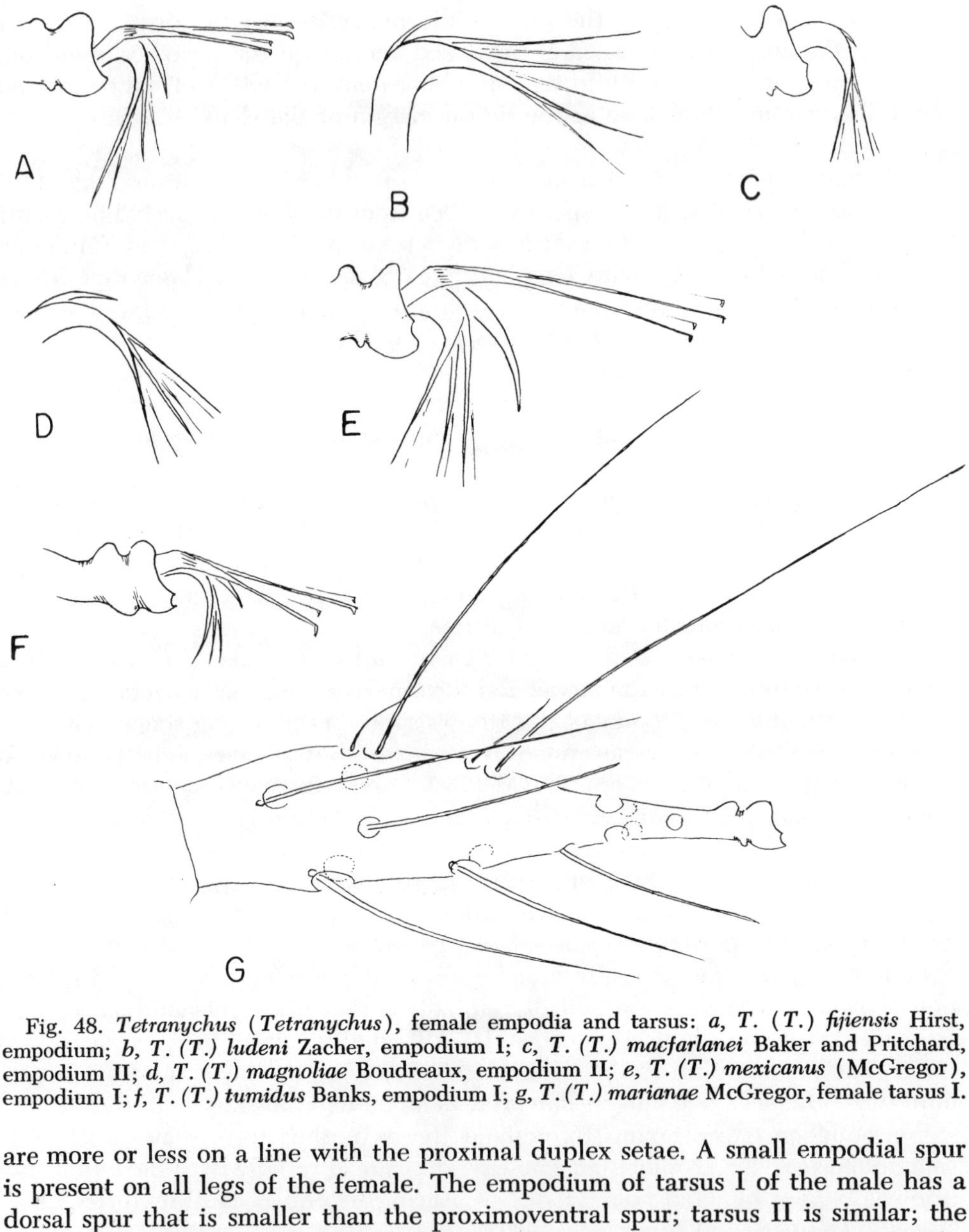

Fig. 48. *Tetranychus* (*Tetranychus*), female empodia and tarsus: *a*, *T.* (*T.*) *fijiensis* Hirst, empodium; *b*, *T.* (*T.*) *ludeni* Zacher, empodium I; *c*, *T.* (*T.*) *macfarlanei* Baker and Pritchard, empodium II; *d*, *T.* (*T.*) *magnoliae* Boudreaux, empodium II; *e*, *T.* (*T.*) *mexicanus* (McGregor), empodium I; *f*, *T.* (*T.*) *tumidus* Banks, empodium I; *g*, *T.* (*T.*) *marianae* McGregor, female tarsus I.

are more or less on a line with the proximal duplex setae. A small empodial spur is present on all legs of the female. The empodium of tarsus I of the male has a dorsal spur that is smaller than the proximoventral spur; tarsus II is similar; the empodia of tarsi III and IV end in three pairs of ventrally directed hairs. The aedeagus consists of a slender shaft; the distal knob forms a strong angle with the axis of the shaft, the knob is small, with a small anterior angulation and an acute posterior tip that is somewhat deflexed.

Tetranychus (Tetranychus) fijiensis Hirst. *Fijiensis* (figs. 46, *d*; 48 *a*) occurs on citrus, coconut, *Seakorthia* palm, and *Dieffenbachia* in India, Fiji and the Philippines (Rimando, 1962).

This species is distinctive in that the empodia are split distally into 2 instead of 3 hairs; the empodia of both sexes have a strong dorsal spur. The tactile setae of tarsus I of the female are on a line with the proximal duplex setae. The aedeagus of the male is very long, slender, tapers distally, and is upcurved.

Tetranychus (Tetranychus) gigas Pritchard and Baker. *T. (T.) gigas* occurs on cotton in Texas and Arizona. This species is similar to *T. desertorum*, differing only in that the aedeagus of the male is about two times as large as that of *T. desertorum.*

Tetranychus (Tetranychus) gloveri Banks. This species (fig. 46, *e*) is sometimes found associated with the tumid spider mite, *T.* (*T.*) *tumidus*, on cotton in Louisiana. The eggs of *T.* (*T.*) *gloveri* are always red. Mated tumid females, however, deposit white eggs and unmated females lay red eggs, but when adult females that develop from red eggs mate, only white eggs are deposited (Boudreaux, 1958).

The empodia of the female of *T.* (*T.*) *gloveri* are rayed and all have a strong dorsal spur; the tactile setae on tarsus I are proximal to the posterior set of duplex setae; the fertilized eggs are red. The aedeagus of the male is indented dorsally; the empodia I and II are clawlike and each has a strong dorsal spur; the other empodia end in ventrally directed hairs and each has a dorsal spur.

Tetranychus (Tetranychus) kanzawai Kishida. *Kanzawai* (fig. 46, *f*) is known to occur in Japan, Philippines, and Okinawa. It is a pest of tea, mulberry, corn, hops, soybeans, apple, pear, grape, citrus, clover, dahlia, peach, elderberry, verbena, and hydrangea.

On tea this red spider mite passes the winter as diapausing females at Kanaya; in the southern districts, however, all stages are found throughout the year including diapausing and nondiapausing females. The population density in the tea plantations fluctuates with the season; high densities are found from March to June and from September to December, but variations occur according to weather conditions.

Cultivated tea varieties vary greatly in their susceptibility to this mite, suggesting that control may be achieved by growing the more resistant varieties (Ehara, 1960; Osakabe, 1965).

The female of *T.* (*T.*) *kanzawai* is carmine; the lobes of the dorsal striae are taller than they are broad. Tiba I has 9 tactile setae and tibia II 7 tactile setae; there are 4 tactile setae proximal to the posterior duplex setae. Tibiae I and II and tarsus I of the male are similar to those of the female. The barb of the aedeagus is similar to that of *T. cinnabarinus* but is larger; the anterior portion is rounded and the posterior is acutely angled.

Tetranychus (Tetranychus) lambi Pritchard and Baker. This species (fig. 46, *g*) is known only from New Zealand and Australia where it infests strawberries. In Australia there are usually 3 crops of berries, but only the second crop is seriously

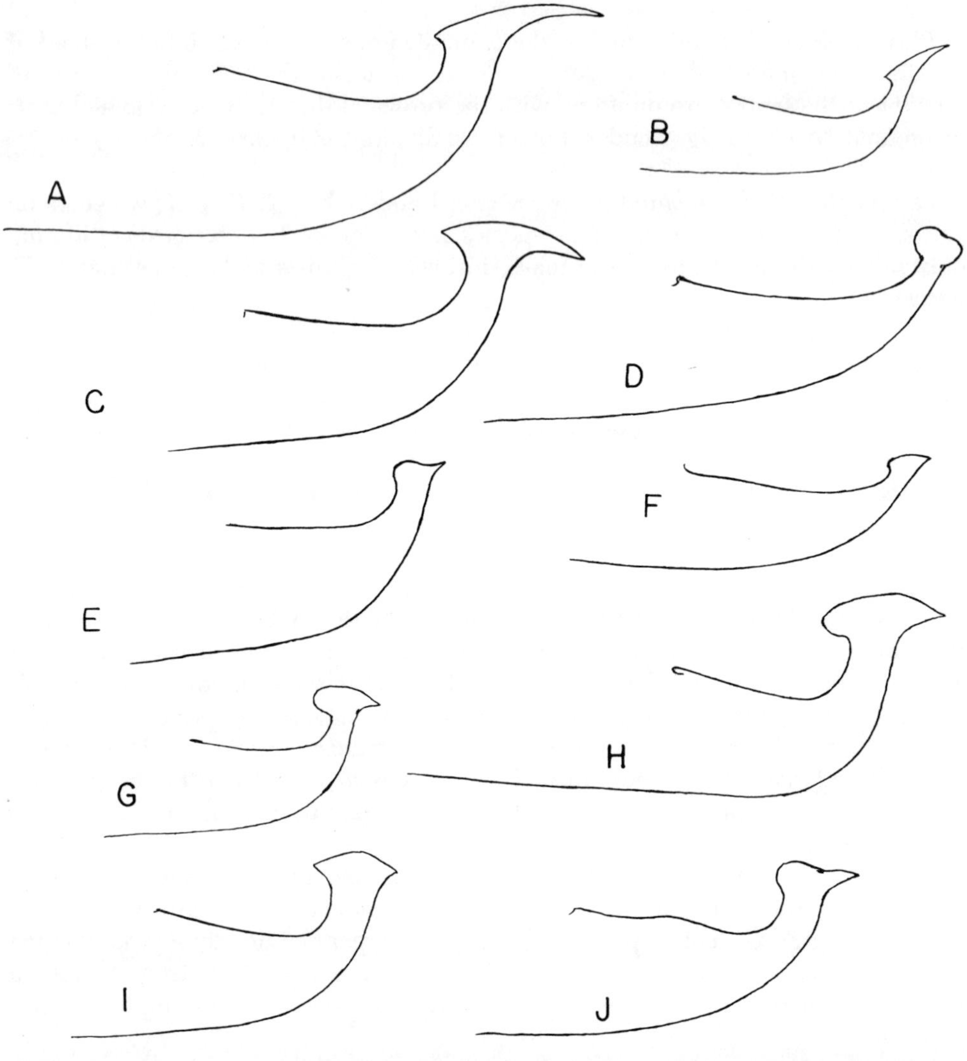

Fig. 49. *Tetranychus* (*Tetranychus*), aedeagi: *a*, *T.* (*T.*) *magnoliae* Boudreaux; *b*, *T.* (*T.*) *marianae* (McGregor); *c*, *T.* (*T.*) *mexicanus* (McGregor); *d*, *T.* (*T.*) *neocalidonicus* Andre; *e*, *T.* (*T.*) *truncatus* Ehara; *f*, *T.* (*T.*) *tumidellus* Pritchard and Baker; *g*, *T.* (*T.*) *tumidus* Banks; *h*, *T.* (*T.*) *turkestani* (Ugarov and Nikolski = *atlanticus*); *i*, *T.* (*T.*) *urticae* Koch; *j*, *T.* (*T.*) *yusti* McGregor.

affected by populations of this mite. This crop is harvested in September and October and is largely used for jam.

Injury to strawberries appears as an abnormal and irregular purple color on the upper leaf surface, usually adjacent to the main veins. When plants are severely damaged the whole undersurface assumes a silvery appearance, leaf edges roll, growth is retarded, and the fruit ripens prematurely (Davis and Heather, 1962).

T. (*T.*) *lambi* females have the diamond-shaped striation pattern between the third and fourth pair of dorsocentral setae, and are similar to other females in the subgenus. The aedeagus of the male is distinctive in that the distal knob is scarcely widened dorsoventrally, but the anterior and particularly the posterior angulation is pronounced. Empodium I of the male consists of 2 almost amalgamated trifid appendages; there is no mediodorsal spur on tarsus I and tarsus II.

Tetranychus (*Tetranychus*) *lombardinii* Baker and Pritchard. This mite (fig. 46, *h*) is widespread in South Africa, Zaire, Rhodesia, and Portugese East Africa where it causes damage to a variety of cultivated plants including cotton, cucurbits, spinach, tomatoes, banana, fig, castor bean, sorghum, *Solanum*, Jasmine sp., storks-bill, *Hibiscus*, and jimsonweed.

The specific injury caused by this mite has not been detailed, nor has its biology. The females are dark red with a dark spot on each side of the body sufficiently large to meet anteromedially. The legs of the young females are cream color, turning dark red in older individuals. The mites usually feed on the underside of the leaves and seem to prefer cotton to other hosts (Goldsmid, 1962; Meyer and Rodrigues, 1966).

The dorsal striation pattern of the female of *T.* (*T.*) *lombardinii* is typical for the subgenus; ventrally the striae have lobes that may extend from the genital area to the gnathosoma. The males are distinctive in that the shaft of the aedeagus narrows gradually until just before the bend dorsad; the knob of the aedeagus has its axis at an angle to the axis of the aedeagus; the anterior development is small, but distinct and rounded, and larger than the small acute posterior angulation; the dorsal margin of the aedeagus is slightly curved.

Tetranychus (Tetranychus) ludeni Zacher. *Ludeni* (figs. 46, *i*; 48, *b*) is widespread outdoors throughout the tropics, across the southern United States, in Mexico, Central and South America, South Africa, and Australia. It is reported as a serious pest of hibiscus, bean, eggplant, pumpkin, and other cucurbitaceous plants on Mauritius Island. It is found on potato, dahlias, and hydrangea in Rhodesia and Nyasaland; on cotton and alfalfa (lucerne) in South Africa, on beans, castor beans, and lantana in the United States; on castor bean in Concordia, Argentina; on Kudzu-vine in San Salvador; on strawberries, apple, plum, and clover in New Zealand; on *Solanum* and cucurbits in France; on muskmelon (cantaloupe), watermelon, tomato, bean, and *Myoporum* in Portugal, and on pyrethrum in Kenya. It causes damage to greenhouse plants in the more temperate climates of northern United States and Europe (Meyer and Rodrigues, 1966; Moutia, 1958).

Infested plants become yellowish, wilt, and droop rapidly, particularly during dry conditions. On pyrethrum the mites initially feed on the underleaf surfaces or in the young curled leaves. Mites become distributed over all leaf surfaces as populations increase and eventually mites produce webbing over the entire plant. The leaves first turn yellow, but soon necrotic patches encompass the leaves.

Moderate populations may greatly influence crop production but heavy infestations cause the plants to die.

All stages of this mite are found on plants throughout the year. The life cycle, during favorable conditions, may be completed in ten days; thus, many overlapping generations per year are possible. The eggs are pallid yellow and more-or-less spherical. The larval color is similar to the eggs, but each successive stage becomes darker red. The adult is carmine with red legs. It differs from *T.* (*T.*) *lombardinii* and *T.* (*T.*) *urticae* by the absence of large lateral spots on its body (Bullock, 1963).

T. (*T.*) *ludeni* tactile setae and the proximal duplex setae of tarsus I of the female are on the same line; both sexes have small dorsal empodial spurs. The aedeagus of the male is distinctive in that there is a small anterior angulation but no posterior angulation.

Tetranychus (Tetranychus) macfarlanei Baker and Pritchard. This species (figs. 46, *j*; 48, c) causes serious damage to okra, eggplant, gourd, pumpkin, and cucumber. It has been collected on bean and okra in India and Mauritius. Infested leaves first show a pronounced yellowish hue, then wilt and drop, especially during dry periods. Several predators feed on this mite in Mauritius, namely *Stethorus*, *Oligota*, and *Feltiella* sp. (Moutia, 1958).

In *T.* (*T.*) *macfarlanei* the tactile setae of tarsus I of the female are on a line with the proximal duplex setae; the empodia bear small dorsal spurs. The empodia I and II of the male are similar, clawlike, with a strong dorsal spur; the shaft of the aedeagus gradually narrows posteriorly, the dorsal margin being nearly straight; the neck of the distal knob is short, and the knob has a slight anterior and posterior angulation; the axis of the knob is parallel to that of the shaft.

Tetranychus (Tetranychus) magnoliae Boudreaux. *Magnoliae* (figs. 48, *d*; 49, *a*) is known from Louisiana on Magnolia tulip trees. The mites spin dense webbing on the upper surface of magnolia leaves and inhabit both surfaces of the tulip tree leaf. The adult females are carmine.

The empodia of the female of this species bear prominent dorsal spurs; the tactile setae of tarsus I are proximal to the duplex setae. The empodium of tarsus I of the male is clawlike; the other empodia are split distally into 3 pairs of hairs; all male empodia possess prominent dorsal spurs. The aedeagus is short, narrows posteriorly, and has a prominent knob that forms an angle with the axis of the shaft; the anterior projection of the knob is short and acutely angulate, and the posterior angulation is considerably longer and acutely angulate.

Tetranychus (Tetranychus) marianae McGregor. *Marianae* (figs. 48, *g*; 49, *b*) has sometimes been confused wth *T.* (*T.*) *evansi* which may be a serious pest of solanaceous plants. *T.* (*T.*) *marianae* is a pest of cotton, but occurs on castor bean, passionflower, and orchid. It is widespread on Pacific islands, being well known from the Marianas and the Marshalls. It occurs in Nicaragua, West Indies, Bahamas, southern Florida, and Argentina. Details of its biology are unknown (Moutia, 1958).

Adult females of *T.* (*T.*) *marianae* are carmine. The tactile setae of tarsus I of the female are proximal to the duplex setae; there are no empodial spurs. Empodium I of the male is clawlike; the other empodia end in 3 pairs of ventrally directed hairs. The aedeagus is similar to that of *T.* (T.) *evansi*, the axis of the terminal knob forming a definite angle with the shaft; there is a small anterior angulation and a longer, dorsocaudally directed posterior angulation; the knob is longer than the width of the stem.

Tetranychus (Tetranychus) mexicanus (McGregor). This species (figs. 48, *e*; 49, *c*) occurs on citrus, Johnson grass, and *Magnolia grandiflora* L. Its known distribution includes Mexico, Texas, Brazil, and Concordia in Argentina (Estebanes and Baker, 1968).

The tactile setae of tarsus I of the female of *T.* (*T.*) *mexicanus* are proximal to the duplex setae; there is a prominent dorsal spur on the empodia of both sexes. The aedeagus of the male is distinctive in that the axis of the knob is parallel to that of the shaft; the anterior angulation is short and acutely angulate, while the posterior angulation is longer and also acutely angulate.

Vegetable mite, *Tetranychus (Tetranychus) neocalidonicus* Andre. This mite (= *Tetranychus cucurbitae* Rahman and Sapra; fig. 49, *d*) is a major mite pest in India, but it is well distributed throughout tropical and subtropical areas of the world including Hawaii, Fiji, Venezuela, Puerto Rico, Mauritius, Bahamas, South America, and southeastern United States. Populations have been reported on more than 110 plants including flowers, peach, coconut, papaya, and many vegetable and field crops (Goldsmid, 1962).

The mites suck the plant sap from the leaves, producing white spots that gradually coalesce as feeding continues. Leaves lose their green color, gradually wilt, dry, and drop. The decreased vitality and leaf drop adversely effect growth, flowering, and fruiting. Damaged portions of the leaves of some plants turn red. The mites web profusely and may form a thick sheath of webbing that covers the entire plant.

The eggs of this mite are spherical and translucent when first laid, but gradually turn brown. The larva is light amber at hatching, but takes on an overall greenish tinge with dark lateral specks. The protonymph and deutonymph are green with dark specks on the dorsum, but adults are carmine.

The vegetable mite overwinters as fertilized females on hollyhock and other weeds. Activity begins when the weather warms in the spring and as populations increase rapidly, the mites move to cultivated plants preferably cucurbits and other vegetables where infestations reach their zenith in May to mid-July. Active populations decline during late July and August, after which only eggs may be found, mainly on the preferred host. Infestations increase again in September and October, primarily on weeds. Gravid females may move to winter crops, such as cauliflower and cabbage, where activity continues until December or February. The gravid females then move to winter crops or weed hosts to hibernate.

The duration of the stages under field condtions are: incubation period, 3 to 9 days; larva, 3 to 5 days; protonymph, 3 to 4 days; and deutonymph 2 to 5 days.

The minimum period from hatching to the adult stage is 10 to 13 days. The oviposition period is 1 to 2 days and the adult female life span varies from 8 to 46 days, averaging about 32 days.

Eggs are laid at random in the web, generally on the lower surface of the leaves. A female mite may lay as many as 13 eggs daily and 60 to 90 during her lifetime. Under ideal conditions the eggs may hatch in 2.5 days, and 32 generations may occur in a year.

Infestations on economic plants may be decreased by destruction of overwintering or summer hosts, leaving the land free of plants between crop plantings. Acaricides used for other tetranychid mites are similarly effective against this mite (Khot and Patil, 1956; Rahman and Sapra, 1945).

The tactile setae of tarsus I of the female of *T.* (*T.*) *neocaledonicus* are in line with the posterior duplex setae; there are no dorsal spurs on the empodia of either sex. The empodium of tarsus I of the male is clawlike; the other empodia are split distally into 3 pairs of ventrally directed hairs. The aedeagus of the male is distinctive in that the knob is berrylike, and the anterior rounded projection is better developed than the rounded posterior projection.

Tetranychus (Tetranychus) piercei McGregor. This species is recorded on sweet potato and palm in the Philippines and Okinawa Island. It is closely related to *T.* (*T.*) *urticae* Koch, but is distinct in that the aedeagal knob is tiny and forms a definite angle with the axis of the shaft, and the posterior projection is short and acute.

Tetranychus (Tetranychus) truncatus Ehara (fig. 49, *e*) is a pest of mulberry and other plants in Japan and the Philippines (Ehara, 1956).

The female is carmine, and the lobes of the dorsal striae are semioblong to semicircular. The tactile setae of tarsus I of the female are proximal to the duplex setae. The aedeagus of the male is stout; the anterior projection is small and rounded, and the posterior projection consists of a tiny barb.

Tetranychus (Tetranychus) tumidellus Pritchard and Baker. *Tumidellus* (fig. 49, *f*) is a host of wild and cultivated peanuts in Brazil and south Turkey and in Georgia and Alabama in the United States.

Young peanut plants may be readily killed by infestations of this species. Information is not available on its biology, but studies indicate that injury by this mite may be prevented through development of resistant varieties (Leuck and Hammons, 1968).

The tactile setae of tarsus I of the female of *T.* (*T.*) *tumidellus* are proximal to the duplex setae; the empodia of both sexes have strong empodial spurs. Empodium I of the male is clawlike; the other empodia end in 3 pairs of ventrally directed hairs. The aedeagus narrows gradually distally, bends dorsally at an obtuse angle, and the anterior and posterior angulations of the knob are small and of equal size.

Tumid spider mite, *Tetranychus (Tetranychus) tumidus* Banks. This mite (fig. 48, *f*; 49, *g*) is commonly found in the southeastern United States, including the states of Florida, Georgia, South Carolina, Louisiana, southeastern Texas, and California. It also occurs in Brazil, Hawaii, Puerto Rico, Canal Zone, Guam, Bermuda, Central America, Mexico, and Trinidad. It is a serious pest of cotton, celery, beans, eggplant, beets, okra, peas, and sweet potato and has been collected from water hyacinth, castor bean, dahlia, morning-glory, palms, *Maranta*, milkweed, mint, avocado, and many ornamental and tropical plants.

Injury to plants by this spider mite appears as a reddening of the upper surface of the leaf. The reddened area may be either a small blotch or many such blotches that often encompass the entire leaf surface, eventually resulting in complete defoliation of affected plants. Moderate infestations on cotton cause a reduction in the number of seeds per ball, a decrease in their weight, viability, and oil content, as well as a reduction in the length and maturity of the fibers and lint index (Roussel et al., 1951).

The adult females of this mite are carmine resembling *T.* (*T.*) *desertorum* and *T.* (*T.*) *cinnabarinus* (Boisduval). The tactile setae of tarsus I of the female of *T.* (*T.*) *tumidus* Banks are proximal to the duplex setae. Both sexes have large empodial spurs. Empodium I of the male is clawlike; the other empodia end in three pairs of ventrally directed hairs. The aedeagus is distinctive in having a broadly rounded anterior angulation and a short, acute angulation posteriorly; the knob is at right angles to the neck and the axis of the knob is parallel to that of the shaft.

Strawberry spider mite, *Tetranychus (Tetranychus) turkestani* (Ugarov and Nikolski). This mite (fig. 49, *h*), best known in American mite literature as *T. atlanticus* McGregor (Baker, 1968), is one of the most widespread and serious mite pests of agricultural crops. It is well distributed throughout the United States, Europe, Russia, Japan, the Near East, and middle eastern countries. It is a serious pest of alfalfa, beans, castor bean, clover, cotton, carrot, cucumber, eggplant, melons, parsley, soybean, squash, strawberry, sunflower, and other low-growing crops. It is also found on apple, hops, maize, mint, peach, peanut, pear, plum, and walnut, but it is not a serious pest on these crops.

The strawberry spider mite feeds in colonies, mainly on the lower surface of the leaf; but the injury shows on the upper surface (pl. 38) as dead areas at the point of feeding on the leaf. High populations cause leaves to drop and the plants soon die. Heavy mite infestations produce sufficient webbing to cause the leaves and stems to become matted together. Infestations on cotton reduce seed yields up to 22 percent, decrease boll size, lower seed viability, decrease lint production, and impair maturity. Feeding by laboratory colonies produces rapid reddening of the leaves and total defoliation within 40 days, regardless of the number of mites used in establishing an infestation. High daily temperatures and limited rainfall favor mite development (Canerday and Arant, 1964).

The interval from infestation until abscission of a cotton cotyledon is directly proportional to the amount of necrotic feeding injury on the upper surface, and abscission always follows when necrosis occupies 70 percent or more of the surface. The strawberry spider mite produces injury twice as rapidly as the carmine

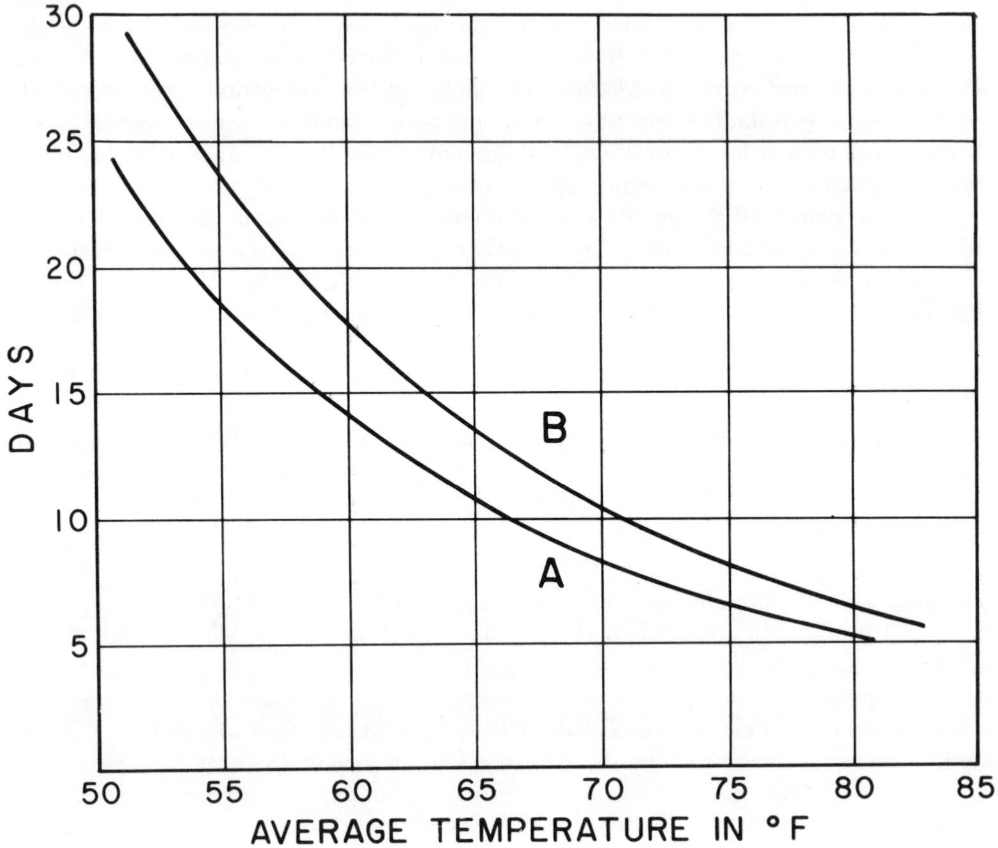

Fig. 50. *Tetranychus* (*Tetranychus*) *turkestani* (Ugarov and Nikolski): *a*, relationship between the incubation period and average temperature in degrees F.; *b*, relationship between the time from hatching to adult and the average temperature.

spider mite. The immature stages of the strawberry spider mite cause as much injury as that caused by adult females, but no visible injury is produced by the males. Noninfested cotyledons opposite heavily infested ones show no more tendency to drop than those on noninfested plants, all of which indicates that a local toxin is injected into the plant by immature and adult female mites during the feeding process (Simons, 1964).

The eggs of the strawberry spider mite are spherical, clear, and colorless at first, becoming opaque and finally ivory just before hatching. The newly hatched larva is pale, almost colorless at first, then it becomes greenish with a black spot on each side of the body. The protonymph and deutonymph are pale straw colored with larger lateral spots. The color of the summer egg-laying female is variable, depending upon the food. It may be amber, green, brownish, or almost black. A massive spot is always present on each side, beginning just behind the eye spots and extending beyond the middle of the body; also 2 distinct black spots may appear, 1 on each side, towards the end of the hysterosoma. Colonies of the straw-

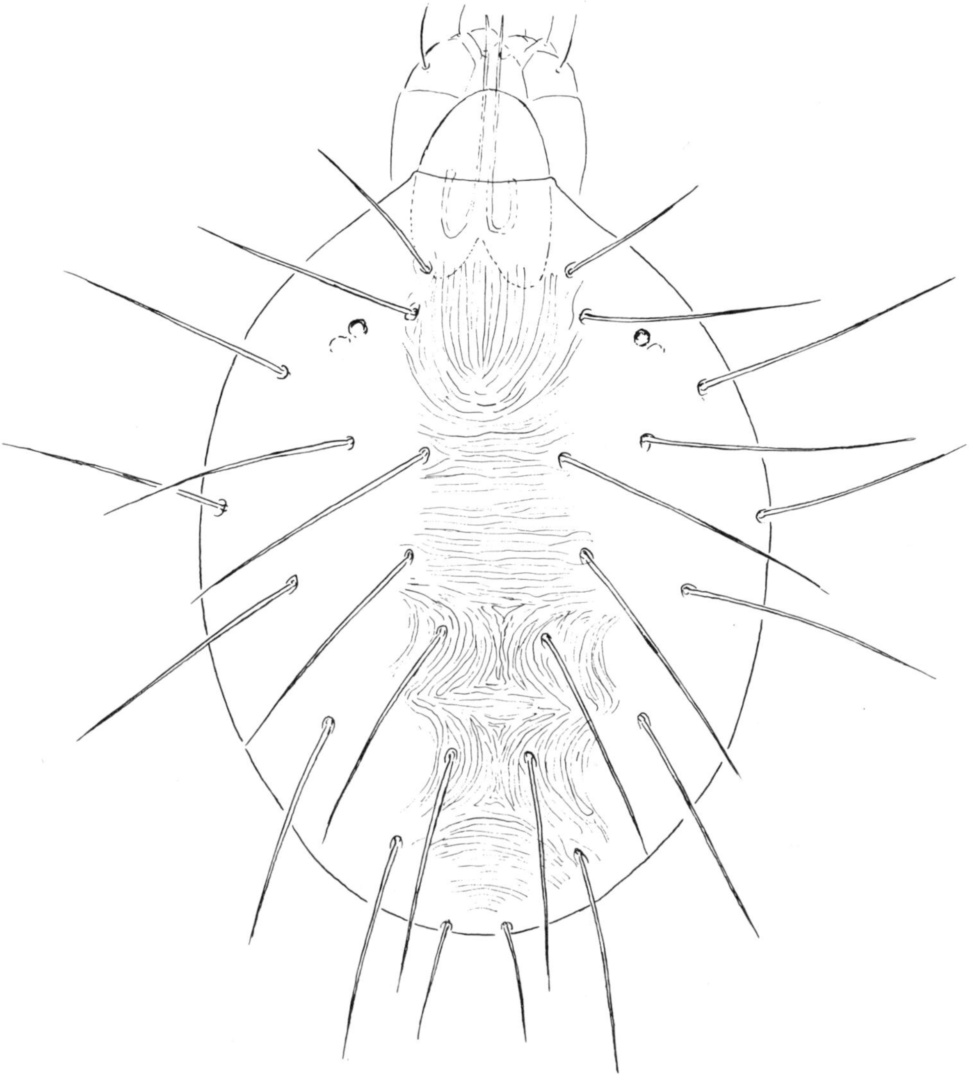

Fig. 51. *Tetranychus (Tetranychus) urticae* Koch, dorsum, showing setae and striations.

berry spider mite may be distinguished from the two-spotted spider mite and Schoenei mite by the number of black spots. The 4-spotted condition usually predominates among the adult females of the strawberry spider mite; the 2-spotted condition on both nymphs and adults is most prevalent in 2-spotted spider mite colnonies, and the 4-spotted condition is prevalent among both the adult females and nymphal stages of *T.* (*P.*) *schoenei.*

Strawberry spider mites hibernate as bright orange females. The hibernating forms first appear near the last of September, and summer forms disappear by the last of October. The hibernating females are first green, but gradually change to bright orange. The black spots persist for a time and then finally disappear.

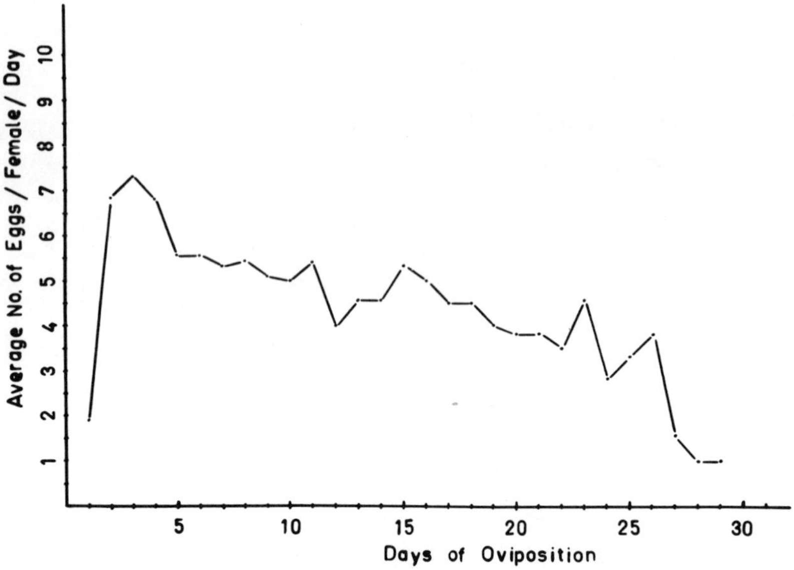

Fig. 52. Average daily oviposition rate of *Tetranychus* (*Tetranychus*) *yusti* McGregor females at indicated days after reaching maturity (after Saba).

Ample shelter for overwintering females is provided in litter, under loose bark scales, and in loose soil. Increased temperature during fall or winter does not affect the hibernating females (Mellott and Connell, 1965).

The length of the incubation period and the time required for development from egg to adult at different temperatures is graphically illustrated in figure 50. The preoviposition period ranges from 1 to 6 days. Females oviposit an average of 7.4 eggs per day during midsummer, but later in the summer and fall only 2 to 3 eggs are laid per female each day. The duration of the adult female is about 8 days during the summer broods and 33 days in the fall. There are 8 to 16 generation per year depending on the length of season (Cagle, 1956).

This mite is generally susceptible to applications of sulfur. Populations may also be reduced by acaricides used for other tetranychid mites. Failures in chemical control generally arise from the difficulties encountered in distributing the toxicant to the leaf undersurfaces of low-growing plants.

In *T.* (*T.*) *turkestani* the tactile setae of tarsus I of the female are proximal to the duplex setae; the empodia end in three pairs of ventrally directed hairs and there is no dorsal spur. Empodium I of the male is clawlike; the other empodia are similar to those of the females. The distal knob of the aedeagus is moderately enlarged, about ¼ as long as the dorsal margin of the shaft; the anterior projection is broad and rounded, and the posterior angulation is small and acute; the axis of the knob forms an angle with that of the shaft.

The two-spotted spider mite, *Tetranychus* (*Tetranychus*) *urticae* Koch. This mite (figs. 49, *i*, 51; pl. 39) has been known as the Glasshouse spider mite, red spider mite, or simple red spider. These common names have referred to a complex that also included *T.* (*T.*) *cinnabarinus* (Boisduval), now considered a separate species (Smith and Baker, 1968). This complex had included about 59 synonyms, each described from different hosts or from different areas of the world. The best known of these are *T. telarius* L., *T. bimaculatus* Harvey, *T. altheae* von Hanstein, *T. multisetus* McGregor, and *Eotetranychus cucurbitacearum* Sayed.

Mites of this species complex have been recorded on more than 150 hosts of some economic value, which includes most of the important agricultural crops and ornamental plants. This species complex is also one of the most destructive to its hosts, often killing them very rapidly (pls. 40, 41). The published host list includes the complex, as well as the synonyms of this and other species, so it is not possible to segregate the hosts of each from the literature.

T. (*T.*) *cinnabarinus* was established as a distinct species from *T.* (*T.*) *urticae* because differences in morphology, biology, distribution, and cross-breeding results. Both are injurious to a wide variety of plants in greenhouses and in the field. *T.* (*T.*) *urticae,* however, is often a serious pest of deciduous fruit and shade trees and shrubs especially in temperate climates, whereas *T.* (*T.*) *cinnabarinus* is most common in semitropical climates, but their distributions overlap.

Two-spotted spider mite populations in the field have an overwintering or diapause form that cannot be broken without an elapsed period of time and is initiated by shortened period of light, decreased temperatures, and unfavorable food supply (see chap. 2). These overwintering females stop feeding and egg laying, leave the host plants, become yellowish orange, and hibernate on the ground under leaves, in cracks and crevices, or other protected places; whereas females of the carmine spider mite may become darker in color during the winter, but they overwinter on their hosts and reproduction continues during warm winter weather, that is, there is no diapause form that requires an elapsed time before activity can be resumed.

The two-spotted spider mite prefers to colonize and lay eggs on the upper leaf surface of some plants and the lower surface of others, but in cases of heavy attack they inhabit all plant surfaces. They prefer the young leaves, but in well-established colonies the older leaves become heavily infested. As the population develops the mites usually spin sufficient webbing to cover the entire plant.

The threshold of development is 12 C (54 F) and the maximum developmental temperature about 40 C (104 F). At 30 to 32 C (85 to 90 F), which is the optimum temperature for development, the incubation period lasts 3 to 5 days. The developmental stages of the female require 4 to 5 days, the preoviposition period only 1 to 2 days, making a total life cycle period of 8 to 12 days. The average life duration of females is about 30 days during which time the average number of eggs laid per female is 90 to 110, but a single female deposited more than 200 eggs (Gasser, 1951; Nuber, 1961).

The life cycle of *T.* (*T.*) *urticae* under a diurnal temperature cycle of 15.0 to 28.3 C (59 to 83 F) was determined by Laing (1969). After a preovipositional period of 2.1 days these mites deposited an average of 2.4 eggs per day for 15.7 days. The duration of each stage of the male and female, respectively was as follows: incubation, 6.7 days; protonymph, 2.7 and 3.0 days; deutonymph stages, 3.1 and 3.5 days. The developmental time for the males was 16.1 days and for females 16.9 days.

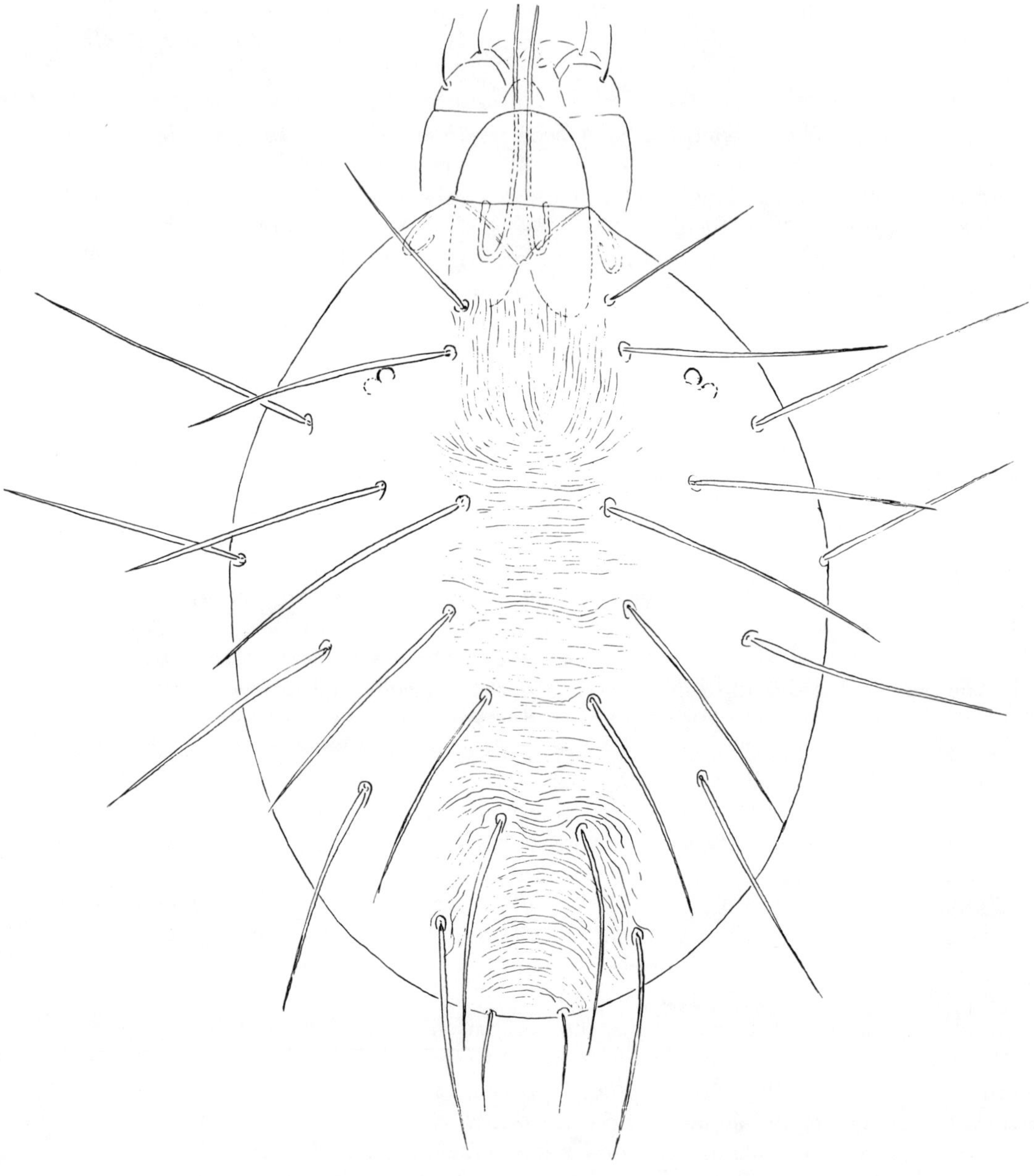

Fig. 53. *Tetranychus* (*Armenychus*) *mcdanieli* McGregor, female dorsum showing setae and striations.

Winter control applications are not effective against this mite species, because their hibernation habits make them unavailable to chemical applications. Many acaricides and application methods have been developed to provide summer control of this mite species, but in many situations populations have rapidly de-

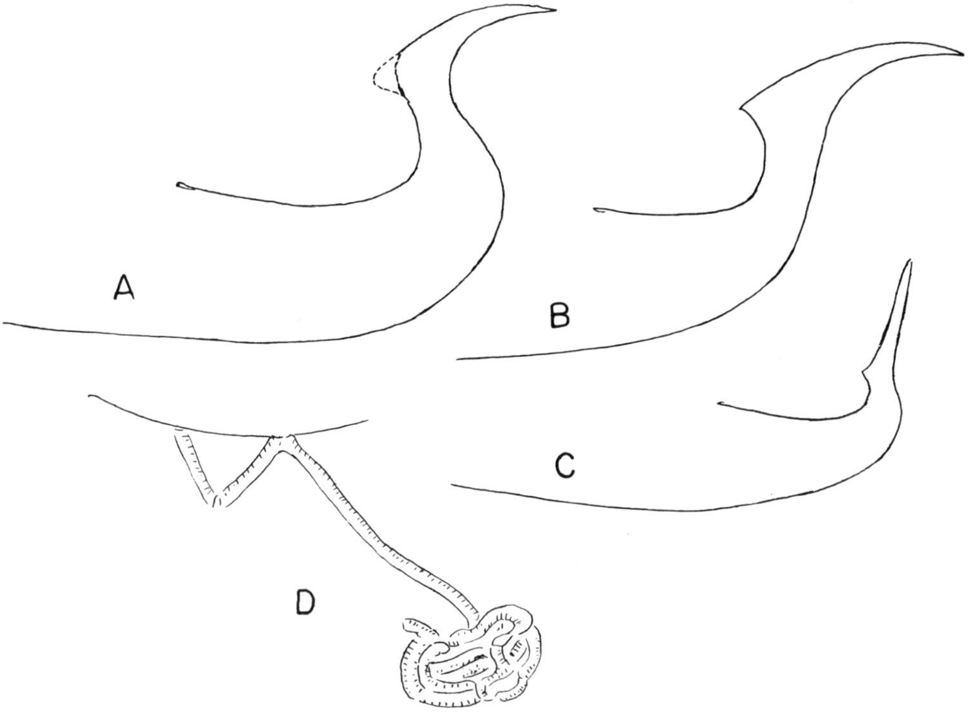

Fig. 54. *Tetranychus* (*Armenychus*): *a*, *T.* (*A.*) *mcdanieli* McGregor, aedeagus; *b*, *T.* (*A.*) *pacificus* McGregor, aedeagus; *c*, *T.* (*A.*) viennensis *Zacher*, aedeagus; *d*, *T.* (*A.*) *viennensis* Zacher, collar trachea.

veloped resistance to available acaricides. The applicable chemical control methods, therefore, depend on the host and its culture, as well as the previous exposure to acaricides. Numerous publications deal with control methods (Bravenboer, 1959).

The tactile setae of tarsus I of *T.* (*T.*) *urticae* are proximal to the duplex setae; there are no or only very tiny dorsal spurs on the empodia. Empodium I of the male is clawlike; the other empodia are similar to those of the female. The aedeagus is distinctive in having a small knob set at right angles to the neck, the anterior and posterior angulations are small and equal.

Tetranychus (Tetranychus) yusti McGregor (fig. 49, *j*) appears to be a subtropical and tropical species, its reported distribution includes Delaware, Louisiana, Mexico, Central America, and Ecuador. It is one of the most abundant and injurious mites to soybean in Delaware where it was reported under the name *T. lobosus* Boudreaux. It lives on a wide variety of plants mainly of no economic importance, belonging to the Compositae, Leguminosae, and Graminae. Economic plant hosts include cotton, roses, okra, sweet potato, sunflower, white clover, marigold, peas, beans, cowpeas and peanuts (Baker and Connell, 1961; Estebanes and Baker, 1968).

At 25.5 C (78 F) females develop in 9 to 10 days, but males require about 1 day less time. The oviposition rate becomes maximum 3 to 5 days after females reach the adult stage, then gradually decreases as the females become older (fig. 52). The average life span is 21 days. Individuals may lay 158 eggs during a life span of 35 days (Saba, 1971).

T. (T.) yusti may be confused in the field with *T. (T.) cinnabarinus, T. (T.) tumidus*, or *T. (T.) turkestani*. They can best be distinguished by the genitalia of the male. The tactile setae of tarsus I of *T. yusti* are proximal to the posterior duplex setae; the transverse striae in the area anterior to the genital opening of the female have lobes. The aedeagus of the male has a relatively small knob rounded anteriorly and angulate posteriorly; the dorsal surface is strongly indented.

Tetranychus (*Tetranychus*) *zambezianus* Meyer and Rodrigues. This mite has been collected on cotton and soybean in South Africa. It is red and indistinguishable in the field from the other tetranychid species. Its biology has not been studied.

It differs from related species by having ventral lobes on an area from the genital shield to the gnathosoma rather than on the posterior hysterosoma and metapodosoma. The empodium has the mediodorsal claw ⅓ to ½ the length of the proximoventral hairs.

Tetranychus (*Armenychus*) Wainstein

The hysterosomal striae of the female are transverse.

McDaniel mite, *Tetranychus (Armenychus) mcdanieli* McGregor. This mite (figs. 53; 54, *a*; pl. 42), first found on raspberry in Michigan in 1931, is now well distributed over northern United States and southern Canada. It is a serious pest of deciduous fruit trees, grapes, berries, and ornamental plants. In addition, it feeds on more than 30 species of weeds and other low-growing plants (Reeves, 1963).

Populations first start on the undersides of leaves, but soon move to both leaf surfaces. Feeding on the terminal leaves causes the leaves to curl upward, providing protection for the mites on the upper leaf surfaces. Mites spin profuse webbing, causing the leaves to become matted together. Feeding injury first appears as stippling owing to removal of chlorophyll (pl. 43), but as feeding continues the leaves turn brown and finally drop. Injury is most severe during hot dry weather. Mites move to the peripheral branches as populations increase, giving a brownish grey appearance to the entire tree.

The bright orange adult females overwinter under bark just below the ground level or in soil and debris at the base of the tree. Females emerging in the spring feed on weed hosts, on buds and on the new growth of water sprouts that develop near the main trunk. Eggs are laid and the overwintering females die. Infestations are not usually present on fruit trees until July and August. Seven to nine generations occur per year depending on temperature conditions. These mites require only eight days to develop from eggs to adults.

The McDaniel mite overwinters in protected niches, so dormant sprays are not effective. Summer treatments are effective but high infestations and profuse webbing make spray coverage difficult, thus preventive rather than corrective control is most effective. Careful observations on mite increase on water sprouts and inside foliage insure early detection and the use of preventive spray applications. This mite has developed resistance to many acaricides and continual evaluation of pesticide effectiveness is necessary to ensure adequate control.

Typhlodromus occidentalis Nesbitt is an effective predator of *T.* (*A.*) *mcdanieli* in the northwestern United States. *Amblyseius fallacis* (Garman) and certain insect predators appear to prevent outbreaks in Ontario (Dondale, 1967; Nielsen, 1958).

The female of *T.* (*A.*) *mcdanieli* is similar to that of *T.* (*A.*) *pacificus* (see below). The male is distinctive in that the dorsally directed bend or neck of the aedeagus is sigmoid and is directed dorsocaudally at almost right angles to the shaft; there sometimes is a small obtuse anterior angulation on the head.

Pacific spider mite, *Tetranychus (Armenychus) pacificus* McGregor. This mite (fig. 54, *b*) is a major pest of a wide variety of crops in the hot interior sections of northern and central California and in Washington, Idaho, Oregon, and Mexico. In these areas it is one of the most injurious pests of cotton, deciduous fruit, walnuts, beans, melons, berries, alfalfa, clover, vetch, and maize. It also attacks citrus, elm, black locust, many ornamental plants, and native weeds, shrubs, and trees.

The Pacific mite feeds on both leaf surfaces of its hosts causing the typical stippling frequently produced by tetranychid mites as they remove chlorophyll from the plant cells. The serious injury produced by low populations of this mite on pear, citrus, and some other plants in California suggests the possibility that a toxin is involved. The leaves, beginning at the tops of the trees, turn brown and die as though they had been scorched by fire. Beans, grapes, and other plants first become stippled; then as populations increase the plants become covered with webbing. Eventually the leaves dry and the plants die. Mite feeding on corn results in premature drying of the foliage which reduces grain yields by increasing stalk breakage, lowering the moisture content of the grain at harvest, and causing kernel shrinkage.

Populations of this mite develop throughout the year in greenhouses and in warmer areas. In more temperate climates the mites hibernate as light orange inactive females beneath the bark and in crevices of the host plant. The mites migrate to the hibernating quarters in late summer and early fall where they remain until the following March. They leave the protective winter quarters, migrate to filaree, shepherds purse, or other weeds growing in the orchards or vineyard, and there deposit eggs. This increases the numbers of mites ready to move onto new growth of grapevines or trees. The first colonies on trees are usually found at the tops or the tips of the branches where they may escape notice until the tree is heavily infested and defoliation occurs. Heavy infestations weaken the trees or vines as a result of defoliation and reduce the quality of the crop through sunburn.

Each female deposits 50 to 100 eggs over a period of 2 to 4 weeks. The cycle from egg to adult requires 10 to 14 days. There are many generations per year, the number depending on the duration of summer temperatures. Populations in July and August of 600 to 1400 mites per grape leaf are common when leaves show severe damage.

Spots on the dorsum may serve as distinguishing characters in the field when the Pacific mite and two-spotted spider mite occur on the same host. The Pacific mite is yellow with 3 irregular dark spots on each side of the hysterosoma; the two-spotted spider mite is generally greenish with larger, almost saddle-shaped, spots on the hysterosoma.

Although the six-spotted thrips, *Scolothrips sexmaculatus* Pergande, predacious mites, and other predators feed on this mite, the natural enemies are usually unable to cope with its reproductive potential (Laminman, 1935). This species is not susceptible to sulfur, and populations have rapidly developed resistance to acaricides. It is therefore one of the most difficult tetranychid mite species to control.

The tactile setae of tarsus I of the female of *T.* (*A.*) *pacificus* are proximal to the duplex setae; the empodia are without dorsal spurs. Tarsus I of the male is clawlike with a strong dorsal spur; the other empodia are similar to those of the female. The aedeagus is distinctive in having a long posterior angulation on the distal knob, its tip reaching well beyond the level of the caudal end of the bend of the neck, forming an obtuse angle.

Hawthorn spider mite, *Tetranychus (Armenychus) viennensis* Zacher. This mite (fig. 54, *c-d*) is well distributed in Europe, Asia, England, and Japan. Its chief hosts are blackthorn, oak, and Rosaceae plants, also fruit trees, including apple, pear, and other stone fruits except cherries.

Initial stages of mite feeding produce yellow spots the size of a hazelnut on the leaves. As feeding continues the foliage becomes yellowish grey. Very few fruit buds are formed and those that do develop remain small.

The females of the hawthorn spider mite occur either as summer or winter forms similar in morphology but differ only in color. Winter females are bright red with light yellow or sandy-colored extremities. Summer females are bluish-violet or carmine red with whitish segments. The male is sandy yellow, but greenish black spots may be seen from under the integument.

Only fertilized females overwinter. Before fall leaf drop the mites seek hibernation quarters under the bark scales or mosses and lichens on the boughs and branches of the trees, but not in the soil or the flora beneath the host plant. Mites emerge in the spring when the mean daily temperatures are about 9 to 10 C (48 to 50 F). The major period of migration from winter hibernation is just before the beginning of blossom time. The leaflets of young shoots are each colonized; at first by a single female that lays its eggs on the under surface of the leaves. Differences in the developmental rate occur during the course of the year. The developmental time for the different stages in the last generation is almost twice as long as the first generation. In Germany, an average of 20 to 30 days are required for the life cycle of females and 8 to 16 days for the males. The females develop in 12 to 25 days in Turkey, so 5 to 6 generations may occur per year in Germany and 9 to 10 in Turkey. At constant temperatures of 22 to 25 C (72.5 to

77 F) in the laboratory, a generation is completed in 12 to 14.5 days; the length of egg, larvae, and nymphal stages require 4 to 5 days, 1 to 2 days, and 0.5 to 3 days respectively. The preoviposition period is 1 to 5 days, and females lay 70 to 82 eggs during their life span.

The change from summer to winter females is stimulated by lack of food and decreasing temperatures. This change becomes evident in the mites as early as the second nymphal stage by the appearance of a brownish color at the intercalations. This begins to occur about mid-August, the time depending on the host condition.

Low temperatures and high humidities are unfavorable for population increase. When such conditions occur in the spring there may be sufficient retardation of population development to preclude the development of destructive populations even during favorable summers. Favorable weather in the spring almost guarantees, if host plants are suitable, a big summer population that, in turn, produces a large number of winter females. This mite is an important pest only in the drier climates in Russia; thus its distribution is closely correlated with that of *Bryobia rubrioculus* (Scheuten) (= *B. redikorzevi* Reck).

Winter applications of dormant spray oils alone, or in combination with acaricides, reduce the spring populations; but a summer application of an acaricide is usually necessary to ensure satisfactory control of this mite pest (Göksu, 1968; Müller, 1957; Beglyarov, 1959).

The *T.* (*A.*) *viennensis* female is similar to the other two species in the subgenus, differing, however, in that the peritreme forms a distal anastomosing pattern. The peritreme of the male is similar to that of the female; the aedeagus is bent sharply dorsad and the distal knob is modified as a small anterior angulation near the base of the bent portion; the caudal angulation is attenuated, straight, and vertical.

Tetranychus (*Polynychus*) Wainstein

The striae of the hysterosoma are transverse except for the longitudinal pattern between the fourth pair of D setae.

Four-spotted spider mite, *Tetranychus (Polynychus) canadensis* (McGregor). This mite (figs. 55; 56, *a*) is a pest of a wide variety of crops, ornamental plants, and trees. It is widely distributed in the United States, Canada, the middle East, Africa, and Poland. The economic plants injured by this mite include peach, plum, apple, crab apple, tomato, beans, sweet potato, okra, cotton, red clover, barley, rye, corn, wheat, ornamentals, rose, elm, linden, horse chestnut, Osage orange, poplar, and umbrella tree.

This mite species inhabits the underleaf surfaces of trees, causing the leaves to turn rusty brown and drop. Defoliation is primarily confined to the tops of the trees. Feeding injury on barley appears as a darkening, followed by yellowing, and withering from the bend of the leaf on to its tip and, occasionally, crinkling of the leaf edges from below.

The four-spotted spider mite overwinters in the adult stage as bright orange females. Eggs and immature stages are not found during the winter, but adult

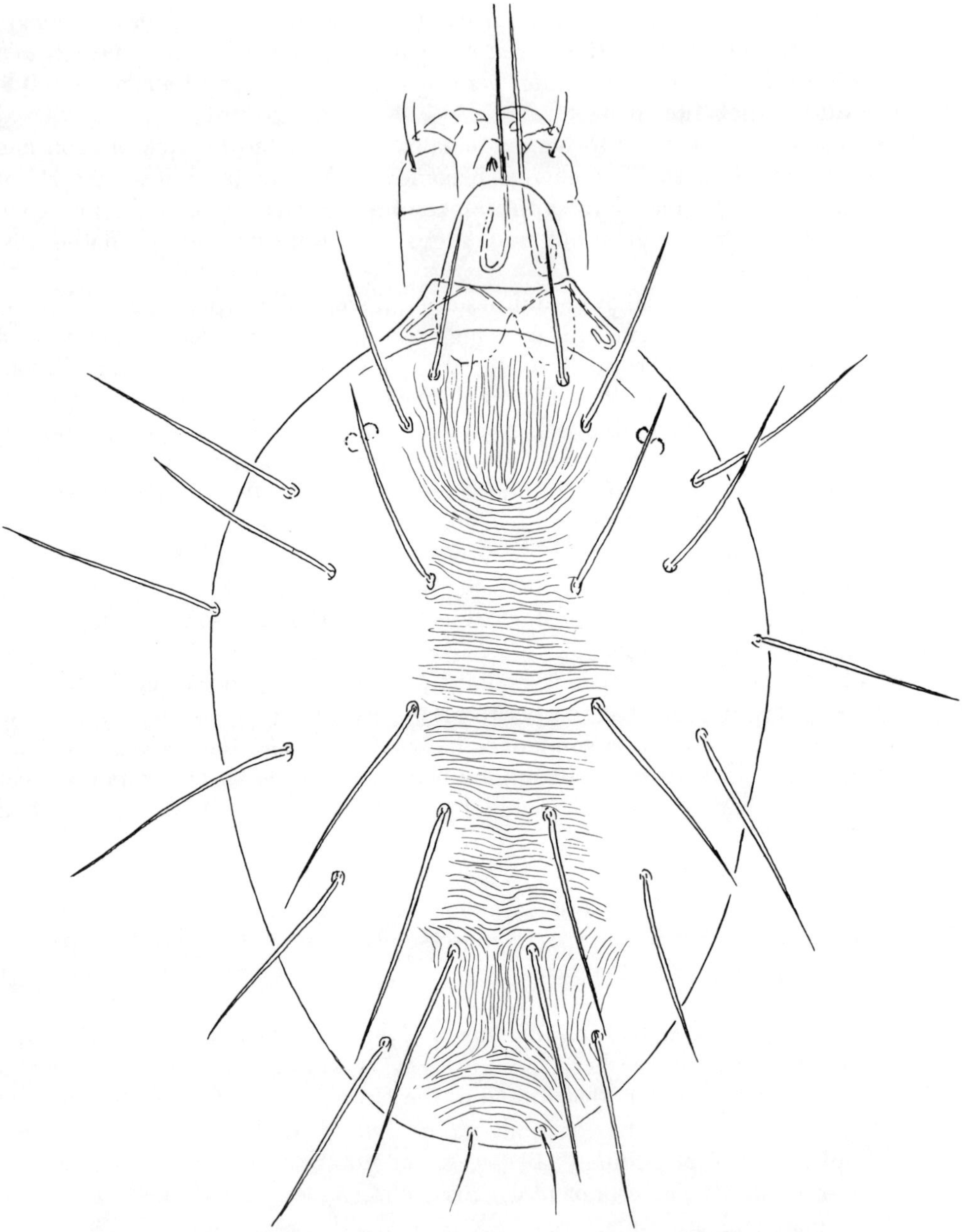

Fig. 55. *Tetranychus* (*Polynychus*) *canadensis* McGregor, female dorsum.

females may be found in the soil and around the trunks of the infested trees.

Summer forms of this species are greenish yellow to orange with dark paired spots on the body, medially and at the posterior end. The mite produces very little webbing. The eggs are spherical and opaque (Atchenson, 1953; Wallace and Sinha, 1961; White, 1966).

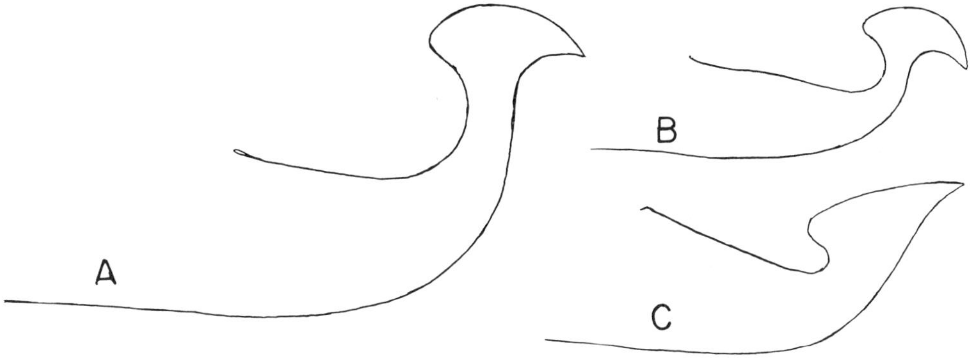

Fig. 56. *Tetranychus (Polynychus)*, aedeagi: *a*, *T. (P.) canadensis* McGregor; *b*, *T. (P.) schoenei* McGregor; *c*, *T. (P.) sinhai* Baker.

The tactile setae of tarsus I of the female of *T. (P.) canadensis* are proximal to the duplex setae; very tiny dorsal spurs may be present on the empodia. Empodium I of the male is clawlike; the other empodia end in three pairs of ventrally directed hairs; empodia I and II have obvious dorsal spurs. The aedeagus is distinctive; the knob is about one-third as long as the dorsal margin of the shaft; the anterior projection is rounded or may be somewhat angulate, and the caudal projection forms an acute angle.

Schoenei spider mite, *Tetranychus (Polynychus) schoenei* McGregor. This mite (fig. 56, *b*) is widely distributed over the eastern and southeastern United States. This species is sometimes a serious pest of apple, cotton, elm, black locust, bean, bramble, raspberry, wild plum, and blackberry. The injury on apple is indistinguishable from that produced by the European red mite, which is often present in the same orchards. The foliage becomes bronzed and the fruit fails to color in the fall, making it necessary to sell the fruit for canning. Heavy infestations cause premature leaf drop. Heavy infestations on staghorn sumac cause leaflets to become yellow near the midveins of both leaf surfaces (Reeves, 1963).

The color of the summer form varies according to the food. It may be green or dark green, or even brown when the mites feed on leaves that have turned brown. Summer adult females have four black spots, one on each side of the propodosoma, and one on each side of the hysterosoma near the posterior end. The front spots are larger than the rear ones and appear to be composed of five small dots compressed into a large triangular spot. The deutonymphs differ from the adult females by having the front spots less triangular and the dots composing them are less distinct. The larvae and protonymphs have only two small black spots. The eggs are almost colorless.

Hibernating Schoenei mites are similar to the summer forms at first, but the color gradually changes through amber to bright orange and the spots disappear. Mating takes place in the fall, but no eggs are laid until spring. There are about nine generations per year; hibernating forms, however, may develop from the fifth to the eighth generation.

The developmental rate of each of the stages is summarized as follows:

Stage	*Days at Indicated Temperatures* 25–28 C (77–82 F)	9–13 C (48–55 F)
Incubation	3	25
Larvae	1	7
Protonymph	1	11
Deutonymph	1	19
Preoviposition	1	5

Females may live up to 38 days and lay 106 eggs during their lifetime. The average egg production is about 3.7 eggs per day.

This species feeds on the lower surface of the leaves, usually close to the veins. It produces a variable amount of webbing (Cagle, 1943).

T. (*P.*) *schoenei* is similar to *T.* (*P.*) *canadensis,* differing only in that the knob or head of the aedeagus is at least twice as large as the shaft.

Tetranychus (Polynychus) sinhai Baker. This species (fig. 56, *c*) was first reported in 1961 as a pest of barley in Manitoba, Canada. It is now known as a pest of barley, wheat, corn, and sunflower in the prairie provinces of Canada and in the midwestern United States as far south as South Dakota. Feeding by this mite causes a darkening of the leaves, followed by yellowing and withering from the bend of the leaf to its tip, with occasional crinkling of the leaf edges from below. Often the entire area along the two sides of the midrib as well as isolated patches in other parts of the leaf are similarly affected. Symptoms resemble those of net blotch. Even when invaded by only a few mites, wheat leaves show a mosaiclike yellowing from the early stages of infestation (Wallace and Sinha, 1961; White, 1966).

T. (*P.*) *sinhai* adults are greenish yellow to orange, with a dark longitudinal band on each side of the dorsum. The eggs are spherical and very little webbing is produced.

Resistant plant varieties appear to offer the most promise for control. Of 165 barley varieties evaluated for their reaction to this mite, 15 were resistant and 47 moderately resistant. In general, the barley varieties grown in arid regions of the world appear to be most resistant to infestations (Sinha and Wallace, 1963).

Females of *T.* (*P.*) *sinhai* are typical for the subgenus in having longitudinal striae between the third pair of dorsocentral setae and are indistinguishable from other members of the subgenus. The male is distinctive in that the aedeagus is upturned distally, the shaft gradually narrowing with a short neck; the knob is relatively large and at a strong angle with the neck, the anterior portion being well rounded. Empodium I of the male is clawlike and has a strong mediodorsal spur; empodium II has a smaller mediodorsal spur and 3 pairs of ventrally directed hairs.

SELECTED BIBLIOGRAPHY

AGARIVAL, R. A. 1957. The sugar cane mite and its control. Indian Sugar 7:394–399.

AMBROSI, M., and R. LENARDUZZI. 1959. Prove di lotta l' "acaro giallo" della vite. Progr. Agr. 5(7):796–807.

ANDERSON, N. H., and C. V. G. MORGAN. 1958. Life histories of the clover mite, *Bryobia praetiosa* Koch and the brown mite, *B. arborea* Morgan and Anderson, in British Columbia. Can. Entomol. 40(1):23–32.

ATCHENSON, W. C. 1953. An ecological study of three species of mites on American Linden. J. Econ. Entomol. 46(4):705.

BAKER, E. W. 1968. Change of name of the strawberry spider mite. Coop. Econ. Inst. Rept. 18(47):1080.

BAKER, E. W., and A. E. PRITCHARD. 1953. A guide to the spider mites of cotton. Hilgardia 22(7):203–234.

———. 1960. The tetranychoid mites of Africa. Hilgardia 29(11):455–574.

BAKER, J. E., and W. A. CONNELL. 1961. Mites on soybeans in Delaware. J. Econ. Entomol. 54(5):1024–1026.

BEGLYAROV, G. A. 1959. On the bionomics of the hawthorne mite *Tetranychus crataegi* Hirst. [In Russian] Rev. Entomol. USSR 38(1):135–144, Moscow.

BINDA, O. S., and G. C. VARMA. 1966. A study in the control of the fig mite, *Eotetranychus hirsti* Pritchard and Baker. J. Res. Punjab Agr. Univ. 3(4):417–420.

BODENHEIMER, F. S. 1951. Citrus Entomology in the Middle East. W. Junk Pub., The Hague. 663 pp.

BOUDREAUX, H. B. 1956. Revision of the two-spotted spider mite (Acarina, Tetranychidae) complex, *Tetranychus telarius* (L.). Ann. Entomol. Soc. Am. 49(1):43–48.

———. 1958. *Tetranychus tumidus* Banks versus *Tetranychus gloveri* Banks. Ann. Entomol. Soc. Am. 51(2):174–177.

BRAVENBOER, J. L. 1959. De chemische en Biologische Bestrijding van de Spentmyt, *Tetranychus urticae* Koch. Landbouwh. Qnderz. 75:1–85.

BROOKING, B. C., JR. 1957. A study of the location and termination of diapause in *Petrobia apicalis* (Banks). Master's Thesis. La. State Univ., Dept. Zool., Entomol., and Physiol. Baton Rouge.

BULLOCK, J. A. 1963. The occurrence, sampling, and control of *Tetranychus ludeni* Zacher on Pyrethrum. East African Agr. and Forest. J. 28(4):252–254.

BUTANI, D. K. 1959. Sugarcane mites—a review. Indian Acad. Sci. Proc. B. 49(2):99–102.

BUTLER, C. D. JR., and M. K. ABID. 1965. The biology of *Oligonychus platani* on Pyracantha. J. Econ. Entomol. 58(4):687–688.

BUXTON, P. A. 1920–1921. Insect pests of dates and the date palm in Mesopotamia and elsewhere. Bull. Entomol. Res. 11(3):287–303.

CAGLE, L. R. 1943. Life history of the spider mite, *Tetranychus schoenei* McGregor. Va. Agr. Expt. Sta. Tech. Bull. 87. 16 pp.

———. 1946. Life history of the European red mite. Va. Agr. Expt. Sta. Tech. Bull. 98. 19 pp.

———. 1956. Life history of the spider mite, *Tetranychus atlanticus* McGregor. Va. Agr. Expt. Sta. Tech. Bull. 124:1–22.

———. 1962. Biology of a red spider mite, *Panonychus* sp. on raspberry in Virginia. Ann. Entomol. Soc. Am. 55(4):373–379.

CALDWELL, S. D. 1967. Cotton, a new host for the spider mite, *Eotetranychus smithi*. J. Econ. Entomol. 60(4):1169.

CALSA, R., and H. F. G. SAUER. 1952. The red spider of coffeae plantations. Biologico 18:201–208.

CANERDAY, T. D., and F. S. ARANT. 1964. The effect of late season infestations of the strawberry spider mite, *Tetranychus atlanticus*, on cotton production. J. Econ. Entomol. 57(6):931–933.

CARMONA, M. M. 1966. Acaros das plantas cultivadas em Portugal IV. Agronomia Lusitana. 26(3):175–203.

CHERIAN, M. C. 1933. The Cholam mite, *Paratetranychus indicus*, on Sorgum. Madras Agr. J. 1:1–6.

CHUNG, C., H. WEI, and Y. TIENG. 1963. The biology of the round wheat mite (*Penthaleus* sp.), and brown wheat mite, *Petrobia latens* (Muller). [In Chinese] Acta phytophyl. sin 2(3):277–284. (From Rev. Appl. Entomol. Ser. A 52:460.)

COLLYER, E., and J. R. GROVES. 1955. Some tetranychid mites on fruit trees. East Malling Eng. Res. Sta. Rept., 1955. 136 pp.

COX, H. C., and F. V. LIEBERMAN. 1960. Biology of the brown wheat mite. J. Econ. Entomol. (53(5):704–705.

CUTRIGHT, C. R. 1963. The European red mite in Ohio. Ohio Agr. Expt. Sta. Bull. 953. 32 pp.

DAS, G. M. 1959. Bionomics of the tea red spider, *Oligonychus coffeae* (Nietner). Bull. Entomol. 50(2):265–274.

———. 1960. Occurrence of the red spider, *Oligonychus coffeae* (Nietner) on tea in northeast India in relation to pruning and defoliation. Bull. Entomol. Res. 51(3):415–426.

DAS, G. M., and S. C. DAS. 1967. Effect of temperature and humidity on the development of the tea red spider mite, *Oligonychus coffeae* (Nietner). Bull. Entomol. Res. 57(3): 433–436.

DAVIS, J. J. 1961. Red spider mites on strawberries. Queensland Agr. J. Ser. 2, 87:619–620.

———. 1969. *Oligonychus araneum* sp. n. and *Oligonychus digitatus* Davis (Acarina: Tetranychidae) as pests of grasses in eastern Australia. J. Australian Entomol. Soc. 7(2):123–129.

DAVIS, J. J., and N. W. HEATHER. 1962. Control of red spiders in strawberries. Queensland J. Agr. Sci. 19(1):143–148.

DEAN, H. A. 1952. Spider mites of citrus and Texas citrus mite control in the lower Rio Grande Valley of Texas. J. Econ. Entomol. 45(6):1051–1056.

———. 1959. Seasonal distribution of mites on Texas grapefruit. J. Econ. Entomol. 52(2): 228–232.

DEPEW, L. J. 1960. Control of *Oligonychus pratensis* attacking winter wheat in western Kansas. J. Econ. Entomol. 53(6):1061–1063.

DONDALE, C. D. 1967. A model outbreak of the mite, *Tetraychus mcdanieli* McGregor in Ontario. Proc. Entomol. Soc. Ont. 98:29–45.

DORESTE, S. E. 1964. Influence of three food plants on the bionomics of *Panonychus ulmi* (Koch). Agr. Trop. 14(2):83–100.

DOSSE, G. 1963. *Bryobia rubricolus* Scheuten in the northern Bekoa region of Lebanon. Z. Pflkrankh. 70:652–666.

———. 1964. Studies on the *Tetranychus cinnabarinus* Boisd. complex in citrus plantations in Lebanon. Z. Angew. Entomol. 53(4):453–461.

DOUCETTE, C. F. 1962. The lewis mite, *Eotetranychus lewisi* on greenhouse poinsettia. J. Econ. Entomol. 55(1):139–140.

DÜZGÜNES, Z. 1965. Preliminary results of rearing two-spotted spider mites in the laboratory. Bol. Zool. Agr. BaChic. Ser. II, V. 7:73–77. (Atti del 5, Simposio Europec D' Acarologic, Milano 23–25, IX.)

EBELING, W. 1959. Subtropical fruit pests. Univ. Calif., Div. Agri. Sci. Los Angeles, Calif.

EHARA, S. 1956. Tetranychoid mites of mulberry in Japan. J. Fac. Sci. Hokkaido Univ. 6(4):499–510.

———. 1959. Mites of the subfamily Bryobiinae from Japan. J. Fac. Sci. Hokkaido Univ. Ser. 6, Zool. 14(3):185–195.

———. 1960. On some Japanese tetranychid mites of economic importance. Jap. J. Appl.

Entomol. Zool. 4(4):234–241.
———. 1963. A new mite of *Oligonychus* from rice, with notes on some Japanese spider mites. Jap. J. Appl. Entomol. Zool. 7(3):228–231.
———. 1964. The tetranychoid mites of Japan. 1st International Congress of Acarology. Fort Collins, Colo. Extrait d'Acarologia 4:409–414.
———. 1966. The tetranychoid mites of Okinawa Island. J. Fac. Sci. Hokkaido Univ. Series 6, Zool. 16(1):1–22.
ELMER, H. S. 1965*a*. Banks grass mite, *Oligonychus pratensis*, on dates in California. J. Econ. Entomol. 58(3):531–534.
———. 1965*b*. The Yuma spider mite, *Eotetranychus yumensis*, on citrus. J. Econ. Entomol. 58(3):534–536.
ENGLISH, L. L., and R. SNETSINGER. 1957*a*. The habits and control of the clover mite in dwellings. J. Econ. Entomol. 50(2):135–141.
———. 1957*b*. The biology and control of *Eotetranychus multidigituli* (Ewing) a spider mite of honey locust. J. Econ. Entomol. 50(6):784–788.
ENGLISH, L. L., and G. F. TURNIPSEED. 1941. The influence of temperature and season on the citrus red mite, *Paratetranychus citri* McGregor. J. Agr. Res. 62(2):65–67.
ESTEBANES, M. L., and E. W. BAKER. 1968. Arañas rojas de México. Ann. Esc. Nac. Cienc. Biol. Méx. 15:61–104.
EVANS, G. O., J. G. SHEALS, and D. MACFARLANE. 1961. The terrestial acari of the British Isles. I, Intro. and Biol. British Museum, London. 219 pp.
FENTON, F. A. 1951. The brown wheat mite, *Petrobia latens*. J. Econ. Entomol. 44(6): 996.
FLECHTMAN, C. H. W., and E. W. BAKER. 1970. A preliminary report on the Tetranychidae of Brazil. Ann. Entomol. Soc. Am. 63(1):156–163.
FRAZIER, N. W., and L. M. SMITH. 1946. The Williamette mite on grapes. Hilgardia 17(4):191–196.
GABELE, M. 1959. Beiträge zur Kenntnis der Guttung Bryobia. Z. Angew. Zool. 46(2): 191–247.
GAHI, M. S. 1968. Three new records of Bryobiinae from India. Indian J. Entomol. 30(1): 88–90.
GARMAN, P. 1923. Notes on the life history of the spruce mite. Conn. Agr. Expt. Sta. Bull. 247:340–342.
GASSER, R. 1951. Zur kenntnis der gemeinen spinnmilbe, *Tetranychus urticae* Koch. 1. Mitteilung: Morphology, Anatomie, Biologie und Oekologie. Mett. der Schweiz. Ges. 24(3):217–262.
GELLATELY, J. G. 1970. N.S.W. Dept. Agr. and Biol. and Chem. Res. Inst. P.M.B. 10, N.S.W., Australia Personal correspondence
GEORGALA, M. B. 1955. Thc biology of orchard mites in the western Cape Province. Union. S. Africa Dept. Agr. Sci. Bull. 360:1–13.
———. 1958. A contribution to the biology of the mite, *Bryobia praetiosa* Koch. Union S. Africa Dept. Agr. Sci. Bull. 367. 41 pp.
GLANCEY, B .M. 1958. Studies on embryonic diapause in the legume mite, *Petrobia apicalis* (Banks). Master's Thesis, La. State Univ., Dept. Zool. Entomol., and Physiol. Baton Rouge.
GÖKSU, M. E. 1968. Studies on the bionomics, control, distribution and food-plants of the hawthorn mite, *Tetranychus viennensis* Zacher, in the Marmara region. Bitke Koruma Bült. 8(3):194–213.
GOLDSMID, J. M. 1962. The mites (Acarina) of the Federation of Rhodesia and Nyasaland. Ministry Agr. P.O. Box 8025 Causeway. Salesburg Rhodesia and Nyasaland. Bull. 2162. 11 pp.
GUTIERREZ, J. P. 1967. Contribution a l'etude morphologique et biologique de *Tetranychus neocaledonicus* Andre. Coton Fibers tropicales 22(2):183–195.
HAMILTON, C. C. 1926. Insect pests of boxwood. N. J. Agr. Expt. Sta. Circ. 179. 14 pp.

HAMILTON, J. T. 1972. *Bryobia repensi* Manson, a new pasture pest and its control. J. Australian Inst. Agr. Sci. p. 62.

HARBANS, S. K., and A. S. SUDHU. 1961. Sugarcane mite and its control. Indian J. Sugarcane Res. 5:92–95.

HERBERT, H. J. 1965. The brown mite, *Bryobia arborea* Morgan and Anderson, on apple in Nova Scotia. Can. Entomol. 97 (12) :1303–1318.

HIGHTOWER, B. G., and D. P. MARTIN. 1956. Ecological studies of spider mites attacking cotton in central Texas. J. Econ. Entomol. 49(3):424–425.

HU, C. C., and L. C. WANG. 1965. A study of the annual life cycle of the tea red spider mite, *Oligonychus coffeae* (Nietner). J. Agr. Assoc. China, 50:1–14.

IGLINSKY, W. E., and C. R. RAINWATER. 1954. Life history and habits of *Tetranychus desertorum* and *bimaculatus* on cotton. J. Econ. Entomol. 47(6):1084–1086.

KANTA, S., B. K. RAI, and L. RATTAN. 1963. Evaluation of the toxicity of some pesticides to the fig mite, *Eotetranychus hirsti* Pritchard and Baker. Indian J. Entomol. 25(1): 26–32.

KHOT, K. N. S., and G. P. PATIL. 1956. Life history of the gladiolas mite, *Tetranychus equatorius* McGregor. Indian J. Sci. 18(2):149–164.

KREMER, FREDRICK-WILHELM. 1956. Studies on the biology, epidemiology, and control of *Bryobia praetiosa* Koch. Höfchenbriefe 9(4):189–256.

LAING, J. E. 1969. Life history and life table of *Tetranychus urticae* Koch. Acarologia 11(1):32–42.

LAMINMAN, J. F. 1935. The pacific mite, *Tetranychus pacificus* McGregor, in California. J. Econ. Entomol. 28(6):900–903.

LEUCK, D. B., and R. O. HAMMONS. 1968. Resistance of wild peanut plants to the mite, *Tetranychus tumidellus*. J. Econ. Entomol. 61(3):687–688.

LO, P. K. C. 1968. (R. O. C. March '57) Tetranychoid mites infesting tea in Taiwan. Taiwan Agr. Res. Inst. Chung-San Academic Cultural Affairs Ser. 1:275–285. Taipei, Taiwan, China.

LO, P. K. C., and D. N. T. HSIA. 1968. (R. O. C. March '57). Tenuipalpid and tetranychid mites infesting citrus in Taiwan, and life history study of the citrus green mite, *Schizotetranychus baltazarae* Rimando. Taiwan Agr. Res. Inst. Chung-San Academic Cultural Affairs Ser. 1:253–274. Taipei, Taiwan, China.

MCGREGOR, E. A. 1956. Mites on citrus trees in southern California. Mem. So. Calif. Acad. Sci. 3:5–42.

MCGREGOR, E. A., and F. L. MCDONOUGH. 1917. Red spider on cotton. U. S. Dept. Agr. Bull. 416:1–72.

MCKENZIE, H. L. 1935. Biology and control of avocado insects and mites. Calif. Agr. Expt. Sta. Bull. 592:1–48.

MCMURTRY, J. A., and H. G. JOHNSON. 1966. An ecological study of the spider mite, *Oligonychus punicae* (Hirst). Hilgardia 37(11):363–402.

MACPHEE, A. W. 1963. The effect of low temperatures on some predacious phytoseiid mites, and on the brown mite, *Bryobia arborea* Morgan and Anderson. Can. Entomol. 95(1):41–44.

MALCOLM, D. R. 1955. Biology and control of the timothy mite, *Paratetranychus pratensis* (Banks). Wash. State Agr. Expt. Sta. Tech. Bull. 17. 35 pp.

MANSON, D. M. C. 1968. The spider mite family Tetranychidae in New Zealand. The genus *Bryobia*. Acarologia 9(1):76–123.

———. 1968. *Panonychus elongatus* Manson. A description and comparison with *P. citri* (McGregor). J. Australian Entomol. Soc. 7(1):6–10.

MANSON, M. G. R., and M. S. GHAI. 1968. Three new records of Bryobiinae from India. Indian J. Entomol. 30(1):88–89.

MATHYS, G. 1954. Contribution e'ethologique a'la résolution du complexe *Bryobia praetiosa* Koch. Mitt Schweiz. Entomol. Gen. 27(2):137–146.

MATTHYSSE, J. G., and J. A. NAEGELE. 1952. Spruce mite and southern red mite control experiments. J. Econ. Entomol. 45(3):383–387.

MELLOTT, J. L. 1968. *Panonychus caglei*, new species. The raspberry red mite. Acarologia 12(2):230–234.

MELLOTT, J. L., and W. A. CONNELL. 1965. Notes on the life history of *Tetranychus atlanticus*. Ann. Entomol. Soc. Am. 58(3):379–383.

MEYER, M. K. P., and M. RODRIGUES. 1966. Acari associated with cotton in southern Africa. Revista da Junta de Inves. do Ultramar. 13(2):1–33.

MOHAMED, I. I. 1963. Acarine mites occurring on cotton plants in Egypt. Bull. Soc. Entomol. Egypte 46:511.

———. 1965. Host preference of the citrus brown mite *Eutetranychus banksi* (McGregor). Bull. Soc. Entomol. Egypte 48:163–170.

MOUTIA, L. A. 1958. Contribution to the study of some phytophagous acarina and their predators in Mauritius. Bull. Entomol. Res. 49:59–75.

MÜLLER, G. F. W. 1957. Morphology, biology, and control of the hawthorne spider mite, *Tetranychus viennensis* Zacher. Höfchenbriefe 10(1):1–60.

MUMA, M. H. 1970. *Oligonychus milleri* on *Pinus caribaea* in Jamaica. Florida Entomol. 53(4):241.

NAGARAJAN, K. R. 1957. A short note on *Paratetranychus oryzae* Hirst. Madras Agr. J. 44 (10) :480.

NICKEL, J. L. 1958. Agricultural insects of the Paraguayan Chaco. J. Econ. Entomol. 51(5):633–637.

NIELSEN, F. 1958. Biology of the McDaniel mite, *Tetranychus mcdanieli* McGregor, in Utah. J. Econ. Entomol. 51(5):588–592.

NUBER, K. 1961. Overwintering of the red spider mite, *Tetranychus urticae* Koch in hop gardens. Höfchenbriefe 14(11):6–15.

OATMAN, E. R., C. A. FLESCHNER, and J. A. MCMURTRY. 1967. New highly destructive spider mite present in California. J. Econ. Entomol. 60(2):477–480.

OSAKABE, M. 1965. Seasonal fluctuations of population density of the tea red spider mite, *Tetranychus kanzawai* Kishida, in the tea plantation. Jap. J. Appl. Entomol. Zool. 9(3): 206–210.

PARENT, B., and A. A. BEALIEU. 1957. Life history of the European red mite. Can. Entomol. 89(7):328–333.

PIERCE, W. C. 1953. Studies of mites and their control on pecan in Louisiana. J. Econ. Entomol. 46(4):561–565.

PRITCHARD, A. E., and E. W. BAKER. 1952. A guide to the spider mites of deciduous fruit trees. Hilgardia 21(9):253–287.

———. 1955. A revision of the spider mite family Tetranychidae. Pac. Coast Entomol. Soc. Mem., Ser. 2. 472 pp.

QUERSHI, A. H., E. R. OATMAN, and C. A. FLESCHNER. 1969. Biology of the spider mite, *Tetranychus evansi* Pritchard and Baker. Ann. Entomol. Soc. Am. 62(4):898–903.

RAHMAN, K. A., and A. N. SAPRA. 1940. Biology of the mite, *Paratetranychus indicus* (Hirst), a pest of sugarcane in the Punjab. Indian J. Entomol. 2:208–212.

———. 1945. On the biology of the vegetable mite, *Tetranychus cucurbitae* Rahman and Sapra. Indian J. Agr. Sci. 15(3):124–130.

RAMAKAEV, KH. KH. 1966. The brown fruit mite (*Bryobia redikorzevi* R.) in the orchard of the experimental teaching farm "Communist" of Khar'kov Agric. Inst. in 1958–1964. [In Russian] (From Rev. Appl. Entomol. A., 56:1680 [1968]).

RAMBIER, A. 1958. Les tetranychus nuisibles à la vigne en France continental. Rev. Zool. Agr. 57(1–3):1–20.

RATTAN, L., B. K. RAI, and S. KANTA. 1962. Evaluation of ovicidal and residual efficacy of some pesticides against the sugarcane mite, *Schizotetranychus andropogoni* Hirst. Indian J. Agr. Sci. 32(4):305–308.

REEVES, R. M. 1963. Tetranychidae infesting woody plants in New York State, and a life history study of the elm spider mite, *Eotetranychus matthyssei* n. sp. Cornell Univ. Agr. Expt. Sta. Mem. 380:99 pp.

REYNOLDS, H. T., and J. E. Swift. 1951. Control of *Petrobia latens* in the Imperial Valley of California. J. Econ. Entomol. 44(5):642–645.

RIES, D. T. 1935. A new mite (*Neotetranychus buxi* Garman) on boxwood. J. Econ. Entomol. 28(1):55–62.

RIMANDO, L. C. 1962. The tetranychoid mites of the Philippines. Univ. Phil. Coll. Agr. Tech. Bull. 11:1–52.

RODRIGUES, M. C. 1967. Mites on cotton in Mozambique, Agronomia Mozamb. 1(3): 141–147.

ROSSI DE SIMONS, N. H. 1966. Déscrípción de *Schizotetranychus oryzae* n. sp. Revista. Invest. Agropec. Ser. 3. Patol. Veg. 3(1):1–10.

ROUSSEL, J. S., J. C. WEBBER, L. D. NEWSOM, and C. E. SMITH. 1951. The effect of infestation by spider mite, *Septanychus tumidus* on growth and yield of cotton. J. Econ. Entomol. 44(4):523–527.

SABA, F. S. 1971. *Tetranychus yusti* McGregor, a spider mite of potential economic importance. J. Econ. Entomol. 64(1):141–144.

SCHUSTER, M. F. 1959. Chemical control of *Tetranychus marianae* McGregor on tomatoes in the lower Rio Grande Valley, J. Econ. Entomol. 52(4):763–764.

SEPASGOSARIAN, H. 1956. Morphologie und Biologie der Gelben Affelspinnmilbe, *Eotetranychus pomi* n. sp. Z. Angew. Zool. 43(4):435–491.

SERAFIMOVSKI, A., and W. THALENHORST. 1962. Biologische und Okologische Beobachtungen an der Fitchenspinnmelbe, *Paratetranychus ununguis*. Anz. Schodlingsk. 36: 37–42.

SHINKAJI, N. 1954. A review of ecological studies on spider mites. Jap. Agr. and Hort. 29:1365:1497–1500.

SIDDIG, M. A., and E. A. ELBADRY. 1971. Biology of the spider mite, *Eutetranychus sudanicus*. Ann. Entomol. Soc. Am. 64(4):806–809.

SIMONS, J. N. 1964. Tetranychid mites as defoliators of cotton. J. Econ. Entomol. 57(1): 145–148.

SINHA, R. N., and H. A. H. WALLACE. 1963. *Tetranychus sinhai* Baker, a new pest of cereals—varietal reaction of barley. Can. Entomol. 95(6):588–596.

SMITH, C. E., and J. C. WEBBER. 1954. The legume mite, *Petrobia apicalis* (Banks), a pest of several winter-growing legumes. La. State Univ. Agr. Expt. Sta. Tech. Bull. 493:1–25.

SMITH, F. F., and E. W. BAKER. 1968. Names of the two-spotted spider mite and the carmine spider mite to be redesignated. U. S. Dept. Agr. Coop. Econ. Ins. Rept. 18(47: 1080.

SMITH, R. H. 1939. Observations on the ilicis mite, *Paratetranychus ilicis*. Bull. Calif. Dept. Agr. 28(5):412–414.

SPECHT, H. B. 1963. *Oligonychus bicolor* in Nova Scotia. Can. Entomol. 95(10):1021–1022.

TOMASEVIC, B. 1964. The yellow popular mite, *Eotetranychus populi* Koch. [In Russian] (Zast Bilja 15(82):687–693. (From Rev. Appl. Entomol. Ser. A, 55:1039.)

———. 1965. On the development and ecology of the brown fruit mite, *Bryobia redikorzevi* Reck. J. Sci. Agr. Res. 18(59):121–132.

TSUGAWA, C. M. Y., and S. SHIRASAKI. 1961. Forecasting the outbreak of destructive insects in apple orchards. III. Forecasting the initial date of hatch in respect of the overwintering eggs of the European red mite, *Panonychus ulmi* (Koch) in Aomori prefecture. Jap. J. Appl. Entomol. Zool. 5(3):167–173.

TUTTLE, D. M., and E. W. BAKER. 1968. Spider mites of southwestern United States and a revision of the family Tetranychidae. Univ. Ariz. Press, Tucson, Arizona. 143 pp.

UBERTALLI, J. A. 1955. Life history of *Eotetranychus uncatus* Garman. J. Econ. Entomol. 48(1):47–49.

VAN EYNDHOVEN, G. L. 1957. Biologische Bemerkungen uber das genus *Bryobia*, Notulae ad Tetranychidae 8. Proc. IV. Intern. Congr. of Crop. Prot. VII, (2):633–634.

WALLACE, H. A. H., and R. N. SINHA. 1961. Note on a new mite disease of barley and other cereals. Can. J. Plant Sci. 41(4):871.

WARD, C. R., E. W. HUDDLESTON, J. C. OWENS, T. M. HILLS, L. G. RICHARDSON, and D. ASHDOWN. 1972. Control of the Banks grass mite attacking grain sorghum and corn in west Texas. J. Econ. Entomol. 65(2):523–529.

WESTIGARD, P. H., and D. W. BERRY. 1970. Life history and control of the yellow spider mite on pear in southern Oregon. J. Econ Entomol. 63(5):1433–1437.

WHITE, L. D. 1966. Plant-feeding mites of South Dakota. So. Dak. Agr. Expt. Sta. Bull. 27 pp.

WILLIAMS, F. X. 1931. The insects and other invertebrates of Hawaiian sugar cane fields. Hawaiian Sugar Planters Assoc. 400 pp.

YOKOYAMA, K., and G. ISHII. 1934. Studies on the mites attacking mulberry leaves. 2. Morphology and biology of *Panonychus mori* Kishida. Jap. Bull. Sericult. Expt. Sta. Japan. 8(9):425–454.

ZEIN-ELDEN, E. A. 1956. Studies on the legume mite, *Petrobia apicalis* (Banks). J. Econ. Entomol. 49(3):291–296.

TAXONOMIC REFERENCES TO THE TETRANYCHIDAE

(including information on hosts and biology)
(indexed by species)

A. Species reviewed by A. E. Pritchard and E. W. Baker, 1955.
(A revision of the spider mite family Tetranychidae. Pac. Coast Entomol. Soc., Mem. Ser. 2.)

B. Species reviewed by D. M. Tuttle and E. W. Baker, 1968.
(Spider mites of southwestern United States and a revision of the family Tetranychidae. Univ. Ariz. Press, Tucson, Ariz.)

C. Species reviewed by E. W. Baker and A. E. Pritchard, 1960.
The tetranychoid mites of Africa. Hilgardia 29(11):526.)

Taxonomic references to other species are in the text.

Chapter 9

Tenuipalpidae Berlese

The family Tenuipalpidae, described by Berlese in 1913 in a private publication, has long been neglected as a family of economic importance. The group has been known by several names—Trichadenidae, Pseudoleptidae, and now Tenuipalpidae. A few generic revisions were published before Pritchard and Baker (1951) surveyed the tenuipalpids of California, and established the modern classification. In 1958 the same authors published on this family on a worldwide basis.

Knowledge of the biology and control of the Tenuipalpidae or false spider mites is fragmentary. Some investigators, however, have made important contributions concerning those species important to agricultural and horticultural crops.

The false spider mites are reddish, slow moving, and usually feed on plant leaves, most commonly on the lower surfaces near the midrib or veins. Some species feed on the bark of plants, some in the floral heads or under leaf sheaths of grasses, and the most specialized members of the family form plant galls within which they feed. Most species are not of economic importance, either because their hosts are not economically important plants or their populations remain below economically injurious levels. This chapter deals with the few species that are injurious to citrus, tea, grapes, and ornamentals.

SYSTEMATICS

General Description of the Family

Generic and specific characters are to be found in the number of marginal hysterosomal setae, the number of dorsocentral setae, the number or lack of mediolateral setae, and the type of setae. Other characteristics include the genital region of the female, the number of palpal segments and their setation, the leg setation, and especially the reticulate pattern to be found on the dorsum of these mites. In most cases the false spider mites are relatively host restricted.

The Tenuipalpidae are tetranychoid mites with both tenent hairs on the tarsal claws and empodia, and long recurved whiplike chelicerae set into a stylophore. The family is distinctive as follows:

There is no palpal claw complex; the distal segment is terminal. The tracheae consist of two anteriorly directed tubes ending in simple bulbs that may be associated with the longitudinal folds of the invagination for the stylophore. The legs are short and wrinkled; there are solenoididia on the distal ends of tarsi I and II of the female which always has two on tarsus I and one or two on tarsus II; the male has a pair of solendidia on both tarsus I and II. The true tarsal claws may be hooked or padlike and with tenent hairs; the empodium consists of a pad with tenent hairs. A rostral shield may or may not be present. The body is divided into a propodosoma and a hysterosoma, and in the male, as well as the females of *Pseudoleptus*, the hysterosoma is further divided into the metapodosoma and opisthosoma.

Basis of Classification with the Family

The dorsal chaetotaxy varies between genera and species. There are 3 pairs of propodosomal setae, but the number of setae on the hysterosoma may vary. There are usually 3 pairs of dorsocentral hysterosomal setae, but 1 or 2 pairs may be lacking. The humeral setae, which are considered to be the first pair of lateral hysterosomal setae, are always present; there are also 5 to 7 pairs of other dorsolateral setae. A series of dorsosublateral setae—1 to 4 pairs—may occur between the lateral and dorsocentral setae. The dorsosublateral setae are usually in a longitudinal line, but may be displaced. The venter of the body has a fairly constant setal pattern, which varies for the genus *Tenuipalpus* and *Dolichotetranychus*. The genitalia of the female consist of a simple trapdoorlike platelet usually with 2 pairs of posterior setae; there may be a ventral platelet anterior to the genital plate; the presence or absence of this plate and the shape of the genital plate is of generic importance. The male genitalia consist of terminal stylets; the aedeagus is long and tapering, the sperm duct enters its funnel-shaped anterior end.

Some of the palpal characters used in the generic classification will probably not be stable, and it is here suggested that the family be separated into groups using setal and genital characters.

For collection and preservation of these mites see chapter 7 on the Tetranychidae.

The Tenuipalpidae, although a plant-feeding family, has only a relatively few species belonging to a few genera that are of economic importance. The genera discussed are marked in the the key by asterisks.

Key to the Genera of Tenuipalpidae

1. Palpus with 4 or 5 segments 2
   Palpus with 3 or less segments 7
2. Hysterosoma with 4 pairs of dorsosublateral setae 3
   Hysterosoma with 2 or less pairs of dorsosublateral setae 4
3. Female with 4 pairs of legs and 3 pairs of anal setae; male with 4 pairs of genitoanal setae . . . . *Aegyptobia* Sayed
   Female with 3 pairs of legs and 2 pairs of anal setae; male with 3 pairs of genitoanal setae . . . . *Phytoptipalpus* Trägårdh
4. Hysterosoma with 2 pairs of dorsosublateral setae; palpus with 4 or 5 segments 5
   Hysterosoma with less than 2 pairs of dorsosublateral setae; palpus with 4 segments . 6
5. Rostral shield, when present and incised, with broad lobes; female with ventral plate . . . . *Pentamerismus* McGregor
   Rostral shield with narrow, acutely pointed lobes; female without ventral plate *Pseudoleptus* Bruyant

6. Hysterosoma with 1 pair of dorsosublateral setae. . *Cenopalpus* Pritchard and Baker*
   Hysterosoma without dorsosublateral setae. . . . . *Brevipalpus* Donnadieu*
7. Podosoma not strongly differentiated from opisthosoma. 8
   Podosoma very broad and opisthosoma narrow. . . . . *Tenuipalpus* Donnadieu*
8. Palpus with 1 or 2 segments. 9
   Palpus with 3 segments; idosoma slender. . . . . *Dolichotetranychus* Sayed*
9. Palpus with 2 segments. 10
   Palpus with a single segment fused with rostrum. 13
10. Dorsosublateral setae present. 11
    Dorsosublateral setae absent. 12
11. Hysterosoma with 4 pairs of dorsosublateral setae; rostral shield absent. . . . . *Raoiella* Hirst*
    Hysterosoma with 3 pairs of dorsosublateral setae; rostral shield present in female . . . . *Phyllotetranychus* Sayed
12. Gnathosoma completely covered by propodosoma. . . . . *Tegopalpus* Womersley
    Gnathosoma not covered by propodosoma; a pair of caudal setae flagellate; genital and anal plates contiguous. . . . . *Colopalpus* Pritchard and Baker
    Gnathosoma not covered by propodosoma; without whiplike setae. . . . . *Priscapalus* DeLeon
13. Adults and nymphs with 4 pairs of legs; hysterosoma with 2 pairs of dorsosublateral setae. . . . . *Obdulia* Pritchard and Baker
    Alults and nymphs with 3 pairs of legs; hysterosoma with 1 pair of dorsosublateral setae. . . . . *Larvacarus* Baker and Pritchard

INJURIOUS TENUIPALPID MITES

Brevipalpus Donnadieu

The palpus consists of four segments. There are 5 or 6 pairs of dorsolateral hysterosomal setae; dorsosublateral setae are not present. The genital plate of the female is usually squarish or rectangular, with a similar anterior ventral plate. The body is oval, and narrows posteriorly. All species are described from the female.

Brevipalpus californicus (Banks) (fig. 57, *a*) at times causes serious injury to a wide variety of ornamental plants and agricultural crops. It is well distributed throughout Asia (except USSR), Africa, Australasia, Pacific Islands, Mexico, Hawaii, North, Central, and South America, and the West Indies. It is the cause of "Leprosis" of citrus in Argentina and Florida, and it is a serious pest of tea in Sri Lanka (Ceylon), India, and Java.

The first observable injury on orchids, for example, appears on the leaves as silvery areas that frequently become sunken and brown. Seriously infested leaves become yellow and drop from the plant. The mites feed on all parts of the plant and, on this host, develop a black pattern on their body (Manglitz and Cory, 1953).

Individuals of *B. californicus* are difficult to see because they lie flat against the leaf surface and are slow to move, but populations can be detected by the light cast skins and full red color of the mites against the green fruit background. Toxic substances injected as mites feed on citrus in Florida result in a condition on leaves and fruits originally known as "nail-head-rust," and symptoms on twigs and branches as "Florida Scaly Bark" (pl. 44, *a*, *c*). It is now known that *Brevi-*

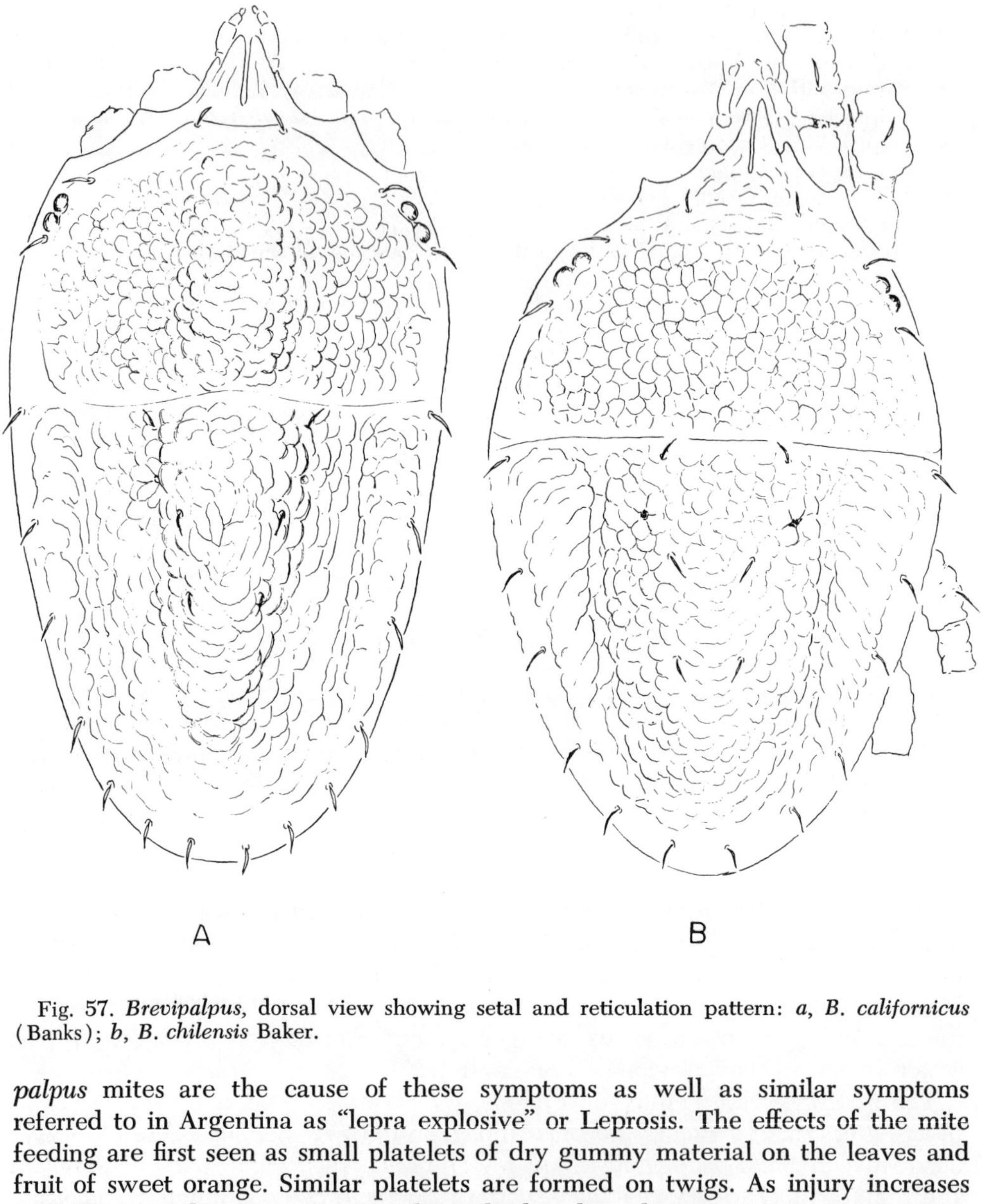

Fig. 57. *Brevipalpus,* dorsal view showing setal and reticulation pattern: *a, B. californicus* (Banks); *b, B. chilensis* Baker.

palpus mites are the cause of these symptoms as well as similar symptoms referred to in Argentina as "lepra explosive" or Leprosis. The effects of the mite feeding are first seen as small platelets of dry gummy material on the leaves and fruit of sweet orange. Similar platelets are formed on twigs. As injury increases and the injured twigs grow into large limbs, these lesions increase proportionately in size resembling scrofulous shelling of bark (Knorr, Webster, and Malaguti, 1960).

This species may become abundant in California, but there are no symptoms that are remotely suggestive of the Leprosis symptoms seen in Florida or Argentina. Injury attributed to *B. californicus* in Spain and in South Africa consists of a brown speckling of the rind on sweet orange fruits. Cell damage is restricted in depth to the thickness of the flavedo. Feeding by *californicus* causes a silvering of the fruit, particularly lemons. In Australia it is known as the silver mite. Feeding by this species and *B. phoenicus* in Texas causes rind spotting of grapefruit (see *B. phoenicus*, below, for description of this injury).

All stages of *B. californicus* develop on the undersides of tea leaves, particularly along the petiole and midrib of mature leaves. A reddish discoloration occurs on the parts attacked. Intensive prolonged attack leads to an extensive darkening of the lower foliage, to a scorched appearance in the basal part of the leaf, and a general reduction in size of the newer foliage. No twisting or distortion occurs, but mature foliage often drops, sometimes resulting in almost complete defoliation. The production of new growth is reduced on partially defoliated bushes, and such bushes often fail to survive the necessary pruning operation.

Under normal conditions these mites do not attain economic proportions for three years subsequent to pruning of the vines; if conditions are particularly favorable, however, injury may occur in 6 months. Population development is favored by shade, light pruning, or a regular succession of dry weather with periods of relatively light rainfall (Baptist and Ranawerra, 1955).

The eggs of *B. californicus* are elliptical, and bright red. They are covered with a sticky substance during oviposition so become tightly glued to the leaf surface. The incubation period lasts about 9 days at 18 to 24 C (65 to 75 F) and 55 percent relative humidity. At first the larvae are dull red; but after feeding a characteristic black pattern usually begins to form. The days required for each developmental stage at 21 to 30 C (70 to 85 F) are: larva 8.6 days, protonymph 6.2 days, and each deutonymph 7.0 days; each quiescent period requiring 3.6 days. The adult female resembles the nymph, but the male does not have a black pattern. Parthenogenesis occurs and there is a high proportion of females to males. Females begin oviposition about 3.8 days after the last molt and lay about 1 egg per day for 25 or more days (Manglitz and Cory, 1953).

This species is susceptible to sulfur, dicofol, and chlorobenzilate, but not to the organophosphorus or carbamate types of acaricides. Control of this mite may be accomplished by providing a host-free period, by discontinuing plucking the tea for a period of time, the length of the period depending on the degree of infestation; and by destroying the mites on other hosts. Branches of infested shade trees of *Albizia* and *Grevillea*, for example, should be removed and burned and acaricides applied to the healthy parts of these trees at weekly intervals. Tea treated with sulfur should not be plucked for 3 to 4 weeks to avoid harvesting tainted tea leaves (Baptist and Ranawerra, 1955).

B. californicus has 6 pairs of short hysterosomal dorsolateral setae plus the humeral setae. Tarsus II has 2 distal solenidia; there are 3 setae on the distal palpal segment; the reticulate pattern usually covers the entire propodosoma.

The Chilean false red mite of grapes, *Brevipalpus chilensis* Baker. This mite (fig. 57, *b*) is a very destructive pest of grapevines in Chile. It also attacks several different species of fruit, forest trees, ornamentals, and even annual weeds. These include citrus, almond, fig, chrysanthemum, geranium, morning-glory, and bindweed.

The mites feed on the lower surface of the leaves. Feeding causes a dark reddish discoloration, leading to dropping of the leaves and a reduction in size of the

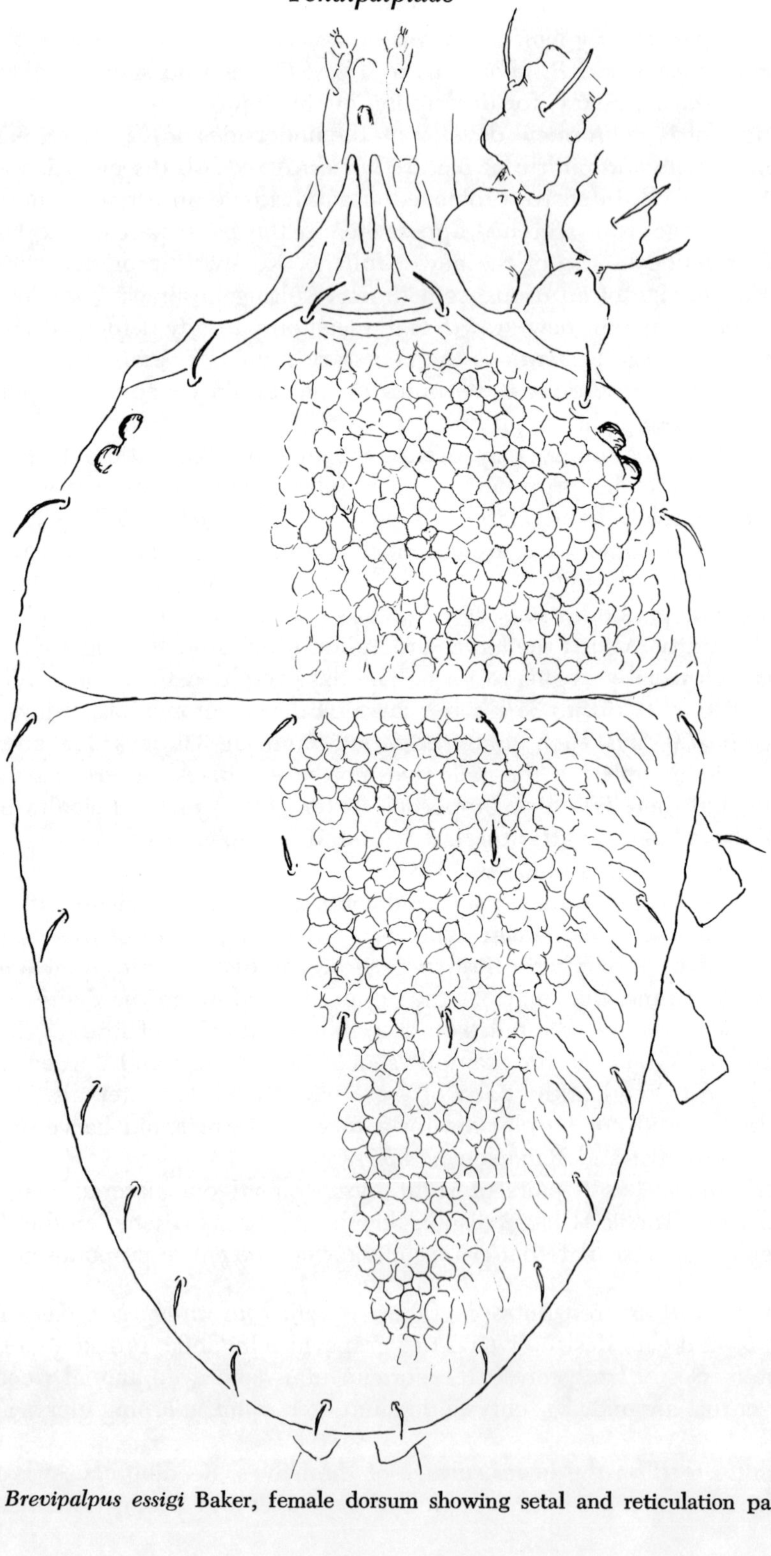

Fig. 58. *Brevipalpus essigi* Baker, female dorsum showing setal and reticulation pattern.

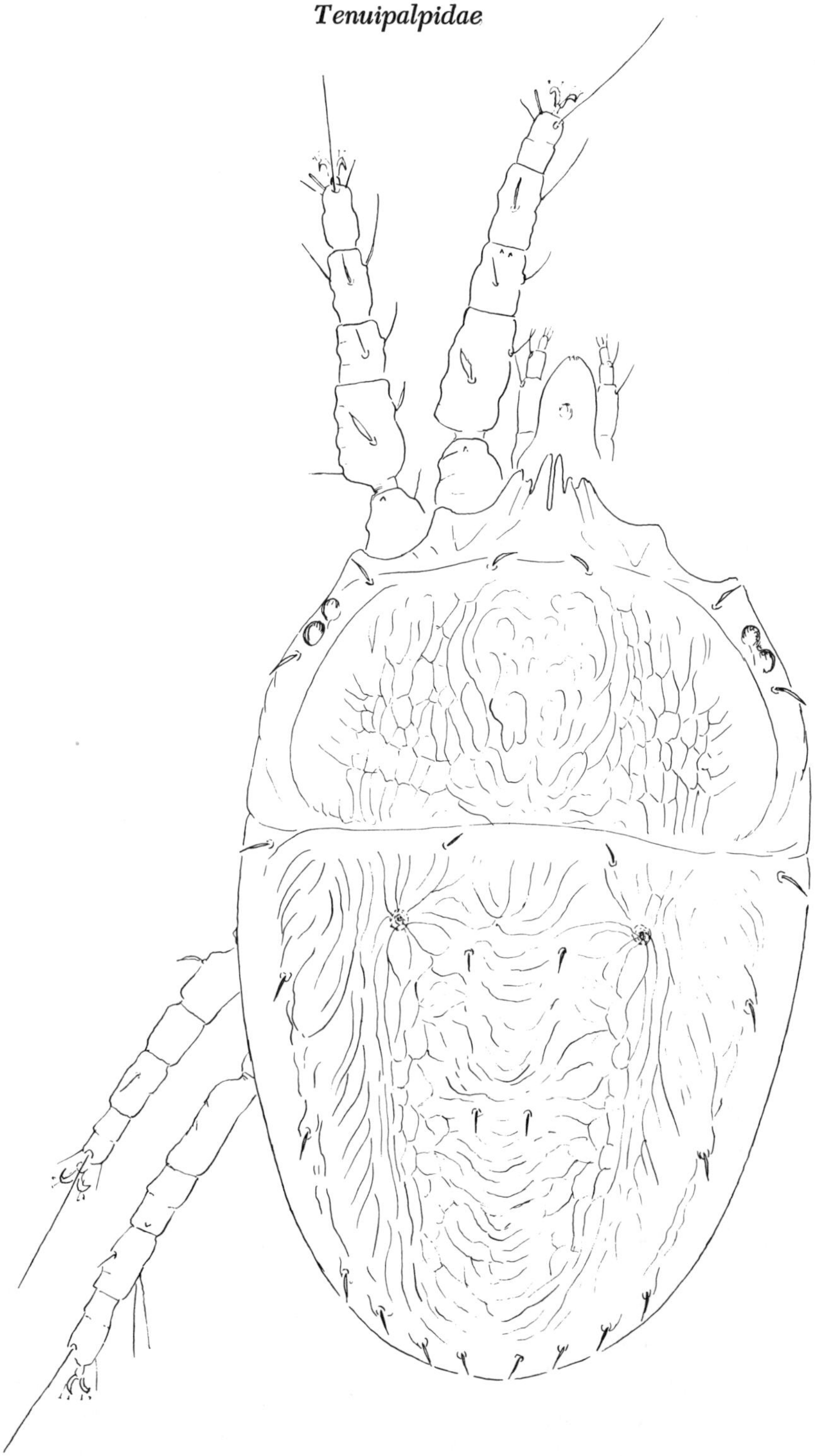

Fig. 59. ***Brevipalpus lewisi*** (McGregor), female dorsum showing setal and reticulation patterns.

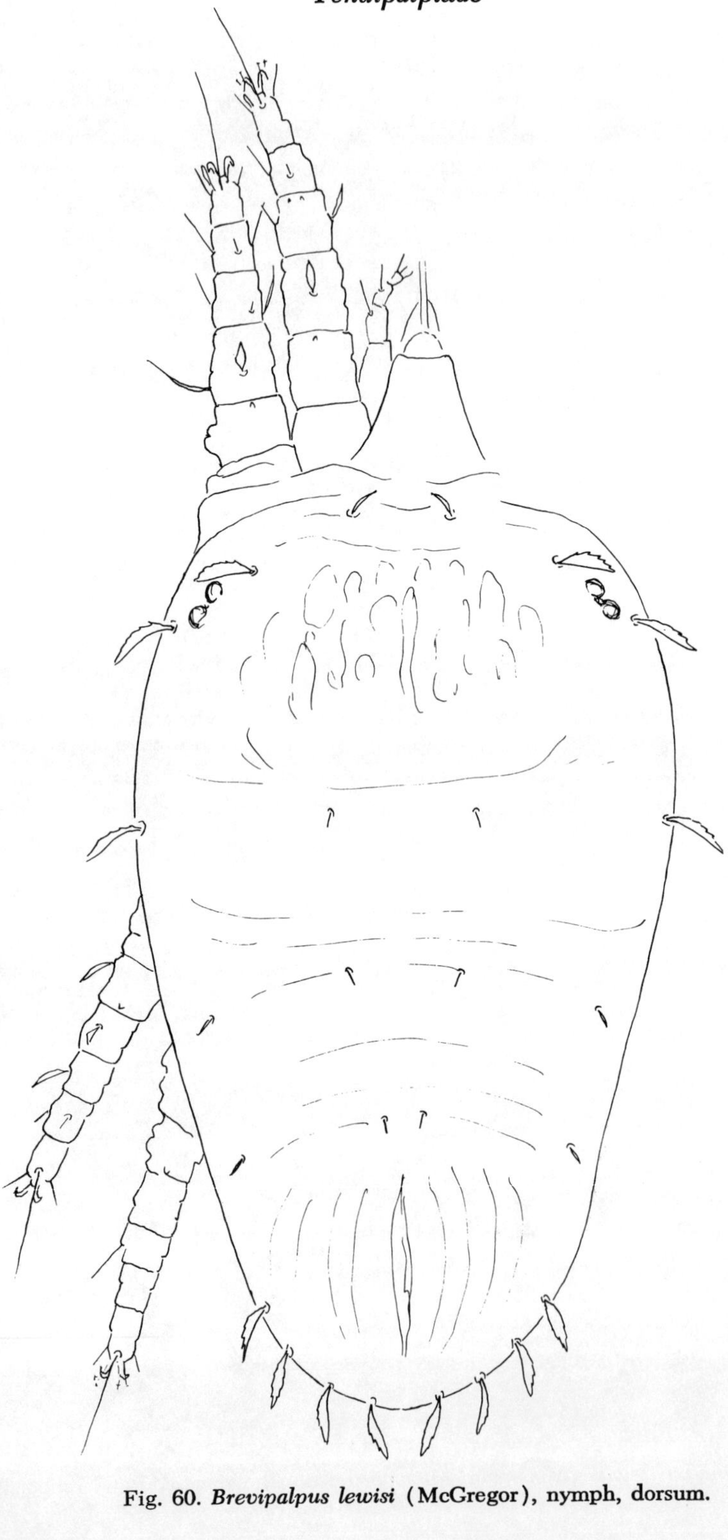

Fig. 60. *Brevipalpus lewisi* (McGregor), nymph, dorsum.

new growth. The most serious damage to vineyards results from spring attacks to young shoots.

The mites overwinter as adult fertilized females, mostly in colonies hidden under bark and crevices of the host plant. In central Chile the mites crawl to the upper part of the plant and begin depositing eggs in October. The average time required to complete a generation is 25.3 days (range is 18 to 59). There are 3 to 6 generations per year depending on the climate (Gonzales, 1968).

Natural enemies are not effective in controlling populations of this mite species. Chlorobenzilate, dicofol, and sulfur have been effective summer treatments. Acaricide applications should be started as early as possible in the spring with additional applications every 15 to 20 days. Dormant type oil sprays are not effective (Hernera-Villamil, 1958).

B. chilensis has 5 pairs of short hysterosomal dorsolateral setae plus the humeral setae; the distal segment of the palpus has 3 setae; tarsus II has a single distal solenidion; the propodosoma is evenly reticulate mediodorsally.

Brevipalpus essigi Baker (fig. 58) lives on *Aucuba, Ficus,* fuchsia, *Howea,* pittosporum, sage, speedwell, butterfly bush, and on orchids in California and Mexico. The biology and the effects of this species on its host are unknown (Baker, 1949).

This species has 6 pairs of short hysterosomal dorsolateral setae plus the humeral setae; the distal segment of the palpus bears 3 setae. There is only a single solenidion on the distal portion of tarsus II; the reticulate pattern is evenly distributed over the entire propodosoma and hysterosoma.

Citrus flat mite, *Brevipalpus lewisi* McGregor. This mite (pl. 45, *a, b*; figs. 59, 60) is a pest of citrus, pomegranates, walnuts, grapes, and many ornamental plants. It is particularly recognized as a pest of citrus in Japan and in the San Joaquin and desert valleys of California, and of grapes in Bulgaria. Distribution additionally includes Arizona, Maryland, North Carolina, Australia, Lebanon, and Egypt. Most of the more than 30 hosts of this species are ornamental plants.

The citrus flat mite is found most abundantly at the stem end of citrus fruit, near or under fruit "button." Eggs are laid singly in cracks and crevices of the fruit, twigs, and leaves. They are oval and pink. The mites prefer green to ripe fruit, but fruit at any stage is preferred to the leaves; thus harvesting the fruit reduces the population. The young and adults are small, flat, slow moving, and light reddish brown to bright red.

The mites overwinter in the adult stage in central California, but in the Imperial and Coachella Valleys they are active throughout the year on citrus, but inactive on grapes. Peak populations occur during the warmest months because periods of extremely high temperatures and low humidities have no deleterious influence on the mite populations, as is the case with the citrus red mite and Yuma mite in these districts. The mites prefer to feed on areas of citrus fruit which have been injured by leafhoppers or citrus thrips. Ordinarily these injured areas tend to become inconspicuous as the fruit changes from green to yellow; conspicuous scablike scars, however, develop when mites feed on such areas (pl. 45, *b*).

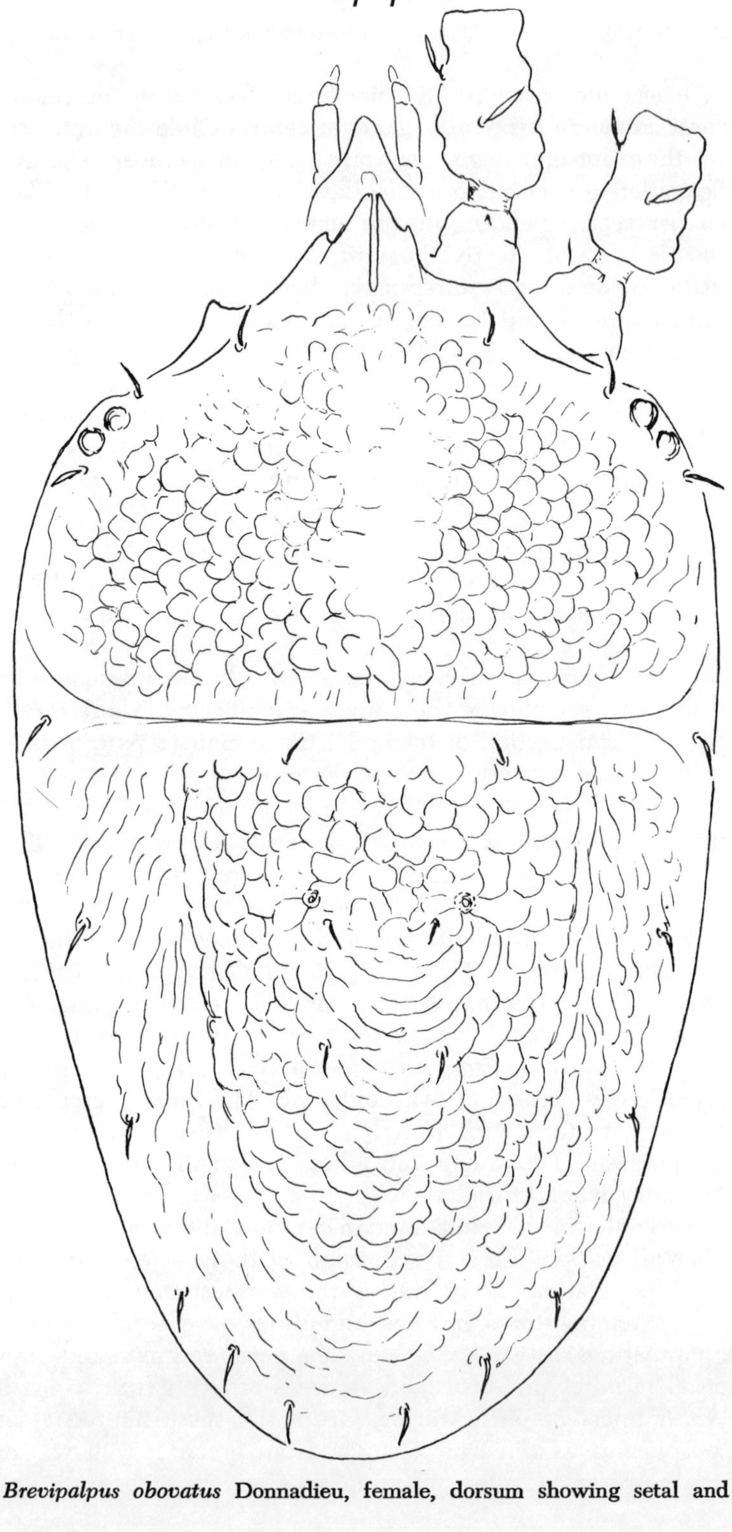

Fig. 61. ***Brevipalpus obovatus*** Donnadieu, female, dorsum showing setal and reticulation pattern.

Heavy populations may cause much of the fruit surface to be scarred. Damage consists almost entirely of reduction in the quality (grade) of fruit. No injury is produced on leaves or wood by infestations of this species. The scablike scars produced by this mite on most varieties of citrus fruit rarely occur on grapefruit (Elmer and Jeppson, 1957).

Feeding by these mites produces a brownish discoloration on pomegranate. Such discoloration usually begins at the stem end and is followed by cracking of the rind and the formation of scab tissue between the cracks.

Infestations of the citrus flat mite on grapes occur on all the green parts of the vine. Feeding injury prevents the development of the berries. There are 4 generations a year in Bulgaria and California where it feeds and reproduces throughout the summer. The mites leave the green parts of the vine in autumn and move to the stems where they overwinter as colonies in cracks in the bark. When spring temperatures reach 20.6 C (69 F), the overwintering females move to the opening buds. The bases of the new shoots become infested, often resulting in death of the shoots (Raikov and Nachev, 1965). *B. lewisi* infests the lower limbs of walnuts and produces scorchlike spots on the leaves often defoliating the infested limbs.

This mite is susceptible to sulfur dusts or sprays. Best results on citrus are obtained when sulfur is applied in late winter and early spring before high temperatures occur. When potentially injurious infestations appear during the summer, after it is too hot to apply sulfur, plant injury by the mite may be prevented by dicofol or chlorobenzilate sprays. Organophosphorus acaricides are not generally effective. Populations on grapes may be controlled by applications of lime-sulfur during the dormant or winter period.

B. lewisi has 6 pairs of short hysterosomal dorsolateral setae plus the humeral setae; obvious hysterosomal pores are present. Tarsus II has only a single solenidion; there are 3 setae on the distal segment of the palpus; the reticulate pattern does not meet dorsally on the propodosoma.

Brevipalpus lilium Baker. This mite occurs in Washington, Oregon, California, Florida, and Hawaii and infests many economic plants including azalea, croton, hibiscus, jasmine, lily, apple, sumac, brambles, berries and grapes; also species of *Acalypha*, *Allamanda*, *Cedrela*, *Dipladenia*, *Ixora*, *Lagerstroemia*, *Sida*, *Thumberqia*, and *Vitex*. This mite is similar to *B. californicus* (Banks), but possesses only a single solenidion on tarsus II of the female.

Privet mite, *Brevipalpus obovatus* Donnadieu. This species (fig. 61; pl. 46, *a*, *b*) is not only a pest of privet and citrus, but of more than 50 genera of ornamental plants. It is distributed throughout the United States, Canada, France, Spain, Cyprus, Israel, Iran, Egypt, Kenya, Sri Lanka (Ceylon), South Africa (Angola), New Zealand, Australia, Japan, Hawaii, Venezuela, and Argentina (Baker and Tuttle, 1964).

The appearance of the plant injury produced by the privet mite varies with the plant species (pls. 46, *c*, *d*, 47). The mites feed on the ventral surface of the leaves and on stems and petioles. The first evidence of injury on fuchsias appears as faint brown flecks. Each fleck is actually a sunken spot where substance of the cells have been removed or the cell has been killed by the toxic saliva injected by the mite while feeding. Additional feeding causes the flecks to merge and leaves to become brownish or bronzy. At this stage a red discoloration occurs on the upper leaf surface and leaf drop begins to occur. Feeding injury on anthuriums produces yellow areas on the upper and brown areas on the lower leaf surfaces. Feeding on sweet orange leaves causes large chlorotic spots, but the privet mite is not a serious pest of commercial citrus in the United States (Morishita, 1954).

Large chlorotic spots with concentric rings of chestnut-colored resinous substance are produced on leaves of sweet orange in Argentina by the privet mite. Such spots may cover ⅓ of the leaf surface. Similar but somewhat smaller spots are produced on the fruit. Ring spots are formed also on the twigs. In later stages these spots erupt to produce scaling of the bark that resembles scaly bark psorosis. Experimental evidence indicates that this injury, known as lepra explosive or Leprosis, is caused by toxins injected by this mite during the feeding process (Vergani, 1945).

Privet mite occurs on citrus in Venezuela where it may be found on trees showing Leprosis symptoms; trees artificially infested with privet mites, however, fail to develop symptoms of Leprosis. In Venezuela, the privet mite is associated with halo scab, which occurs on leaves where sour orange seedlings are grown in humid situations. When both *B. obovatus* and *B. phoenicis* (Geijskes) occur on the same plant, more serious effects on the leaves and stems occur. These symptoms are described under the latter species (Knorr et al., 1960).

The egg of the privet mite is elliptical, bright orange red when first laid, becoming darker until just before hatching, at which time the chorion assumes an opaque, white appearance. The larvae and nymphs are orange red with dark areas on the hysterosoma. The adult females vary in color, ranging from light orange to dark red with various patterns of dark pigmentation. The extent of such pigment is correlated with the host and amount of feeding. The body of the female is broadest at the suture between the propodosoma and hysterosoma.

The winter is passed mostly in the adult stage. The mites congregate around the base of the plants or in sheltered places on the ventral surface of leaves. Eggs and immature stages may occasionally be found during the winter months. The mites reproduce and develop throughout the year in greenhouses kept under favorable temperature and humidity. Females of this mite produce female offspring parthenogenetically and males are rarely found.

The length of the life cycle increases as the average temperature is decreased, as shown below.

Temperature, C	*Average number days in each life stage at different temperatures*					
	Larva	Protonymph	Deutonymph	Total	Life span of adult	Eggs laid
32	3.9	2.7	4.5	11.1	3.0	0.5
30	3.5	4.1	2.7	10.3	23.4	32.1
27	5.3	4.0	4.0	13.3	38.1	54.3
25	7.1	5.0	7.1	19.2	52.5	44.1
20	9.5	7.9	9.2	26.6	67.0	3.0

Like other *Brevipalpus* mites the privet mite is susceptible to sulfur, but not to the organophosphorus insecticides except diazinon. Besides sulfur this species appears to be most susceptible to dicofol, although other acaricides such as chlorobenzilate and ovex may also result in effective control of the privet mite (Morishita, 1954).

B. obovatus has 5 pairs of short hysterosomal dorsolateral setae plus the humeral setae; hysterosomal pores are present. There is a single solenidion on tarsus II; there are three setae on the distal segment of the palpus; the body striae are irregular dorsomedially on the propodosoma and hysterosoma.

Brevipalpus oleae Baker. *B. oleae* occurs on olive leaves in Morocco and *B. olearius* Sayed occurs in Egypt where it feeds on the bark of olive trees. These species are potential pests of olive trees (Baker, 1949).

B. oleae and *B. olearius* are similar, differing only in the type of dorsal body setae. The 2 species may be described as follows: There are 6 pairs of dorsolateral setae plus the humeral setae; the distal segment of the palpus has three terminal setae; tarsus II has only one rodlike seta; the rostrum reaches to the distal end of genu I; and the dorsal body setae are shorter than the length between their bases. In *B. oleae* the dorsal body setae taper distally; in *B. olearius* these setae are subclavate.

Brevipalpus oncidii Baker (fig. 62, *a, b,*) is an occasional pest of orchids, *Oncidium* spp., and *Odontoglossum* spp., in greenhouses in California and England (Baker, 1949). There are 6 pairs of short hysterosomal dorsolateral setae plus the humeral setae. There is a single solenidion on tarsus II; there are 3 terminal setae on the distal segment of the palpus; the reticulate pattern covers the entire dorsum of the body; the anterior and posterior podosomal ventral setae are short and of equal length.

Brevipalpus phoenicis (Geijskes). This mite (fig. 63, *a*) is an important pest of citrus and tea, but it has also been found on coffee, peach, papaya, loquat, coconut, apple, pear, guava, olive, fig, walnut, grape, and more than 50 genera of ornamental plants. It is evidently well distributed throughout the world and has been reported from Holland, Spain, Portugal, Sicily, Italy, Kenya, Tanganyika, Ethiopia, Mauritius, India, Malaya, Taiwan, Syria, Hawaii, California, Texas, District of Columbia, Florida, Cuba, Trinidad, Argentina, Brazil, Venezuela, Okinawa, Philippines, and Australia.

The toxic saliva injected into the host as a result of feeding by this species results in unusually severe injury to citrus in certain areas of the world, but it does not appear to produce leprosis symptoms. High populations in Florida occasionally cause diffuse chlorotic spotting of foliage of orange trees, symptoms that have been called *phoenicis blotch*. The spots resemble early stages of Leprosis, but there is no gumming of affected areas. The symptoms may be produced by either nymphs or adults. Damage seems to be limited to plants previously defoliated by various stress factors.

Populations of *B. phoenicis* on sour orange seedlings in Venezuela cause a disorder known as *halo scab*. These symptoms occur when the mites infest scabs

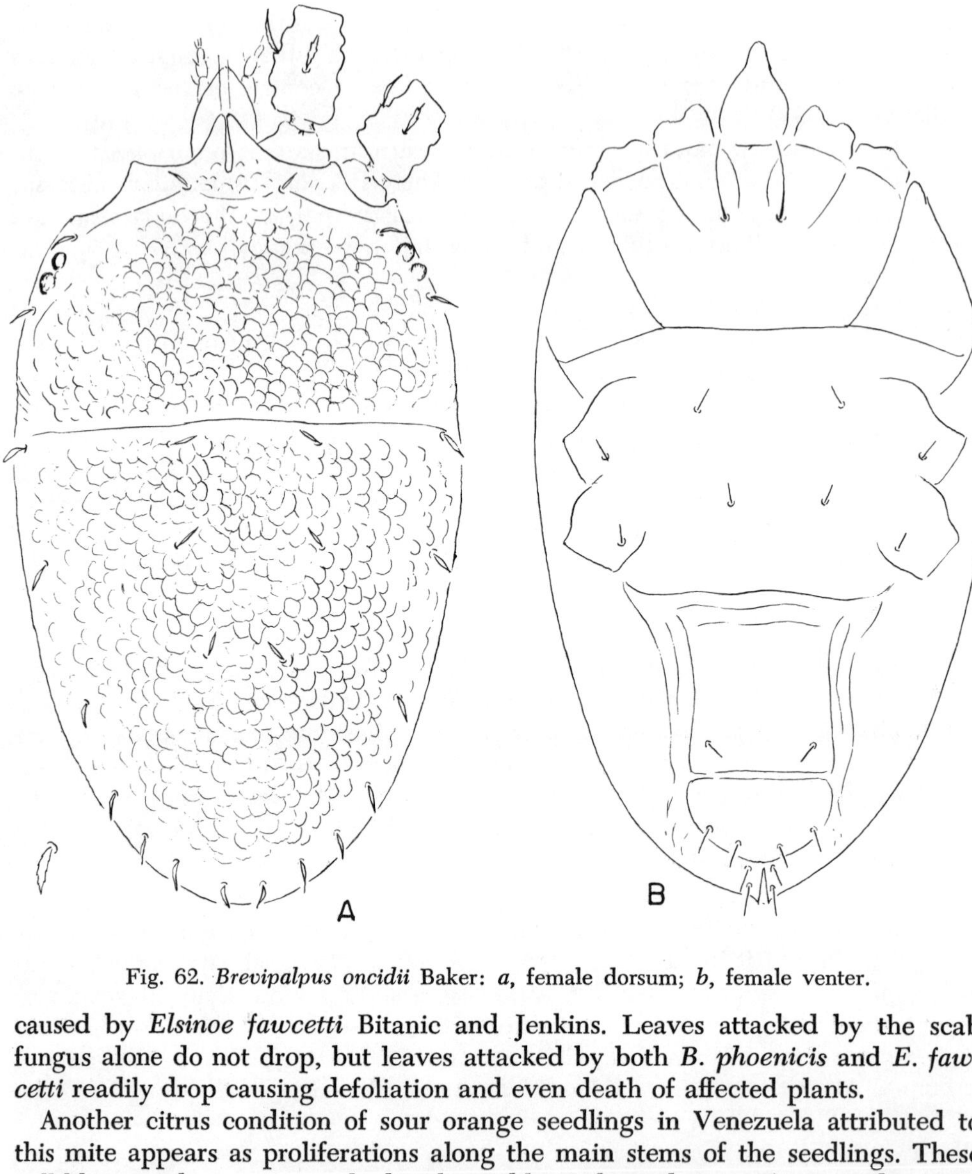

Fig. 62. *Brevipalpus oncidii* Baker: *a*, female dorsum; *b*, female venter.

caused by *Elsinoe fawcetti* Bitanic and Jenkins. Leaves attacked by the scab fungus alone do not drop, but leaves attacked by both *B. phoenicis* and *E. fawcetti* readily drop causing defoliation and even death of affected plants.

Another citrus condition of sour orange seedlings in Venezuela attributed to this mite appears as proliferations along the main stems of the seedlings. These gall-like protuberances may be barely visible to those that are 5 mm in diameter. The galls are woody and appear like axils that have proliferated to the point where bud-studded cushions are developed. No leaves develop at the axils occupied by these cushions, and when all the buds are replaced by cushions the trees become devoid of leaves and soon die. Gall formation follows the abscission of the initial leaves; adventitious buds sprout but are successively killed, producing hypertrophies at the bud loci (Knorr, 1964; Knorr, et al., 1960; pl. 44, *a*, *b*).

B. phoenicis and *B. californicus* have been associated with a rind spotting of grapefruit in Texas. The leprosislike spotting occurs only when populations are high. Spots appear as irregular-shaped brownish blemishes that vary in size from

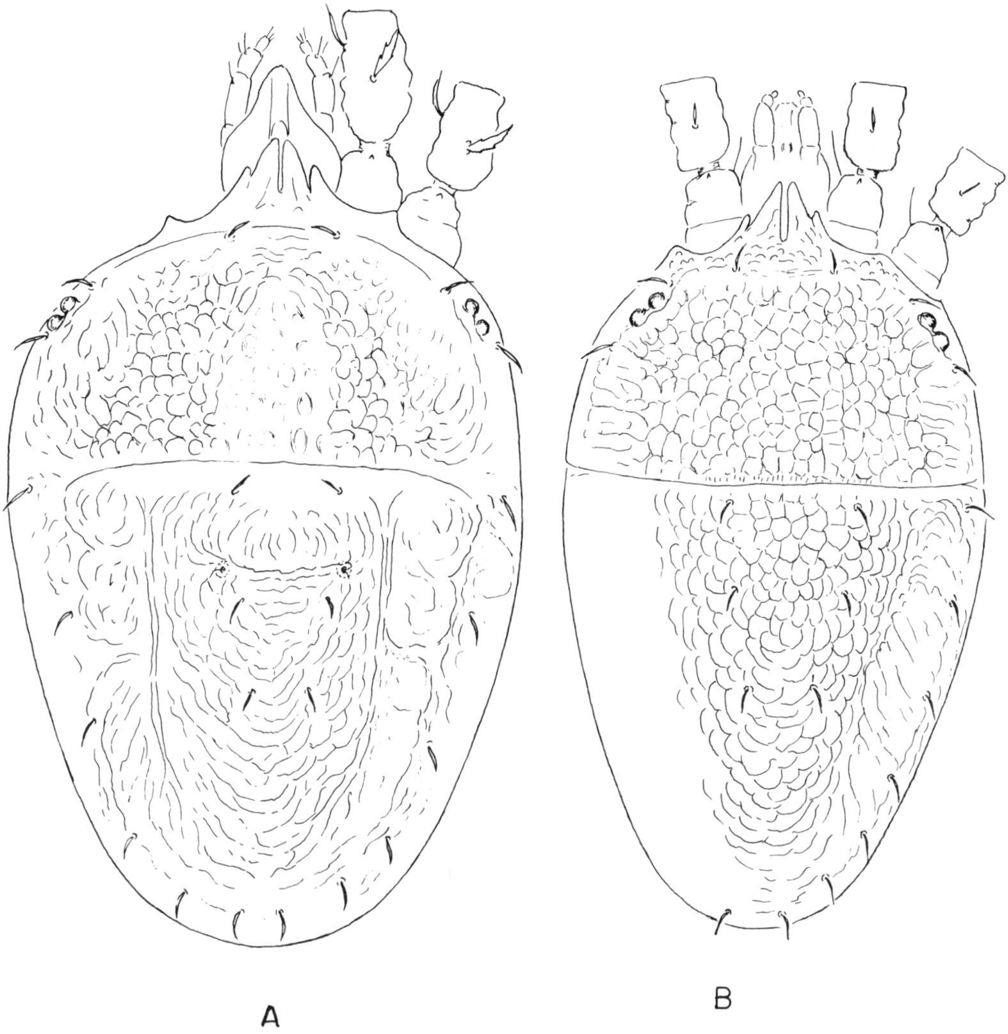

Fig. 63. *a*, *Brevipalpus phoenicis* (Geijskes), female dorsum; *b*, *Brevipalpus russulus* (Boisduval), female dorsum.

1 to 30 mm or larger. Each spot contains an aggregation of small spots sometimes seemingly fused together. Spots are most common first on the sides and stylar end of the fruit, but as feeding continues the spots may cover the fruit. Initially the spots are level with the surface of the peel, but on storage the spots tend to become raised and darker in color (Dean and Maxwell, 1967). *B. phoenicis* causes bronze colored scablike spots on citrus in South Africa (Schwartz, 1970).

The biology and general appearance of this mite are very similar to *B. obovatus* and *B. californicus*. As these species are often found on the same plants it has been difficult to determine the species responsible for the symptoms.

As with other *Brevipalpus* mites this species is susceptible to sulfur, dicofol, and chlorobenzilate and not to the organophosphorus or carbamate compounds.

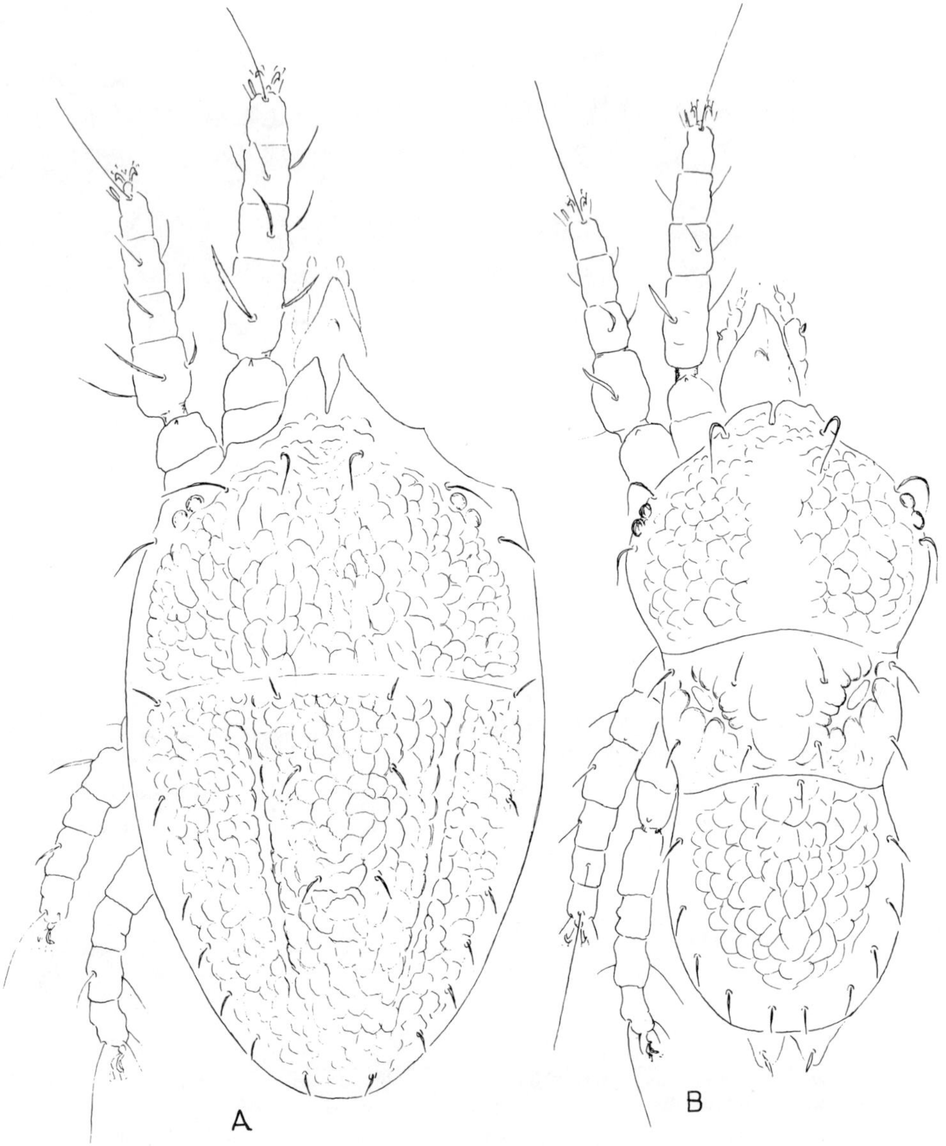

Fig. 64. *Brevipalpus sayedi* Baker: *a*, female dorsum; *b*, male dorsum (not typical).

B. phoenicis has 5 pairs of short hysterosomal dorsolateral setae plus the humeral setae. There are two solenidia on tarsus II: hysterosomal pores are present; the dorsocentral area of the propodosoma is covered with a scallop pattern and the dorsocentral area of the hysterosoma has irregular striae.

Brevipalpus russulus (Boisduval). *B. russulus* (fig. 63, *b*) is a pest of cactus and succulents in France, Germany, Belgium, Netherlands, California, Mexico, Peru, Argentina, and Japan. It not only is a pest of ornamental cactus, but causes dam-

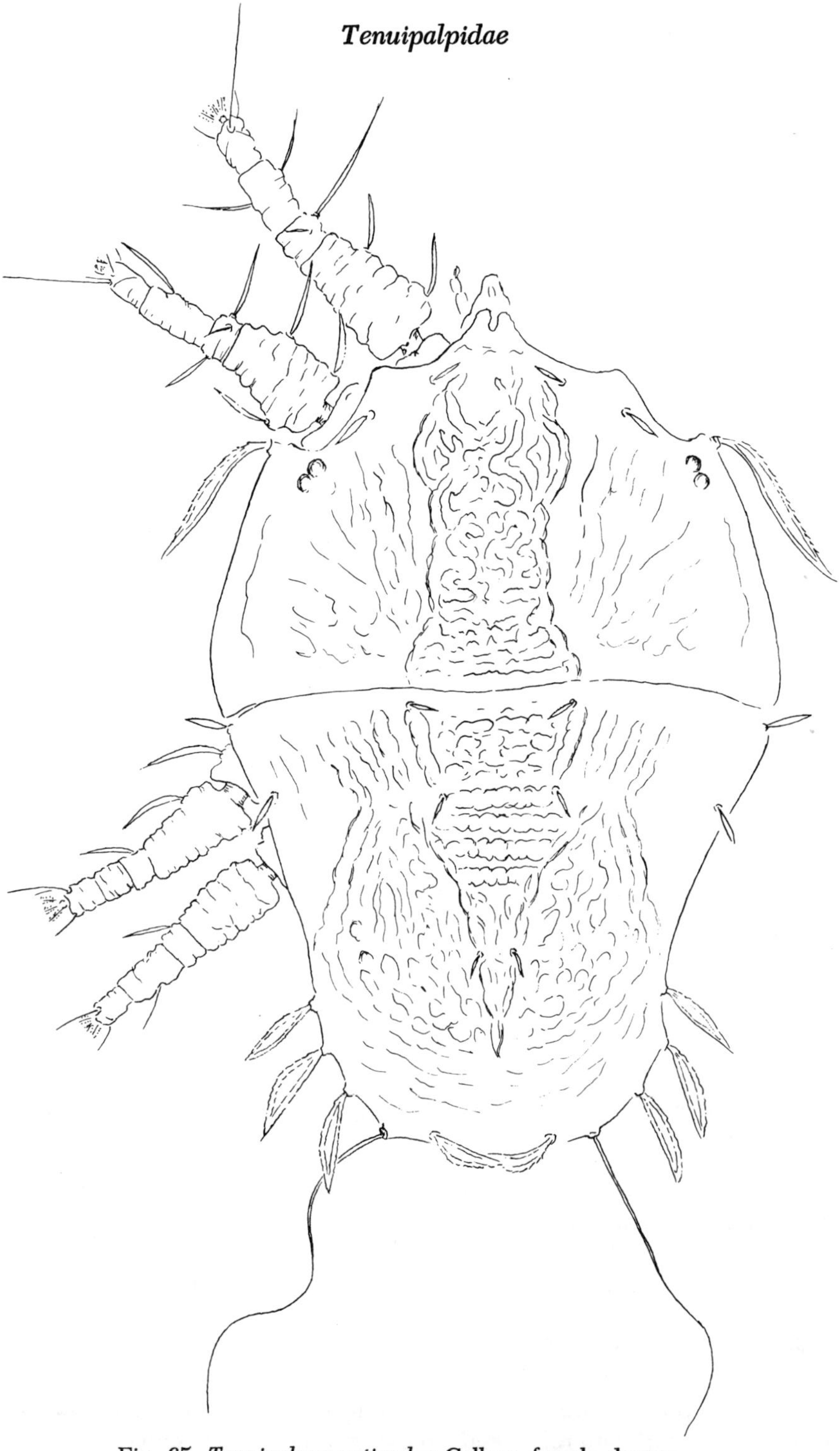

Fig. 65. *Tenuipalpus antipodus* Collyer, female dorsum.

age to range plants. Infested plants are a uniform reddish grey (Evans, Sheals, and MacFarlane, 1961). It has recently been causing much damage to cactus in Brazil.

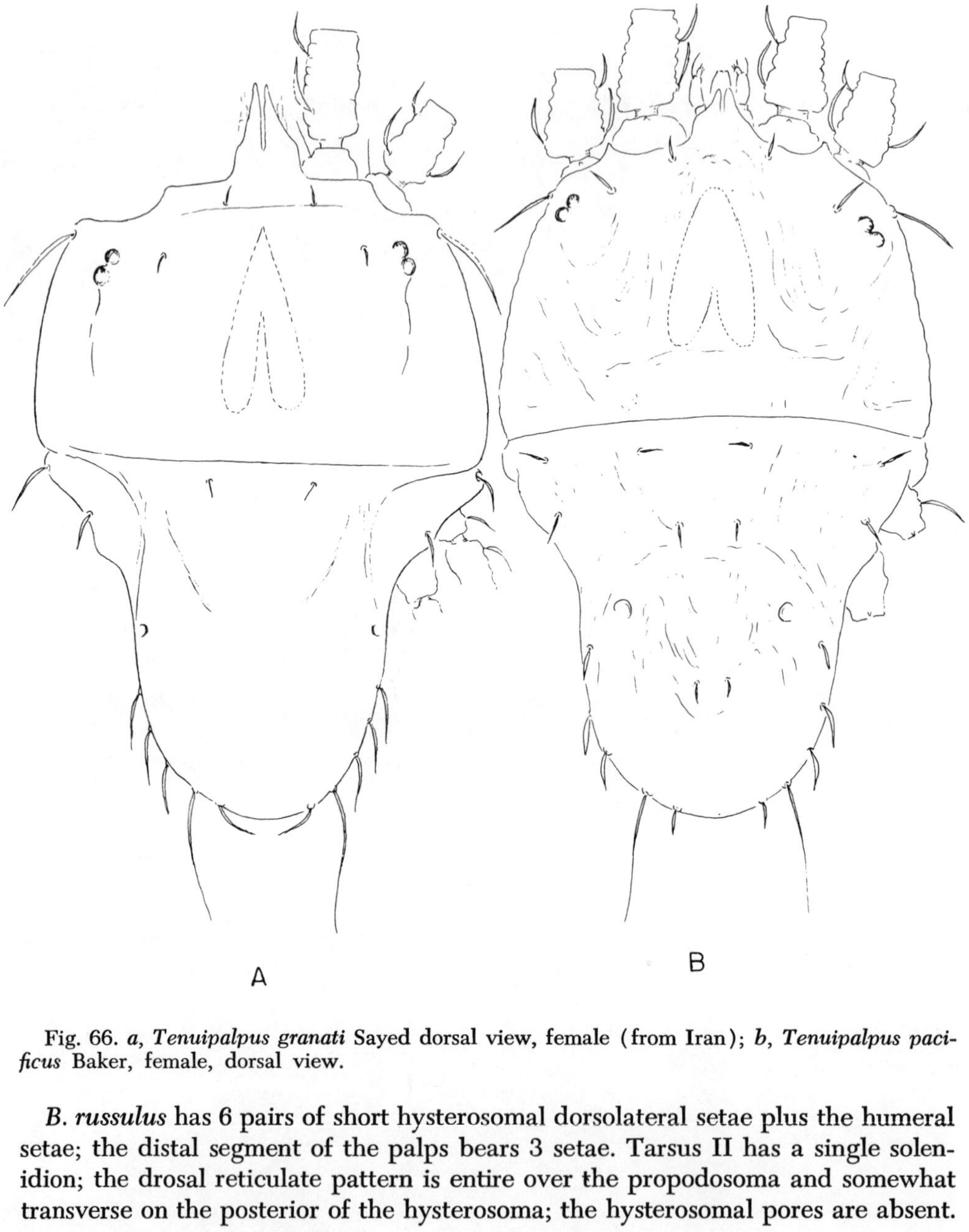

Fig. 66. *a, Tenuipalpus granati* Sayed dorsal view, female (from Iran); *b, Tenuipalpus pacificus* Baker, female, dorsal view.

B. russulus has 6 pairs of short hysterosomal dorsolateral setae plus the humeral setae; the distal segment of the palps bears 3 setae. Tarsus II has a single solenidion; the drosal reticulate pattern is entire over the propodosoma and somewhat transverse on the posterior of the hysterosoma; the hysterosomal pores are absent.

Brevipalpus sayedi Baker. This species (fig. 64, *a, b*) occurs on hickory and pecan (*Carya*) in Florida, Maryland, and Indiana. It has 6 pairs of short hysterosomal dorsolateral setae plus the humeral setae; the hysterosomal pores are not present; there is a narrow mediolateral dorsal groove on the hysterosoma; the reticulate pattern is large and entire dorsally; and there are three setae on the distal palpal segment.

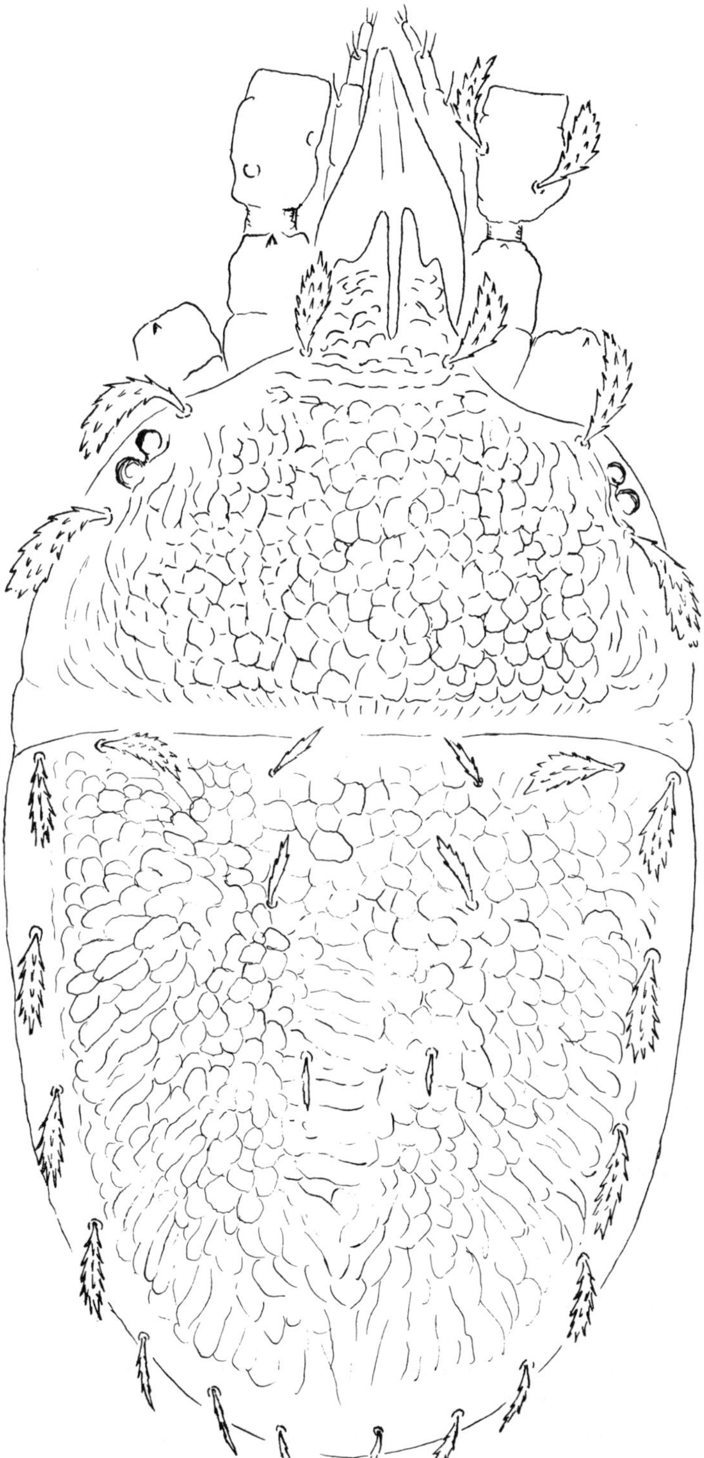

Fig. 67. *Cenopalpus lanceolatisetae* (Attiah), female dorsum.

Tenuipalpus Donnadieu

Tenuipalpid mites have a varying number of palpal segments; the posterior margin of the hysterosoma usually possesses a pair of long, flagellate setae; there are 5 or 6 pairs of dorsolateral setae plus the humeral setae; the podosoma is distinctly broad and the opisthosoma is narrow; the genital and anterior ventral plates are fused and not distinct from one another.

Tenuipalpus antipodus Collyer (fig. 65) infests tea on Taiwan Island, and occurs on several bush plants belonging to the genera *Coprosma* and *Polyscias* on the North Island of New Zealand. These mites are reddish yellow and deposit flattened orange eggs (Lo, 1968).

T. antipodus has a single pair of simple anterior medioventral setae on the venter of the podosoma, and a pair of simple posterior medioventral setae; the paired anterior ventral plate setae are pilose as are the genital plate and anal setae. Dorsally the mite is distinct in the rugose striation pattern, in that the first 2 pairs of propodosomal setae are very short, and in that the third pair of propodosomal setae and posterior caudal setae of the hysterosoma are large and lanceolate serrate.

Tenuipalpus granati Sayed. This mite (fig. 66, *a*) is an occasional pest in vineyards in both lower and upper Egypt. It also infests leaves, small branches, and sometimes the fruit of pomegranate trees. In addition, this species has been reported from Iran, Greece, and from Georgia, Kazakhstan, and Azerbaidzhan in the USSR. It prefers the lower surface of pomegranate leaves. Populations increase during the spring, reaching the highest peak in July, followed by a gradual decrease until December (Zaher and Elbadry, 1964).

T. granati has a single pair of short simple setae on the venter of the podosoma, and 2 pairs of simple posterior medioventral setae; the paired ventral plate setae and 2 pairs of genital plate setae are short and simple. The dorsal body sculpture consists of a few longitudinal markings; there is only a single pair, the anterior, of dorsocentral hysterosomal setae; the 4 pairs of posterior caudal setae are slender and lanceolate; the posterior flagellate setae are short.

Tenuipalpus pacificus Baker. This species (fig. 66, *b*) is a pest of orchids (Orchidaceae) in California, Florida, Panama, Australia, Siam, Java, England, Germany, and Holland. Feeding by these mites causes dark spots on the leaves and eventual necrosis of the tissue (pl. 48).

The development of this species is slow. The incubation period requires 18 to 23 days, the developmental stages each 14 to 15 days. The duration of the life cycle is at least 64 days (Dosse, 1954).

T. pacificus has 2 pairs of anterior medioventral podosomal setae, the inner pair being short; there are 2 pairs of posterior medioventral setae, the inner pair being much longer than the outer pair. There is a single pair of anterior plate setae, and 2 pairs of genital plate setae; all setae are nude. The first 2 pairs of propodosomal setae are about ½ as long as the third pair; the posterior marginal setae are about equal in size to the first 2 pairs of propodosomal setae; the flagellate

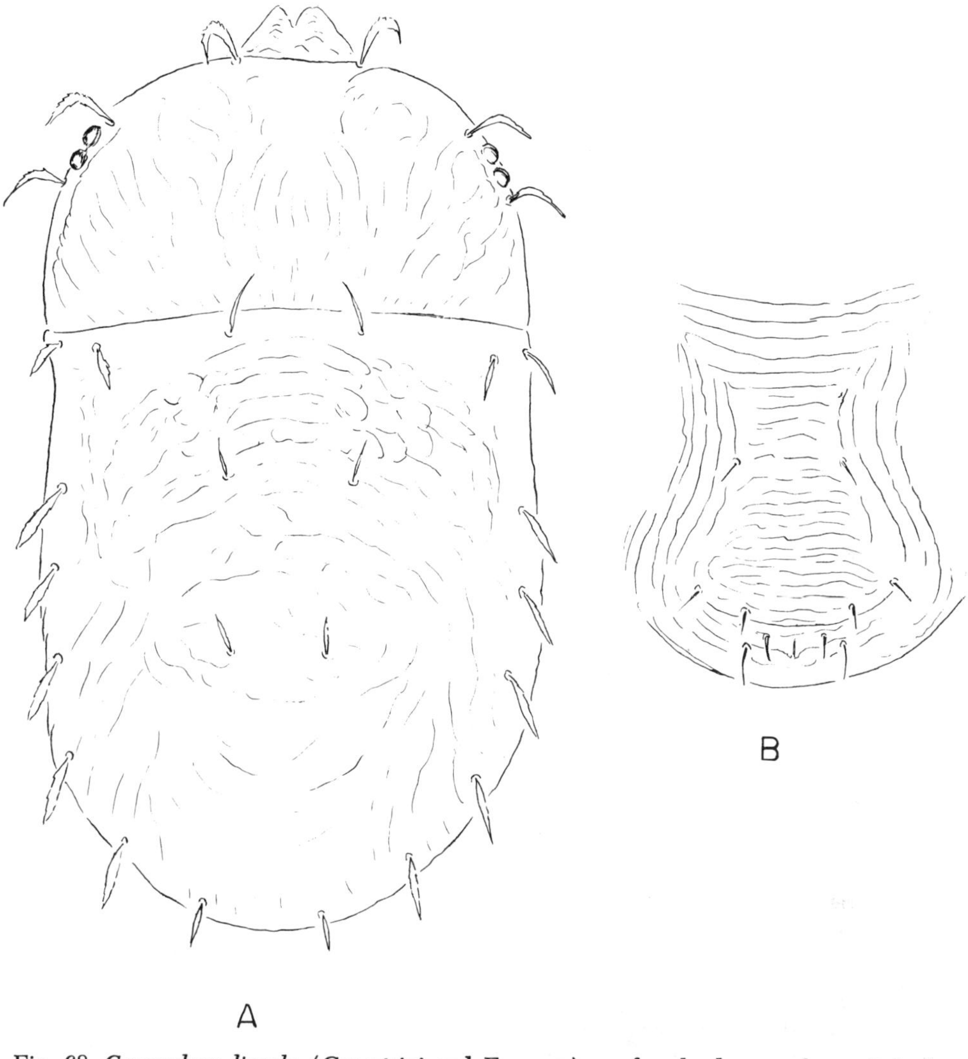

Fig. 68. *Cenopalpus lineola* (Canestrini and Fanzago): *a*, female dorsum; b, genital plate area.

setae are present but not long. The dorsal body ornamentation consists, in general, of a few longitudinal striae.

Tenuipalpus punicae Pritchard and Baker. *T. punicae* occurs in Spain on pomegranates where it causes defoliation and sometimes spots on fruit, and may be responsible for some fruit cracking. This species hibernates in the adult form. It may be controlled by dicofol and tetradifon (J. M. del Rivero by correspondence).

Cenopalpus Pritchard and Baker

The palpus has 4 segments; there are 5 or 6 pairs of hysterosomal dorsolateral setae, and 1 pair of dorsosublateral setae. Three pairs of dorsocentral hysteroso-

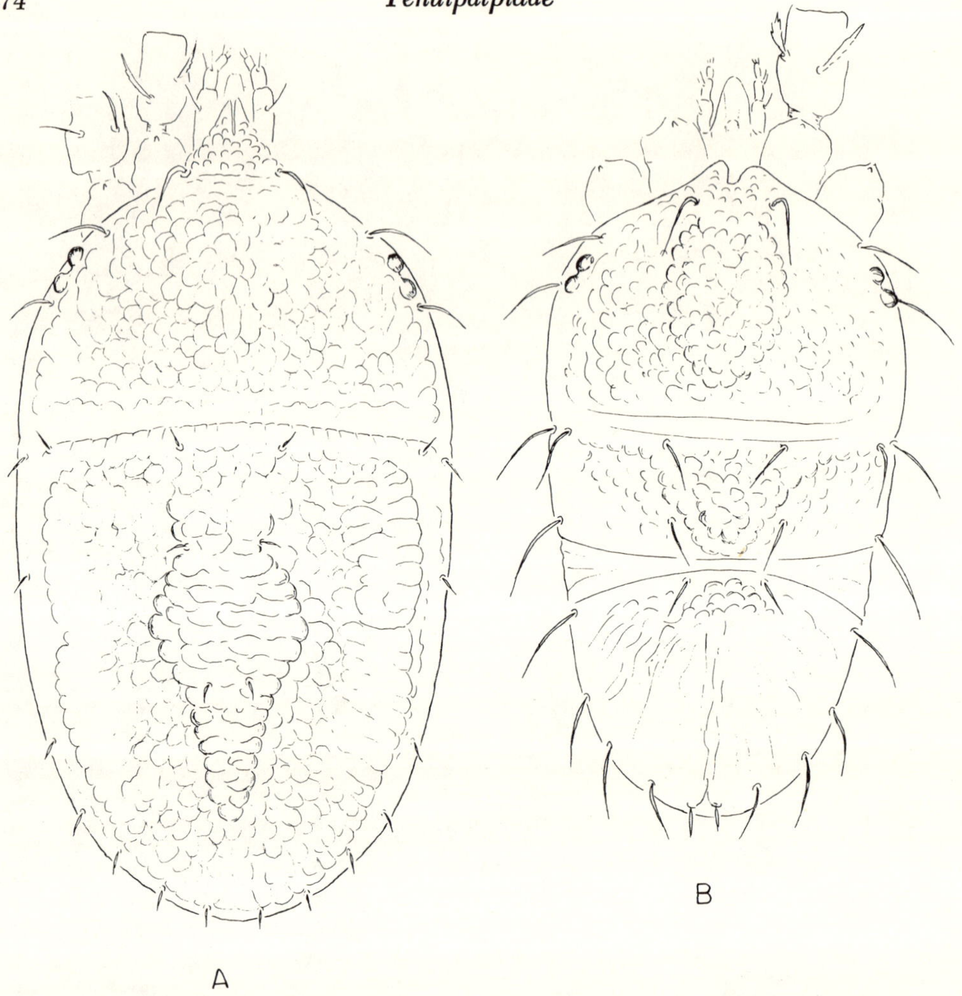

Fig. 69. *Cenopalpus pulcher* (Canestrini and Fanzago), showing setal and striation patterns: *a*, female dorsum; *b*, male dorsum.

mal setae are always present; the body is broadly rounded. The genital plate is broader than the anterior ventral plate, which widens anteriorly, and is medially constricted.

Cenopalpus lanceolatisetae (Attiah). This species (fig. 67), along with *C. pulcher* Canestrini and Fanzago, is a major pest of deciduous fruit trees in Egypt and Cyprus. In Egypt, *C. lanceolatisetae* prefers apricot, peach, pomegranate, plum, and pear while *C. pulcher* infests only quince. *C. lanceolatisetae* prefers the buds. It may be found on lower leaf surfaces only during favorable summer weather conditions.

These mites appear later in the spring than red spider mites; they reach their highest density during July, August, and September. Populations normally remain high through autumn. *C. lanceolatisetae* and *C. pulcher* are similar in appearance, differing slightly only in biology. Therefore they have often been confused in the field (Zaher and Elbadry, 1964).

C. lanceolatisetae has 6 pairs of dorsolateral hysterosomal setae; the dorsal body setae are strongly lanceaolate; the propodosoma is evenly reticulate dorsomedially; and the rostrum tapers distally and reaches well past femur I.

Cenopalpus lineola (Canestrini and Fanzago) (fig. 68, *a*, *b*) is a pest of pine trees in Holland, Italy, and in Georgia of the USSR. It has 6 pairs of dorsolateral hysterosomal setae that are strong, lanceolate, and serrate; the propodosoma has a few irregular striae dorsomedially (Baker, 1949).

Flat scarlet mite, *Cenopalpus pulcher* (Canestrini and Fanzago). *C. pulcher* (figs. 69, *a*, *b*; 70) has been reported from England, Denmark, Holland, Portugal, Austria, Bulgaria, Libya, Iran, Syria, Germany, Italy, Sicily, Cyprus, Lebanon, Algeria, Egypt, Israel, Turkey, Afghanistan, Georgia, SSR, Crimea, Transcaucasia, and Soviet Central Asia. It is an occasional pest of neglected apple, pear, prune, and walnut trees in England and the European countries. In Egypt and Turkey it is primarily a pest of quince (Pritchard and Baker, 1958).

The flat scarlet mite is small and noticeable only because of the intense scarlet color. Females are about 0.32 mm long and 0.16 mm wide. The male is shorter and paler than the female; its abdomen is almost transparent and curves upwards. The eggs are bright red, oval, and measure 0.11 mm by 0.07 mm.

The mites are relatively sedentary and normally live in groups on the undersurface of the leaves along the midrib and leaf veins. The adult females overwinter in cracks of the bark of their host. They become active early in the spring, from April onwards, although they do not move completely to the leaves until May. The first eggs are laid on wood late in April, but subsequently eggs are laid along the midrib, buried beneath the leaf hairs. In England and Europe egg laying continues until mid-July. Populations in Egypt reach their peak in August and remain high until the end of December. Mites prefer the lower leaf surface and move to buds during the winter when the trees are bare. The incubation period averages 55 days; thus eggs begin to hatch near the end of June. The next generation of adults appears in late July. At first there is a proponderance of males, but later the sexes reach virtual equality. Mating occurs in August and September, after which the males die and the females go into hibernation; there is only 1 generation per year in Europe (Dosse, 1953). Sepasgosarian (1970) reports 3 generations per year in Iran. He also found that overwintering mites could survive temperatures as low as –30 C (–22 F).

The hysterosoma of *C. pulcher* possesses 6 pairs of dorsolateral setae and a pair of humeral setae; the propodosoma is evenly reticulate dorsomedially; the rostral shield tapers distally and is pebbled dorsally; the dorsal body setae are tapering, setiform; the rostrum is short and does not reach the end of femur I. The nymph its distinctive in that there are six pairs of dorsolateral setiform setae, the third, fifth, and sixth pairs being minute. The male is similar to the female except that the male has much longer dorsolateral setae than the female and the posterior of the body is striate rather than reticulate.

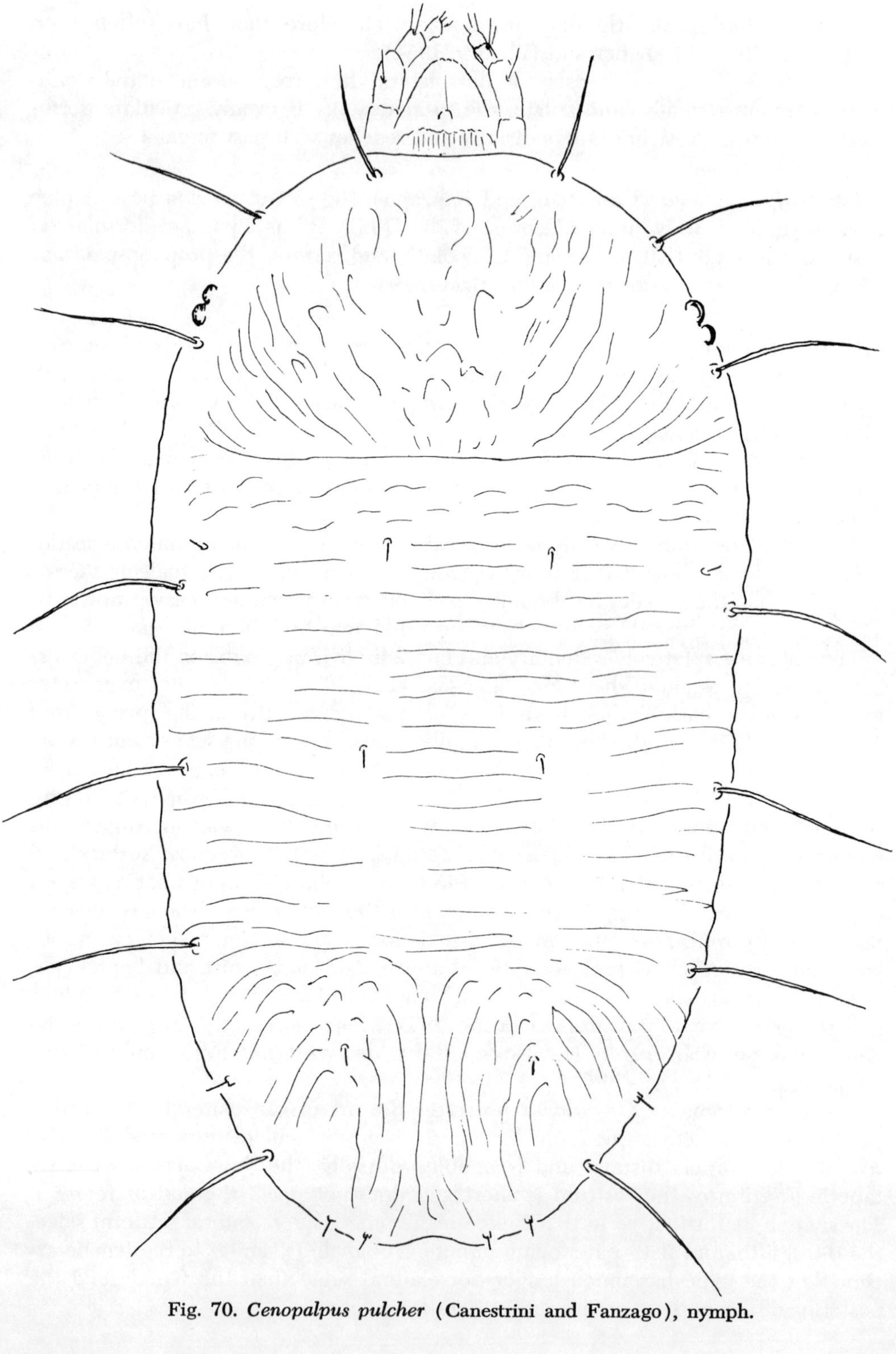

Fig. 70. *Cenopalpus pulcher* (Canestrini and Fanzago), nymph.

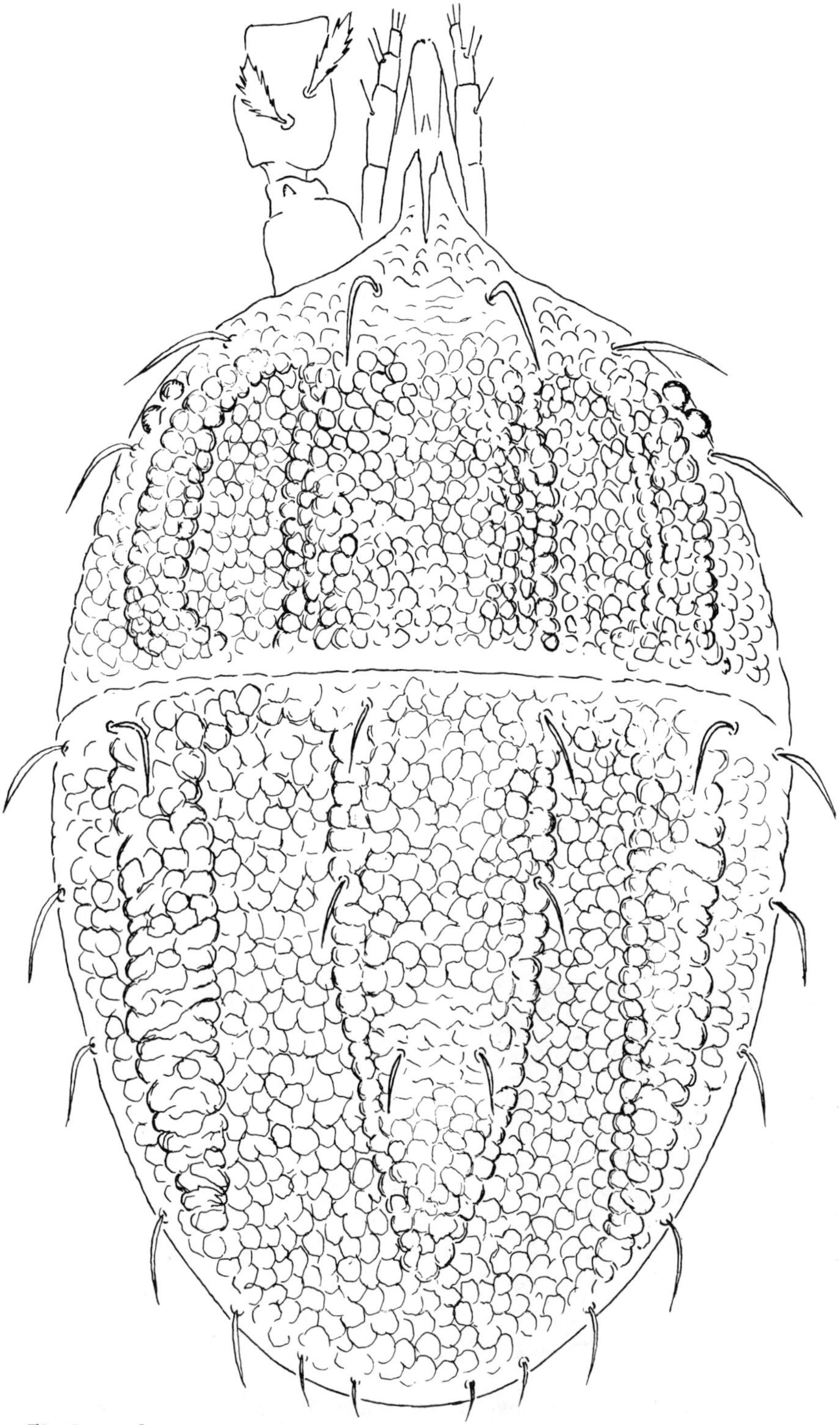

Fig. 71. *Cenopalpus spinosus* (Donnadieu), female dorsal view showing setal and striation patterns.

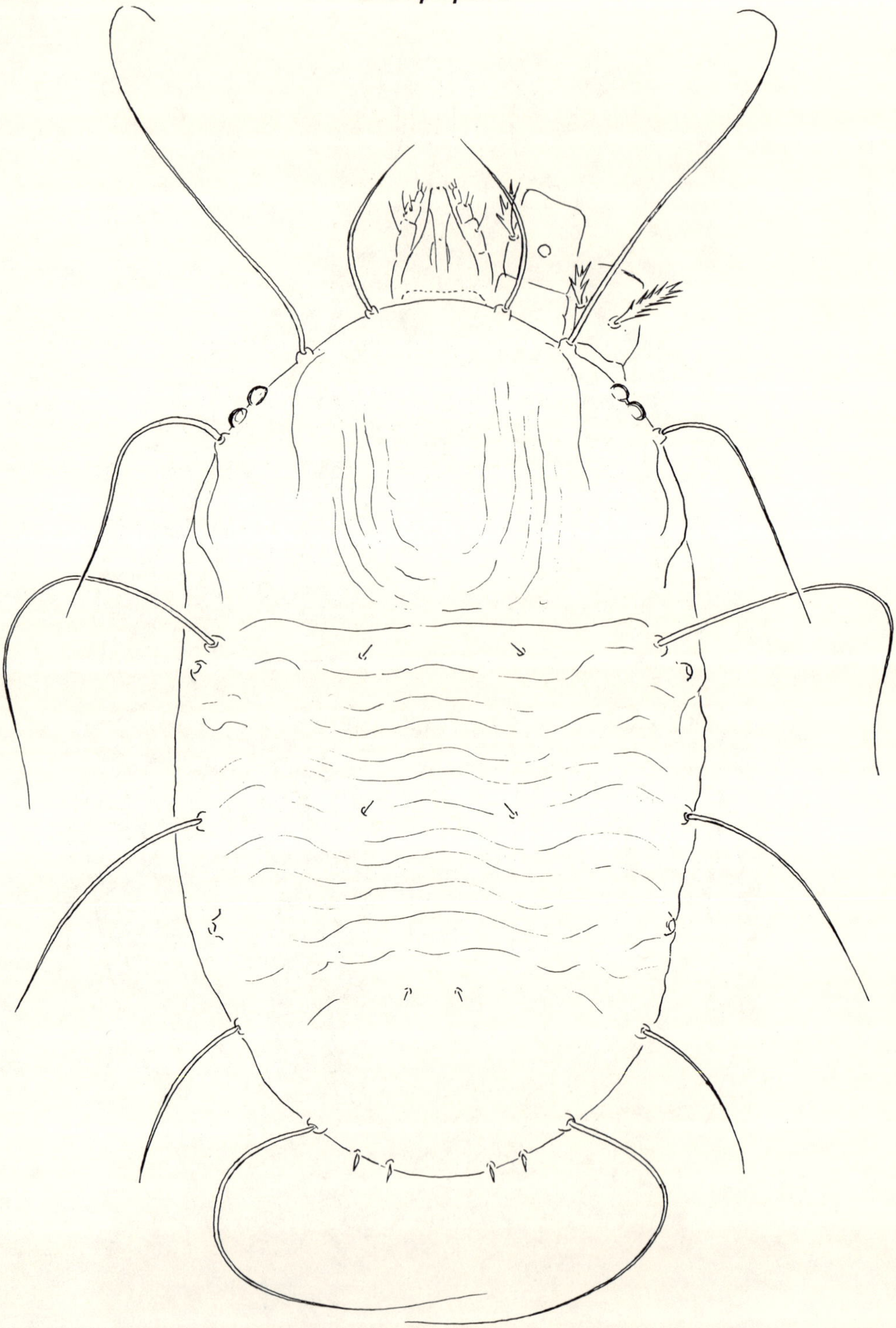

Fig. 72. *Cenopalpus spinosus* (Donnadieu), nymph, dorsal view.

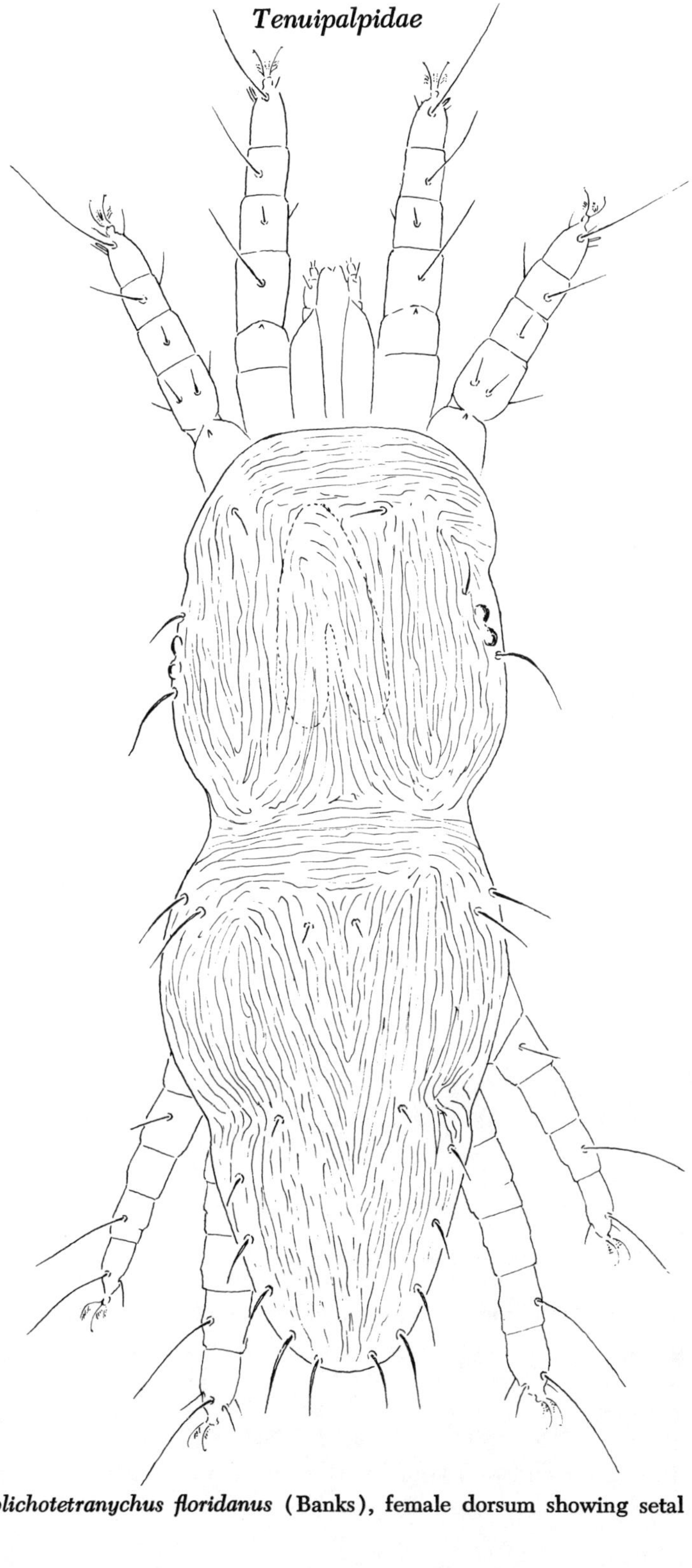

Fig. 73. ***Dolichotetranychus floridanus*** **(Banks), female dorsum showing setal and striation pattern.**

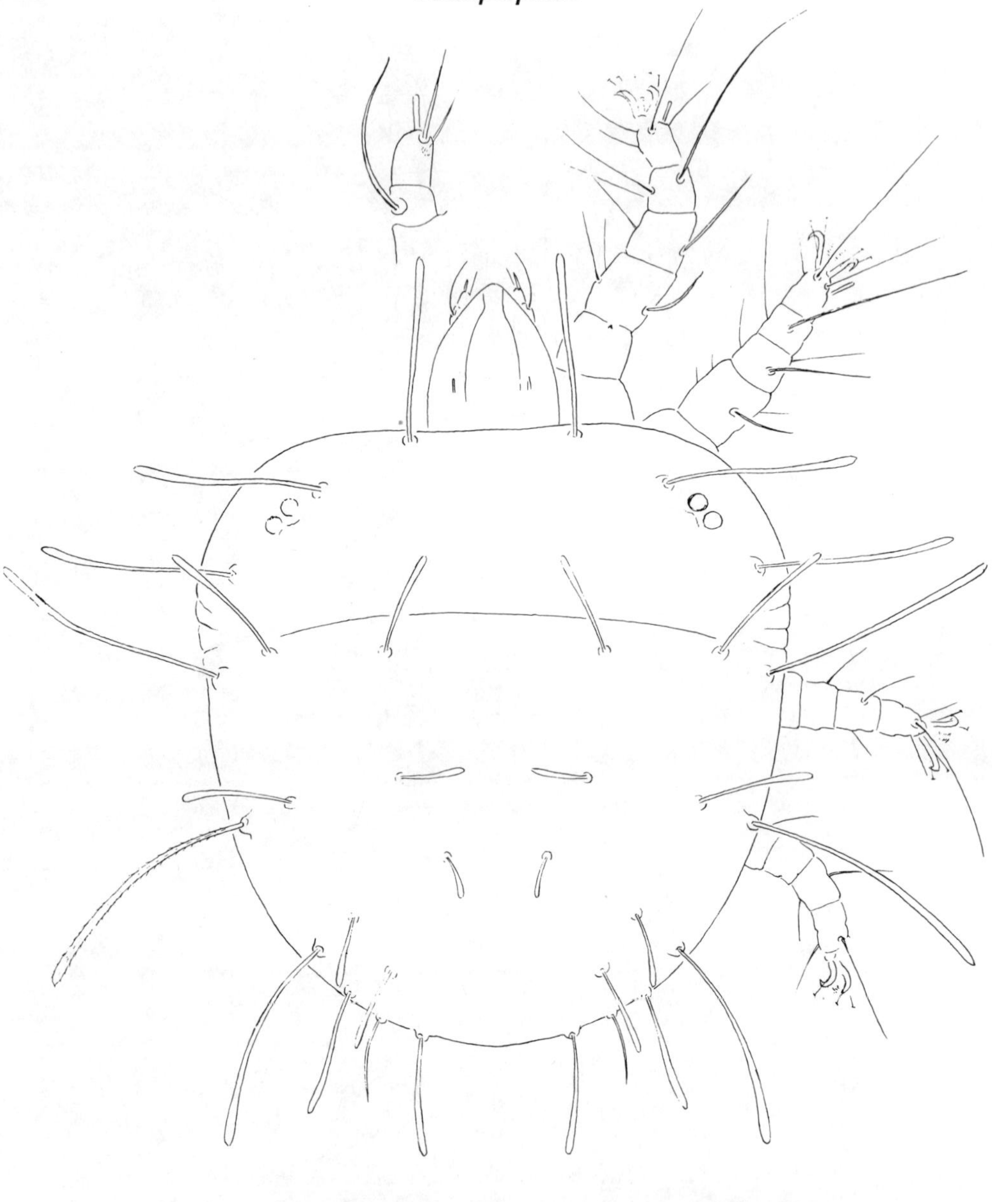

Fig. 74. *Raoiella indica* Hirst, showing dorsal setae and palpus.

Cenopalpus spinosus (Donnadieu). This species (figs. 71, 72) is a pest of primroses, blackberries, dewberries, raspberries (Rubus), roses, and dogwood (*Cornus*) in France, Germany, and Monaco (Baker, 1949).

The hysterosoma of *C. spinosus* has 6 pairs of dorsolateral setae; the propodosoma is evenly reticulate mediodorsally; the rostrum tapers distally and does not quite reach the distal end of femur II; the dorsal body setae are long, tapering, and setiform. The male is not known. The nymph is distinctive in that the dorsolateral setae are exceptionally long; the second dorsopropodosomal, the humeral, and the fourth dorsolateral hysterosomal setae are much longer than the others.

Dolichotetranychus Sayed

Dolichotetranychus species have 3 palpal segments; there is no rostral shield. The hysterosoma has 2 pairs of dorsocentral setae, 1 pair of dorsosublateral setae, and 5 pairs of dorsolateral setae. The body of each sex is characteristically shaped: that of the female is somewhat oval; and that of the male is extremely pointed posteriorly.

Dolichotetranychus floridanus (Banks). *Floridanus* (fig. 73) is found only on pineapple. It occurs in Florida, Cuba, Puerto Rico, Panama, Honduras, Mexico, Central America, Hawaii, Philippine Islands, Japan, Okinawa, and Java. These mites feed on the tender white tissue at the base of pineapple plants. Feeding injury appears to be insignificant, but the injury produced affords an entrance for bacteria and fungi, which cause the bud to rot (Pritchard and Baker, 1958). Control measures consist of applying an insecticide and a fungicide to the buds, depending on rain and dew to carry the pesticides down into the buds.

D. floridanus is characterized in having short dorsal setae on femur II; the female has 2 pairs of anal setae; the male has 2 solenidia each on tarsi I and II. The striae of the genital plate of the female are smooth and without lobes.

Raoiella Hirst

Raoiella species have only 2 palpal segments and no propodosomal shield over the rostrum; there are 4 pairs of dorsosublateral hysterosomal setae; there is no anterior ventral plate in the female, and the body of the female is strongly rounded.

Raoiella indica Hirst. This mite (fig. 74) is a pest of coconut in Mauritius. It also lives on the date palm and other palms. Young coconut plants are the most severely injured. In the summer the plants show a sickly yellowish appearance, a condition that may be the combined result of mite feeding and dry season conditions or even a virus disease.

R. indica lives on the undersurfaces of the coconut leaves where the eggs are deposited in colonies ranging in number from 110 to 330. The eggs are red, oblong (111 μ long by 88 μ wide), smooth, and shiny with a stipe. The larvae are reddish and sluggish. The preoviposition period is 3 days in summer and 7 days during the winter. Females lay an average of 2 eggs per day over an average oviposition period of 27 days. The average time for development of each life stage is: egg, 6.5 days; larva, 9.5 days; protonymph, 6.5 days; deutonymph 10.5 days. Thus, the average time required to complete the life cycle is 33 days.

These mites are generally abundant in Mauritius on coconut from September to March, except when heavy rains occur during November to January. Starting in April, there is normally a decline in populations, which continues through August (Moutia, 1958).

The body of *R. indica* does not have striae. The first pair of dorsocentral hysterosomal setae are longer than the others; the fourth pair of dorsosublateral setae are shorter than the first pair. All dorsal body setae are slightly clublike and serrate.

SELECTED BIBLIOGRAPHY

BAKER, E. W. 1949. The genus *Brevipalpus* (Acarina: Pseudoleptidae) Amer. Midland Naturalist 42(2):350–402.

BAKER,E. W. and D. M. TUTTLE. 1964. The false spider mites of Arizona. Ariz. Agr. Expt. Sta. Tech. Bull. 163:1–80.

BAPTIST, B. A., and D. J. W. RANAWERRA. 1955. The scarlet mites of the genus *Brevipalpus* as pests of tea in Ceylon. The Tea Quarterly 26(4):127–136.

COLLYER, E., and J. R. GROVES. 1955. Some tetranychid mites on fruit trees. East. Malling Eng. Res. Sta. Rept., 1955. 136 pp.

DEAN, H. A., and N. P. MAXWELL. 1967. Spotting of grapefruit as associated with false spider mites. Rio Grande Valley Hort. Soc. 21:35–45.

DOSSE, G. 1953 *Tenuipalpus oudemansi* Geijskes, eine fur Deutschland neue spinnmilbenart. P. Angew. Entomol. 34(4):587–597.

———. 1954. *Tenuipalpus orchidarum* Parfitt num auch in duetschen Gewachschausern. Zeitschr. Z. Angew. Entomol. 36(3):304.

ELMER, H. S., and L. R. JEPPSON. 1957. Biology and control of the citrus flat mite. J. Econ. Entomol. 50(5):566–570.

ESTEBANES, G. M. L., and E. W. BAKER. 1968. Arañas rojas de México. An. Esc. Nac. Cienc. Biol. Méx. 15:61–104.

EVANS, G. O., J. G. SHEALS, and D. MACFARLANE. 1961. The terrestrial acari of the British Isles. An introduction to their morphology, biology, and classification. British Museum, London. 219 pp.

GONZALES, R. H. 1968. Biologia y control de la falsa aranita de la vid. *Brevipalpus chilensis* Baker. Univ. Chile Estac. Expt. Agr. Bol. Tech. 1:1–31.

HERNERA-VILLAMIL, G. 1958. Biologia y control de la falsa aranita roja de la vid (*Brevipalpus chilensis* Baker). Agr. Tech. Chile. 18:35–42.

KNORR, L. C. 1964. World citrus problems. V. Venezuela. F. A. O. Plant. Prot. Bull. 12(6):125–126.

KNORR, L. C., B. N. WEBSTER, and G. MALAGUTI. 1960. Injuries in citrus attributed to brevipalpus mites, including brevipalpus gall, a newly reported disorder in sour-orange seedlings. Plant Prot. Bull. 8(12):35–42.

LO, P. K. 1968. Tetranchoid mites infesting tea in Taiwan. Chung-San Academic Cultural Affairs Ser. 1 :275–286. Taipei, Taiwan, China.

MCGREGOR, E. A. 1949. Nearctic mites of the family Pseudoleptidae, Mem. So. Calif. Acad. Sci. 3(2):1–45.

MANGLITZ, G. R. and E. N. CORY. 1953. Biology of *Brevipalpus australis*. J. Econ. Entomol. 46(1):116–119.

MORISHITA, F. S. 1954. Biology and control of *Brevipalpus inornatus* (Banks). J. Econ. Entomol. 47(3):449–456.

MOUTIA, L. A. 1958. Contribution to the study of some phytophagous acarina and their predators in Mauritius. Bull. Entomol. Res. 49(1):59–75.

PRITCHARD, A. E., and E. W. BAKER. 1951. The false spider mites of California. Univ. Calif. Pub. Entomol. 9:1–94. Univ. Calif. Press, Berkeley and Los Angeles.

———. 1958. The false spider mites. Univ. Calif. Pub. Entomol. 14(3):1–274. Univ. Calif. Press, Berkeley and Los Angeles.

RAIKOV, E., and P. NACHEV. 1965. Winter control of the vine phytoptipalpid. Past. Zash. 13 pt. 8:6–8. (From Rev. Appl. Entomol. Ser. A. 56:1912.)

ROLFS, P. H. 1899. Pineapple fertilizers: Red spider, *Stigmaeus* sp. Fla. Agr. Expt. Sta. Bull. 50:99–100.

SCHWARTZ, A. 1970. Brevipalpusmyt on citrus, S. Afr. Citrus J. 438:27–28.

SEPASGOSARIAN, H. 1970. The red false spider mite of apple, *Cenopalpus pulcher* (Can. and Fanz.), Agr. College Tehran Univ. Bull. 115. 36 pp. Karadj, Iran.

VERGANI, A. R. 1945. Transmission y naturalez de la "Lepra explosiva" del Naranjo. Argentina Inst. Sanidad. Veg. 1 Series A (3):10 pp.

ZAHER, M. A., and E. A. ELBADRY. 1964. Survey and population studies of red and false spider mites. First Intern. Cong. Acarology Proc., Fort Collins, Colo. Extrait d'Acarologia 6:425–435.

Chapter 10

Tarsonemidae Kramer

HISTORY OF KNOWLEDGE OF TARSONEMIDS

The mite family Tarsonemidae was erected in 1877 by Kramer based upon the type genus *Tarsonemus* Canestrini and Fanzago, 1876. The family has been variously placed in the order Acari by acarologists since that time.

The first comprehensive taxonomic work on mites belonging to the family Tarsonemidae was published by Ewing (1939). He considered the Tarsonemidae as one of the families that compose the superfamily Tarsonemoidea. He divided the family into three subfamilies: the Tarsopolipinae, the Podapolipinae, and the Tarsoneminae. In the subfamily Tarsopolipinae, he placed certain mite parasites of insects, such as *Acarapis woodi* (Rennie), the tracheal mite of honeybee, and *Locustacarus trachealis* Ewing, the tracheal mite of grasshoppers. He stated that the subfamily Podopolipinae are entirely parasitic on insects. In the subfamily Tarsoneminae he included some species parasitic on insects, others that are necrophagous, still others that attack living plants.

Beer (1954) transferred certain of the genera formerly included in the family to other families. He redefined the family to include two of Ewing's three genera comprising the subfamily Tarsoneminae. He recognized three additional genera: the *Steneotarsonemus*, *Xenotarsonemus*, and *Rhynchotarsonemus*. Much of the taxonomy, morphology, and general biology used in this chapter have been summarized from his monograph.

The importance of tarsonemid mites to agriculture has long been established. The first definite record of tarsonemid damage to an agricultural crop was made in 1877, when *Steneotarsonemus bancrofti* (Michael) was noted as a pest of sugar cane in Queensland. Several other species of tarsonemid mites have been incriminated as agricultural pests since that early date. Certain species are recognized as parasites of scale insects; others are suspected as having a parasitic relationship to higher animals, including man; and some are indicated as fungus feeders. The validity of some of these records should be investigated.

The species of undisputed agricultural significance include *Steneotarsonemus ananas* (Tryon), *S. bancrofti*, *S. furcatus* (DeLeon), *S. laticeps* (Halbert), *S. pallidus* (Banks), *S. spiriflex* (Marchal), and *Polyphagotarsonemus latus* (Banks).

S. pallidus causes the greatest amount of damage, followed closely by *P. latus.* Both have enormous host ranges, which include many commercially grown crops and plants. The biologies of these two species have been rather extensively investigated to develop means of biological and chemical control. Another species, *Lupotarsonemus randsi* (Ewing), may occasionally be a pest of importance in the commercial growing of mushrooms. *L. randsi* and certain species of the genus *Tarsonemus* may become quite destructive to fungus cultures in research laboratories. Control of such invaders without damaging the fungi or culture media has been studied, but no general methods of control are available.

MORPHOLOGY AND SYSTEMATICS

General Description

Tarsonemid mites are very small, ranging in length from 100 to 300 μ. Mature forms have a relatively hard and shiny integument. The body and posterior legs are rather sparsely beset with setae. The anterior pairs of legs, especially their terminal segments, are more densely clothed with setae and are often equipped with specialized sensory setae of various configurations and sizes.

Pronounced sexual dimorphism is characteristic. The males are not only much smaller in size than females of the same species, but the general body contour is markedly different. The usual shape of the body of the female is ovoid with the dorsum convex. The interior pairs of legs are separated from the posterior pairs by a distinct interval. A group of species in the genus *Steneotarsonemus* has apparently undergone much modification in respect to general body contour, no doubt related to adaptations for their particular habitats. Females of these species are quite elongate, and the anterior and posterior pairs of legs are widely separated. Both sexes are dorsoventrally depressed—a configuration quite suitable for mite activities in the confined spaces between the sheaths and stems of the grasslike hosts. Extreme convexity of the female dorsum is characteristic of the genus *Polyphagotarsonemus*; the carapacelike dorsal idiosoma conceals the capitulum when living specimens are viewed from above. This condition is approached by some species in the genus *Tarsonemus.*

The body of a tarsonemid mite is divided into three well-defined portions, but may be further divided by the use of established terminology defined below. The mouthparts are contained in a distinct capsular head called the capitulum. The remainder of the body comprises the idiosoma, which is transected by a distinct suture. This main body suture is located between the anterior and posterior pairs of legs. The unsegmented area anterior to the main body suture is called the propodosoma, and the portion of the idiosoma behind the main body suture is the hysterosoma. The propodosoma is a single, more-or-less continuous body region, which in some species has the dorsum prolonged anteriorly forming what is called a cephalothoracic or rostral shield. This shield is sometimes separated from the remainder of the dorsal propodosoma by a suture. Such a prolongation or forward extension of the dorsum of the propodosoma, which occurs in many families of mites, has also been designated by acarologists as the cephalothoracic hood or rostral hood. The hysterosoma may be further divided into anterior and posterior portions; namely, the portion from the main body suture to the hind margins of coxae IV and the portion behind the legs. Names applied to the anterior and posterior regions of hysterosoma are the metapodosoma and opisthosoma, respectively.

The mouthparts consist of stout, paired palpi of indistinct segmentation inserted on the apical portion of the capitulum. Also the mouthparts include the slender, styliform, paired chelicerae, the bases of which are inserted just medial to the bases of the palpi.

Tarsonemids are characterized by the pronounced development of apodemes on the ventral portion of the body. These apodemes are called epimera by some acarologists. Each has been assigned a number to aid in identifying them in taxonomic descriptions.

The males are equipped caudally with a unique structure known as the genital papilla or genital plate. The papilla is situated terminally on the opisthosoma in living specimens. In microslide preparations it usually tilts so as to be visible in a dorsal position with the dorsal margin; thus it appears as the anterior margin, and the ventral margin projects caudad beyond the apex of the opisthosoma. This genital plate contains within its clearly defined limits the paired, styliform aedeagus, as well as other accessory genital organs and appendages, the exact identity of which are as yet unknown. Another structure, the anal plate, is often quite conspicuous in slide-mounted specimens. Its position in living males is subterminal on the ventral opisthosoma just anterior to the ventral margin on the genital papilla, although in microslide mounts this normal position is not often apparent. The plate lacks clearly defined lateral limitations. The most conspicuous portion of the structure is a central disc or aperture from which fingerlike apodemes radiate. The usual number of anal apodemes is 3; 2 extend anterolaterally for a short distance from the anterolateral margins of the disc and the third projects caudally from the posteromesal margin of the disc. In some species, however, there may be 4 anal apodemes, 2 projecting from the anterior margin of the disc and 2 from the posterior margin. One species has 1 apodeme projecting forward from the disc and 2 projecting caudad.

Females are characterized by the possession of specialized organs located dorsolaterally between coxae I and II. These organs, which vary somewhat in size and shape, are of uncertain function and have been called clavate sense organs or pseudostigmatic organs. Probably these paired structures are highly modified sensilla trichodea, and are more properly referred to as specialized sense organs since they seem to have no relationship to the tracheal system.

There appears to be little or no evidence of a tracheal system in male tarsonemid mites. Stigmatal or tracheal openings are quite distinct in the females, being situated dorsolaterally near the anterior margin of the propodosoma. The tracheae extend internally from these external orifices, converge medially in the region of legs II, diverge posteriorly from this point and disappear inconspicuously in the opisthosomal region. The tracheae in some species connect with paired, heavily sclerotized, elongated structures situated medially, near the point of convergence of the tracheae, or in subspheroidal, dilated pouches similarly situated.

BASIS OF CLASSIFICATION

Classification of the family Tarsonemidae has been based largely upon characters on the posterior pair of legs of the males. This appears to be a logical method of separation because of the variability of these appendages. Legs IV of the males may be regarded as accessory copulatory appendages because of their function in premating behavior and the mating process. These appendages appear to be highly modified in various ways, presumably representing adaptations developed for better performance of copulatory functions. A "plastic character," such as this, lends itself well in distinguishing the lower categories. Legs IV of the male are generally 4-segmented, but in some species a fusion of the tibia and tarsus has reduced the segmentation. The terminal claw of these appendages also varies considerably from the condition of prominence to that of nearly complete degeneration. Modifications of the femur are quite evident in some species groups, ranging from a thin, membranelike inner flange to a spurlike projection of the inner margin. This type of specialization is probably related to the requirements of the various species in the copulatory process, as this pair of legs functions only passively in locomotion.

Most body setae are identified by size, shape, and location when used to distinguish species. There are several types of specialized setae ranging from the normal conditions, that is, those with a rather narrow base from which the seta tapers to threadlike apex, to modified types, that may appear clavate, lanceolate, peglike, or of various shapes. Leg segments are referred to by name except where names cannot be applied with any degree of certainty owing to a lack of knowledge of the homologies involved.

Chaetotaxy of the legs is described on the basis of normal orientation of the mite on a microslide. In this position legs I and II extend forward and the posterior pairs extend to the rear. By reference to this "normal" position the geographic location of the seta is readily and easily indicated by the simple statement of dorsal or ventral position or the outer or inner margin of the segment; that is, the margin is either away from or close to a hypothetical line bisecting the mite along the longitudinal axis.

BIONOMICS

Life History

Comprehensive biological studies have been conducted on only 2 species of tarsonemid mites, but studies so far indicate that tarsonemids have 4 distinct stages in their life history. Eggs are laid singly by the gravid female. They are white, ovoid, opaque, and large in comparison to the size of the adult. In some species the smooth surface is dotted with small tubercular swellings while the surface of eggs of other species is broken by numerous pitlike depressions. The egg hatches into 6-legged larva that is opaque white, with the 2 anterior pairs of legs situated as in the adult, but the posterior pair in the position of legs III of the adult. Larvae are further characterized by the presence of a peculiar enlargement of the opisthosoma into a triangular platelike development that is most prominent in males. Male larvae are considerably smaller than female larvae. From this active stage the mites of both sexes enter a quiescent "pupal" stage in which transformation to the adult takes place. This pupal stage is sessile and the larval integument appears inflated or bloated with the cuticle tightly stretched. Transformation to the adult takes place within the larval skin through successive stages of withdrawing the appendages from the old integument, formation of legs IV behind legs III, as well as development of genitalic structures. The pupal skin splits dorsally at the completion of the transformation to the adult and the mature individual emerges. Usually the integument darkens somewhat following emergence.

The color of many species seems to vary according to the food ingested; thus body color is generally not a reliable diagnostic character. Some of the phytophagous species are commonly green, and fungivorous species may assume a body color comparable to the color of the fungus upon which they have been feeding.

Locomotion

Locomotion by females is accomplished by the use of all 4 pairs of legs. The mites walk on the ventral subterminal setae of legs IV. The hind legs of the males are rarely used in locomotion, being most frequently carried in a semierect position above and behind the body. These appendages are reportedly used by the males in transporting pupae and adult females, both of which are carried on the male's back. Pupae are held in the grip of legs IV and fastened to the male by structures or appendages of the genital papilla. Larvae are not carried by the males and rarely are male pupae so carried, but nearly mature female pupae are more commonly the portage than the adult females. In most species unfertilized eggs produce only males, although it is reported that *Steneotarsonemus pallidus* resulting from parthenogenetic reproduction are invariably females.

Influence of Environment

Optimum environmental conditions for the various species studied appear to involve a combination of warm temperatures, high humidity, and low light intensity. The adult stage of tarsonemid mites is known to survive through prolonged exposure to freezing temperatures, but seems sensitive to temperatures above 35 C (95 F).

Relation of Mouthparts to Food Sources

Species of tarsonemids that are known to feed on higher plants are restricted to three genera. The remaining mites in this family probably feed on fungi or possibly algae. Apparently, tarsonemid mouthpart appendages are unsuitable for effective penetration of renitent tissues. Palpi are simple and quite reduced and probably most often serve a tactile function or, in some instances, may function in guiding solid food particles to the mouth. The oral opening and anterior digestive tract, in at least some species, do accomodate ingestion of whole fungus spores. Chelicerae are simple, styliform, and are only slightly eversible. Such feeding appendages are quite suitable for penetrating thin-walled mycelial strands

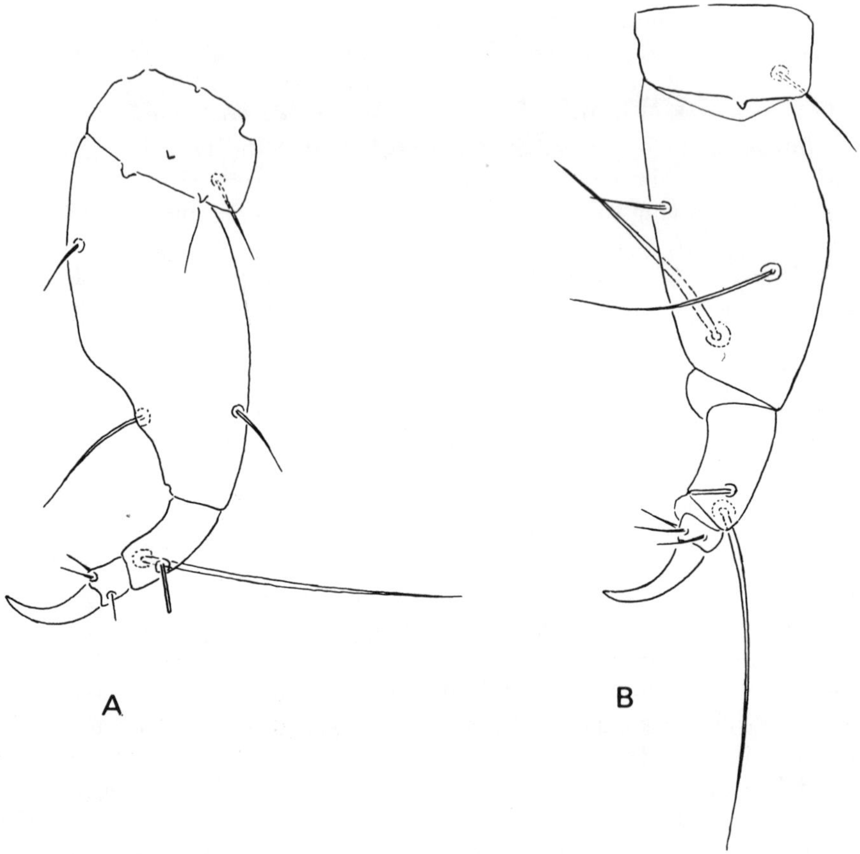

Fig. 75. *a*, *Tarsonemus smithi* Ewing, male, Leg IV, ventral view; *b*, *Tarsonemus setifer* Ewing, male, Leg IV, ventral view.

and highly succulent tissues; but they are quite incapable of penetrating thick-walled, lignified, and often varnished tissues, such as are found in mature stems and leaves. Occasionally, however, toxins injected during feeding, presumably of salivary gland origin, cause alteration of normal tissue ontogeny in the host plant. In at least some cases continuous proliferation of thin-walled cells, in a localized area around the feeding site on mature leaves, produces material available to the mites for food. Often this type of deformity begins when the leaf is proximal to the apical meristem—a time when its epidermal cell walls are thin and vulnerable to attack. Species that are known to be able to feed successfully on old leaves, probably by virtue of their ability to produce such toxemia, are *Steneotarsonemus pallidus* (Banks) and *Polyphagotarsonemus latus* (Banks).

INJURIOUS TARSONEMID MITES

The females have a typical treacheal system posterior to the gnathosoma; rudimentary stigma may be present in the males; the females usually possess a distally expanded pseudostigmatic organ between coxae I and II; and the mouthparts are

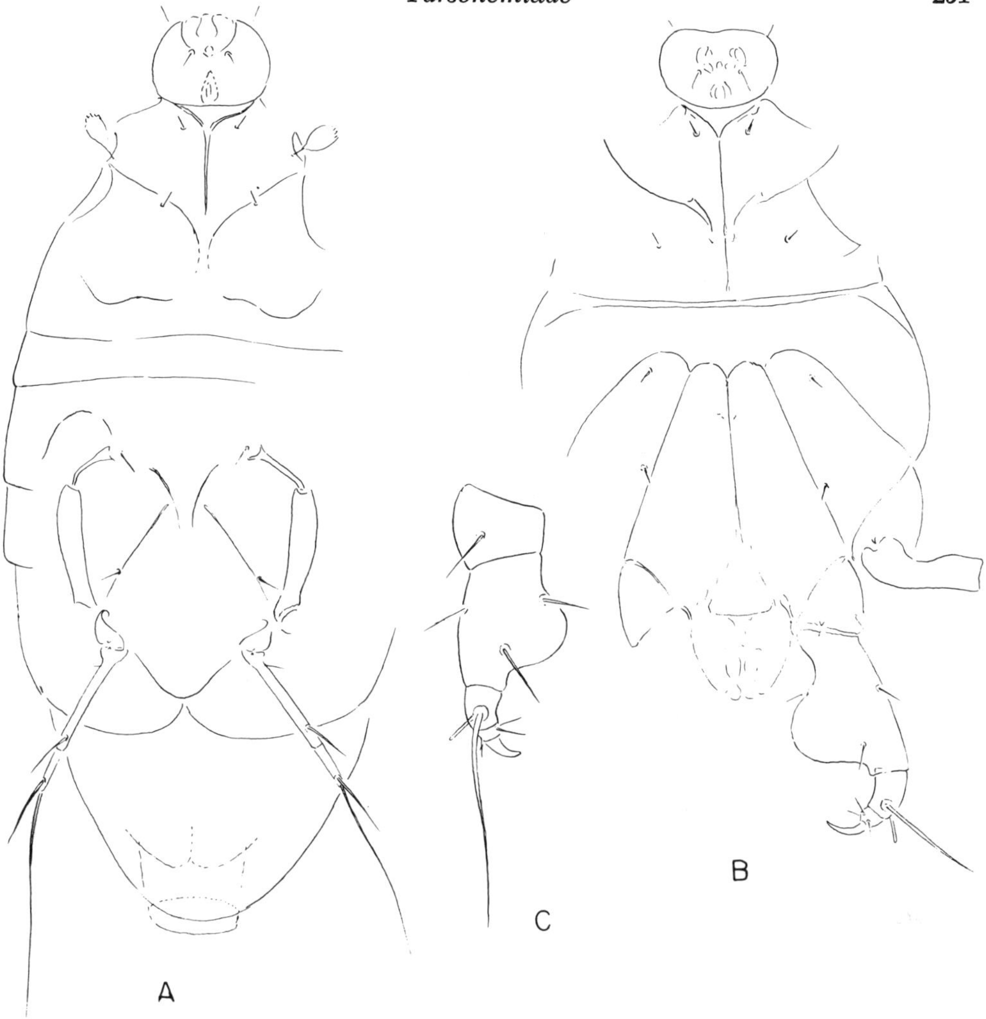

Fig. 76. *Steneotarsonemus ananas* (Tryon): *a*, female, ventral view; *b*, male, ventral view; *c*, male, leg IV.

reduced. Both males and females have 4 pairs of legs; leg IV of the female ends in a terminal and subterminal seta; legs IV of the male are highly specialized for clasping the female during copulation; the leg segments vary in number. In most cases specific identifications are made by studying the males.

Tarsonemus Canestrini and Fanzago

The pseudostigmatic organs of the female are expanded distally. The propodosomal setae of the female are widely separated; the second pair is on the posterior half of the propodosoma. Leg IV of the male has a distinct tibia and tarsus, their combined length being less than ½ the femur and less than 3 times the basal width of the femur IV. There is no flange on femur IV of the male.

White-tailed mite, *Tarsonemus setifer* Ewing. This mite (fig. 75, *b*) is wide-

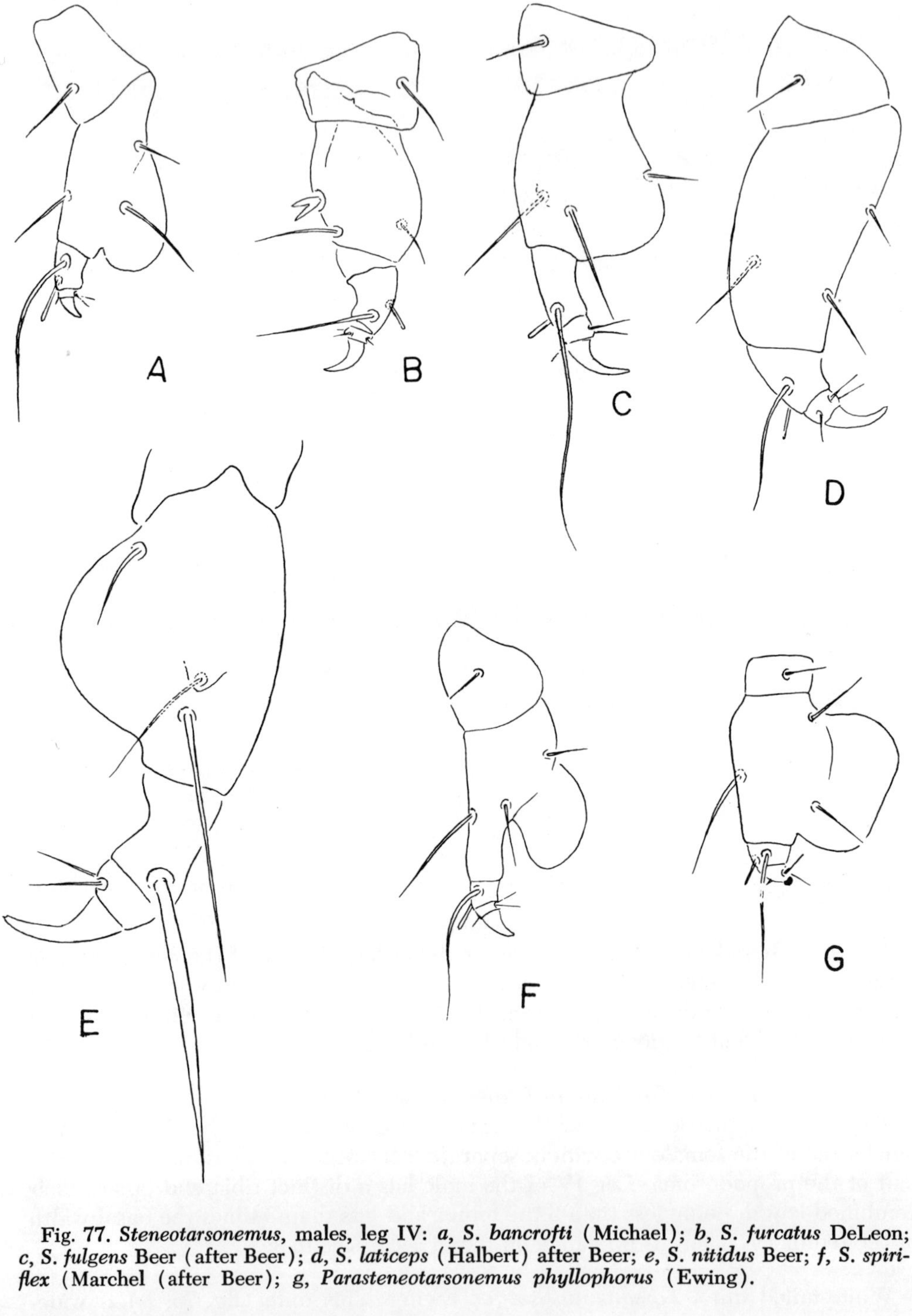

Fig. 77. ***Steneotarsonemus***, males, leg IV: *a*, *S.* ***bancrofti*** (Michael); *b*, *S.* ***furcatus*** DeLeon; *c*, *S.* ***fulgens*** Beer (after Beer); *d*, *S.* ***laticeps*** (Halbert) after Beer; *e*, *S.* ***nitidus*** Beer; *f*, *S.* ***spiriflex*** (Marchel (after Beer); *g*, ***Parasteneotarsonemus phyllophorus*** (Ewing).

ly distributed in Europe and in North America, having been reported in British Columbia in Canada; and in California, Florida, Louisiana, New York, Virginia, and Washington in the United States.

This species lives on a wide variety of agricultural plants, including citrus, pomegranate, strawberry, peach, grape, blackberry, dewberry, and raspberry. Its hosts among the ornamental plants include gerbera, verbena, *Penstemon*, (copper leaf,) rose, *Daphne*, spiderwort, and bamboo.

The mites probably live mainly on fungi. Those living on citrus are commonly associated with the fungi living on honeydew secreted from scale insects or aphids (Beer, 1954). Daniel Klimker (personal interview) attributes certain scarring of citrus fruits in Morocco to *T. setifer*. The first evidence of injury is the disappearance of stomata; next the rind cracks. A white liquid is extruded at the cracked areas, resulting in light-colored scar tissue which later becomes dark brown.

Tarsonemus smithi Ewing. This species (fig. 75, *a*) occurs on apple, citrus, peach, plum, tomato, elm, chrysanthemum, and hollyhock. It has been reported from Portugal and from Virginia and Washington, D.C. in the United States. Plant injury by this species has not been reported (Beer, 1954).

Steneotarsonemus Beer

The males usually possess 4 pairs of propodosomal setae; femur IV usually has a flangelike process on the inner margin, but never a spurlike process. The gnathosoma is usually subcircular and often as broad or broader than long and has short palpi; the body is dorsoventrally depressed. The female body is often elongate; the legs I and II are widely separated as are legs III and IV. There are 2 pairs of propodosomal setae.

Pineapple tarsonemid mite, *Steneotarsonemus ananas* (Tryon). This species (fig. 76, *a-c*) is host specific on pineapple. It may initiate fungal infection as well as injure fruit by its feeding activities. The total effect is to cause some segments of the fruit to remain green and become rotten inside. The mites may also cause injury to young plants. This species occurs on the island of Oahu, Hawaii, and in South Queensland, Australia; but is has not been found on the continental Western Hemisphere (Beer, 1954; Hughes, 1959).

The inner flange on tarsus IV of the male of S. *ananas* is evenly rounded; all 3 setae of the tarsus are short and subequal in length. The propodosoma has 4 pairs of dorsal setae: the third pair is longer than the others, and the first pair is the shortest. The female is elongate, legs I and II, and III and IV, are well separated; genu (patella) II has a stout seta and 2 normal setae; the first pair of ventral propodosomal setae lie behind apodemes I; the terminal seta of tarsus IV is much longer than the subterminal seta. The pseudostigmatic organs are rounded; apodemes III and IV are weak and are not contiguous.

West Indian sugar can mite, *Steneotarsonemus bancrofti* (Michael). This mite (fig. 77, *a*) appears to be host specific and occurs wherever sugar cane is grown, except in southern India. The mite colonies live in cane tops underneath the green leaf sheaths where they produce peculiar transparent brown craterlike depressions (0.5 to 2.0 mm in diameter) on the young cane stalk, giving the plant tissue a scabby or scarred appearance. As the cane grows older, these areas become opaque, brown, and rough; thus, at cutting time, the canes may be covered with roughened areas. Mites are never found on the older parts of the cane and only a few may be found on the youngest portions where the lesions are produced. Feeding injury allows the entrance of spores of red rot, *Colletotrichum*. The fungus alone produces red spots resulting in a reddened appearance of the leaves. These mites are small, opalescent light green. The young are soft, yellow, transparent and shiny (Butani, 1959; Holloway, 1936).

The female of S. *bancrofti* is typical for the genus. The first pair of dorsal propodosomal setae of the male are shorter than the second and fourth pairs. The gnathosoma and legs IV are small in comparison with the body. Coxae III are not ornamented. The flange on femur IV is baglike and reaches to the distal margin of the femur, and is indented distally between the femur and the flange; and the tibial seta is strong and about as long as the length of the femur.

Steneotarsonemus furcatus De Leon. *Furcatus* (fig. 77, *b*) feeds on an ornamental grass, *Paspalum* spp., and greenhouse-grown maranta plants, *Maranta leuconeura* Morr. Injury is expressed as severe distortion of maranta leaves and shortened internodes, which result in stunted plants.

S. *furcatus* is a typical member of the genus, but it is distinguished from all others in that in the male there is a short, strong bifurcate seta on the inner margin of the femur. In all other species this seta is simple.

Steneotarsonemus fulgens Beer. This mite (fig. 77, *c*) is not injurious to plants, but is included here because it utilizes as a food source the tissue inside galls engendered by an eriophyid mite, *Phyllocoptes didelphis* Keifer, which feeds on *Populus*. The tarsonemid mites invade the erinea or galls caused by the eriophyid mites, after which the eriophyid mites leave the erinea to the tarsonemids. The *Populus* erinea provide an environment very smilar in physical structure to mycelial mats. The tissue inside the erinea is sufficiently succulent for the weak mouthparts of these tarsonemids to penetrate and obtain food. The erinea also provides an environment characterized by tightly packed, treadlike material that not only causes reduction of the ambulatory movements characteristic of all tarsonemid mites on exposed and smooth surfaces, but also elicits a feeding response. This indicates how a fungus-feeding group might develop species that feed on tissues of higher plants that resemble mycelial mats (Beer, 1954).

S. *fulgens* has four pairs of propodosomal setae in linear arrangement on the male. Setae I are longer than IV, which are equal in length; setae III are at least 3 times longer than the others. Coxae III are not ornamented or punctate. The tactile setae of tiba IV are about as long as femur IV; femur IV has an inner flange that is baglike and broadest at the posterior distal level of femur IV.

Steneotarsonemus nitidus Beer has a similar social parasitism with the eriophyoid mite, *Phytoptus laevis* (Nalepa) on birch (Beer, 1963).

The female of S. *nitidus* Beer (fig. 77, *e*) is typical for the genus. The male is distinctive in having strong dorsal body setae, and in the setal and flange pattern of leg IV. The femur is widest at the midpoint; and the proximal seta is the shortest, the mid seta next in length, and the distal seta about as long as the segment. The tibial seta is as long as the femur; it is strong, and lanceolate.

Bulb scale mite, *Steneotarsonemus laticeps* (Halbert). This mite (fig. 77, *d*), previously known as *Tarsonemus approximatus* Banks, is a pest of narcissus bulbs and bulbs of other plants in the family Amaryllidaceae. This species is found in Ireland, England, Holland, and Sweden, and on the West Coast of the United States.

The mites feed principally on the epidermal surfaces of the scales of the bulbs. The focal center is at the neck area, extending upward from the point where the leaves separate, which is usually at about the soil surface; at this location space is available for the mites to work their way down into the shoot primordia.

Feeding by this mite on developing shoots causes distortion, stunting, and often mortality of the leaves and flowers. Longitudinal bronze streaks and transverse cracks may occur on the foliage and flower stems as a result of scarlike tissue produced by mite feeding. Infestations in the bulbs produce yellowish brown areas where mites feed, but this injury is seen only by cutting the bulb open from the neck to the base and separating the scales and shoots.

The mites are unable to penetrate into the vigorously growing bulbs in the spring, but during August and September the bulbs lose moisture and shrink, permitting the mites to work their way into the soft tissues. When root action begins in the spring, the bulbs swell and large numbers of mites are crushed by the resulting pressure. In vigorous bulbs the mortality may be nearly complete. As the foliage develops in February, the mites move to the leaves and flowers where they produce the typical streak type patterns. Migration from plant to plant takes place in June.

Females lay from 5 to 28 eggs during their life span. Eggs hatch in about 11 days and the first larvae develop in about 15 days. The life cycle from egg to adult in the field requires about 7 weeks.

This species, like other bulb mites, is spread onto new bulbs while bulbs are in storage; therefore prevention of infestations may be accomplished by a hot water treatment at 43 C (110 F) for one hour, or vapor heat treatment at 43 C for 2 hours (Carmona, 1966; Doucette, 1936).

Control on narcissus plants in greenhouses has been achieved by 0.1 percent drench sprays of either endrin, endosulfan, or azinphos-methyl (Winfield, 1964).

The female of S. *laticeps* is typical for the genus. The femur IV of the male does not have a flange. The tactile seta of tiba IV is short and not as long as the femoral segment; the third pair of propodosomal setae are ⅔ as long as the width of the gnathosoma; the other setae are in linear arrangement.

Cyclamen mite, *Steneotarsonemus pallidus* (Banks). This mite (fig. 78, *a*, *b*) is a destructive pest of strawberries, watercress, and many ornamental flowers and shrubs, such as cyclamen, gerbera, begonia, African violet, ivy, and

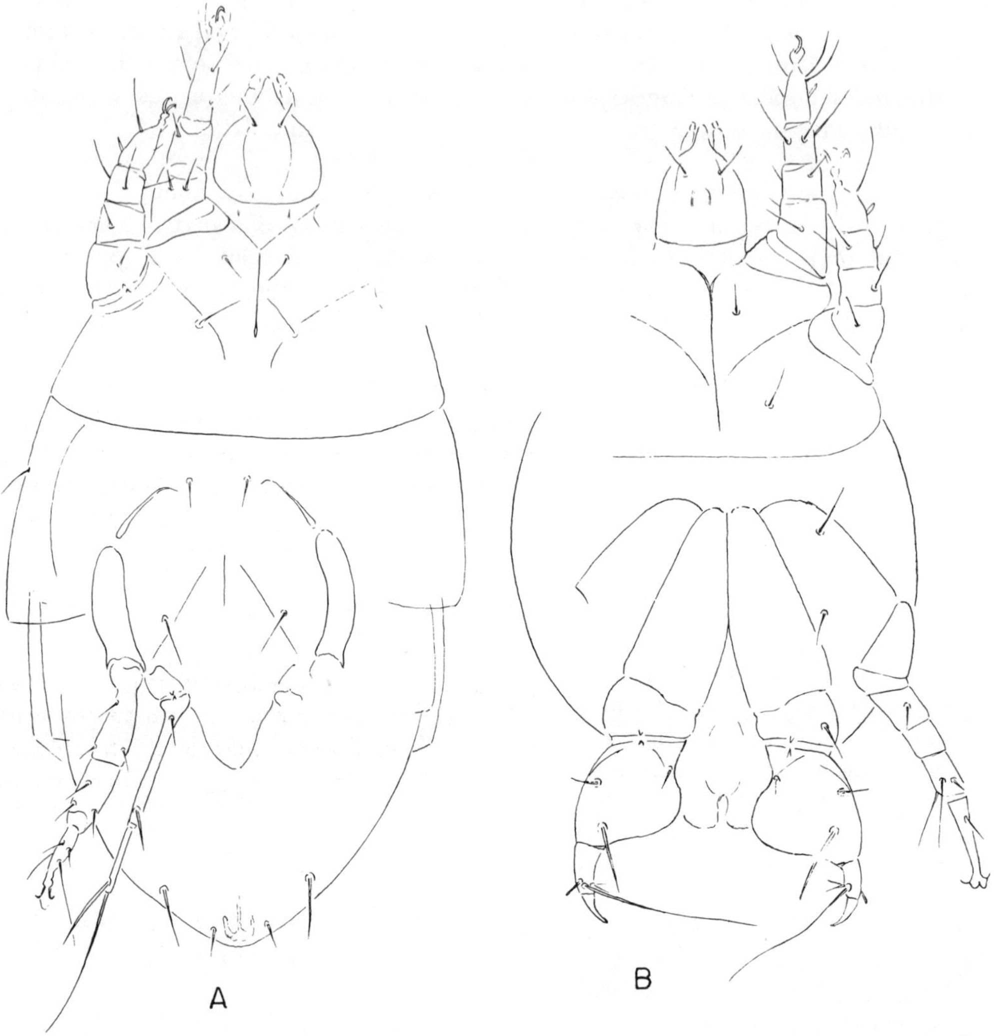

Fig. 78. *Steneotarsonemus pallidus* (Banks): *a*, female, ventral view; *b*, male, ventral view.

pikake (van Eyndhoven and Groenewold, 1959). It appears to be widely distributed throughout the world, having been reported on flowers and shrubs throughout North America, Hawaii, Europe, and Asia (pls. 49, a, *b*; 50, *a-c*).

Infestations of this mite on pikake, for example, result in fewer flowers per cluster and often complete abortion of the clusters. Buds and flowers that do appear are often poorly formed. Mature as well as young leaves are twisted, distorted, and usually smaller than normal. Infested shoots commonly have elongated internodes. Also, the leaf bud is generally smaller at the shoot apex than that of an uninfested shoot. Leaf buds of badly infested shoots are sometimes completely destroyed. The leaves on such a shoot are progressively smaller and more distorted toward the tip. Mite populations on gerbera foliage produce bronzed patches along the midribs and slight curling of the foliage. Flowers are attacked in the bud stage causing the flower rays to become deformed, and not fit for market (Boyle and Haramoto, 1956).

Infestations of cyclamen mite on strawberries produce a roughened, wrinkled appearance of the upper leaf surfaces, irregular folding and fluting of the leaf margins, and veins that bulge upward like blisters. Plants with mild injury assume a dense appearance because petioles fail to elongate. Those with severe symptoms are dwarfed at the crown and have small leaflets that fail to unfold completely. The smaller leaflets become pale yellowish green with a hard brittle texture and finally turn brown or silvered when the undersurfaces are exposed to the sun. Infested flowers and young fruit are brown near the inner bases of the sepals, and in severe cases may turn black and dry (Dustan and Mattewman, 1932).

Cyclamen mites avoid light and require humidity near saturation; thus they occur in unopened leaflets in the crown of their hosts, between tightly packed young leaves in the leaf bud, or in the cuplike cavities of embryonic flower buds. Only adults overwinter in temperate climates, but oviposition and development may take place during mild winters or in mild climates. In Canada, the adult females hibernate in the crowns within the leaf sheaths of strawberry plants. The overwintering females begin depositing eggs by mid-April and populations peak from the middle of June to the end of July; at that time there may be as many as 300 mites per plant. Population levels vary inversely in proportion to the rate of new leaf growth. Oviposition decreases in August and ends in October.

Distribution in strawberry fields may occur as mites crawl along runners, by transplanting daughter plants, or by wind. Migration over soil surface is unlikely. Mites exposed to direct rays of the sun die in a few hours or succumb in a few days in dry soil. Dispersion from field to field or from one area to another is by transporting infested plants by pickers and their equipment, by irrigation, or by bees or other flying insects.

The adult female cyclamen mite is yellowish brown, 250 to 260 μ long, with hind legs reduced to slender threadlike structures. The eggs are relatively large, (125 x 75 μ), at least half as large as the adult female. They are elliptical, opaque, smooth, and half again longer than they are wide, with 2 ends equally rounded. The shell is sufficiently thin for the embryo to be clearly visible in the egg for some time before the eggs hatch. The larvae are opaque white, with a peculiar triangle enlargement at the posterior end of the body. The larvae molt to the pupal stage, which is without means of locomotion, and adults emerge from the pupal stage. The males are able to pick up and transport the pupae or even adult females with their modified fourth pair of legs. This species commonly reproduces parthenogenetically.

The eggs are usually laid in clusters and sometimes in masses two or three eggs deep located between the young leaves of the bud. Each adult female may lay 1 to 3 eggs per day and a total of 12 to 16 during her lifetime. The life cycle requires 1 to 3 weeks. The duration of the egg stage is 3 to 7 days; the larva, 1 to 4 days; and the resting or pupa, 2 to 7 days (Smith and Goldsmith, 1936).

The cyclamen mite is difficult to control by acaricidal dusts or sprays because the mites are so well protected in the leaf buds. Spread from one area to another may be prevented by fumigating the planting stock in closed chambers with methyl bromide or ethylene dibromide. Field populations on strawberries may be controlled by using polyethylene (plastic) covers over the beds during fumigation. A complete kill of mites on planting stock can be obtained by immersion of the plants in water at 43.5 C (110 F) for 30 minutes. A more efficient method is to treat the plants with saturated air at 43.5 C for 1 hour. Plants must be loose on screen or slat boxes or stacked to permit penetration of the gas. After treatment the plants should be carefully dried (but not in excess) before packing, and they should be planted as soon as feasible (Munger, 1933; Smith, 1939). Mites on plants that cannot practically be fumigated may be controlled by repeated and thorough spray applications of endrin, endosulfan, or dicofol.

The first ventral propodosomal seta of the female of S. *pallidus* is well behind apodeme I. Femur I has a stout seta and three slender setae; there are no setae between coxae IV. The pseudostigmatic organ is expanded distally and is subcircular in shape. There are 2 pairs of setae on the propodosoma: the first pair is as long as genu I, and the second pair is about ¾ the length of the body width at the sejugal sutre between the propodosoma and the hysterosoma. The flange on femur IV is longer than femur IV. The fourth pair of propodosomal setae are shorter than the third pair and are slightly laterad of the linear line between the first and third pair; the first dorsopropodosomal setae are longer than the second pair but are shorter than the third pair of setae.

Steneotarsonemus spirifex (Marchal). *Spirifex* (fig. 77, *f*) occurs in England and continental Europe. Its feeding causes considerable damage to cereals during dry seasons, especially to late-ripening varieties of oats. The mites are found within the upper part of the leaf sheath where their feeding activities result in a spiral malformation of the rachis, blind spikelets, and an incompletely emerged inflorescence (Evans, Sheals, and Macfarlane, 1961).

The female of S. *spirifex* is typical for the genus. Femur IV of the male has a strong flange that reaches to the distal end of the segment; it is strongly indented between the segment and flange area distally. The distal seta of tibia IV is strong, but it is not as long as the femur. There are 4 pairs of propodosomal setae; the first pair of propodosomal setae are longer than the second and fourth pair; the second pair of setae are shorter than the other pairs; the first pair of ventral propodosomal setae are set well behind apodemes I; and coxae III are not ornamented.

Polyphagotarsonemus Beer and Nucifora

Legs II and IV of the female do not have claws; the female gnathosoma is rounded; there are 4 pairs of ventral metapodosomal setae; and the pseudostigmatic organs are expanded distally. There are 3 to 4 pairs of propodosomal setae on the male; there are 4 pairs of ventral metapodosomal setae; tibia and tarsus IV of the male are fused to form a tibiotarsal segment bearing a buttonlike claw, and coxae III and IV of the male are distinctively contiguous.

Broad mite, *Polyphagotarsonemus latus* (Banks). This species (fig. 79, *a*, *b*; pl. 51, *a-c*), also known as the yellow tea mite and the tropical mite, is distributed

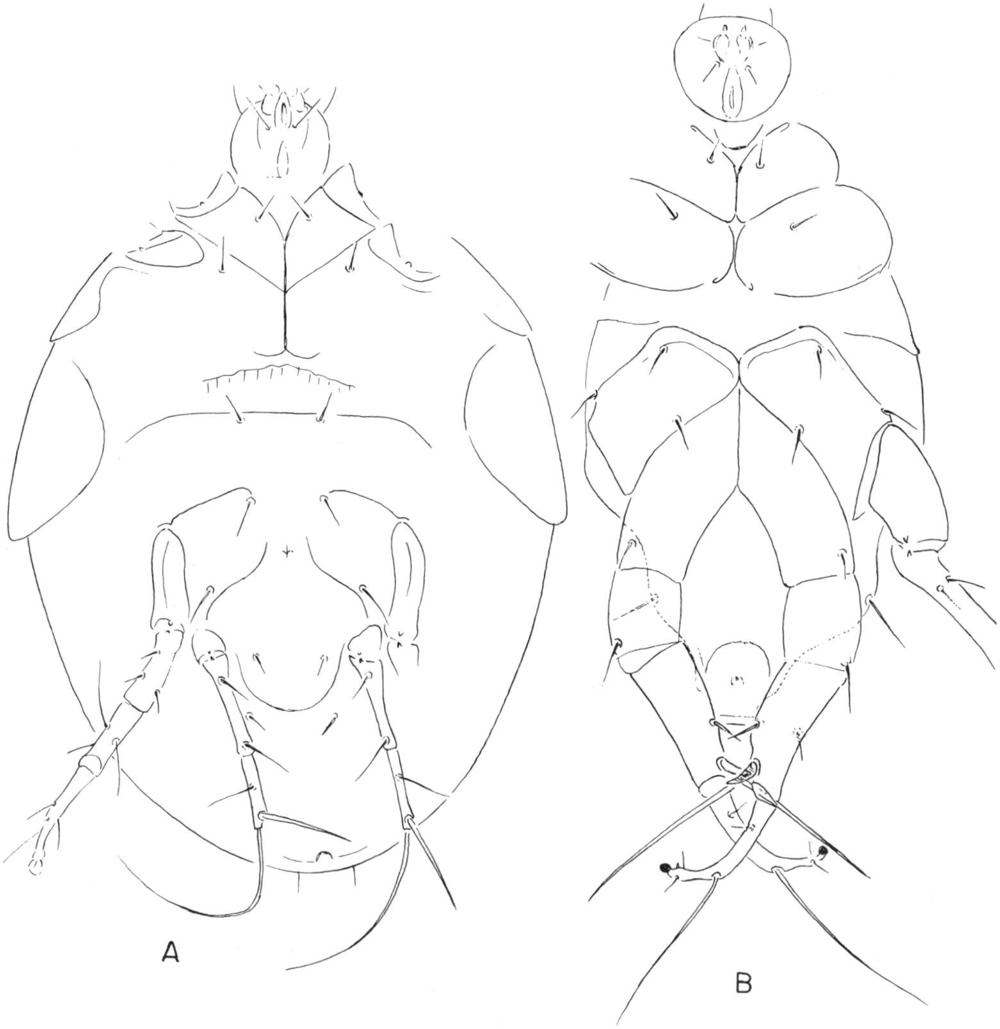

Fig. 79. *Polyphagotarsonemus latus* (Banks), ventral views: *a*, female; *b*, male.

in the field throughout the tropics and in greenhouses in temperate regions on a wide variety (about 50) of agricultural crops, ornamental, and wild plants. It has been known in Africa since 1890 where symptoms on cotton are known as "acariose." Although injury is produced to many crops, it causes most concern to man by attacking such major crops as cotton, tea, rubber, citrus, tobacco, potatoes, beans, peppers, gerberas, dahlias, zinnias, and chrysanthemums (Hambleton, 1938).

The mites feed almost entirely on the lower leaf surface causing the leaves of gerberas to become rigid and rolled under at the edges. Feeding injury is confined to young foliage or flower parts. As the leaves age they may spilt or crack open, producing a ragged appearance of various shapes. The lower leaf surfaces become bronzed, and injured flowers have part or all the rays distorted or discolored. Feeding injury is usually expressed on many hosts as sudden curling and crinkling of leaves followed by blister patches. Plants severely attacked stop growing and die (Smith, 1939).

Injury to potato appears first as oily blackish spots on the undersurfaces of young leaves. The undersides of the leaves turn reddish, the plants become rosetted, the leaf hairs become prominent and the leaves become wrinkled at their edges. Starting at the tip, the plants wither and auxiliary buds are produced which in turn are killed ("Tambera" disease in India). Similar effects are produced on red pepper where attacks in the seedling state prevent flower and fruit developemnt while later infestations cause flower drop ("Murda" disease of chili). Besides producing multiple buds on citrus seedlings, this species causes discoloration of the skin of lemons (Kulkarni, 1923; Mann, Nagpurkar, and Kulkarni, 1920).

The first sign of the disease on tomato is a superficial shiny bronze or brownish discoloration on the surface of the succulent stems of terminal shoots and on the undersurface of the young leaves. The injury is first limited to the browning of epidermal cells, but later the cells of the rapidly growing tissues collapse and die. At the same time the young expanding leaves become narrow, stiff, twisted, or crumpled, fail to elongate, and finally may wilt and dry up rapidly, as if the top of the plant were scorched by flame. As soon as the top appears scorched, the succulent part of the stem of the young plant may be slightly swollen, roughened, or russetted and take on a greyish green color.

Feeding on cotton by the mites is confined almost entirely to the lower leaf surface where they are most commonly found. Their feeding causes cotton leaves to become rigid and rolled under at the edges. With age, the leaves may split or crack open, and assume a ragged appearance of various shapes (Hambleton, 1938).

The eggs of the broad mite are oval and elongate, but when attached to the leaf surface they have a flat unornamented base. The upper surface is studded with longitudinal rows of tubercles that have high refractive index and are white, in contrast to the general body of the egg, which is transparent. The larvae, except for size, resemble the adult in appearance, and the absence of the pseudostigmatic organ and skin annulations. The nymphs remain enclosed within the skins of the quiescent larvae until the adult is formed. The adult female is large, oval, and broad, rich amber or dark green, the color depending on the host and food supply. Young females differ from full-grown adults in being subcircular, less deeply pigmented and in having a pair of posterior lateral setae situated on the last body segment. The adult male is short, broad, tapers at the posterior end, has long legs, and is colorless when young, but rich amber when fully developed. At the apex, on the ventral side of the male is a suckerlike organ used to hold and carry the pupa. Lavoipierre (1940) indicates that it is only the pupa that appears to be held by the hind legs while being carried.

This species multiplies rapidly so only 4 to 5 days are required to complete a generation in the summer and seven to 10 days in the winter. The average egg deposition during the summer is 3.6 eggs per day, but activity and reproduction continues throughout the year, even though productivity decreases during the winter. Eggs are laid in almost imperceptible depressions on the leaf or fruit sur-

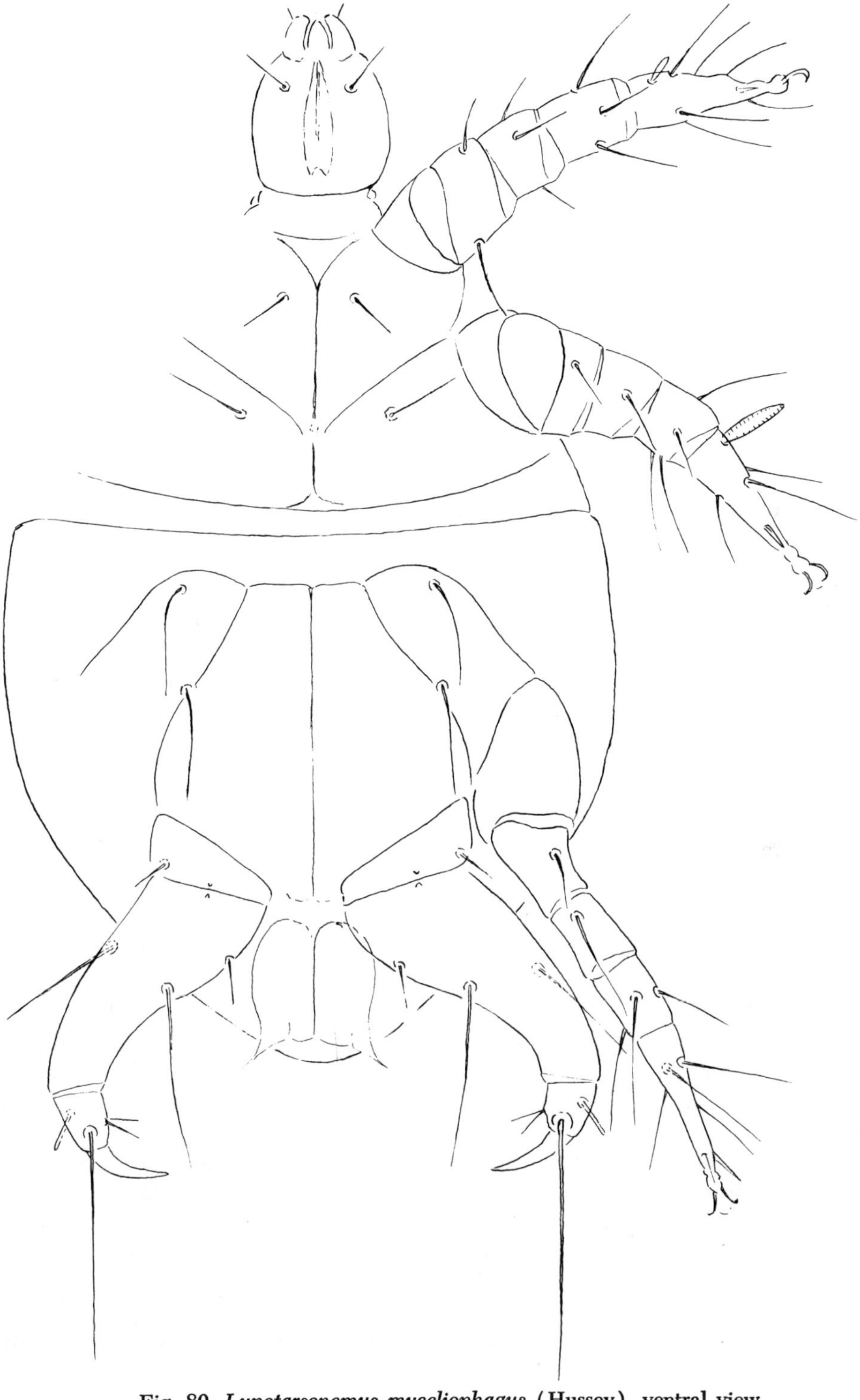

Fig. 80. *Lupotarsonemus myceliophagus* (Hussey), ventral view.

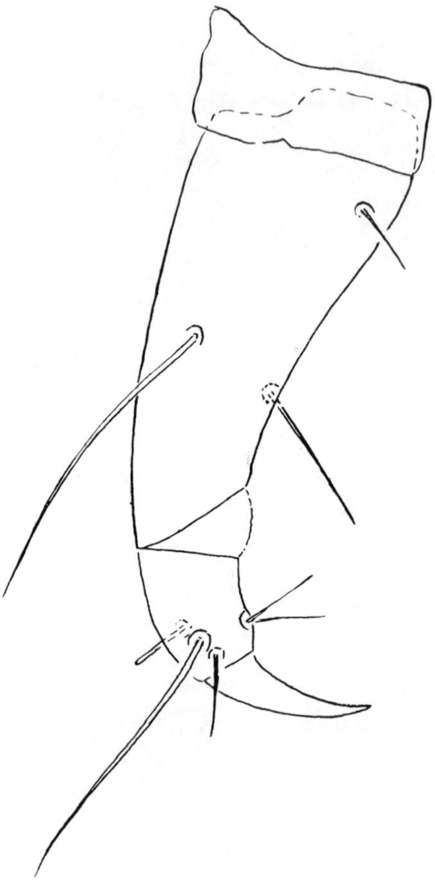

Fig. 81. *Lupotarsonemus randsi* (Ewing), male, Leg IV, dorsal view.

face. They are firmly attached at the base. Mites are most numerous in damp, shady places.

The females and larvae do not tend to wander from the leaf where they are located, yet the colonies disappear from the matured leaves and become established on the terminal or younger leaves. The males may transfer the colony from mature to young leaves by carrying female pupae as they aimlessly wander about (Lavoipierre, 1940; Moutia, 1958).

Control of broad mite may be achieved by dusting the plants with sulfur, by fumigation with calcium cyanide or naphthalene, or by heat treatment such as is used for control of the cyclamen mite. The egg stage is not very susceptible to sulfur, but the larvae and adults succumb to sulfur residues. Two or 3 applications at 5-day intervals may be necessary. An application of dicofol at the beginning of September followed by another in 35 days is effective on fruit trees. Successive applications of dicofol, chlorobenzilate, and carbophenothion should control this mite (Ingram, 1960).

The dorsal body setae of both sexes of *P. latus* are relatively short and there are 4 pairs of propodosomal setae in the male. Otherwise, the species is typical for the genus.

Parasteneotarsonemus Beer and Nucifora

The pseudostigmatic organs of the female are expanded distally and there are 2 pairs of ventral metapodosomal setae. The male has 4 pairs of propodosomal setae; the tarsal IV claw is reduced and buttonlike, and the inner flange is about as long as the femur. Leg IV has tibia and tarsus; and the gnathsoma is broader than it is long.

Parasteneotarsonemus phyllophorus (Ewing). This mite (fig. 77, *g*) is an occasional pest of bamboo in Florida and Georgia in the United States and in Japan (Beer, 1954). Femur IV of the male of *P. phyllophorus* has a long, broad flange. There are only 3 pairs of propodosomal setae, the "true" second pair is lacking. The first pair of propodosomal setae are the shortest and the other two are longer, the second being twice as long as the third pair; the tibial IV seta is not as long as femur IV; and the tarsal IV claw is knoblike.

Lupotarsonemus Beer and Nucifora

The pseudostigmatic organs of the female are expanded distally; the propodosomal setae of the female are widely separated, the second pair being on the posterior half of the propodosoma; leg IV of the male has the tibia and tarsus fused to form a tibiotarsal segment; an unmodified claw is present on leg IV of the male.

Lupotarsonemus myceliophagus (Hussey). This species (fig. 80) is a serious pest in mushroom culture. Significant damage occurs when the initial infestation takes place soon after spawning. The life cycle requires only 8 days at 24 C (75 F) and 12 days at 16 C (61 F). Effective control depends on clean culture between crops. Methyl bromide fumigation at 66 to 71 C (150 to 160 F) for several hours followed by 0.01% dicofol sprays directed toward the internal surfaces of the empty building usually prevents the development of injurious infestations in the compost (Hussey and Furney, 1967).

L. myceliophagus is a typical *Lupotarsonemus* in which the flange on femur IV on the male is missing and the fourth pair of dorsal propodosomal setae are set laterad from the third pair. The female is typical for the genus and the male is here used for specific identificatinon. The obvious specific character is the greatly enlarged solenidion on tarsus II and the design of the coxal apodemes; femur IV is rounded at the base. The tactile seta of tibia IV is about as long as femur IV; the third dorsal propodosomal setae are distinctly longer than the others; the tactile setae of tibia IV are longer than the others on leg IV; and femur IV is 2 times as long as it is broad.

Lupotarsonemus randsi (Ewing). *L. randsi* (fig. 81) occurs on the European plum, sweet cherry, citrus, rose and sugar cane. The reported distribution includes Portugal and Florida and Virginia in the United States (Beer, 1954).

The ventral transverse apodeme of the female *L. randsi* is entire; otherwise the female is typical for the genus. Femur IV of the male is broadest near its base and is more than twice as long as broad at the base. The tactile seta of the tibia IV is not longer than the longest femoral seta. There is no flange on femur IV; the fourth pair of propodosomal setae of the male are usually displaced laterally.

SELECTED BIBLIOGRAPHY

Beer, R. E. 1954. A revision of the Tarsonemidae of the Western Hemisphere (Order, Acarina). Univ. Kans. Sci. Bull. 36, pt. 2, 16:1091–1387.

———. 1958. A new species of *Steneotarsonemus*, and additional information on the plant feeding habits of *Steneotarsonemus furcatus* de Leon. J. N. Y. Entomol. Soc. 66:153–159.

———. 1963. Social parasitism in the Tarsonemidae, with description of a new species of tarsonemid mite involved. Ann. Entomol. Soc. Am. 56(2):153–160.

Beer, R. E., and A. Nucifora. 1965. Revisionedei generi della famiglia Tarsonemidae (Acarina). Boll. agrariaedi Bachicoltura per. 11, 7 :19–43.

Boyle, W. W., and F. H. Haramoto. 1956. Cyclamen mite on pikake. Hawaii Farm Sci. 5(2):1, 7, 8.

Butani, D. K. 1959. Sugarcane mites—a review. Indian Acad. Sci. Proc. B. 49(2):99–102.

Carmona, M. M. 1966. Acaros das plantas cultivadas IV, Agronomic Lusitana. 26(3): 175–203.

Doucette, C. F. 1936. Observations on bulb scale mite as a major pest of narcissus. J. Econ. Entomol. 29(6):1103–1105.

Dustan, A. G., and W. G. Mattewman. 1932. Some notes on the cyclamen mite, *Tarsonemus pallidus* (Banks), a pest of strawberry plants. Entomol. Soc. Ont., 62d Ann. Rept. 1931:34–37.

Evans, G. O., J. G. Sheals, and D. Macfarlane. 1961. The terrestrial acari of the British Isles. I. Introduction and biology. 219 pp. British Museum, London.

Ewing, H. E. 1939. A revision of the mites of the subfamily Tarsoneminae of North America, the West Indies, and the Hawaiian Islands, U. S. Dept. Agr. Tech. Bull. 653. 63 pp.

Fajardo, T. G., and G. C. Bellosillo. 1934. A mite disease of tomato, tobacco, potato, and other plants in the Philippines. Philippine J. Sci. 54(4):523–543.

Gadd, C. H. 1946. Observations on the yellow tea mite, *Hemitarsonemus latus* (Banks). Bull. Entomol. Res. 37(2):157–162.

Garman, P. 1917. *Tarsonemus pallidus* Banks, a pest of geraniums. Md. Agr. Expt. Sta. Bull. 208:327–342.

Hambleton, E. J. 1938. A occorrencia do acarotrobical *Tarsonemus latus* Banks. Causador da rasgadura das folhas nos algodoais de S. Paulo. Instituto Biologico. 9:201–209.

Hodson, W. E. H. 1934. The bionomics of the bulb-scale mite, *Tarsonemus approximatus* Banks var. *narcissi* Ewing. Bull. Entomol. Res. 25 :177–184.

Holloway, T. E. 1936. Insect pests of sugarcane XIV. Facts about sugar 31(6):216–217.

Hughes, T. E. 1959. Mites or the Acari. The Athlone Press, Univ. London. 225 pp.

Hussey, N. W., and B. Furney. 1967. Bionomics and control of *Tarsonemus myceliophagus* Hussey in mushroom composts Entomol. Expt. and Appl. 10(3–4):287–294.

Ingram. W. R. 1960. The control of yellow tea mite, *Hemitarsonemus latus* (Banks) with DDT on cotton in Uganda. Bull. Entomol. Res. 51(3):575–582.

Kulkarni, G. S. 1923. The "murda" disease of chili (Capsicum). Agr. J. India 17:51–54.

LAVOIPIERRE, M. M. J. 1940. *Hemitarsonemus latus* (Banks), (Acarina) a mite of economic importance to South Africa. Entomol. Soc. So. Africa 3:116–123.

MANN, H. H., S. D. NAGPURKAR, and G. S. KULKARNI. 1920. The "tambera" disease of potato. Agr. J. India 15:282–288.

MOUTIA, L. A. 1958. Contribution to the study of some phytophagous acarina and their predators in Mauritius, Bull. Entomol. Res. 49:59–75.

MOZNETTE, G. F. 1917. The cyclamen mite. J. Agr. Res. 10(8):373–390.

———. 1925 A pest in the mango nursery. Quart. Bull. State Plant Bd. Fla. 9(3):121–211.

MUNGER, F. 1933. Investigations in the control of the cyclamen mite (*Tarsonemus pallidus* Banks). Minn. Agr. Exp. Sta. Tech. Bull. 93:1–20.

SMITH, F. F. 1939. Control of cyclamen and broad mites on Gerbera. U. S. Dept. Agr. Circ. 516:1–14.

SMITH, F. F., and E. V. GOLDSMITH. 1936. The cyclamen mite, *Tarsonemus pallidus*, and its control on field strawberries. Hilgardia 10(3):53–54.

VAN EYNDHOVEN, G. L., and H. GROENEWOLD. 1959. On the morphology of *Steneotarsonemus pallidus* and *S. fragariae* (Acarina, Tarsonemidae). Entomol. Ber. 19:123–124.

WINFIELD, A. L. 1964. Chemical control of bulb scale mite on forced narcissus. Exp. Hort. 11:69–77.

Chapter 11

Tydeidae, Tuckerellidae, Pyemotidae, Penthaleidae, Astigmata, and Cryptostigmata

THE TYDEIDAE KRAMER

The family has been described by Baker (1965:96–97) as follows:

The family is difficult to characterize, although easily recognized. They are small to very small mites, the adults ranging in size from 150 to 500 μ, with a weakly sclerotized or non-sclerotized body. The palpus is four-segmented and typically shaped, the setal count varying among the genera; there are five setae on the distal (tarsus) segment, and at times one of these segments may possess a small basal seta (duplex setae), a solenidion may also be present ventrally and proximally on this segment; the penultimate (tibia) segment has either one or two simple setae; the strong (femur-genu) segment proximal to this always has two setae; and the short basal (trochanter) segment has none. The movable chelae of the chelicerae are needlelike and unopposed. Body setation is simple. There are three pairs of dorsal setae plus one pair of sensory setae on the propodosoma; in two species a pair has dropped out [Fig. 82]. The sensory setae are usually distinctive and set in large pseudostigmata, although in a few species all the propodosomal setae and their bases are similar; these sensory setae are usually set inside the second pair of propodosomal setae, but this position can vary. The propodosomal setae are here designated as P_1 and P_2 (the anterior row), and as the sensory setae and P_3 (the second row). The hysterosomal setal pattern is also simple, consisting of either five transverse rows of four setae each, or four and one-half rows, the posterior lateral pair lacking. These setae are labeled D_1 to D_5 (the dorsal setae), and L_1 to L_5 (or to L_4) (the lateral setae). D_4 and L_4, and D_5 and L_5 may be much longer and different from the others. Dorsal body setae may be simple and nude, pilose or serrate lanceolate, clublike, and so so on, with many combinations possible. Setae L_2 in the *Pronematus-Triophtydeus* group have migrated dorsally to be in line with the D setae. There are always three pairs of ventral body setae, but the genital and anal setae may vary in number according to the genus and species; the setae represented are: anal, genital, paragenital, and ventral setae.

The dorsal striation pattern is important both generically and specifically. The simplest type is that found in the genus *Tydeus*, in which the striae are longitudinal on the propodosoma and transverse dorsally on the hysterosoma; the lobes of the striae may vary in height and width. In *Pronematus* the striae are longitudinal on the dorsal anterior region of the hysterosoma. In *Lorryia* the striae form a reticulate pattern, either in whole or in part, which also may be found ventrally; the reticulations usually possess a few sharp tubercles.

The hysterosoma may be strongly lobed dorsally and laterally, but this structure is found throughout many genera, and is not here considered to be of generic value. Mounting may cause this condition to disappear.

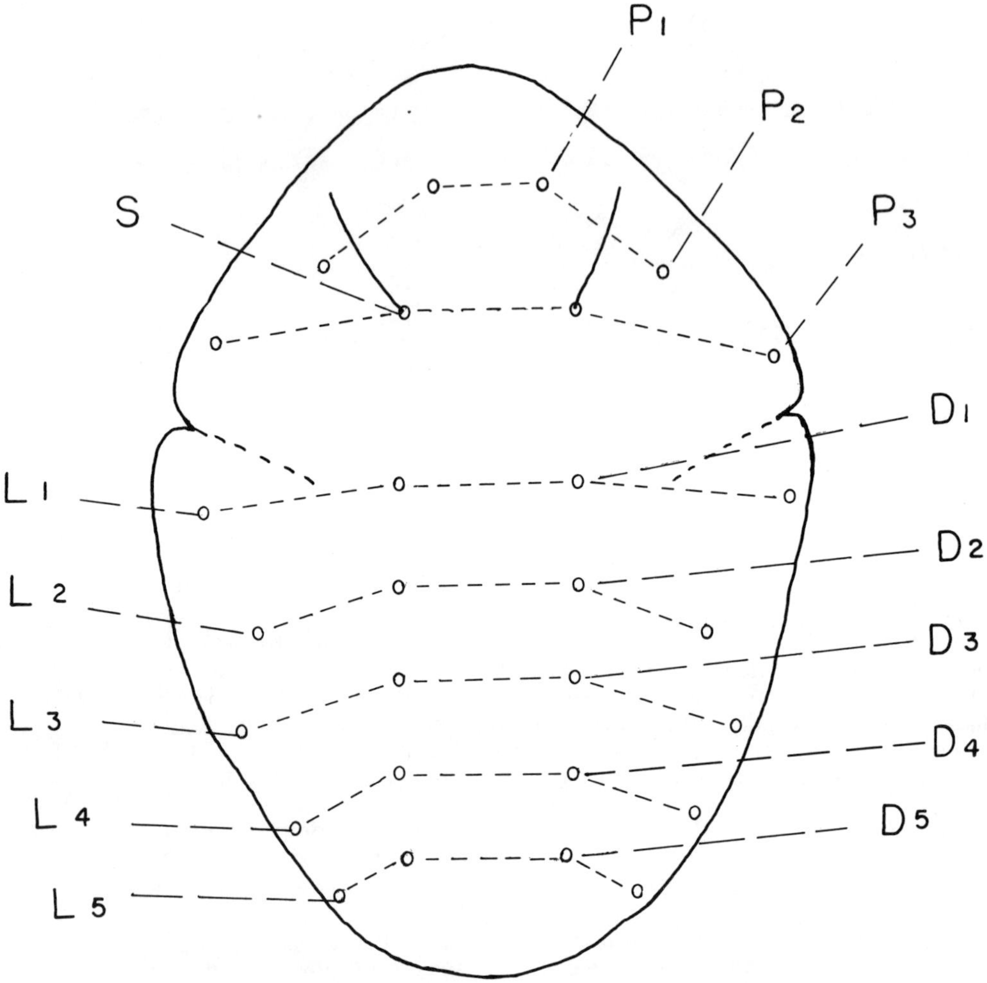

Fig. 82. Tydeidae setation, dorsal view (after Baker, 1965).

Eyes, or pigmented area, may be present on the prodosoma. Usually there is a single pair, but in *Triophtydeus* a third or anterior median eye is present. Since these eye "spots" may disappear upon mounting, they are not here used for either generic or specific identification. The genitalia vary with the genera.

Males and females are similar except that the male is smaller and possesses a much smaller genital opening. Mating has not been observed so it is presumed that the males are spermatophore layers. Most of the species are of no economic importance, probably feeding on fungi, honeydew, etc. *Tydeus californicus* (Banks) is a recognized plant feeder, as probably is the closely related species *T. caudatus* (Dugès). *Lorryia formosa* Cooreman has been found damaging citrus. *Pronematus ubiquitus* (McGregor) has been observed preying on the fig mite, *Eriophyes ficus* Cotte (Baker, 1965).

Tydeus Koch

Tarsus I has empodium and claws. The striae are longitudinal on the propodos-

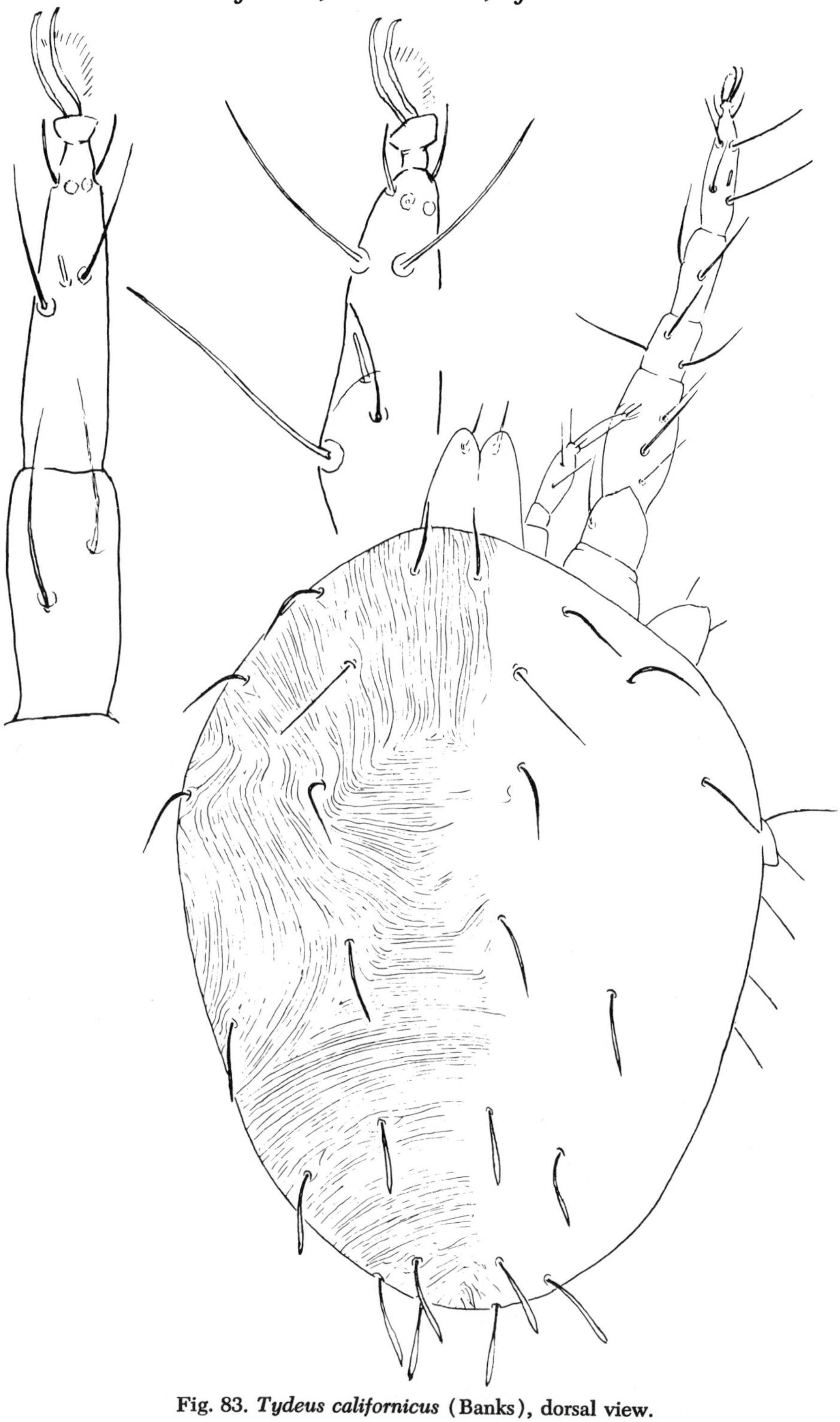

Fig. 83. *Tydeus californicus* (Banks), dorsal view.

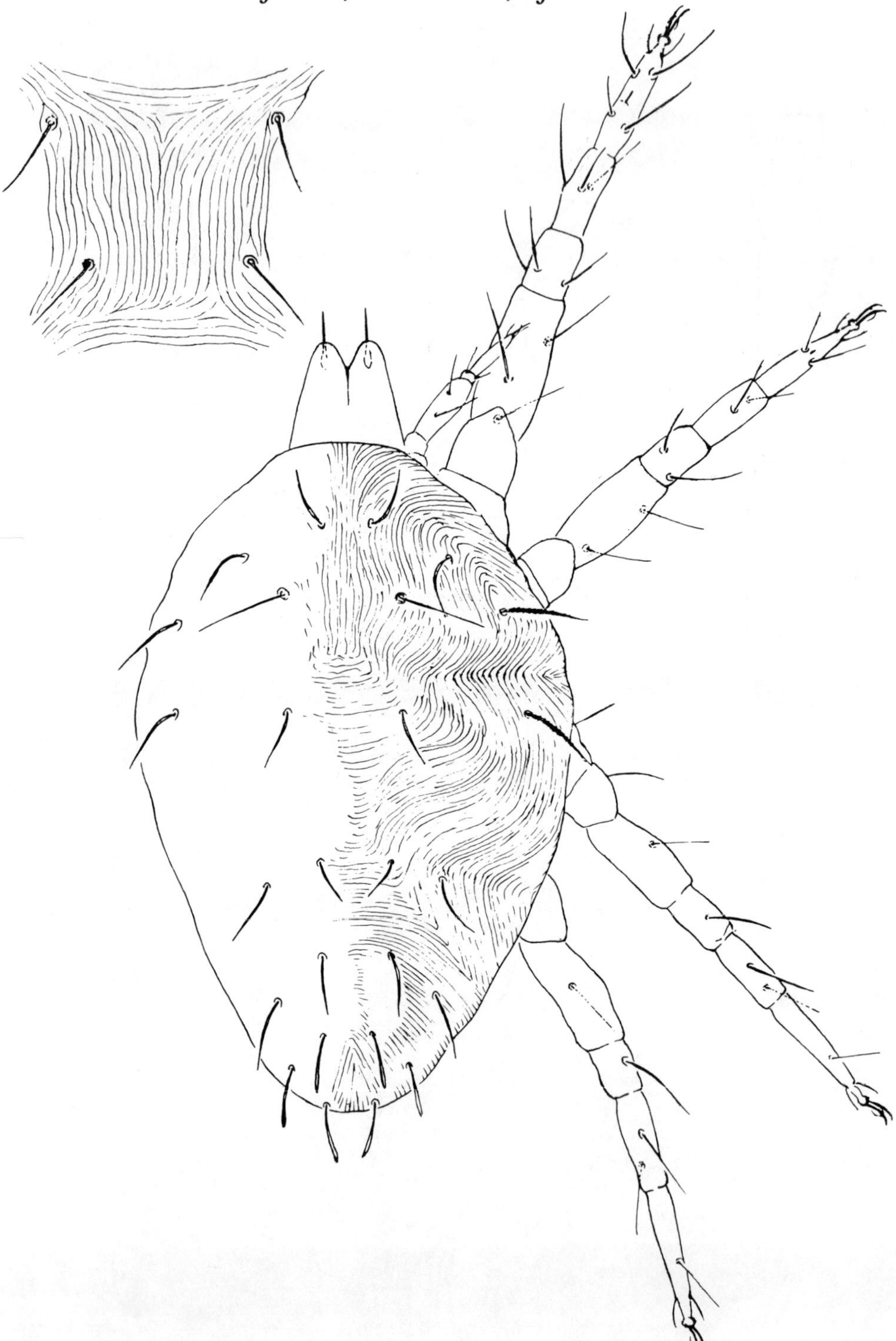

Fig. 84. ***Tydeus caudatus*** (Dugès). ***a***, female; ***b***, ventral striation pattern.

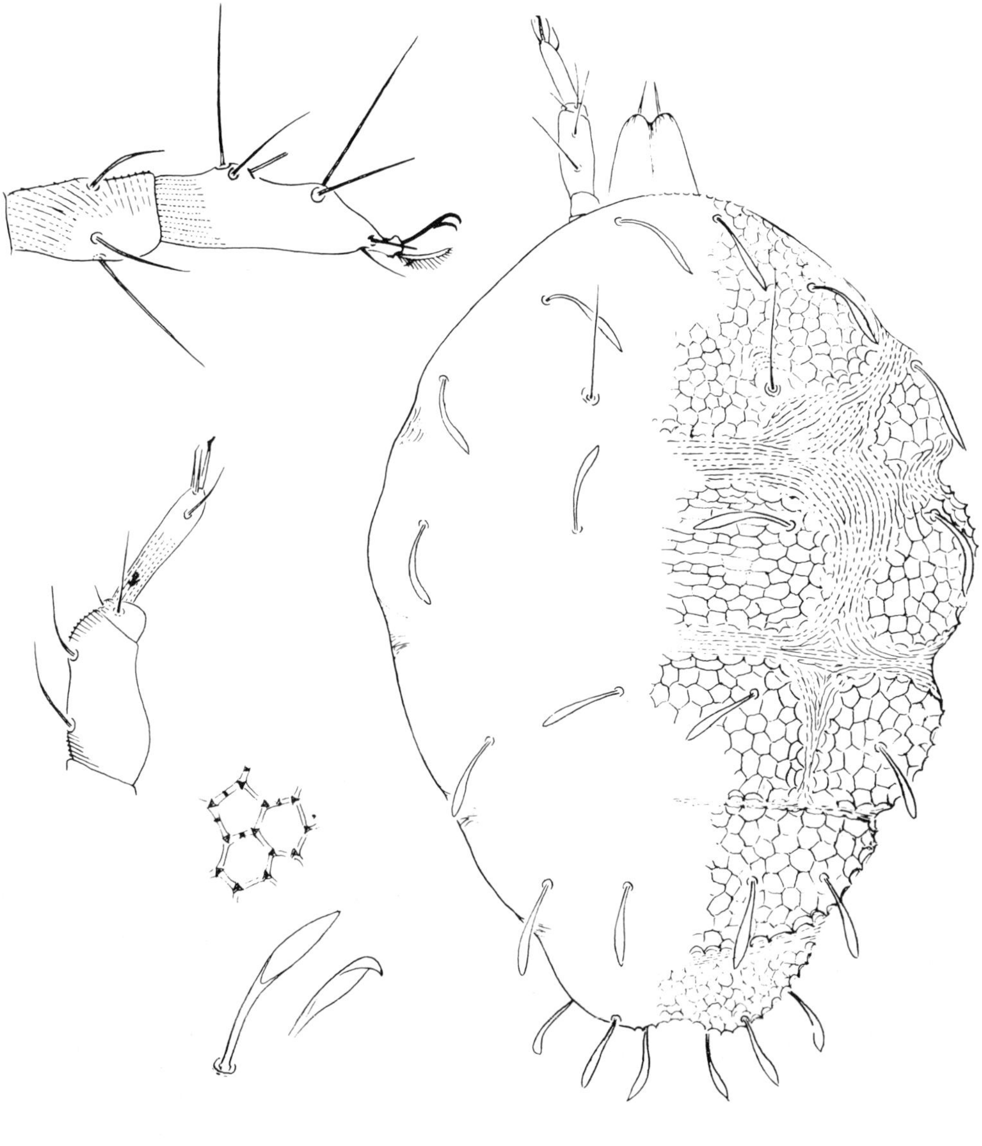

Fig. 85. *Lorryia formosa* Cooreman. *a*, dorsum of female; *b*, tarsus I; *c*, palpus; *d*, reticulate pattern; *e*, dorsal setae.

oma and transverse on the hysterosoma; the D and L setae of the hysterosoma lie in transverse rows. The number of genital setae varies within the genus, as does the ventral striation pattern.

Tydeus californicus (Banks). This mite (fig. 83) is common on citrus in coastal districts of Southern California and in other southern climates; it often lives in dense groups, chiefly on the undersides of leaves, where cast skins occur in compact masses. It is pear-shaped, small, and slower in movement than *Pronematus ubiquitus* McGregor. It has been reported both as a predator of the citrus bud mite and as a cause of injury to the citrus. Its habits and biology have not been studied (McGregor, 1956).

T. californicus is distinctive in that there are 5 pairs of spatulate setae on the posterior dorsal area of the body—the D_3, D_4, D_5, and L_4 and L_5 setae. The dorsal striae are typical for the genus; the ventral striae are longitudinal between setae V_2 and V_3. There are 6 pairs of genital setae. Empodial claws are not present.

Tydeus caudatus (Dugès). *Caudatus* (fig. 84) is found in temperate climates (Baker, 1965). It is similar to *T. californicus* (Banks), differing only in having 3 pairs of hysterosomal spatulate setae—the D_4, D_5 and L_4 setae.

Lorryia Oudemans

In this genus the dorsum of the body is entirely or partially covered with a reticulate pattern; if striae are present they do not form a longitudinal pattern between the third (D_3) pair of dorsal hysterosomal setae. There are usually 6 pairs of genital and 4 pairs of paragenital setae. The dorsal body setae are in a transverse pattern.

Lorryia formosa Cooreman. This mite (fig. 85) is a widespread species, originally described as a pest of citrus in Morocco. It has been reported on citrus in Spain, Argentina, Brazil, Chile, Uruguay, and Portugal, on avocado in Ecuador, and on gardenia from Mexico.

Populations of this mite on citrus produce premature sclerification of the green branches followed by desquamation at the areas where mites are concentrated. The injury results in a ring of dead brown tissue which enlarges as the fruit grows. The damage is similar to that produced by thrips under the same conditions. The mites are attracted to honeydew excretions of scale insects and the accompanying fungi upon which this mite probably feeds.

Females lay their oval, white, faintly translucent eggs close together vertically, the narrow end directed downward, and sometimes in two or three layers. The incubation period is 3 to 4 days. Larvae and nymphs are whitish becoming yellowish as they approach the molts. Adults are pale lemon yellow becoming slightly darker on the dorsum. The life cycle lasts from 12 to 41 days.

The mites congregate at the base of the twigs, petioles of flowers, and various rough areas on the branches. Yellow stains, formed by congregations of adults and the white of larvae and their exuvia, show very clearly on the green background of the bark. In the winter the mites congregate on top of the fruit stems and forming fruit. As mite colonies become more dense in the spring, the mites settle on the lower part of the leaf near the center rib where the larvae remain, but produce no injury until the first molt. After molting they abandon the leaves, and by the end of July the mites have moved to the young fruit where females begin to lay eggs under the sepals and fruit peduncles. There the mites are protected as they feed and produce injury to the young fruit tissue.

This mite species is susceptible to sulfur sprays and dusts or to the specific acari-

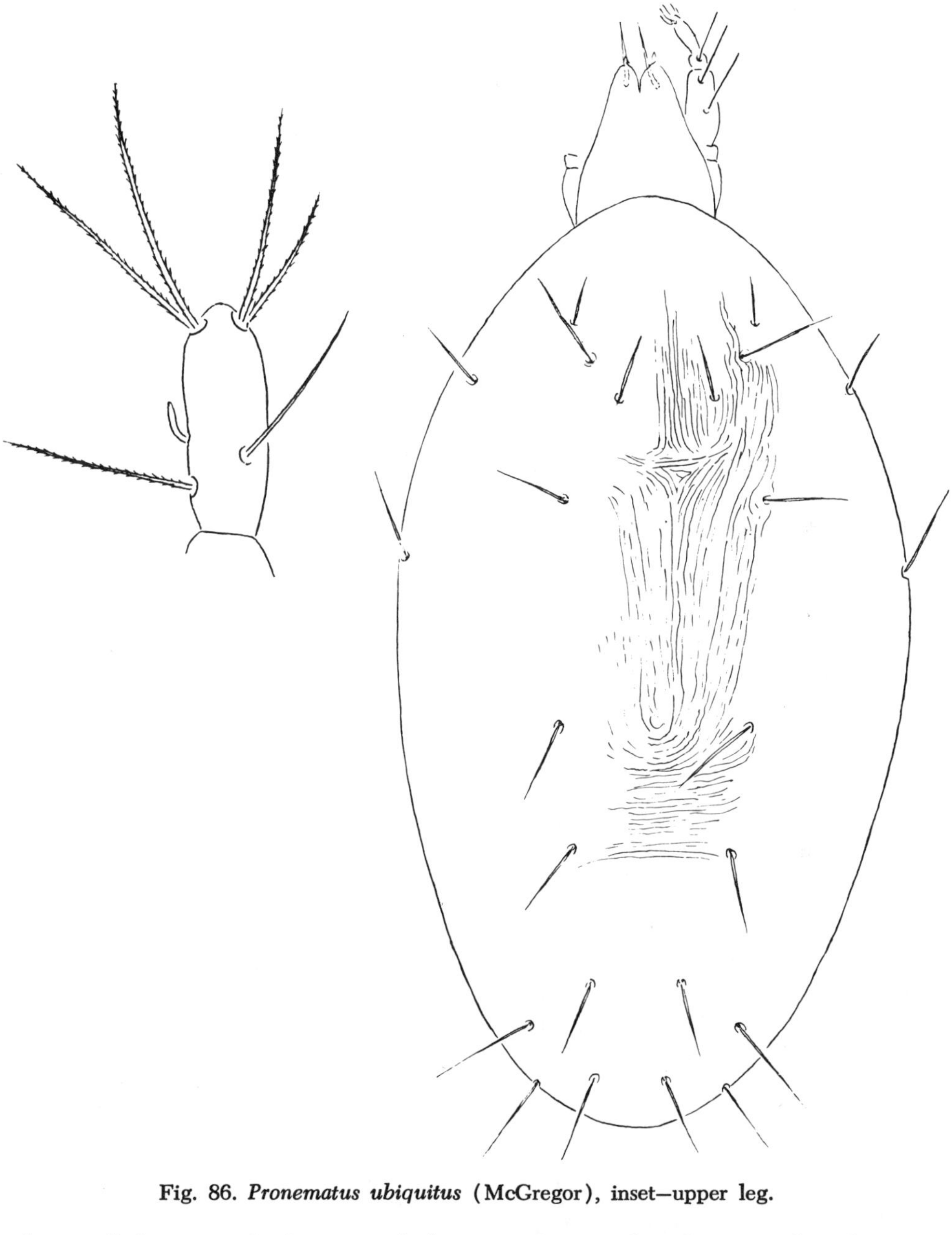

Fig. 86. *Pronematus ubiquitus* (McGregor), inset—upper leg.

cides used for control of tetranychid mites. Organophosphorus and carbamate acaricides are not generally effective (Smirnoff, 1957).

The reticulate pattern of *L. formosa* nearly covers the entire dorsal portion of the body, but it is broken into discrete areas; the dorsal body setae, except for the propodosomal trichoboths or sensory setae, are smooth and lanceolate and curved distally, all more or less of equal size.

Fig. 87. *Tuckerella pavoniformis* (Ewing).

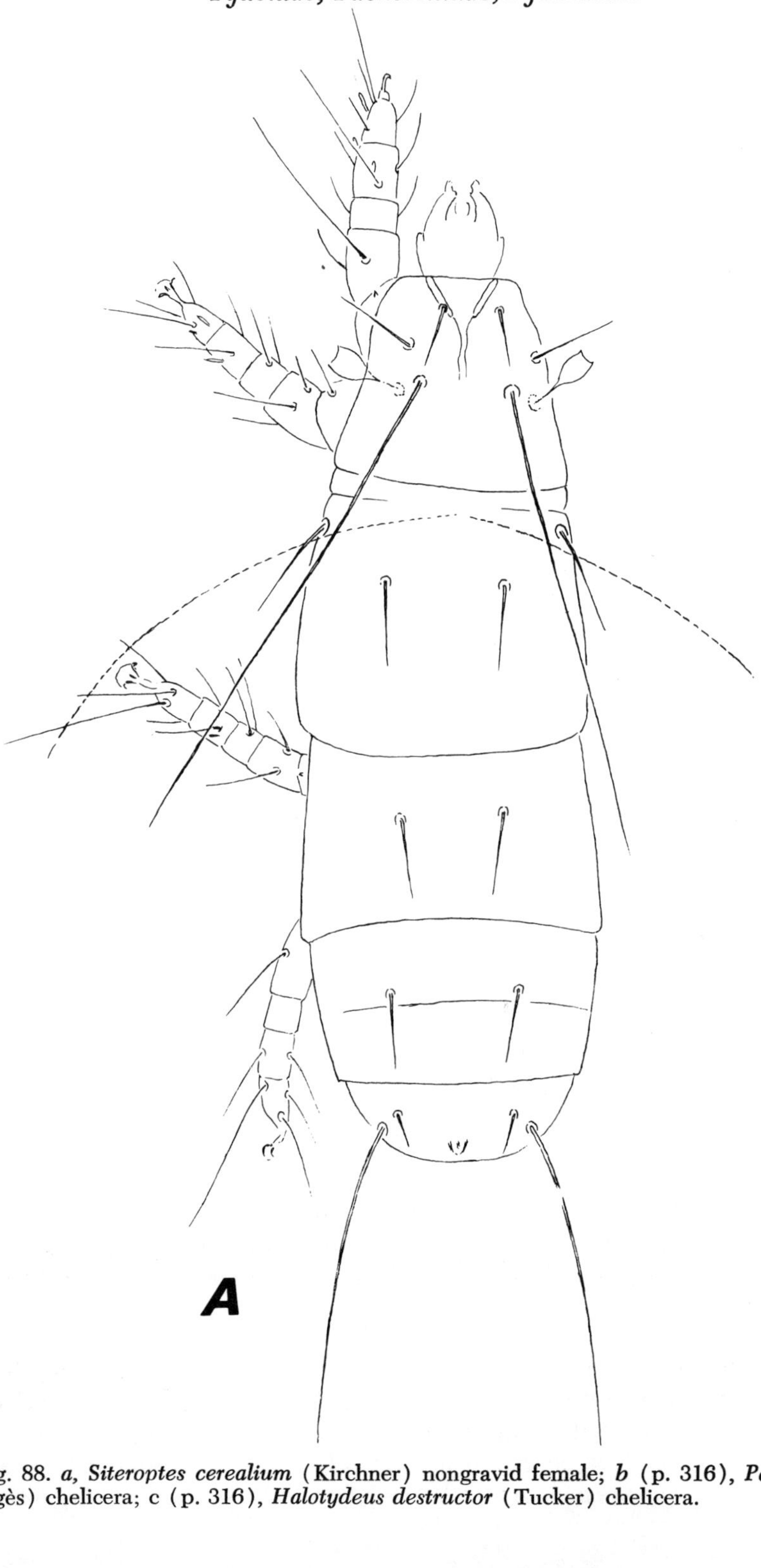

Fig. 88. *a, Siteroptes cerealium* (Kirchner) nongravid female; *b* (p. 316), *Penthaleus major* (Dugès) chelicera; c (p. 316), *Halotydeus destructor* (Tucker) chelicera.

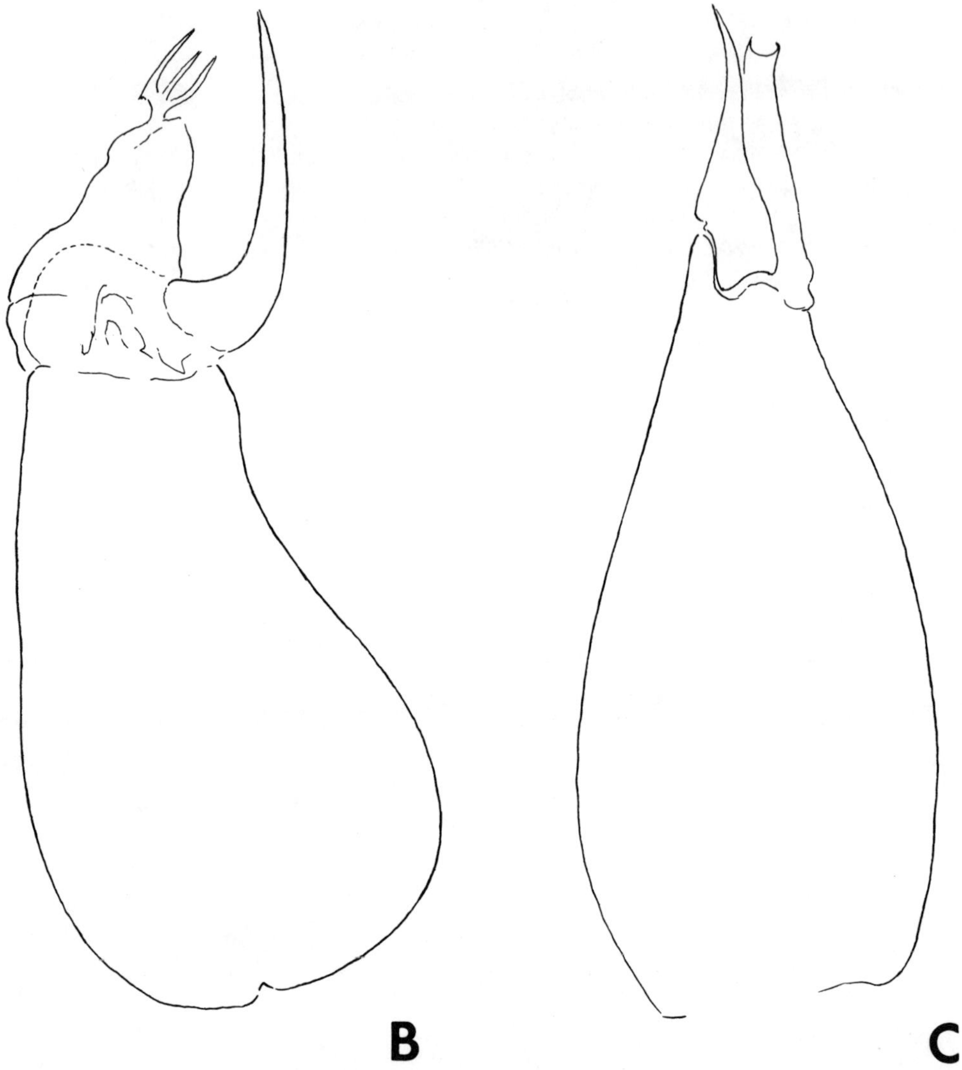

Fig. 88. *b*, and *c*.

Pronematus Canestrini

This genus is distinctive in that tarsus I ends bluntly, bearing 4 distal setae. Femur IV is not divided; and the L_2 setae of the hysterosoma are on a longitudinal line with the D_2 setae.

Pronematus ubiquitus (McGregor). This species (fig. 86) often becomes sufficiently abundant on citrus trees and other plants that growers consider it a pest, but it probably feeds on honeydew, fungi, dead insects, and mite fragments. It has also been observed feeding on *Eriophyes ficus* Cotte. It is common in western United States, Florida, Mexico, Egypt, Portugal, South Africa, and Argentina.

This mite is easily distinguished from most mites by its agility and speed of movement. It is light pinkish with a white line front to rear. The egg is borne on a fiilamentous stalk; often attached to, or near, dead or living scale insects (Mal' chenkova, 1967; McGregor, 1956).

Tarsus I of *P. ubiquitus* is longer than tibia I; the distal setae of tarsus I are as long as or longer than the segment and serrate along most of the entire length. The solenidion of tarsus I is located near the middle of the segment. The ventral body setae are about ½ as long as the distance between their bases and are placed in a longitudinal line. The longitudinal striae of the propodosoma reach to the D_2 setae on the hysterosoma.

THE TUCKERELLIDAE BAKER AND PRITCHARD

The dorsal chaetotaxy of the body is distinctive for the family. There are 4 pairs of dorsal palmate propodosomal setae and 36 pairs of dorsal palmate hysterosomal setae, as well as 5 or 6 pairs of flagellate setae caudally. The palpus is elongate and has a thumb-claw complex. The genitoanal region of the female is distinctive.

Tuckerella Womersley

There is only a single genus in the family and it has the familial characters.

Tuckerella pavoniformis (Ewing). This mite (fig. 87) is tropical and semitropical in distribution having been found in Hawaii, California, Florida, Georgian USSR, Mauritius, and Okinawa. The hosts include citrus, hibiscus, papaya, tea, and a wide variety of noneconomical plants. It may be recognized by the 6 pairs of whiplike caudal setae; the lateral pair of the palmate setae in the last row of hysterosomal setae are larger than the inner pair; and the anterior pair of propodosomal setae are about as broad as they are long (Ehara, 1966; Baker and Pritchard, 1955).

THE PYEMOTIDAE OUDEMANS

The gnathosoma is small and free; the palpi are small and simple; the chelicera are minute and needlelike. The peritremes are rarely absent, are dorsal, and are directed anteriorly on the "shoulders" of the propodosoma; a pseudostigmatic organ is usually present in the females. The body is segmented; leg I is usually without an empodium but with a claw; other legs bear empodia and claws. The males may be strongly heteromorphic.

Siteroptes Amerling

Leg I has 5 segments; the single claw of tarsus I is simple, pedicellate and lacks a thumb. The chelicerae are small and indistinct; the palpi are minute and free. The gravid female distends behind the second pair of legs.

Siteroptes cerealium (Kirchner). This species (fig. 88, *a*), known in literature also as *Siteroptes graminum* (Reuter), is a pest of grasses and cereals in England and continental Europe. Over 30 species of grasses as well as wheat, barley, and rye serve as hosts. The mites feed within the sheaths on the upper part of the stem, causing growth to be retarded sufficiently that the partially emerged inflorescences become malformed and silvered—a condition known as "silver tip." The mites are often associated with, and may aid in the distribution of the fungus of *Nigrospora* cob rot, *Nigrospora oryzae*. The mites feed not only on healthy plant tissue, but on the fungus and parts of the plants destroyed by the fungus.

S. cerealium is also associated with a central bud rot of carnations. The rot is caused primarily by a fungus, *Fusarium poae*, but the mites actively spread the spores (Cooper, 1940).

After the female mites become attached to the host, their hysterosoma swells with fluid and attains a size of up to 500 times the original body volume. Eggs form, hatch, and develop to mature adults within the hysterosomal sac. A mass birth then occurs, resulting in the breakdown of the hysterosoma and death of the parent. Copulation takes place within the body of the mother, but this species is faculatively parthenogenetic, that is, virgin females give rise only to males (Cooper, 1937).

The peritremes of *S. cerealium* are elongate; the femurgenu has 1 to 3 setae; tarsi II and III each bear 7 tactil setae. The anterior ventral plate of the propodosoma has 5 pairs of setae; the ventrites II always have 2 pairs of setae. The tergum of the hysterosoma III often bears only a single pair of setae, the lateral setae being absent. The posterior ventral plate is entire; the apodemes III are incomplete medially; the apodemes IV are weak, complete or incompete, but distinct along the midline and laterally in front of coxae IV; the apodemes V are usually absent; the posterior median apodeme is weak or absent anterior to its union with apodemes IV.

THE PENTHALEIDAE OUDEMANS

These are soft-bodied mites, without external peritremes and with the tracheae orignating at the base of the chelicerae. "Rhagidial organs" (solenidia) lie flat in specialized areas on tarsi I and II; the tarsal claws are hooked, and the empodia are padlike and have radiating hairs. The anterior medial portion of the propodosoma bears a tubercle with a pair of setae.

Penthaleus Koch

This genus illustrates the character of the family. *Penthaleus* is distinctive generically in having a dorsal anal opening on the posterior portion of the hysterosoma; the movable chela is straight.

Winter grain mite, *Penthaleus major* (Dugès). This mite (fig. 88, *b*), also known as the blue oat or pea mite, is widely distributed, particularly throughout the North Temperate Zone, and from Australia, New Zealand, South America,

South Africa, mainland China, and Taiwan, and in the South Temperate Zone (Goldsmid, 1962).

Small grains and grasses are favored hosts of the winter grain mite; but the mite infests and damages legumes, vegetables, ornamental flowers, cotton, peanut, and various weeds.

Heavily infested fields appear grayish or silvery, a result of the removal of plant chlorophyll by mite feeding. When high infestations feed on the plants for several days, the tips of the leaves exhibit a scorched appearance, then turn brown, and the entire plant may die. These mites do not cause the yellowing characteristic of tetranychid mite feeding. Many of the infested plants do not die, but become stunted and produce little forage or grain; damage on young plants, however, is more severe than on large, healthy ones. There are two types of damage to the small grains, namely, reduced amount of forage throughout the winter and reduced yields of grain in the spring and summer. Reduction in forage probably is of the greater importance where winter grazing is practiced because farmers rely on the small grains for fall and winter pasturing of livestock.

The life cycle of the winter grain mite consists of the following stages; egg, deutovum (prelarva), larva, protonymph, deutonymph, and adult. The ovoid, bright pink to orange red eggs that are attached to the substratum at their somewhat flattened end become wrinkled, pale, and straw-colored as soon as they lose contact with moisture. There are two physiologically different types of eggs laid, namely, the winter eggs that have a short incubation period, and the aestivating eggs that have an extended incubation period. The only external difference is the size, which is because of the thicker membrane covering of the aestivating eggs.

Attempts to break diapause (aestivation) in the fall by subjecting the eggs to low temperatures and high humidities have failed; the eggs swell, but do not hatch. The incubation period of these eggs is 110 to 140 days; winter eggs hatch in 25 to 35 days. Exposure of aestivating eggs to winter conditions for 14 to 18 days causes them to lose the bright pink color; also 6 to 8 longitudinal stripes become visible. The appearance of a longitudinal split along 1 side of the egg indicates the beginning of the deutovum stage. The mite remains inside an inner membrane while the rudiments of appendages undergo segmentation. For the larvae to emerge, adequate moisture is essential during the entire prelarval period. Conversely, if excessive moisture and fluctuating temperatures occur, the eggs swell enormously, the inner membrane splits, and the larvae lose protection and fail to emerge. This response of the egg to moisture helps explain why this species is largely restricted to loose sandy or loamy rather than hard clay soils.

Larvae emerging from the deutovum are bright pink or orange. Soon after feeding they become orange red and later dull brown with a mixture of green. The area around the dorsal anus becomes lighter and clearly visible, and a small drop of excretory material may be seen outside the anal opening. The protonymph is at first pale brown with green legs and mouthparts, but after feeding 3 days the protonymphs turn green. The deutonymph has a pale green body with red gnathosoma and legs. The adult is pale green, but becomes darker as it feeds on more mature leaves.

The length of the life cycle varies with the weather. Under optimum conditions the average time from oviposition to death of the adult is 98 days, not including the period of aestivation of the eggs. The incubation period averages 25 days. The average number of days in the immature stages are: larval, 12.2 days; first nymphal, 8.6 days; second nymphal 7.8 days; and third nymphal, 7.0 days; the average life span as adults is 37.5 days.

There are 2 generations per year. The first develops from oversummering eggs. Development begins after the onset of favorable temperature and moisture conditions in late September and October, and populations peak in December and January. The second generation develops from eggs laid by the first generation reaching maximum infestation density in March and April. Populations then decrease as temperatures exceed the range of tolerance. The females of this generation lay aestivating or oversummering eggs.

The larvae become very active soon after hatching and begin to feed on the sheath leaves or tender shoots near the ground. The larvae as well as the adults feed higher up on the plants at night or on cloudy days. As the sun rises the mites descend the plants and seek protection during the hot part of the day on the moist soil surface under foliage. If the soil is dry and there is little foliage cover, they dig into the soil in search of moisture and cooler temperatures. At sunset and thereafter the plants become covered with feeding mites where, with the aid of a searchlight, they can be observed feeding at all hours of the night. They drop to the ground upon being disturbed.

The female mites deposit their eggs on the sheath leaves and stems and on and in soil near the base of the plants. Those on the leaves are usually fastened by mucilaginous substance secreted by the female to the inner surface next to the stem and on the stems. They may be deposited singly, but usually large numbers are found close together.

Temperature and moisture are the most important factors influencing mite development and abundance. Cool rather than warm temperatures favor their development. Oviposition is heaviest between 10 and 15.5 C (50 and 60 F); the optimum conditions for hatching are between 7 and 13 C (45 and 55 F); and adult activities are greatest between 4.5 and 23.5 C (40 and 75 F). When temperatures drop below or rise above these extremes, the mites stop feeding, descend to the ground, or burrow into the soil. Mite activity in the spring drops rapidly and the eggs fail to hatch when the daily temperature exceeds 23.5 C (75 F). Aestivating eggs do not hatch in the fall until rains provide adequate moisture. On hot, dry days it may be necessary to dig into the soil to a depth of 4 or 5 inches to find mites. The mites are not harmed by short periods of sleet or ice cover or by ground frozen to a depth of several inches.

Wallace and Mahon (1971) indicated that the winter grain mite can tolerate a drier climate than the red-legged earth mite, reaching a May-October isohyet of about 190 mm, and a higher proportion of summer rain. This enables this species to survive on the south coast of New South Wales and on the northern tablelands of New South Wales as far as southern Queensland, but not in north coastal areas where the midsummer rainfall exceeds about 500 mm. These authors suggest that the winter grain mite could probably survive in most areas in both hemispheres between the latitudes 25° and 50 to 55°. In the tropics the mite is unlikely to be found except perhaps at high altitudes.

Strew and Gingrich (1972) found that mites increase on turf grass in New Jersey through November and remain high during the winter, decreasing in April. These mites survive longer and deposit more eggs on blue grass than on rye grass or fescue. Rye grass and fescue support mites for about a month and mites die in a day or two when placed on bent grass.

Dispersion from field to field may occur by transportation of aestivating eggs or mites on grain stubble or leaves, on soil adhering to implements that are moved about, or on forage or straw carried from infested fields in livestock feeding operations. Aestivating eggs may also be transported on debris by wind, and local distribution may occur by adult migration. Such migrations to grain fields may take place from fence rows or other uncultivated areas. Fescue grass appears to be the favored host in such situations in Texas.

Cropping practices have a marked effect upon the occurrence and damage caused by the winter grain mite. Injury by this mite may be prevented by crop rotation, that is, by not planting more than two years in succession. Crops such as cotton, corn, clover, or sorghum may be used in such rotations. Although *Chrysopa* larvae and the predatory mite, *Balaustium* spp., prey on the winter grain mite, these predators are not of importance in reducing populations (Chung, Wei, and Tieng, 1963; Greenup, 1967; Narayan, 1962; Swan, 1934).

Chemical control of the winter grain mite may be achieved by sprays or dust applications of organophosphorus acaricides such as malathion, parathion, and dimethioate; and also by the organic sulfur acaricides, Imidan, and DDT (Chada, 1956).

P. major possesses the characters of the genus and family. The movable chela is long, slender, and nearly straight; the fixed chela is membranous and has a tridentate fork distally. The body setae are short and serrate; the body itself is large and globular and difficult to mount.

Penthaleus minor (Canestrini) is similar to *P. major* (Dugès), but it can be distinguished by its very small movable chelae; there is no fixed chela.

Halotydeus Berlese

Halotydeus is similar to *Penthaleus*, but differs in that the anal opening is posterior rather than dorsal. Also, the chelicerae are more specialized. *Halotydeus* has slender rather than stout palpi.

Red-legged earth mite, *Halotydeus destructor* (Tucker). This species (fig. 88, *c*) is a serious pest of many crops in Australia and Cyprus, and in Rhodesia and Nyasaland of South Africa, where it is known as the "black sand mite." Feeding of this mite causes silvery or whitish blemishes along the main veins of host plant leaves. The whole surface of the leaves become bleached progressively, rapidly wilt, and shrivel such that the infested plants appear scorched. Seedlings may be killed outright as though from frost damage. Sections of silvered *Sonchus olera-*

ceus leaves show that the cells of the palisade parenchyma are empty, including chloroplasts. In these areas the injured epidermis collapses, leaving fine breaks in the cuticle. The silvery appearance results from the replacement of the cell contents by air.

This mite feeds principally on annual, broad-leaved plants and grasses such as permanent pastures containing clovers, young wheat, potatoes, tobacco, peas, beets, tomatoes; it feeds especially on seedlings in hotbeds and greenhouses and on ornamental annuals.

The eggs of this species are of two different kinds. Winter eggs are bright yellow or orange and are laid in masses in a single layer. They are mainly deposited on the undersides of the leaves located in damp places or in contact with the soil. The egg surface is smooth and glossy but may assume a whitish bloom when dry, owing to a secretion used to stick the eggs to the plant surface. When temperatures exceed 18 C (65 F), mite activity decreases and the oversummering eggs are formed. These eggs are not laid, but are retained in the body of the mite when it dies—an adaptation that serves to protect the eggs from desiccation. As summer eggs lose moisture they become flattened or concave on one side and have a thicker shell than the winter eggs—an additional protection against hot dry summer conditions. On exposure to a water surface after the dry period, the eggs assimilate moisture and assume the appearance of winter eggs. Adequate, but not excessive, moisture is essential for hatching. Excessive moisture causes the eggs to continue to swell and eventually to burst. The habitat of this species is therefore limited to light sandy soils. The eggs hatch in 2.5 to 8.5 days. There are four to five nymphal stages, each requiring 5 days for development; thus 20 or more days may elapse before the immature stages are completed and adults start laying eggs. Adult mites live 25 to 50 days.

Temperature and humidity are important factors in the distribution and population fluctuations of this mite. The thin cuticle of the mite appears to be permeable to water, because mites exposed to dry air lose their moisture rapidly. Actively feeding mites are less affected by air humidity, because moisture lost by evaporation is replaced by the juices imbibed from the leaf tissues, but nonfeeding mites must remain in damp sheltered situations to survive. Survival at a given temperature is longer as humidity is increased, the optimum humidity being near saturation. In atmospheres having the same saturation deficit, but different temperatures, the survival period decreases with rise in temperature. Within the range of 18 to 26 C (65 to 79 F), however, temperature changes have no effect. Above 30 C (86 F), temperature becomes the major factor influencing survival rate. At 34 C (93 F) the mites die within a few hours, and at 50 C (122 F) they die in a fraction of a minute (Solomon, 1937).

The occurrence of an aestivating diapause in the eggs of *Halotydeus destructor* was estabilshed by Wallace (1970, *a*, 1970, b) who found that the production of diapause eggs in females was influenced by increasing maturity of the food plants in the spring. Morphogenesis was resumed following exposure to summer conditions for four to six weeks in the field. In the laboratory, development was achieved by exposing the eggs to various combinations of temperature and humidity. Most effective was 32 days at 50 C (122 F) and 50 percent relative humidity, although diapause development occurred within the temperature range of 15 to 70 C (59 to 158 F) and the relative humidity range of 10 to 100 percent, even when eggs were completely immersed in water. Development was most successfully promoted in atmospheres having vapor pressures of about 40 mm of mercury at all combinations of temperature and relative humidity. At 20 percent humidity diapause development was relatively slower than at higher or lower humidities, whereas tolerance to high temperature was greatest at 20 percent. Diapause eggs were not harmed either by complete immersion in water or by periodical desiccation and remoistening.

After diapause development is complete, moistened aestivating eggs of *H. destructor* develop within the temperature range of 5 to 20.5 C (41 to 69 F). The rate of development up to the deutovum stage increases with temperatures up to 20 C (68 F) and then rapidly decreases. Eggs may tolerate temperatures between 25 and 31.5 C (77 and 87 F) for up to 53 days without undergoing morphogenesis, but this treatment retards subsequent development at 16 C. Exposure to alternating temperatures accelerates the rate of development, but, if the higher temperature exceeds 20.5 C, development is retarded. Air-dry eggs may tolerate temperatures up to 75 C (167 F), but moist eggs are killed by exposures of one day at 45 C (113 F). This high temperature tolerance, together with the limitations on development imposed by temperature, ensures that aestivating eggs do not hatch before early autumn in the event of unseasonal summer rain (Wallace, 1970, *a*).

Wallace and Mahon (1971) indicate that the red-legged earth mite is restricted to the southern parts of Australia having a Mediterranean-type climate, with a warm dry summer and a cool wet winter. The northern or inland distribution agrees closely with the 205 mm isohyet for the growing season May to October inclusive. The eastern distribution is limited by the quantity of midsummer rainfall and agrees closely with the 225 mm isohyet for this period. Other limitations to distribution within these isohyets are imposed by high summer temperatures (mean monthly maximum of hottest month 33 C [91 F]). The same climatic limits also fit the South African distribution. Other areas of the world that might have suitable climate include the central western coast of Chile, the central western coast of the North American continent from Vancouver to San Diego, most of the coastal Mediterranean regions, the Atlantic coast of Spain, Portugal, and Morocco, the southern coastal regions of the Baltic and the Caspian seas, and a narrow strip between the northeastern end of the Persian Gulf and eastern Turkey.

The nymphs and adults are velvety black; the legs are red and the body setae plumose. This species may be distinguished in the field from the pea or blue oat mite, *Penthaleus major*, by the lack of a reddish area on its dorsum and by the differences in habits. The larvae have distinct segmentation and are more elongate than adults. The mites feed at any time during the day or night depending on the moisture. They are gregarious when feeding and when not feeding they congregate in hollows on the ground or beneath leaves. Although they prefer moist soil, free water is avoided. They are sensitive to touch stimuli, therefore select their habitat in response to texture, food, and other tactile surroundings. Webbing is not produced by this species. Injurious mite populations may be present in the fields from the onset of winter rains until early summer when they succumb to high temperatures and desiccation (Goldsmid, 1962; Norris, 1948; Swan, 1934).

Chemical control measures against the red-legged earth mite should be initiated as soon as possible after egg hatching is complete in the autumn because the egg stage is the most difficult to control by acaricides. Also, plants that are small, delicate, and often desiccated, are particularly susceptible to injury by a few mites. To achieve early control with nonpersistent acaricides such as the organophosphorus compounds, the time of application must be carefully determined. The more persistent types of acaricides may remain sufficiently toxic to the young mites as they emerge from the egg or molt stage and thus provide adequate control over an extended period. Injurious infestations of agricultural crops may be prevented by clean cultivation, especially during autumn and early winter; weed hosts serve as breeding areas (Wright, 1961).

The body of *H. destructor* (Tucker) is dark and without lighter markings. The legs are reddish; the palpal segments are relatively slender. The dorsal body setae are short; the chelicerae are distinctive in that the fixed chela ends in a membranous section opposing the short movable chela.

THE ASTIGMATA

Rhizoglyphus, Tyrophagus, and *Caloglyphus* mites live on vegetable and flower bulbs, roots, and tubers; but the root, bulb, and potato mite, *R. echinopus* (Fumouse and Robin) is the most injurious to living plants. Soil-inhabiting species in the other genera feed primarily on seeds or dead or decaying plant parts.

Rhizoglyphus Claparède and *Tyrophagus* Oudemans

R. echinopus and *R. callae* Oudemans live on onion, lily, narcissus, hyacinth, tulip, orchid, and other bulbs; on potato and dahlia tubers, beetroot, and other vegetables, on the roots of vines, wheat, oats, and a whole series of cultivated plants; either in storage or in the field. They have also been found on fallen fruit, in forest litter, and on mushrooms, and sometimes on grain having high moisture content. These mites are recognized as serious pests of decorative bulbs, onions, potatoes, and other tuberous plants; they are particularly destructive in the field and cause serious losses during storage.

These species develop and reproduce most rapidly at high humidities or in moist situations. Optimum temperatures are 23 to 26 C (74 to 79 F). The life history is completed in 17 to 27 days at 18 to 24 C (65 to 75 F) and in 9 to 13 days at 20 to 27 C (68 to 82 F). The hypopi are usually carried by diptera, which develop in the same habitat as the mite, and by some beetles (Zakhvatkin, 1941).

Caloglyphus Berlese

Caloglyphus spp. live in damp places on grain, nuts, and decaying fruit. Some species will attack insect cultures; including eggs, larvae, and adults. Hughes (1961) indicates that the life cycle under saturated conditions takes 8 to 9 days at 22 C (72 F). A female may lay 213 eggs in 24 hours; one laid 1,034 eggs in 39 days.

Tyrophagus species are often found within the leaf sheaths of grasses and cereals, or they may occur in immense numbers on plants damaged by insects or phytophagous mites. For example, *Tyrophagus dimiatus* (Hermann) is commonly found on plants damaged by tarsonemids, and it almost invariably occurs on grasses and cereals damaged by the frit fly, *Oscinella frit* L. (Zakhvatkin, 1941).

THE CRYPTOSTIGMATA

The cryptostigmata or oribatids are best known as inhabitants of litter and upper soil strata. Several species are associated with plants, however, and evidence indicates that some soil-dwelling species spend part of their life cycle on plants. The soil-dwelling cryptostigmatid *Perlohmannia dissimilis* (Hewitt) has been reported damaging the root systems of potato, strawberry, and tulip. *Minunthozetes semirufus* (Koch), a very common soil-dwelling species, lays its eggs on grasses, and the larvae and nymphs burrow in the stems (Zakhvatkin, 1941).

Several oribatids are arboreal in habit; of these the best known are *Camisia segnis* (Hermann) and *Humerobates rostrolamellatus* Grandjean. The latter is particularly common on fruit trees, where it may feed mainly on algae growing on the bark. This species may, under certain conditions, become harmful to cherries. During June and July, especially in wet seasons, the mites feed on the rinds of split fruits where they may become so numerous that the fruit must be washed before packing. The mites apparently do not feed on sound fruit (Evans, Sheals, and Macfarlane, 1961).

SELECTED BIBLIOGRAPHY

BAKER, E. W. 1965. A review of the genera of the family Tydeidae (Acarina). Advances in Acarology 2:95–133. Cornell Univ. Press. Ithaca, N.Y.

BAKER, E. W., and A. E. PRITCHARD. 1953. The family categories of the new families Linotetranidae and Tuckerellidae. Ann. Entomol. Soc. Amer. 46(3):243–258.

CHADA, H. L. 1956. Biology of the winter grain mite and its control in small grains. J. Econ. Entomol. 49(4):515–520.

CHUNG, C., H. WEI, and Y. TIENG. 1963. The biology of the round wheat mite, *Penthaleus* sp., and brown wheat mite *Petrobia latens* (Müller). [In Chinese with English summary] Acta Phytophyl. 2(3):277–284.

COOPER, K. W. 1937. Reproductive behavior and haploid parthenogenesis in the grass mite, *Pediculopsis graminum* (Reut.). Proc. Nat. Acad. Sci. Wash. 23(2):41–44.

———. 1940. Relations of *Pediculopsis graminum* and *Fusarium poae* to central bud rot of carnations. Phytopathology 30(10):853–859.

EHARA, SHOZO. 1966. The tetranychoid mites of Okinawa Island. J. Fac. Sci. Hokkaido Univ. Ser. IV. Zool. 16(1):1–22.

EVANS, G. O., J. G. SHEALS, and D. MACFARLANE. 1961. The terrestrial acari of the British Isles. An introduction to their morphology, biology and classification. Vol. I. Introduction and biology. British Museum, London. 219 pp.

GOLDSMID, J. M. 1962. The mites of the Federation of Rhodesia and Nyasaland. Bull. Fed. Minist. Rhod. Nyasad. No. 2162. 11 pp.

GREENUP, L. R. 1967. The blue oat mite, *Penthaleus major* (Dugès). Agr. Gaz. N. S. W., 78(7):410–411.

Hughes, A. M. 1961. The mites of stored food. Ministry of Agr., Fisheries, and Food. Tech. Bull. 9, 287 pp. Her Majesty's Stationary Office, London.

Lo, Paul Kong-chen. 1968. Tetranychoid mites infesting tea in Taiwan. Chung-San Academic Cultural Series 1:1–11.

McGregor, E. A. 1956. Mites on citrus trees in southern California. Mem. So. Calif. Acad. Sci. 3(3):13.

Mal'chenkova, N. I. 1967. A mite of the genus Tydeus (Acariformes Tydeidae) that is a grape pest in Moldavia. Entomol. Rev. (Entomologischeskoe Obozrenie) 46(1):66–68.

Narayan, D. S. 1962. Morphological, biological and ecological studies of the winter grain mite, *Penthaleus major* (Dugès) Pt. 1, J. Zool. Soc. India 14(1):45–63.

Norris, K. R. 1948. Seasonal severity of the attack of the red-legged earth mite (*Halotydeus destructor*) on subterranean clover. Australia Council Sci. and Indust. Res. J. 21:7–15.

Smirnoff, W. W. 1957. An undescribed species of *Lorryia* (Acarina, Tydeidae) causing injury to citrus trees in Morocco. J. Econ. Entomol. 56(3):361–362.

Solomon, M. E. 1937. Experiments on the effects of temperature and humidity on the survival of *Halotydeus destructor* (Dugès). S. Australia J. Agr. 38:353–367.

Strew, H. T., and J. B. Gingrich. 1972. Seasonal activity of the winter grain mite in turfgrass in New Jersey. J. Econ. Entomol. 65(2):427–430.

Swan, D. C. 1934. The red-legged earth mite, *Halotydeus destructor* (Tucker) in south Australia; with remarks upon *Penthaleus major* (Dugès). S. Australia J. Agr. 38:365–367.

Wallace, M. M. H. 1970*a*. Diapause in the aestivating egg of *Halotydeus destructor*. Australia J. Zool. 18(3):295–313.

———. 1970*b*. The influence of temperature on post-diapause development and survival in the aestivating eggs of *Halotydeus destructor*. Australia J. Zool. 18(3):315–329.

Wallace, M. M. H., and J. A. Mahon. 1971. The distribution of *Halotydeus destructor* and *Penthaleus major* in Australia in relation to climate and land use. Australia J. Zool. 19(1):65–76.

Wright, W. E. 1961. Red-legged earth mites. The Agr. Gaz. N. S. W. 72(4):213–215.

Zakhvatkin, A. A. 1941. Fauna of USSR Arachnoidea, Tyroglyphoidea (Acari) 6(1): 1–573. Trans. and ed. by A. Ratcliffe and A. Hughes.

Chapter 12

Eriophyoidea Nalepa

These mites are known as gall, bud, blister, and rust mites. They are basically wormlike, and are from 1/10 to ⅓ mm long. They are essentially invisible to the unaided eye. This small size originally made them difficult to understand, and still challenges attempts to disclose their secrets.

EVOLUTIONARY DEVELOPMENT

Eriophyoids are a distinct and very successful type of acarine, which seems to indicate that they evolved long ago. Since a fossil rust mite has been found in 37,000,000 year-old North Maslin Sands in Australia, and since this mite is essentially the same as present-day leaf vagrant eriophyoids, it is not overemphasizing to estimate that this mite group originated more than 50,000,000 years ago.[1]

Present information indicates that eriophyoids are entirely phytophagous. They usually travel by drifting in wind, but they may also ride on insects that feed on their respective host plants, and on birds. Eriophyoids feed on many plant species from the tropics to beyond the arctic circle.

Most eriophyoids seek microenvironments in which to live, feed, and reproduce. Many insert themselves into crevices under bud scales or at petiole bases on their host plant, and in that way gain cover and food. Others induce growth of galls on plants, these galls being one of the most characteristic manifestations of these mites (pl, 52, *a-c*; figs. 98–101, below). Many species, however, are open leaf vagrants or rust mites for at least part of their seasonal cycles, although those living exposed often seek leaf depressions or hairs for protection.

Erophyoids are not only microscopic in size, but also show great reduction in body structure. They have lost certain parts that are typical of most Acari, such as the rear 4 legs and nearly all body setae. The mouthparts have retained basic structures (see morphology of gnathosoma) but all parts are modified. The cephalothorax is but a remnant of the typical leg-bearing section. Eriophyoid genitalia are proximal, as compared to terminal tetranychoid genitalia.

A peculiar family of mites, Nematalycidae (Baker et al., 1958; Krantz, 1970), may offer suggestions as to the evolutionary loss of the rear legs by eriophyoids. The rear legs on nematalycids have drifted back on the extended wormlike body. If, on eriophyoids, the rear leg pair moved back and finally disappeared, then part of the abdomen is a composit structure. The nematalycids have proximal genitalia, as do eriophyoids, suggesting that the eriophyoid genitalia continued to move ahead, as the rear pairs of coxae and legs disappeared.

The theory adopted here is that the disappearance of body parts was originally an adaptation for life in microenvironments on plants. Galls seem to be the most suitable place for this sort of evolution to take place, but nalepellids, which are the most primitive of the 3 families (Nalepellidae, Eriophyidae, Rhyncaphytoptidae), are not strong gall makers (Farkas 1966).

Existence of open leaf eriophyoids is explained in part by the dispersal activities of these mites, during which times they are forced to expose themselves. Thus part of the species in most of the groups have independently developed the ability, and acquired structures, enabling them to live in the open. An intermediate form between species that live under cover, and open leaf mites, is exemplified by *Paraphytoptus* spp. (fig. 121, *a* below). Species in this genus live in natural plant hair masses. They keep the front of the body immersed in the hairs, but the rear is exposed. The softer front part of the abdomen has retained the narrow and heavily microtuberculate rings of those species that live under cover, and is not exposed directly to desiccating influences. The exposed rear of the abdomen has developed the tergite-sternite relationship possessed by rust mites, which resists drying effectively.

If we compare structural groups in the eriophyoids with life habits of the species in these groups, the only possible conclusion about the evolutionary development of these tiny mites is that some species have been able to discard gall making, and then independently to reacquire it; also species can independently become deuterogynous when moving to, and adapting to, a deciduous host.

HISTORY OF KNOWLEDGE ON ERIOPHYOIDS

In the pre-Linnean period, M. de Reaumur, 1737, in his *History of Insects*, commented on various eriophyid galls and erinea, but did not associate the organisms in the galls with the Acari. He speculated that the tiny white worms in these galls were maggots of very small flies, and used the term "nail gall" for pointed linden leaf galls (caused by *Phytoptus tiliae* Pagenstecher (figs. 100, *a*; 111, *e*). He also used the term "mold galls" for erinea. In associating erinea with arthropod action, he was much more correct than the first post-Linnean taxonomists, who considered them fungi.

Linnaeus to Nalepa

Development of knowledge with the advent of binomial nomenclature is divisible into several periods. One is the mycological or "fungus" period. It began in the 1790s and extended to the 1830s. During this time students of fungi collected erinea and developed a binomial name system for them. They integrated these names with the general mycological system of the time.

Generic names applied to erinea were: *Erineum*, *Phyllerium*, and *Taphrina*. These names still have no other use, except as Nalepa employed them to define the erinea in his 1929 host list. Nalepa did not use them as names for the mites making the erinea, and they have not been used for these mites for well over 100 years. (The name *Mucor* was also used as a name for erinea, but it has another and valid application in mycology.)

The literature of the "fungus" period contains disputes among mycologists as to the correct taxonomic assignments for various erinea. There were even claims that erineal strands were seen to develop spores, but gradually workers came to notice the wormlike creatures present in these hairy masses on leaves and in galls. Improved microscopes enabled observers to associate these worms with the Acari.

A book on botany and mycology by Fée, dated 1834, signaled the beginning of the next advance in knowledge on eriophyoids. Fée noted that the erinea had no spores, contained wormlike creatures, attributed the growth of these hairy masses to animal action, and suggested that they be transferred to the animal kingdom.

The tiny wormlike mites in erinea and galls were difficult to see so workers groped for means of adequately characterizing them. Those who proposed names for these mites usually did it by describing the galls and basing the names on these gall formations, plus the host. As late as the 1880s, Garman, in Illinois, believed that there could be no way really to describe the mites themselves.

Two basic names for these mites came into use during this period. These are: *Eriophyes* von Siebold, dated March 1851, and *Phytoptus* Dujardin, dated July 1851. They were originally proposed without accompanying specific names.

The first binomials for *Phytoptus* were by Pagenstecher (1857). He listed the specific names *tiliae, pyri, vitis,* and *rhamni* under that genus. The first binomial under *Eriophyes* was *Eriophyes labiatiflorae* Thomas (1872). This established *labiatiflorae* as the genotype of *Eriophyes* by subsequent monotypy.

One type of thought concerning these mites in this second period is illustrated by Donnadieu (1875). He published a paper on Tetranychidae and used the genus name *Phytocoptes* for certain tetranychids that he postulated came from four-legged wormlike larvae in galls. He recognized increase in numbers of mites in the galls, but attributed it to larval reproduction. Donnadieu was the first to use *Phytocoptes* binomially. He listed three species under it: *epidermi, gallarum,* and *nervorum.* Since he indicated 3 or 4 host plants under each of these specific names, the so-called species are composites as far as eriophyoids are concerned. Donnadieu assigned *Phytocoptes* to the Tetranychidae so it really belongs to that family, but tetranychidologists continue to ignore it.

The Nalepa Period

Alfred Nalepa began publishing on eriophyoids in 1886, and this period ended with his death in 1929. He lived in the vicinity of Vienna, Austria. Various workers in addition to Nalepa contributed to eriophyoid taxonomy during this time, but since Nalepa was by far the dominating worker, a general appraisal of his activities defines the period.

Nalepa's first publication was an anatomical study of the pine twig knot mite, *Trisetacus pini* (Nal.)[2] (figs. 103, *a, b*). The part dealing with internal anatomy stood as the only such study for many years.

Nalepa was the first to recognize the differences between male and female eriophyoids. He was the first adequately to describe species on the basis of their structure. (Canestrini and Massalongo, in Italy, soon followed Nalepa with good descriptions and figures). Nalepa averaged about 10 new species a year for the 40 years he studied these mites. He published papers on eriophyoid bionomics. While most of his species are European, he did receive collections from elsewhere, notably Java.

Nalepa's microscope equipment was probably equal to the best available at the time, but he seems to have been unable to discern featherclaws sufficiently. He was equivocal about the number of featherclaw rays on some species. Certain species named in his early work he evidently never reviewed carefully, with resulting perpetuation of minor errors. Examples are featherclaw rays on *Eriophyes tenuis* (Nal.), and the hosts of *Aculus schlechtendali* (Nal.). Schlechtendal reported that the apple rust mite was collected from both apple and pear. Nalepa always reported both as hosts, but the rust mite does not occur on pear at least in North America.

Most of Nalepa's earlier publications were illustrated by artistic drawings and half-tones. These illustrations show general dorsal and ventral views (often with female genitalia exploded), rarely show side views, and they do not include details of leg setation and of ring granules. He discontinued illustrating before 1910.

One of the more enlightening parts of Nalepa's 1911 *Zoologica* publication is the inclusion of his four postulates. They give clues to his taxonomic activities in later years. A translation of them is:

1. Structurally similar galls on hosts that are not particularly closely related to each other are caused by different species of mites. Examples are big buds on filbert, and birch.

2. Structurally similar galls on closely related hosts are made by the same species of eriophyid, or by varieties of that species. Examples are the leaf blisters on various pomaceous hosts, such as apple and pear.

3. Structurally different galls on the same host, such as erineum, bead galls, bark galls, on a host such as maple, are caused by different mite species, or subspecies. Examples are the various types of galls and erinea on hedge maple and sycamore maple.

4. Structurally different galls on unrelated hosts are the work of different species of mites. Nalepa added that experimentation would prove or disprove these points.

These postulates emphasize activity and host, rather than precise mite anatomy. They tend to direct attention away from the mite itself. An examination of Nalepa's 1929 host list will disclose some of the results of these postulates. They led him into the development of numerous multi-nomials for various complexes. It can be said about these long names that they do call attention to certain morphology that the student can then reinterpret if he feels he has sufficient basic data to support revisions.

One reinterpretation has to do with what we now know about alternating forms that occur in many eriophyoid species (deuterogyny). While it had been suggested that something of this sort existed, and Nalepa published a paper discussing polymorphism (1924), his 1929 host list does not disclose that he took the concept very seriously, or really knew what to do about it. Many of Nalepa's subspecific and varietal names in this list stand for alternate forms, and are therefore synonyms.

Nalepa's generic and subfamily assignments indicate he allowed the habits of a species to influence his taxonomic conclusions. For example, he referred most gall formers to *Eriophyes*, whereas gall making is a bionomical trait that has been adopted through the ages by actually diverse structural groups in the superfamily.

The theory previously mentioned, based on comparative structure, is opposed to giving habit a primary place in taxonomy and assumes that various structural types have independently moved into, out of, and then back into available places on plants. An example is *Vasates quadripedes* Shimer, which makes bladder galls (fig. 98, *b*) on silver maple leaves. This bladder gall mite has rust mite structures indicating it has returned to a microenvironment (in galls protected from weather variations), after having been a free-living species.

Another example of this is the willow leaf gall mite, *Aculops tetanothrix* (Nal.), which has rust mite form (Keifer, 1969). Examples of groups that have independently occupied about all available eriophyoid habitats on plants are the Nalepellidae, the Nothopodinae, and the Cecidophyinae. Parenthetically it should be noted here that the principal group of eriophyoids that occupy all available habitats on plants open to this type of mite are the combined subfamilies Eriophyinae and Phyllocoptinae, of the Eriophyidae (strict sense). Among species of these two subfamilies the division is mostly along habitat lines as far as anatomical adaptations have provided structural characters for separation. The species in the Eriophyinae are principally the wormlike ones occupying microenvironments, while the species in the Phyllocoptinae are primarily leaf vagrants and rust mites, which are less wormlike and have dorsal adaptations to compensate for open habitats. Actually these two subfamilies have much in common, which makes drawing distinctions between them for the benefit of keys rather difficult.

Nalepa's use of binomial nomenclature does not altogether coincide with certain provisions in the *International Code of Zoological Nomenclature*. Nalepa was somewhat of a free agent in his disposition of eriophyoid names; he was always the master, names were servants. While this is an enviable position, it creates problems. But those who have followed Nalepa into the study of these mites use his name system, and wish to preserve it as far as possible.

Nalepa's next-to-last publication was his 1929 general host list of eriophyoids of the world, a publication that could not possibly have been achieved, even yet, without his contributions. In this list one can find all of the then-known species. subspecies, and varieties he considered worth recognizing. He did correlate early fungus names with his mite names, not for the names themselves, but as definitive of erinea. Amerling's and Donnadieu's names are omitted.

Regardless of reappraisals, Nalepa's accomplishments far overshadow the categories in which he fell short. When he started there was little information about these mites and misconceptions were rampant. Eriophyoid names at the time were on the nude name plane. He left the science with numerous examples of adequate species descriptions, much information about life histories and hosts, reproduceable illustrations, and the host list. As a result of what he did, the study of these mites has been able to proceed constructively.

Recent Period

Since the time of Nalepa, the agricultural importance of eriophyoids has proved to be much greater than was formerly suspected. Discovery of new pests proceeds apace. That eriophyoids carry plant viruses was first shown in connecttion with currant reversion disease (Carter, 1966; see chap. 6). This period has also seen the explanation of dimorphism or polymorphism, also known as deuterogyny which is the existence of heteromorphic females.

It has been known for some time that males of certain Trombidiiform mites (to which group eriophyoids belong) deposit spermatophores on stalks, to be picked up later by females. Several acarologists have suggested that eriophyoids may accomplish fertilization by means of spermatophores (Hall, 1967, *b*), but actual evidence of the presence of spermatophores was not available until Oldfield, Hobza, and Wilson (1970), announced the discovery of deposited spermatophores (pls. 53, *a*, *b*; 54; 55, *a*, *b*), and carried out some preliminary breeding experiments. At this writing additional papers are in preparation on this general subject, including eriophyid sperm.

Tetranychoids differ markedly from the process described above in respect to the fertilization of the females. Tetranychoid males have aedeagi and copulate with the females, and do not use the spermatophore method. In addition, the internal chitinized frames inside female eriophyoid genitalia have no counterpart in the tetranychoid females.

Unsolved Problems

Nothing is known at present about eriophyoid chromosomes. Neither is anything known about the biochemical nature of the salivary ingredients that these mites have to induce gall growth.

Another category remaining for investigation is the life histories of eriophyoids in tropical regions, especially in areas that have alternate wet and dry cycles, with compensating host adjustments. Many temperate and boreal host plants of eriophyoids are periodically deciduous, and these mites cope with this by the production of alternating generations. How do tropical species of these mites react to wet and dry cycles? (See *Dicrothrix anacardii* K.)

The homogeneity displayed by the eriophyoid group suggests that these mites may have originated from a single successful primordial ancestor in an ancient geological era. While there has been speculation that tropical areas on earth that have never been disturbed by glaciation might still harbor such an ancestor, it seems unlikely that such a mite species, or something like it, would be surviving by now. So the evolutionary route that eriophyoids took to attain their present structural state may probably have to await the discovery of a fossil older than the rust mite referred to at the beginning of this chapter. The place hopefully to look for such an ancient species would be in a fossil twig or leaf gall, if such is ever found, which might have recognizable contents.

At present most of the land surface on the world has not been adequately explored for eriophyoids, so it seems reasonable to estimate that only a minority of the actual species and genera in existence have been found. This situation, along with all of the other possibilities for research in this mite group, offers the student untold opportunities to find the study of these tiny creatures very rewarding.

ERIOPHYOID ANATOMY AND MORPHOLOGY

General Anatomy

Whitmoyer, Nault, and Bradfute (1972) studied the internal structure of an eriophyid mite, *Eriophyes tulipae*, using transmission and scanning electron microscopes. From these studies and literature reviews they have illustrated the general anatomy of eriophyoid mites (fig. 89). The fore- and hindgut appeared as small cuticle-lined tubes. The foregut passes from the mouth ventrodorsally through the lower portion of the neuroganglion, connecting to the anterior opening of the midgut in the area of the genital flap but dorsal to the reproductive canal. The large saclike midgut is composed of a single layer of epithelial cells from which numerous microvilli project into the lumen of the gut. The midgut extends to the posterior quarter of the mite where it joins the hindgut.

The hindgut, which appears to follow an irregular path in a dorsoventral direc-

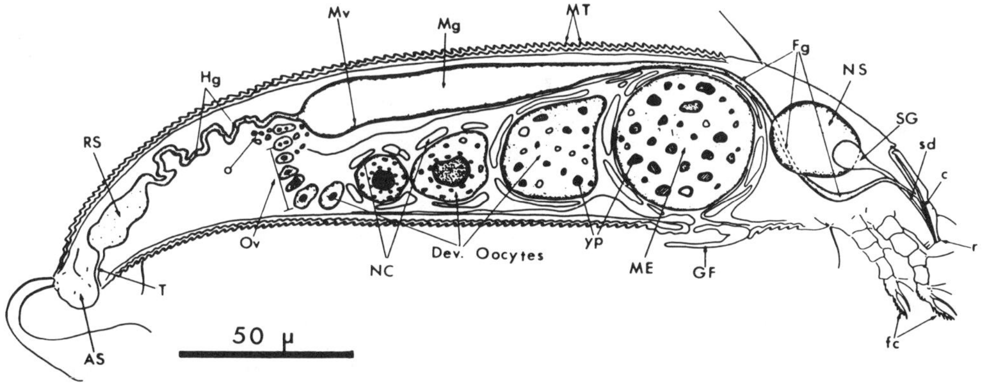

Fig. 89. *Eriophyes tulipae* Keifer: Diagram of female adult; fc, featherclaws; r, rostrum; sd, salivary duct; c, chelicera; SG, salivary gland; NS, neurosynganglion; Fg, foregut; GF, genital flap; ME, mature egg; Yp, yolk platelets; MT, microtubercles; Mg, midgut; Mv, microvilli; Dev. Oocytes, developing oocytes; NC, nurse cells; o, ovariole; Ov, oviduct region; Hg, hindgut; RS, rectal sac; T, rectal tube; AS, anal sucker (after Whitmoyer, Nault, and Bradfute).

tion posteriorly, connects to the rectal sac. The rectal sac is then connected by a short tube to the rectal orifice located within the foldings of the anal sucker. The presence of microvilli in only the midgut regions and the cuticlelike lining of the fore- and hindguts would indicate that the midgut is the prime area of active absorption.

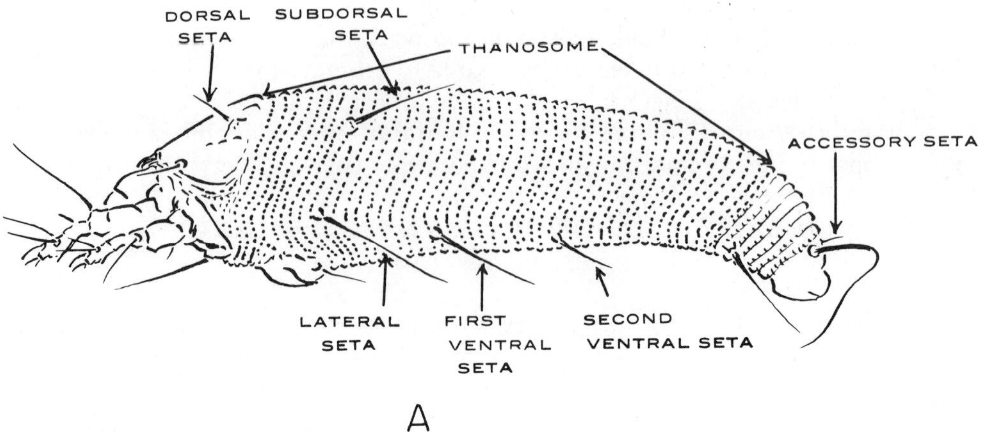

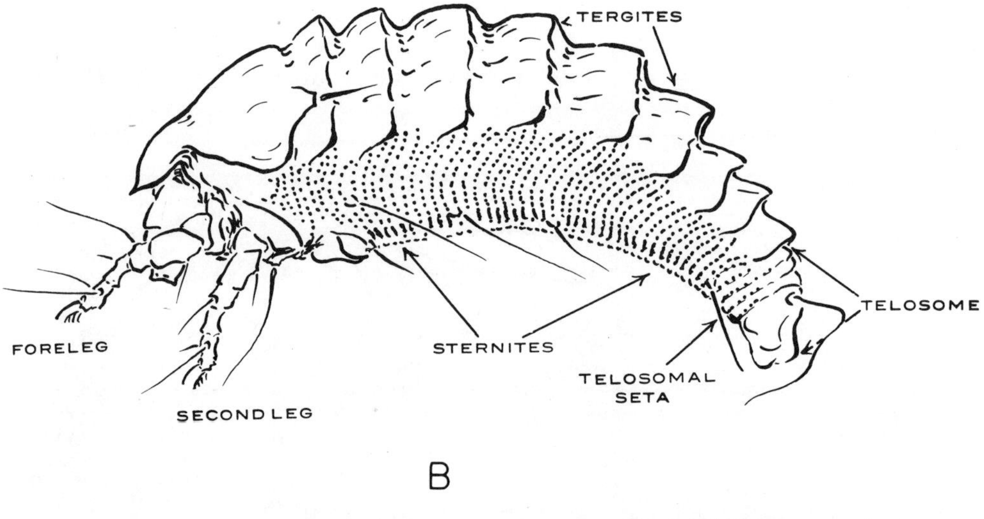

Fig. 90. Lateral diagrams of adult females, with designations of body sections and structures: *a, Phytocoptella leucothonis* (K.) (Nalepellidae); *b, Anthocoptes helianthella* K. (Eriophyidae).

The nervous system is composed of a single neurosynganglion located in the posterior cephalothoracic region. The neurosynganglion consists of a central membranous neuropile surrounded by numerous small nuclei. There is some evidence for lateral mesacon processes, particularly in the area of the extremities, but there is little evidence for a well-developed neural tube or canal extending posteriorly from the synganglion.

The reproductive system consists of a pair of ovarioles located in the posterior dorsal portion of the mite. Developing oöcytes pass downward through bilateral

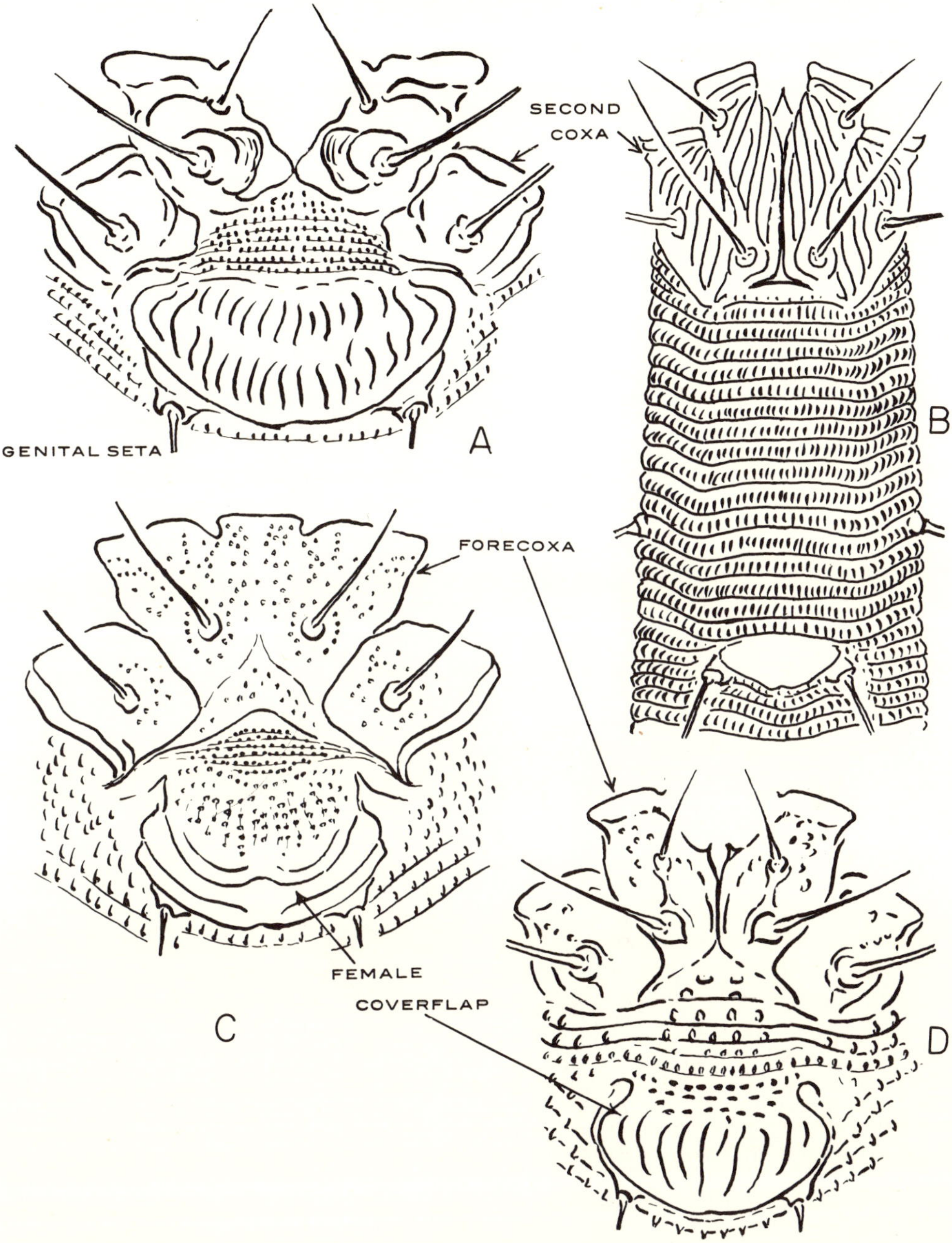

Fig. 91. Ventral drawings of coxae and external female genitalia: *a, Cecidophyes* sp. (Eriophyidae, Cecidophyinae); *b, Novophytoptus* sp. (Nalepellidae, Novophytoptinae); *c, Floracarus* sp. (Eriophyidae, Nothopodinae); *d, Aculops* sp. (Eriophyidae, Phyllocoptinae).

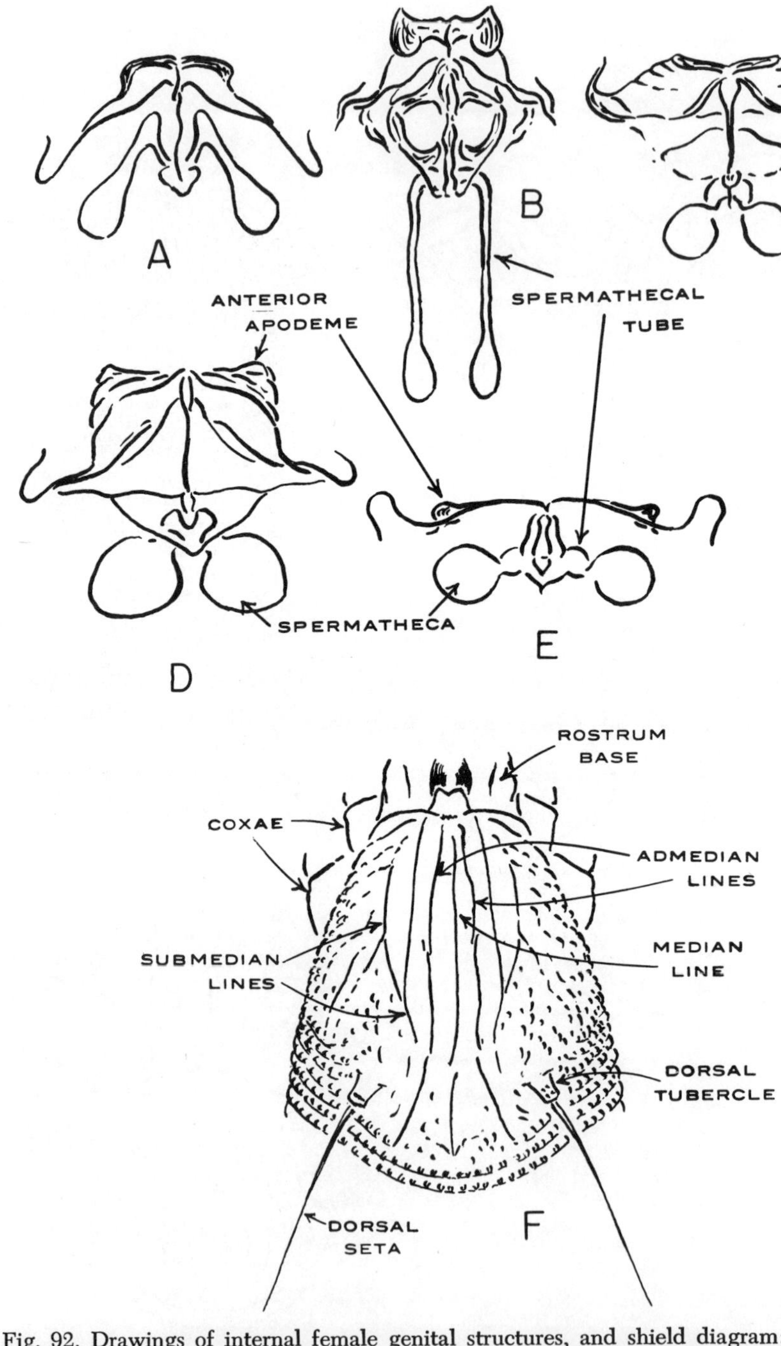

Fig. 92. Drawings of internal female genital structures, and shield diagram: *a*, *Acathrix* sp. (Nalepellidae, Phytocoptellinae), showing recurved spermathecal tubes; *b*, *Trisetacus* sp. (Nalepellidae, Nalepellinae), with extra long spermathecal tubes characteristic of the Nalepellinae; *c*, *Diptilomiopus* sp. (Rhyncaphytoptidae, Diptilomiopinae); *d*, *Anthocoptes* sp. (Eriophyidae, Phyllocoptinae); *e*, *Cecidophyes* sp. (Eriophyidae, Cecidophyinae) showing shortened internal anterior apodeme characteristic of the Cecidophyinae; *f*, *Eriophyes muhlenbergiae* (K.) (Eriophyidae, Eriophyinae) with designations of lines in shield pattern.

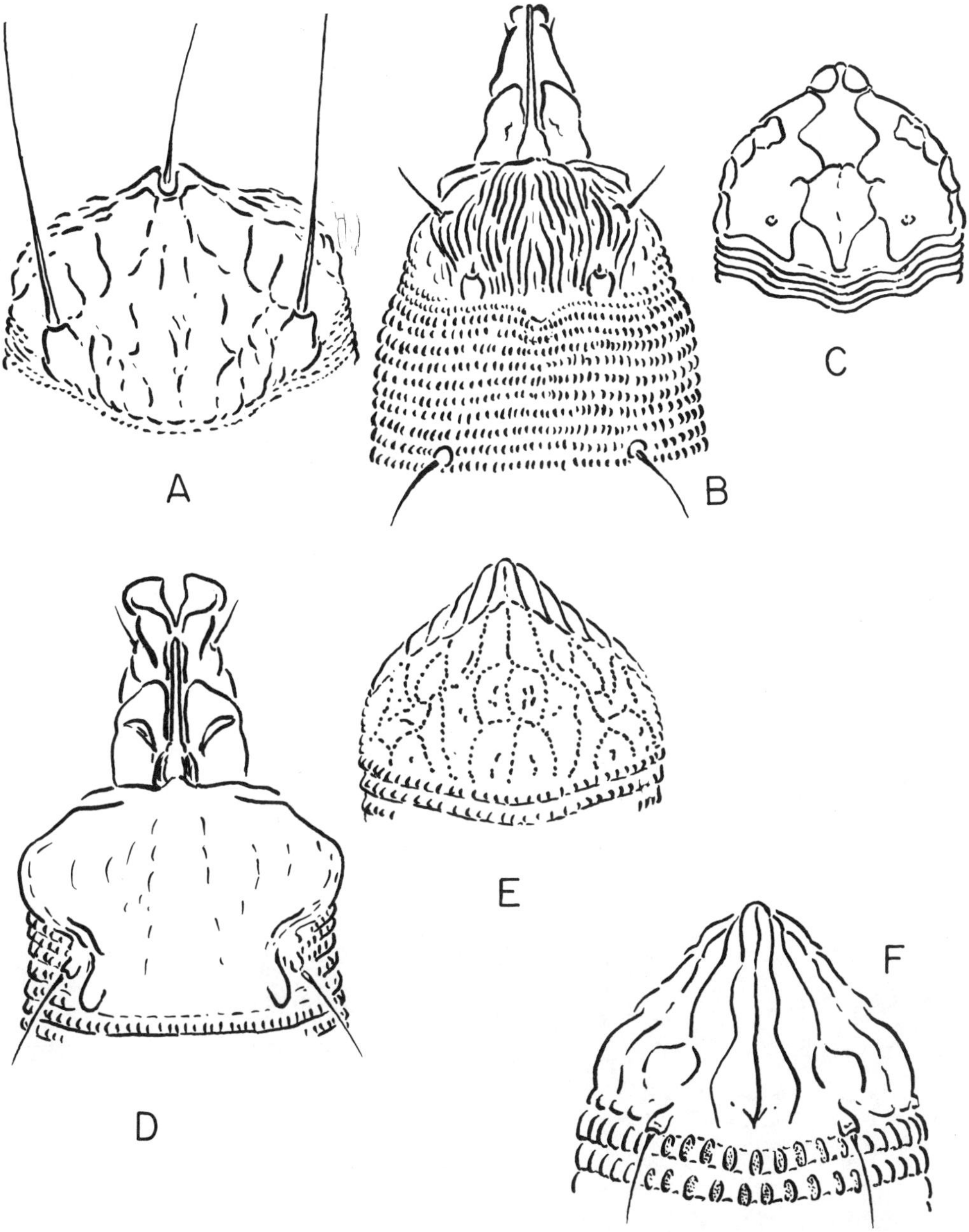

Fig. 93. Drawings of various shield formations: *a, Nalepella* (Nalepellidae, Nalepellinae) showing single anterior shield seta characteristic of the Nalepellinae; *b, Acathrix* (Nalepellidae, Phytocoptellinae) showing anterior shield seta pair, and minute dorsal setae; *c, Calacarus* (Eriophyidae, Phyllocoptinae) showing dorsal tubercles with minute setae; *d, Cisaberoptus* (Eriophyidae, Aberoptinae) showing terminal rostral spatulae; *e, Cecidophyes* (Eriophyidae, Cecidophyinae) illustrating absence of dorsal setae; *f, Aculops* (Eriophyidae, Phyllocoptinae) showing lack of spinules on anterior shield lobe.

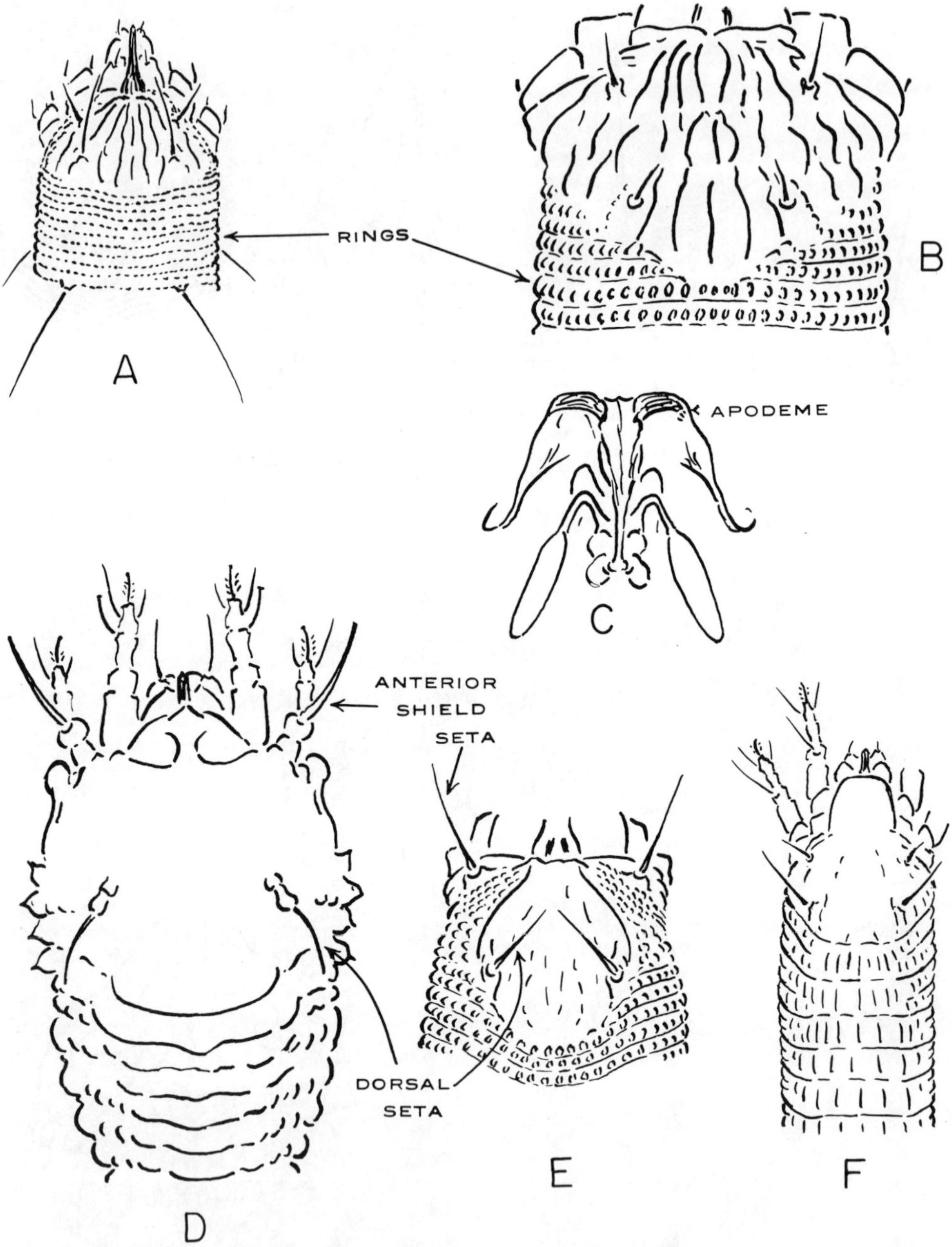

Fig. 94. Drawings of shield patterns and anterior abdominal structures in the Nalepellidae: *a*, *Phytocoptella hedericola* (K) (Phytocoptellinae) showing anterior seta pair and subdorsal abdominal setae; *b*, *Phytocoptella yuccae* (K.) (Phytocoptellinae) shield; *c*, *P. yuccae* internal female genital structures showing recurved spermathecal tubes; *d*, *Retracrus johnstoni* K. (Sierraphytoptinae) showing backward direction of dorsal setae, and lack of subdorsal abdominal setae; *e*, *Phytocoptella garryana* (K.) (Phytocoptellinae) shield; *f*, *Mackiella phoenicis* K. (Sierraphytoptinae, Mackiellini) illustrating lack of subdorsal abdominal seta pair.

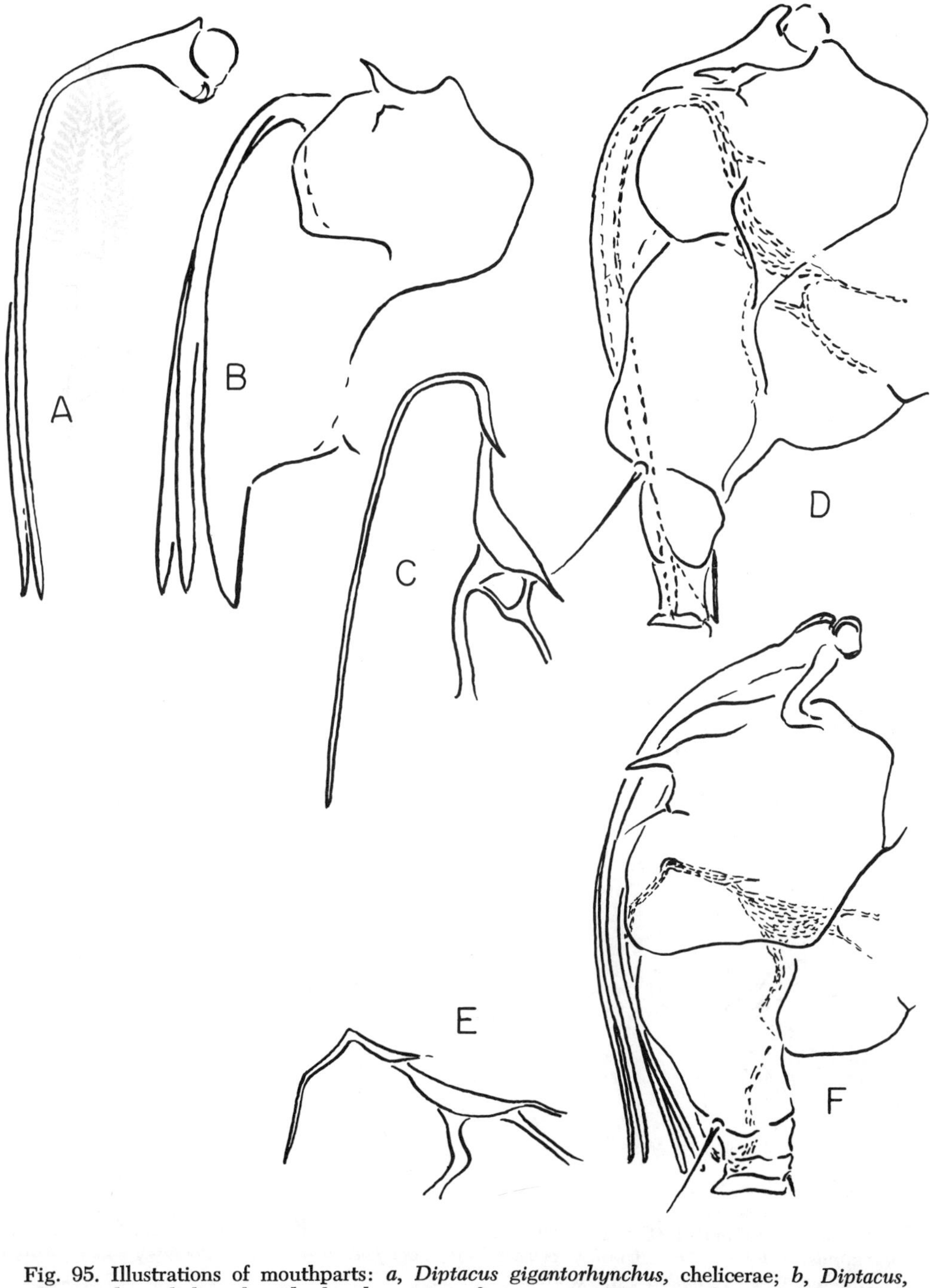

Fig. 95. Illustrations of mouthparts: *a, Diptacus gigantorhynchus,* chelicerae; *b, Diptacus,* auxiliary stylets, cheliceral guide, basal segment of rostrum; *c, Diptacus,* pharyngeal pump and oral stylet; *d, Diptacus,* side view of mouthparts; *e, Calacarus carinatus* (Green), pharyngeal pump and oral stylet; *f, Calacarus,* side view of mouthparts.

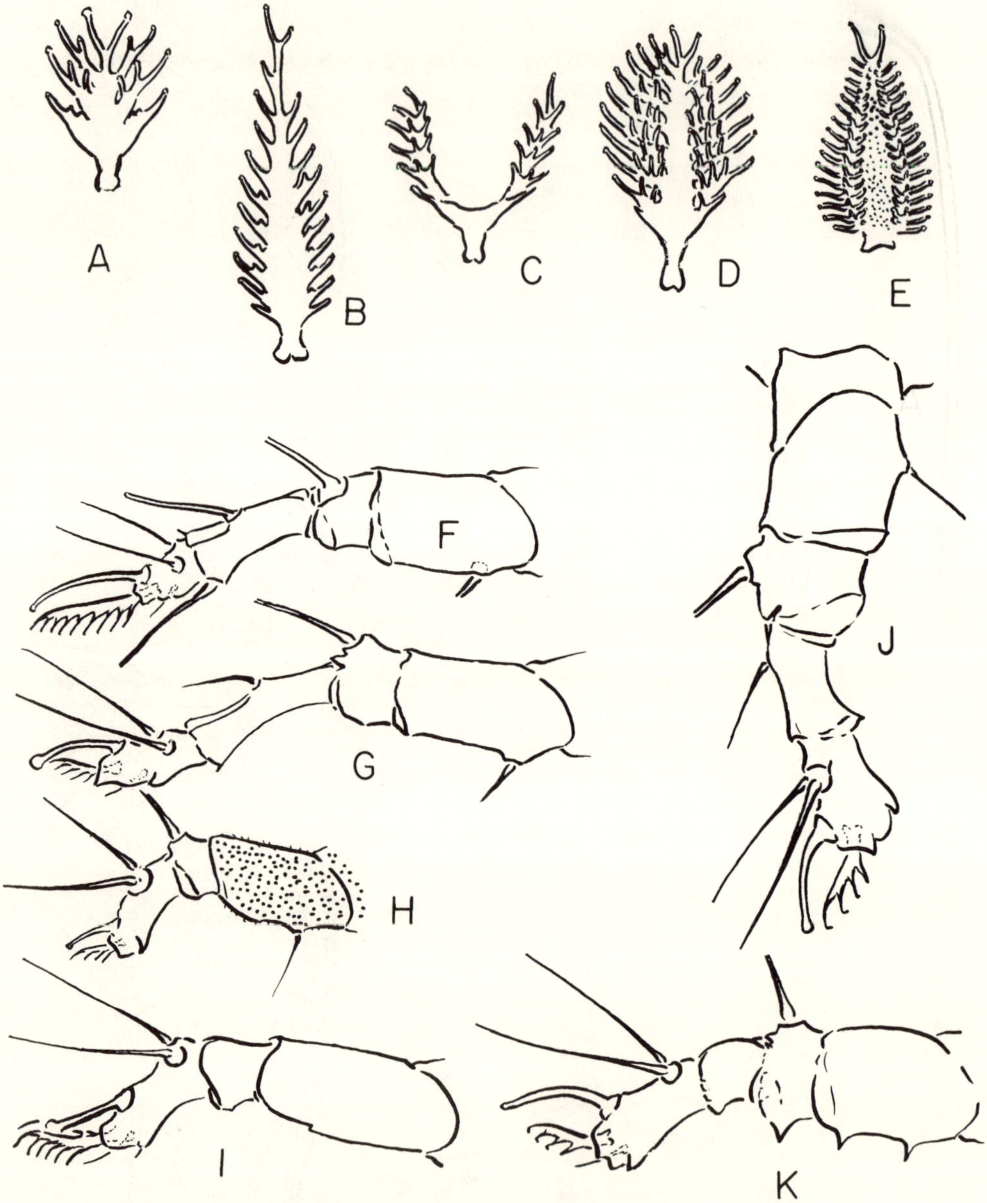

Fig. 96. Legs and featherclaws: Featherclaws of: *a, Anthocoptes; b, Acathrix; c, Diptilomiopus; d, Nalepella; e, Cisaberoptus*. Forelegs of: *f, Nalepella; g, Calacarus; h, Floracarus; i, Diptilomiopus; j, Aculops; k, Acalitus*.

ovarioles to the oviducts and into the main genital canal. Some nutrient material is supplied to the oöcytes even before they reach the genital canal. Before reaching the genital flap the chorion and vitilline membrane becomes evident and the eggs are filled with yolk platelets and lipidlike bodies. Nurse cells along the genital canal contribute yolk protein and lipid material for storage in the developing oöcytes.

The circulatory system is simply composed of various cells or of cellular organs lying in the hemocoele. The hemolymph appears to be circulated in the mite by body movements. The excretory mechanisms are discussed later under Waste Disposal.

Little is known concerning the eriophyoid muscular system, but Whitmoyer et al. have found the muscles nonstriated. Tetranychid muscles are supposed to be striated.

General Morphology

This is a morphological prospectus of the more visible, that is, chitinized parts of the subcyclindrical eriophyoid body, and of their mechanisms (figs. 90–97).

The specialized and reduced eriophyoid body still has the three standard acarine sections: (1) rostrum or ganathosoma; (2) propodosoma (cephalothorax); (3) hysterosoma (abdomen). The rostrum, while greatly modified, still has most of the original parts. The gnathosoma plus propodosoma with shield above and coxae below, is but a remnant. Embryological studies might indicate what has happened to the cephalothorax rear part, and 2 rear leg pairs, but postembryonic stages offer no information on this. The abdomen, which is probably a composite structure, undoubtedly has elements of the cephalothorax incorporated into it which are now indistinguishable.

Rostrum features, and number and location of residual setae, furnish structures for segregating 3 major groups or families of the Eriophyoidea (see also the taxonomic definitions in appendix 3). No forms possessing intermediate structures between these groups are now known.

Absence of Ocelli and Tracheae

There are no definite ocelli on these mites, but slight subglobular projections on the rear lateral shield angles on many of them suggest that these projections may be light receptors.

As in the case of many of the smaller Acari, eriophyoids have no tracheae. Since there are no spiracles or tracheae in these mites the association of them with the Prostigmata must rely on more tenuous characters such as transverse genitalia, needlelike mouth stylets, brushlike empodia, and the plant-feeding habit.

Oxygen absorption by these mites has to come through the epidermis, but the chelating compounds for absorping oxygen are unknown. That sulfur is extremely poisonous to eriophyoids suggests that this element is enough like oxygen to destroy the oxygen-receiving apparatus.

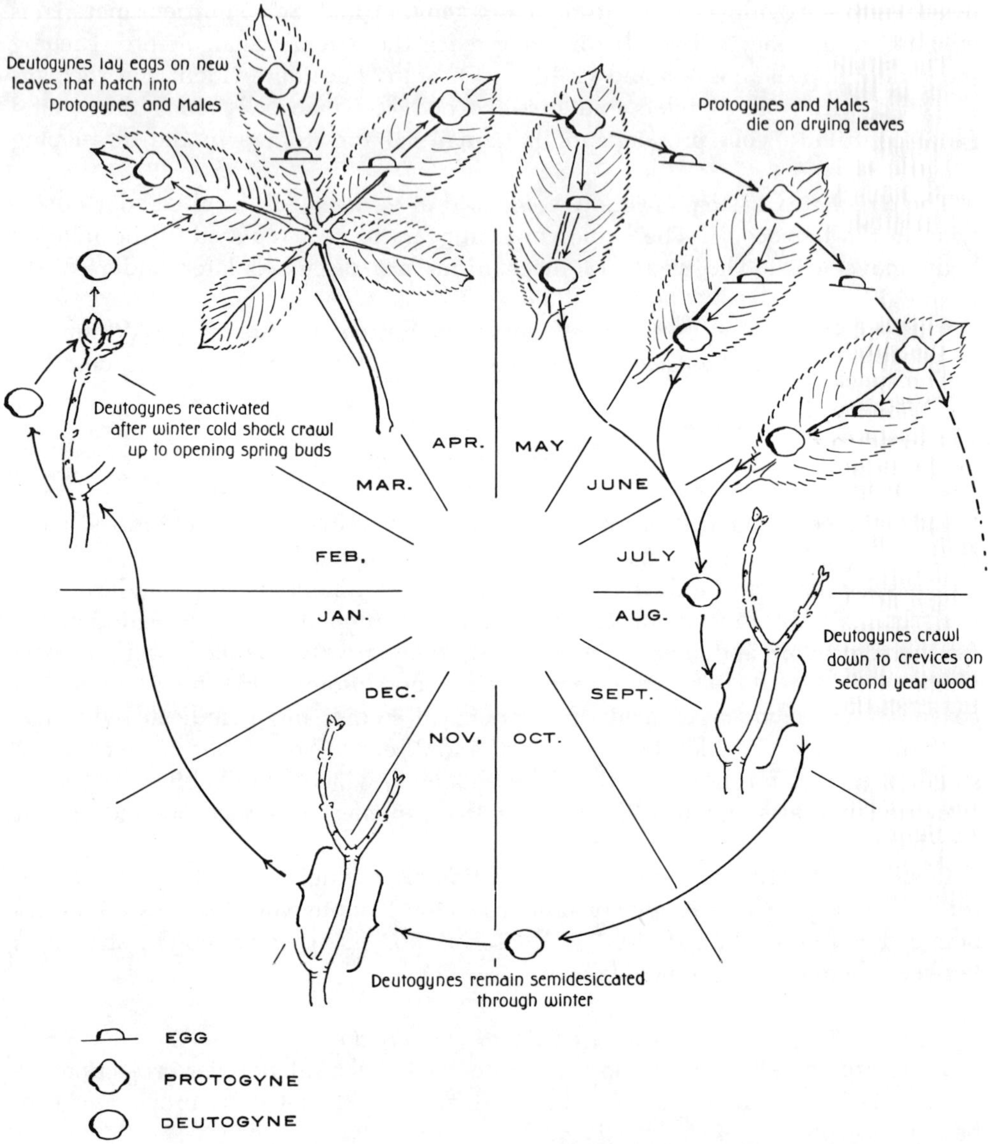

Fig. 97. Diagram of life history of *Tegonotus aesculifoliae* (K.).

Familial Differences

Nalepellidae. The first group, here called the family Nalepellidae, has the most residual setae and may be the most primitive. The nalepellid shield, unlike the shields in the other two families, has rarely lost dorsal tubercles and dorsal setae. These dorsal tubercles, with few exceptions, are located well ahead of rear shield margins, and dorsal setae nearly always point forward. A diagnostic feature of the family is the invariable presence of 1 or 2 anterior shield setae (fig. 93, *a*, *b*), regardless of the position or fate of dorsal tubercles and setae. Additional nalepellid defining structures are: the frequent presence of anterior pair of subdorsal abdominal setae (fig. 90, *a*); usual presence of lateral foretibial spur (fig. 96, *f*); the female genital coverflap never ribbed; the internal female spermathecal tubes extending diagonally ahead of the spermathecal pore at rear end of genital slit (fig. 92, *a*), and then recurving to rear. In the subfamily Nalepellinae, which is the group with a single anterior shield seta, the spermathecal tubes offer an exceptionally useful character by being extra long (figs. 92, *a*; 104, *g*; 106, *e*; 107, *g*). Nalepellids have the short form oral stylet as do the eriophyids (fig. 95, *e*).

First stage *Trisetacus* nymphs (Nalepellinae) (fig. 105), have the same setae as the adult, but unlike first nymphs in the Eriophyidae (strict sense), these *Trisetacus* nymphs have a well-defined rear shield margin.

Eriophyidae sensu stricto (s. str). The great majority of gall and erineum making species occur in this family. These mites also have the short form oral stylet, but differ from nalepellids by lacking anterior shield setae, never having subdorsal abdominal setae or foretibial spur. Eriophyid spermathecal tubes are short and extend laterally or diagonally to the rear from the spermathecal pore (fig. 92, *d*, *e*). The female genital coverflap is usually ribbed.

Some genera and species of eriophyids lack dorsal tubercles and setae, or they are much reduced, but the majority have these structures. There is more variation in the position of dorsal tubercles within this family than in the other two families. Part of the genera and species have these tubercles set ahead of the rear shield margin, or the dorsal setae are pointed forward, (figs. 111, 127), but the only eriophyoids with dorsal tubercles set on the rear shield margin, with setae pointing to the rear, are in this family (figs. 115, 131). Dorsal tubercles set ahead of the rear shield margin, when not cyclindrical, have the tubercle axis either longitudinal to the body, or at an angle to the body (figs. 128, 129). Dorsal tubercles when set on the rear shield margin, and when not cyclindrical, have axes transverse to the body length (figs. 120, *c*; 130, *h*; 131, *a*, *b*, *d*), or nearly so. First nymphs of the species with dorsal tubercles situated on the rear shield margin on the adult have these tubercles located ahead of the rear margin, presumably recapitulating an ancestral trait.

Rhyncaphytoptidae. This family seems to have affinities to the Eriophyidae rather than to the Nalepellidae, but these mites are aberrant in rostral structure, and in habit. The rostrum on mites in this family is always large in comparison to the body, averaging between 50 and 70 μ long. The chelicerae are long, so these mites can feed deeper in plant tissue than can most species in the other 2 families. The oral stylet is of the long form (figs. 95, *a-d*).

Setation on rhyncaphytoptids is essentially as it is on eriophyids, but the dorsal setae, which are present on most species project up when short, and forward in some degree when long. First stage nymphs in this family have poorly defined rear shield margins, as do these nymphs in the eriophyids (s. str).

In habit the rhyncaphytoptids have proved so far to be all leaf vagrants that do little damage to their host. This indicates that no injurious chemical is injected while feeding. That these big-beaked mites, except for some hibernating forms, have completely moved away from microenvironments, and that they have the unusually long oral stylet, presents a problem in theorizing as to their precise relationship. Perhaps further discoveries of new species may help answer this question.

Eriophyoid Gnathosoma

The rostrum is enclosed on each side by the pedipalps, or mouth feelers (pl. 56). Observation of struggling mites indicates that eriophyoid pedipalps are capable of some independent action, and even of crawling movements. They also telescope, or fold back to allow feeding penetration of the stylets, as discussed later. The pedipalps hold the mouth stylets in an anterior groove or sheath. This groove projects from the base of the pedipalps, and since the cheliceral stylets are able to lift out of it upon slide-mounting preparation, the groove is open in front (fig. 95, *b*).

Identification of palpal structures depends in part on how they are associated with interior anatomy. The *basal palp* segment lies ventrally just ahead of the fore-coxae, and contains the pharyngeal pump. The second or *proximal segment* in front, projects forward or diagonally down from the anterior part of the shield (fig. 95, *d*, *f*). On species with an anterior shield lobe the proximal segment is just under this lobe. At about the first ⅓, or ½, on this proximal segment, there is a short seta, and just before it is a centrally directed spine. This spine crosses the cheliceral groove, and evidently helps direct cheliceral stylets. This spine is here called the *cheliceral retainer*. A third, or intermediate *palpal segment*, bears a terminal frontal seta, which is the antapical palpal seta of descriptions (fig. 95, *d*, *f*).

Terminally the pedipalp is a more-or-less telescoped series of segments, the basal one being the fourth. This fourth one is longer on rhyncaphytoptids than on species in the other two families. The remaining two palpal segments are more distinguishable on rhyncaphytoptids than on species in the other two families. A sensory peg arises from the rear of the next-to-last segment; this peg is longer on rhyncaphytoptids.

Cheliceral guides. As stated above, these are stiffer interior parts of the pedipalps that project from the base. These guides have rounded ends on nalepellids and eriophyids, and are sharper apically on rhyncaphytoptids.

Scanning electron microscope pictures of the front of the eriophyoid rostrum show the terminal part of the pedipalps to be in the form of a suction cup. Cupping action would furnish the mechanical support for penetration of the mouth stylets, which are described in the next section.

Oral Stylets. Eriophyoids feed by means of 5 stylets in the gnathosoma (fig. 95), which are 3 more than the number of stylets in the tetranychoid gnathosoma. The anterior 2 stylets in the eriophyoid mouth are presumably the chelicerae. These cheliceral stylets lie close together, but are separate for their entire length, and move alternately when penetrating plant tissue. The alternate movement is achieved by the lateral vibrations of a basal knob called the *motivator*. The stylets themselves can move at most but a short distance ahead of their basal attachment, or perhaps not at all. Penetration of plant tissue by these stylets is permitted by telescoping, or folding back, of the pedipalps. These cheliceral stylets are forced into plant tissue slowly, evidently cause little or no mechanical damage, and remain in position as long as the mite feeds at the point of penetration.

The homologies of these cheliceral stylets in eriophyoids, as compared to more generalized acarine chelicerae, are principally on a speculative basis. A limp filament is frequently seen to hang down from the cheliceral stylets on slide-mounted eriophyoids. Schevtshenko (1968) calls this limp filament a stylet. If that is the case it might be that it is the movable digit with the stiff part the fixed digit.

A comparison of these eriophyoid cheliceral stylets with the stylophore and 2 appendant stylets in the tetranychoid gnathosoma, indicates that the 2 structures are quite different. The stylophore on tetranychoids is the basally fused fixed digit and shaft, with the hanging needles the movable digits. If there is some fundamental relationship between eriophyoids and tetranychoids the chelicerae then suggest that the two groups evolutionarily parted before there was a basal fusion of the chelicerae.

Most of the eriophyoids have chelicerae that average between 15 and 40 μ long, which limits the penetration depth of these needles. Such mites occur mostly in the Eriophyidae, and especially among gall formers. Some nalepellids have chelicerae as long as 60 μ, and most rhyncaphytoptids have cheliceral stylets that can penetrate from 50 to 70 μ into plant tissue. In contrast to this, common tetranychoids can push their stylets in about 100 μ.

Auxiliary stylets. Just behind the eriophyoid cheliceral stylets is a second pair of stylets about as long as the first pair. The precise function of this second pair is unknown since the stylets have yet to be connected with body structures basal to them. The supposition is that they are salivary ducts.

Tetranychoids, with the thrust and return of the stylophore, and consequent in-and-out action of the stylets, continually keep their stylets lubricated with fluid that issues from salivary duct openings in the groove that guides these stylets, and which groove is located on the upper side of the fused pedipalp base.

In contrast to this, the eriophyoid stylets bore into plant tissue slowly since there is no thrusting stylophore. So, in the absence of in-and-out action of the chelicerae, fluid cannot be carried by them into plant tissue. If the auxiliary stylets are salivary ducts, they then furnish an efficient way to deliver saliva to cheliceral tips.

The oral stylet that sucks up nutrient liquids is identifiable because it hinges off of the upper, anterior part of the pharyngeal pump. It is able to move down into plant tissue behind the anterior four stylets. Eriophyoids therefore feed within plant tissue.

There are two forms of eriophyoid oral stylets. The short form (fig. 95, *e*) is in nalepellids and eriophyids. The long form oral stylet (fig. 95, *c*) is found only in the rhyncaphytoptid rostra. This long form projects up higher than the short stylet and extends from the hinge up to just below the cheliceral base, and then recurves down to the cheliceral tips.

Unlike all of the tetranychoid mites except the gall-making forms, eriophyoids depend for the most part upon keeping plant tissue alive so that it will continue to feed them. This fact is illustrated in various ways, such as the growth of galls and of erinea, which could not develop from dead tissue. Within galls the several months required for the reproduction of the summer mite brood, and the continual succulence of the nurse cells lining these galls, shows that the mites stimulate the life and succulence of these cells. Also, virus transmission by eriophyoids requires that the virus be injected into the presence of living tissue.

The spatulate, or shovel-shaped rostrum on the Aberoptinae, which is a subfamily of the Eriophyidae, deserves special mention. The burrowing structures these mites have are discussed under the heading: Relationships of Eriophyid Mites to their Host Plants (below).

Propodosomal (Cephalothoracic) Shield

Cephalothoracic shields on eriophyoids (figs. 93, 94) are essentially triangular dorsal-covering plates. The setae on shields have been mentioned above. On bud and on gall mites that have always lived in sequestered environments this shield usually has little or no anterior extension over the rostrum. A relatively few exceptional bud and gall mites have slender forward extensions over the chelicerae. Such extensions are narrow-based and basally flexible.

Rust mites and leaf vagrants, adapted to exposed conditions, usually have anterior shield extensions and these are nearly always broad-based and rigidly attached. All rhyncaphytoptids are leaf vagrants and they have both broad and narrow-based anterior shield extensions, or in some cases no anterior lobe.

Broad-based anterior shield extensions, or lobes, are, as indicated above, usually coupled with open life, and are also usually associated with broad transverse back plates or tergites on the abdominal dorsum (as discussed below). When gall mites have broad-based anterior shield lobes it is an indication that they have returned to gall making from a formerly exterior life habit (Keifer, 1969).

Legs of Eriophyoids

Eriophyoid legs, as already pointed out, attach to coxae on the cephalothoracic underside. Anterior coxae embrace the basal pedipalp segment anteriorly and are usually connate behind it. The contact line between these coxae, when present, is the *sternal line*. When coxae are either separate or united this line fades or is absent. The sternal line is essentially an internal apodeme. A similar internal apodeme surrounds and defines the hind coxae, and has an extension from the rear hind coxal angle which attaches to the lateral parts of the interior genital structures.

Most eriophyoid legs have the usual acarine segments. Segments that are absent, or fused together, are of taxonomic importance. The legs of all eriophyoid species so far seen show that the second legs have the same segmentation as the forelegs. This includes segment reductions. Thus, if the tibiae and tarsi are fused on the forelegs, the hindlegs have the same fusion. If the patella is absent from the foreleg, it is also absent from the hindleg. Illustrations of eriophyoids which show different segmentation on the forelegs than are shown on the hindlegs are probably in error.

Normal leg segments from the coxa are trochanter, femur, patella or genu, tibia, tarsus. Leg segment reductions are present on the Nothopodinae (fig. 96, *h*), the tibia and tarsus are fused, or the tibia is but slightly indicated. The legs of the Aberoptinae are thickened and the tibia is reduced in length. The fore- and hindlegs of members of the genus *Diphlomiopus* (fig. 96, *i*), lack the patellae.

Leg Setation. Coxal setae are typically 3 pairs (fig. 91, *a*, *b*, *d*), with the forecoxae having 2 setal pairs and the hind coxae 1 pair. Setae on the hind coxae are relatively long and quite similar from species to species. They are important tactile setae. Forecoxal setae vary as to occurrence and position, but occasionally the first pair are absent (fig. 91, *c*).

No setae occur on the trochanter, but the femur typically has a ventral seta that on some species is absent, thereby offering a taxonomic feature. The patella typically has a large upper seta, but it may move to a lateral position on the hindleg of some nalepellids.

There is usually a foretibial seta, the absence of which is of taxonomic importance (figs. 96, *k*; 124, *d*). There is never a hind tibial seta.

There are usually two subbasal tarsal setae on the upper surface which project diagonally forward from their sockets. When tibial or patellar setae are absent, the tarsal setae are usually large and long. Some species have but one upper tarsal seta.

Claw and Empodium. Unlike most mites, eriophyoids have no true claw. The curved, often knobbed, and inarticulate tarsal appendage that projects forward from the upper front of the tarsus, while called "claw," is probably an adapted sensory club.

The eriophyoid featherclaw (fig. 96, *a-e*—also note the terminal position of this empodium on the illustrated legs) is, however, the true empodium. It is set in the end of the tarsus in a ball and socket joint or attachment. This empodium is essentially a clinging hairbrush, and in that respect is similar to empodia on many acari. Branches from the central empodium shaft are called *rays*. All but the terminal ray pair are secondarily branched, with few exceptions. Each ray ends in a bulb-like tip, which usually appears bent down in side view, a characteristic that resembles tetranychid tenent hairs. Electron microscope photographs of featherclaw rays disclose that there are types with small terminal bulb endings, and other types with expanded endings. These differences may indicate group characters (pls. 62, *a-f*; 68, *b*; 71, *b*; 72, *b*; 73, *b*, *c*; 74, *d*).

The number of rays on featherclaws varies from 2 on each side to more than 10 on the central stem. The most common number of rays is 4, with 5 nearly as common (figs. 96, *a*; 124, *c*). The number of species with 6 or more rays diminishes as the ray number increases. The Aberoptine species with spatulate attachments have more than ten rays but it has been difficult to determine the exact number.

Most featherclaws are simple, that is, the rays branch from a central stem (fig. 96, *a*, *b*, *d*). A minority of species, usually in genera separated on that basis, have divided featherclaws (figs. 96, *c*; 125, *b*; 138, *b*). The most conspicuously divided featherclaws are in the Rhyncaphytoptidae, with *Diptilomiopus* as the most outstanding example. In one species in the Eriophyidae (*Notostrix jamaciae* K.), part of the featherclaws on the same individual may show a shallow division.

Eriophyoid Hysterosoma (Abdomen)

The so-called abdomen on these mites is the principal part of the body that gives these mites their wormlike aspect. This is especially true on species living in microenvironments (figs. 90, *a*; 117, *a*). Free-living species tend to be less wormlike, having flattened abdomens with ridged or grooved dorsums (figs. 126, 127). Since the rear part of the original cephalothorax has disappeared along with the rear 4 legs, some of it has undoubtedly become incorporated into this wormlike section. So the term "abdomen" in the case of eriophyoids is more a term of convenience than a name for a definite entity.

One of the more notable features on eriophyoid abdomens is the presence of transverse surface rings. The invariably transverse character of these rings may perhaps be in part the result of life in sequestered habitats, since the 8-legged demodecids, which are not closely related to eriophyoids and live in mammalian hair follicles, also have similar rings.

Tetranychid mites also have epidermal lines, or ridges, but examination shows that they describe much more diverse patterns than do the eriophyoid rings. Surface striations on tetranychid mites are often beautifully elaborate, and are both transverse and longitudinal on the same individual. Many tenuipalpid flat mites have dorsal ocellarlike epidermal designs. There is therefore considerable difference displayed between epidermal structures on eriophyoids and on tetranychoids.

Gall and bud mites in the eriophyoids, which live under cover, usually have narrow rings that are similar dorsoventrally. Leaf vagrants and rust mites often show broad back plates that are probably a defense against dehydration. These broad back plates are also known as dorsal half rings or *tergites*. They contrast with ventral half rings, which remain narrow.

Rust mites and leaf vagrants of the genus *Tegonotus*, which have broad tergites, also have lateral extensions of these tergites on the thanosome (see below). These lateral extensions vary from conspicuous lobes, to sharp processes, and to spectacular and pointed spinelike projections (fig. 126, *c*; also Keifer, ES, X:pl. 149; Nalepa, *Zoologica*, 1911, taf. VI, 16 *a*). A few phyllocoptine mites have anterior lateral abdominal expansions just behind the shield that in part recapitulate the shield (fig. 126, *f*; Keifer, ES, V:pl. 81).

Thanosome and Telosome. For descriptions it has been convenient to divide the abdomen into two sections. The anterior part, from the rear shield edge to the third or last ventral seta, is the *thanosome* (fig. 90, *a*). The part from this last ventral seta (now the telosomal seta) to the terminal lobes is the *telosome* (fig. 90, *b*). On wormlike bud and gall mites the telosome is much like the thanosome. On rust mites and leaf vagrants with broad back plates, the telosome usually maintains its more conservative structure with narrow rings.

Microtubercles or Ring Granules. These small humps (Nalepa's *puncthöcker*) in the great majority of species stud the abdominal rings (pl. 57; figs. 103, *g*, *j*; 104, *b*; 107, *b*; 108, *k*). Many species of Acari have minute lobes, or granules, on skin striations, but it is not possible here to draw precise comparisons between what they have and the eriophyoid microtubercles. The basic outline of the eriophyoid microtubercle is either elliptical, elongate, or oval (pl. 57). Outwardly these humps are flattened, rounded off, pointed, or even produced as a spinule or short hairlike projection. Eriophyoids living in sticky habitats tend to have coarse, or spinulate microtubercles, which may help them avoid becoming caught in viscous adhesive liquids.

In connection with abdominal ring microtubercles it is possible, on the basis of available evidence, to theorize that their presence or absence has a relation to water exchange between the body and its surroundings (see Water Loss, chap. 2). Mites in galls, and those that live under cover, where they can feed, are usually microtuberculate. Species that live where they are partially or entirely exposed, and where modifications to aid water conservation are more necessary, tend to have these ring granules reduced or eliminated, at least dorsally. Many of the more wormlike species that tend to expose the rear parts of their abdomens have either obscure microtubercles on the dorsal abdominal rear, or the dorsum on the rear thanosome and on the telosome completely lacks these granules. Regional levels of atmospheric humidity may also be a factor in the amount of microtuberculation species that live in any respective area may display.

Leaf vagrants and rust mites, which are often fusiform and flattened, are entirely exposed to such water loss through their integuments as the regional atmospheric humidity exacts. These exposed species show considerably more variation in their microtuberculation than do species living under cover. Individuals that are feeding actively on fresh succulent tissue may offset water loss through ring granules. Many of these exposed species have broad, hard, back plates, however, with the microtubercles either fainter and fused with the surface, or these granules are entirely gone. The differences in microtuberculation between protogynes and deutogynes in the same species offer thought-provoking comparisons, and these are discussed in the section on deuterogyny (see chap. 2).[3]

Eriophyoid species that have dorsal white wax stripes, or cover themselves with flocculent wax, have what are probably microtubercles that are wax-producing organs. This is especially so with granules on dorsal ridges, where they often join laterally to form thick transverse wax-making bases on the ridge apex. Flocculent wax may aid in water conservation or offer some protection against predators.

The microtubercles on the telesomes are usually finer than on the thanosome. Often these small humps project over the ring margins as small sharp points. Ventrally the microtubercles on most telosomes are peculiarly elongate and may be the only ones that persist on deutogynes that have lost all thanosome microtubercles.

Many actively feeding tetranychoids have small lobes attached to the epidermal striations. These lobes may have the same function in regard to regulation of body water equilibrium with surroundings as do microtubercles on eriophyoids. This relationship is strikingly illustrated by hibernating forms in the Eriophyoidea and in the Tetranychidae, which usually have rings and striations respectively, that possess reduced microtuberculation and reduced strial lobes, or microtubercules and lobes may be entirely absent from rings and striations. Since hibernating forms that lack these minute granules retire to locations where they cannot feed or take in water, the reason for absence of these structures seems clearly to be that of rendering the epidermis impervious to water so the mites can conserve water. For further treatment of this question see the section on eriophyoid deuterogyny (chap. 12).

Abdominal Setae. All eriophyoids have a basic set of abdominal setae. As discussed, many nalepellid species have a special anterior subdorsal adominal seta pair (fig. 90, *a*). The standard thanosomal seta number on each side, including genital setae, is 4. The first abdominal seta, the *lateral seta*, is located slightly below the midlateral line, about 5 to 12 rings behind the shield, and directly above or slightly behind the genital seta. That the lateral seta occurs below tergites on leaf vagrants shows that it is essentially ventral. The *first ventral seta*, which is usually longest, occurs a little ahead of the thanosomal midpoint and lower than the lateral seta. The *second ventral seta* lies still nearer the midventral line than the preceding seta and near the 3/5 point on the thanosome. This second ventral seta is usually the shortest seta of the series. The third ventral or *telosomal seta* (fig. 90, *b*) is on the first telosomal ring. It is almost always moderately long and frequently stiff.

Above the terminal lobes is a long curved seta, the *caudal seta*. It is useful in helping the mite to rear up, or leap. Close to this caudal seta there is usually a small seta known as the *accessory seta*. This accessory seta is usually largest on nalepellids (fig. 104, *c*).

Genitalia

Eriophyoid genitalia are proximal, that is, located on the anterior end of the abdomen just behind the coxae (fig. 90, pl. 58, *a, b, d*; and other figures showing the genitals just behind the coxae). A pair of genital setae always accompany these structures; each seta arises just behind the lateral genital angle (fig. 91). The displaced condition of the body rings indicates that external eriophyoid genitals have been forced through the body wall during the resting stage of the second nymph and have crowded the rings both ahead and to the rear.

Neither nymphal stage has external genitals, but the seta pair is always present (figs. 105, *d*; 117, c). On these nymphs the genital setae may be located near the midventral line 6 to 12 rings behind the coxae. In contrast, the adult genital location usually shows but 4 to 6 or 7 displaced or crowded rings between coxae and genitals. On species with genitals that are appressed to the coxae (Cecidophyinae, figs. 91, *a*; 108, *f*), the ring number appears to be even less. Nymphs of the species whose adults have relatively appressed genitalia do not indicate by the position of their genital setae the close position found on the adults.

The major exception to the usual genital position on adult eriophyoids occurs on the novophytoptines (fig. 91, *b*). In this group the adult genitalia may be from 12 to 16 uncrowded rings behind the coxae. While novophytoptine nymphs have not been available for study, such other nymphs as have been available minimize this novophytoptine difference.

In the Acari in general the reproductive organs may be situated anywhere along the midventral line from just back of the gnathosoma, to terminal locations on or near the abdominal rear. The impression is, however, that the location of these organs is in part a feature of mite groups, and to some extent defines the respective groups.

The genitalia on many mites, and perhaps the majority, show the principal dimension and opening to be longitudinal. Baker and Wharton (1952) note that the genital plates on both eriophyoids and on tetranychoids are transverse. They deduce a relationship between the groups on that basis. While this may very well indicate affinity, the evolutionary parting of the groups may still be far back in geological time. There are certain genitalic differences between the eriophyoids and tetranychoids which await resolution. A list of these differences is: (1) anterior or proximal genitalia on the eriophyoid abdomen; (2) terminal or subterminal genitalia on the tetranychoid abdomen (note: no intermediate positions of the genitals in either group have so far appeared); (3) male tetranychoids with aedeagus for copulation; (4) male eriophyoids lacking aedeagus and fertilizing females by means of spermatophores; (5) chitinous frame (as described below) within eriophyoid female genitalia, which has no counterpart in the female genitalia in tetranychoids. It is not possible to state at this writing what other mite group possesses an internal female genital chitinous frame at all comparable to that present in eriophyoids.

Eriophyoid Male Genitalia

Male genitalia on these mites consist of a somewhat produced, anteriorly convex, transverse opening, just behind the coxae, and ahead of the genital setae (fig. 120, *g*; 138, *d*). A pair of sensory pegs, one on each side of the midline, occurs just behind the transverse opening. Internally there is little or no chitinization in the male genital apparatus.

Eriophyoid Female Genitalia

Eriophyoid female genitalia (fig. 91) also protrude somewhat from the ventral surface, and are distinguishable in lateral view on slides by the anterior downward projecting coverflap. They are outwardly covered by the anteriorly hinged coverflap, which is in the form of a scoop, and probably aids in squeezing sperm from the spermatophores (pl. 58). Examination of female genitalia under an ordinary microscope gives the distinct impression of a central longitudinal genital slit, with the spermathecal pore at the rear end (fig. 92, *a-e*; also illustrations such as figs. 104, *g*; 106, *e*; 118, *g*). The function of the pore is indicated by the spermathecal attachment.

A scanning electron microscope picture of one female genital structure indicates that there is a transverse opening just behind the coverflap base (pl. 58). How this relates to what is visible under an ordinary microscope is not yet clear.

A principal feature of the eriophyoid female genitalia is the chitinous frame frame located just within these organs. As stated above, the frame shows a central longitudinal slit, or line, extending back from the broad anterior apodeme, which apodeme is the anterior part of the frame. This apodeme varies in shape, and usually extends a short distance forward. The apodeme is of particular significance in defining the Cecidophyinae, as in that subfamily the crowding of the genitalia up against the coxae causes the apodeme to bend up and appear short in ventral view.

As with the internal male genitalia, the female oviduct extends forward from the abdominal rear to the abdominal anterior. This is, of course, opposite from the direction of the alimentary canal. Eggs of various sizes are often visible in the oviduct in slide-mounted specimens. Much less often there is a first stage nymph, or even two nymphs, internally in the female. Eriophyoid egg diameter is nearly as great as female body width, so eggs must undergo considerable squeezing upon extrusion. While the usual egg shape, after deposition, is ellipsoidal, some rust mite eggs lie almost flat on the leaf surface.

Spermathecae. The paired spermathecae are usually visible at the rear end of the chitinous frame within the female genital apparatus, if the internal structures have not been destroyed during preparation. These structures are bulbous sacks, appended to the ends of the spermathecal tubes that project laterally from the spermathecal pore. The form and direction of these tubes help define families and subfamilies in the Eriophyoidea, and this is discussed to some extent at the beginning of this section. Illustrations of these tubes and spermathecae are on figure 92.

As stated elsewhere, transfer of sperm from males to females is accomplished by spermatophores, deposited by males, and then visited by females (pls. 53, 54, 55). This process is further described in the section on life histories.

Eriophyoid Nymphs

The 2 nymphal stages in the species in this superfamily differ from adults not only in size (the second nymph reaches adult size before molting), but also by the amount of microtuberculation. Nymphs also lack external genitalia. Genital setae are always present on the venter of these nymphs, and lie on each side of the midventral line a few rings back from the coxae.

Second nymphs are very similar to adults, if not identical, in respect to setal direction and to microtuberculation.

First stage nymphs in the Eriophyidae (strict sense) and in the Rhyncaphytoptidae, show certain peculiar ring and setal differences from the adults, principally on the anterior abdominal dorsa. All first stage nymphs so far studied in the family Eriophyidae, and in the Rhyncaphytoptidae, have a peculiar dorsal discontinuity in the first 4 to 6 abdominal rings just behind the cephalothoracic shield (figs. 110, 117, 119). Heavily microtuberculate first nymphs have these tubercles spaced irregularly over this postshield area.

Judging from the lateral ends of these anterior discontinuous rings on examples of the first nymphs studied, the first dorsal shield tubercles in the Eriophyidae are always situated ahead of the rear shield margin on these nymphs and the dorsal setae that arise from these tubercles are nearly always directed dorsally or anteriorly, regardless of the directions these setae may take on the second nymph and on adults. The adjustment from first nymph to adult is not a major change for species that have adults with forward-directed dorsal setae. The rearrangement of these tubercles and setae during the molt from the first to the second stage, however, is considerable for species possessing second nymphs and adults with dorsal tubercles on rear shield margins, and with dorsal setae directed in some degree to the rear. Eriophyoid adults with dorsal tubercles located on the rear shield margin, and with setae directed from them to the rear, are only found in the Eriophyidae.

A survey of these dorsal tubercles and setae throughout the eriophyoid group indicates that the great majority of species have forward-directed setae that arise from tubercles set more-or-less ahead of the rear shield margin. For the genera and species that have adults with rear-directed dorsal setae, the first nymphs are evidently recapitulating phylogeny.

Some of the genera in the Eriophyidae which are characterized in part by dorsal tubercles located on the rear shield margin, and with dorsal setae directed to the rear, are *Eriophyes*, *Paraphytoptus*, *Acalitus*, *Anthocoptes*, *Aculops*, and *Tegolophus*. In these genera the change in dorsal tubercle location, and in dorsal seta direction, between the first and second instars, is a major alteration.

The first nymph of blackberry-infesting species of *Acalitus*, with dorsal tubercles situated ahead of the rear shield margin, and also with scattered microtubercles just behind the shield dorsally, has dorsal setae projecting to the rear, but the tubercles are anterior to the position these tubercles will take on the second nymphs and on the adults (fig. 117, *d*).

First stage eriophyoid nymphs are variable as to microtuberculation. These abdominal granules are absent from the rings on the Persian walnut erineum mite, *Eriophyes erineus* (Nal.), (fig. 117, *a*); they are present only on the rear rings on the holly bud mite, *Cecidophyopsis verilicis* (K.) (fig. 110, *a*); the first nymph of *Acalitus orthomera* (K.) (fig. 117, *d*, *e*) is entirely microtuberculate. First nymphs are entirely microtuberculate on the filbert big bud mite, *Phytocoptella avellanae* (Nal.) (fig. 102, *e*), and the pine needle sheath mite, *Trisetacus ehmanni* K. (fig. 105, a). The latter 2 species belong to the Nalepellidae.

As discussed elsewhere, these microtubercles seem to have a bearing on water exchange through the integument, perhaps in both directions in microenvironments. Adult eriophyoids show definite adaptations to their particular environments by reduced numbers or absence of microtuberculation. As immature stages are of short duration, the problem of nymphal body water equilibrium with surroundings may not be nearly as critical as it is for adults.

The two-spotted mite, *Tetranychus* (*T.*) *urticae* Koch, has actively feeding summer-type females with epidermal striae thickly set with lobes; the lobes are comparable to eriophyoid microtubercles, and evidently have the same function. Hibernating adult females of these two-spotted mites, which move to places where they will be unable to feed for a long period, have striae completely without lobes, which appear to be for water conservation. First and second nymphs of this two-spotted mite, however, also have striae lacking lobes, although they are actively feeding. This may be an indication that these red spider nymphs are but ephemeral stages that do not have the water conservation problems of the adults.

First stage eriophyoid nymphs have larger rostra in comparison to body size than do adults. The first stage nymph of *Trisetacus ehmanni* K. (Nalepellinae) has a rostrum measuring 22 μ long and the body measures 95 μ long. This is a ratio of about 1 to 4.5 (fig. 105). Second stage nymphs of this species have nearly adult dimensions. This second nymph has a 25 μ long rostrum, and the adult rostrum is 27 μ long. The adult *Trisetacus* body is about 150 μ long, and thus a ratio of rostral to body length is 1 to 9.

Comparison of first stage nymphal rostra to lengths of adult rostra in the Eriophyidae show that the first nymphal rostra vary from 4/5 to 5/6 of adult rostral length. But the length of the first stage nymphal body is less than 1/2 the adult body length.

A first stage rhyncaphytoptid nymph has a rostrum 40 μ long; the second stage has a rostrum 48 μ long; the adult has rostra varying from 55 to 69 μ long. But the adult body is again 2 or 3 times the length of the first stage nymphal body.

Some eriophyoid species can be most easily separated from each other on the basis of differences in microtuberculation on first stage nymphs. Wilson (1965), has shown that *Phytoptus pseudoinsidiosus* (Wilson) is one of these. This mite is implicated in pear leaf blistering, and either occurs alone in these blisters, or in company with *Phytoptus pyri* Pagenstecher. *Phytoptus insidiosus* (Wilson and Keifer) is the peach mosaic vector mite; it has very similar shield lines to those on *pseudoinsidiosus* adults. Wilson was able best to define the differences between *insidiosus* and *pseudoinsidiosus* by showing that the first nymph of *insidiosus* has much sparser microtuberculation on its abdominal rings than does the first nymph of *pseudoinsidiosus*.

RELATIONSHIPS OF ERIOPHYOID MITES TO THEIR HOST PLANTS

A knowledge of how these tiny creatures live on their hosts and what they do to these plants is necessary to comprehend them, whether one approaches eriophyidology from the economic or from the noneconomic viewpoint. Since every eriophyoid species has, in its way, a peculiar association with its host or hosts, the bionomics of these mites will be never-ending research projects. While this account is necessarily superficial, it may help those who are interested in discerning some of the general patterns displayed by eriophyoid life. More precise information on economic species is in chapter 13.

Plants that harbor these mites show the entire range from complete toleration, through various types of tissue modifications, to actual host killing. This latter result is unnatural and will be discussed later (figs. 98–101).

Eriophyoids feed only on succulent plant tissue, usually green areas. Galls are special sites that the mites develop for themselves that not only provide protection for the brood, but also supply food in extrasucculent cells or papillae. Some of these mites appear to stay in contact with soft tissue throughout the year, living at the bases of outer bud scales, under petiole bases, or at leaf bases on evergreen hosts. Deciduous hosts have forced some kinds of mites to undergo alteration of form (deuterogyny). The resulting specialized females are able to aestivate or hibernate in dry bark crevices. All kinds of eriophyoids suffer considerable mortality through less favorable seasons, but always some mites find survival niches and give rise to populations the next year.

Specialization on Host Preference and Plant Response

A general characteristic of most eriophyoids, especially those on broad-leaved plants, is their narrow host ranges. This may be in part because of the ancient age of the group, or to intimacy of gall formation. Actually few species have been tested in this respect, so conclusions can only be based at present on the interpretation of mite structure.

Species that serve as examples of single host mites are: (1) the petiole gall mite of American black walnut in eastern North America, *Eriophyes caulis* Cook; (2) the camphor leaf gland mite, *Gammaphytoptus camphorae* K.

Species that live on plants belonging to one botanical genus are: (1) the alder bead gall mite, *Phytoptus laevis* Nal., holarctic in range and known from 4 to 5 alder species; (2) the alder erineum mite, *Acalitus brevitarsus* (Fockeu), with about the same range and hosts as *P. laevis*; (3) a plum finger gall mite, *Phytoptus emarginatae* (K.), which occurs in western North America on 3 plum species. (If *P. emarginatae* proves to be synonymous with the European *P. padi* Nal. then the range and hosts are extended.)

Some grass mites show less host specialization, and are able to range across several monocotyledonous genera. One is the grass rust mite, *Abacarus hystrix* (Nal.), which lives on hosts belonging to *Bromus, Poa, Elymus, Lolium, Festuca,* and so on. The wheat curl mite, *Eriophyes tulipae* K., is able to breed on both graminaceous and liliaceous hosts. (This latter apparent jumping of family lines by *tulipae* is disputed by some.)

A most unusual example of an eriophyid that can live on several species of broad-leaved plants in the Solanaceae is the tomato russet mite, *Aculops lycopersici* (Massee). This mite feeds on various species in the genus *Solanum*, breeds actively on petunia, and kills most tomato varieties. This latter association is disastrous to both host and mite. It removes the indispensable perennial basis the mite must have for survival.

Many eriophyoids cause no detectable alteration of, or damage to, their hosts. To find such mites it is necessary to hunt randomly in likely places on plants. But as these mites are often numerous, such searching is usually rewarding.

Presumably most benign eriophyoids either lack salivary growth regulators, or the host plant is resistant. An example of host resistance occurred on a grape trellis, where two grape varieties were intertwined. Both varieties had grape bud and erineum mite, *Colomerus vitus* (Pagenstecher), in their buds, but only one grape variety had leaf erineum.

A Suspected Case of Mechanical Damage by Eriophyids

While ordinary eriophyoids cause no mechanical damage to plant tissue by their feeding, Aberoptine mites, with spatulate or shovel-shaped appendages, which live on mango leaves, have been suspected of leaf mining. The mite species in this group are: *Aberoptus samoae* K. with spatulae on the foretibiae; *Cisaberoptus kenyae* K. with spatulae on the rostral termen (fig. 93, *d*). Dr. El Fatih Osman Hassan, of the University of Khartoum, has investigated the action of *C. kenyae* on mango leaves in the Sudan. He has found that these mites cause a spreading silvery layer to develop mainly on the upper surface of the leaves, but that the actual leaf tissue is not part of this layer. The nature of this silvery layer, or coating, and its origin, are now under investigation. The mites live under this coating.

Growth Regulators

Growth regulators are salivary chemicals possessed by some eriophyoids that when injected into plants either discolor leaves or change growth patterns of the affected cells. Composition of these growth regulators in mites is unknown, but

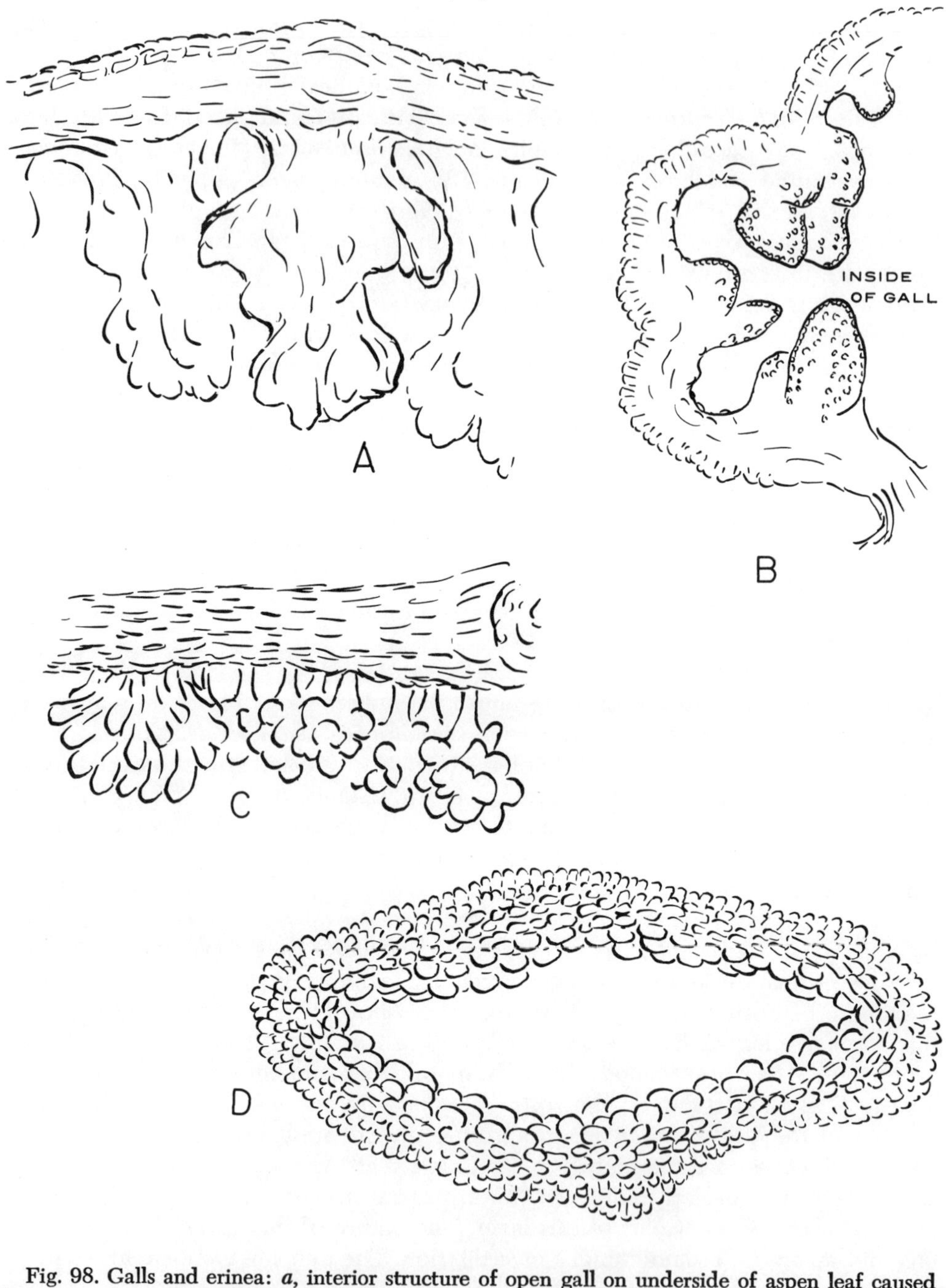

Fig. 98. Galls and erinea: *a*, interior structure of open gall on underside of aspen leaf caused by: *a, Phyllocoptes didelphis* K.; *b*, interior structure of bladder gall on silver maple caused by *Vasates quadripedes* Shimer; *c*, compound capitate erineal growths on a *Dicranopteris* fern frond caused by an *Eriophyes*; *d*, cross section of plum finger gall caused by *Phytoptus padi* Nal.

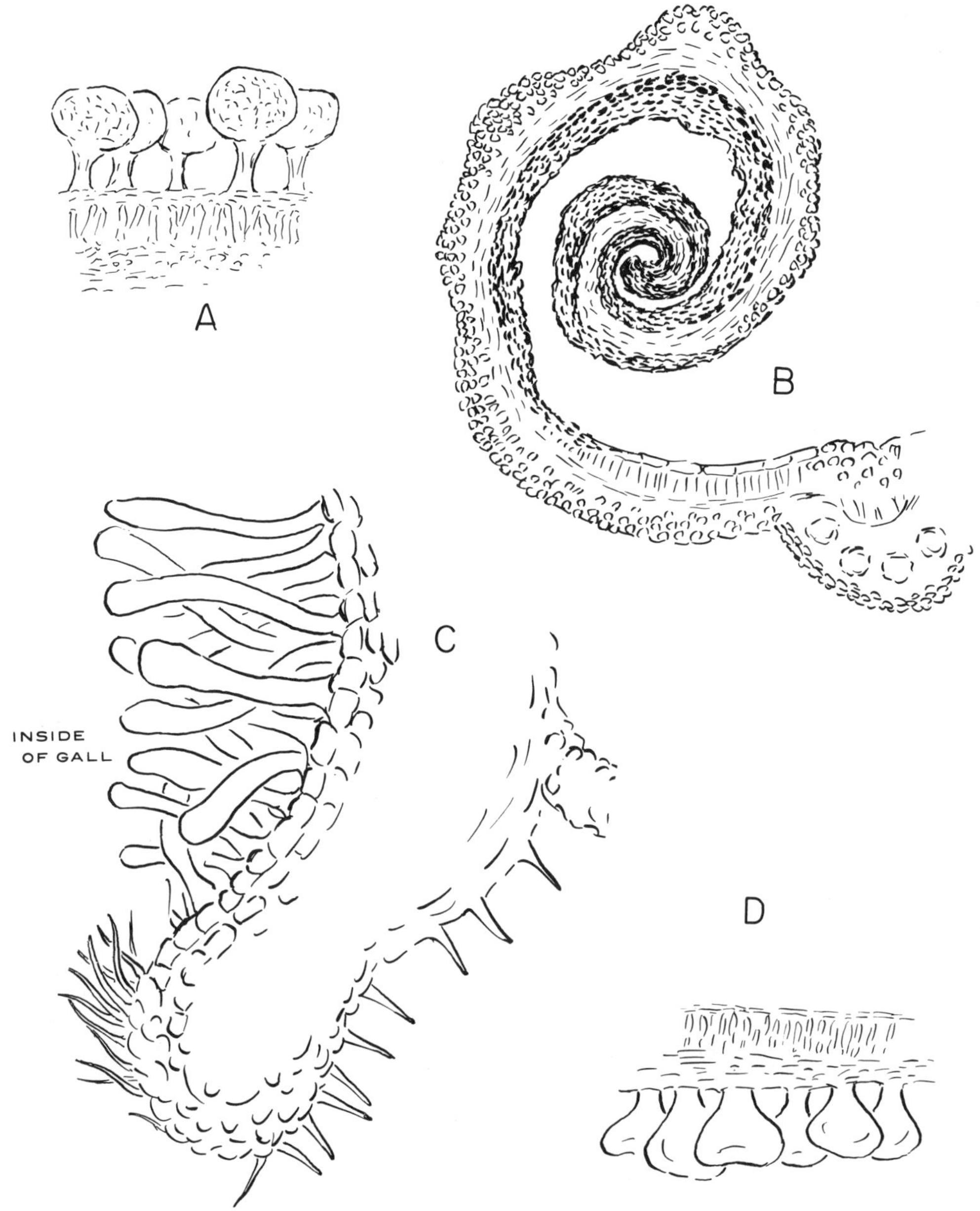

Fig. 99. Galls and erinea; *a*, capitate erineal papillae in crimson erinea on upper side of sugar maple leaves caused by ***Eriophyes elongatus*** Hodgkiss; *b*, cross section of leaf roll on pistacio leaves caused by ***Eriophyes stefanii*** Nal.; *c*, part of leaf gall on ***Prunus spinosa*** L., with interior papillae on left side, caused by ***Eriophyes similis*** (Nal.); *d*, capitate erineal papillae in green erinea on underside of sugar maple leaf, causd by ***Eriophyes modestus*** Hodgkiss.

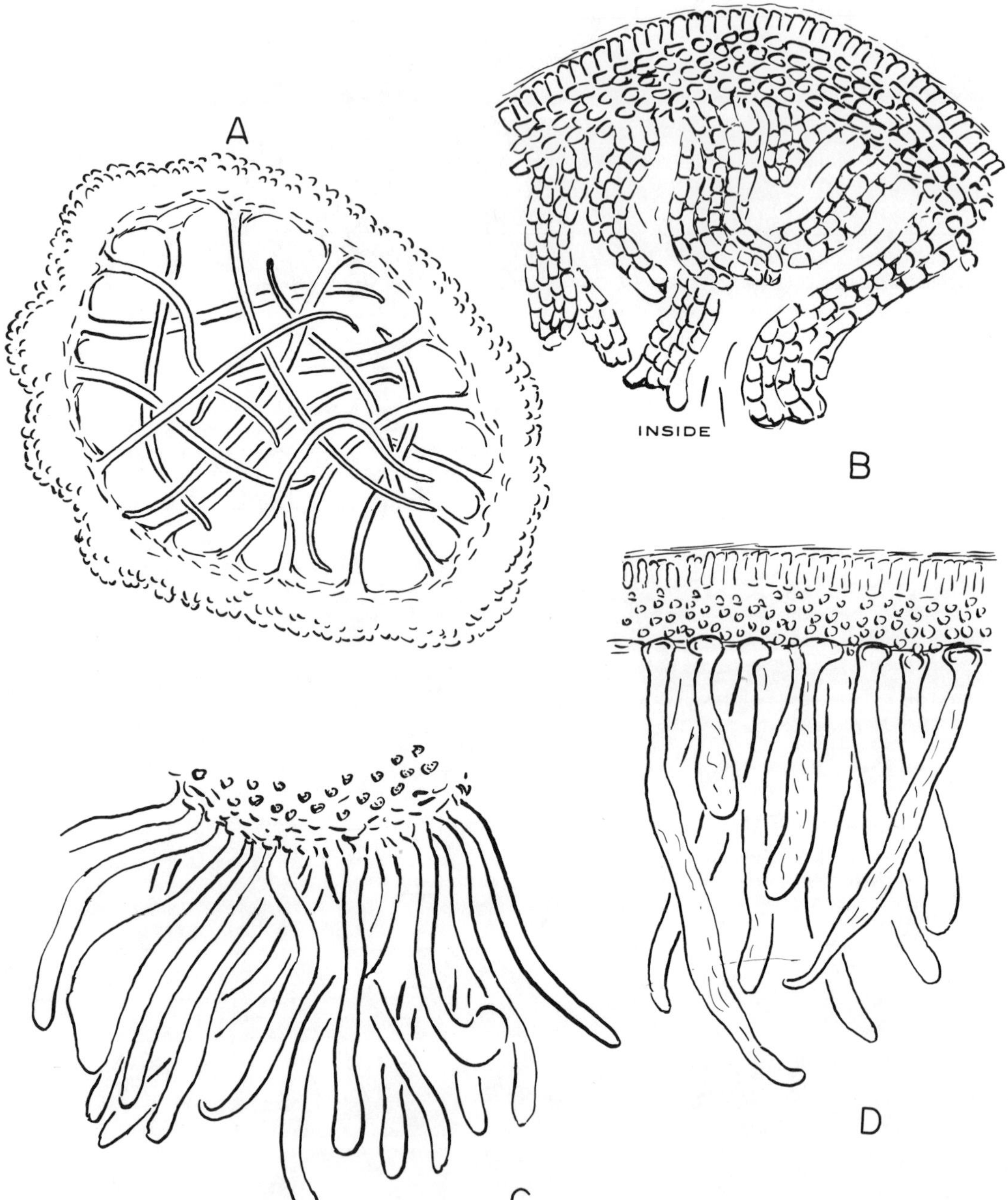

Fig. 100. Galls and erinea: *a*, cross section of nail gall on linden leaf caused by ***Phytoptus tiliae*** Pgst.; *b*, interior structures in pouch gall on leaf of ***Juglans hindsii*** Jepson, and caused by ***Eriophyes brachytarsus*** K.; *c*, erineal papillae in petiole erineum gall caused by ***Eriophyes caulis*** Cook, on ***Juglans nigra*** L.; *d*, erineal papillae on underside of Sorbus leaves, caused by ***Phyllocoptes calisorbi*** K.

these salivary chemicals are diverse. As already mentioned, one of the more simple kinds of such regulators produces kernel red streak in corn. These red-streak growth regulating agents behave as if water soluble.

Another type of injury is russeting on leaves and fruit by rust mites. Plant tissue is not deformed, but badly russeted leaves may shrivel somewhat. Russeting proceeds from heavy feeding, usually on leaf undersurfaces. Whether or not possible entrance of air is involved is unknown. Probably digestive enzymes are the main russeting factor. Variability of russet damage from species to species of mite, suggests that some kinds, notably members of *Aculus*, have more damaging digestive factors than mites in many genera. Mites of the genus *Acarelliptus* settle on undersides of oak leaves after the manner of scale crawlers, but make no appreciable leaf spots or cause other injury.

Leaf galls made by these growth regulators remain localized, as in the case of discrete projections through leaves. But erineal producing factors usually spread laterally on leaf surfaces. These two considerations show that these chemicals vary in structure. The student can find repeated examples showing that different mite species attacking one host plant make strikingly different alterations, respectively. The undersurface erineum leaf pocket maker on California live oak is *Eriophyes mackiei* K. A closely related species, *Eriophyes paramackiei* K., causes brooming and bud clustering on the same host.

Pomaceous shrubs and trees suffer from leaf blistering by mites belonging to the *Phytoptus pyri* Pgst. complex. Another eriophyid complex, on these same hosts, collectively referable to *Phyllocoptes goniothorax* (Nal.), make undersurface papillar leaf erineum patches.

A striking illustration indicating that each eriophyid species has its own specific type of salivary growth regulator occurs on eastern sugar maple. On this host upper surface capitate papillar crimson erineum on the leaves is the work of *Eriophyes elongatus* Hodgkiss. Green underside leaf erineum on sugar maple is made by *Eriophyes modestus* Hodgkiss. *Vasates aceris-crumena* (Riley) causes upper surface finger galls on sugar maple leaves.

The opposite question—how do different plant species respond to the salivary injections of one eriophyid species?—has fewer answers as yet. A mite attacking a variety of umbelliferous plants, *Eriophyes peucedani* (Canestrini) (fig. 113, *a*), causes deformation of leaves and umbels, and yellows these plants. *Eriophyes tulipae* K. (fig. 115, *h*), which is becoming one of the most investigated species, as well as carrying viruses, produces kernel red streak on corn, curls wheat leaves, dries onion leaves, and, as a storage pest causes premature garlic clove drying.

The tomato russet mite, *Aculops lycopersici* (Massee), lives on several solanaceous plants referred to several different genera within that family. It kills tomato, thrives on potato and on petunia, but can only develop sparse populations on nightshade.

Growth modifications are initiated only on embryonic plant tissue. Russeting of leaves, which is not a growth reaction, occurs on mature leaves. There is no evidence that galls, or other mite induced growth abnormalities, develop on mature plant tissue. One example of this concerns pear leaf blisters. Such blisters appear in the first leaves coming out of the bud, since the overwintering mites fed on

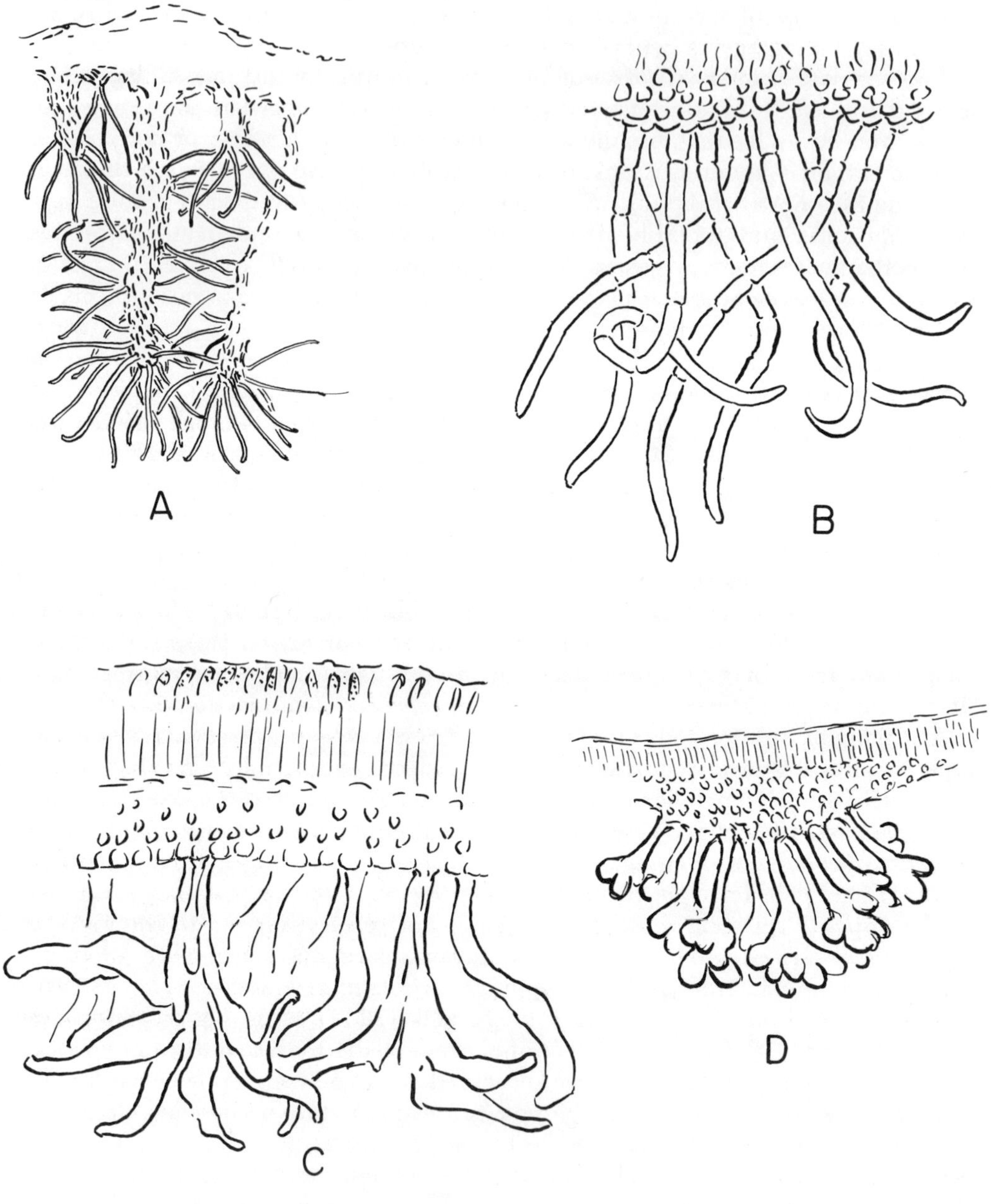

Fig. 101. Erinea: *a*, structure of leaf erineum on Persian walnut caused by *Eriophyes erineus* (Can.), showing partitions; *b*, erineal papillae with cellular partitions on litchi leaves, caused by *Eriophyes litchii* K.; *c*, treelike papillae in erineum on *Quercus ilex* L., caused by *Eriophyes ilicis* Nal.; *d*, compound-capitate erineal tuft in erineum patch on underside of alder leaf, caused by *Acalitus brevitarsus* (Fockeu).

these leaves when they were still embryonic. All of the mites remain in these blisters to start colonies, so the next leaves to emerge from buds lack blisters. Later, when new mites have become numerous enough to migrate, they again enter the terminal bud on the shoot. If new leaves are still developing in this bud, these leaves are then blistered. The result is that the first leaves have blisters, the next leaves lack blisters, and the following leaves have blisters.

This same sort of condition occurs on California black walnut. Early leaves carry pouch galls made by *Eriophyes brachytarsus* K. These galled leaves are followed by leaves without galls, and if the shoot continues growing long enough the terminal leaves will be found to bear a few pouch galls.

Preferred Locations on Plants

As the reader can now perceive, every eriophyoid species has it own living site on the host to which it is adapted. Eriophyids that live under cover, even though they make no galls, require surfaces both above and below them. The spaces between these approximate surfaces where bud mites live can be called *mite space.* Often the mites cause tissue shrinkage in these retreats and thus provide their own mite space. Students rearing bud mites must provide such space for the mites successfully to found colonies.

Some rust mites live on the underside of leaves, some on the upper side only. Others, such as the peach silver mite, occupy both leaf surfaces. One vagrant, *Anthocoptes pickeringiae* K., lives both on leaves and in hairs on green stems.

Embryonic plant parts, as they emerge from buds, often determine the position of growth abnormalities or galls by the way they are folded (Nalepa, 1928). The undersides of immature leaves are most available for mite feeding as buds loosen in the spring, with the upper surfaces appressed. This explains why most gall openings and erinea are on the undersurface. That some erinea are upper surface developments can only be explained by assuming either that some leaves are not as tightly folded as others, or that some mites are proficient at penetrating folds.

Boczek (1961) has published copies of figures of preferred sites for leaf galls. He credits this work to Schevtshenko. Each gall-making species is supposed to select the leaf area it prefers. Heavy infestations would obscure these preferences.

An instructive example of a preferred leaf site for the placement of erinea occurs on Caribbean satin leaf, *Chrysophyllum oliviforme* L. The mite making the erineum patches, *Eriophyes chrysophylli* Cook (1906), starts them on the upper surface, and at the time of the initial feeding is able to penetrate only part way across the unfolding leaf. When the leaf matures these erineum patches have become deeply invaginated out of the undersurface, into gall-like upper surface projections. These protuberances are fairly uniformly subparallel to the outer leaf edges, but about ⅓ of the way toward the midvein.

Eriophyoid mites have exceedingly thin and short mouth stylets. The smallest of such stylets, occur in mites with short form oral stylets in the Nalepellidae and the Eriophyidae. These are principally bud mites, gall mites, rust mites, and leaf vagrants. A few rust mites in these families, especially species on conifers, have larger rostra and longer chelicerae; they still have short form oral stylets and are in the minority in these groups.

Short chelicerae can only penetrate plant tissue 25 μ to 50 μ. Longer chelicerae in the short form oral stylet groups can go as deep as 65 μ, such as species in the genus *Nalepella.*

Rhyncaphytoptids with long form oral stylets have rostra ranging from 40 μ to 70 μ long. These rhyncaphytoptids consistently do little damage to plants. This indicates that stylet penetration is not significant from the standpoint of plant injury.

No information is available to affirm that these tiny mites actually suck entire cell contents out and thus kill the cell. Air spaces do not appear to develop in eriophyoid feeding sites, unless russeting has something to do with intrusion of air.

Indications are that these mites not only do not kill the tissue they feed on by inserting their mouthparts, but the saliva they inject frequently promotes succulence and even prolongs the life of some plant appendages. With large cottonwood male catkin galls, caused by *Eriophyes neoessigi* K., the catkin evidently first succeeds in discharging pollen, but then becomes a false stem, replete with leaflike growths that have curled edges for the mites. This whole structure may become a hanging mass 20 cm or more long; it hangs on the tree until early summer, long after normal catkins have dried and dropped.

The interior of galls remains succulent with turgid nurse cells, papillae, and nutritive lobes that are constantly fed upon by the growing mite colony. In other words, eriophyoid stylets are not of significance in plant injury; only the salivary chemicals cause alterations in plant tissue.

Another facet of this lack of injury is shown by virus transmissions. These mites are efficient virus carriers. This requires that cells in penetrated tissue remain alive to receive the introduced pathogen.

TYPES OF PLANT ABNORMALITIES CAUSED BY ERIOPHYOIDS

Bud mites that cause no galls or are in such a position that they can make no deformities on the host, as under fruit bottoms, frequently injure or brown tissue on which they are making colonies. An example of such a mite making brown spots under fruit buttons is the citrus bud mite, *Eriophyes sheldoni* Ewing. Redberry disease of blackberry is delayed drupelet ripening caused by the feeding of *Acalitus essigi* (Hassan) (fig. 122, *a, b*) colonies on the berries. Peach silver mite, *Aculus cornutus* (Banks) (fig. 131, *a*) makes yellow spots on spring leaves, and also causes upper longitudinal leaf curling on trees that lack basal leaf glands.

Leaf Edgerolling

A fairly common induced deformity is edgerolling. Mite colonies then develop inside this tight roll. One example is edgerolling on pomegranate, caused by *Eriophyes granati* (Canestrini). Heavy infestations of this mite may stunt all leaves along a shoot. An edgeroller that makes tight upper surface leaf rolls on pistacio is *Eriophyes stefanii* Nal. (fig. 114, *a-c*). Another edegroller is *Eriophyes caryae* K. (fig. 116, *q-s*). It rolls the leaf edge over the upper surface.

Leaf furrowing (Bremi's *Legnon*, see Nalepa's 1929 host list) occurs on European hornbeam. The furrows start from the upper surface, follow lateral leaf veins, and have the sides tightly appressed with the outfolded part sinuate. The mite causing this is *Eriophyes macrotrichus* (Nal.). In California a similar leaf furrowing occurs on barberry and results from *Eriophyes caliberberis* (K.). Leaf crumpling on globe mallow is the result of work by *Acalitus sphaeralceae* K. This occurs in southwestern Arizona and adjacent areas.

Effects of Bud Damage

The simplest type of bud damage is the actual killing of the bud as it begins spring growth. *Phytoptus pyri* Pagenstecher, the pear leaf blister mite, causes this when in heavy infestation.

A second type of bud damage is known as "Big Bud." In typical big bud damage the interior embryonic parts of the bud swell, become succulent from mite feeding, and after the brood leaves the bud dies. Such big buds develop on filbert and are engendered by the filbert big bud mite, *Phytocoptella avellanae* (Nal.) (fig. 102).

Big bud development on Douglas fir is caused by *Trisetacus pseudotsugae* K.; salivary growth inhibitors do not cause permanent genetic alteration within host cells. Most of the big buds die owing to prolonged mite feeding. Such cells in big buds as do manage to survive after the mites leave then resume normal twig growth. This is contrasted with growth directors that cause leaf gall formation. Cells that make galls remain alive, but they proceed to grow into a gall whether or not mites are present. Note: *T. pseudotsugae* causes terminal bud proliferation on potted seedling Douglas fir, rather than big buds.

Peach mosaic vector mite, *Phytoptus insidiosus* (Wilson and Keifer), influences development of lateral buds on peach twigs, causes them to enlarge, and founds colonies that may persist for several years at the site. The bud may be able to put out a rosette of small leaves, but it remains somewhat enlarged and produces no twig.

Woody bud galls occur on plum and are caused by the plum bud gall mite *Acalitus phloeocoptes* (Nal.) (fig. 124). In this instance several galls may form around a bud, stunting growth of the twig.

On various poplars, including cottonwoods, large woody bud galls may appear on most of the buds along a twig. The mite making this is *Eriophyes parapopuli* K. These galls may grow to an inch or more in diameter over a period of several years; the mites continually move to fresh folds in the gall as the new folds develop, stopping cottonwood growth. Such injury occurs in places where these poplars are used as a principal shelter tree.

Tissue Breakdown—Blisters

Most eriophyoids do not penetrate into plant tissue, except for leaf blisters and a few other cases where they actually invade plant tissue. The most common type of this action is engendered by the pear leaf blister mite, *Phytoptus pyri* Pagenstecher (fig. 111, *a*), and other members of this group complex that live on leaves of pomaceous trees and shrubs. Mite feeding on the undersurface of embryonic folded leaves in loosening buds causes vacancies to develop in the leaf tissue, and a minute hole on the epidermal undersurface necessary for mites to escape. In heavy infestations these blisters coalesce damaging most of the leaf. At first the blisters are green with few mites present, but as the mite population increases there is a continual movement of these eriophyids from blister to blister and even into buds until early summer. During the summer the blisters dry, leaving dead leaf areas at which time the blisters are largely abandoned. Some mites can take advantage of early summer leaf developmnt to make new blisters, if there is a late growth flush. Eventually all surviving mites will be found in the terminal buds. These eriophyids also blister fruit, but growth soon obliterates these blisters, leaving only the scar. Mites on drying blistered leaves will line the edges of the leaves, rear up on their hind lobes, and paw the air, looking for an escape means.

Members of the pomaceous genus *Sorbus*, known by the common names of mountain ash or rowan, are Holarctic in distribution, but are heavily attacked by blister mites of *P. pyri* complex. The life history of these rowan mites may not be precisely the same as the mites on pear and apple (this question is discussed in the section on life histories). The *Sorbus* mite bears the name *sorbi* Canestrini.

Various Eurasian elms have a complex of leaf blister mites belonging to the threadlike *Eriophyes filiformis* (Nal.) group. These mites, which have curious 3-rayed featherclaws, often make baggy blisters, with undersurface holes. Boxwood leaf blisters, covering the entire leaf undersurface, are frequently intercepted in quarantine from the British Isles. The mite involved is *Phytoptus parabuxi* (K.). All the leaves seen were dry and torn, obscuring the exit hole.

A type of tissue breakdown in juniper seeds is the work of *Trisetacus quadrisetus* (Thomas). These mites penetrate the very young berry, change each seed into a brood cavity, but do not prevent the berry from enlarging. The escape hole is terminal with either the seed end protruding or the shriveled end of the seed projecting out. The mite has many juniper hosts in its Holarctic range. In California the western or Sierra juniper and the California juniper are hosts. Infestations are frequently so heavy that nearly every berry on a bush contains a mite colony with destroyed seeds in it.

A gall-like growth, pine twig knot, evidently stands between tissue breakdown and the growth regulator type diversion of cell growth. The mite causing this is *Trisetacus pini* (Nal.) and the knots are typically on scots pine. The knot is said to last for more than a year and produce sucessive mite broods.

Brooming

Brooming appears as twig elongation, or bud proliferation, accompanied by either absence of leaves, or stunted leaves, and often internode shortening. Twigs and flower clusters may show this effect. The salivary factor that induces brooming seems to travel in conductive tissue, being more of an inhibitor than one that diverts cell growth. At this writing it is not possible to state whether or not brooming is a temporary condition dependent on mite feeding, or if stunted twigs would continue to grow poorly even after mite removal.

Various eucalyptus species in Australia have twig brooming caused by members of the eriophyid genus *Acadicrus* (fig. 129, *a*). *Pistacio terebinthus* L., in its uncultivated condition in Turkey, suffers from flower stalk brooming. The mite responsible for this is *Eriophyes pistaciae* Nal. (fig. 114, *d, e*). The brooms on this pistacio are prominent reddish proliferations in staminate flower heads.

In California, coast live oak infested with broods of *Eriophyes paramackiei* K. develop short shoots that are covered with buds, but have no leaves.

Structure and Development of Mite Galls

Galls per se

Mite galls develop from epidermal cells that are diverted by injected growth regulators. As already seen, each mite has particular regulators that cause galls to grow which are of specific benefit to that mite. After the induced change has altered the behavior of the affected cell or cells, the mite does not have to remain on the site to insure continuation of gall growth. It is even possible to demonstrate that early season galls start development before they are inhabited by mites. Later, when the brood has increased, the mites move into the waiting galls.

A series of names describe and define mite galls. Examples are: pouch or purse galls, bladder galls, nail galls, finger galls, bead galls. Many galls become hairy on the outside; the hairs are rather similar to natural leaf hairs but denser. In some cases galls intergrade with erinea, especially invaginated erineal patches. All eriophyid galls retain the escape holes, usually on the leaf underside first available to feeding females. These escape holes are a necessary feature of all eriophyid galls because the mites are incapable of forcing an exit.

Mite-induced galls, especially on leaves, tend to be localized, that is, each discrete gall has definite limits. The common type consists of a projection that presumably grows through the leaf, with the underside hole, but the principal development is out of the upper surface.

With the exception of roots, eriphyoid galls occur on all soft plant parts, usually on green tissue that received the growth director when embryonic. While most galls are on leaves, there are also flower galls, green stem galls, petiole galls, and the like. Some galls are tumorlike; some cause semireversion of flower heads back to false leaf growth.

Galls provide interior turgid cells or papillae on which the brood feeds. The size and shape of these nurse organs suggests that, in at least some of the papillae without definite cell walls have more than one nucleus floating inside. In some galls, such as bead galls on North American alder leaves engendered by *Phytoptus laevis* Nal. (fig. 98, *d*), the walls have merely enlarged surface cells. Bead galls on alder in Scotland, made by *P. laevis*, do have papillae inside. In elongate bead galls on blackthorn formed by *Phytoptus similis* Nal. (fig. 99, *c*), the inner walls have turgid clavate papillae. California black walnut pouch galls have complicated internal lobes that are united papillae and show numerous cell-like divisions. These walnut galls are the work of *Eriophyes brachytarsus* K. (fig. 100, *b*). (For further reading on mite galls, and other deformities see Mani, 1964, and Felt, 1940. These books contain long discussions and excellent pictures.)

Erinea

Mite-induced growth of surface hairs or erinea is here called *erineum* (singular), *erinea* (plural). This is the characteristic result of the feeding of many mite species, all in the family Eriophyidae. Unlike the galls with escape holes, erinea are not closed growths, but enable mites to find shelter within the hair masses or "forests." Erinea vary from very localized patches to those that cover much of the leaf or petiole surface. This suggests that the growth regulators engendering erineal development differ from gall regulators by being able in most instances to become translocated laterally. Examples of very localized erineal patches occur in the American tropics on *Eupatorium* spp. Here species of the genus *Acalitus* produce separate erineal tufts on leaf undersides (Keifer, 1969).[4]

Eriophyids seem to intergrade into the development of erinea, from those that live in natural plant hairs to such highly organized growths as the crimson maple leaf erineum. As already mentioned, members of the genus *Paraphytoptus* live half in and half out of natural plant hair masses. In some instances the precise relation to surface hair is uncertain. In California, some twigs of arroyo willow bear leaves with more undersurface hairs than normal. *Rhyncaphytoptus acilius* K. favors these leaves. Rhyncaphytoptids are not known to engender any sort of gall, so the supposition is that this vagrant is merely taking advantage of a condition it did not make.

A plum leaf mite, *Phyllocoptes abaenus* K., favors basal hairy areas along the midrib toward leaf bases of underleaf surfaces. The part played by *abaenus* in developing these hair masses is unknown. But a complex of close relatives to *P. abaenus* which collectively go under the name of *P. goniothorax* (Nal.) do make undersurface papillar erinea on leaves of pomaceous trees and shrubs. The California representative is *P. calisorbi* K., on mountain ash.

Felty erineum pads on undersides of Persian walnut leaves are the work of *Eriophyes erineus* (Nal.) (figs. 101, *a*; 116, *k-m*). These pads have fairly thin hairs, but are thickened by vertical partitions.

An instructive range of erineal form occurs on undersides of leaves of various oaks. White oak in eastern North America has erineal pockets made up of fairly thin papillae that show no divisions. These erineum pocket developments are stimulated by *Eriophyes triplacis* (K.). On coast liveoak in California, *Eriophyes mackiei* K. causes erineal pockets with papillae showing some partitions.

An oak erineum found on the island of Cyprus consists of subcapitate papillae, and on holly oak in Portugal, *Eriophyes ilicis* (Can.), produces erineal pockets with the papillae partially fused basally, but branching treelike apically.

Undersurface grape erineum, made by *Colomerus vitis* (Pagenstecher), the grape erineum mite, consists of papillae. Litchi undersurface erineum, which makes large brown patches that twist leaves, consists of papillar strands that have some cell-like partitions, and are the result of infestations of *Eriophyes litchii* K.

Erineum sometimes develops on petioles of compound leaves. One example is on *Acacia leucophylla* Willdenow and is instigated by *Eriophyes acaciae* Nal. In North America the best-known petiole erineum has the peculiarity of growing on a swelling. It consists of a papillar mass covering the swelling, and is caused by *Eriophyes caulis* Cook on black walnut.

Capitate erineal papillae are often the most spectacular and colorful of all such growths. There are two general types of these papillae, capitate, and compound-capitate. Upper surface crimson erineum on sugar maple in the eastern United States caused by *Eriophyes elongatus* Hodgkiss (Hodgkiss, 1913) has simple capitate papillae. In western North America a very similar crimson or purple growth occurs on leaf edges and on upper surfaces on western mountain maple and is engendered by *Eriophyes calaceris* (K.).

Mites in the genus *Acalitus* offer some of the better examples of compound-papillar erinea. Yellow beech erineum in eastern North America is caused by *Acalitus fagerinea* (K.), and is of this compound type. Erineum on the undersurface of alder leaves in the northern hemisphere offers a series of erineal types. Erinea containing mites that bear the name *Acalitus brevitarsus* (Fockeu) vary from white to orange and from simple to compound papillae. In North America the only kind seen is the simple entwined type. In Europe these orange erinea are of capitate papillae. Examples received from Turkey, of *brevitarsus* alder leaf erinea, are composed of compound-capitate papillae.

ADDITIONAL OBSERVATIONS

Waste Disposal

Scrutiny of eriophyoid galls, erinea, sequestered depressions, and of other places occupied by eriophyoid colonies, never discloses evidence of excrement masses, or of other types of fecal matter. W. P. Styer points out that eriophyoids, with their very small mouthparts, can only feed near the surface tissue in their host plant. For this reason these mites do not have a carbohydrate disposal problem, such as certain homopterous insects have that feed in phloem.

Whitmoyer et al. (1972) state: "As in many arthropods, the excretory mechanisms are not well developed in the grass mite, *Eriophyes tulipae*, the metabolic waste products appearing to remain within the hemolymph. The hemolymph contains many concentrically layered ovoid granules, comparable with spherical granules found in insects and generally associated with waste products from the hemolymph." They also add the suggestion that these granules may result from a low-grade pathogenic condition.

A possible further step in elimination of otherwise unexcreted material comes from an article on demodecids, by Nutting (1964). Demodecids are wormlike follicle mites, living in mammalian skin. These demodecids are supposed to void excreta in part as an ingredient of egg coverings, or shells. If eriophyoids remove excreta via eggshells, perhaps the process could include spermatophores.

The oral stylet in the eriophyoid rostrum is exceedingly thin and small. It could not suck up anything but liquified plant tissue that had been predigested by salivary injections. This contrasts with tetranychid food, which consists in part of torn out fragments. Such restriction of eriophyoid food to liquids must have considerable bearing on prevention of fouling of breeding sites.

Annual versus Perennial Hosts for Eriophyoids

Eriophyoids depend on forces beyond their control to travel among their host plants. Such travel is very uncertain and evidently only a minor percentage ever succeed in reaching new members of their specific host. For this reason the overwhelming number of eriophyoid species live on perennial hosts so that failure to migrate successfully any one year can be overcome another year.

A few eriophyoids, however, can manage to breed on annual host plants, especially if the annual is closely related to a perennial plant that also serves as a regular host. One such mite is the wheat curl mite, *Eriophyes tulipae* K., which lives on perennial grasses along fence rows, but attacks wheat that is grown on an annual basis. This mite moves actively around wheat fields in summer, greatly aided by the solid expanses of wheat as it is commercially grown.

The tomato russet mite, *Aculops lycopersici* (Massee), kills tomato plants that are grown annually, but this mite overwinters on such perennial hosts as petunia and nightshade.

One California eriophyoid is only known as living on an annual host. How it succeeds in doing this remains to be explained. The mite is *Eriophyes bowlesiae* (Wilson). Its host is *Bowlesia incana* Ruis Lopec and Pavon, an umbellifer. The only possible explanation for this host relationship is that the mite also lives on a perennial unmbellifer in the same general area where *Bowlesia* grows, and is able to transfer when the annual host plant is available.

BIOLOGY AND HABITS

Types of Life Histories

Most eriophyoids, as far as is known, develop from the egg, through two nymphal instars, to the adult. This direct type of life history pertains apparently to bud mites that live under bud scales, to mites that live on evergreens, and to eriophyoids that live in warm or tropical areas. Yet a minority of species inhabiting deciduous hosts has acquired a revealing form of alternation of generations. This alternation enables the species to have a primary female that resembles the male, and is capable of rapid reproduction during favorable times of the year. The species can then change, upon approach of change of food, and often weather, to the secondary female that has no male counterpart, and the function of which is to carry the species through unfavorable periods. The presence of two different structural types of females within a species is called *deuterogyny*. The ramifications of deuterogyny are treated in a section below. As far as known no eriophyoid has alternation of hosts.

During the past thirty years several important papers have given new data on life histories and environmental responses of these mites. Some of the contributors are as follows: Putman (1939) reported on experiments on the plum rust mite, *Aculus fockeui* (Nal.), showing that males hatch from unfertilized eggs, females hatch from fertilized eggs, and certain females are specialized for hibernation. Keifer (1942) described rearing experiments on the California buckeye rust mite, *Tegonotus aesculifoliae* (K.), which disclosed that the two structurally different females, closely associated on buckeye leaves, are the primary and secondary females of the same species. Schevtchenko (1957) reported deuterogyny in the alder bead gall mite, *Phytoptus laevis* Nal., thus adding gall mites to known deuterogynous species. Davis (1964) described the life history of *Rhynacus breitlowi* Davis, and its temperature requirements. Hall (1967 *a*) reviewed simple and deuterogynous life cycles of various eriophyoids, suggesting some may be capable of ovovivipary. Schevtchenko had also suggested ovovivipary among these mites. Oldfield (1969) disclosed that the chokecherry finger gall mite, *Phytoptus emarginatae* (K.), in southern California mountains, possessed only the secondary female, or deutogyne, having discarded the primary female, or protogyne. Oldfield, Hobza, and Wilson (1970) announced the discovery of eriophyoid spermatophores. Sternlicht and Goldenberg (1971) published on the bionomics of the citrus bud mite, *Eriophyes sheldoni* Ewing, discussing male spermatophore production. These latter two papers complete our understanding of eriophyoid reproduction.

Reproduction

Fertilization of the Female

There are statements in literature that many eriophyoids are parthenogenic. These statements now require reconsideration. While there may be parthenogenic species, that is thelytokous eriophyoids, recent information that has been accumulating about these mites does not support that idea. It is not possible to name a single, proven, parthenogenic species at this writing. The basis for these statements to the effect that there are many parthenogenic eriophyoids may rest on several errors in observation. Many eriophyoid colonies have not been investigated thoroughly enough. Perhaps the worker has not known how to distinguish between males and females. With the recent discovery of spermatophores, and the observation that all females properly studied have been found to possess functional spermathecae, the existence of parthenogenic species has considerably diminished as a possibility.

Like most tetranychoids, and species in other mite groups, all male eriophyoids presumably hatch from haploid or unfertilized eggs. That is, the eggs producing males have only a half series of chromosomes. Females, however, hatch from fertilized eggs, and have the complete set of chromosomes characteristic of the respective species.

Some workers had suggested that eriophyoid reproduction was accomplished by males depositing spermatophores that were then picked up by females. This speculation resulted from the observation that certain prostigmatid (trombidiiform) mites have males that deposit spermatophores (see Lipovsky, Byers, and Kardos, 1957; Krantz 1970). Also, no male eriophyoid had ever been seen to copulate with a female. Hall (1967, *b*) illustrated a mushroom-shaped spermatophore within a slide-mounted male body and indicated that E. W. Baker had speculated (unpublished) that males deposit spermatophores and females pick them up.

The discoveries of actual male eriophyoid spermatophores by Oldfield, Hobza, and Wilson (1970; pls. 53, 54, 55) and the finding of spermatophores by Sternlicht and Goldenberg (1971) were approximately simultaneous.

The report by Oldfield et al. (1970) was based primarily on studies of the citrus rust mite, *Phyllocoptruta oleivora* (Ashmead), and of the peach silver mite, *Aculus cornutus* (Banks). They also found spermatophores of the peach mosaic vector mite, *Phytoptus insidiosus* (Wilson and Keifer), the pear leaf blister mite, *Phytoptus pyri* Pagenstecher, the big-beaked plum mite, *Diptacus gigantorhynchus*

(Nal.), and a Nalepellid *Novophytoptus* sp. from *Vaccinium* leaf blisters.

For *Aculus cornutus*, Oldfield et al. (1970) showed that uninseminated females produce only males whereas those inseminated from fresh spermatophores produce both sexes, with the number of female progeny outnumbering that of male progeny by several times. Whether or not males are produced also from fertilized eggs remains to be investigated. Also, Oldfield et al. (1970) observed that previously uninseminated females exposed to 3-day-old spermatophores failed to produce any female progeny. Later, Oldfield and Newell (1973 *a*) observed that protogynes of *A. cornutus* would visit newly deposited or 2-day-old spermatophores but would not visit 4-, 6-, or 8-day-old spermatophores. The production of females as a result of a visitation of either newly deposited or 2-day-old spermatophores indicated that sperm remained viable for at least 2 days in the spermatophore.

The reports by Oldfield and Newell (1973, *b*) indicate that sperm also remained viable in the spermatheca of the reproducing protogyne for several days and in that of the overwintering deutogyne for several months.

Male Spermatophore Production

Some of the *A. cornutus* males, in experiments by Oldfield, Newell, and Reid (1972), started depositing spermatophores within 24 hours after becoming adults. They averaged between 20 and 30 a day. One male lived 20 days and deposited 614 of these sperm packets. Males averaged about twelve times as many spermatophore as the number of eggs laid per female. Preferred sites for depositing spermatophores coincide with areas frequented by females.

Sternlicht and Goldenberg (1971) found that males of the citrus bud mite *E. sheldoni* deposited from 2 to 15 of the packets a day. The authors' table shows that the maximum number a male bud mite deposited in about 39 days was 88 spermatophores; but conditions for spermatophore production may not have been ideal. Citrus bud mite males were about 7 percent of a colony in winter, with the percentage rising to 25 percent in summer.

Oldfield and Newell (1973, *a*) found that eriophyid sperm are spheroidal and insemination in the peach silver mite involves the deposition of the sperm from one spermatophore into one of the two spermathecae located internally in the posterior region of the female genitalia. The sperm nucleus is large and elipsoid. Each spermatophore contains from 40 to 60 sperm (Oldfield and Newell, 1973, *a*).

A female peach silver mite when sensing the proximity of a spermatophore will move up over it and then apparently squeeze the sperm from the sack on top of it. The hinged female coverflap is probably the squeezing mechanism.

The discovery of spermatophores discloses the reproductive cycle of these tiny mites, and the importance of colony founding stands revealed. These mites cannot transfer sperm from male to female by a chance meeting. Males must feed and lay down sperm sacks where females can find them. Males each deposit many spermatophores, which shows that enough of these sperm-bearing packets must be present to guarantee that females will find some. This is one reason why experimental transfer of these mites from one host plant species to another is so important in determining species limits among these mites. If any given species of eriophyoid cannot found a colony on a plant other than its natural host, males will have no chance to lay down spermatophores, and thus no further generations will develop.

Developmental Stages

Eriophyoid mites have four stages in their growth: egg, first nymph, second nymph, adult. Sternlicht and Goldenberg (1971) interpolate the term "nymphochrysalis" between the first and second nymphs, for the resting period, and "imagochrysalis" between the second nymph and adult. During this second resting period the genitalia protrude through the epidermis of the forming adult.

Schevtchenko (1957), while studying the alder bead gall mite, *Phytoptus laevis* Nal., found some females containing a first nymph in their oviducts. He considered this ovovivipary. Hall (1967, *a*) reported these internal nymphs in protogynes of the bladder gall mite on maple, *Vasates quadripedes* Shimer.

Those who make many slides of eriophyoid mites will find that females with these internal nymphs, while almost rare, occur in a wide spectrum of species. Such internal nymphs can be found alone, or behind 1 fully formed egg in the oviduct, or behind 2 or 3 eggs. As well as the supposition of ovovivipary for this condition, there are 2 additional suppositions. First, the eggs from which these internal nymphs have developed have likely not moved far enough toward the spermathecae to have been fertilized. If so, the nymphs[5] are haploid males. Second, the retention of the egg in the oviduct long enough to allow nymphal development, would seem to be a haphazard occurrence. A possible explanation for internal nymphs might be that this happens at the end of the female's life, when she has become unable to extrude the egg. If these internal nymphs actually represent the beginning of ovovivipary, then it is simultaneously, but weakly appearing, in many unrelated species. At any rate the fate of such nymphs remains to be discovered.

Schevtchenko believes that a membrane he observed in an eriophyid eggshell is the remnant of a previously existing instar.

Direct Development

This comprises the 4 stages just recited. Perhaps the majority of eriophyoid species have this simple growth pattern. Examples are mites that live under cover, such as in buds or within petiole bases, and all species inhabiting evergreen hosts. Presumably all tropical species have this development, but these mites in tropical areas remain for investigation (see *Dicrothrix anacardii* K.).

An example of a simple life history is illustrated by the privet bud mite, *Eriophyes ligustri* (K.). This mite always lives under cover in terminal buds, feeding and breeding on green bases of scales. Both sexes overwinter under bud scales, and egg laying can occur any time of year in the Sacramento Valley of California, when the temperature is high enough. While there is no information on any possible difference between resident females and females that seek to migrate, all females of this species are presumably similar as regards external structure and microtubercles.

Deuterogyny

Not all species of eriophyoids have a simple development. Some have alternation of generations, and such species, as far as known at present, are confined to deciduous host plants in northern temperate and arctic latitudes.

The term *deuterogyny* applied to a species means that it has 2 different kinds of females within its organization. While this account of deuterogyny, as it applies to eriophyoid mites, is necessarily a preliminary discussion of the subject, we do know of many species in this superfamily which are deuterogynous. A recognition of deuterogyny, and how the life histories of the species are modified, by it, is not only essential to taxonomic and bionomic studies, but is also important in mite control. The seasonal activities of these mites, especially as they relate to agricultural hosts, and their positions upon these plants at any particular time of year, have a direct bearing not only on timing of acaricide applications, but also upon the composition of these acaracides.

The terms *protogyne* and *deutogyne* apply only to adults. The first and second stage nymphs of deuterogynous species that have been studied do not indicate whether they will produce protogynes, deutogynes, or males.

Deuterogyny occurs in all three families of the Eriophyoidea. The Eriophyidae, which is the largest of the three families, has the most deuterogynous species. Various rhyncaphytoptid species are deuterogynous. So far only one deuterogynous species is known in the Nalepellidae. This latter group, the most primitive of the three, also has two other types of form change, one of which is the existence in a *Trisetacus* sp. of two forms of both males and females, the other being a radical change in the second nymph (*Phytocoptella avellanae* (Nal.)).

While there may be nonstructural deuterogyny, the easily recognized type has two anatomically distinct females within a species. In such a species there are three kinds of adults: (1) males; (2) primary females or protogynes; (3) secondary females or deutogynes. The male, with few exceptions, has the full compliment of abdominal ring microtubercles found in the species. Typical males of wormlike species have microtubercles around the entire abdomen. On rust mites with broad tergites, the narrower sternites on males are often the only area with microtubercles. Males of deuterogynous species exist only during favorable times of the year. Evidence so far accumulated indicates that male spermatophores supply sperm to both protogynes and to deutogynes.

The protogyne is typically the female with exactly the same exterior anatomy as the male, except the genitalia. Protogynes are coeval only with males, and are active egg layers. The probable life span of an active protogyne is about a month or five weeks. New protgynes continue to develop in an active colony as long as conditions are favorable.

The protogyne and male are the primary forms of a deuterogynous species. Classification, with few exceptions, must rest upon these two forms because: (1) the male and protogyne are the perfect generation, showing the highest development of the species; (2) the protogyne in a deuterogynous species is directly comparable to females in all other species, with few exceptions, whether deuterogynous or not. In a sense all eriophyoid species have protogynes, except when discarded.

The deutogyne, or secondary female, is the imperfect generation since it has no male counterpart, anatomically. Deutogynes may differ from protogynes in the same species in several ways. Most conspicuous differences are apparent in microtuberculation. While deutogynes most often have reduced microtuberculation, compared to their protogynes, yet in some species deutogynes have complete microtuberculation that differs in shape from microtubercles on their protogynes. As discussed below, these microtubercle modifications on deutogynes seem to have a bearing not only on hibernation locations, but also on the time of year the deutogynes move to resting positions on the plant.

The cephalothoracic shield on rust mite deutogynes is generally less plainly ornamented than on protogynes, and the deutogyne anterior shield lobe tends to be thinner and more downturned. Rust mite deutogynes usually have narrower tergites than do protogynes, and ridges, furrows, and projections displayed by protogynes are nearly all absent from the respective deutogyne.

The presence, position, and direction of body setae, however, are about identical on both protogynes and deutogynes of the same species. Deutogynes have about the same genital structure as the respective protogynes, including functional spermathecae.

While direct evidence is available at present on only three or four species showing that deutogynes of these species carry sperm with them through hibernation, the presence of functional appearing spermathecae in all deutogynes examined suggests they carry viable sperm.

No deutogyne so far studied can reproduce the year it grows to maturity. Deutogynes must be subjected to a period of winter chilling, followed by rising spring temperatures, before they can lay eggs. Deutogynes emerging from winter quarters crawl to embryonic leaves in opening buds, feed, and lay eggs in habitual places. Primary forms hatch from these eggs, and soon the old deutogynes disappear. For a period after the death of the old deutogynes the only members of a species present in a growing colony are immature forms and males and protogynes.

The appearance of new deutogynes in active colonies of deuterogynous species in spring and summer occurs at various times during these periods as a result of interaction between growth habits of the respective host, and character of the mite species present. For rust mites and leaf vagrants the development of new deutogynes seems related to leaf hardening. Leaf hardening may make feeding more difficult and result in diminution of moisture intake. When eriophyoids living in the open meet such unfavorable conditions they must develop deutogynes, which promptly move off of leaves to habitual resting quarters. In this way deutogynes carry out their function of preserving the species during less favorable times of year when food is removed from the species.

Erineum-making and gall-making eriophyoids can control the intertistices of these microenvironments by continued growth regulator injections, a process seemingly unavailable to rust mites. So these gall species do not have the immediate urgency to move out of their quarters which is shown by rust mites and leaf vagrants. New deutogynes show up in erinea and in galls evidently in response to gall maturity, but the interiors of these galls remain succulent as long as the respective leaf is functional. Thus the gall species can leisurely fill these microenvironments with deutogynes over a period of 2 or 3 months. Urgency to abandon galls only comes when the plant prepares to shed its leaves at the end of the season.

Variations in life history of deuterogynous forms. Species of rust mites and leaf vagrants that produce deutogynes during the summer come under the following categories according to the time and conditions of their appearance: (1) deutogynes appear and move off of the leaves owing to late spring temperature rise; (2) deutogynes develop ahead of early summer leaf drop; (3) the appearance of deutogynes results from early or midsummer leaf hardening; (4) the species changes to deutogynes as the result of a late summer leaf change; (5) deutogynes develop because fall temperature drop causes the host plant to stop producing fresh leaves.

The privet rust mite, *Aculus ligustri* (K.) in the Sacramento Valley of California, moves off leaves when late spring temperatures rise. At this time there is a variety of intergrades of females as regards microtuberculation, but all crawl to aestivation quarters under loose lateral bud scales or behind buds. None of the mites return to privet leaves during the current season. Observations indicate that only the typical deutogyne types of females, with reduced microtuberculation, have been able to survive until late winter. At that time they move around among lateral buds, evidently able to start feeding on green bud scale bases. A slide mount of such a deutogyne showed what are probably sperm nuclei in the spermatheca, indicative of insemination the previous spring. At Riverside, California, this species may be found all winter on privet leaves.

The California buckeye rust mite, *Tegonotus aesculifoliae* (K.), develops deutogynes in late spring or early summer to keep ahead of impending leaf drop. California buckeye ranges a considerable lateral distance across California, with the inland range being drier than it is in the coast range. In the coast range buckeye keeps its leaves until late summer, whereas inland in the Sierra Nevada foothills, the leaves may drop in early summer (see fig. 97). The mites always keep ahead of leaf drop by producing deutogynes wherever they infest the shrub. This buckeye rust mite displays maximum differences between the protogyne and deutogyne. Protogynes have a prominent longitudinal middorsal ridge and lateral projecting tergal lobes. Deutogynes not only lack this ridge and lobes, but have no abdominal microtubercles. These deutogynes aestivate in dry bark crevices at the apex of the previous year's growth. This aestivation is followed by hibernation in the same crevices.

Both males and protogynes of the pear leaf rust mite, *Epitrimerus pyri* (Nal.), inhabit spring leaves of the host. These primary forms are rather flattened-fusiform, with a slight subdorsal longitudinal thanosomal furrow on each side. The males and protogynes are completely set with fine microtubercles on the rings. New deutogynes appear in June owing to leaf hardening. Deutogynes are less flattened than protogynes, lack the subdorsal furrow, and have no microtubercles. Deutogynes proceed off of the hardening leaves in early summer and seek crevices at the apex of the previous year's growth.

One reason why leaf hardening is known as the cause of the appearance of deutogynes is that late spring or early summer growth of fresh twigs promptly stimulates a vigorous development of primary forms on this new foliage, forms that persist until the late growth has in its turn hardened. These rust mites also persist later in the summer on growing pear fruit, because it presents fresh surface tissue longer than ordinary leaves. Those seeking to control this rust mite should be aware that the pear rust mite by habit moves off of the leaves during early summer; they should not attribute that move to acaricide applications. To learn how well a late spring treatment has actually done, it is necessary to examine the dry crevices in locations referred to above.

The only member of the Nalepellidae so far known to be deuterogynous, that is, to have but one male and two female forms, is the alder rust mite, *Sierraphytoptus alnivagrans* K. This mite lives on alder at the 6,000 foot level in the mountains east of the Sacramento Valley in California. This deutogyne has reduced numbers of sternal microtubercles, which are more pointed than on the protogyne. Deutogynes move off of the leaves during July and August, and go to resting positions behind lateral buds of the current season's growth. Before snowdrifts have entirely melted away in spring, these nalepellid deutogynes may be found moving slowly along twigs, feeding on bud parts that are somewhat succulent and exposed. While this mite is not related to rust mites of the genus *Aculus*, it has acquired characters very similar to members of that genus. These include the flattened-fusiform body, and the pair of small spines projecting ahead from the anterior male and protogyne shield lobe. *Sierraphytoptus alnivagrans* differs from *Aculus* spp. by having the extra nalepellid shield and subdorsal abdominal setae, as well as the characteristic spermathecal tube structures.

European elm in California has several leaf vagrant mites that were imported with it. This elm leafs out promptly in spring, but the deuterogynous rhyncaphytoptid, *Rhyncaphytoptus ulmivagrans* K., does not appear on the underside of elm leaves in noticeable numbers until late June. During July, *R. ulmivagrans* protogynes and males become extremely common on leaf undersides, but by early August the mite turns to the production of the peculiar cross-ribbed, flat back deutogynes, and promptly moves off of the leaves. Elm leaves would seem to have hardened by early June, but there is evidently a secondary hardening of the leaves in late July which triggers the deutogyne appearance. Here again, late succulent sprouts on these elms become infested with *R. ulmivagrans* primary forms that persist past the period when the main population of the mite has moved from the leaves. The flat back deutogynes of *R. ulmivagrans* hibernate in exposed depressions along twigs, and evidently revive late the following spring.

The peach silver mite, *Aculus cornutus* (Banks), feeds on a continual supply of fresh leaves throughout the summer, so it is only when fall temperatures drop and the peach trees cease to put out new leaves that the silver mite turns to deutogyne production. Deutogynes move down twigs to look for survival niches behind lateral buds on the current year's growth. Silver mite deutogynes have microtuberculation over the entire abdomen, but abdominal tergites are narrower than on protogynes and the microtubercles are more obscure, especially dorsally. Oldfield (verbal communication) states that he has recovered sperm from silver mite deutogynes, and that he has broken the winter diapause ahead of time by appropriate refrigeration.

From the standpoint of its life history the chokecherry (*Prunus demissa* Walpers) finger gall mite in California is one of the most remarkable of the eriophyids.

The mite is *Phytoptus emarginatae* (K.), a very close relative of the European *Phytoptus padi* Nal. In the California mountains this mite has but one annual generation because it has eliminated the protogyne, and has only deutogyniform females with no thanosomal microtubercles. These females, after becoming full-fed and inseminated, leave the finger galls. Since they cannot reproduce before going through the winter cold shock, they move to dry crevices at the apex of the previous year's growth for hibernation. Males of *P. emarginatae* show the original form of the species, being completely microtuberculate, with granular lines on the shield. Dorsal male microtubercles are weaker than the ventral ones. That spring-emerging deutogyniform females lay eggs that hatch into both sexes in finger galls shows that these females have carried sperm with them through the winter.

The red erineum-making mite on mountain maple in California is *Eriophyes calaceris* (K.). This mite engenders a leaf edge brilliant red erineum that extends out over upper leaf surfaces in heavy infestations. Males and protogynes have rather coarse, unevenly distributed ring granules on the abdomen (pl. 58), with deutogynes lacking all dorsal microtuberculation (pl. 59,*a*). Deutogyne dorsal half rings show a central transverse ring crease across the back. Deutogyne development continues during the summer months until leaves begin to dry in September, and the erineum deteriorates. At that time males, protogynes, and deutogynes crawl down the maple twigs as if all would hibernate. Examination of the twigs on these same maple shrubs in October, after night frosts have begun and after the first snowfall, however, discloses that only the deutogynes, which lack dorsal microtubercles, are present in dry twig crevices. Males and protogynes had perished.

A deuterogynous species that has deutogynes with abdominal rings completely microtuberculate can be tentatively called *Phytoptus sorbi* Canestrini. The mite population studied in this connection makes leaf blisters on mountain ash in the California mountains. In May, before snow has completely melted, deutogynes of this mountain ash mite lurk under edges of bud scales on active terminal and lateral buds. The buds cover themselves with very sticky syrup, but the mites avoid miring in it by remaining at scale edges where the surfaces are drier. The abdomen of these overwintering deutogynes is entirely studded with round, prominent microtubercles. The shape of these ring granules evidently helps the mite to remain free from entrapment in the sticky liquid. In July the leaf blisters on mountain ash bushes are inhabited by males and females (protogynes) that have elongate and closely packed microtubercles. The protogynes often have eggs in the oviducts. New deutogynes, with the round microtubercles as described, begin to appear in the blisters in August. The new deutogynes do not display eggs in their oviducts.

Some of the mites found in leaf blisters on mountain ash have a diagonal line across in front of the dorsal shield tubercle, after the manner of *Phytoptus pseudoinsidiosus* (Wilson). While they probably have the same sort of deuterogynous development as described for the species that lacks this line, the diagonal-lined mites were not studied.

Pear leaf blister mite, *Phytoptus pyri* Pagenstecher, in the Sacramento Valley,

California, displays no such deuterogynous changes as does *P. sorbi.* Pear leaf blister mites hibernate under scales of terminal fruit buds, where both sexes occur during the winter. These pear mites will feed and lay eggs in buds when winter temperatures rise high enough to allow such action. Apple bud mites, with about the same exterior structure as the pear leaf blister mites, also do not show a deuterogynous change of exterior features.

A second eriophyid with deutogynes that are entirely microtuberculate is the California black walnut leaf pouch gall mite, *Eriophyes brachytarsus* K. on black walnut. The males and protogynes of *brachytarsus* are yellowish, with microtubercles that are projected into points, giving these stages a spiny appearance. Deutogynes, however, are red, and have coarse, round microtubercles, distributed over the entire abdomen (pl. 60). In late winter in the Sacramento Valley, only red deutogynes are present in the terminal buds. These buds have considerable viscous syrup in them, and the coarse microtubercles may aid in keeping the mites from becoming entrapped. New leaf galls begin to grow out of upper leaf surfaces in early April, and at this time the only mites in the galls are one or two red deutogynes. Gradually eggs and nymphs appear in these galls, and by late April the red mites are gone. During May the only inhabitants of the galls, aside from eggs and nymphs, are yellow, spiny, males and protogynes. New red deutogynes begin to appear in early June, and gradually fill the galls through July and August. But males and spiny protogynes also continue to be present in galls. When walnut pouch galls begin to deteriorate in September, both yellow protogynes and the more numerous red deutogynes start migrating down petioles, then up to terminal buds. Individuals die along the way, especially at petiole bases. Spiny protogynes do reach terminal buds, but none overwinter.

The red deutogynes of this walnut pouch gall mite were named *E. brachytarsus* first in 1939. In 1940, before the significance of the life cycle of this mite was properly understood, the protogynes were named *E. amiculus* K. These pouch gall mites do not have the produced setiferous genital tubercle characteristic of many of the walnut and *Carya* mites with 3-rayed featherclaws.

The distribution of these walnut pouch galls on leaflets along a walnut twig illustrates an action of various gall-making eriophyids which takes place on vigorous twigs. With mites present in opening buds the first leaflets are galled, since the mites have to feed upon embryonic tissue for the growth regulators to effect growth. These first leaflets occupy the energies of the overwintering mites, and additional leaves are gall-free for a time. Later, when protogynes become sufficiently numerous, they evidently move to terminal buds, and by feeding on embryonic leaflets give rise to a new set of galls on more terminal leaves. It is assumed that deutogynes would not be able to make these additional galls, since the secondaries cannot reproduce the year they develop.

Two other walnut mites, which are of some economic importance, are also deuterogynous much after the manner described for *E. brachytarsus.* One is the black walnut petiole knot mite of eastern North America, *Eriophyes caulis* Cook. Galls made by this mite are comparatively large warts covered with exposed erineum. They are located just above the petiole bases. These large knots often severely twist the petioles and stunt leaflets. The deutogynes of *E. caulis* are very similar to those of *E. brachytarsus* because they evidently have to contend with sticky terminal buds much the same as on the western walnut. But the protogynes of *E. caulis* have shorter spines on the microtubercles, showing that the exposed erineum is not as viscious as the interior of the western pouch galls.

The Persian walnut leaf blister mite, *Eriophyes tristriatus* (Nal.), has males and protogynes with slightly pointed microtubercles, but the deutogynes lack dorsal microtubercles (fig. 116, *g*). This lack of dorsal microtubercles suggests that the deutogyne hibernates in dry twig crevices. *E. tristriatus* differs from *E. caulis* and *E. brachytarsus* by having produced setiferous genital tubercles.

Eriophyes caryae Keifer rolls leaf edges on pecan trees. This mite has produced setiferous genital tubercles, and the mites described, which lack dorsal microtubercles, suggest that they are deutogynes. The name for a spiny pecan mite, *E. vaga* Keifer, may actually apply to the protogyne of the species.

Another deuterogynous gall mite is *Eriophyes theospyri* Keifer, for which the deutogyne is the stage described. This deutogyne has flattened microtubercles, suggesting it hibernates behind lateral buds on current season's growth. The protogyne has slightly pointed microtubercles, indicating that the persimmon leaf bead galls are not particularly moist inside.

Eriophyids can become deuterogynous independently, if the evidence presented by two closely related species of *Aculus* is correct. The first species, *Aculus rhamnivagrans* (K.), lives on the hairy variety of *Rhamnus californicus* Escholtz, an evergreen shrub. This *rhamnivagrans* has four spinules projecting forward from just under the front edge of the anterior shield lobe. It is not deuterogynous, so far as is known. A closely related species, *Aculus amandae* (K.), has the same four anterior spinules, and lives on the deciduous *Rhamnus purshiana* de Candolle. The deutogyne differs from the protogyne in *R. amandae* by having narrower thanosomal rings and less prominent microtubercles. The deutogyne is entirely microtuberculate, however, indicating that the hairy buds where it hibernates do furnish it some moisture during the resting period. The deutogyne also lacks the anterior lobe spinules.

A possible deuterogyny in grass mites occurs in what we can assume is *Eriophyes tenuis* (Nal.). The mites were first found in drying seed heads of sheeps fescue, growing near Acherslaben, East Germany. The males have 5-rayed featherclaws, but the females have 6-rayed featherclaws, a condition never mentioned by Nalepa. All specimens from the drying seed heads were similar in outward anatomy, except for genitalia and size. Two sizes of females came from these seed heads. The larger females had internal eggs, or 1 or 2 internal nymphs. These larger females are, therefore, the resident reproductives. Smaller females show no internal indication of gravidity, and likely are the migrating form that is blown around. New infestations are started if they land in Festucas.

There has been but brief mention of possible existence of purely functional deuterogyny. Deuterogyny that lacks structural manifestations is more difficult to detect, and requires careful study of life habits of the species under investigation. The only eriophyid known which might possibly suggest this sort of deuterogyny is the Persian walnut leaf erineum mite, *Eriophyes erineus* (Nal.). This mite moves out of the thick undersurface erineum pads during the summer, and females crawl down petioles and up to terminal buds in the same manner as the deuterogynous *E. brachytarsus.* Many individuals die and dry up on the way. Those that survive have penetrated into softer parts of the buds. Examination of the females in erinea, and in terminal buds in early fall discloses no structural differences, but the habits of the mite, and the deciduous nature of its host, suggest that deuterogyny is actually present.

Schevtshenko (1967) states that *Trisetacus kirghisorum* Schev. has two forms of both males and females that overwinter. If this is in any way a manifestation of deuterogyny it is not the same as the types described above. Another reason for not considering this deuterogyny is that the host is evergreen. Other species of *Trisetacus* have overwintering males, and *T. ehmanni* K., a pine needle sheath mite in California, is one of these. It is not possible to state here whether or not the overwintering forms differ from summer forms.

Rules for Detecting the Presence of Deutogynes. Many deutogynes in the Eriophyoidea, especially in rust mite species, are so radically different from protogynes of the same species that they have in the past been viewed not only as distinct species, but have often been placed in genera different from the protogynes of the respective species. The following is a series of rules, learned after considerable trial and error, that the classifier or bionomist can observe to avoid making this sort of mistake. (1) If two closely associated types of adult females are on a deciduous host, suspect them of being but two phases of the same species. (2) Find males in the colony and define protogynes on that basis. Deutogynes will not have the same microtuberculation as the males. (3) Note setation on the associated forms. Protogynes and deutogynes of the same species should have precisely the same setal makeup, but setal lengths may differ. (4) Disregard dorsal abdominal structures. The dorsal anatomy of males and protogynes is of great use in defining genera, and will always be so. But deutogynes do not follow the rules in this respect, and allowances must be made for them. Deutogynes have more generalized features than protogynes, as a rule, and often show intergeneric relationships. (5) Note microtuberculation on the abdomens of the two types of closely associated females. If one type has considerably reduced ring granules, or if it lacks them, it could be the deutogyne of the species. Some deutogynes, as indicated above, have stronger microtubercles than the protogynes have, but these ring granules are then of a different shape. (6) The anterior shield lobe on rust mite deutogynes is usually thinner than this lobe on protogynes, and lacks spinules. (7) Experimentation and observation will disclose the relationships between two closely associated types on a host, and will contribute to further understanding of the bionomics of the species.

Deutogyne Microtuberculation Correlates with Aestivation-Hibernation Retreats. A survey of epidermal structures on eriophyoid deutogynes that inhabit deciduous hosts in the Northern Hemisphere discloses a considerable correlation between the microtuberculation patterns on secondary females and the locations where they aestivate and/or hibernate. Both those studying the bionomics of these tiny mites and those who devise control methods for pests will find these structural hints valuable in giving clues as to where deutogynes conceal themselves for the resting period. While each eriophyoid species has its own peculiarities, it is possible to divide these hibernating types into four rough groups, as follows, with examples illustrating respective habits.

First Group: The deutogynes in this group are strongly microtuberculate on all sides of the abdomen. The deutogyne microtubercles are more widely spaced in general than on protogynes, but these deutogyne granules are usually more prominent than on protogynes.

Eriophyinae. These mites leave leaf galls in late summer or early fall and retire to terminal buds. Survival depends on ability of the individuals to penetrate far enough into the buds. Mites going in far enough to reach embryonic bud tissue encounter viscous coatings that entrap some, but this syrupy fluid holds enough moisture to supply the mites with water, both through the mouth, and perhaps through the microtubercles. The California black walnut leaf pouch gall mite, *Eriophyes brachytarsus* K., is one of these. The protogyne of *brachytarsus* has pointed microtubercles, and the deutogyne has prominent rounded microtubercles. The eastern North American black walnut petiole gall (erineum) mite, *Eriophyes caulis* Cook, is closely related to *brachytarsus* and undoubtedly hibernates in terminal buds, but has not been investigated.

Another eriophyid belonging here is the western *Sorbus* leaf blister mite, *Phytoptus sorbi* Canestrini, or a closely similar species in the *Sorbus* Holarctic complex. This protogyne has narrow, rather elongate microtubercles, whereas the deutogyne has prominent rounded microtubercles.

Phyllocoptinae. The example, which is not a typical species in the group as regards habits, is *Aculops glabri* (K.). The deutogyne is *paraglabri* (K.), which name becomes a synonym. This mite is a rust mite type that has abandoned open leaf for an inquiline existance in red mountain maple erinea. Its hibernation reactions closely follow the erineum maker, *Eriophyes calaceris*, as treated under group three. The deutogynes of *A. glabri* abandon deteriorating erinea in September and move to dry bark crevices along with the deutogynes of the erineum maker.

Second Group: Deutogynes in this group have microtubercles entirely around the abdomen, but these granules are weakened and somewhat thinned out, especially dorsally. These mites typically retire to lateral buds, or to retreats within bud clusters. Survival depends on the mites being able to crowd in as far as possible into deep niches. These deutogynes probably begin to crawl about in early spring before bud opening.

Phyllocoptinae. The best example of this type is the common peach silver mite, *Aculus cornutus* (Banks). This species does not need to develop deutogynes until fall temperatures drop, as peach continually grows new leaves all summer.

A second mite of this type is the cascara buckthorn leaf mite, *Aculus amandae* (K.). The name *amandae* originally applied to the deutogyne, and the protogyne was named *purshivagrans* K. The host, *Rhamnus purshiana* DC, grows on the coast of Oregon and Washington and is a deciduous bush or small tree. Deutogynes have thinned and weakened microtubercles and retire to hairs around bud clusters.

The sugar maple finger gall mite, *Vasates aceris-crumena* (Riley), may have deutogynes that fit into this group.

Third Group: Deutogynes in this group have microtubercles only on the lower parts of the abdominal rings. Rust mite types may or may not have microtubercles on protogyne tergites. These mites, as far as known, leave the feeding and breeding sites rather late in summer, or in the early fall, and probably all retire to dry bark crevices.

Nalepellidae. An alder rust mite on the west coast of North America, *Sierraphytoptus alnivagrans* K., is the only deuterogynous species so far known in this family. It occurs on its hosts at sea level on the Olympic Peninsula, Washington, and at 6,000 to 7,000 feet elevation in California. Protogynes and deutogynes lack tergal microtubercles. On primary females the rounded microtubercles are evenly distributed ventrally. On deutogynes these ring granules are sparsely distributed and pointed. Deutogynes may hibernate in dry bark crevices or behind buds, but they do reactivate in spring before bud opening and wander along twigs.

Phyllocoptinae. The protogyne of the privet rust mite, *Aculus ligustri* (K.), has elongated tergal microtubercles. Deutogynes lack dorsal microtubercles, and have prominent but sparse and rounded microtubercles on the sternites. All females move off of privet leaves upon late spring temperature rise, and seek loose, dead buds low on the twigs; but only the deutogyne types survive to the following spring. They reactivate before spring bud opening.

Eriophyinae. As already described, photogynes of the red erineum mountain maple mite, *Eriophyes calaceris* (K.), have coarse unevenly distributed microtubercles on all parts of the abdomen. Deutogynes possess microtubercles only on lower sides and ventrally on the abdomen. Dorsally the deutogyne half rings show a central crease across the back when viewed under the ordinary microscope, but appear doubled as compared to ventral half rings in stereoscan pictures. Males, protogynes, and deutogynes abandon deteriorating red leaf erinea in September, but the only ones that survive are the deutogynes that find dry bark crevices.

Eastern North American species of maple erineum mites that probably have habits much like *E. calaceris* are *E. elongatus* Hodgkiss on sugar maple and *E. major* Hodgkiss on red maple.

The Persian walnut leaf blister or gall mite, *Eriophyes tristriatus* (Nal.), has protogynes with evenly distributed and pointed microtubercles. Deutogynes of this species have rather flattened abdominal microtubercles that are only present laterally and ventrally. The structure of this deutogyne suggests that it hibernates in dry crevices on twigs.

The alder leaf bead gall mite, *Phytoptus laevis* Nal., has deutogyne structures that place it between this third group and the fourth group. The deutogyne microtubercles are present only along the midventer of the abdomen and are somewhat scattered. Protogynes of this species have sparse microtubercles generally distributed on the abdomen. While it is not possible to state with certainty where *laevis* spends the winter, it must retire to a type of location that is dry.

Fourth Group: Deutogynes in this group move out of galls, or off of leaves, during hot summer months. This means they first aestivate, then hibernate, before they become reactivated. They locate in either dry masses of bud hairs, or crawl into dry bark crevices on the previous year's wood, where they undergo some desiccation. This is the most distinct group of deutogynes and their chief feature is the total absence, or almost total absence of microtubercles from the abdominal rings. This elimination of microtubercles is obviously to make the cuticle as impervious to the passage of water as possible so that mites can maintain some body fluidity, and not completely dry up. Some elongate microtubercles may remain on the telosome venter.

Eriophyinae. The example is a unique species (or two species) that makes finger galls on *Prunus* spp. leaves. *Phytoptus emarginatae* (K.) (synonym of *padi* Nal.?) has divested itself of the protogyne and the only female remaining is the deutogyne (at least in California latitudes). This restriction of the species to the deutogyne limits the yearly generations to one. Males, as recounted elsewhere, retain the completely microtuberculate abdomen and show what the species was originally. Females hibernate in dry crevices at the apex of the previous year's wood.

Phyllocoptinae. The first two species here treated in this subfamily with deutogynes of this type aestivate or hibernate in dry bud hairs. One is *Calepitrimerus vitis* (Nal.), the grape rust mite, and the other is *Calepitrimerus baileyi* K., the interior apple rust mite.

Two species that aestivate or hibernate in dry bark crevices at the apexes of the previous year's wood are the California buckeye rust mite, *Tegonotus aesculifoliae* (K.), which moves off of the leaves as early as late spring in interior areas, and the pear rust mite, *Epitrimerus pyri* (Nal.). The pear rust mite moves off of pear leaves when they harden in early summer.

Another species that probably fits in with this fourth series is the European horse chestnut rust mite *Tegonotus carinatus* Nal.

On eastern silver maple the bladder gall mite, *Vasates quadripedes* Shimer, produces deutogynes that lack ring microtubercles. While it is not possible now to state where these deutogynes go, they must seek some sort of dry retreat.

A comparison of eriophyoid deutogynes with certain tetranychid hibernating female forms (which could also be called deutogynes) discloses that there is close parallelism with regard to epidermal alterations to prepare the adult females in both mite superfamilies to resist water loss through the epidermis. For example, hibernating females of *Tetranychus urticae* Koch, and of *T. pacificus* McGregor, not only change color to pinkish, but also are entirely divested of strial lobes. Strial lobes on tetranychids are entirely comparable in function to eriophyid microtubercles. The two tetranychid female types quoted above would fit exactly into group four, except that the tetranychid females retire later in the season. This parallelism strengthens the theory that absence of microtubercles from rings on certain eriophyids, and absence of strial lobes from some hibernating tetranychid females, enables resting mites to conserve body water by making the cuticle more resistent to the passage of water. This general question has also been discussed in chapter 2.

Influences of Weather on Biology

California, with its sharply defined life zones, is a good area in which to observe eriophyoid environmental preferences. There are many examples as to how these zones limit the range of some of these mites. The citrus rust mite, *Phyllocoptruta oleivora* (Ashmead), and citrus bud mite, *Eriophyes sheldoni* Ewing, infest their hosts only on the immediate southern California coast, although citrus is grown in many inland areas. (Rust mite also occurs on citrus at the head of the Gulf of Baja California, where there is a prevailing moist wind off of the gulf water.) In contrast to the high atmospheric humidity requirement of these citrus mites, the peach mosaic vector mite, *Phytoptus insidiosus* (Wilson and Keifer), remains back of the ocean frontages, preferring lower interior atmospheric humidities behind the barrier ranges.

The common apple rust mite, *Aculus schlechtendali* (Nal.), in California, occurs only on planted apples in coast range localities, in the higher atmospheric humidities. Bailey's rust mite, *Calepitrimerus baileyi* K., replaces *A. schlechtendali* in interior California apple orchards and in lower mountain foothills to the east of the central valley, where atmospheric humidities are lower than in coastal mountains. An unexpected aspect of the ranges of these two apple mites, after observing them in California, is that both species are found in Brookings, South Dakota (Personal communications by Magdlena L. Briones). In this South Dakota area, *A. schlechtendali* lives on crabapple, and *C. baileyi* lives on regular commercial varieties of apple.

The tomato russet mite, *Aculops lycopersici* (Massee), flourishes on its solanaceous hosts in the central California valleys during the hottest part of summer. Rice and Strong (1962) reported that experiments showed this mite to have its shortest life cycle at 27.6 C (80 F) and at a relative humidity of 30 percent.

Davis (1964) reported that a magnolia leaf vagrant, *Rhynacus breitlowi* Davis, which he tested in the higher atmospheric humidities of Georgia, required about 30 C (83 F) temperature and a 4 mm saturation deficit for optimum growth. This is high atmospheric humidity, and indicates that the only California area in which this mite could survive on its host would be the southern coast.

Some eriophyid species can range through these wide western variations in temperature and atmospheric humidities. One of these is the privet rust mite, *Aculus ligustri* (K.); its habits in the central valleys are known, but its life cycle on the coast is unknown. A mite living in knotweed buds, *Eriophyes sawatchense* (K.), prefers the lower humidities of the central valleys in California, also lives in Arizona, and attacks knotweeds in both areas.

The above accounts would imply that warm weather is the most favorable breeding time for all eriophyids. But at least one species in temperate climate habitats has its optimum growth in winter. This is a grass mite, *Aculodes mckenziei* (K.), which lives on a variety of hosts placed in several separate grass genera by botanists. Beardless wild rye grass frequently harbors this mite in the Sacramento Valley, California. The mite develops in large numbers in upper surface blade grooves during winter, at which time all stages of the mite are present. *A. mckenziei* also breeds during Ohio winters as shown by collections from quack grass on December 27, 1967, by W. R. Styer.

Another grass mite, *Abacarus hystrix* (Nal.), bears a reciprocal relationship to *A. mckenziei* in the Sacramento Valley. This mite is plentiful in summer when the *A. mckenziei* population is depressed. Since males of both types of mites are present, the possibility that the two kinds are alternate forms does not exist.

Life Span of Individual Eriophyoids

The length of life of an individual eriophyoid depends on the species, on the time of year the mite hatches, on the place the individual occupies in the reproductive cycle of the species, on dispersal activities, and on survival of the species through unfavorable periods. Immature stages, as far as is known, always complete their growth as rapidly as possible. Males, and females that resemble males, usually have the shorter life spans, depending on the time of year and species survival. In nondeuterogynous species both males and females may live overwinter.

Putman (1939) found that under optimum conditions actively breeding plum rust mites, *Aculus fockeui* (Nal. and Trt.), were capable of starting to lay eggs as soon as 6 days after hatching, and could live between 20 and 30 days. Davis (1964) conducted starvation tests on *Rhynacus breitlowi* and discovered the mites would die within 72 hours without food.

There is no information as to the action or fate of migrating eriophyoids. Some are known to reach new plants of their primary host, but how long the journey took for an individual mite, or whether or not it was able to feed on nonprimary hosts, are as yet unanswered questions.

Deutogynes that remain on their primary host can serve as a positive basis for measure of their length of life. Peach silver mite deutogynes (*Aculus cornutus*) that move to lateral buds in October do not find new leaves until the following April, so live six or seven months. California buckeye rust mite deutogynes (*Tegonotus aesculifoliae*) that retire in May or June, before leaf drop, do not emerge from their bark crevices until the next April, a life span of about ten months.

Egg Deposition

The number of eggs a female eriophyoid will lay when actively breeding can vary from 1 to 5 a day, with between 3 and 4 the probable average. Oldfield (1969) noted that a female of the 1 generation plum finger gall mite (*Phytoptus emarginatae*) laid about 50 eggs. Putman (1939) observed a female plum rust mite (*Aculus fockeui*) lay 79 eggs over a period of 30 days. As far as known no eriophyoid species overwinters in the egg stage.

Dispersal

Since eriophyoids cannot travel any distance under their own power, they must rely on wind, insects, and birds to carry them. The adult females and a few males are the dispersal stages. Presumably flying insects are the most efficient means of aiding dispersal, especially those that prefer the same host plant as the eriophyoid, for they are likely to fly more directly from one favored host plant to another. But wind is probably the principal dispersal means used by these mites. They can be seen, under a hand lens or dissecting microscope, to rear up on their terminal lobes and long caudal setae and to paw the air, apparently a habit that aids dispersal. On occasion they form chains by crawling up on each other. They also leap or drop from whatever plant part they wish to abandon. Davis (1964) noted that mites facing the wind become airborne more easily than those facing another direction.

The small size of eriophyoids contributes to air flotation for distribution. All plant mites use wind for traveling, but larger, 8-legged types, such as tetranychids, have much greater running and crawling ability than eriophyoids, and can move from one plant to another much more easily. Since these tiny eriophyoids can only crawl from one plant to another if the plants touch, a vital necessity is a size small enough to enable traveling mites to be suspended in slight air currents.

Nault and Styer (1969) report on experiments in trapping traveling eriophyids, using standard 1 x 3 inch glass slides coated with silicone grease, hung near the top of 6-foot high and 1-foot square frames, one slide on each side. The frames, set out in a wheat fiield, caught wheat curl mites, *Eriophyes tulipae* K., on each side. Glass slides smeared with grease and set on top of high buildings will catch these mites. Species trapped on these glass slides show what eriophyoids are actively traveling at any particular time of year. Single, traveling mites, obviously not on their primary host, have been observed to rear up and blow away in the next light gust of wind. But wind travel is hazardous and probably well over 90 percent of these mites perish without reaching their primary host.

It is, however, easy to show that seedling plants, growing more-or-less isolated from older and infested members of the same plant species, always develop a population of eriophyids that live on that plant. Seedlings of California white oak, which have sprung up nearly a quarter of a mile from the nearest old white oak, soon become infested with *Rhyncaphytoptus megarostris* (K.), a white oak leaf vagrant. Seedlings of holly, growing more than 200 feet from nearest older hollies, and separated by houses and bushes, became infested with three holly mites: *Cecidophyopsis verilicis* (K.), a bud mite, and by two leaf vagrants, *Diptacus swensoni* K. and *Acaricalus hydrophylli* K.

THE COLLECTION, PRESERVATION, SLIDE MOUNTING, AND ILLUSTRATING OF ERIOPHYOID MITES

Collecting

Those wishing to collect and study these mites may find them on growing plants, in collections of plant galls caused by mites, and to a lesser extent in herbaria. Mites in dry galls and on herbarium leaves will be mummified. It is usually possible to cook dry-infested plant parts in chloral hydrate-water media (see the following formulas for these media), and to bring the mites back to nearly their original shape. The success of this recovery depends on several factors: ingredients in the medium and the way it is handled, the treatment of the mites before they were placed in the collection, and the age of the specimens. There is room for experimentation in improving recovery methods.

While it may take considerable searching to discover mite indications in herbarium specimens, galls and outgrowths may usually be removed without appreciable damage to plant specimens.

For the collection of eriophyids on living plants (field collecting) first ascertain mite habitats. Galls or other mite-induced deformities are relatively easy to see. Leaf vagrants and rust mites are usually not too small for successful collecting with a 10X hand lens. Plant parts may also be brought into the laboratory for microscopic examination.

Eriophyoid species are normally most abundant on their host plants during the summer and early fall in temperate and cold latitudes, but they may also be found during less favorable weather. In the wintertime, hibernating deutogynes are apt to be present; they do not, however, completely represent the species. Mite abundance in the tropics is likely to be modified by wet and dry cycles.

Fast-growing or vigorous spring shoots usually have fewer eriophyids because the growth has not allowed the establishment of new colonies. Annual plants will likely not harbor these mites although this question remains for further investigation.

The collector will discover many exceptions to the above general rules on timing, depending on the mite species. Certain juniper mites are most abundant on vigorous tips. Two kinds of grass mites on perennial grasses are most abundant in leaf grooves during the winter. They are not represented by hibernating forms.

Useful equipment for field collecting is: a 10X hand lens, long letter envelopes, plastic bags, capsule vials 1 inch in diameter by 4 or 5 inches long, clippers and a small icebox or some means of keeping picked plant material cool. If the collector limits his search to looking for galls or plant deformities he will miss most of the eriophyoid species present in the area. Random searching, with the places where these mites live on their hosts in mind, will nearly always amply reward the collector.

Although a 10X hand lens will usually disclose the mites on plants, some kinds still remain hidden. If it is suspected that species are being missed, bring plant parts into the laboratory and examine them at higher magnifications under a stereoscopic binocular microscope.

The collector will usually pick up far more material than is possible to study immediately. Therefore much of this material should be stored for future examination. Mites in galls, buds, and in other sequestered places usually remain on the plant parts after they have mummified. Dry mites are usually recoverable if the plant parts were originally well infested.

The simplest way to handle fresh plant parts bearing plenty of mites is to put them into paper envelopes. Write the data concerning host, locality, collection date, and collector, on the outside, and file for future reference. Allow drying to begin immediately, but do not heat the specimens. Plastic bags are undesirable for storage as they promote mold. While out collecting, shield the specimens from the sun's heat. A small icebox is essential on hot days.

Leaf vagrants and rust mites, as opposed to mites in galls, tend to fall off of drying leaves and may be lost. For collecting leaf vagrants a liquid preservative is desirable, but do not carry the liquid into the field. Push infested plant parts into a vial (a capsule vial 1 inch in diameter by 4 inches long) and pour in the liquid after return to the laboratory. Write data on an outside sticker. Some people prefer to pick mites off of plant parts laboriously and to needle them into, or place them into, small vials.

Preservation

Do not use ordinary liquid preservatives for eriophyoids because they modify the bodies and make preparation in chloral hydrate media unsatisfactory or impossible. Preservatives to avoid include 70% alcohol in water, formaldehyde, weak solutions of chloral hydrate, ketones, and so on.

A liquid that has so far proved to be satisfactory for holding eriophyoids, and in which they remain preparable (as opposed to 70% alcohol), is a thin sorbitol syrup in a 25% solution of isopropyl alcohol. Make the 25% alcohol solution in a jar and then add sorbitol, allowing the sugar to dissolve until the fluid becomes a thin syrup. Add a small amount of iodine crystals to aid preservation and prevent mold. As previously directed, keep this syrup in the laboratory and pour into vials onto infested plant parts after they are brought in.

Recover mites from this liquid by pouring it into a watch glass and examining with transmitted light. Thick syrup is a good adhesive for holding mites on needles when moving them during preparation.

Recovery of mites from dry infested plant parts consists in placing small amounts of these parts, galls, and the like, in a thick slide concavity, dropping on preparatory media, and continuing as directed. Heat the mites in smallest casseroles to produce the best cleared specimens, then pour out onto slides to find the mites.

Experience shows that mummified mites vary greatly in their response to heating, some becoming only partly expanded. Usually the mites that have been dry for 8 to 10 years respond well. An example of older eriophyid mummies are specimens recoverable from Trotter's *Cecidotheca Italica.* Some of these are at least 70 years old and many do not expand much. They usually show diagnostic characters, but are worthless for delineating. Perhaps variability in response to preparation on the part of these mumies is the result of different methods of handling when the mites were collected.

Slide Mounting

Ingredients in Mounting Media for Eriophyoid Mites

The slide-mounting media recommended here for study and preservation of eriophyoid mites are water media, based primarily on *chloral hydrate.* Chloral hydrate is a relatively simple and almost unique chemical. It is basically an aldehyde, with two carbon atoms in each molecule. Three chlorine atoms attach to one carbon, enabling the molecule to hold a water molecule in chemical combination with the other carbon.

As chloral hydrate is strongly crystalline, it is necessary to prevent media containing it from recrystallizing after the specimens are under coverslips. A suggestion toward preventing recrystallization is in the paragraph on sorbitol. Chloral hydrate is a strong clearing agent, dissolving soft body tissue when heated. While chitin is at first resistant to this action, it will tend to dissolve over long periods of subjection to chloral hydrate. A suggestion toward preventing this dissolution is in the paragraph on formaldehyde.

Besides helping to remove opaque soft tissue, chloral hydrate softens, expands, and plasticizes exoskeletons of delicate arthropods. This expansion enlarges specimens and displays structural features.

Some of the chemicals in media, with water as the solvent, are as follows: gum arabic or acacia, sorbitol, carbolic acid or phenol, resorcinol, formaldehyde, glycerin, and iodine.

Gum arabic is a natural product picked as sap droplets from various Asiatic and African acacias. No water-soluble synthetic polymer so far fabricated is anything like it. It is a giant molecule substance, and in common with large soluble molecules, tends to dissolve slowly in its solvent. It is compatible with some water-soluble chemicals, usually simple ones with small molecules, but is very intolerant of many water-soluble types with large molecules. It would be a considerable advantage to blend gum arabic with water-soluble sythetics, and still retain a completely transparent solution, but none are known that do not either cause instant separation, or evenutual clouding, so none can be recommended here.

Ordinary alcohols quickly precipitate gum arabic from water solution. Butanediol produces a cloudy solution, and full strength formaldehyde may also make a cloud.

The size of the gum arabic molecule is supposed to be on the order of 240,000 carbon atoms. Unlike giant molecules of linear synthetics, gum arabic does not make particularly rubbery solutions. Linear rubbery giants require excessive amounts of solvent to reach suitably low viscosities. This excessive solvent results in considerable collapse, when, under the coverslip, the polymer releases its solvent. Specimens in such collapsing media are inevitably excessively flattened or crushed, which is particularly disadvantageous in connection with very soft mites, such as eriophyoids. But in contrast, when gum arabic is the noncrystalline polymer in the medium, in combination with simple sugars, and a moderate amount of water, the medium is thin enough to be entirely satisfactory. This suggests that the gum arabic molecule is more bunched than linear.

Gum arabic's principal value is in giving substance to the medium and helping it to become sufficiently viscous upon losing some water. Gum arabic tends to resist humidity in the atmosphere and as such an ingredient its percentage can be varied to adjust the medium to presistently moist or dry airs. Because it is a natural product, gum arabic carries some dirt. Removal of minute particles of bark and sand is possible, but the methods, such as precipitation or filtering, are time consuming. The so-called "chemically pure" powder usually contains a minimum of foreign material, and in the amounts recommended in the formulas, dirt is rarely very annoying.

Sorbitol is apparently the best of the mono- and disaccharides to blend with gum arabic because it is a completely hydrogenated monosaccharide that not only resists chemical change much better than unsaturated forms, but is also a humectant. Sorbitol occurs naturally in small amounts, but is made in large quantities by hydrogenating glucose. It holds chloral hydrate in solution better than gum arabic, and, along with glycerin, resists recrystallization of the chloral hydrate. Sorbitol is useful as a preventor of overreaction by preparatory chemicals, depending on the percentage present.

Phenol or carbolic acid is monohydroxy benzene. It is hygroscopic, and forms a clear solution with a minimum amount of water, but clouds with excess water. To liquefy phenol, add small amounts of water; this liquid phenol is used in the recipes below. Phenol assists chloral hydrate in removing soft tissue, thereby clearing specimens, and in chitin expansion. The worker will have to experiment with the amount of phenol, drop by drop, to add to the preparatory steps, to suit his particular desires.

Resorcinol is 1, 3 dihydroxy benzene, or metadihydroxy benzene. It is quite water soluble. There are two other dihydroxy benzenes, but the 1, 3 is most effective for mite preparation. Formulas for eriophyoid-mounting media which recommend resorcinol are in *Eriophyid Studies* (Keifer XIX, 1952; XXI, 1953). But resorcinol is difficult to handle, and problems in connection with its use often overshadow its value. The principal peculiarity that resorcinol has is that it must be in solution for several weeks before it becomes effective, perhaps owing to action of oxygen. Resorcinol that has had dry exposure to air for several years is better than the fresh chemical. When resorcinol is working properly it is a strong tissue dissolver, and at times is strikingly effective in making transparent specimens. In the presence of formaldehyde it almost instantly forms an insoluble wine red polymer, and with hydrochloric acid may cause the solution to become opaque black.

Formaldehyde is the simplest aldehyde, and by fastening onto nitrogen radicals in organic compounds, hardens tissue. It tends to harden mounted eriophyoids, and its presence is intended to slow the dissolving action of chloral hydrate on chitin. Media with formaldehyde are stable, and will remain unchanged in a vial for long periods. For the medium, cut the 37% commercial solution down to 5% or less by water addition, and use this for the slovent. Full strength formaldehyde as the medium solvent not only tends to cloud gum arabic, but shrivels mites. *Never include formaldehyde in liquid mite preservatives*, for these reasons. Strong alcohol solutions used as preservatives will also harden mites beyond recovery.

Glycerin or glycerol is trihydroxy propane. It is always a thick liquid at ordinary temperatures, and one purpose for adding it to the media is to prevent mites from becoming completely hard. Probably diethylene glycol, or propylene glycol, are suitable substitutes for glycerin. Those interested should experiment with higher glycols, but may discover that some, such as butylene glycol, will cloud in the presence of gum arabic. (*Do not use* lactic acid, lacto-phenol, or acetic acid, in any liquid preservatives or media for eriophyoids. These chemicals are either astringent, or will ruin mite bodies for preparatory purposes.)

Iodine will sublime, so improperly closed iodine bottles containing crystals will lose their contents. The use of iodine in mite-mounting media is as a stain, but it is admittedly poor as such. At present no other substance is known to be as good for staining mites in water media. To speed the solution of iodine in water media "salt" it in with small quantities of potassium iodide.

Gum arabic, like starch, has an affinity for iodine, and will take it away from mite bodies. Sorbitol and formaldehyde prevent this action.

Iodine-colored media should be fairly dark in the vial. If specimens disappear in the preparatory process, when iodine proves to be in too small concentration, add more finely divided crystals and continue heating. The worker can experiment with amounts of iodine to suit his particular methods. Too intense iodine color causes reflcetions, and disturbs phase illumination. Potassium iodide will hold iodine to a considerable extent from subliming from slides. Ringing with non-water-soluble material will also help hold the chemical.

Oil Soluble Resins as Mite-Mounting Media. Unfortunately at present no recommendation is possible as to the use of oil soluble resins, either natural or synthetic, in eriophyoid study. In general the natural resins have smaller molecules that make them more suitable for handling, especially as to amounts of solvent.

Natural resins in the form of fossil amber have long since proved their ability to preserve arthropods through the millenia. Comparison of this preservative power against water media with chloral hydrate, immediately indicates the precariousness of these water media from a longevity standpoint. But there are other advantages to resin media. Staining possibilities to make mites visible are greatly increased. Perhaps experimentation will eventually disclose ways to use resins in this study, but these possible oil solvent techniques must not sacrifice the specimen clarity or expansion achieved in chloral hydrate media. But the search for resin techniques should continue.

Mite-Mounting Media Formulas

The unlimited ways to compound water media allows the worker to discover what he likes best. Formulas given here are only suggestions. Those making slides of eriophyoids, and of other mites, should try additional chemicals besides the ones listed. The condition and visibility of the mites on the final slides, and their longevity, are the final tests of the excellence of any medium.

Hoyer's Medium. A similar formulation has long been known as Berlese Fluid. The formula for Hoyer's medium in the tetranychid section of this book is: distilled water 40 cc; gum arabic 30 grams; chloral hydrate 200 grams; glycerin 20 grams. The reader should compare preparation suggestions in the tetranychid section with those in this eriophyoid section. Tetranychids and other mites larger than eriophyoids are more resistant to body form change, and therefore chemicals useful for them are not necessarily useful for eriophyoids.

Singer (1967) discusses various water media at some length and lists various useful chemicals. He gives Hoyer's medium as: distilled water 50 cc; gum arabic (clear flakes) 50 grams; chloral hydrate 125 grams; glycerin 30 cc. He mentions recrystallization as a problem in this medium. If the formula is modified to read: gum arabic 20 grams; sorbitol 30 grams; recrystallization will be eliminated to a considerable degree.

While the writer has had no firsthand experience with Hoyer's medium, the understanding is that it is usually a one-step medium. That is, the specimen is needled onto the slide into a drop of the medium, the slide is labeled, then put into a low temperature oven to remain until desired clearing is reached.

Tokuwo Kono, mite identifier for the California Department of Agriculture, recommends use of a preliminary medium with Hoyer's. This preliminary medium is: chloral hydrate 100 grams; glycerin 10 grams; water 50 cc; concentrated hydrochloric acid 1 cc. Gently cook the mites in Kono's preparatory mixture. When they are cleared, needle them over into a wash of Hoyer's and then to the final slide. Examples of eriophyoids prepared and mounted in this way show them to be in excellent condition. Do not heat the final slide mounts.

A Formaldehyde Medium. Another approach to slide mounting of eriophyoids utilizes the following recipes. The solvent for the hydrochloric acid and phenol solutions is distilled water. The solvent for the formaldehyde medium is 4% dilution of the original formaldehyde solution.

F. Medium

Sorbitol	3.0 grams
Gum arabic powder	1.0 gram
Iodine crystals	0.02 gram
4% F solution	5 cc approx.

allow to disolve, with agitation, for 24 hrs. or more, then add the following:

Chloral hydrate crystals	14.0 grams
Glycerin	20 drops (1 cc, or more)
KI	0.1 or 0.2 gram
Iodine	0.1 or 0.2 gram

add more 4% F if necessary

Booster mixture

Sorbitol	1.0 gram
Chloral hydrate	2.5 grams
Iodine crystals	small amount
Water	5 cc or more
HCI, reagent grade	7 or 8 drops

Phenol solution .... 1 drop to 3 drops of above solution of phenol

Keep each of these media in separate vials, tightly corked or stoppered. Dispense with a dropper pipette, each pipette used only for the particular medium. Do not heat to speed up solution as that will cut effectiveness. Keep the F medium quite fluid by occasional addition of a drop or two of either 4% or water.

Slide Preparation

Equipment in addition to slides and thin covers should include: thick two-depression slides; fine needles (a size 00 insect pin stuck into a match stick, with point out); very small casserole for cooking pieces of leaves and reclaiming dry material; kapok fibers for holding covers from crushing mites.

Place two drops of the F medium in the depression on one end of a slide. Needle mites into it. Add a drop of HC1 solution, and a drop of phenol solution. Heat gently on hot plate, set just below the water boiling point, until gentle boiling is reached. Examine under low power binocular and continue heating, and adding more media—a drop more of each, for example—until mites reach required transparency. Then transfer to drop of fresh F medium, stir gently to wash; transfer to drop of medium on final slide.

Before placing cover arrange kapok fibers around specimens to prevent the cover from pressing too heavily on the mites. Mites are most easily controlled if the cover touches them lightly. They are then turnable, and after a little manipulation, will remain fairly well positioned.

Do not heat the final slide preparation. The medium is not designed for heating after the mites are on the final slide. Actually, this medium should not become hard, only viscous, and permanent preservation is achieved by ringing.

The above proportions are basic. If larger quantities are necessary multiply by whatever factor is suitable. It is usually best to mix small quantities at a time. The HC1 medium will need occasional additional drops of the acid.

To use resorcinol in these media, substitute it for HC1 in the booster medium. The main medium should then be mixed, minus formaldehyde. Proportion resorcinol in the booster at the rate of 1/10 gram in place of each HCl drop. Wash out the resorcinol in the medium minus 4% F, and then transfer the mites onto the final slide into the formaldehyde medium.

To keep from destroying eriophyoid genitalia during preparation of the mites, it is necessary to cut into some specimens so that internal pressure will escape through the cut. Cutting mites in two is effective. With dry mummies it is necessary to start cooking slowly, then work over examples that are partially softened.

To gain a proper perspective of any eriophyoid species, to associate all mature and immature forms, it is necessary to study at least 40 or 50 examples. For that reason this account recommends that each study begin by placing 5 to 10 specimens on each of 5 or 10 slides. This procedure makes a thorough survey of the various aspects much easier.

The alternative, that is, restricting each slide to one example, will obscure the relationships of the various forms, make it necessary to shuffle a maximum number of slides, inevitably result in a number of worthless slides, and rapidly fill up available space.

Admittedly, the placing of several mites on each slide will now and then result in the presence of two or three species under one coverslip. The segregation of species so associated will sharpen the perception of the taxonomist.

If, for reasons of policy or to prevent the presence of more than one species on a slide so as not to confuse type designations, it is necessary to end the study with one mite to a slide, then good examples can be remounted or new slides made.

The writer judges from some slides received for study, which have but one unoriented, flattened, and rather opaque specimen to a slide, that the ability to recognize what constitutes a good slide mount has not been too well distributed. Published delineations also often indicate that they were drawn from poor preparations.

Illustrating Eriophyoid Mite Publications

Pictures of eriophyoid mites have appeared in articles on control, in publications on life histories, in many recent taxonomic papers, and in various books. Diagrammatic drawings, and photographs showing general views, are probably of some value in control publications. But even these must not be totally uninformative, or misleading. Eriophyoids vary in form from group to group, and some sort of definitive generic and specific features are useful in all depictions. Bare diagrams are a waste of time and space. Earlier pictures of the citrus rust mite, printed with control studies, give the reader no useful information; many details are incorrect, and lack any indication of the broad dorsal longitudinal trough so characteristic of this mite.

All taxonomic articles require illustrations that show generic and specific structures. Good photographs, especially electron scanning microscope pictures, will convey this information. As to drawings, those who can do no better than make bare diagrams that do not follow precise form of the parts delineated should use photographs, or should not attempt this type of taxonomy. For whatever type of illustration is used in taxonomy, the following text figures suggest minimum structural categories to include with each new species description: see figures 104, 106, 109, 118, 120, 134, and others.

Ways to Illustrate Eriophyoids

There are four general methods of illustration: (1) line sketches or drawings; (2) photographs through ordinary microscopes (including phase illumination); (3) photographs through electron microscopes; (4) distribution of actual mite galls in packets, with explanatory literature. This latter method is the *Cecidotheca Italica*, Trotter and Cecconi, issued before 1915.

Recommendations for Line Drawings. These recommendations are presented for those who have the judgment and ability to draw pictures showing precise forms of parts, and which convey generic and specific characters of these tiny mites. Perhaps the best microscope accessory for transferring size and shape to the board is the camera lucida. Rough in views with a blue pencil. A zoom lens in connection with high magnification is exceedingly helpful, when using the camera lucida.

Of the pen points available, the writer finds flexible crowquills will enable the delineator to show depth and character to the drawings much better than rigid points.

Additional recommendations are: (1) prepare well cleared and transparent specimens; opaque inside tissue makes correct drawings impossible; (2) use only well-formed examples of the mites, with legs projecting ahead and diagonally down (while it is preferable to place the head to the left, the delineator should at least be consistent); (3) prop coverslips up with fiber to prevent crushing the mites; if cover rests lightly on specimens they tend to remain suitably stable; (4) push cover to turn mites for lateral, dorsal, and ventral views; defer ringing until illustrating is complete; (5) use partially dissected mites for genitalia delineations; avoid specimens with female coverflap turned up toward the coxae; (6) since these mites show important characteristics by the presence or absence of microtubercles, do not leave these structures off of rings when they are actually present on the specimen being drawn; (7) carefully note presence or absence of all setae on body and legs; show precise locations of these setae; (8) follow body and leg form carefully, especially in depicting ridges, furrows, and thickenings.

The person making these drawings can improve his technique by noting which of his pen strokes are steadiest, and turning the board to take advantage of these strokes.

Photography through Ordinary Light Microscopes. Photographs through ordinary microscopes, including phase illumination, offer authors of economic articles fairly satisfactory means of showing general views of eriophoids. But for taxonomy these pictures must be sharp enough to define specific characters. A major problem inherent in this photography is that the higher the magnification, the shallower the depth of focus. The result is that while a focus is possible on some of the critical features, other features are blurred. Flattened mites are more suitable for higher magnifications, but the flattening itself may destroy characters.

Electron Microscope Photography. The present rapid development of the electron microscope results from the clear, greatly magnified pictures it produces of very minute objects. Those publishing economic articles on eriophyoids will find these machines to be excellent means for obtaining illustrations. It will only be possible here to mention some of the procedures necessary for preparing specimens for this photography. Those wishing pictures should consult experts. There are two types of electron microscopes. The older one uses electrons that pass through the specimen, and this fact makes thin objects necessary. Eisbein and Proeseler (1969) in East Germany, who photograph virus-carrying eriophyids with this type of machine have published highly magnified views of body rings and of featherclaws. These pictures show characters not visible under ordinary light microscope magnification. They describe methods for preparing specimens.

Scanning Electron Microscope Photography. This second kind of electron microscope utilizes reflected electrons. As these electrons issue from the gun, magnets divert part into angular paths, which gives depth to the photographs. Magnifications from about 500 diameters, up to over 2 million, are possible (Crewe, 1971). Because air interferes with viewing the specimen, the machine first draws a vacuum around the object to be photographed.

It is necessary to harden soft-bodied specimens before placing them into the scanning electron microscope. It is possible to harden eriophyids by pouring hot anhydrous isopropyl alcohol onto them. It is usually advantageous to coat mites with metal before photograhy; the process is accomplished by evaporating the metal onto the specimens in a vacuum chamber. Gold proves to be a less obscuring coating than aluminum. These metal coatings also have the advantage of carrying away condensing electrons that would otherwise dull the view. Another way of removing condensing electrons is the glycerol-KC1 method (Brody and Wharton, 1971).

Plates 1, 4, 5, 53–60, 65, 68, 71, 72, 73*a*, 74 are scanning electron microscope pictures of eriophyids.

SELECTED BIBLIOGRAPHY

Baker, E. W., J. H. Camin, F. Cunliffe, T. A. Woolley, and C. E. Yunker. 1958. Guide to the families of mites. Institute of Acarology Contrib. no. 3. 116 pp.

Baker, E. W., and G. W. Wharton. 1952. An introduction to acarology. The Macmillan Co., N. Y. 465 pp.

Boczek, J. 1961. Studies on eriophyid mites in Poland. [In Polish, with short English summary] Prace Nauk. Inst. Och. Roslin 3(2):5.

Brody, A. R., and G. W. Wharton. 1971. The use of Glycerol-KC1 in scanning microscopy of acari. Ann. Entomol. Soc. Am. 64(2):528–530.

Browne, F. G. 1968. Pests and diseases of forest plantation trees. p. 19. Oxford, Claredon Press.

Carter, Walter. 1966. Insects in relation to plant diseases. Interscience Publ. 705 pp.

Crewe, A. A. 1971. A high-resolution scanning electron microscope. Sci. Am. 224(4): 26–35.

DAVIS, R. 1964. Autecological studies of *Rhynacus breitlowi* Davis. Fla. Entomol. 47(2): 113.

DONNADIEU, A. L. 1875. Recherches pour servir a l'histoire des Tetranyques. Ann. Soc. Lin. Lyon Ser. 2.25:153–155.

EISBEIN, K., and G. PROESELER. 1969. Weitere Untersuchungen uber einige Morphologische Merkmale bei Eriophyiden Institut. fur Phytopath. Aschersleben Deutschen Akademie Landwirtschafteswissenschaften zu Berlin 11(11/12):900–909.

FARKAS, H. K. 1960. Uber die Eriophyiden Ungarns I. Acta Zool. Akad. Sci. Hungariciae, 6(3–4):315–339.

———. 1963*a*. Eriophyids of Hungary III. Ann. Hist-Nat. Mus. Hungarici, Pars. Zool. 52:429.

———. 1963*b*. A new genus and three new species of eriophyid mites from Africa and Java. Ann. Hist-Nat. Mus. Hungarici, Pars. Zool. 55:509–511.

———. 1965*a*. *Boczekella laricis* n. g. and n. sp., Aceria szepligeti n. Ann. Hist-Nat. Mus. Hungarici, Pars. Zool. 57:467.

———. 1965*b*. Die Tierwelt Mittleeuropas, Lief 3, III Band.

———. 1966. Some problems in eriophyid systematics. Zes. Prob. Nauk. Roln. 65 :195–198.

———. 1967. Eriophyids collected by Dr. T. Pocs in Vietnam. Ann. Hist-Nat. Mus. Hungarici, Pars. Zool. 59 :385–388.

FEE, A. L. A. 1834. Mémoire sur le groupe des Phylleriees de Fries. 75 pp.

FELT, E. P. 1940. Plant galls and gall makers. Comstock Publ. Co. 364 pp.

HALL, C. C., JR. 1967*a*. Eriophyidae of Kansas. Univ. Kans. Sci. Bull. (Oct.) 47(9):601–675.

———. 1967*b*. A look at eriophyoid life cycles (Acarina Eriophyoidea). Ann. Entomol. Soc. Am. 60(1):91.

HODGKISS, H. E. 1913. New species of maple mites. J. Econ. Entomol. 6:420–424.

———. 1930. The Eriophyidae of New York. Tech. Bull. N. Y. Agr. Expt. Sta. 163 pp.

KEIFER, H. H. 1942. Eriophyid Studies XII (Experiments proving structural deuterogyny). Bull. Calif. Dept. Agr. 31(3):117.

———. 1944. Eriophyid Studies XIV (Account of rearing *Rhyncaphytoptus ulmivagrans* K. to prove *Abacoptes platynus* K. was deutogyne). Bull. Calif. Dept. Agr. 33(1):30.

———. 1969. Eriophyid Studies. C-3.

KRANTZ, G. W. 1970. A manual of acarology. 335 pp. O.S.U. Book Stores, Corvallis, Oregon.

LIPOVSKY, L. J., G. W. BYERS, and E. H. KARDOS. 1957. Spermatophores, the mode of insemination of chiggers (Acarina: Trombiculidae). J. Parasitol. 43:256–262.

MANI, M. S. 1964. Ecology of plant galls in E. P. Felts' plant galls and gall makers. Comstock Publ. Co., Ithaca, N. Y. 434 pp.

NALEPA, A. 1886. Die Anatomie der Phytoptiden. Sitz. Akad. Wiss. Vienna 99:115.

———. 1911. Eriophyiden. Zool. Stuttgart, Heft 61, 169.

———. 1922. Zur kenntnis der Milbengallen einiger Ahornarten und iher Erzeuger. Marcellia 19:3–33.

———. 1924. Polymorphe Eriophyiden. Marcellia 20:87.

———. 1926. Dr. Jegens Eriophyidenstudien in Kritischer Beleuchtung. Marcellia 22:120.

———. 1928. Zur Phänologie und Entwicklungsgeschichte der Milbengallen. Marcellia 24:87.

———. 1929. Neuer Katalog der Bisher Bescrieben Gallmilben. Marcellia 25:67 et seq. (host list).

NAULT, L. R., and W. E. STYER. 1969. The dispersal of *Aceria tulipae* and three other grass-infesting eriophyid mites in Ohio. Ann Entomol. Soc. Am. 62(6):1446.

NUTTING, W. B. 1964. Demodeicidae-status and prognostics. Acarologia 6(3):441–454.

OLDFIELD, G. N. 1969. The biology and morphology of *Eriophyes emarginatae*, a Prunus gall mite, and notes on *Eriophyes prunidemissae*. Ann. Entomol. Soc. Am. 62(2):269.

OLDFIELD, G. N., R. F. HOBZA, and N. S. WILSON. 1970. Discovery and characterization of spermatophores in the eriphyoidea (Acari). Ann. Entomol. Soc. Am. 63(2):520 526.

OLDFIELD, G. N., and I. N. NEWELL. 1973*a*. The role of the spermatophore in the reproductive biology of protogynes of *Aculus cornutus* (Banks) Ann. Entomol. Soc. Am. 66 (1):160–163.

———. 1973*b*. The spermatophore as the source of sperm for deutogynes of *Aculus cornutus* Ann. Entomol. Soc. Am. 66(1):223–225.

OLDFIELD, G. N., I. N. NEWELL, and D. K. REID. 1972. Insemination of protogynes of *Aculus cornutus* from spermatophores and description of the sperm cell. Ann. Entomol. Soc. Am. 65(5):1080–1084.

PAGENSTECHER, H. A. 1857. Uber Milben besonders die Gattung *Phytoptus*. Verh. Ver. Heidelberg 1:46.

PUTMAN, W. L. 1939. The plum nursery mite (*Phyllocoptes fockeui* N. & T.). Seventh Ann. Rept. Entomol. Soc. Ontario, p. 33.

REAMUR, M. de. 1737. Mémoires pour servir a l'histoire des insects tome troisième, de l'imprimerie royal. Academie Royale des Sciences, Paris, 421–423; 511–515.

RICE, R. E., and F. E. STRONG. 1962. Bionomics of the tomato russet mite, *Vasates lycopersici* (Massee). Ann. Entomol. Soc. Am. 55(4):431–435.

ROIVAINEN, H. E. 1947. Eriophyid news from Finland. Acta Entomol. Fen. 3:8.

SCHEVTSHENKO, V. G. 1957. The life history of alder gall mite, *Eriophyes laevis* (Nalepa). Zool. J. 36(3):598–618.

———. 1961. Peculiarities of the postembryonic development of gall-mites and some notes on the classification of *Eriophyes laevis* (Nal.). Zool. J. 40(8):1143–1158.

———. 1967. Account of dimorphism in *Trisetacus kirghisorum* Schevtshenko. Leningrad Univ. Vestnik. Serian. Biolosu. 3:60–67.

———. 1968. The feeding organs of the four-legged mites. [In Russian with English summary] Biologia, Estonian Inst. Expt. Biol. 17(3):248–265.

SINGER, G. 1967. A comparison between different mounting techniques commonly employed in acarology. Acarologia 9(3):475–484.

SOMSEN, H. W. 1966. Development of migratory form of wheat curl mite. J. Econ. Entomol. 59(5):1283.

STERNLICHT, M. 1969. Effect of different wave length of light on behavior of an eriophyid mite, *Aceria sheldoni*. Entomol. Expt. and Appl. 12:377.

STERNLICHT, M., and S. GOLDENBERG. 1971. Fertilization, sex ratio, and postembryonic stages of the citrus bud mite, *Aceria sheldoni* (Ewing), Bull. Entomol. Res. 60(3): 391–397.

THOMAS, F. A. W. 1869 et seq. Uber *Phytoptus* Duj. etc. Zeits. für Naturw. Halle, Stuttgart. 33:312.

———. 1872. Schweizerische Milbengallen Zeitschr. für die Gesamten Naturwis n.s. 5, 39:459–572.

WHITMOYER, R. E., L. R. NAULT, and O. E. BRADFUTE. 1972. Fine structure of *Aceria tulipae*. Ann. Entomol. Soc. Am. 65(1):201–215.

WILSON, N. S. 1965. A new species of blister-forming mite on pear. Ann. Entomol. Soc. Am. 58(3):327–330.

WILSON, N. S., and G. N. OLDFIELD. 1966. New species of eriophyid mites from western North America, with a discussion of eriophyid mites on *Populus*. Ann. Entomol. Soc. Am. 59(3):585–599.

Chapter 13

Injurious Eriophyoid Mites

THE NALEPELLIDAE NEWKIRK AND KEIFER

Phytocoptella Newkirk and Keifer

Filbert big-bud mite, or nut gall mite, *Phytocoptella avellanae* (Nal.).[1] This mite (fig. 102) (= *Acarus pseudogallarum* Vallot, and *Phytoptus coryligallarum* Targioni-Tozzetti) is a serious pest of hazel, filbert, and cob nuts, and has therefore attracted attention for a long time. It evidently occurs wherever these nuts are grown, which includes Eurasia, North America, and Australia. It makes "big buds" on native *Corylus* in North America. It has an unusual life history that is as yet imperfectly known.

Bud injury is the most serious result of the attack of this mite. Terminal buds are the favorite site of invasion, and infested buds become swollen, deformed, fleshy, and pinkish. Damaged male catkins become rigid, brittle, and produce little pollen. Light bud infestations cause injury to external bracts, with resulting deformed, weak, and sickly shoots. Weakened buds produce no nuts. Hazelnut bushes are injured to a greater extent than filbert bushes.

The life cycle of *P. avellanae* is relatively complex. The two nymphal instars that develop in overwintering "big buds" presumably have narrow abdominal rings, as do the adults. Females emerging in spring from "big buds" lay eggs on leaf undersurfaces. These eggs hatch into first stage nymphs that have narrow abdominal rings. But the second nymph on the leaf undersurface displays a most unexpected external anatomy. This second nymph is flat and has broad tergites with laterally projecting fleshy points. These "*Tegonotus*-like" nymphs lie next to veins in a semistationary manner. It would not be possible to connect these peculiar nymphs to the *avellanae* adult were one not able to find adults developing inside of them. Adults that issue from these "*Tegonotus*-like" second nymphs migrate to terminal buds, where they begin making new "big buds."

The common inquiline in *Corylus* "big buds" is *Cecidophyopsis vermiformis* (Nal.), which is easy to confuse with *P. avellanae*, but careful examination of the mites will disclose definite differences between the two species. *C. vermiformis* differs from *P. avellanae* by lacking all shield setae, lacking subdorsal abdominal setae, and by having the female genital coverflap heavily ribbed. *P. avellanae* has no ribs on the female coverflap; also there are internal genital differences between the two species.

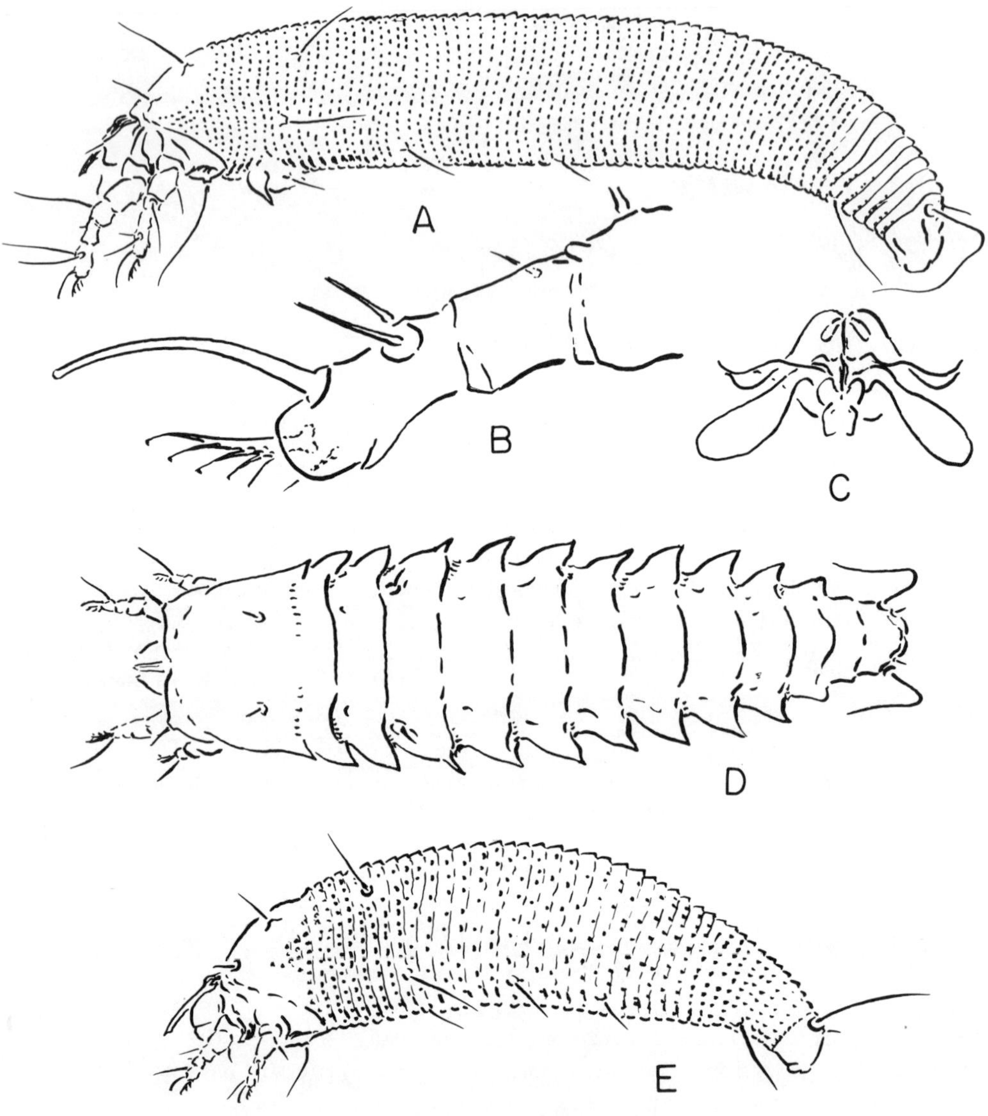

Fig. 102. *Phytocoptella avellanae* (Nal.): *a*, side view of adult female; *b*, foreleg; *c*, internal female genital structures; *d*, dorsal view of second nymph; *e*, side view of first nymph.

The inside walls and partitions in "big buds" have densely matted hairs and thick warty excrescences, or yellow-colored fleshy developments on the inner parts. These projections increase the surface and provide extensively protected habitats for the mites. External bud surfaces become hairy, but lack protuberances.

Mites living within the buds find protection from adverse weather conditions, but during migration to leaves, they are subject to desiccation by warm, dry air,

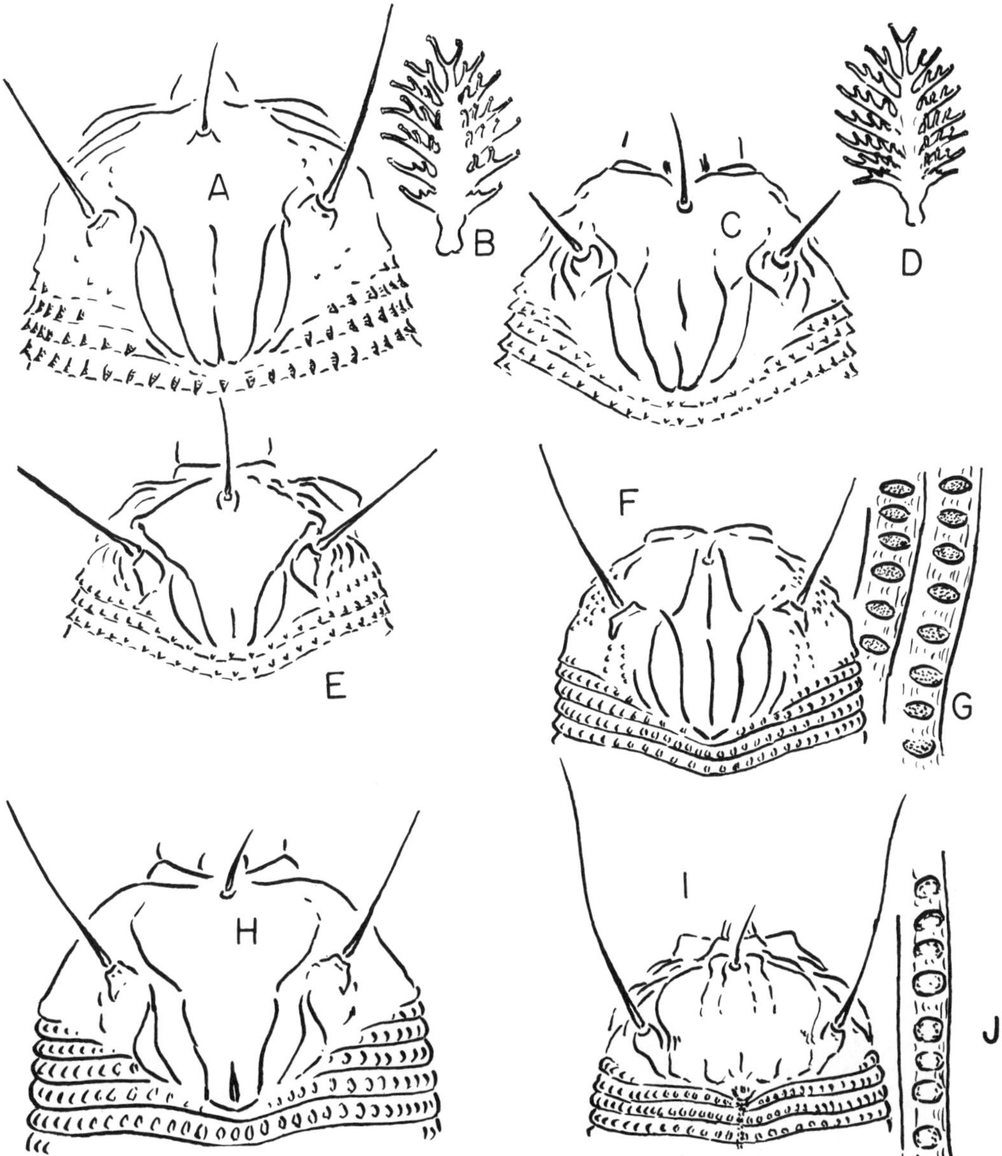

Fig. 103. Shield pattern of *Trisetacus* spp.: *a*, shield of *pini* (Nal.); *b*, featherclaw, *pini*; *c*, shield of *pseudotsugae*; *d*, featherclaw of *pseudotsugae*; *e*, shield of *grosmanni* K.; *f*, shield of *ehmanni* K.; *g*, side skin of *ehmanni*; *h*, shield of *alborum* K.; *i*, shield of *sequoiae* K.; *j*, detail of microtubercles on *sequoiae*.

and some wash off when it rains. It is estimated that during migration periods there is more than 90 percent mortality.

Eriophyid mites commonly associated with *P. avellanae* on the undersides of leaves are: *Tegonotus depressus* Nal., *Aculus comatus* (Nal.), *Coptophylla lami-*

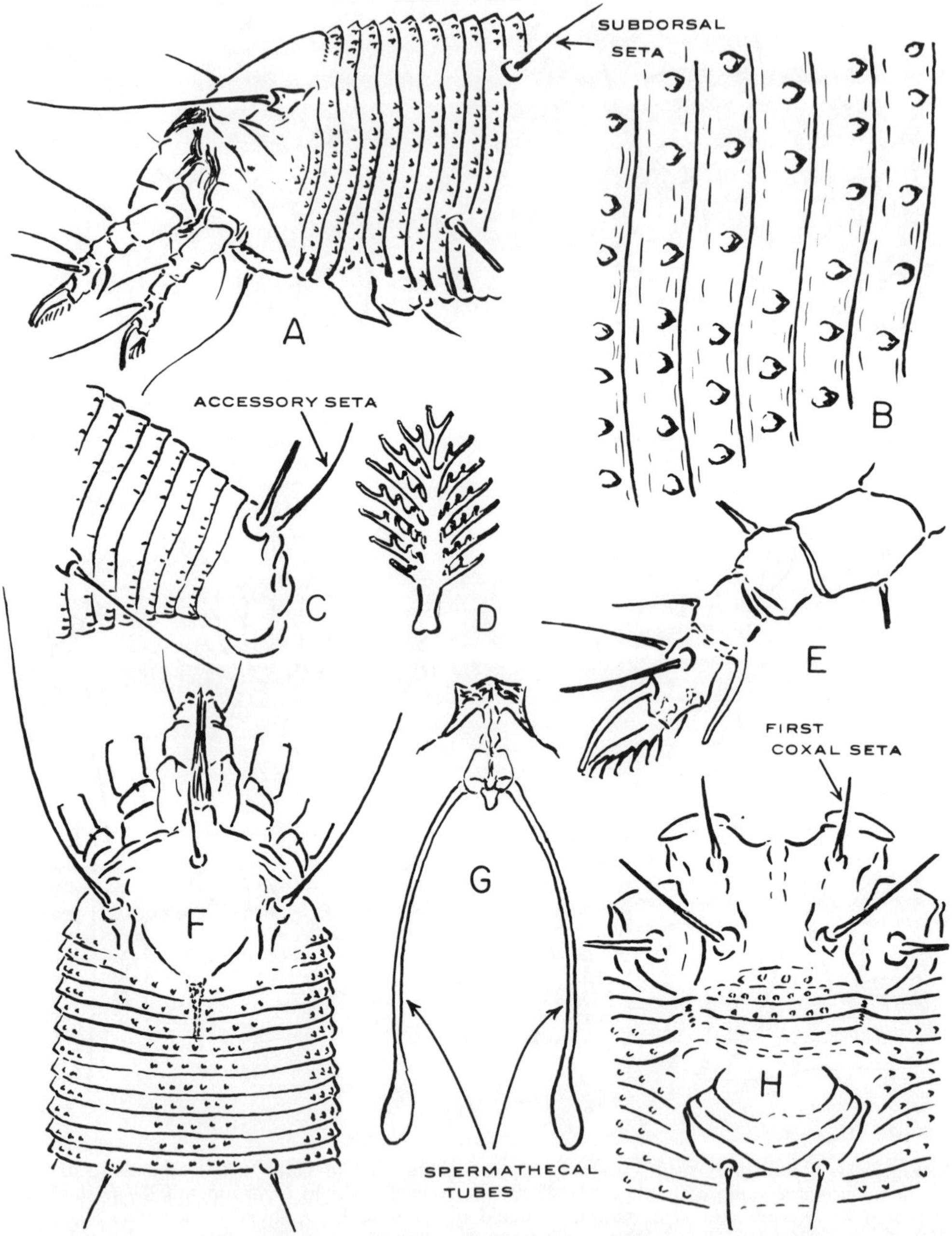

Fig. 104. *Trisetacus juniperinus* (Nal.): *a*, side view of anterior section; *b*, side skin; *c*, caudal section of telosome; *d*, featherclaw; *e*, foreleg; *f*, dorsal view of anterior section; *g*, internal female genital structures, showing extra long spermathecal tubes; *h*, coxae and female genitalia.

mani K., and *Anthocoptes loricatus* Nal. These species either cause leaf rusting, or are innocuous.

Various predators attack exposed mites, including *Anthrocnodax coryligallarum* (Targarth) and *Typhlodromus aberrans* (Oudemans).

Spray applications for this mite are most effective during the migration period, which usually occurs when terminal shoots begin to form the fifth and sixth buds (Massee, 1930; Planes et al., 1965; Vidal-Barraquer et al., 1966).

The female of *Phytocoptella avellanae* (fig. 102, *a, b, c*) is 220 to 260 μ long. The shield has 4 setae and the admedian lines extend from the chelicera base to the rear shield margin between these setae. The featherclaws are 4- or 5-rayed and the foretibia, unlike in most *Phytocoptella* species, lacks the lateral apical spur. The ring microtubercles (granules) are elliptical and slightly pointed. There is a strong sternal line between the forecoxae, and the female genital coverflap is ribless (fig. 102).

Phytocoptella abnormis (Garman). This mite makes leaf galls on North American linden in the eastern United States. Nalepa refers this mite to his *Eriophyes tetratrichus*, calling it a subspecies. But the name *abnormis* has seven years priority over Nalepa's name, so it cannot be a subspecies or variety. The precise taxonomic relationship between *abnormis* and *tetratrichus* remains to be investigated. Nalepa's *tetratrichus* inhabits European linden, or leafed lime. *P. abnormis* is similar to *avellanae* by lacking the foretibial lateral spur, but the shield differs by having lateral lines, and the microtubercles have no appreciable point.

Phytocoptella hedericola (K.). This species (fig. 94, *a*) occurs in buds and at petiole bases of English ivy along the California coast, causing leaf stunting and deformation, particularly on dwarf potted ivy. That this mite lives near the ocean shows that it prefers high atmospheric humidity. This mite never appears on the same host growing in interior California valleys.

P. hedericola is 150 to 180 μ long. Its distinguishing features are the 5-rayed featherclaw, the foretibial lateral apical spur, and strongly lined shield surface. The dorsal seta pair are longer than the anterior shield setae. *P. hedericola* has elongate microtubercles.

Phytocoptella yuccae (K.) (fig. 94, *b, c*) occurs deep in the central more succulent leaf bases on yucca in the southwestern deserts. It is 300 μ and more long and is similar to grass-infesting *Phytocoptella* spp. It illustrates the tendency for these grass infesting mites to have dorsal setae that are shorter than the anterior setae.

Additional species of *Phytocoptella* in California are: *montanus* (K.) 1954, on sedge; *corniseminis* (K.) on dogwood; *garryana* (K.) (fig. 94, *e*) on *Garrya elliptica* Douglas; and *leucothonis* (K.) (fig. 90, *a*), on sand myrtle. All of these species have the foretibial lateral spur.

Trisetacus Keifer (figs. 103, 104)

Trisetacus quadrisetus (Thomas). This typical mite is the juniper berry mite and appears to be limited to that host. It occurs in the Northern Hemisphere. During the summer it lives inside juniper berries on several different species of that genus. Western juniper and California juniper are regularly found to have heavy infestations, with nearly every berry on the shrub or tree formed into a mite gall. Part of the apex of the seed or seeds of attacked berries is so exposed that the small escape hole is outside. The inside of the seeds is entirely destroyed, having been changed into a mite brood chamber (Morgan and Hedlin, 1960).

T. quadrisetus, as it exists in California juniper berries, has a 9-rayed featherclaw. The shield has no central lines and there is no gland pit at the central rear margin. The abdominal microtubercles are conical and evenly distributed. Accessory setae on these mites are especially prominent, being 20 to 22μ long, stout, and upcurved. We consider *"Eriophyes" ramosus* Hodgkiss, 1918, to be a synonym of *T. quadrisetus.*

Trisetacus juniperinus (Nal.) (fig. 104) is the species that Nalepa reported as causing needles of ornamental junipers to become basally swollen and crowded; also terminal buds of low, ornamental junipers with pointed projecting needles, are often killed. *T. juniperinus* is similar to *quadrisetus*, but it has a longer frontal shield seta. Perhaps the most characteristic feature of *juniperinus* is the microtubercle pattern on the front part of the abdominal dorsum. These microtubercles are absent from two subdorsal areas that extend longitudinally back from the rear shield margin. Another character of *juniperinus* is the presence of a gland pit at the rear central shield margin.

Trisetacus pseudotsugae. This mite (fig. 103, *c*, *d*) attacks Douglas fir in western North America and causes both "big bud" and terminal bud proliferation. Forest trees show a range of resistance to "big bud" development. Part of the buds affected with "big bud" die, but some recover and resume twig growth. Bud proliferation has so far been confined to potted seedlings on which the mite prevented further growth (Buxton, 1969; Coop. Econ. Ins. Rept., 1969).

T. pseudotsugae is 300 to 400 μ long, and wormlike. Body color is yellowish white. The featherclaw is 8-rayed. The central anterior shield seta is 13 μ long. Shield lines are confined to the rear shield half, but the median line, admedian lines, and submedian lines are present. Abdominal microtubercles are pointed as on *T. pini* (Nal.), but *pini* has coarser microtubercles, and a 7-rayed featherclaw.

Trisetacus laricis (Tubeuf). This species occurs on European larch and evidently causes "big bud" on that host. *T. cembrae* (Tubeuf causes bud proliferation on Swiss stone pine in Europe. Until these species are carefully studied on their type hosts, the precise placement of them in the genus is uncertain. Nalepa placed both as subspecies of *pini*, but that relationship is probably not tenable.

Trisetacus abietis Postner, which is characterized in the description as having 5-rayed featherclaws, attacks *Abies alba* Miller in Europe. Damaged needles are said to have brownish cross stripes.

Trisetacus grosmanni K. This mite is the Sitka spruce bud mite (fig. 103, *e*). It damages Sitka spruce buds in Europe, and probably elsewhere. It is a yellowish wormlike mite, 225 to 245 μ long, with an 8-rayed featherclaw. The anterior shield seta is 20 μ long. In company with other species in the genus which have pointed microtubercles, the shield lines are mostly confined to the rear part of the shield; here the median line is very short.

Trisetacus ehmanni K. This species (fig. 105) infests three-needle pitch pines in western North America. It has been reported (under the name of *pini*) as having caused needle yellowing on Monterey pine on the California coast (Walther, 1925). It probably contributes to gumming at needle bases. This species is a yellowish wormlike mite 215 μ to 300 μ long. A principal character is the very short anterior median shield seta, which is 6 μ long. The shield lines differ from species in the genus which have pointed microtubercles because of the fact that these lines extend well onto the anterior half of the shield. The featherclaws are 7-rayed, and the ring microtubercles are elliptical and flat.

Trisetacus alborum K. This mite (fig. 103, *h*) infests white pines, principally in western North America. The type host is mountain white pine. The mites live in needle sheaths, but the sheaths tend to shatter early on white pines so mites may be difficult to find on this host. This mite has been reported to be the cause of bud proliferation of sugar pines; but the specimens received for identification had shorter anterior shield seta than those on mountain white pine.

T. alborum is 310 to 410 μ long, wormlike. The anterior median shield seta on examples from the type host is 10 μ long. The front half of the shield lacks middle lines, the flared admedians running forward just inside the dorsal tubercles. The ring microtubercles on *alborum* are elliptical-flattened.

Trisetacus sequoiae K. (fig. 103, *i, j*) occurs on coast redwood in California. It damages buds, causes browning under bud scales, and prevents further growth, principally on lateral buds along twigs. Occasionally large twig galls develop on redwood, but as yet it is not possible to connect directly the galls with this mite.

T. sequoiae has an 8-rayed featherclaw. The anterior shield seta is 4 μ long. Curved shield lines frame the center of the shield, but there arc no appreciable central lines. There is a central rear gland on the caudal shield margin. Ring microtubercles are more subcircular than elliptical, and but gently humped.

Trisetacus pini (Nal.) This species (fig. 103, *a, b*) is the twig gall or twig knot mite of Scotch pine in Europe (pl. 61). The twig swellings, or knots, are supposed to persist several years and turn out successive broods of mites (Kruel, 1963).

T. pini is a member of the genus that has a 7-rayed featherclaw, a 13 μ long central anterior shield seta, and ring microtubercles that are pointed over the entire abdomen. The shield lines are confined to the rear half of the shield, but the median line extends forward to a point even with the dorsal tubercles.

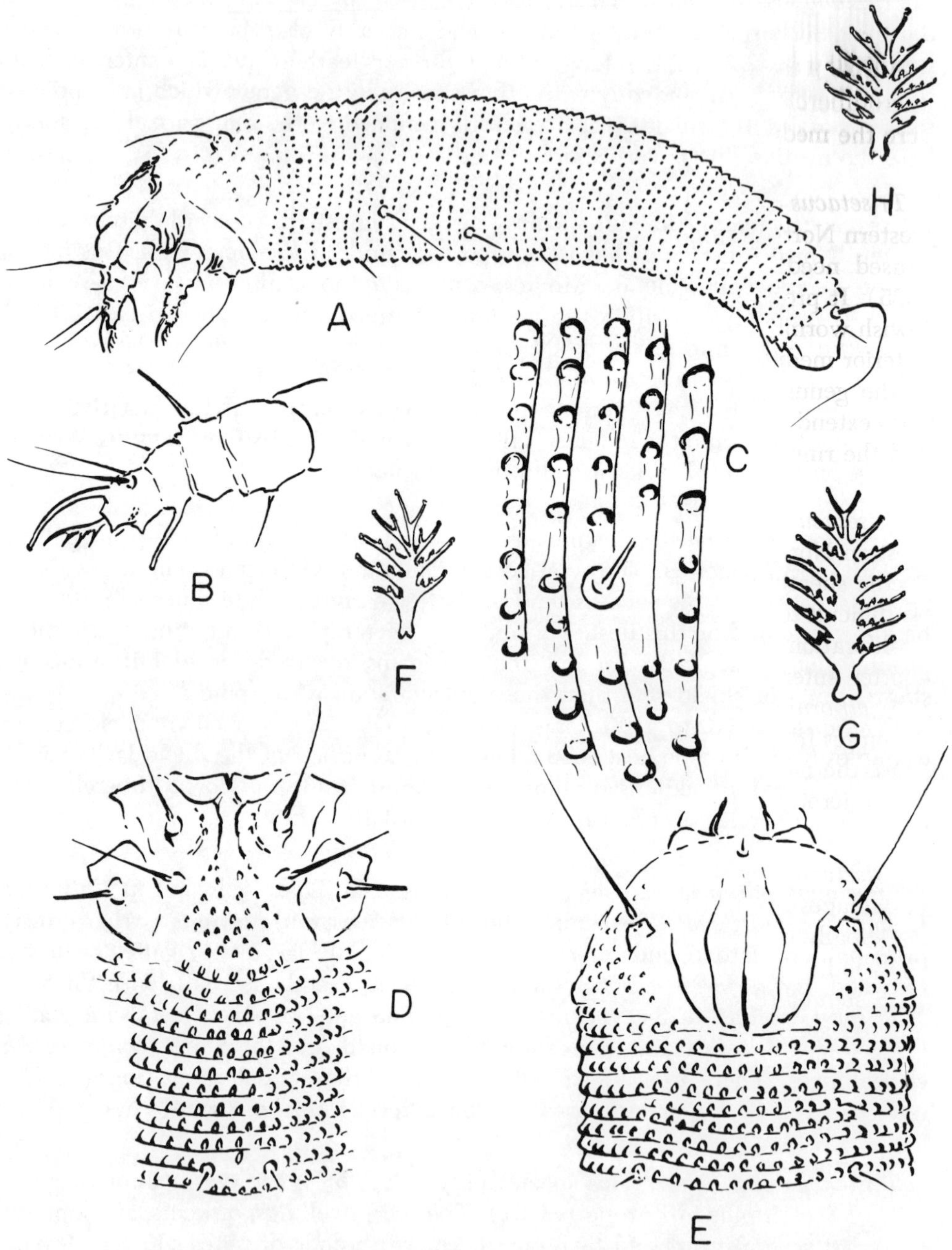

Fig. 105. ***Trisetacus ehmanni*** K., first nymph: *a*, side view; *b*, foreleg; *c*, side skin structures; *d*, coxae and genital setae; *e*, shield and subdorsal setae; *f*, featherclaws of, first nymph; *g*, adult; *h*, second nymph featherclaw.

Trisetacus kirghisorum Shevtchenko. This mite lives on *Juniperus semiglobosa* Regel in Siberia. According to Schevtchenko and Demillo (1968) both males and females hibernate (as does *ehmanni* in California). But there are supposed to be two forms of both males and females in this juniper species.

Trisetacus capnodus Keifer and Saunders. *Capnodus* attacks Scotch pine in the state of Washington, feeding within needle sheaths, and typically causes chlorosis, stunting, and needle distortion. The distinguishing features of this species are the shape of the microtubercles. These structures are basally elliptical, and each bears a backward projecting point dorsally and ventrally on the abdomen; but laterally the microtubercles are unpointed.

Acathrix Keifer

Acathrix trymatus K. *Trymatus* (figs. 92, *a*; 93, *b*) was originally collected on coconut in the Philippines by Magdalena L. Briones in connection with investigations on coconut blight, or cadang-cadang. The mites live on the emerging leaflets as they protrude folded from the bud. Mites also occur under brownish floss material on the third, fourth, and fifth fronds above the base of the crown. Eggs are oval, translucent, or whitish, with dimensions of 50 to 72 μ. Eggs are able to withstand immersion in water for more than 3 hours without adverse effect (Briones and Sill, 1963).

A. trymatus is similar to the genus *Phytocoptella*, but differs mainly in the extremely reduced dorsal setae that occur on dorsal tubercles near the rear shield margin. The featherclaws have eleven rays, and are unusually long and slender; the hind claw is especially long. The shield design consists of many subparallel fine longitudinal lines. Ring microtubercles on *A. trymatus* are elongate-elliptical.

Nalepella Keifer

The members of this genus are needle vagrant mites on fir, spruce and hemlock. They are robust species with large rostrums and long chelicerae. As presently constituted the genus seems to have two major divisions. One of these includes the genotype *triceras* Börner and has wide tergites; the other group has tergites that are nearly as narrow and the sternites and resemble members of the genus *Setoptus* K.

Nalepella tsugifoliae K. (fig. 106) is an example of the group with tergites that are similar to the sternites. In eastern North America, it browns and discolors needles on Canadian hemlock.

Setoptus Keifer

Setoptus strobacus K. This species (fig. 107) attacks white pine in eastern North America where it stunts the needles. *S. strobacus* is wormlike, yellowish, and is from 200 to 225 μ long. The featherclaw is 10-rayed. The shield surface is set with scattered small granules. The anterior shield seta is 21 to 25 μ long. The rear edge of the foretibia and hind tibia bears spines. The microtubercles on the abdominal rings are spinelike. The coxae have some spinules distributed over the sur-

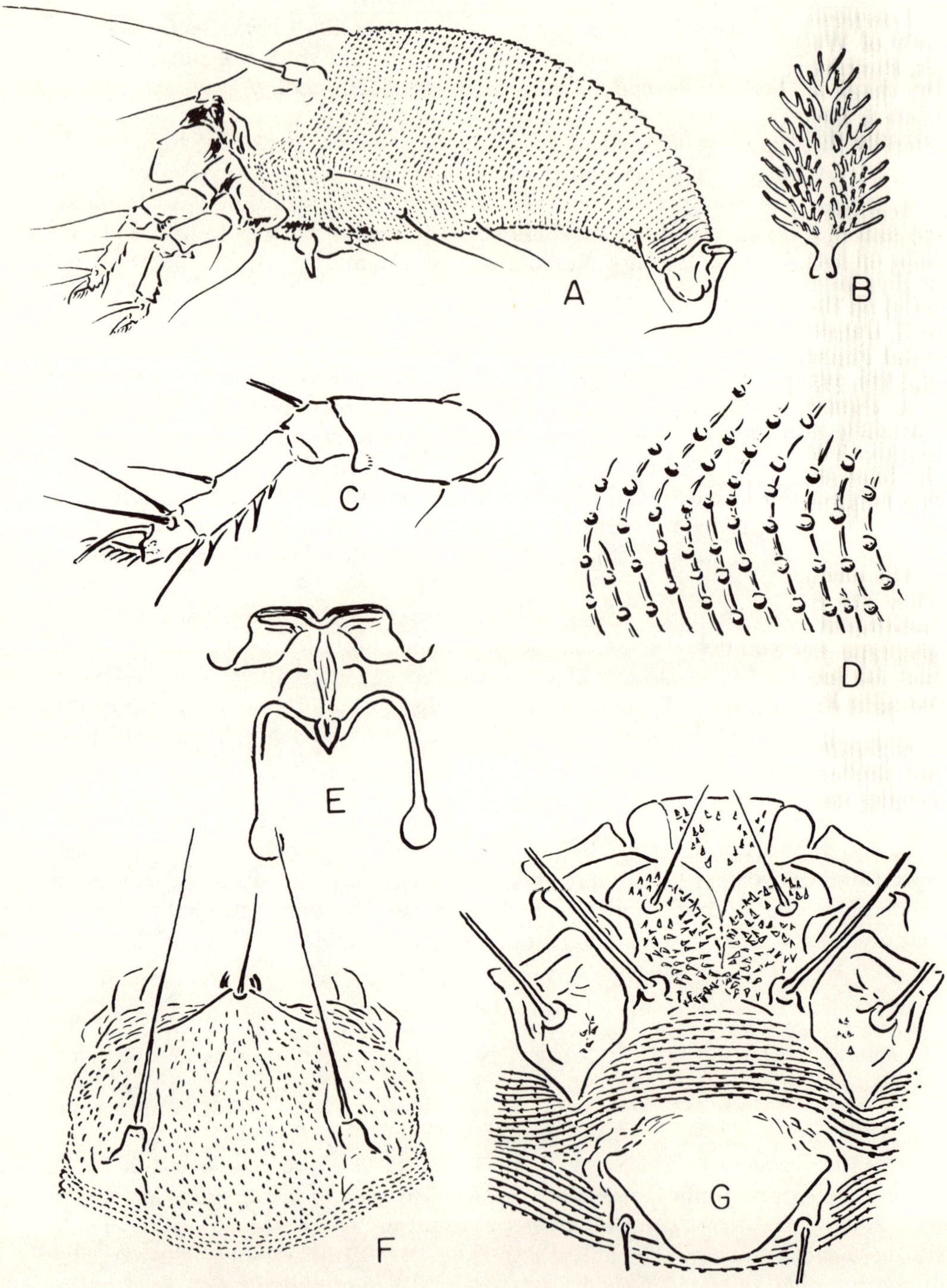

Fig. 106. *Nalepella tsugifoliae* K.: *a*, side view of mite; *b*, featherclaw; *c*, foreleg; *d*, side skin; *e*, internal female genital structure; *f*, shield; *g*, coxae and external female genitalia.

face. *S. strobacus* differs from *S. jonesi* by having a longer anterior shield seta, more prominent spines on the tibiae, and more pointed dorsal ring granules.

Mackiella Keifer

Mackiella phoenicis K. This mite (fig. 94, *f*), found on date palm, belongs to the tribe Mackellini, because it lacks subdorsal abdominal setae. It occurs on its host in folds in emerging fronds. So far no damage has been connected with it. The mite occurs in California, but undoubtedly lives on date palm in the regions where the palm is native. *M. phoenicis* is 180 μ long, with a 7-rayed featherclaw. There are 4 shield setae and the anterior shield lobe is broad. The tergites on this species are much broader than the sternites, and have longitudinal lines.

Retracrus johnstoni K. This mite (fig. 94, *d*) makes black blotches on the undersides of fronds of *Chamaedorea* palms in Central America, including southern Mexico. The mite secretes copious amounts of flocculent waxy covering, which is usually scattered on the black blotches of the host. Since the mites deface the fronds of a palm used as an ornamental, the species is an important pest. These palms are constantly intercepted at the Texas-Mexican border. As the colonies of the mite continue to flourish on black areas, the color of these areas does not indicate that the tissue has died.

The female of *R. johnstoni* is 170 to 185 μ long. The mite, minus wax, is a whitish species, partially translucent. It belongs to the Mackellini tribe because it has 4 shield setae, but it lacks the subdorsal abdominal setae. The featherclaw is 6-rayed. There is no shield design, but the dorsal tubercles are unusual in the Nalepellidae, and direct the dorsal setae caudad. Shield setiferous tubercles are produced and bulbous. The foretibial lateral spur and tarsal claws have much the same form. Partial tergites below the dorsal tubercles end in points. Abdominal tergites are somewhat broader than the sternites.

THE ERIOPHYIDAE NALEPA

THE NOTHOPODINAE KEIFER

In the subfamily Nothopodinae, the tibia and tarsus on both front and rear legs are fused, or nearly fused, into a tibiotarsus. The result is that the foretibial seta, which is present on the great majority of eriophyoids, is absent; and the tarsal seta pair are strengthened to accomodate for the loss of this seta. A characteristic that most often goes with this leg segment reduction is the fusion of the forecoxae across the midline, and the accompanying absence of the first coxal setiferous tubercles.

Colopodacus Keifer

Colopodacus africanus K. This species (fig. 108, *d, e*) occurs on coffee in the Congo where it joins several other rust mites in browning the underside of coffee bush leaves. *C. africanus* differs from most Nothopodines by having the forecoxal first setiferous tubercle pair, and the coxae are somewhat separate. The legs are

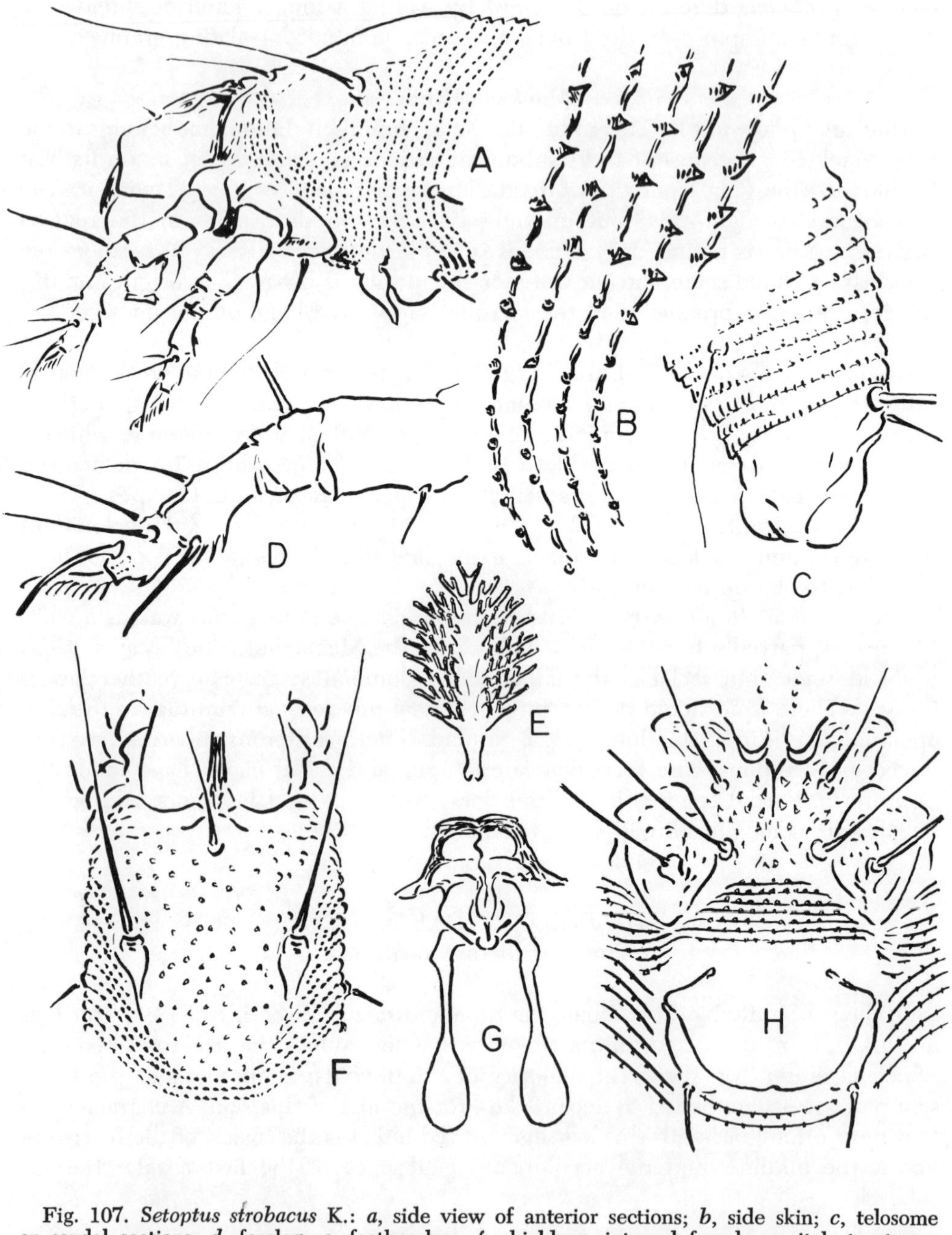

Fig. 107. *Setoptus strobacus* K.: *a*, side view of anterior sections; *b*, side skin; *c*, telosome or caudal sections; *d*, foreleg; *e*, featherclaw; *f*, shield; *g*, internal female genital structures; *h*, coxae and female genitalia.

typically Nothopodine in segmentation. The shield is ornamented with a complex network of lines. It has a short anterior lobe and is subcircular dorsally in lateral view. The dorsal tubercles are located ahead of the rear shield margin, with the setae pointing up and toward each other. This species has a 5-rayed featherclaw. The abdomen tapers with little difference between the rings on the dorsum and venter. The microtubercles are longitudinal lines that touch the rear ring margins, making a slightly sharp projecting point.

Cosella Newkirk and Keifer

Cosella fleschneri (K.) This mite (fig. 108, *g, h*) is a rust mite, or leaf vagrant, on several kinds of citrus in India. This mite has almost entirely fused and irrorated forecoxae, which lack the first setiferous forecoxal tubercles. The shield has a slight anterior lobe, and is ornamented with an open network of lines. Shield dorsal tubercules are a little ahead of the rear margin and direct the setae up and inward. This mite has a 4-rayed featherclaw. The abdominal rings are moderately wide, with fine elongate microtubercles. The forecoxae are generally irrorated with fine granules that extend onto the suboral plate. The female coverflap is also irrorated with granules. The species was originally placed in the genus *Floracarus*, but in that genus the forecoxae are more separate. For a reference to *Cosella*, see the generic key in the appendix 3, below.

THE CECIDOPHYINAE KEIFER

In members of this subfamily, the genitalia are appressed to the coxae, which tends so to spread these coxae that the forecoxae, instead of having the usual broad contact along the sternal line, are narrowly tangential centrally on most species. This forward push of the genitalia tends to project them down below the body surface more than usual. Another fairly consistent characteristic is the heavily ribbed coverflap, with ribs in 2 uneven ranks. The forecoxae of the Cecidophyinae usually have prominent curved lines around the setiferous tubercles.

The most notable Cecidophyine characteristic is the shortened female internal anterior apodeme on the female chitinous frame under the coverflap. The forward crowding of the genitalia causes this apodeme to bend up, so in ventral view it appears in the great majority of these species as a thick transverse band or line, either curving, or with lateral ends bulbous. Members of the Cecidophyine tribe Cecidophyini Newkirk and Keifer (see appendix 3) lack dorsal tubercles and dorsal setae. This tribe includes *Cecidophyopsis* K. (1966). Members of the Cecidophyine tribe Colomerini Newkirk and Keifer (see appendix 3) have dorsal setae on the shield.

As with the Nothopodinae, the cecidophyine species have independently moved into most of the plant habitats available to eriophyoid mites. Thus, within both of these subfamilies are rust mites, bud mites, "big bud" mites, erineum makers, and gall mites.

Cecidophyopsis Keifer

Cecidophyopsis ribis (Westwood). *C. ribis* (fig. 109) is the black currant big bud mite. Throughout its history *C. ribis* has appeared in literature under various generic names, including *Phytoptus* and *Eriophyes*. It is a pest of black currants in Europe and occurs in British Columbia. This mite not only damages buds, but

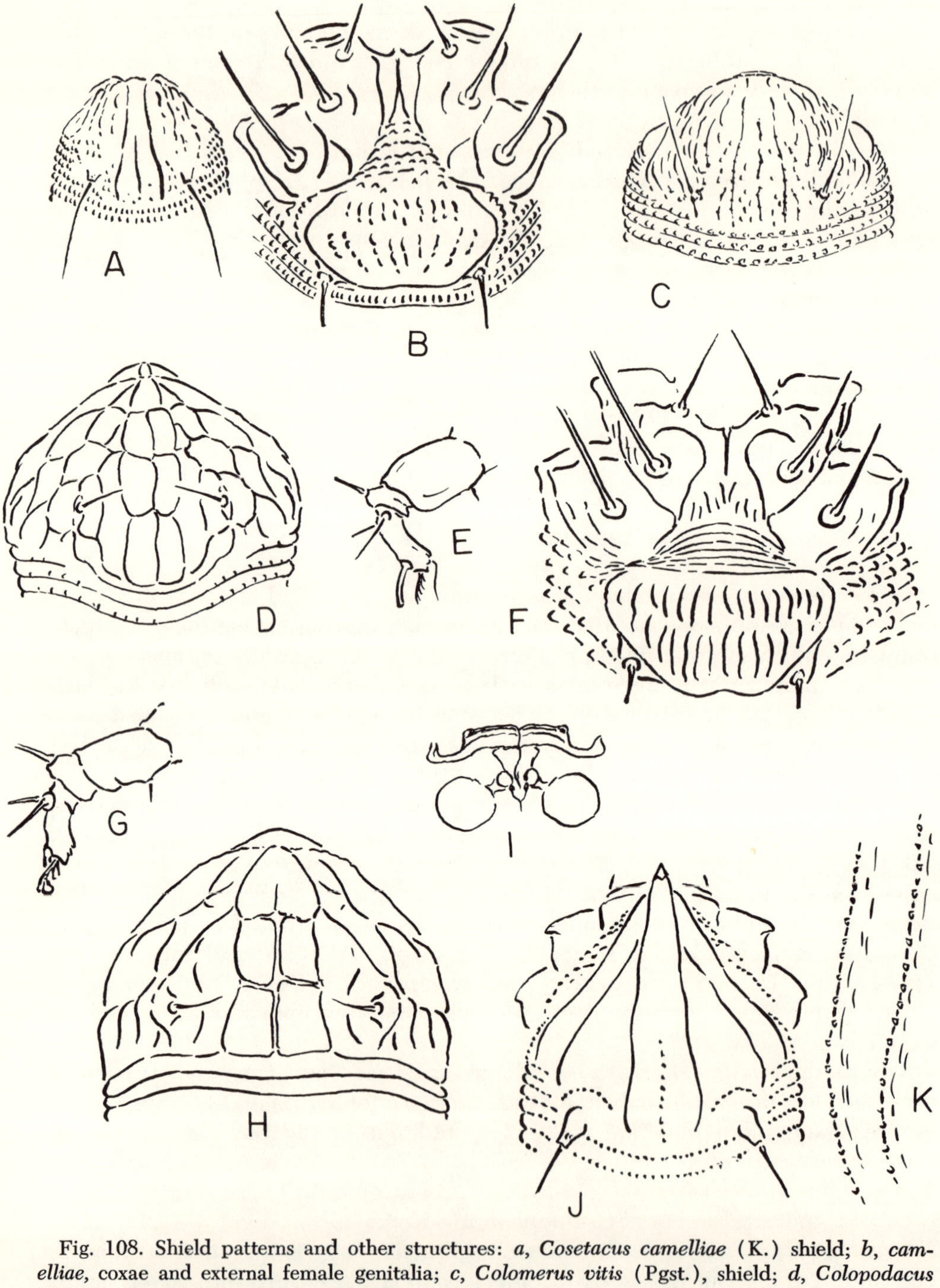

Fig. 108. Shield patterns and other structures: *a*, *Cosetacus camelliae* (K.) shield; *b*, *camelliae*, coxae and external female genitalia; *c*, *Colomerus vitis* (Pgst.), shield; *d*, *Colopodacus africanus* K., shield; *e*, *africanus* foreleg; *f*, *vitis*, coxae and external female genitalia; *g*, *Cosella fleschneri* (K.), foreleg; *h*, *fleschneri*, shield; *i*, *vitis*, internal female genitalia; *j*, *Aculodes dubuis* (Nal.), shield; *k*, *dubuis*, lateral microtubercles.

also carries the virus of currant reversion.

Medium to high infestations of this mite cause the young buds to become galled, to fail to flower, and frequently to turn into "big buds" that fail to differentiate into flowers or true leaves. Buds less seriously injured may develop one or more secondary buds that give rise to grossly distorted foliage and abnormal flowers. A general reduction in vigor occurs when infestations are high and leader growth is replaced by the production of many short twiggy side shoots.

The injury caused by mite feeding and the symptoms of the reversion virus are sometimes confused, but each show distinct characteristics. When mites are controlled the injury does not appear the following year, but virus symptoms continue the second and third year after infection. Segregation of symptoms has been difficult, not only because the mites transmit the virus, but the virus makes the plant more susceptible to infestation by the mites; that is, fewer mites reach or enter healthy buds than buds on bushes with reversion virus. An obvious difference between healthy and "reverted" black currants is the hairiness of the leaves and young stems. Mites are impeded by the hairs during the spring dispersal period, especially those on the stipules and at the base of the petioles. As hairs are almost absent on reversion bushes, the mites can readily migrate and infest the young buds. Reversion virus appears to break down a natural resistance mechanism, which increases colonization and reinfestation by mites.

The several currant varieties differ considerably in the hairiness of their leaves and stems. More resistant varieties to mite infestations, and thus to the virus, are those with the most hair (Thresh, 1964*c*).

Bushes infected with the reversion virus can be detected earliest from leaf characteristics arising from buds on the previous year's growth wood. The reverted leaf tends to be narrower and smaller than normal, and flattened at the base, owing to a reduction in the depth of the basal sinus. Infected leaves from extension growth are flatter than healthy ones. Healthy leaves of most varieties have many pointed serrations along the leaf margin and 13 or more serrations on one side and 5 or more submain veins from the midrib. By comparison, reverted leaves have usually less than twelve broad, rounded serrations with less than 5 submain veins; but diagnosis is often complicated by varietal differences and the effects of damage by aphids, capsids, or other insects, as well as the effects of pruning. Diseased bushes soon cease to bear fruit when attacked by the most severe virus strains; whereas less injurious virus strains merely cause a decrease in fruit production.

The flower sepals of healthy black currants are actually magenta but appear grey because of numerous hairs. By comparison, sepals on reverted bushes show the magenta color because the hairs are almost entirely absent. The difference is best seen just before the first flowers open. Similar symptoms occur in all currant varieties but most clearly in hairy types, such as Baldwin.

C. ribis mites feed and reproduce only on black currant. Migration from buds begins in early March and continues until mid-June, but the migration peak occurs during the latter part of April or in May. During this migration period the mites live unprotected and feed on the outer bud scales, shoots, and leaves. Some find their way to the petiole and seek protection between its upper surface

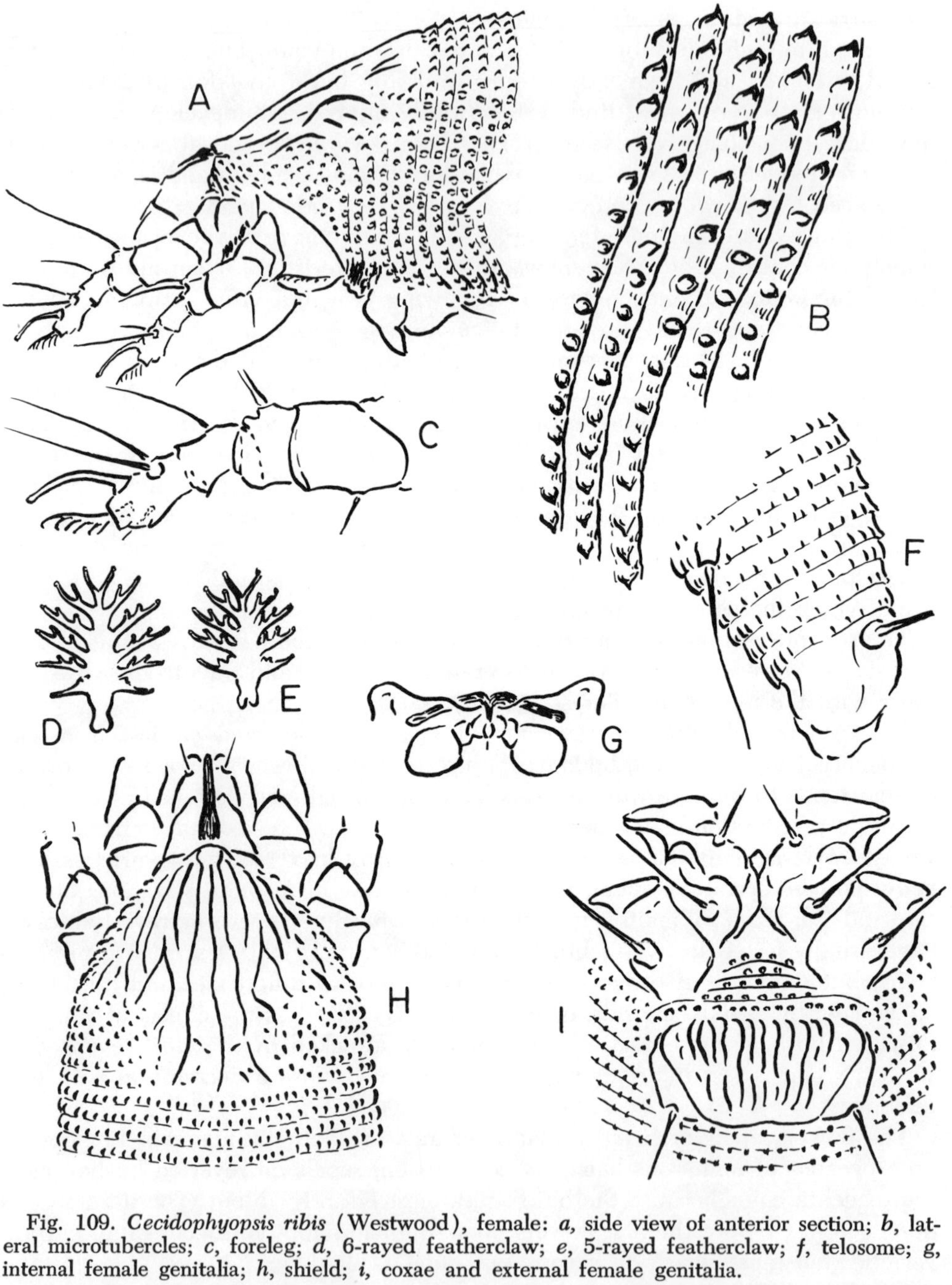

Fig. 109. *Cecidophyopsis ribis* (Westwood), female: *a*, side view of anterior section; *b*, lateral microtubercles; *c*, foreleg; *d*, 6-rayed featherclaw; *e*, 5-rayed featherclaw; *f*, telosome; *g*, internal female genitalia; *h*, shield; *i*, coxae and external female genitalia.

and the main stem; there they feed, reproduce, and eventually enter the small, almost invisible buds. As the lower leaves of the new shoots harden, the mites migrate toward the apex of the newly developing leaves where most of the infestation finally develops. They enter new buds from late May until late July where they lay their eggs and multiply rapidly. "Big buds" may appear as early as July, become abundant from September to November, and continue to develop sometimes until March. Egg laying occurs throughout the year, but mainly during the spring and autumn months. The mites can resist temperatures of –7 to –2 C (20 to 29 F) and become active as soon as the temperature rises. Egg-laying starts 3 to 4 days after entry into newly formed buds. The eggs hatch in about 3 days, and many generations of mites develop each year.

In the fall the mites crawl out of the "big buds" and some distribute themselves around the extremities of the deformed leaves, others make their way down the stems where they may fasten themselves to the plant surface by means of their suckerlike abdominal appendage and wave their body and legs in the air. In such position they may readily be picked up by the wind or moving objects or they may actually leap into the air and drift to a new location on leaves, then make their way to a new bud. They have been observed to leap up to 400 μ or about 16 times their own length (Kanervo and Makien, 1967; Massee, 1928; Smith, 1960; Thresh, 1963, 1964*a*, 1964*b*; Vrie, 1967).

It is possible to depress the prevalency of reversion disease by using certain nursery and field practices. One method is to propagate only from plants that have no "big bud" or reversion symptoms, then treat cuttings in warm water (40 C for 40 minutes). Plant these cuttings away from currants that have disease and mites, and spray thoroughly with an acaricide several times during the period of spring mite dispersal. Inspect bushes for reversion at the end of July in the first year, in May, June, and in July the second year and for galled buds during the first winter on lifting.

Fruiting plantations established with bushes of a reliable certified stock should be planted away from sources of infestation and sprayed intensively to eliminate mites, particularly before the bushes start to crop. Inspections made at the appropriate times followed by roguing of the bushes should control the spread of the mites and disease.

Treat nursery plants with an appropriate acaricide when growth starts; follow by four more applications 10 to 14 days apart. Mature plants producing berries for the market may be treated at first flower, 3 weeks after first flower, or as last flowers open and again 2 weeks later. Lime-sulfur, endrin, and endosulfan have been effective treatments (Thresh, 1964*b*).

In Finland, the larvae of *Tetrastichus eriophyes* Taylor destroys a considerable number of mites but it does not appear to be of great practical significance. (Callingwood, 1959).

The female of *C. ribis* is wormlike and 230 to 240 μ long; the color is light yellowish. Featherclaws on *C. ribis* are 5- or 6-rayed (pl. 62). The shield has but a slight central lobe over the rostrum base. The shield design consists of strong longitudinal lines, with submedians forked just laterally to the admedians. The sides of the shield are granular. The shield bears no dorsal tubercles or dorsal setae, a characteristic of *Cecidophyopsis.* The ring microtubercles cover the whole abdomen and are pointed. There is no accessory seta. The female genitalia are appressed to the coxae and the female coverflap has about fourteen longitudinal ribs in two uneven ranks.

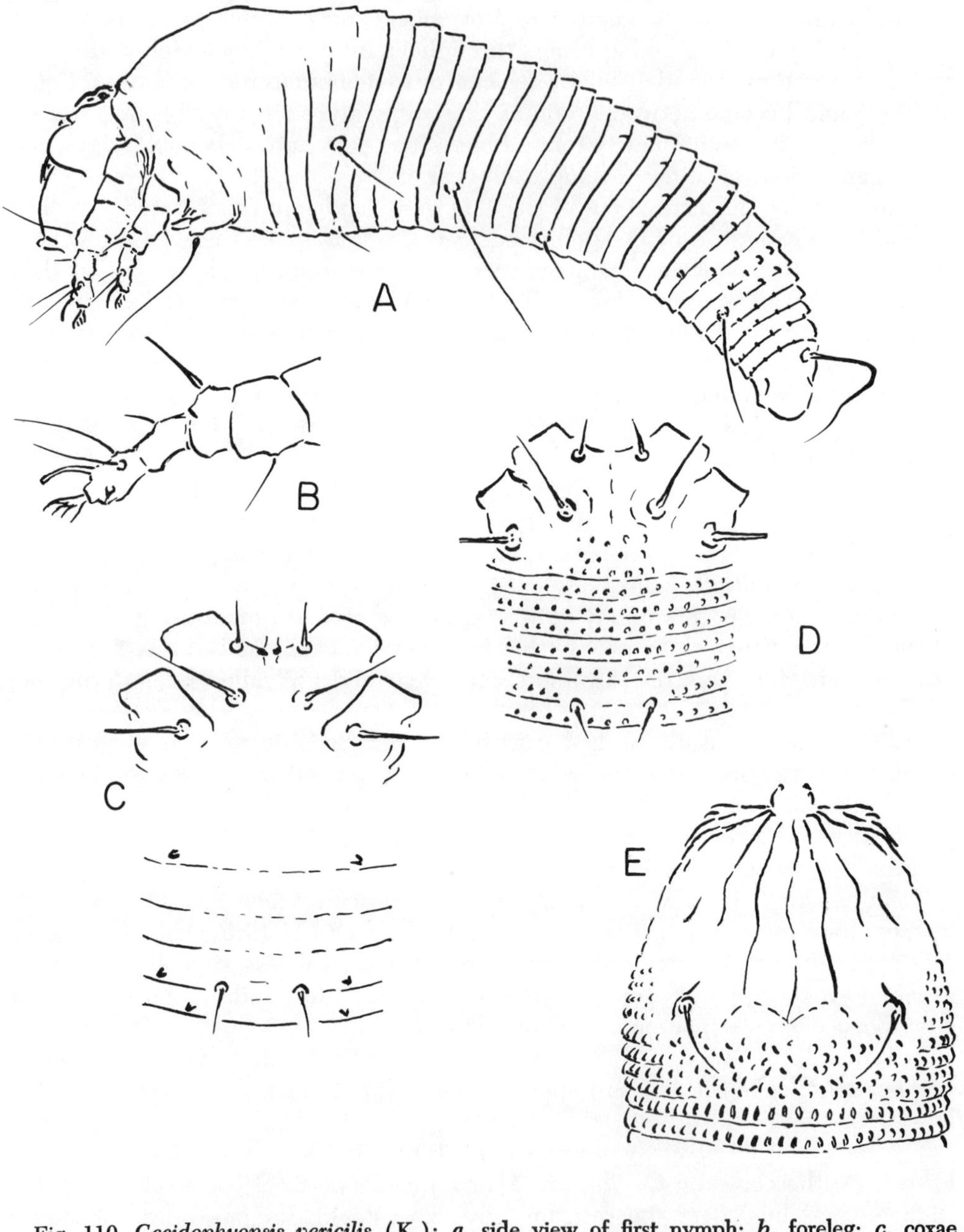

Fig. 110. *Cecidophyopsis vericilis* (K.): *a*, side view of first nymph; *b*, foreleg; *c*, coxae and genital setae; *Gammaphytoptus comphorae* K.: *d*, coxae and genital setae, first nymph; *e*, shield, first nymph.

Cecidophyopsis selachodon van Eyndhoven. This is the red currant big bud mite (fig. 133, *h*). It is structurally very similar to *C. ribis* but may have slight differences in the submedian shield line arrangements. The female genital coverflap has fewer longitudinal ribs. Van Eyndhoven states that experiments on transferring these mites from red to black currant, and vice versa, have always failed. This experiment provides important knowledge in connection with these mite populations. Few eriophyids have so far received this sort of experimentation, which is so badly needed (van Eyndhoven, 1967).

Additional eriophyids that Nalepa indicates are on black currant, and which are in his 1929 host list are: *Anthocoptes ribis* Massee, *Tegonotus neglectus* (Massee), and *Phyllocoptes masseei* Nal.

A mite, which Nalepa considered to be an inquiline in filbert big buds, is *Cecidophyopsis vermiformis* (Nal.). This species occurs along with *Phytocoptella avellanae* (Nal.). The precise relationship of *vermiformis* to *avellanae* in the development of the big buds remains to be discovered. Nalepa also listed *vermiformis* (*betulae* Nal.) as an inquiline in certain birch buds, the birch buds having been damaged by *Acalitus rudis* (Can.).

The yew big bud mite, *Cecidophyopsis psilaspis* (Nal.), was discovered originally in Europe, but has spread to the Western Hemisphere. It now occurs on its host in North America from New England to British Columbia.

Cosetacus Keifer

Cosetacus camelliae (K.). This species (fig. 108, *a, b*) occurs under scales of both vegetative and flower buds on camellia. It probably contributes to premature flower drop. Camellia is a native of Asia, but this mite has never been reported from that area. *C. camelliae* is a whitish wormlike mite, 180 to 200 μ long. The featherclaw is 6-rayed. The median shield line is present on the rear half of the shield. The admedian shield lines are complete. The shield laterally has confused short lines and some granules. The dorsal shield tubercles direct the setae divergently to the rear. Abdominal microtubercles are elongate-elliptical. The female genitalia are appressed to the coxae and the female genital coverflap has light longitudinal lines in 2 ranks; there are a few lines present on the coxae. This mite was originally named in the genus *Aceria*, which is now a synonym of *Eriophyes*. It has genital characters that place the mite species in the subfamily Cecidophyinae, hence the erection of *Cosetacus* to receive the species.

Colomerus Newkirk and Keifer

Colomerus vitis (Pagenstecher). *C. vitis* (fig. 108, *c, f*) is the grape bud mite and grape erineum mite. It has previously been placed under the name *Eriophyes vitis* (Pagenstecher), but the genital characters place it in the subfamily Cecidophyinae. It is widespread throughout the world, being a pest of grapevines in nearly every area where the vine grows.

Three strains of this mite are reported, each characterized by the type of injury produced: the bud strain, the erineum strain, and the leaf-curling strain. The last strain produces distinct leaf curl, few abnormal plant hairs on spots chosen for colonizing, but no typical erinose areas. The strains seem to be physiologically distinct as the bud form is never on the leaves, but confines its activities entirely to the buds.

Feeding by this mite on the buds of susceptible grape varieties causes deformation of the primordial bud cluster, distortion of the basal leaves, stunting of the main growing point of the buds and often death of the overwintering buds. Highest infestation and thus the most injury occurs in the first ten basal buds. There is a tristichous arrangement of the leaf primordia with the basal leaves forming the base of the bud and associated with the outer stipular scales. Mites normally enter the bud and establish colonies under the bud scales or the outermost stipular. As fruit primordia and the more apical leaves are progressively distal and deeper in the bud, the greatest amount of feeding and damage occurs on the basal leaves of the embryonic shoot. As populations become more dense, the mites work progressively deeper in the bud; thus the depth of the penetration as well as the density of the population is reflected in the damage to new growth shoots (Kido and Stafford, 1948).

The effects of mite feeding on leaves first appears to the eye as whitish areas where the mites feed, then feltlike yellow patches develop, followed by blisterlike formations on the uppr leaf surfaces, and finally the areas appear reddish brown. The blistered areas are incapable of resuming normal development, which results in disruption or coalescence of venation, irregular leaf margins, and generally misshapen leaves at maturity (Batchelor, 1952; Slepyan, Landsberg, and Mal' chenkova, 1969).

Badly damaged shoot apexes are short and thick in contrast to normal shoot apexes that exhibit a more open growth habit. If the scar tissue is extensive, one or more lateral bud primordia initiate growth. If these are already damaged, a secondary or tertiary growing point may produce a cane. This may often result in witches' broom. Mite populations sufficient to kill the shoot or retard shoot growth usually occur before the shoot is two or three inches long. The smaller secondary and tertiary growing points in the overwintering bud are infested infrequently and may give rise to a normal cane, even though the shoot from the primary growing point is killed.

Mites feeding in flower primordia first produce a slight enlargement of the cells. Typical polyplike formation is most noticeable along the peduncles and bracts. Severe injury to flower buds leaves only a shell formed by the calyx and, at times, the anthers. Dead cluster branches appear to be dry and discolored, and, when damage is extensive, the entire cluster usually will abscise before bloom.

The mites prefer to attack the lower epidermis, particularly on the apical part of the leaf comprising the lobe above the upper lateral incisions. The next most favored area is the narrow area situated slightly below the upper lateral incision. When dense infestations occur, the mites may infest the remainder of the lower and even the upper leaf surfaces (Batchelor, 1952).

Slepyan et al. (1969) have detailed the development and appearance of these erinea. The undersides of juvenile grape leaves are covered with dense, long, narrow, bent, thin-walled mainly unicellular trichomes plaited together. But trichomes on the fully formed leaves are usually 2–6–cellular and hornlike. Erineous trichomes on mature leaves are most commonly thin-walled, elongate, unicellular, and rounded at the apex, having the appearance more nearly like those found on the juvenile leaves.

The trichomes in the erineum are sharply distinguishable from those of the uninjured epidermis, but at the boundaries of the erinea they retain the structure characteristic of uninjured leaves. This is evidence of the localized action of the mites on leaf morphogenesis. Erineum trichomes are mostly thin-walled, elongate, unpartitioned, with convexly round apexes in contrast to the acuminate apexes of normal trichomes. The characteristics of the majority of erineous trichomes show resemblance to the trichomes on leaves that are just beginning to form, suggesting that the trichomatous integument of the erinea is of juvenile origin.

After the leaves have ceased to unfold, fat granules appear in the trichomes of the erinea and then, towards the end of the growth period, there is a deposit of tannic substances. Their accumulation imparts a rusty brown color to the erinea. The gradual browning results from the oxidizing of the tannic matter. Toward the end of the growth period the browning trichomes dry out and are shed.

In addition to the transformation of the trichomatous integument, the epidermal cells are also modified in the process of erineum formation. The cuticle either does not develop fully, or is completely wanting, while the walls are usually thickened. The epidermal stomata opposite the erinea remain normal or are slightly deformed. Significant changes occur in the mesophyll chlorenchyma at the points of erineum development. A characteristic feature of juvenile grape leaves is the tight closure of the mesophyll cells, without the formation of large intercellular spaces. In the very first stages of leaf development, the palisade chlorenchyma has not yet separated and its cells have the same columnar appearance and meristematic character as the cells of the upper and lower epidermis. Like the cells of the spongy chlorenchyma, those of the columnar chlorenchyma are distinguished by large nuclei and a large column of cytoplasm. In the process of normal leaf growth the structure of its mesophyll changes and the palisade and spongy chlorenchyma become morphologically separate; numerous intercellular spaces develop; and the chlorenchymal cells themselves, like the epidermal cells, enlarge, losing their rectangular outlines.

In the course of erinuem formation, the mesophyll acquires a different structure. The cells of the columnar chlorenchyma undergo one or two periclinal divisions in the affected parts, resulting in a local invagination of the leaf blade towards the upper side of the leaf. Two, three, or even more divisions occur also in the cells of the spongy chlorenchyma. These divisions, however, are not oriented in the same way as the division of the palisade cells. Comparison of all the structural and growth features of the erinea caused by *Colomerus vitis* indicates that these are examples of the most simply organized cataplasmic galls.

The nondevelopment or even absence of a cuticle, and also the various structural disorders of the leaf at the sites of erineum formation facilitate their infestation by *Plasmopara viticola* Berlese and de Toni, with the result that the mycelium of this fungus is more abundant among erineous trichomas than in the adjacent parts of the leaf blades (Smith, 1960; Smith and Schuster, 1963; Zeck, 1955).

Necrotic spotting, uniform crinkling of leaves, or open petiolar sinuses are not normally expressions of mite damage. Also several diseases and other conditions may cause symptoms that are easily confused with bud mite injury. A virus disease known as "Fanleaf" is characterized by early season dwarfing, shoot and leaf deformities, slight mottling in young leaves, and irregular zigzag growth of shoots. Pierce's disease of grapevines is characterized by delayed foliation. Armillaria root rot causes a decline in vigor. Effects produced by feeding by the grape phyloxera, nematodes, and root infestations of grape rootworm all can cause sparse leaves, stunted and slow growth, and reduced yields. The time of year vines are pruned may result in rather typical bud mite symptoms (Barnes, 1958).

The mites develop from egg to adult in about 10 to 14 days. Eggs may be laid, about 1 per day, or in groups of 2 to 10 or more. Immatures appear in 5 to 9 days after eggs are deposited in early April and become adults in 14 days after hatching. Spring activity begins when temperatures reach about 15.5 C (60 F) (Collingwood and Brock, 1959; Kido and Stafford, 1948).

Not all grape varieties are susceptible. Vines of two varieties may be intertwined with one vine consistently infested with erinea while the other is free from injury. American varieties are usually resistant to injury.

The growth period of the mites in the erinea is the most suitable time for control measures. After the mites have migrated into the buds, they are more protected from the action of the acaricides than are mites still on the leaves. In particular, the mites are most accessible to poisons, whether of the systemic or the contact type, when they are just beginning to disperse, to leave their winter refuges, and to attack the leaves and cause the formation of erinea. Erinea start developing on the leaves as soon as the invaded buds begin to swell and constitute the first ecological niches in the annual growth cycle of the mite.

The female of *Colomerus vitis* (Pagenstecher)[2] is a wormlike mite, light yellowish white, and from 160 to 200 μ long. The shield is set with numerous longitudinal lines. The median shield line is present on the rear ⅔ of the surface. Admedian lines are complete and sinuate. There are several submedian longitudinal lines: the inner ones end in front of the dorsal tubercle, and the outer ones run below the tubercle to the rear shield margin. There are fine lines above the coxae, and an ocellar spot or bulge near the rear lateral shield angle. The dorsal tubercles are set a little ahead of the rear shield margin. These tubercles have longitudinal axes and direct the dorsal setae ahead and diagonally centrad. The featherclaws are 5-rayed. The coxae are somewhat spread owing to the cecidophyine characteristic of the appressed genitalia. The first setiferous coxal tubercles on the forecoxae are set slightly ahead of the anterior end of the sternal line. There are prominent curved lines around the second setiferous coxal tubercles. Ring granules, or microtubercles, are elongate-elliptical and touch ring margins. The female genitalia protrude somewhat, and the coverflap is strongly ribbed, with the ribs being in 2 uneven ranks. The internal female genital anterior apodeme is shortened, but not as short as in many cecidophyine mite females.

Cecidophyes Nalepa

Cecidophyes malifoliae (Parrott). This mite was originally assigned by Parrott, Hodgkiss, and Schoene (1906) to the genus *Eriophyes.* They found it in company with the common apple rust mite, *Aculus schlechtendali* (Nal.), on the underside of apple leaves in the vicinity of Geneva and Canandaigue, New York, and noted that *C. malifoliae* was never as common as *A. schlechtendali.*

While it may be possible by diligent searching to recover this mite from apple leaves in these localities, apparently no one has attempted to find the mite again. Parrott's statements in his original publication clearly place this species in the genus *Cecidophyes.* The characteristics that put *malifoliae* in *Cecidophyes* are: large rostrum; absence of dorsal setae on the cephlothorax; large shield with broad anterior lobe over rostrum, and with pattern of strong shield lines; robust tapering body; lack of accessory setae; large protruding genitalia. Additional definitive characters as indicated by Parrott are: 4-rayed featherclaw; from 60 to 70 abdominal rings.

There seems to be no other leaf vagrant eriophyid listed as occurring on pomaceous trees or shrubs which is referable to the Cecidophyinae. To explain this occurrence it is necessary to resort to the supposition that this *malifoliae* had, in reality, been blown over onto apple form its primary host, which was growing in the immediate vicinity.

As regards primary hosts for this mite there are several possibilities. Oak trees in eastern North America are regularly infested with heavy populations of leaf vagrants that belong to the genus *Cecidophyes.* Examples of this are: *Cecidophyes pusilla* K., (1962), which lives on *Quercus falcate* Michaux; *C. lyrata* K. (1959), on *Quercus lyrata* Walter; a mite structurally identical to *C. lyrata* on *Quercus phellos* L.; *C. caryvagrans* K. (1964), on *Carya cordiformis* Wangerin; *C. collegiatus* K. (1961), on *Acer platanoides* L. Undoubtedly some of these hosts are growing in the localities specified by Parrott. Figure 133, *j*, is a diagram of the shield pattern of *C. lyrata.*

The Eriophyinae Nalepa

Phytoptus Dujardin: Species Inhabiting Fruit Trees

Phytoptus pyri Pagenstecher. This species (fig. 111, *a*), commonly known as pear leaf blister mite, or pear bud mites, produces tissue disintegration blisters in leaves of many pomaceous hosts, cause bud deterioration, or actually kill buds. Apple and pear are prime examples of agricultural plants that regularly suffer these blisters on both leaves and fruit, or show bud damage.

Pomaceous blister and bud mites have long been known as *Eriophyes pyri* (Pgst.), but in conformance with Nomenclature Code directives, the technical name is now *Phytoptus pyri* Pgst.

Nalepa, as indicated by his 1929 host list (Marcellia, 1929), considered the *pyri* complex on the various hosts to represent a series of subspecies and varieties. For example: he designated the pear blister mite as *Eriophyes pyri typicus* (Pgst.), the apple blister mite as *Eriophyes pyri* var. *mali* Nal., and a rowan (*Sorbus*) blister mite as *Eriophyes pyri* var. *sorbi* (Canestrini). Studies in progress indicate that there are actually several species within this complex. A blister mite on apple and pear is now named *Phytoptus pseudoinsidiosus* (Wilson) on the basis of shield pattern features.

A blister mite on California *Sorbus* (mountain ash, rowan) is definitely deuterogynous; both are described below. While it has not been possible to find

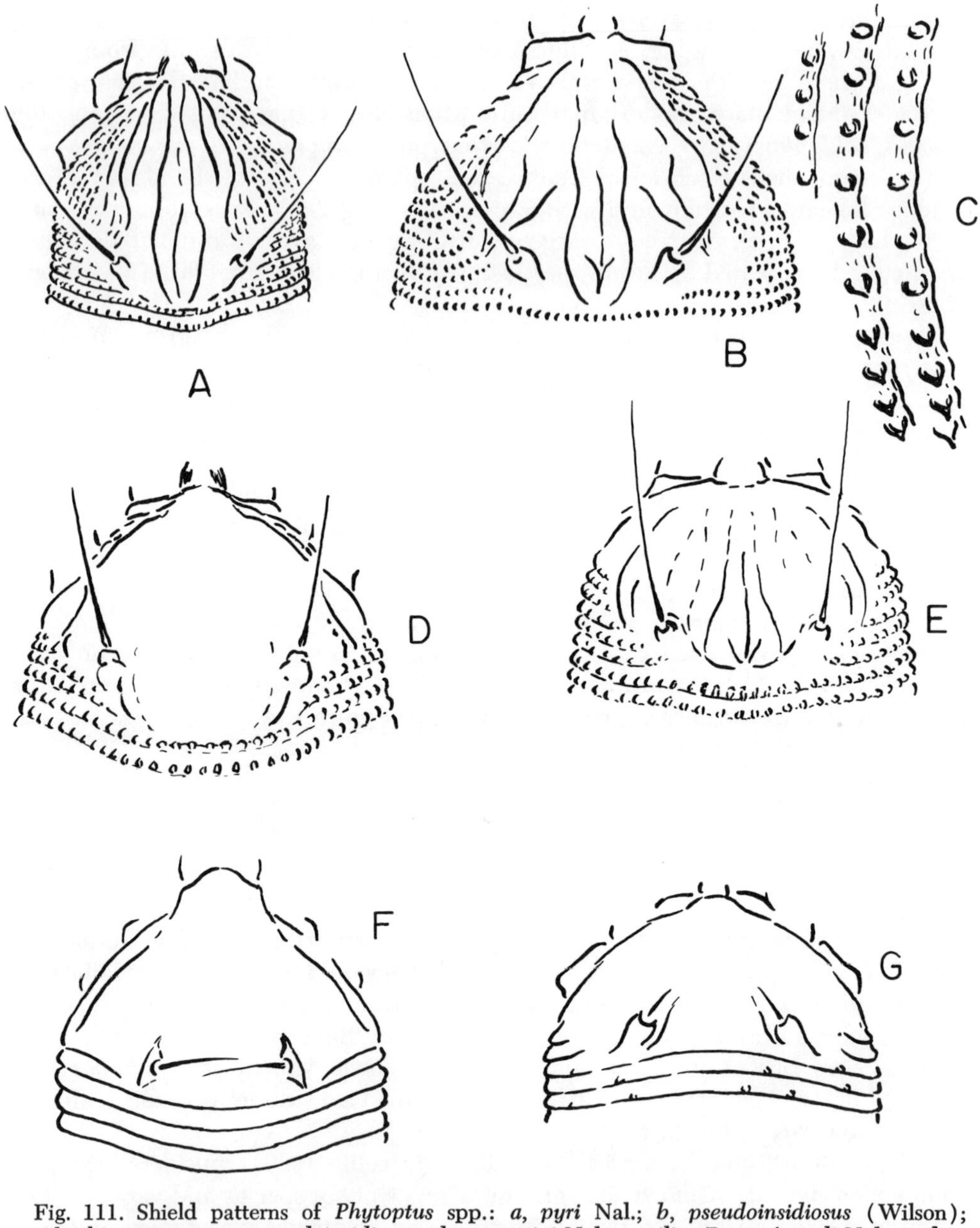

Fig. 111. Shield patterns of *Phytoptus* spp.: *a, pyri* Nal.; *b, pseudoinsidiosus* (Wilson); *c*, side skin structure on *pseudoinsidiosus; d, canestrini* Nal.; *e, tiliae* Pgst.; *f, padi* Nal.; *g, laevis* Nal.

evidence of deuterogyny in blister mites on pear and apple in California, deutogynes do occur in the pear leaf blister mites, *Phytoptus pyri* Pagenstecher, in Iran.

The hosts of these mites making blisters on pomaceous shrubs and trees, as given by Nalepa, are pear, apple, rowan, (*Sorbus*), quince, serviceberry, hawthorne, cotoneaster, and other Pomae.

It is probable that many workers who have been previously involved with the biology of the pear leaf blister mite were not aware that several species might be involved. Therefore, the following accounts of biology and life histories are based on what might have been populations of mixed species.

The most well known type of damage caused by mites in the *P. pyri* complex consists of blisters or galls on the undersides of pear and apple leaves. These first appear as minute greenish pimples with reddish tinge, at first barely distinguishable from the leaf (pl. 63). With increase in size the galls become reddish, and often cause brilliantly colored spots on the leaves. Later, with the drying up of the diseased tissue, the spots turn to a dark brown or black. On the undersurface of the leaf, the affected areas are corky and raised, with the epidermis rather loose and wrinkled, suggestive of a blister. Leading into each spot is a tiny opening that permits the mites to enter or leave the blister. Blisters are irregular and vary in size according to age, the largest being about ⅛ inch in diameter. Severe mite infestations on some varieties cause the leaves to become completely reddened, swollen, and blistered; leaves, when unfolded, become wrinkled and unshapely.

The blisters are usually arranged in a row on each side of the midrib, a grouping apparently determined by the pubescence and degree of fold of the leaf at the time of migration of the mites from under bud scales. As infestations increase, the spots coalesce, forming dark brown patches that may cause areas of the leaf to rupture, especially along the margins. Feeding injury may cause light-colored pimples on stems and around the calyx of the fruit.

In midsummer the mite feeding causes much yellowing of the foliage with light brown or dark green spots on the upper leaf surfaces and brown on the undersurface. The dark bands along each side of the midrib, contrasting with the yellow of the remainder of the leaf, is suggestive of variegated foliage of certain ornamental plants. The mites also feed on fruits and stems, producing injury that in rare cases causes fruit drop.

These mites require a highly humid atmosphere; in drier weather mites will leave the blister through the minute opening and seek a suitable environment elsewhere. They can survive outside the blister up to 4 days.

In South Africa the mites are active day or night during spring bud development. They swarm into leaf axils and crevices around the base of blossom stems and on the calyx of the blossoms, crippling or killing buds and causing bud and blossom drop. Severe damage results in dwarf leaves and trees with sparse foliage. Damage may cause a 25-percent loss of crop production (Morgan, Yee and Brinton, 1962).

The biology of the pear leaf blister mite complex varies in different areas probably according to the identity of the populations included. In the eastern United States, feeding injury results primarily in the production of blisters on the young leaves.

Throughout its distribution, this mite spends the winter in the buds, preferably under the second and third layers of the bud scales or in depressions or grooves more-or-less concealed by the pubescence of the bud scales. Spring and summer activities vary in different areas.

The life cycle described for populations in the Pacific Northwestern United

States (probably Nalepa's *E. pyri typicus*) appears to be typical of the more northern areas of the mite's range. The overwintering mites undergo a diapause during which they do not feed or lay eggs even though they may be active during warm winter weather. As the pear buds begin to swell, the mites start to feed and deposit eggs on the succulent basal portion of the bud scales. This bud enlarging enables the mites to penetrate deeper into the buds and soon mites begin to feed and oviposit on the young leaves and blossoms.

Blisters are initiated on the young leaves within the buds at the cluster bud stage by feeding activities mainly of overwintering adults, soon after which the overwintering mites attach themselves to the young leaves and leaves ⅕ to 1/20 inches long. Most of the blisters develop during this time. The time required for blister development after the initial feeding ranges from 3 to 7 days. The first small blisters produced have no opening and do not contain mites. After a feeding period of 4 to 6 days, 1 to 4 epidermal cells, usually near the center of the blisters, show hypertrophy. After about 6 to 15 days the hypertrophied cells die, producing small holes in the epidermis. Many mites insert their rostrums into the holes but cannot enter. As feeding progresses, more epidermal cells die and allow entrance of an adult mite. Usually only one female occupies the blister originally and begins to reproduce. Mites feeding at the periphery of the blister enlarge the cavity.

Since the early 1920s, workers on apple and pear pests have recognized that a form of *P. pyri* caused severe damage to fruit buds, caused abnormal falling of blossoms and young fruits, but rarely if ever produced leaf blisters. Stubbings (1947) studied the bud form in South Africa and published data indicating that the bud form of the mites differed from the blister form in the ratio of body width to body length, the number of body rings and placement of abdominal setae. Wilson (unpublished data) found no morphological differences between the bud or blister populations from northwestern states in North America.

In previous literature various morphological and biological differences have been indicated between the bud form and the blister form of *P. pyri*. A summary of biological differences is here presented, but it should be recognized that a more precise determination of the populations involved in different geographical regions may result in modification of these data.

In California, as in South Africa, there appear to be two strains—the blister form and the bud form—with different biologies. The biology of the blister form is similar to that described for the cooler climates. Heavy infestations of the bud form cause the fruit buds to turn brown early in the winter and frequently to flare open. During the winter the buds drop readily when touched and are so dry that they may be rubbed to a fine chaff between the thumb and first finger. At blooming time many of the fruit buds fail to open; others open with only one or two blossoms or a weak bloom that often drops or fails to set. Mite feeding around the floral parts may result in misshapen, badly russeted fruit. Fruit produced from buds on old spurs is more likely to be "round" (Morgan et al., 1962).

Mites enter the outer bud scales during the period from August to early October. During the migratory period they are pinkish and cluster about the leaf buds behind the bases of the leaf stems on new growth. Once hidden under the fruit bud scales they are white and tend to collect in groups. Egg deposition soon starts within the bud and as the mites mature and continue to multiply throughout the winter, they work deeper and deeper into the bud. At first they feed on and destroy the white tissue at the base of bud scales but later they attack the floral parts enclosed in the bud. When the remaining live buds open in the spring the mites leave the cluster buds and feed upon the young tender new growth and soon establish colonies in leaf axils. They feed on new growth as it develops until June, at which time oviposition and most other activities stop. During the warm period of June, July, and August there is little or no activity; not until the evenings are cool again do mites cluster about the leaf buds in their fall migration to fruit buds (Heriot, 1935; Parrott et al., 1906; Prescott, 1934; Venables and Heriot, 1937).

A summary of the principal differences between habits of blister and bud forms of *P. pyri* (after Borden, 1932) is as follows:

Time of activity	*Bud Form*	*Blister Form*
Oviposition period	September to June	March to October
Winter months	Actively feeding and reproducing	Hibernating—no activity until buds begin to open
Attack at bloom time	May cause resseting and short-length fruits	Fruit and leaves blistered
Early spring	First on new foliage; later in axils of new growth; no blisters	Blistering spreads to all new growth; often defoliation of old leaves
Early fall	Activity starts in August: migration to fruit bud	May drop with foliage; migration to buds September to November

Each female of the pear leaf blister mite lays 7 to 21 eggs that require 18 days to hatch at 10 to 17 C (50 to 63 F) or 5 to 8 days at 18 to 24 C (64 to 76 F). Sexual development requires 18 days in May and 10 to 12 days in June. The life cycle thus requires 23 to 36 days, and 2 or 3 generations may develop in a year. There may be up to 14 eggs laid in a single gall. The egg is translucent and whitish, oval with bluntly rounded ends. The larvae are 70 μ long, have smooth shields, visible striae on the abdomen with fine punctuation.

As the mites live within the blisters throughout the summer and spring and are free living for only a brief period in the spring before they move into the blister, populations have been most effectively controlled by a dormant treatment of either lime-sulfur, lime-sulfur plus dormant oil, the sodium salt of dinitro-o-cresol, or dinitro-o-cresol plus dormant oil. The bud form has been best treated before the mites enter the buds in early fall. Some of the more recently developed acaricides, however, are able to penetrate through the leaf or bud and effectively kill the mites in their protected habitat. Pesticides such as Sevin and diazinon applied during the summer months have resulted in effective control of both the blister form and the bud form of this injurious eriophyid mite.

Early illustrations of *Phytoptus pyri* Pgst.—the 1910 issue of *Zoologica* for example—showed figures of *P. pyri* as having subparallel and rather close admedian shield lines running back to the rear shield margin and turning inward there to meet centrally. This pattern is then adopted as the basic indicator of the original *P. pyri*, and is shared with other characters by members of the complex.

All of Nalepa's varieties of *pyri* have certain structures in common. These include 4-rayed featherclaws; wormlike rather elongate bodies; prominent central longitudinal shield lines; lateral shield lines of granules curving up just below the dorsal tubercles; moderately prominent ocellarlike lateral lobe at rear shield angles; well microtuberculated abdominal rings, with adult microtubercles fairly evenly distributed; prominent accessory setae on rear lobes. The female genitalia are not appressed to coxae; female coverflap has 10 to 12 longitudinal ribs. The significance of an outwardly directed diagonal line across in front of the dorsal tubercles is discussed under *Phytoptus pseudoinsidiosus* (Wilson).

Phytoptus pyri blister mites on pear with rear-meeting admedian lines are 200 to 240 μ long, slender, and wormlike. The featherclaws are 6 to 7 μ long. The median shield line is variable, but usually present only on the rear ½ of the shield. Submedian shield lines are variable, but the longer ones run back and curve out in front of the dorsal tubercle. Lateral lines of granules run back subparallel to the lower shield margin; some then join the portions of body rings below the dorsal tubercles. The lower shield angle ocellar lobes have granules. Dorsal tubercles are situated not far ahead of rear shield margin, and they direct the dorsal setae up and divergently ahead. The legs are moderately long; the foretibia is as long as the tarsus and the seta arises on the foretibia at ⅓ out from the base. Coxal ornamentation consists of lines of granules. Abdominal ring granules are elongate-elliptical and usually reach ring margins. These granules are unpointed.

Wilson (1965) published the description of *Phytoptus pseudoinsidiosus* new species (fig. 111, *b*, *c*). He found it in pear leaf blisters. He separated it from the well-known designation of pear leaf blister mite, *P. pyri* Pgst. Both species have the same dimensions. Nalepa's figures show a shield pattern with admedian shield lines curving strongly centrad at the rear margin. On Wilson's species these lines remain variably separate at the rear, but the most definitive character is the caudolaterally directed diagonal line across in front of the dorsal tubercle. This diagonal line is often the inner part of a dart-shaped mark lying just below the dorsal tubercle. More prominent lateral shield lines that curve up and forward below the dorsal tubercle seem to be another feature of these mites with this diagonal line. Females with both shield patterns often have pointed ventral ring granules. Both first stage nymphs and adults of *P. pseudoinsidiosus* have evenly distributed ring microtubercles. Blister mites with both shield patterns are often recoverable from the same blisters on apple, pear, *Sorbus*, and probably other pomaceous hosts. Both kinds seem able to make leaf blisters. This places a serious question on the validity of the rule that the description of the work of an animal establishes the name of that animal. See *Code* Article 16 (a) (viii).

Wilson's name, *pseudoinsidiosus*, refers to his comparison of it with *Phytoptus insidiosus* (Wilson and Keifer), the peach mosaic vector mite. Wilson states that the first nymph of the vector mite has unevenly distributed microtubercles. Wilson tried to transfer *P. pseudoinsidiosus* from pear to peach without success.

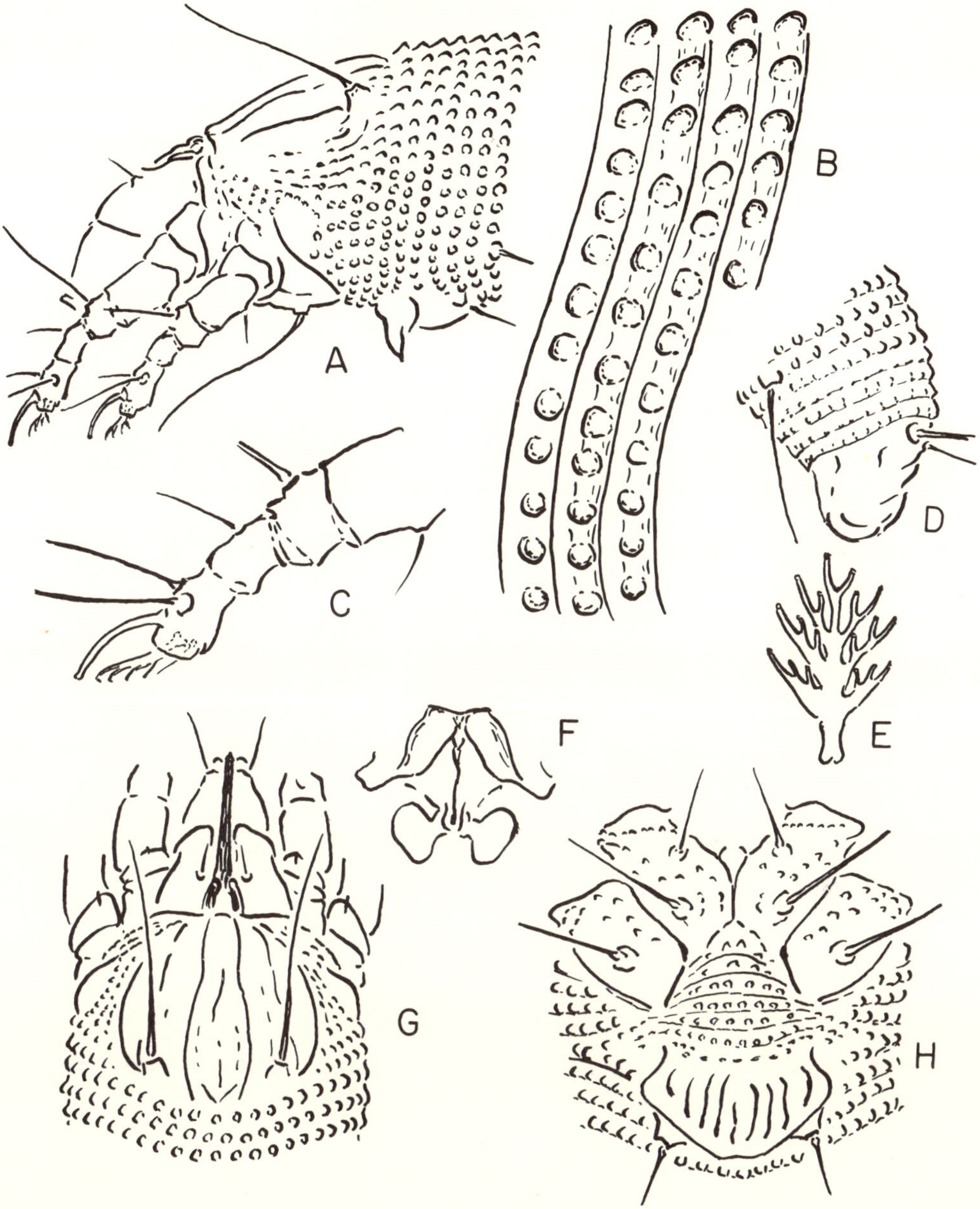

Fig. 112. *Phytoptus marginemtorquens* (Nal.): *a*, side view of anterior section of female; *b*, side skin structure; *c*, foreleg; *d*, telosome; *e*, featherclaw; *f*, internal female genital structures; *g*, shield diagram; *h*, coxae and external female genitalia.

There is obviously much more experimentation necessary to establish the precise role these mites bear in relation to each other, and to blister formation.

Pear and apple leaf edgeroller, *Phytoptus marginemtorquens* (Nal.). This mite (fig. 112) rolls the leaf edges on its pomaceous hosts and produces misshapen leaves. The rolling in this case is onto the upper surface of the leaf. Structurally this edgeroller is very similar to *P. pyri* Pgst., but the admedian shield lines are distinctly more widespread, especially to the rear, the ocellar spot on the lateral rear shield angle is clear, with few or no granules, and the ring microtubercles are larger, at least on one type of female. Two kinds of females come out of these edgerolls, similar to the kinds described for the *Sorbus* blister mites; the species is therefore likely deuterogynous. So far this edgeroller has not come to North America. Its hibernation quarters are unknown.

Nalepa, in his 1929 host list (Marcellia, 25:105, 106), lists this edgeroller as it occurs on pear as *Eriophyes pyri marginemtorquens* Nal. The one on apple he calls *Eriophyes pyri marginemtorquens* var. *mali* Nal.

Peach mosaic vector mite, *Phytoptus insidiosus* (Wilson and Keifer). This mite lives on drupacious rosaceous hosts; it is similar in shield design to *P. pseudoinsidiosus* (Wilson, 1965). This mite (and others like it) lives on several hosts belonging to the genus *Prunus* (including the genus *Amygdalus*), but none of these mites make leaf blisters. The peach mosaic vector mite lives in buds, and usually forms colonies in lateral twig or limb buds, where it makes a sort of perennially stunted bud for the mite colony that flourishes there. Peach trees with mosaic and with mite colonies thriving in these lateral buds serve as much more effective virus sources than do similar trees which have not had mites on them long enough to make large colonies.

The female *insidiosus* is 160 to 190 μ long, elongate, and wormlike. The featherclaw is 4-rayed. The shield design consists of long central lines with a complete median line. Beside this median line the admedians gradually diverge to rear and slightly recurve centrally at margin. The submedian line runs back to the dart-shaped mark on the outer side of the tubercle, the inner line of which forms the diagonal line across in front of the tubercle. Laterally the shield lines curve up and forward above the rear coxa. The foretibia is shorter than the foretarsus and not much longer than it is thick. Elliptical abdominal ring granules touch the ring margins. The coxae are coarsely granulate. The female genital coverflap has basal granules and about 12 longitudinal ribs. The first nymph is said to have unevenly distributed microtubercles. The adult has evenly distributed microtubercles.

Additional eriophyids in western North America are similar to *insidiosus* and live on *Prunus* spp. *Phytoptus inaequalis* (Wilson and Oldfield) (1966) lives in lateral buds on *Prunus emarginata* (Douglas), but will move to sweet cherry; the adult keeps the uneven nymphal distribution of abdominal microtubercles and is thus differentiated. *Phytoptus prunandersoni* (K.) makes erineum in baggy leaf deformations on desert peach; it has somewhat broken shield lines and pointed microtubercles. Perhaps the mite on desert peach, which was named *Phytoptus breechii* (K.) and which has a simpler shield pattern but much more spiny microtubercles than *P. prunandersoni*, is the alternate phase of this erineum mite. *Phytoptus geratopus* (K.) lives in buds on Pacific or Sierra plum and has micro-

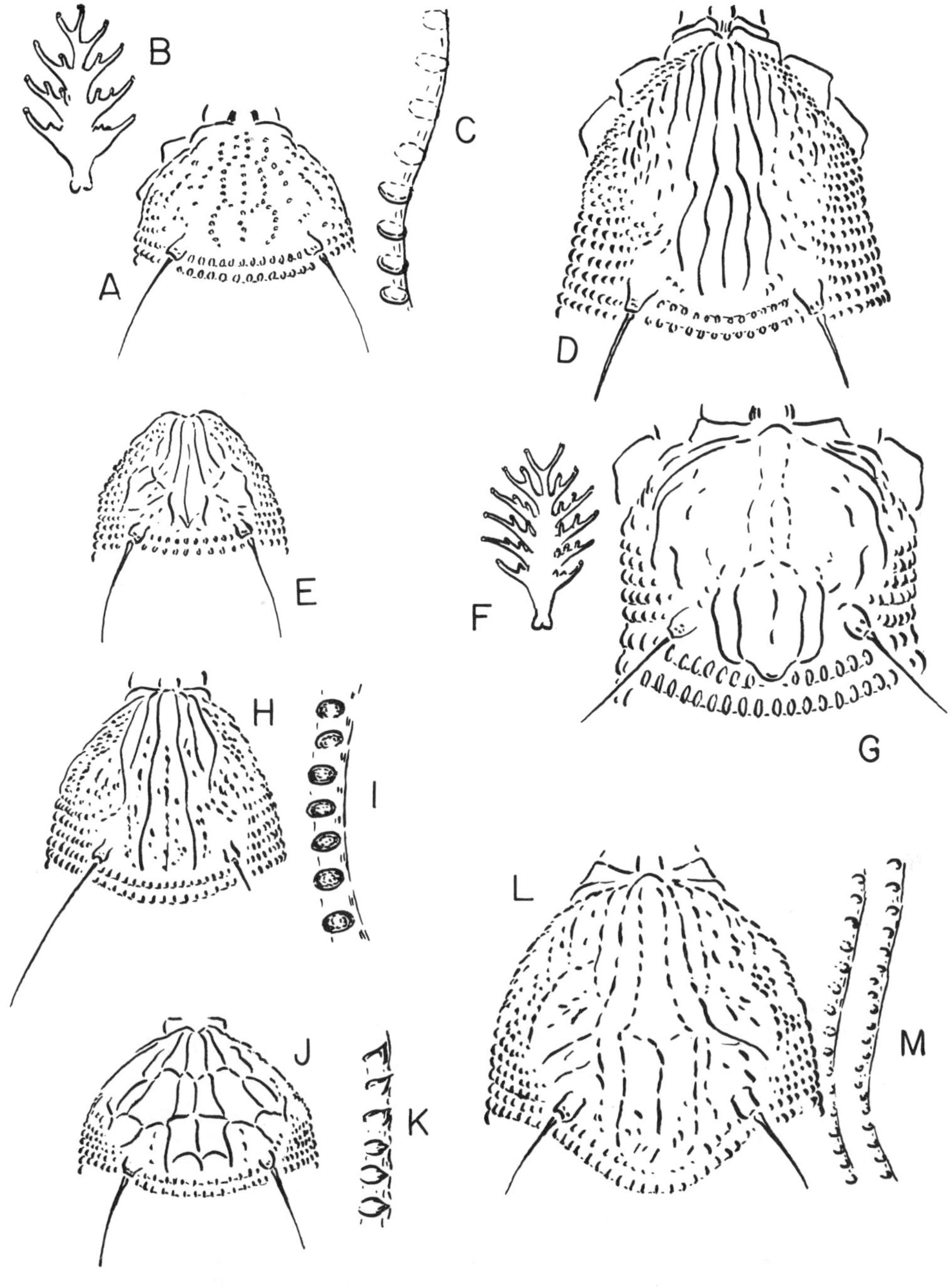

Fig. 113. Shield patterns of *Eriophyes* spp.: *a, peucedani* (Can.); *b, peucedani,* featherclaw; *c, peucedani,* microtubercles; *d, rossettonis* K., shield; *e, litchii* K., shield; *f, mangiferae* (Sayed), featherclaw; *g, mangiferae,* shield; *h, cynodoniensis* (Sayed), shield; *i, cynodoniensis,* microtubercles; j, *paradianthi* (K.), shield; *k, paradianthi,* microtubercles; *l, bakkeri* (K.), shield; *m, bakkeri,* microtubercles.

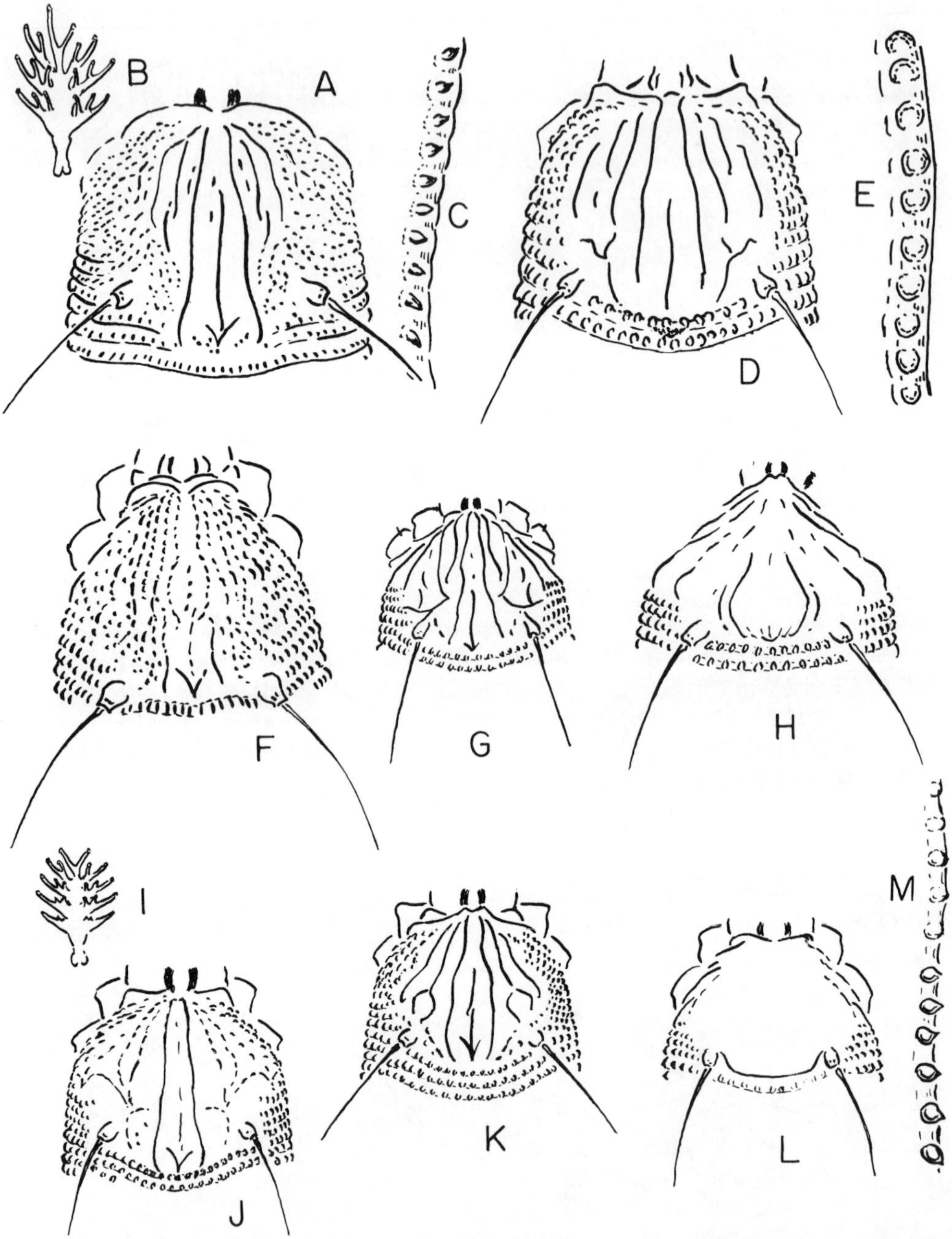

Fig. 114. Shield patterns, featherclaws, and microtubercles of *Eriophyes* spp.: *a, stefanii* Nal., shield; *b, stefanii,* featherclaw; *c, stefanii,* microtubercles; *d, pistaciae* Nal., shield; *e, pistaciae,* microtubercles; *f, biopsidia* (K.); *g, granati* (Can.), shield; *h, hibisci* Nal., shield; *i, sheldoni* Ewing, featherclaw; *j, sheldoni* shield; *k, ficus* Cotte, shield; *l, medicaginis* K., shield; *m, medicaginis,* microtubercles on lateral rings.

tubercles set a little ahead of ring margins; the female genital coverflap has wide ribs.

Phytoptus hybridicola (K.) lives in buds on orchard plums in California. It has so far appeared on trees growing near San Luis Obispo, and at Winters, California. Its chief feature is the series of longitudinal short lines just to the side of and below the dorsal tubercle. No damage has so far been connected with *P. hybridicola.*

Species Inhabiting Ornamental Plants and Shade Trees

Alder bead gall mite, *Phytoptus laevis* Nal. (fig. 111, g), is a pest on alder in various parts of Europe. It attacks seedlings and makes bead galls on leaves, the galls projecting through the leaf evenly on both surfaces, but the opening is usually on the undersurface. This underside opening may very well be owing to the fact that the embryonic leaf in the bud is folded with the upper surfaces appressed to each other, allowing the mite to make its initial feeding punctures only on the exposed lower surfaces. Nalepa suggested this in connection with elm leaf gall mites.

This species exhibits deuterogyny. After hibernation the deutogynes induce the formation of galls and deposit eggs that develop into protogynes and males of the first generation. The duration of the embryonic and postembryonic development is 23 to 30 days, which allows the development of 2 or 3 generations before leaf changes and decreasing day length stimulates the development of deutogynes that survive through the winter.

Two or 3 generations occur during the vegetative period of the host plant. Galls may contain more than 400 mites including all stages. The emergence of the mites from galls is facilitated by the pronounced positive phototaxis of the deutogynes. Dispersal occurs as they migrate to typical winter habitats on old twigs. This summer and fall migration lasts from 1½ to 3 months (Schevtchenko, 1957, 1961).

Phytoptus laevis is a wormlike, rather yellowish mite in the protogyne form. It has a 4-rayed featherclaw with the rays at least partially thickened basally. Length of *P. laevis* is 225 to 250 μ. The shield pattern is absent, or the admedians slightly indicated. The dorsal tubercles are set a little ahead of rear shield margin, and direct short setae up and diagonally centrad. On the protogyne these setae seem pointed more toward the front. Legs are slender; the tibia and tarsus are about the same length, with the foretibial setae at ⅓. Between unornamented coxae the sternal line is moderately long. Abdominal rings on the protogyne have sparse granules; the distribution varies from general to being limited to sides and venter. The female genitalia are shallow, bowl-shaped, with the coverflap unmarked. The male is similar to the protogyne, but examples seen have ring granules limited to sides and venter.

Deutogynes of *P. laevis* are reddish and differ from the protogyne, not only in the setting of the dorsal setae, as noted above, but also by the greater limitation of the ring granules. These granules (microtubercles) are limited to the venter, and sometimes to the rear venter.

Phytoptus laevis Nal. was the subject of research by Schevtchenko, who, in 1957,

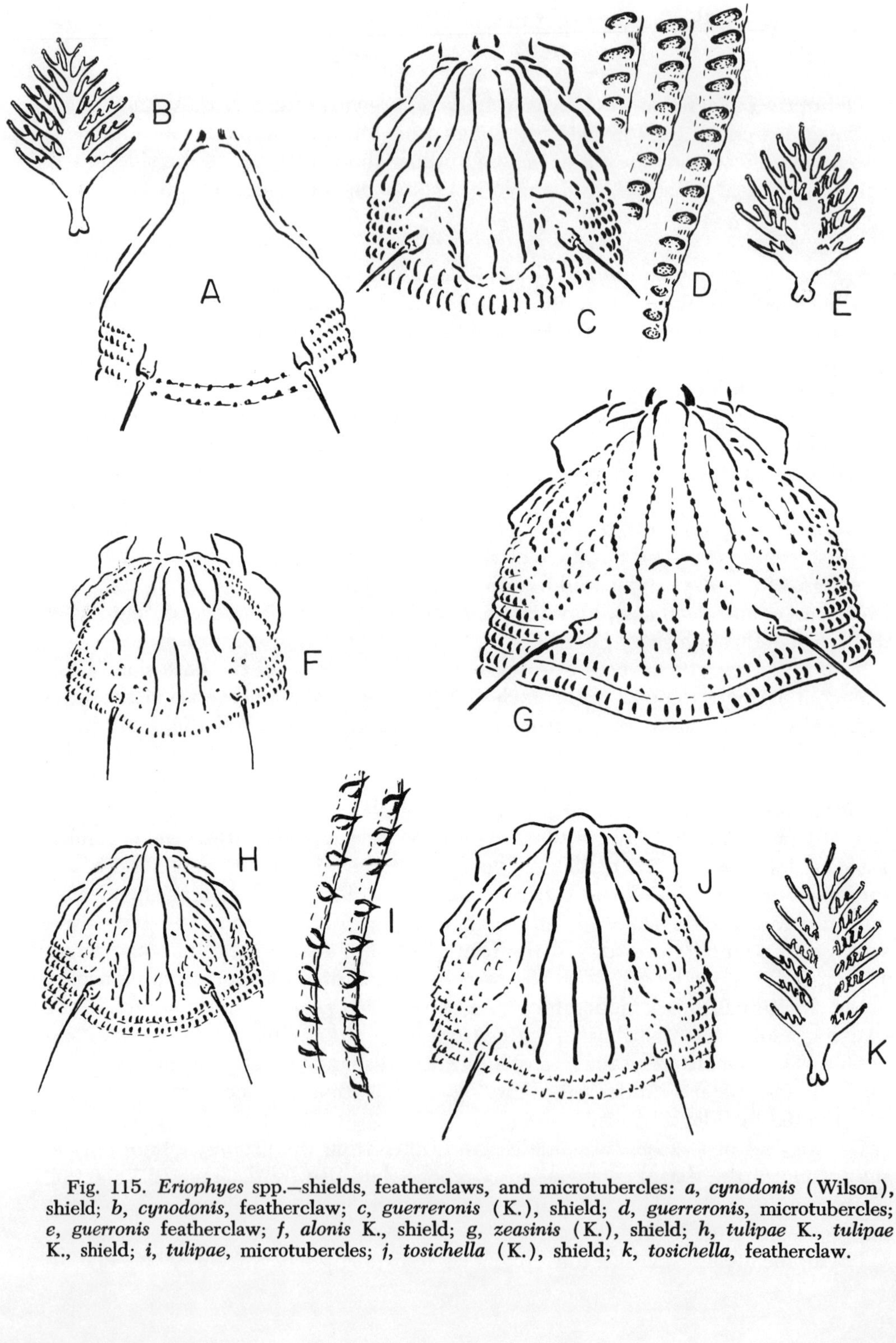

Fig. 115. *Eriophyes* spp.—shields, featherclaws, and microtubercles: *a*, *cynodonis* (Wilson), shield; *b*, *cynodonis*, featherclaw; *c*, *guerreronis* (K.), shield; *d*, *guerreronis*, microtubercles; *e*, *guerronis* featherclaw; *f*, *alonis* K., shield; *g*, *zeasinis* (K.), shield; *h*, *tulipae* K., *tulipae* K., shield; *i*, *tulipae*, microtubercles; *j*, *tosichella* (K.), shield; *k*, *tosichella*, featherclaw.

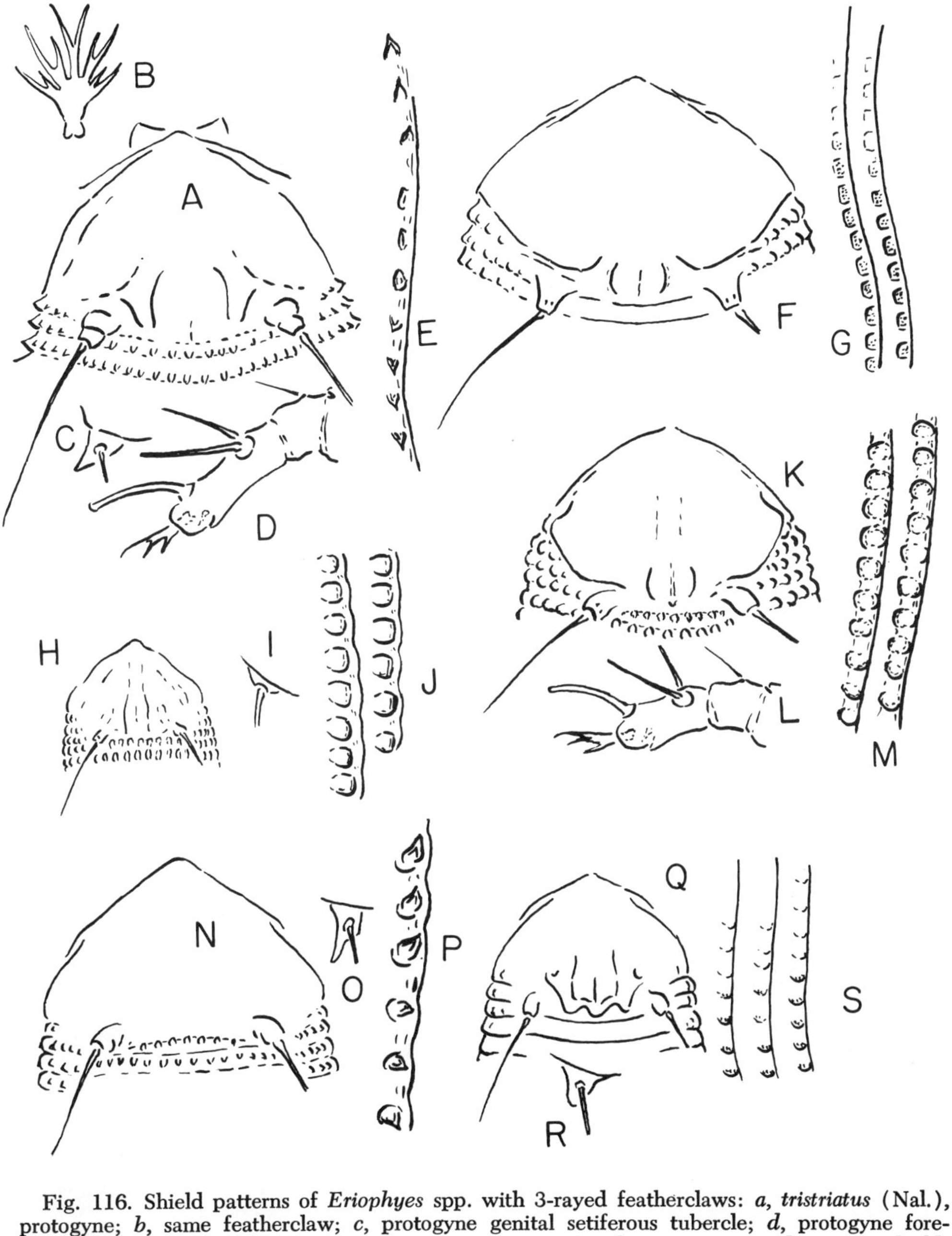

Fig. 116. Shield patterns of *Eriophyes* spp. with 3-rayed featherclaws: *a*, *tristriatus* (Nal.), protogyne; *b*, same featherclaw; *c*, protogyne genital setiferous tubercle; *d*, protogyne foretibia and tarsus; *e*, detail of *tristriatus* protogyne microtubercles; *f*, *tristriatus* deutogyne shield; *g*, deutogyne of *tristriatus* showing microtubercles on rings. *h*, shield of *caulis* Cook; *i*, genital setiferous tubercle of *caulis*; *j*, microtubercles on caulis; *k*, *erineus* (Nal.), shield pattern; *l*, foreleg of *erineus*; *m*, microtubercles of *erineus*; *n*, shield of *vaga* (K.); *o*, genital tubercle of *vaga*; *p*, microtubercles on *vaga*; *q*, shield of *caryae* K.; *r*, genital setiferous tubercle on *caryae*; *s*, lateral microtubercles on *caryae*.

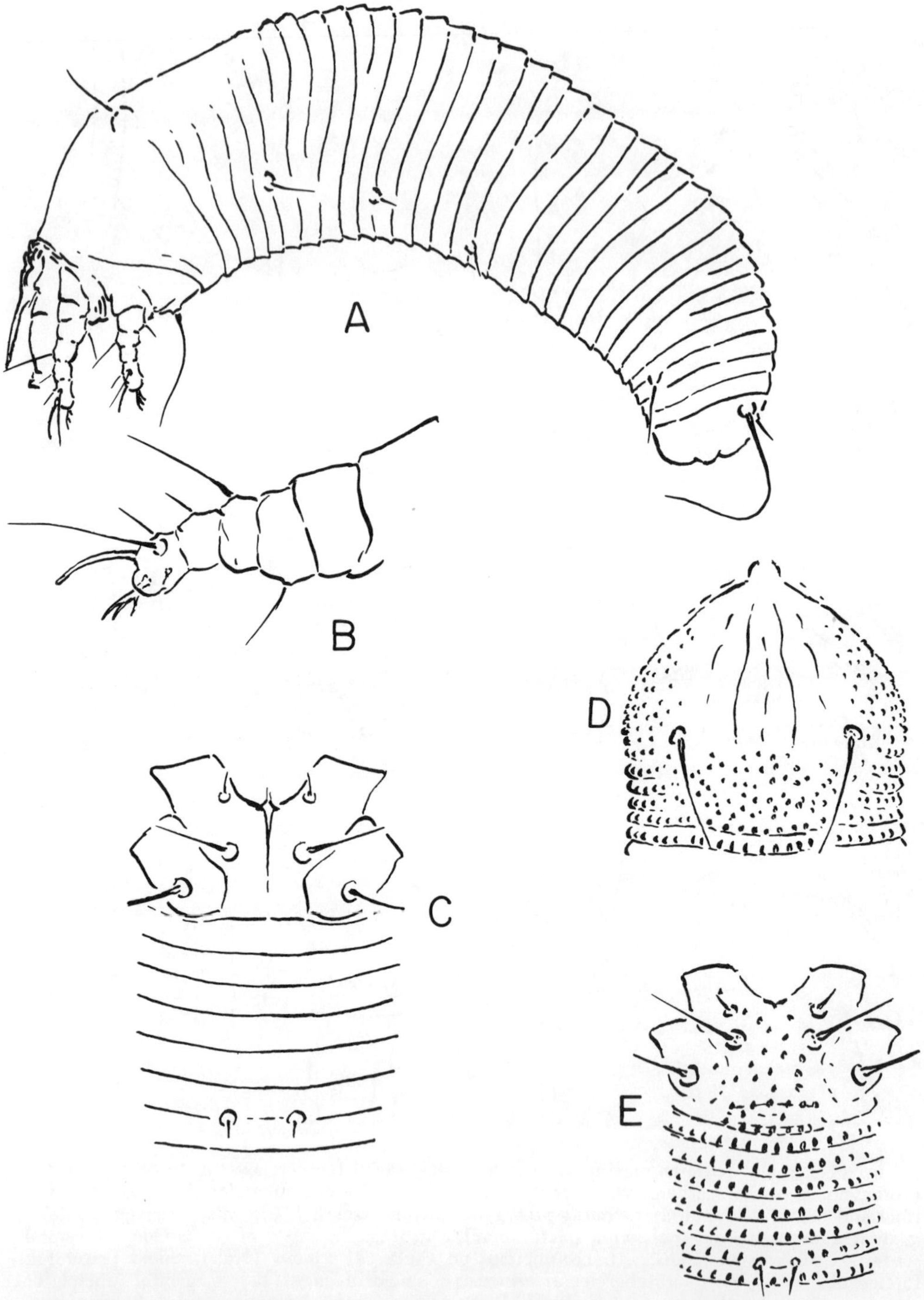

Fig. 117. *Eriophyes erineus* (Nal.), first nymph: *a*, side view of nymph; *b*, foreleg; *c*, coxae and genital setae; *Acalitus orthomera* (K.), first nymph; *d*, shield from above; *e*, coxae and genital setae.

published proof of deuterogyny in the species. It was the first leaf gall mite proven to be deuterogynous. As pointed out by Oldfield (1969), *Phytoptus laevis* is related to *Phytoptus padi* Nal., on *Prunus*, to *Phytoptus emarginatae* (K.), also on *Prunus*, and to *P. lionotus* Nal., on birch.

Lilac bud mite, *Phytoptus löwi* Nal. This mite feeds on lilac buds in Europe, causing drying of the buds. Shoots become dwarfed, and an excess of lateral growth develops, suggesting a witches'-broom effect (Hughes, 1959). This mite is 160 to 170 μ long. It has a 4-rayed featherclaw. The shield is somewhat acuminate anteriorly and has a faint design that seems to be of longitudinal lines centrally, with submedian lines arching back toward the dorsal tubercles. The dorsal tubercles sit near the rear shield margin, but direct the dorsal setae toward the front. The body of this mite is represented as completely microtuberculate. The female genitalia have coverflaps with longitudinal ribs.

Eriophyes von Siebold: Walnut and Oak Group

Walnut, oak, and hickory mites (fig. 116, *a-c*; 117, *a*, *b*, *c*) with 3-rayed featherclaws are a homogeneous group. These mites cause damage to their hosts evident by twig brooming, leaf galls, leaf erinea, and leaf edgerolling. Persian walnut leaf gall and erineum mites, as they occur in Europe, were the first 3-rayed featherclaw species to be named. Representatives of the group are now known to occur throughout the Northern Hemisphere.

Persian walnut leaf gall or blister mite, *Eriophyes tristriatus* (Nal.). This mite (fig. 116, *a-g*) is widespread in Europe and parts of Asia, but has so far not appeared in North America. In leaves it makes small, brown, hard pustules that are about 1¼ mm in diameter. The mites place these galls along midribs and larger lateral veins, but in heavy infestations blisters occur elsewhere. Badly galled leaves are twisted and misshapen. Careful examination of the mite populations in these galls discloses males, smaller than females, and which have pointed thanosomal microtubercles. These males then define the protogynes of *tristriatus*, which also have pointed microtubercles. By August the individuals with pointed microtubercles are giving way to deutogynes that have abdominal rings without dorsal microtubercles, and flattened lateral and ventral microtubercles.

The protogyne of *E. tristriatus* is 200 to 240 μ long; the shield design consists of short lines just ahead of the rear margin. Abdominal microtubercles are pointed, more so ventrally. The tarsus of *E. tristriatus* is three times as long as the tibia, a feature definitive for the species. The anterior coxae are elongate, with a well developed sternal line between them, bifurcate to the rear. The female genital coverflap has no markings and the genital setiferous tubercles extend into a point. The *tristriatus* deutogyne, as noted above, has no dorsal microtubercles and flattened ventral ones. No male has these features.

California black walnut pouch gall mite *Eriophyes brachytarsus* K. This mite (pl. 60; fig. 100, *b* of the gall interior), causes rather large upper surface leaf galls to grow on its host, the California black walnut. The galls are semiglobular and about 4 mm in diameter. Some hairs grow around the lower gall orifice; but the interior of this gall contains unusual lobes that are consolidations of papilla groups, and have numerous cross divisions in each of the joined papillae. The lobes and surrounding interior tissue become red as the gall matures during the summer, and this gall color determines the deutogyne color.

The protogyne of *E. brachytarsus* (originally named *amiculus*; pl. 60, *b*) is a spiny mite 195 to 210 μ long, wormlike, and tends to be light reddish. The shield has a short median line just ahead of the rear margin. Admedian shield lines are moderately close to each other, faint anteriorly, but plainly visible on the rear shield third where they diverge at the rear margin. There is a lateral ocellar spot on the rear shield angle. The sternal line between the coxae is strong. The foretarsus is 1½ times the tibia length. Abdominal microtubercles extend into short spinules, giving the mite a spiny appearance. The female genitalia have an unmarked coverflap, and, unlike *E. tristriatus* and *E. erineus*, the genital setiferous tubercles are unproduced. Male *E. brachytarsus* are spiny like the female, but smaller, and more easily overlooked.

Deutogynes of *E. brachytarsus* (pl. 60, *a*) are about the size of the protogynes, and deep red. They differ also from protogynes in the shape of the coarse, rounded off, and evenly distributed microtubercles on the thanosome (pl. 60).

For an account of the life history of *E. brachytarsus* refer back to the section on deuterogyny (chap. 12). In that section are notes on the differences between the black walnut petiole knot mite of eastern North America, *Eriophyes caulis* Cook, and *E. brachytarsus.*

Persian walnut erineum mite, *Eriophyes erineus* (Nal.). This species (figs. 116, *k*, *l*, *m*; 117, *a-c*) makes felty erineum pads on the undersides of the leaves. While this mite is listed in many publications as a variety of *E. tristriatus*, its shorter foretarsus in comparison to tibial length, its different life history, and the different type of injury it makes on walnut leaves, mark it as a very distinct species. *E. erineus* is discussed in the section on deuterogyny as a mite that might possibly have nonstructural deuterogyny. At any rate, no matter where the individuals of this species are collected, from erinea, from petiole bases, or under terminal bud scales, they all have elliptical and rather flattened microtubercles. *Erineus* females range in size up to 240 μ long. They have but slight shield lines at the rear margin. The genital setiferous tubercles are produced, but unlike those on *tristriatus*, are rounded off.

Butternut erineum mite, *Eriophyes cinereae* K. This mite makes extensive undersurface leaf erineum patches on its host; patches tend to run together. This species is very similar to *E. erineus*, but perhaps differs most by having subtriangular microtubercles, and more smooth dorsal rings caudally. The butternut mite ranges up to 240 μ long. The genital tubercles are produced as they are on *E. erineus.*

The black walnut petiole gall mite, *Eriophyes caulis* Cook. This mite (figs. 100, *c*; 116, *h-j* of the deutogyne) makes curious prominent erineum-covered swellings on black walnut petioles in eastern North America. The erineal papillae on the swellings show no divisions or cell boundaries. These erineal galls often cause deformation in the petiole: the petiole curls back onto itself.

The protogyne of *E. caulis*, which may be found in the erinea in early summer, is wormlike, 240 μ long, and has the 3-rayed featherclaw of the group. This protogyne has pointed microtubercles on the rings. Later in the season, deutogynes which have rounded and humped microtubercles, are the form recoverable from the erinea. The eastern *E. caulis* differs from the western *E. brachytarsus* mainly in the protogyne stage; the *E. caulis* protogyne has shorter points on the microtubercles. Both *E. caulis* and *E. brachytarsus* have flat, or very slightly produced genital setiferous tubercles. In that respect *E. caulis* differs markedly from *E. caryae* K., *E. vaga* (K.), *E. tristriatus* (Nal.), and *E. erineus* (Nal.).

Pecan leaf edgeroll mite, *Eriophyes caryae* K. This mite (fig. 116, *q-s*) causes leaf spots on pecan and rolls leaf edges. The females of this species are 190 to 245 μ long, wormlike, with 3-rayed featherclaws. The shield design is limited to 3 short lines just ahead of the rear shield margin, followed by a sinuate cross line. The original description of *E. caryae* is evidently of the deutogyne form since microtubercles are only ventral on the abdomen. The foretibia is 3 μ long, and the tarsus is 7.5 μ long. There is a well-defined sternal line between the forecoxae. The female coverflap has no ribs, and the setiferous tubercles project, being somewhat acute apically.

Eriophyes vaga (K.) (fig. 116, *n-p*) occurs on pecan in Brazil. This mite was found making discolored spots along veins on the leaf underside. It has pointed microtubercles, and in analogy with *E. brachytarsus*, this form, *E. vaga*, can possibly be the protogyne of *E. caryae*. Recently mites with the structure of *E. caryae* have been found associated with pecan bunch disease.

California black walnut catkin gall mite, *Eriophyes neobeevori* (K.). This mite alters male catkins on southern California black walnut. The catkins enlarge, become succulent, and remain hanging most of the season. *E. neobeevori* females are 175 to 250 μ long, wormlike, and yellowish. The shield design on this mite is obscure, with three short lines present at rear center. The foretarsi are about twice the tibial length, and there is a strong sternal line between the forecoxae. The most distinctive feature on *E. neobeevori* is the finely pointed microtubercles. Projected and rather pointed genital setiferous tubercles indicate that this species is not close structurally to the pouch gall mite, *E. brachytarsus*, which is common on California black walnut in southern California.

Oak erineum mites. Several species of mites in the genus *Eriophyes* have 3-rayed featherclaws and cause various alterations on oaks in the Northern Hemisphere. Some are bud mites, some live around acorns and sequester themselves under cap scales or at the nut base. One or 2 species cause male catkin damage. But the most prominent types are those making undersurface leaf erinea. All of these 3-rayed featherclaw oak mites are divided into two groups on the basis of the rear margin of the female genitalia. In 1 group this margin curves posteriorly, even with the edge of the female coverflap. In the other group this margin is con-

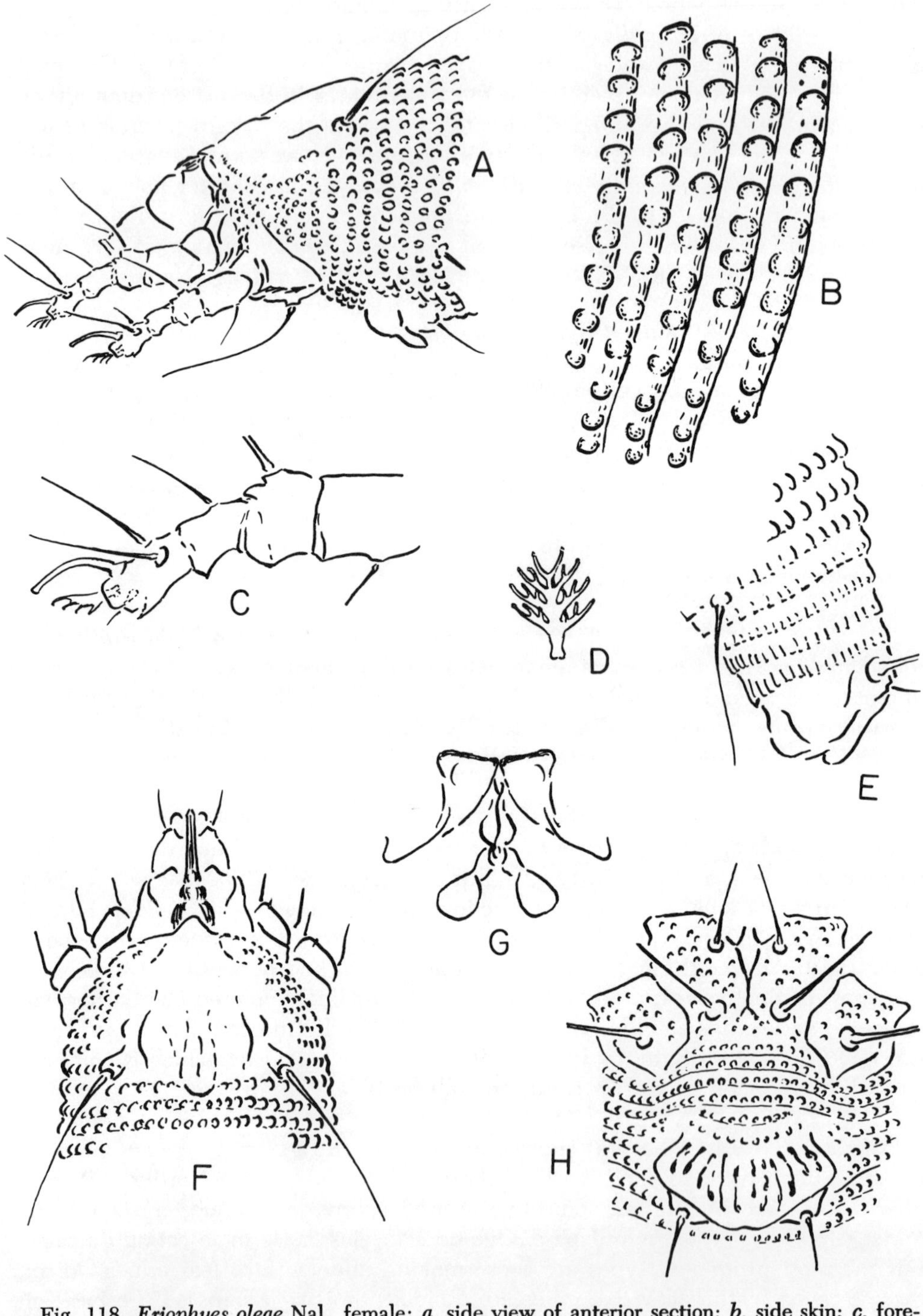

Fig. 118. ***Eriophyes oleae*** Nal., female: *a*, side view of anterior section; *b*, side skin; *c*, foreleg; *d*, featherclaw; *e*, telosome; *f*, shield; *g*, internal female genital structures; *h*, coxae and external genitalia.

vex anteriorly, and does not follow the coverflap edge. *Eriophyes ilicis* (Canestrini), the erineum mite on holly oak in Europe, is an example of the group with the posterior curve, or convexity. *Eriophyes ecantyx* (K.), a bud mite on holly oak, and *E. triplacis* (K.), are examples with the edge convex anteriorly. *E. triplacis* makes leaf erinea on North American white oak.

These oak erinea are basically constructed of the finger type of papillae. Some are clavate apically and others are united basally, branching apically.

Eriophyes Inhabiting Fruit and Nut Trees other than Walnut

Eriophyes biopsidia (K.) (fig. 114, *f*) inhabits guava buds in Venezuela, causing bud damage. It is a wormlike species with 4-rayed featherclaws. Females are 105 to 120 μ long. The shield design is of granular lines; the median line is present on the posterior half. The admedian lines are complete, sinuate, and subparallel. Submedian lines are about four in number; the first 3 are nearly complete and ending in the area before the dorsal tubercle. The coxae are ornamented with granules. Thanosomal microtubercles are rounded off, oval above, and rounder basically below. The genital coverflap is basally granular and with about 14 longitudinal ribs.

Persimmon bud mite, *Eriophyes diospyri* (K.). This species occurs under fruit buttons and under buds scales on cultivated persimmon in southern California. Mites blacken the fruit surface and stem under fruit buttons and evidently are implicated in premature dropping of fruit. This mite is wormlike, 140 to 170 μ long, with a 5-rayed featherclaw. The shield design is mostly lines of granules. The median shield line is fairly complete. The admedian lines are subparallel to the median and gradually diverge to the rear. The first submedian line projects back toward the dorsal tubercle and suddenly angles outward in front of the tubercle. The second submedian is shorter than the first. Legs of *E. diospyri* are rather stocky, with the foretibial seta arising at ⅓. The coxae are ornamented with granules and the sternal line is somewhat obscured. Abdominal microtubercles are elliptical. The female genital coverflap has basal coarse granules and 12 to 14 longitudinal ribs.

Fig bud mite, *Eriophyes ficus* Cotte. This mite (fig. 114, *k*) not only injures fig buds, but it also transmits fig mosaic virus. The mite ranges everywhere figs are grown. Feeding by mites that carry no virus produces variable symptoms such as russeting or surface browning, bud blasting, impedence of new growth, bud distortion, leaf chlorosis, and in severe cases the result can be defoliation of branches or of whole trees. These mites make no galls. In the greenhouse, feeding by this mite may entirely prevent new growth. In the field it is not possible to separate virus symptoms from feeding effects because the virus is universally present in all varieties of figs. Calculation of loss in fruit production in the field is not possible because virus-free trees are not available.

All stages of *E. ficus* are present in and around buds in winter. As buds break in spring the mites move to stems and leaves and begin laying eggs. Most eggs are deposited among lower leaf surface hairs. Development of this mite from egg to adult requires 5 to 7 days. Females are usually found on buds and young leaves, but they may be abundant among undersurface leaf hairs on older leaves, and under apical fruit scales. Virus carrying females can produce virus symptoms on figs in less than 10 days of feeding, but males are unable to transmit the mosaic virus.

Populations of *E. ficus* increase more rapidly on plants supplied with optimum nutrients than on plants deficient in nitrogen and phosphorus; also populations of this mite develop more rapidly on nonviruliferous plants than on infected ones. Predators of the fig bud mite include *Typhlodromus* mites, *Leptothrips mali* Fitch, Cheyletid mites, and Chrysopidae (Baker, 1939; Flock and Wallace, 1955).

E. ficus is a wormlike species 160 to 170 μ long, with a 5-rayed featherclaw. The median shield line is present on the rear ⅔ and ends in a dart-shaped mark. Admedian lines are complete and recurve slightly at the rear. The most distinctive feature of this fig mite is the fork in the first submedian line just in front of the dorsal tubercle. Abdominal microtubercles are elliptcial and rounded off, the coxae are ornamented with fine granules, and the female genital coverflap has about 8 longitudinal ribs.

Olive bud mite, *Eriophyes oleae* Nal. This mite (fig. 118) is a serious pest of all varieties of olive in the Mediterranean area, and is especially injurious to young trees. It prefers warm areas. This mite twists and deforms leaves, causes the fruit to be stunted and misshapen, and seriously reduces the amount and quality of olives available for pickling. Feeding by *E. oleae* on and around growing points makes developing leaves twisted and stunted, and may even kill shoots. Healthy leaves have a silvery sheen. As the result of mite infestations, however, mature leaves may show subcircular, irregular greenish patches, owing to destruction of the normal silvery stellate hairs. These greenish patches may bulge out as small chlorotic areas above the general leaf surface, giving the leaf an embossed appearance. These patches turn brown as necrosis progresses. In heavy infestations, the mites may extend their feeding to leaf margins, which results in deformities.

Injured or rusty-colored leaf tissue results from the presence of dead cells scattered throughout the upper tissue layers. Such cells are shriveled, often triangular; the walls are irregular, and the cell contents are coarsely granular and shrunken from walls. Only upper layer cells are affected. Damage on young fruit first appears as silvering, then browing, and ends in fruit deformation. Silvered areas on fruit are speckled with tiny brown droplets of exuded gum from the skin. Mites congregate in large numbers at the stem end of fruit, and find shelter under sepal rudiments. Most damage to fruit is confined to the stem. Fruits that are fully developed before the mites become abundant do not show much damage (Harpaz, 1955).

E. oleae is a small, wormlike species with a 4-rayed featherclaw. The central shield pattern is rather obscure with lines only appearing on the rear half. There is the suggestion of a median and two admedian lines at the rear shield margin, and a curved line extends back just centrad of the dorsal tubercle. Coxae are ornamented with granules, and the sternal line is short and bifurcate at both ends. Thanosomal microtubercles are subcircular and rounded off. The female genital coverflap has two transverse lines of granules across the base; the longitudinal ribs are irregular and tend to be broad. There are about 12 ribs.

Citrus bud mite, *Eriophyes sheldoni* Ewing. This species (fig. 114, *i, j*) probably occurs in all lemon-growing areas in the world where the humidity is high enough for it to develop. In California, *E. sheldoni* only attacks citrus along the immediate coast, and does not live in citrus orchards that are behind even the first high ranges of hills. Under the right conditions for its growth, it lives on all varieties of citrus, but lemon is probably the most favored host, owing to its larger buds that offer good protection.

Citrus bud mites inhabit protected locations under bud bracts, at petiole bases next to buds, under buds scales and within buds, inside developing blossoms, and under fruit buttons. Mites will even live where two growing lemons touch, but they do not favor open surfaces. Mites infest all ages of buds, including dormant buds on old wood. They continually move up to new growth, and sequester themselves under friut buttons as soon as the buttons touch the growing fruit. All of this indicates that the mites continually disperse themselves throughout the trees.

Mite feeding within lemon leaf axil buds results in the principal injury caused by *E. sheldoni.* Feeding blackens buds scales and often kills the entire bud. As adventitious buds appear owing to this injury the mites proceed to blast them also. Thus extensive multiple budding occurs on infested twigs, which either stops twig growth and blossom development, or results in an inferior type of growth. Even small amounts of bud injury depress proper growth of the various plant parts, and twigs that grow from infested and injured buds may be shortened, or thickened and flattened, or both. Rosetted growths sometimes come out of bud proliferation (pls. 10, 64) on lemon trees.

Leaves from damaged buds assume curious shapes, and blossoms are malformed, with floral parts misshapen or aborted. Fruits from damaged flowers may drop prematurely, or assume grotesque shapes (pls. 10, *a*; 64, *b*). Lemons sometimes grow round, rather than ellipsoidal, depending on the amount of blossom injury.

Symptoms on orange trees are similar to those on lemon except deformities on twigs, leaves, and fruits are not as pronounced. Feeding by *E. sheldoni* mites may flatten oranges vertically, and may also cause ridges and seams in the rind.

Biological studies by Sternlicht (1970) indicate that hatching is most successful at 25 C (77 F) and 98 percent relative humidity (RH) but is reduced by low humidities (35 to 40%) when abnormal dwarf larvae emerge. The eggs hatch in 3 to 14 days; the length of a generation (egg to egg) is 12 to 33 days. The threshold of embryonic development is 9 C (48 F) and that for completion of the life cycle is 12.5 C (73 F). The average number of eggs laid per female is 6 (4 to 8). It increases to 12 (5 to 19) if the mite was fed on buds during its larval stages. The mites fail to survive at high or low temperatures; 50 percent died in 30 minutes at 30 C (86 F) and at –15 C (5 F).

As indicated for other eriophyoid mites, the males of the citrus bud mite deposit stalked spermatophores from which the sperm masses are picked up by virgin females. Males may deposit 25 to 100 spermatophores (2 to 15 daily) during their adult life. Spermatophores exposed for one hour to relatively extreme temperatures of 5 C (41 F) or 38 C (100 F) do not fertilize eggs. Mite populations from lemon trees usually comprised a low percentage of males in winter and a high percentage in spring and autumn (Sternlicht and Goldenberg 1971).

Eggs of this bud mite are present wherever the mites live. Each female lays about 50 pearly white, subspherical eggs. Incubation periods, according to Boyce and Korsmeier (1941), vary from 2 to 6 days, depending on temperature. After hatching, the mites develop through the first nymphal stage, first molt, second nymphal stage, second molt, and then into the final stage or adult. The second molt, which is the longest resting period of the 2 molts, is in part for the purpose of producing a sexually mature individual. Neither nymphal stage has external genitals, and as the adult forms within the second nymphal skin the genitals are extruded between the coxae and the genital setae on the anterior part of the thanosomal venter. The devlopmental cycle for *E. sheldoni,* from egg to adult, requires about 15 days during warmer summer weather, and from 20 to 30 days in winter.

Lime sulfur applied at 2% plus 4 lbs wettable sulfur per U.S. 100 gals of spray is effective and once was used for control of this mite. A spring and a fall application of 1¾% summer petroleum oil has usually prevented serious damage to lemon trees by this mite. More recently one application per year of 0.5 to 1 lb of 25% chlorobenzilate has been more effective in preventing development of injurious populations of this citrus bud mite in California (Boyce and Korsmeier, 1941; Boyce, Korsmeier, and Persing, 1942; Jeppson, Jesser, and Complin, 1955).

Eriophyes sheldoni is a wormlike species, 170 to 180 μ long. It has a 5-rayed featherclaw. Shield patterns are variable, ranging from indistinct lines to rather clearly marked designs. On better-defined shields the median line is broken and ends as a dart-shaped mark ahead of the rear shield margin. Admedian lines are complete; submedian lines extend back toward the dorsal tubercles and meet transversely curved lines in front of this tubercle. The coxae are granular, and there is a short sternal line between the forecoxae. Abdominal microtubercles are elliptical, touching ring margins. On some mites the microtubercles tend to be pointed over the telosomal ring margins. The female genital coverflap has 10 to 12 longitudinal ribs.

Litchi erineum mite, *Eriophyes litchii* K. This mite (figs. 101, *b*; 113, *e*) (*E. cordai* Oudemans) is the most important pest of litchi nut in Hawaii and Pakistan. The tree is both a commercially grown plant, and an ornamental. Mites attack young leaves, fruits, shoots, and flower buds. Erineum developing on leaf undersurfaces causes the leaves to curl, dry, and fall, ultimately resulting in stunted plants. Mite feeding also depresses flower bud development. Greatest damage occurs during intermittent flush growth periods. Erineum, or erinose, is the first symptom of mite damage, and begins to develop as leaflets are unfolding. At first the erineum is greener than normal plant color. Within 2 days the erineum becomes thick silvery white, but in a span of 3 or 4 days this hairy growth changes from light brown to deep reddish brown. By that time the leaves are twisted. The deep brown color indicates that the erineum is drying, and the mites begin to abandon it for newer parts of their host.

These mites deposit eggs singly at the bases of the erineal papillae. Incubation periods average 2.5 days. The first nymph grows for 2 or 3 days before molting; the second instar, including the pseudopupa, lasts about 6 days. The preoviposition period is 1.5 days. This mite completes a life cycle in from 13 to 19 days. Population peaks occur at each new flush of growth of the host, but high temperatures, high humidity, and heavy rainfall are unfavorable for mite development (Alan and Wadud, 1963; Nishida and Holdoway, 1955).

Eriophyes litchii[3] is a wormlike species, 110 to 135 μ long. It has 5-rayed featherclaws. The shield design is of fairly heavy central lines and a broad band of lateral granules. The median shield line ends in a dart-shaped mark. Admedian lines are subparallel to the median and gradually diverge, slightly recurving at rear shield margin. Submedian lines extend back from rostrum base and join in a series of quadrate designs in front of the dorsal tubercle.

The coxae are heavily granular and there is a strong sternal line between the forecoxae. Thanosomal microtubercles are elliptical and touch ring margins. The female genital coverflap has a series of basal transverse lines and about 16 crowded longitudinal ribs.

Mango bud mite, *Eriophyes mangiferae* (Sayed). This species (fig. 113, *f*, *g*) is wormlike, ranges up to 225 μ long, and is whitish. This mite stunts and brooms twigs, causing bud proliferation. Young trees are most seriously affected. It is said to be slower in its action on the terminal growth than the rust mite *Metaculus mangiferae* (Attiah). The bud mite lives under cover and is more difficult to control than the rust mite. While the bud mite was originally named from Egypt, it also occurs on its host in southern Asia, and in Brazil (Sayed, 1946).[4] Its recognition features are: the featherclaws are 6-rayed; the shield design is indistinct anteriorly but between the dorsal tubercles there is a broken part of the median line, strong admedian lines that curve to the center at the rear shield margin, and a submedian line on each side of the admedians that curves to unite with them at the rear margin. Among the more characteristic features are the noticeably elongate dorsal microtubercles behind the shield. The female genital coverflap has longitudinal ribs in two uneven ranks, the rear rank stronger.

Cashew bud mite, *Eriophyes rossettonis* (K.). This mite (fig. 113, *d*) lives on cashew in South America, feeding on buds and flowers and decreasing nut production. It is a wormlike species, 150 to 185 μ long. The featherclaw is 5- and 6-rayed. The shield design centrally is of close-set subparallel lines: the median line is present on the rear half; the admedian lines are complete, subparallel, and sinuate, diverging to the rear and slightly recurving. The first submedian lines are completely sinuate, but not quite as long as the admedians. Laterally the shield is broadly granular. The coxae are granular, with a short sternal line between the forecoxae. The first setiferous coxal tubercles are pushed ahead of the usual position. Microtubercles on thanosomal rings are beadlike, with slight points, and set ahead of the ring margins. The female genital coverflap has about 10 weak longitudinal ribs.

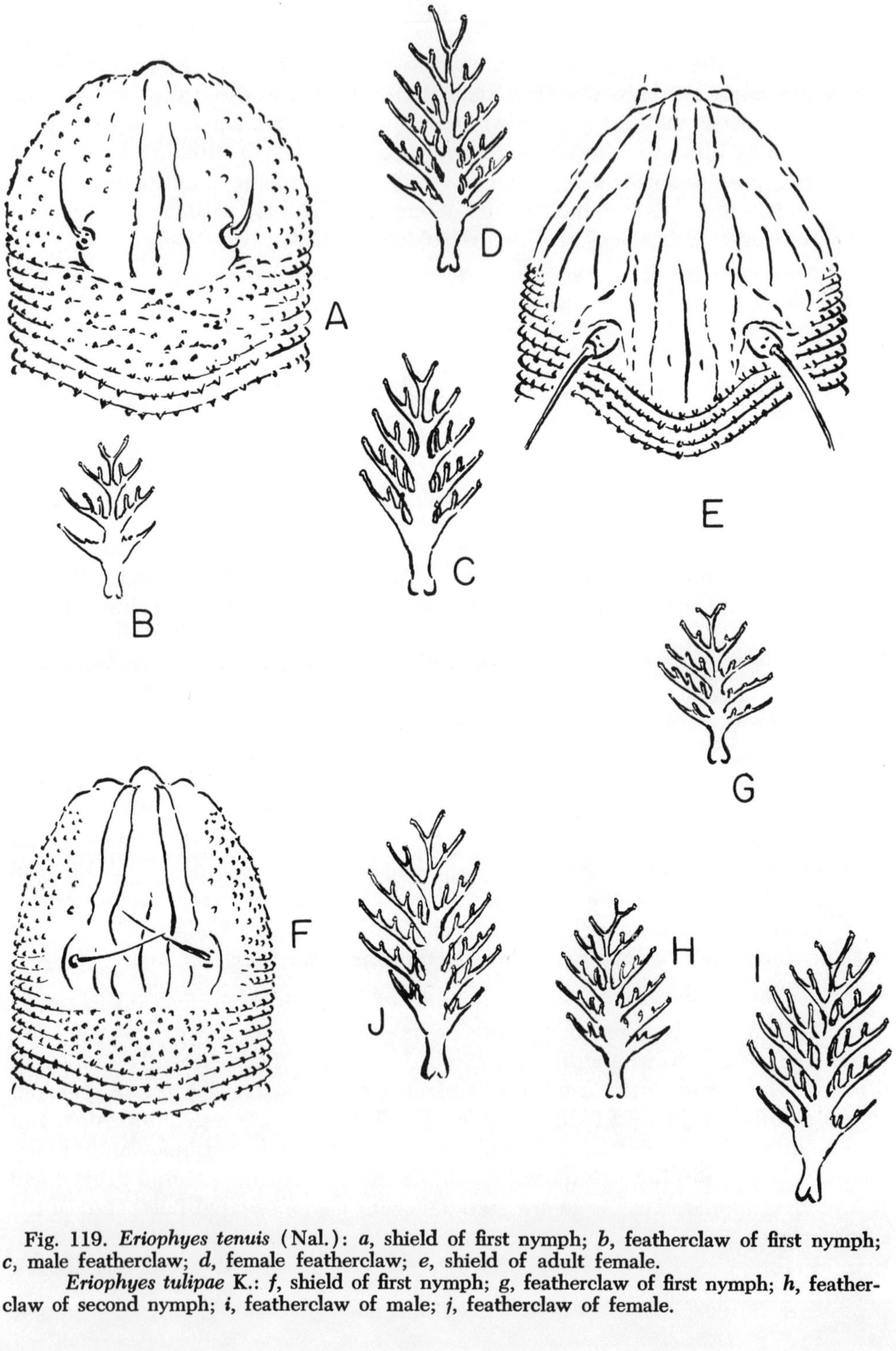

Fig. 119. *Eriophyes tenuis* (Nal.): *a*, shield of first nymph; *b*, featherclaw of first nymph; *c*, male featherclaw; *d*, female featherclaw; *e*, shield of adult female.

Eriophyes tulipae K.: *f*, shield of first nymph; *g*, featherclaw of first nymph; *h*, featherclaw of second nymph; *i*, featherclaw of male; *j*, featherclaw of female.

Pistacio bud mite, *Eriophyes pistaciae* Nal. *Pistaciae* (fig. 114, *d, e*) causes flower stalk brooming and some leaf deformation on certain pistacio species. The brooms are reddish and noticeable. The mite ranges up to 170 μ long, is wormlike, and has 4-rayed featherclaws. The central shield design is of strong longitudinal lines. The median line is present on the rear ½. Admedian lines are nearly complete and gradually diverge. The first submedian line runs back toward the dorsal tubercle and ends in a diagonal line extending inwardly across in front of the dorsal tubercle. There are two more short submedian lines, and lateral granules. The abdominal microtubercles are subcircular and slightly ahead of ring margins, which distinguishes this species from *stefanii*; as *stefanii* has pointed microtubercles. The coxae are ornamented with coarse granules. The female genital coverflap has about 6 rather broad longitudinal ribs.

Pistacio leaf edge roll mite, *Eriophyes stefanii* Nal. This species (figs. 99, *b*; 114, *a-c*) rolls pistacio leaflets over the upper surfaces into tight rolls, and transforms these leaflets into thin sticklike structures. A whole branch may at times be so badly infested as to appear almost leafless. The mite is a wormlike species, with females ranging up to 160 μ long. The featherclaws are 4-rayed. The central shield design has strong longitudinal lines, but laterally the shield is broadly granular. The median shield line is present on the rear ⅔ and ends in a dart-shaped mark at the rear margin. Admedian lines are slightly sinuate, complete, and gradually diverging. Short submedian lines are present anteriorly. The coxae have lines of granules. *E. stefanii* has pointed microtubercles on the thanosome that lie just ahead of ring margins. The female coverflap has about 10 longitudinal ribs. The two most distinctive features of this species, which separate it from the flower broom mite, are the broad granular areas on each side of the shield, and the pointed microtubercles.

Jasmine erineum mite, *Eriophyes jasmini* (Channabasavanna), infests jasmine in India. It causes feltlike hairy outgrowths on leaf surfaces, on tender stems, and on flower buds. This injury results in stunted growth and flower suppression. *E. jasmini* has a 4-rayed featherclaw and lacks lines or design on the shield.

Eriophyes Inhabiting Grasses and other Monocotyledonous Crops

Wheat curl mite or dry bulb mite, *Eriophyes tulipae* K. This mite (figs. 115, *h, i*, 119, *h-j*; pl. 65) is one of the most injurious eriophyid pests because it not only damages important agricultural plants by feeding, but also because it is able to carry plant viruses. Kernel red streak of corn behaves like the result of an injected toxin. Known viruses that *E. tulipae* transmits are wheat streak mosaic, wheat spot mosaic, and a third virus similar to the streak mosaic virus. These viruses have already been treated in chapter 6. These viruses, and evidently others related to them, are widespread, and transmitted by *E. tulipae* and by additional species of eriophyids that are structurally similar to *E. tulipae*. For these reasons *E. tulipae*, and some of its relatives, have been receiving intensive research attention (Slykhius, 1959; 1967).

Tulip bulbs, which are members of the Liliaceae, were the first known host of *Eriophyes tulipae*. *Ornithogallum* is another liliaceous plant host of the mite, but onion and garlic are the most commonly infested host in this plant family. All mites of this type so far seen (i.e., wormlike eriophyids with females having 7-rayed featherclaws, plainly lined shields, and spiny microtubercles) that were taken from garlic or onions, whether from North America or Europe, have the structure possessed by the original specimens from tulip.

In addition to liliaceous hosts, this mite infests grasses throughout North America (but not necessarily in Europe). Grass species that harbor *E. tulipae* fall for the most part into the following genera: *Agropyron*, *Elymus*, *Hilaria*, *Hordeum*, *Lolium*, *Muhlenbergia*, *Triticum*, and *Zea*. Nault and associates have been able to transfer these mites experimentally from corn to wheat, one of the more instructive developments in research on this mite.

In the field, mites inhabiting the upper parts of garlic concentrate on growing tips and to some extent on white bracts. Infested foliage shows stunting, twisting, curling, and yellow mottling, especially at margins. Mites in all stages of development may be found along the midrib on upper surface of leaf blades. From there they move into protected leaf folds. Tips of some leaves never unfurl, owing to mite feeding but remain within the next older leaves, giving the plants an arched appearance. Severe infestations can permanently disfigure a plant, but plants can outgrow light infestations.

From folds on aerial parts of garlic plants the mites move down into bulbs, feeding and breeding between bulb layers and causing drying or decay. The wheat curl mite, together with *Rhizoglyphus callae* Oudemans, a root mite, is implicated in spread of fungi causing bulb rot in storage. *E. tulipae* also causes clove drying of garlic in storage (Lange, 1955).

Perhaps the major corn disease made by this mite is known as *kernel red streak of corn*. Twisting and looping of corn leaves are other troubles produced by *E. tulipae*. This red streak occurs on corn throughout middlewestern area of Canada and the United States. Similar corn problems are reported from European areas, but the identity of the European mite making the discoloration is as yet uncertain. Kernel red streak is not a virus disease, but is probably the result of injected mite salivary toxins.

Kernel red streak occurs on sweet corn, flint corn, and on popcorn, but yellow dent corn is discolored the most, and white corn receives the least damage. Badly streaked corn sells at depressed prices as the discoloration forces it into mixed grade classifications. While streaked grains appear at random on ears, they are most often near ear tips. Loosened husks give mites easier entry and ears in such husks show more damage. Red streak colors range from deep red on yellow kernels, to pink or purple on white kernels. Streaks from the pericarp extend from kernel bases toward, and over the kernel crown. Pericarp streaking appears to result from formation and deposition of red pigment in these irregular streaks; endosperm discoloration is rare.

Wheat curl mites first attack corn plants by settling in leaf grooves. The mites favor areas where cells, high in water content, extend from bottom of the groove to the stomata. These cells rapidly carry mite toxin upward in the leaves. The mites move from vegetative portions of corn stalks early in the season to developing ears. They crawl along silk strands and into husks through the tip. As mentioned above, loosened husks facilitate this entry (Nault et al., 1967).

Wheat streak mosaic virus occurs in large wheat-producing areas of North America, Europe, India, North Africa, Australia, and New Zealand. *E. tulipae* successfully transmitted wheat streak mosaic virus to all wheat varieties tested (27) and from wheat to corn and back to wheat. A wide difference in vector efficiency occurs when transfers are attempted from other host plants to wheat depending on the species of the infected host. The virus is not transmitted transovarially, but by the adult and both nymphal stages. Wheat plants were successfully inoculated with this disease using macerates of nymphs and adults. Older adults were found to be less efficient than newly emerged adults.

The mites are not generally infective after only a 10-minute feeding period. A linear relationship appears to exist between length of acquisition–feeding period and the percentage of plants that become infected.

The virus appears to persist in *E. tulipae* for at least 6 days and even through the molt. Mites immersed in a 1 percent formaldehyde bath for 2 minutes did not lose ability to transmit the virus. Viruslike particles were found in the gut of mites fed on infected wheat, the geratest concentration occurring in the lumens of the hindgut and posterior part of the midgut. There appears to be disagreement as to whether the virus occurs in mite tissue (Oldfield, 1970).

Wheat spot mosaic is known only in the midwestern United States and Canada. Wheat plants infected with this virus show chlorotic spots, severe chlorosis, stunting, and necrosis. Virus symptoms continue to develop even after mite removal. Damage by nonviruliferous mites is not the same as that made by the virus. Wheat spot mosaic virus is not sap transmissable. Two types of symptoms may appear. One starts as light green spots on expanding leaves. The spots intensify to yellow and may become necrotic and eventually render the whole leaf chlorotic or necrotic. A second type starts as light green mottle and develops into yellowing that involves the entire leaf. Loss of leaf function and often death of the plant results.

Wheat is the principal host involved in the epidemiology of both wheat streak and wheat spot mosaic viruses. Immature diseased wheat infested by mites during the summer becomes a potent source of virus-carrying mites that spread to adjacent wheat by wind, or on flying insects. Winter wheat sown near infected oversummered wheat is subject to virus infection. One way to avoid infection of winter wheat is to plant it away from fields of immature infected wheat. Late seeding achieves virus-free winter wheat more effectively than early seeding. Minimizing of virus infection of spring sown wheat may be achieved by locating it away from infected winter wheat stands (Orlob, 1966; Nault, 1970).

All stages of *E. tulipae* are present in rather inactive condition on their hosts during the winter in the north-central United States. Mites overwinter on winter wheat, on volunteer wheat, and on nearby perennial grasses. As the mite population increases upon spring temperature rise, the mites develop under leaf sheaths, inside newly emerged unfurled blades, and later within wheat head glumes. High populations of *E. tulipae* in Ohio are found on wheat in spring and early summer, and on corn in late summer and early fall. Mites prefer terminal wheat blades, from which they move to each new leaf as it emerges, and ascend on the plant in this manner. They migrate into wheat heads and remain while the heads are succulent.

As long as the host tissue is suitable for feeding, the mites are negatively phototropic; but after plant decline, mites crawl in large numbers to uppermost parts of their hosts and rear up on their terminal lobes. They crawl up on each other and make chains. Mites seeking transportation may become so thick on top plant parts as to produce so-called "fuzzy masses."

Dispersal of *E. tulipae* is influenced by wind, temperature, and light. The mites seem to prefer daytime rather than nighttime travel, but this photopositive preference decreases in magnitude as temperatures rise. Most wind dispersal occurs at temperatures of about 18 C (65 F) and wind velocities above 15 mph.

Just what adaptations *E. tulipae* has for traveling are unknown. Precise structural studies in this connection remain for future investigation. Somsen (1966) has reported what he believed to be a migratory form of *E. tulipae* that is larger than the average colony female.

Eggs of *E. tulipae* are deposited or fastened in straight lines along parallel leaf veins. Favored oviposition times are early morning or late afternoon. Individual females lay from 3 to 25 eggs, averaging nearly an egg a day during their lives. Incubation lasts from 3 to 5 days at 9 C (45 F), but hatching stops at freezing. A complete life cycle covers 8 to 10 days during most favorable conditions. First nymphal growth periods, including the molt, are about 2.25 days long. The second nymph grows and molts in 2.75 days. Adult females have a preoviposition period of from 1 to 3 days.

The wheat curl mite can survive without food or water less than eight hours at 24 C (75 F) and 30 to 40 hours at 3 C (36 F). These survival times indicate that these mites have poor defenses against drying. On sterile media they have survived about 80 days.

Control of the wheat curl mite on garlic and onion is best done by eliminating mites on planting stock. Fumigation with methyl bromide at 2.5 lbs per 1000 cu. ft. for two hours at 27 C (80 F) kills all mites with no bulb damage. In the field, dusting the foliage with sulfur or with sulfur plus insecticide combinations (as commonly used against thrips) results in reduction of these mites. But mites in leaf folds protected from the toxicant are difficult to reach. Acaricides applied to wheat have not resulted in effective control of wheat streak mosaic, or of wheat spot mosaic (Gibson, 1957; Kantach and Knutson, 1958; Rosario and Webster, 1964).

The female of *Eriophyes tulipae* K. is a wormlike mite 210 to 250 μ long. The female featherclaws nearly always have 7 rays, rarely 8. Males have 6-rayed featherclaws and are usually shorter than the female. The antapical rostral female seta is 12 μ long. The shield is subtriangular, with a small lobe over the chelicera base. The shield pattern is of strong longitudinal lines, mostly solid. The median line is present on the rear ¼. Admedian lines are complete, sinuate, diverging gradually to the rear margin. The first submedian lines run from the anterior lobe divergently back toward the dorsal tubercles, becoming lines of granules to the rear and turning out in front of the tubercles. An inner submedian branch runs inside the dorsal tubercle position. There are two laterally directed lines from anterior positions on the submedian; lateral shield areas are granular and have partial rings. The dorsal setae are moderately long, and diverge to the rear. The coxae have coarse granulations. The thanosome has 70 to 85 rings; the telosome has 5 rings. Thanosomal microtubercles reach ring margins and are pointed (these spiny microtubercles are especially well depicted by stereoscan photographs), the points extending over ring margins. The accessory seta is 4 μ long. Female genitalia are deep, bowl-shaped, and the coverflap has 10 to 12 longitudinal ribs.[5]

Aloe wart mite, *Eriophyes aloinis* K. (fig. 115, *f*), attacks aloes and *Haworthia*, which are hosts belonging to the Asphodeloideae of the Liliaceae. The mite makes large warty deformations on leaf bases. It is not known to attack other types of monocotyledons. *Eriophyes aloinis* has 6-rayed featherclaws. It is generally similar to *E. tulipae* structurally, but differs most noticeably by having the submedian shield lines first uniting and then double-forking in front of the dorsal tubercles. *E. aloinis* is purple.

The Kenya rice mite, *Eriophyes bakkeri* (K.). This mite (fig. 113, *l, m*) lives on rice and on an unidentified grass in Kenya. The female is 170 to 185 μ long, wormlike, with 6- to 7-rayed featherclaws. The male has 5-rayed featherclaws. Over the rostrum base is a slight anterior shield lobe. Shield lines are composed of granules and short dashes, but otherwise are similar to *E. tenuis* series patterns. The median shield line is present on the rear ⅓. Admedian lines are complete, rather close anteriorly, farther apart when enclosing the median. The first submedian line divides once or twice in front of the dorsal tubercle. There are lines of granules on the coxae and on each side of the sternal line. Foretibiae are shorter and stockier than on *E. tulipae*, with the seta arising from ⅔. Thanosomal microtubercles are small, somewhat crowded, and rounded off. These microtubercles or ring granules are elongate dorsally and touch ring margins. Laterally the granules are ahead of margins. The female genitalia are not as deeply bowl-shaped as those on *E. tulipae*, and the coverflap has 6 ribs.

Coconut flower and nut mite, *Eriophyes guerreronis* (K.). This mite (fig. 115, *c-e*) is a serious pest of coconut palm trees along the west coast of Mexico, in Venezuela and Brazil of South America, and in Dahomey of West Africa. Feeding by these mites produces brown blemishes over the fruits, which begins at the separation of the floral bracts, and extends towards the free part of the nut. These areas become gangrenous, hard, cracked, and sufficiently extensive to deform fruit. Nuts that develop from infested flowers are smaller than normal.

Eriophyes guerreronis lives in colonies beneath bracts of young nuts, but mites do not seem to attack nuts smaller than 6 cm in diameter. The presence of but a few mites on smaller nuts indicates that the colonies have not had opportunity to develop higher populations up to that point. The injury described appears to be initiated entirely by the mites, but disease pathogens or plant deficiencies may also contribute to total injury. Most severe infestations may damage 50 to 100 percent of the nuts. No coconut virus transmission has as yet been linked to *E.*

guerreronis, although it could possibly carry the cadang-cadang disease of coconut trees, which is present in the Phlippines (Ortega, Rodriquez, and Garibay, 1965).

Eriophyes guerreronis is another member of the *tenuis* group in the genus *Eriophyes*. It differs from *E. tulipae* by having rounded-off ring granules, by the transverse line across in front of the dorsal tubercles, and by having short dashes on the coxae, but no coxal granules. The *E. guerreronis* female is 205 to 255 μ long, wormlike, with 6-rayed featherclaws. There is the usual slight anterior shield lobe over the rostrum base. The shield design is mostly of strong lines centrally; the admedian line extends back from the anterior ⅓. Admedian lines are complete and diverge gradually. The first submedian line curves back from the side of the base of the anterior lobe, extends toward the dorsal tubercle, and turns centrad as it meets the inner end of a transverse line across in front of the tubercle. The second submedian line runs divergently toward the partial rings below the dorsal tubercle. The coxae are ornamented with short longitudinal lines and the first coxal setiferous tubercles are well behind the anterior end of the sternal line. Thanosomal microtubercles or ring granules are elliptical, longer dorsally, but ahead of ring margins. Female genitalia are somewhat bowl-shaped, and the coverflap has about 12 longitudinal ribs.

Sugar cane blister mite, *Eriophyes sacchari* (Channabasavanna). This species causes injury to sugar cane in India and other parts of Asia, but will not grow on corn or sorghum. The mites live on the inner surface of leaf sheaths. Their feeding makes spots about 1 or 2 cm in diameter on both sides of the sheath. These spots are first watery in appearance, but later turn dark red. Hypertrophied growth results from spherical-like leaf blisters, with mite colonies inside (Wang, 1964). Eggs of *E. sacchari* are about 45 μ in diameter and are subspherical and colorless.

There are two female forms reported for *E. sacchari*. One is slender with 6-rayed featherclaws; the thicker type usually has 7-rayed featherclaws but the number is variable. These mites are light yellowish and wormlike. The length varies from 180 to 260 μ. The shield is represented as somewhat truncate anteriorly and longer than it is wide. Centrally the shield has long lines with the median and admedian lines complete. The submedian line is close to the admedian but represented as irregular. The coxae have numerous lines. Abdominal microtubercles are drawn as elongate-elliptical, touching ring margins. There are 8 to 12 longitudinal female coverflap ribs.

A sugar cane eriophyid living in South Africa is in Nalepa's 1929 host list as *Eriophyes merwei* (Tucker). It is supposed to damage the buds and cause dwarfing. The description of *E. merwei* states it to be narrow-cylindrical in shape, the females being 200 μ long. There are about 80 abdominal rings, and the featherclaw is stated to be 6-rayed. The claw is represented as being very long.

Corn sheath mite, *Eriophyes zeasinis* (K.) (fig. 115, *g*), causes distortion of growth and stunting of corn grown in southern greenhouses. It differs from *E. tulipae* by having ring granules that are rounded off, by the granular shield lines, and by having a 6-rayed featherclaw. Wormlike females of *E. zeasinis* range from 255 to 270 μ long. The shield has the usual slight projection over the rostrum base. The granular median shield line occupies the rear ⅔ of the shield length. There is a prominent cross line, transversing the median, at just behind ½. Admedian shield lines are complete, sinuate, farthest apart at the cross line. The first submedian line curves back from the front end of the admedian and extends toward the dorsal tubercle, forking in front of the tubercle, with the inner line broken and running toward rear inside the tubercle position, the outer fork abruptly angling past the tubercle on its outer side. The coxae have lines of granules and a moderately prominent sternal line, flanked by subparallel granular lines. The abdominal microtubercles are rounded off, ahead of ring margins, and more elongate-elliptical dorsally; dorsally some reach margins. Female genitalia are shallow, bowl-shaped, with about 10 irregular ribs.

Buffalo grass mite, *Eriophyes slykhuisi* (Hall). This species is known only from buffalo grass. The host and the mite occur in midwestern areas of North America. It differs from *E. tulipae* mainly by not having the dorsal and lateral microtubercles pointed. *E. slykhuisi* is wormlike and about 220 μ long. The featherclaw rays are usually 6, less often 7. It has a shield design much like *E. tulipae.* The coxae are granular. There are about 65 to 70 thanosomal rings. The ring granules are elongate dorsally, more beadlike laterally, slightly pointed ventrally, and ahead of ring margins. The female genitalia are shallow, bowl-shaped, with about 7 to 8 longitudinal ribs on the coverflap.

Bermuda or couch grass mite, *Eriophyes cynodoniensis* (Sayed) (fig. 113, *h, i*) (= *neocynodonis* Keifer), infests Bermuda grass nodes, causing bunching and stunting (pl. 66). The mites congregate in these thickened shoots, living under broadened leaf bases. The mite prefers climates where atmospheric humidity is fairly low. In the western United States it damages its host grass in central and southern California, but especially in the even drier Arizona area. This mite is probably native to Australia, where its host is native; but now is widespread, occurring in New Zealand, North Africa, and western North America. Undoubtedly other locations will soon be found (Wilson, 1959).

E. cynodoniensis Sayed is a wormlike mite, 165 to 210 μ long. Females have 6-rayed featherclaws, the males having 5-rayed ones. The shield has no appreciable anterior lobe. Shield lines are fairly solid, but many granules and short dashes accompany them. The median line is present on the rear ⅔. Admedian lines are complete and gradually diverge. The first two submedian lines join at ½, but give way to a large fork, the inner line of which extends toward, and curves out past, the dorsal tubercle; the outer line runs laterally toward the partial rings below the tubercle. The legs are stocky with the foretibia slightly shorter than the tarsus. The coxae are granular with a moderate sternal line, unforked to rear, and flanked by subparallel sinuate granular lines. The abdominal thanosome has about 60 rings; the ring granules are elliptical and set ahead of margins. The granules are unpointed. The female genitalia are shallow, bowl-shaped, with 10 to 12 longitudinal ribs.

The first nymph of *E. cynodoniensis* has dorsal setae pointing up, but more inclined to the rear than many first nymphs in the *tenuis* grass series.

Eriophyes cynodonis (Wilson). A second Bermuda grass mite in North America is *E. cynodonis* (Wilson) (fig. 115, *a, b*), which occurs in the central valleys of California, in southern California, and has been collected in Kansas and Arkansas. It lives in the grass heads and does not cause stunting. The grass plants are able to extend their height, but heavy infestations of *E. cynodonis* cause terminal blades to stick in partially unfolded lower blades, and the result is conspicuous terminal arching. *E. cynodonis* is not similar to other grass mites, and does not belong in the *tenuis* series of grass species. The *E. cynodonis* shield not only lacks markings, but the anterior lobe is broad, rigid, and extends well over the rostrum base. It is a wormlike species that ranges up to 210 μ long.

Eriophyes Inhabiting Vegetables, Field Crops, and Berries

Alfalfa broom mite, or Lucerne mite, *Eriophyes medicaginis* K. (fig. 114, *l, m*), causes witches'-broom on alfalfa (lucerne) in Australia and North America. While the alfalfa symptoms caused by this mite are similar to certain virus diseases (for example the tomato big-bud disease carried by a leafhopper), the alfalfa symptoms prove to be entirely the result of mite feeding. The alfalfa witches'-broom symptoms are several. The plants become dwarfed; there is considerable increase in crown shoots. Stems become fine with erect bunchy foliage; the leaves remain small, rounded, puckered, and slightly chlorotic. The intensity of the disease is directly related to mite population density, and recovery of affected plants is slow even after removal of the mites. While little has been reported on the habits of *E. medicaginis*, one test showed that the mites could remain alive for 2 weeks at 38 C (100 F) (Stubbs and Meagher, 1965).

E. medicaginis is wormlike, with females from 150 to 180 μ long. The featherclaws are 5-rayed. There is no shield pattern, the shield surface being without lines or granules. The most distinctive feature of *E. medicaginis* is the fainter development of the dorsal microtubercles. The microtubercles, or ring granules, are pointed over the ring margins, but the ventral ones are stronger than the dorsal granules. Dorsal parts of the rings on the thanosome rear, and telesome, are smooth. The female genital coverflap has scattered basal granules, and 13 to 16 longitudinal ribs.

Mites with the exact external structures of *medicaginis*, as described above, occur on *Hedysarum coronarium* L., a leguminaceous plant growing in Italy.

Eriophyes plicator (Nal.). This mite occurs on alfalfa, *Medicago falcata* L., in Europe. It is similar to *medicaginis* by having a 5-rayed featherclaw, but *plicator* differs by having strong longitudinal shield lines. The microtubercles on rings on the European legume-infesting bud mites have not been sufficiently characterized as yet.

The carrot bud mite, *Eriophyes peucedani* (Canestrini). This mite (fig. 113, *a-c*) occurs in Europe and in parts of western North America on carrots and apparently on other umbelliferous plants. It is an important pest of Idaho field carrots. It causes deformation of leaves and umbels, resulting in greenish dwarfed plants with mosaiclike symptoms. Subsequently umbel parts turn brown early in their development and fail to have the normal creamy white color.

Damaged plants make little or no seed and infested seedlings may develop undersized roots. In storage, infested carrot stecklings (roots) offer the mites the chance to feed and multiply. Infested roots with storage damage perform poorly when planted the following spring. In some of the carrots, the growing point on these roots may have been killed, thus preventing or delaying growth. Or, if the plants do grow, they produce stunted tops, with poor blooms, or none. During bloom or petal fall these mites congregate in large numbers on top parts of the plants from where they travel to new hosts by wind, or on flying insects.

While there is no significant control of the carrot bud mite, it is known that diazinon, demeton, di-syston, and phorate are toxic to the mite. Growers should isolate steckling fields from seed carrot fields (Keifer, 1952; Scott, 1965).

The *peucedani* female is wormlike, about 210 μ long, and has 4- to 5-rayed featherclaws. Shield lines are of granules. The median line extends back to the rear shield margin from ⅓. Admedian lines are close to each other before the median line, but then diverge as two convex arcs in extending back to the rear margin. The shield is granular laterally. The coxae are ornamented with coarse granules. A particularly important character useful in recognizing this mite species is the weak dorsal ring granules in combination with the lines of shield granules. The coxae have coarse granules. There are about 12 female genital coverflap longitudinal ribs.

Hibiscus erineum and leaf crumpling mite, *Eriophyes hibisci* Nal. This mite (fig. 114, *h*) makes a leaf erineum on rose of China, or Chinese hibiscus, and perhaps on other species of *Hibiscus*. On okra the mites form irregular erineum pockets, which result in misshapen leaves. This mite has been collected in South Sea islands such as Tonga, and has also appeared in Brazil, but its full range is not known. Females of *hibisci* are 160 to 185 μ long, robust, wormlike, with 4-rayed featherclaws. The shield is subtriangular in dorsal view and the lines are only strong to the rear. The median line is hardly indicated. The admedians appear on the rear shield half as outwardly convex arches, ending on the rear margin. There are some lateral shield lines, and there is an ocellar spot at the rear shield angles. The thanosome is divided laterally into stronger rings above, and weaker rings below. Ring granules are elongate above, reaching ring margins, and more beadlike below, tending to be acute to the rear. The abdominal dorsum lacks granules on the rear ¼. Coxal ornamentation is absent. Female genital coverflap longitudinal ribs number 8 to 10.

Nalepa's species *hibiscitellus*, genus unassigned, is listed as making leaf galls on *Hibiscus similis* B.

Sweet potato leaf gall mite, *Eriophyes gastrotrichus* Nal. This species causes the development of extensive irregularly shaped outgrowths on leaves and stems of sweet potato. This mite ranges across southern Asia and into the East Indies. The activities of the mite result in severe deformation of sweet potato leaves and stems. Channabasavanna (1966) gives a description of this mite, as well as figures, from which the following characterization comes. The wormlike females are 145 to 160 μ long, with featherclaws varying from 4- to 5-rayed. The shield is somewhat concave on the sides toward the front, and the front edge is rather broad and slightly emarginate. There are so many fine longitudinal shield lines as to completely obscure the usual type of pattern. Elliptical ring granules are depicted on this mite. Dorsal telosome ring halves are smooth, with numerous elongate microtubercles on the telosome venter. The female genital coverflap has an enclosed triangular design on each side of the base, and there are numerous fine longitudinal ribs. The coxae are drawn with coarsely transverse lines.

Tomato erineum mite, *Eriophyes lycopersici* (Wolffenstein). This mite has an almost holotropical range on various solanaceous plants. It may occur in more temperate areas in greenhouses. The damage it causes consists of excessive hairlike growths on leaves and stems, which sometimes gives the impression of white mold. Attacks by this mite are more likely to occur in dry seasons. Tomato and eggplant are among the principal hosts. Channabasavanna (1966) has not only given a good description of this mite, but his drawings are unusually informative. (Note—do not mistake *E. lycopersici* (W.) for *Aculops lycopersici* (Massee), which see. The *Aculops* is the tomato russet mite.)

E. lycopersici (W.) is a robust wormlike species with females 150 to 200 μ long. The species has a 4-rayed featherclaw. The subtriangular shield has an almost complete median line, and the admedians are also almost complete, though widespread. There is a broad lateral granular band on the shield sides. Coxae are represented as without markings, and with a moderate sternal line, unforked posteriorly. Abdominal microtubercles are elliptical, thick posteriorly, and touch ring margins. The telosomal rings have either suppressed granules, or are smooth dorsally. There are 10 to 12 longitudinal ribs on the female coverflap.

Pomegranate leaf curl mite, *Eriophyes granati* (Canestrini and Massalongo). *E. granati* (fig. 114, *g*) occurs throughout the Mediterranean region and has followed pomegranate plantings elsewhere in the world. The mite tightly rolls the leaves from the sides down onto the undersurface; these leaves may be so tightly rolled as to produce a nearly leafless appearance to the twig. But the twigs continue to elongate, indicating the twig terminal is not damaged. A description of the mite is taken from California specimens.

Somewhat robust, wormlike females range up to 180 μ long. The featherclaw is 4-rayed. The median shield line is faint anteriorly, but ends in a dart-shaped mark at rear shield margin. The admedians are complete and gradually diverge. The first submedians are short. The second submedians curve back toward the admedians before the dorsal tubercles and then abruptly angle out past the front of the tubercles. Heavy granules on the coxae obscure the sternal line between the forecoxae. Thanosomal ring granules are somewhat elliptical and situated just ahead of the ring margins. The female genital coverflap has about 5 heavily granular transverse posteriorly convex lines basally, with 14 to 16 longitudinal ribs present on the coverflap itself. (Note—*Eriophyes granati* must not be confused with *Phyllocoptes graniti* K. The latter species, known only in Italy, makes undersurface erineum patches on almond leaves.)

Eriophyes Inhabiting Ornamental Plants and Forest and Shade Trees

Three species assignable to the genus *Eriophyes* attack carnations. These species are: *E. paradianthi* (K.) a European carnation mite, *E. dianthi* Lindroth, and *E. georghioui* (K.). The precise ranges of each of these species are unknown, but *E. paradianthi* occurs in Europe, Argentina, and in Maryland and California. The mites live between leaf bases and stems, and are especially numerous on lower portions of the plants. Their presence is evident by the somewhat greasy, distorted, and stunted growth they make. The plants also lose color.

Eriophyes paradianthi (K.). The female of this mite (fig. 113, *j*, *k*)is wormlike, with a 6-rayed featherclaw. The shield has a heavy network design on it. The median line is on the rear ¾. Admedian lines are complete, with 3 cross lines joining them to the median, and to submedian lines. The first cross line joins the front end of the median, and joins the two submedians. The second cross line, at just behind ½, also joins both submedians, with a backward projection just outside the dorsal tubercle. The third cross line, at beyond the ¾ position, involves only the median and admedians. Coxae are ornamented with numerous short dashes, and the sternal line is absent owing to separation of the forecoxae. Abdominal microtubercles are pointed. The female genitalia have about 5 basal cross lines of dashes, and there are about 14 longitudinal coverflap ribs.

A European carnation mite, *Eriophyes dianthi* Lindthroth, has a 5-rayed feather claw and is said to be free living on its host.

Eriophyes georghioui (K.) 1959, on the island of Cyprus, makes similar carnation injury to that described for *E. paradianthi.* Diazinon, chlorobenzilate, dicofol, and parathion effectively reduce exposed populations of these mites, but repeated applications are necessary to prevent injury (Pritchard, 1957).

Eriophyes georghioui (K.) is considerably different from *E. paradianthi,* not only in shield design, but also in having a 7-rayed featherclaw. Females of *E. georghioui* are wormlike and 210 to 220 μ long. The shield is primarily composed of longitudinal lines. The median line is faint anteriorly, but ends close to the rear margin in a dart-shaped mark. The admedians are complete, sinuate, gradually diverging. The first submedian runs back from the shield front to about the ⅔ point, then abruptly angles out into a cross line well ahead of the dorsal tubercle. The second submedian forks at the outer end of the cross line. Forecoxae touch narrowly, centrally, the sternal line forks ahead and behind this short contiguity, the ornamentation on the coxae is a series of granules more-or-less arranged in curved lines. Abdominal ring granules are rounded off and ahead of the ring margins. Female genitalia have 3 or 4 basal cross lines of granules, and 14 to 16 longitudinal ribs.

Eriophyes cajan (Channabasavanna). This mite lives on *pigeon pea* in India. It may cause slight leaf yellowing, but also carries pigeon pea sterility virus. The female ranges from 120 to 130 μ long and is represented as having a robust, wormlike shape. Shield lines are light centrally, with complete median and admedian lines. There are numerous short longitudinal shield lines laterally. Coxal ornamentation is not represented, and the sternal line is moderately long. Abdominal ring granules are elliptical and touch ring margins. The female coverflap has about 12 longitudinal lines.

Flax leaf edgeroller, *Eriophyes peyerimhoffi* Nal. This species occurs in North Africa, where it causes flax leaves to roll upward. The mite is about 160 μ long, wormlike, and has very fine featherclaws that are probably 4-rayed. Central shield lines are present with the median line on the rear 1/2. Admedian lines are said to be broken centrally. Dorsal abdominal rings are wider than those on the venter, and more strongly granular. The telosome lacks granules on the dorsal parts of the rings. The female coverflap has fine lines.

Lantana flower gall mite, *Eriophyes lantanae* Cook. This mite was originally named from deformed flowers on *Lantana camara* L., in Cuba. The mite causes flowers to change to a mass of small leaflike outgrowths. As well as occurring in Cuba, the mite is present in southern Florida, and in the general Caribbean area. The mite has 4-rayed featherclaws, rounded microtubercles on the abdominal rings, and the cephalothoracic shield lacks a definite pattern. A very similar mite makes simple leaf galls on this same host throughout Central and South America.

Ash spangle gall mite, *Eriophyes fraxinivorus* Nal. *Fraxinivorus* has a wide distribution and causes development of multilobed hanging galls on ash seed clusters. It may also produce galls that are somewhat like the spangles, but are firmly attached, and issue from lateral twig buds. The female is wormlike and about 180 μ long. The featherclaw has 4 rays. There are no longitudinal shield lines, or other surface markings.

Bead gall mite on ash, *Eriophyes chrondriphora* (K.). This mite causes bead galls on ash leaves. Its precise range is unknown, but the mite is evidently widespread. This mite is deuterogynous. The protogyne is wormlike, with 3-rayed featherclaws, and an involved network of granular lines on the shield. Abdominal ring granules are set on the margins, beadlike, and rounded off. The female coverflap has no ribs. The length of the protogyne is about 210 to 260 μ. The deutogyne has no shield granulations and no ring granules.

Cottonwood or poplar bud gall mite, *Eriophyes parapopuli* K. This species ranges throughout the western part of North America, principally on the great plains and in the intermountain region. These mites make woody galls from buds, and stop twig growth at that point. The gall continues to enlarge for several years and the mites move into fresh surface creases as they develop. Eventually the gall may reach a diameter of an inch or more, and die. But as it is firmly attached, this gall hangs on. Galls may erupt on most buds along twigs, resulting in growth retardation of the tree and twig deformation. Branches may die prematurely, apparently as the result of the mite attacks. Owing to the presence of the mites, ornamental poplars often fail to develop into satisfactory trees. This mite has a reddish wormlike body, which ranges up to 200 μ or more long. The featherclaw is 4-rayed. The shield design is faint, but the median line is present on the rear ¼ behind a cross line between the admedians. Coxae bear no ornamentations, and there is a short sternal line. Abdominal microtubercles are almost subquadrate and slightly ahead of the ring margins. The terminal lobes have a distinct upward extension between the caudal setae. The female coverflap is unmarked.

A similar mite, *Eriophyes populi* (Nal.), occurs in Europe. It is not possible to make a statement as to the importance of this mite on poplars, but H. K. Wagnon reports that in Turkey the poplars do not show noticeable, if any damage.

Eriophyes Infesting Maple Trees

Solitary maple leaf gall mite, *Eriophyes macrochelus* (Nal.). This mite is a European species that makes upper surface solitary, subspherical leaf galls, mostly along veins, on hedge maple. The galls engendered by this mite are conspicuously larger than the much more numerous tiny bead galls common on maple leaves in Europe. Nalepa designated this mite *Eriophyes macrochelus typicus* Nal. in his 1929 host list. The wormlike females are about 190 μ long, with 4-rayed featherclaws. The dorsal setae are slightly longer than the shield. The second legs have claws longer than the featherclaws. The abdomen has about 58 rings, generally granular, but the granules are faint caudally, especially above. Examination of mites from this type of gall, on this host, shows that there are two types of females present. One female has fainter microtubercles (ring granules) than the other, and could be the deutogyne form. If so, this form is not similar to the deutogynes in leaf erinea on these trees, which entirely lack dorsal granules. Considering the structure of the apparent deutogyne, and the galls this mite makes, it seems inappropriate to use the name *E. macrochelus* for mites that make European maple leaf erinea. This book therefore uses *E. macrochelus* only for the solitary gall maker.

Purple erineum mite, *Eriophyes eriobius* Nal. This is a yellowish-white species that makes undersurface purple erinea on European hedge maple leaves. The erineum is composed of capitate papillae. Nalepa's name for the mite is *Eriophyes macrochelus eriobius* Nal. He described the females in part as being 185 μ long and wormlike. The shield design is faint, with the dorsal setae slightly longer than the shield. The featherclaws are 4-rayed. There are about 68 narrow thanosomal rings that are microtuberculate, with the microtubercles being faint above caudally. A mite that Nalepa called an inquiline in this purple erineum was named by him as *Eriophyes macrochelus psilomerus* var. *aceris campestris* Nal. Nalepa defines *E. psilomerus* as lacking dorsal microtuberculation. This lack of dorsal microtubercles allies this mite to several erineum mites on other maples, on linden,

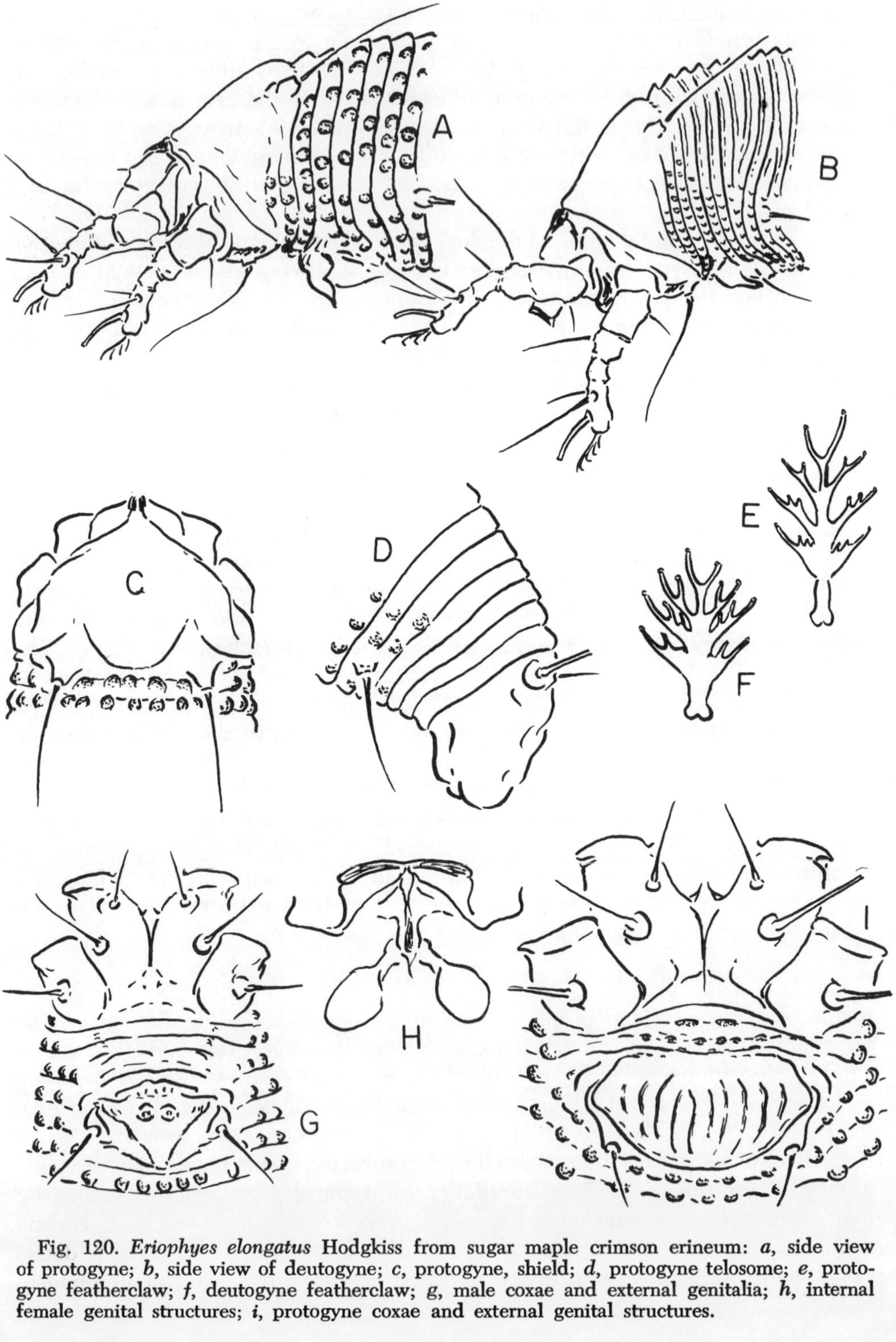

Fig. 120. *Eriophyes elongatus* Hodgkiss from sugar maple crimson erineum: *a*, side view of protogyne; *b*, side view of deutogyne; *c*, protogyne, shield; *d*, protogyne telosome; *e*, protogyne featherclaw; *f*, deutogyne featherclaw; *g*, male coxae and external genitalia; *h*, internal female genital structures; *i*, protogyne coxae and external genital structures.

and on beech. The conclusion here is that the name *psilomerus,* as applied to the mite with smooth dorsal rings, which on this host, in reality designates the deutogyne. Therefore the name for it in this case is a synonym of eriobius.

Eriophyes psilomerus Nal. This mite makes purple erinea on sycamore maple leaves in Europe. Since sycamore maple is the type host for *E. psilomerus,* the name *psilomerus* has a reciprocal relationship to the use of *eriobius* on hedge maple. So Nalepa's use of the name *eriobius* on sycamore maple becomes a synonym of *psilomerus.* Nalepa's name for the purple leaf erineum sycamore maple mite was *Eriophyes macrochelus psilomerus* Nal. He referred the actual erineum making to what he called *Eriophyes macrochelus eriobius* var. *pseudoplatani* (Corti). Nalepa differentiated *eriobius* on sycamore maple, from *eriobius* on hedge maple, by noting that the females on hedge maple are larger, 200 μ long, and with stronger legs than those on sycamore maple. Microtuberculation is also stronger. Since there are these distinctions, and since there is lack of data on experimental transfers, the conclusion here is that these purple erineum mites on the two different hosts are closely related but specifically distinct.

Crimson erineum mite, *Eriophyes elongatus* Hodgkiss (fig. 120), is a yellowish-white species that makes upper surface crimson erinea on North American sugar maples. It occurs along the East Coast. The protogyne is 216 μ long, wormlike, and has a 4-rayed featherclaw. The shield lacks much marking, but there is an inverted V-shaped cross line in front of each dorsal tubercle, the inner extension of which attaches to the V on the other side. The thanosomal ring granules are large and unevenly distributed both dorsally and ventrally. These granules are absent from the dorsal abdominal rear. The female coverflap has about 10 longitudinal ribs. Deutogynes lack dorsal abdominal microtubercles, and there is a fine central transverse crease in each dorsal half-ring. The deutogyne featherclaw is longer than that on the protogyne. While *E. elongatus* has page priority, Hodgkiss has added a second name to the mites in this crimson erineum. The second name is *regulus* and the conclusion here is that one of these names applies to the protogyne and the other to the deutogyne, so *regulus* becomes a synonym.

Western erineum mite, *Eriophyes calaceris* (K.). This species is quite similar structurally to *E. elongatus* but the protogyne ring granules are smaller, and the deutogyne ring creases are stronger than on *E. elongatus. E. calaceris* makes leaf edge, or upper surface bright red erinea on western mountain maples.

Box elder erineum mite, *Eriophyes negundi* Hodgkiss. The erineum pockets this mite makes on the undersurface of box elder leaves protrude out above the upper surface. The erineum is dull tan or gray. The mite species is characterized by unusual deutogynes, which are identified by the strong tergal-sternal ring differentiation. The shield does not overhang the rostrum.

Eriophyes pseudoplatani (Corti). This mite is, according to Nalepa, a form lacking dorsal ring granules, which makes it a deutogyne. Mites of this type come from rather small yellowish undersurface erineum patches near leaf veins, usually located basally, on European sycamore maple. Nalepa always expressed doubt as to the precise maker of this erineum. The form from this erineum which fits the protogyne definition, was named by Nalepa as *Eriophyes phyllerites.* This *phyllerites* was described by Nalepa as being wormlike and 190 μ long. The shield has three central lines. Those extending back fork in front of the dorsal tubercles. The dorsal setae are as long as the shield. The abdomen has about 65 rings, strongly granular, the rings broader and smoother caudally. The accessory seta is short. Such deutogyne forms as have been available from this yellow erineum show dorsal transverse ring creases that ally the species to such North American mites as *E. elongatus* Hodgkiss, on sugar maple, and to *E. calaceris* in the west.

Eriophyes major Hodgkiss. This mite makes white to red undersurface leaf erinea on red maple leaves. The range of the mite in eastern North America is presumably the same as the range of the host. Hodgkiss's name *ornatus*, for another supposed species in this erineum, and on this host, presumably really refers to the deuterogynous condition of the species. The protogyne of *E. major* has stronger shield lines than *E. elongatus* on sugar maple, but the deutogyne of *E. major* has less conspicuous transverse dorsal ring creases. The name *major* has priority.

Eriophyes aceris Hodgkiss. This species engenders greenish or reddish undersurface leaf erinea on silver maple. No deutogyne is known. *Eriophyes modestus* Hodgkiss causes undersurface green erinea on sugar maple leaves in eastern North America.

Maple Mites in Other Genera

Two maple rust mites that Hodgkiss placed in the genus *Oxypleurites* (now *Tegonotus*) occur on certain eastern North American maples and are deutogynous. One is *Tegonotus dentilobis* (Hodgkiss) on sugar maple, and the deutogyne is *Phyllocoptes variabilis* Hodgkiss. The other is *Tegonotus dentatus* (Hodgkiss), on Norway maple, and the deutogyne is *Phyllocoptes trilobis* Hodgkiss. Both specific names with *Tegonotus* take precedence.

Artacris macrorhynchus (Nal.) (fig. 123, *a-c*) makes tiny, slightly elongate, very numerous bead galls, on upper surfaces of sycamore maple leaves in Europe. Nalepa's name for this mite is *Eriophyes macrorhynchus typicus* Nal. The female is 160 μ long, wormlike, with 4-rayed featherclaws. The shield design has central longitudinal lines, with the admedians flaring at rear. The most distinguishing specific feature, and the one defining the genus *Artacris*, is the narrow, basally hinged, anterior projection over the rostrum.

Three different forms of this anterior shield projection are now recognized on maple mites. The species these forms designate are:

1. *A. macrorhynchus* (Nal.) has a narrow, sharp-pointed shield projection over the rostrum. It makes tiny, slightly elongate upper surface leaf galls on sycamore maple in Europe.

2. *A. cephaloneus* (Nal.) (fig. 123, *d, e*) has a slightly broad, apically rounded anterior suprarostral shield projection, and makes tiny bead galls on sycamore maple. (Nalepa's name for this mite is *Eriophyes macrorhynchus cephaloneus* Nal.)

3. *A. antonimus* (K.) is a bud mite on California boxelder. It makes no galls. This species has a rounded anterior projection over the shield, which projection is narrower than on *A. cephaloneus.* Another character of *A. antonimus* is the dorsal rear ridge on the abdomen. Mites with pointed anterior shield projections have recently been found in Martinez, and on Angel Island, in California.

Aculops maculatus (Hodgkiss) (fig. 93, *f*) is a leaf vagrant or inquiline in erinea on sugar maple. It is recognized by the short, rounded shield lobe, and by the elongate-elliptical dorsal microtubercles.

Aculops aceris (Nal.) is an inquiline in purple erinea on hedge maple in Europe. Nalepa described it as having a female about 140 μ long. The dorsal setae are about shield length; the abdomen has 28 tergites. *Aculops acericola* (Nal.) is an inquiline in purple erinea on sycamore maple. This latter species, according to Nalepa, has an obscure shield design, with the dorsal setae shorter than the shield. There are about 23 abdominal tergites. The female of *A. acericola* is described as being 120 μ long, which is smaller than *A. aceris.* Nalepa synonymized these two names under *aceris* in later publications, but the opinion here is that they are distinct.

In recent literature (Browne, 1968), the species *acericola* has been implicated in maple leaf damage in Europe. As far as can be determined at present, both from literature and from observations on western North American eriophyids similar to *acericola,* this type of erineum inquiline is harmless to maple.

A mite on sycamore maple in Europe which Nalepa stated caused extra hair growth on leaf undersides, was named by him as *Phyllocoptes gymnaspis* Nal. This mite lacks dorsal shield setae, and the genital characteristics, which are the appressed genitalia with internal shortened anterior apodeme in the female, all combine to put this species in the Cecidophyinae (see key in appendix 3). Roivainen (1947) put the species *gymnaspis* into the cecidophyine genus *Coptophylla.* A similar species, *Cecidophyes collegiatus* K., functions as a leaf vagrant on the underside of Norway maple leaves in various parts of the United States (such as Maryland and California).

Maple bladder gall mite, *Vasates quadripedes* Shimer. This mite (fig. 130, *a-c*) is native on eastern North American silver maple. This species is treated below. The mite is strongly deuterogynous. Deutogynes are especially active and appear during the latter part of the season and wander into many locations. Hodgkiss's name *confusus* would seem to be the description of the protogyne of *V. quadripedes.* The mite makes bladder galls on red maple in eastern North America, but they tend to be more solitary than those on silver maple.

Sugar maple finger gall mite, *Vasates aceris-crumena* (Riley) (fig. 130, *d, e*),

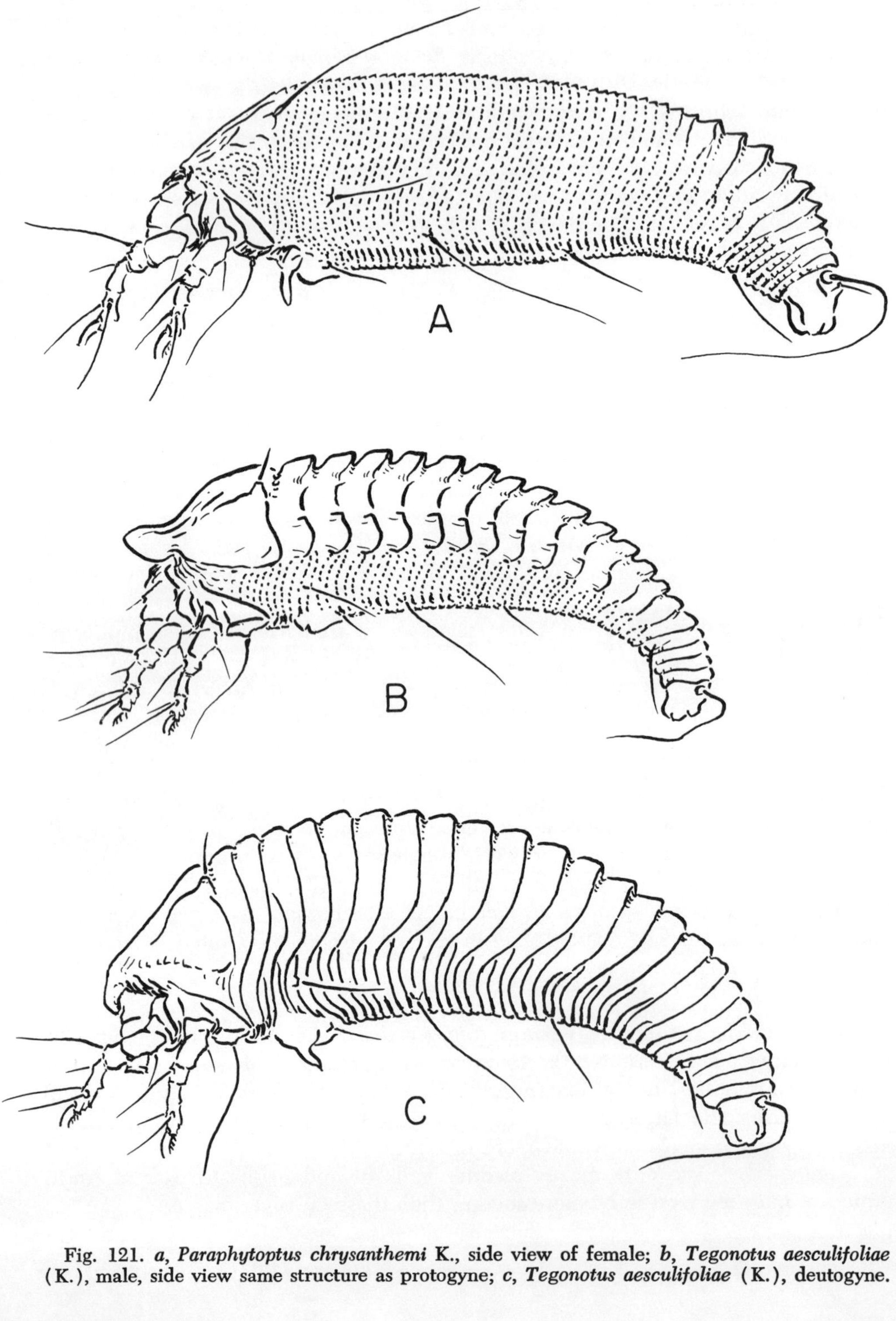

Fig. 121. *a, Paraphytoptus chrysanthemi* K., side view of female; *b, Tegonotus aesculifoliae* (K.), male, side view same structure as protogyne; *c, Tegonotus aesculifoliae* (K.), deutogyne.

makes elongate galls on sugar maple upper leaf surfaces. It has certain similarities to *V. quadripedes*, such as the position of the dorsal tubercles, but the shield pattern is more involved. Protogynes of *V. aceris-crumena*, and males, show a pair of points on the anterior shield lobe, which are not present on *V. quadripedes*. Garman's name, *acericola*, is a synonym of *aceris-crumena*.

Some of the rust mites on eastern North American trees often do considerable damage to their hosts, but there has been little economic notice of such injuries. An eriophyid that can seriously damage the underside of silver maple leaves (in Maryland) is a rust mite, *Notalox rubigator* K. It is a small, flat species, with short dorsal setae that point up and inward. The abdominal dorsum is principally a broad longitudinal trough.

Paraphytoptus Nalepa

The species of eriophyids referable to *Paraphytoptus* have adapted themselves to a life in natural plant hairs. The features peculiar to the genus are: (1) The lack of an anterior shield projection over the rostrum allies these species to *Eriophyes*. (2) The anterior end of the *Paraphytoptus* abdomen is wormlike, and with narrow rings that are subequal dorsoventrally. (3) The rear abdomen differs from the front part by having a tergite-sternite development, on which a tergite covers three or four narrow sternites. The extent to which the tergite-sternite structures extend forward often defines particular species.

The life mode for these mites is that they project their anterior softer end down into the plant hairs, with the broader tergites exposed above the hairs. Variations in shield patterns possessed by these *Paraphytoptus* species, plus the different types and extents of the posterior tergite-sternite relationships, indicate that the species have come into this genus independently at various times from *Eriophyes*, whenever the opportunity to evolve in this manner became available. Plants belonging to the Compositae often are infested with mites belonging to this genus.

Chrysanthemum rust mite, *Paraphytoptus chrysanthemi* K. This species (fig. 121, *a*) lives and feeds among hairs on lower leaf surfaces and on green stems on chrysanthemums, in parts of North America and in the British Isles. Undoubtedly it has a wide range. Feeding by the mites brings about stunting and backward curling of apical leaves, with resulting reddishness. The leaves also become brittle. Stem internodes shorten, and bud clustering results in some brooming. The mites also engender flower phyllody with only partial petal coloring; the flower parts tend to revert to leaflike structures (Breakey and Batchelor, 1950; Miles, 1964).

P. chrysanthemi is a wormlike mite, the females ranging from 130 to 150 μ long. The featherclaws are 5-rayed. Mites are light yellow. The shield has strong central lines, with a complete median line ending at the rear margin in a dart-shaped mark. The admedian lines are complete, gradually diverging. The first submedian line runs back only half way, but the second submedian forks twice in front of the dorsal tubercle. The dorsal setae are quite long. The coxae are ornamented with curved lines and dashes; the sternal line is forked both front and rear. The

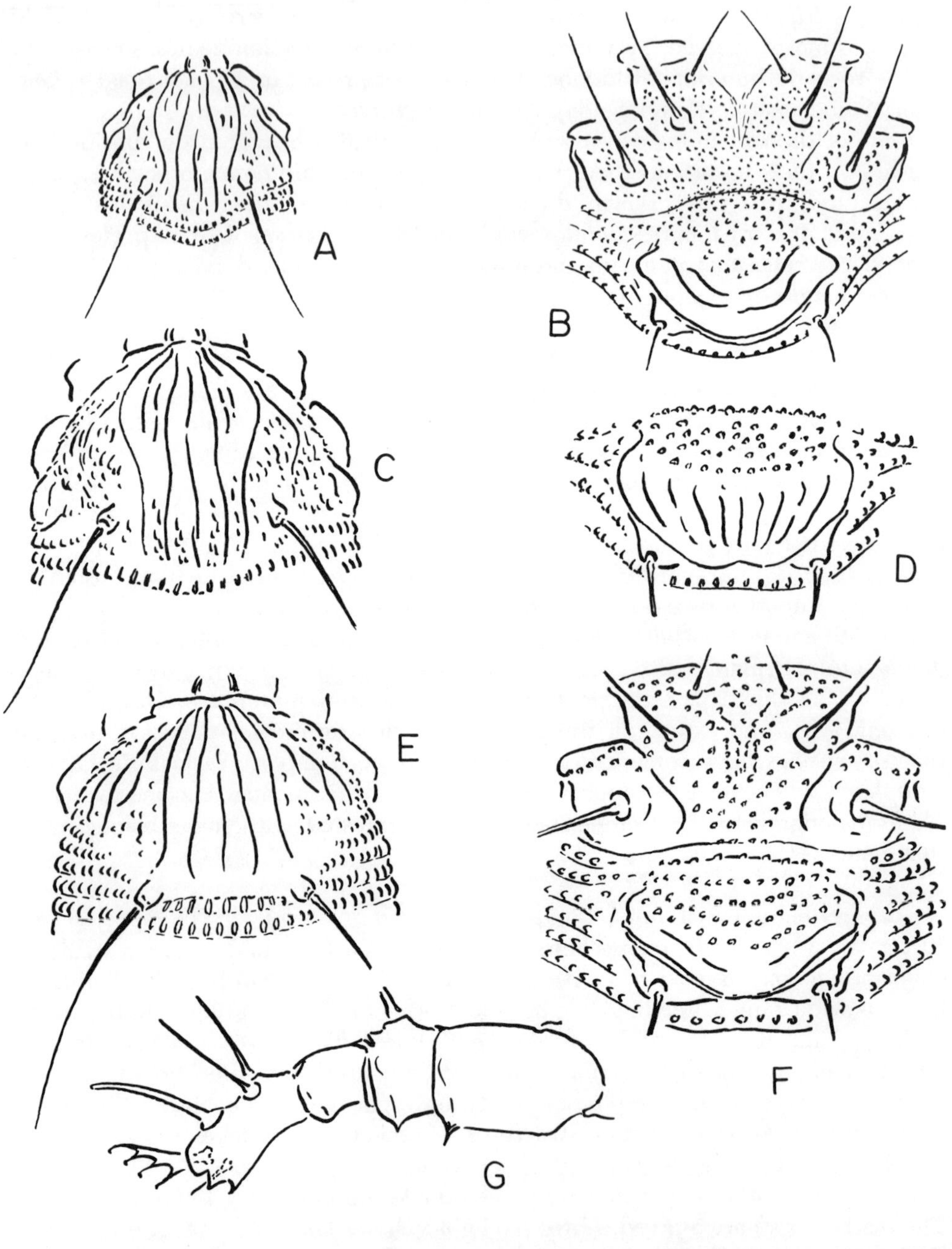

Fig. 122. *Acalitus* spp., shields and other structures: *a*, *essigi* (Hassan), shield; *b*, *essigi*, coxae and external female genitalia. *c*, *orthomera* (K.), shield; *d*, *orthomera* external female genitalia; *e*, *sphaeralceae* K., shield; *f*, *sphaeralceae*, coxae and external female genitalia; *g*, *sphaeralceae*, foreleg.

abdomen becomes differentiated into tergites and sternites at about the rear 1/4; each of these tergites covers about 2 sternites. The microtubercles on the rear tergites are elongate and the tergites are ridged longitudinally. The average microtubercles reach ring margins, but are not particularly pointed. The female genital coverflap has about 3 basal transverse lines of granules, and about 16 longitudinal ribs.

Another chrysanthemum mite, *Epitrimerus alinae* Liro, occurs in Finland.

Acalitus Keifer

The genus *Acalitus* Keifer has the following features: (1) absence of the forefemoral seta and of the foretibial seta; (2) setiferous forecoxal tubercles moved farther ahead than is usual; (3) forecoxae are often granular and are usually either poorly separated, or are combined centrally. On females of *Acalitus* the female genital coverflap usually lacks longitudinal ribs, with extensive granules replacing them. A very few species referable to *Eriophyes* lack the foretibial seta, but in that case the forefemoral seta is present.

Alder erineum mite, *Acalitus brevitarsus* (Fockeu). This mite (fig. 123, *f, g, i, j*) makes white or orange erinea on leaf undersurfaces on various alders in Eurasia and in North America. European erinea are composed of various forms of capitate papillae ranging from simple capitate types in western Europe to compound-capitate papillae in Turkey. In North America the alder erinea are made up of simple, tangled filaments. A species of *Eriophyes* is implicated in North American alder erinea.

A. brevitarsus is a deuterogynous species. Males have fine, rather close-set elongate microtubercles; the ventral ones are shorter and with slight points over ring margins. Protogynes resemble males as regards ring granules. On deutogynes the dorsal ring granules, or microtubercles, are rounded and on slide-mounted examples are lighter in color. Ventral deutogyne microtubercles are either rounded off, or slightly pointed, but unlike on protogynes, the deutogyne ring granules tend to be ahead of margins.

Redberry mite, *Acalitus essigi* (Hassan) (fig. 122, *a, b*). The type locality named by Hassan is Berkeley, California. While he implicated loganberry and raspberry as hosts for this mite, the only precise indication of a species of *Rubus* as a host for *A. essigi* is "Himalaya blackberry," *Rubus thyrsanthus* Focke. Other hosts of this species, in literature, are evergreen blackberry, dewberry, wild dewberry, and *Rubus leucodermis* Douglas.

As *A. essigi* can be confused by investigators with *Acalitus orthomera* (K.) and with *Phyllocoptes gracilis* (Nal.), the hosts and worldwide range of *A. essigi* are somewhat uncertain at this writing. Locations where *A. essigi* is supposed to occur are western North America, England, and New Zealand. The species undoubtedly has a wider range than is known as yet.

While *A. essigi* may cause some cane bud damage, the principal drawback to commercial berry growing engendered by this mite is delayed drupelet ripening, known as redberry disease." Infested berries have mite colonies at drupelet bases and around the core and stem end. Mite toxin prevents proper drupelet coloring, and the berries show uneven greenish red or red drupelets, with only parts of infested berries having normal dark maroon or black drupelets. Intensity of damage seems proportional to numbers of mites present.

Late-maturing berry varieties show the most damaged berries because the mites are late in migrating from winter cane buds to the fruit, and consequently early varieties of berries are harvested before mites become numerous. Fruit discoloration of the early maturing Texas berry variety, for example, occurs only on a few berries late in August, after harvest of the main part of the crop. In New Zealand the red berries appear about 2 weeks before the main crop ripens, and at that time are distinguished from normal fruit by their swollen condition and bright red color.

In the state of Washington mites occur throughout the year in and around buds. There is no general fall migratory period from berries, and many mites remain in them until berry decomposition. Mites that have left berries, and those remaining along canes, move up to new axillary cane buds in fall. They remain in and around these buds until the development of berries the next spring, many then moving to the berries. Fall and winter mortality among these migrating mites is high as the majority are not successful in finding suitable hibernating quarters. Predatory mites destroy many exposed redberry mites during these movements from one place to another. Redberry mites are able to survive field temperatures down to –6 C (21 F). Egg laying by *A. essigi* starts around buds in early March, increases to September, and then decreases until it stops in early December. There are many generations of the mites each year.

Sulfur compounds are particularly effective against the redberry mite. The toxic action of sulfur on this mite is both by contact, and by sulfur vapor. Sulfur vapor kills mites that are otherwise protected from direct contact. Lime-sulfur applications made in the fall (October and November), lime-sulfur and oil combinations in February, or a delayed dormant application (March 15 to April 1) are effective. It is advisable to apply lime-sulfur twice when combating high infestations from the previous season. Remove old canes before the fall applications of acaricides. Do not treat with sulfur during hot weather periods, and use sulfur and oil only during February (Edwards et al., 1935; Essig, 1925; Hamilton, 1950; Hanson, 1933; Massee, 1932).

A. essigi (Hassan) is a whitish, wormlike mite, 160 to 175 μ long in the female stage. The featherclaw is 4-rayed. The median shield line is broken, but the admedian lines are solid, sinuate to some degree, and gradually diverging. The first submedian line arches outwardly from the rostrum base, then recurves back toward the dorsal tubercle, and ends with a slight outward curve just on the inner side of the dorsal tubercle. The abdominal microtubercles are elliptical and slightly pointed. The granular forecoxae are almost fused centrally. The coxal granules extend down onto the base of the female coverflap. Across the coverflap are two uneven transverse curved lines, usually broken centrally.

Boysenberry bud mite, *Acalitus orthomera* (K.). This mite (fig. 122, *c, d*) is a pest of boysenberry, native blackberry, and dewberry in California. This species is very similar to *A. essigi*, but differs mainly by having longitudinal ribs on the coverflap. It can be easily confused with *A. essigi*. Feeding by *A. orthomera* on the inside of bud scales causes growth of warty blisters inside the buds. Heavily infested buds become entirely penetrated by the mites, the tissue becomes swelled, twisted, the whole bud swells to some extent, and the bud becomes a gall that cannot grow into a twig. The mite also causes some shoot stunting and blistering at drupelet bases within berries.

Cotton blister mite, *Acalitus gossypii* (Banks). This species (see *A. sphaeralceae* fig. 122, *e-g, k*) is widely distributed on cotton in the American tropics. It occurs on the Florida Keys. The earliest West Indian report of this mite is dated 1903. It has also been recorded in India, but because another and somewhat different mite, *Eriophyes puttarudriahi* (Channabasavanna), attacks cotton in southern Asia, it will be necessary to recheck all Indian records of *A. gossypii*. As far as known, cultivated and wild cottons, including Levant cotton, are the only hosts of *A. gossypii*, but within the genus *Gossypium* there is a wide range of susceptibility. For example, Sea-island cotton is very susceptible to this mite, whereas Upland cotton suffers little damage. There may also be other native malvaceous hosts in the American tropics.

The cotton blister mite is destructive in most countries where it occurs, and can completely destroy the cotton plants on fields where the infestations are heavy. While the mite thrives best in humid tropical and subtropical areas, its occurrence in Peru shows that it does adapt to semiarid conditions.

This mite attacks all parts of the cotton plant except larger stems and roots. It makes hairy deformations (erinea) on leaves and flowers, in which the mites live. Injury begins on young fruiting buds in leaf axils, and this leads to distortion and crumpling of the attacked parts. Silvery spots and extensive growth of fine short white hair spread out on leaves, especially lower surfaces. Palisade cells on damaged leaves grow to only one-third their normal length. Most conspicuous injuries on cotton plants are crumpled leaves, distorted growth, and paucity of fruiting branches. Later, where fruiting branches should normally form, small leaves with short petioles grow out from the main stem. Severely affected squares do not open into normal flowers because the petals become tightly bound with overgrowth and cannot separate. Another result of mite injury is abnormal shedding of squares (Fife, 1937; Thaker and Desa, 1929).

Since the cotton blister mite does not hibernate, it is able to reproduce throughout the year, and the life cycle is short.

One recommended control for cotton blister mite is to combine field clean-up with cotton-free periods. Host free periods from 2 to 4 weeks, during which time growers uproot and burn old plants and destroy wild cotton, have proved to suppress mite populations. Another suppression method is to select mite resistant cotton varieties for planting. Sulfur applications against this mite are more successful in dry climates, such as Peru. Sulfur use in India has not proved as suc-

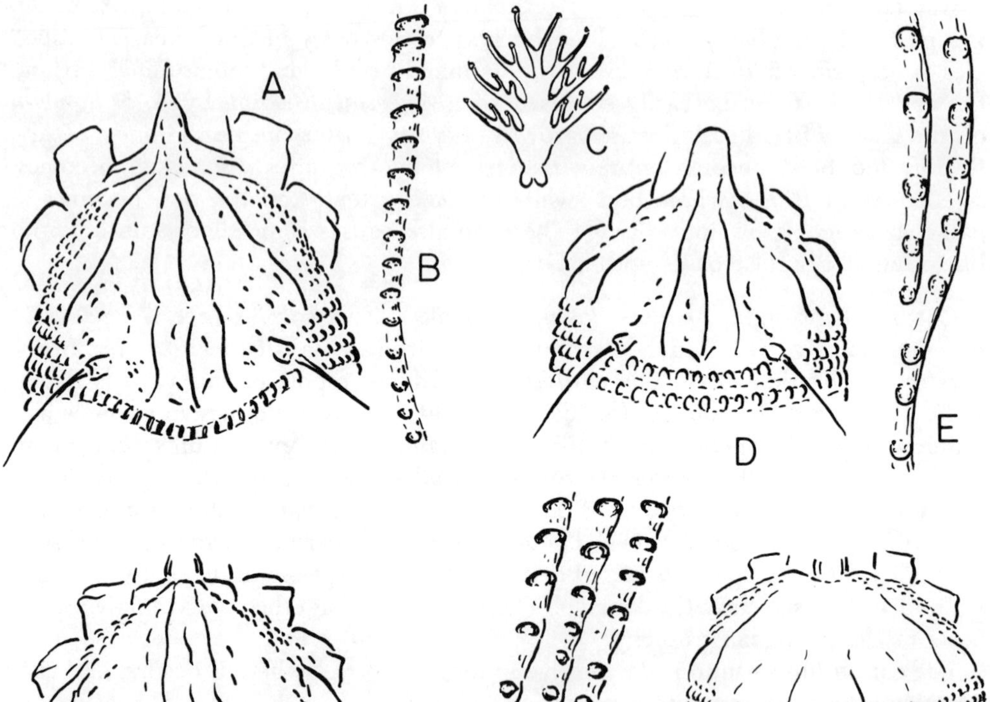

Fig. 123. Shield patterns and other structures: *a*, *Artacris macrorhynchus* (Nal.), shield; *b*, *macrorhynchus*, microtubercles; *c*, same, featherclaw; *d*, *Artacris cephaloneus* (Nal.), shield; *e*, *cephaloneus* microtubercles; *f*, *Acalitus brevitarsus* (Fockeu), shield; *g*, *brevitarsus*, microtubercles; *h*, *Acalitus vaccinii* (K.), shield; *i*, *brevitarsus*, foreleg; *j*, *brevitarsus*, coxae and external female genitalia.

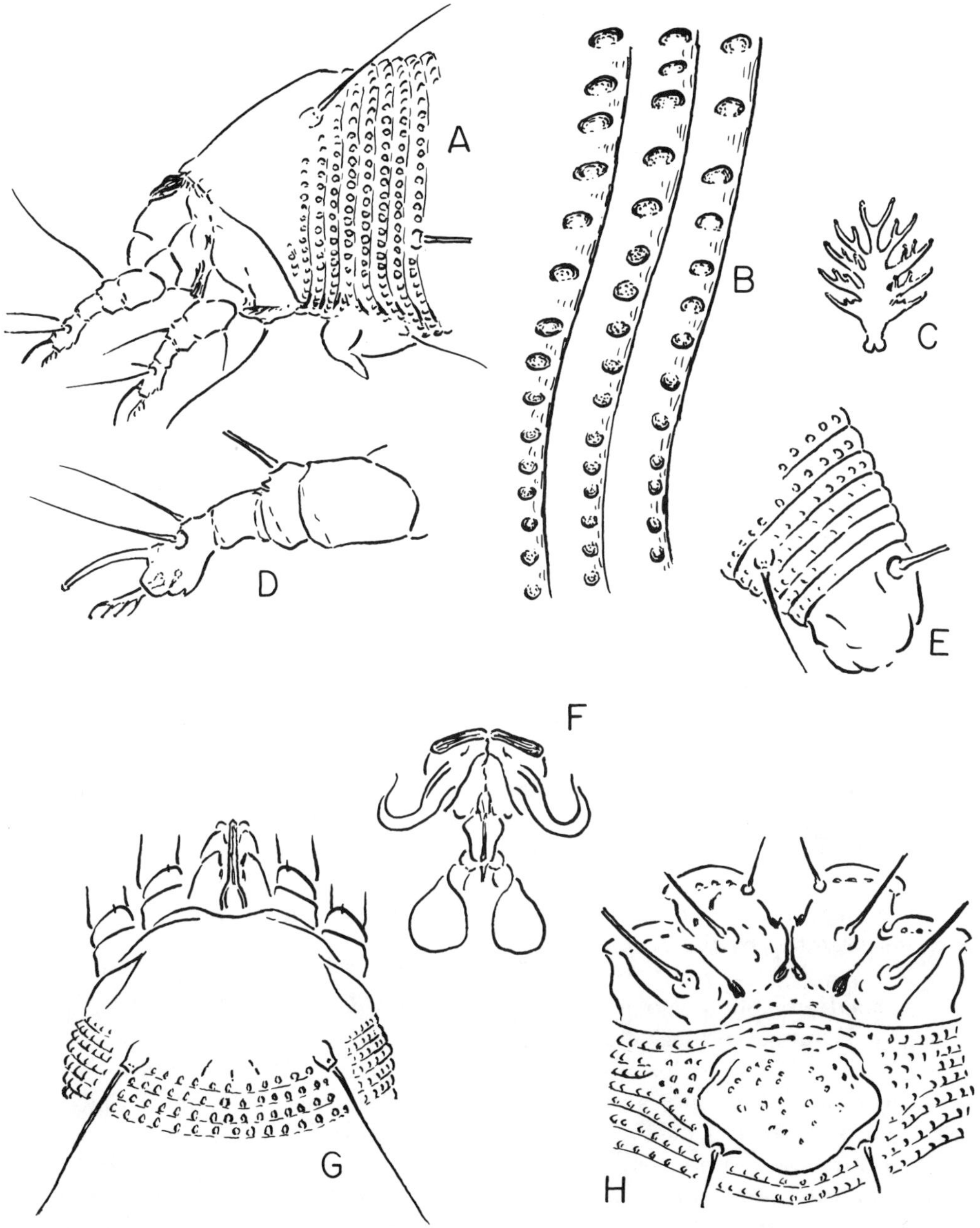

Fig. 124. *Acalitus phloeocoptes* (Nal.), female: *a*, side view of anterior section; *b*, lateral microtubercles; *c*, featherclaw; *d*, foreleg; *e*, telosome; *f*, internal female genitalia; *g*, shield; *h*, coxae and external female genitalia.

cessful as in Peru, perhaps because of higher atmospheric humidity, but it also suggests that the mite, against which sulfur was used in India, has greater resistance to sulfur than the American cotton blister mite (see below for the Indian cotton mite species; Wilson, 1923).

Acalitus gossypii (Banks), the American species, is whitish, wormlike, and females reach 225 to 250 μ lengths. The featherclaw is 4-rayed. The median line is strong, but only the rear ¾ of it is present. Admedian lines are complete, gradually diverging, and slightly recurved to rear. The first submedian line is short, but the second gently curves back, almost reaching the dorsal tubercle. Abdominal microtubercles are rounded off and beadlike, set ahead of ring margins. The coxae generally bear granules, with forecoxae centrally joined. Granules from coxae extend back onto the female genital coverflap, on which the granules are in transverse rows. A slight line runs across the flap just inside the rear margin of the flap.

A mite very closely related to *A. gossypii*, and at present structurally indistinguishable from it, is *Acalitus sphaeralceae* K. (fig. 122, *e-g*). *A. sphaeralceae* attacks globe mallow, *Sphaeralcea orcutti* Rose, in the southwestern United States. This mallow mite has never attacked cotton, although cotton is cultivated extensively in the vicinity of the native habitat of the mallow and its mite.The damage to mallow is that of severely crumpled leaves, with mites living down in the creases where the tissue is succulent.

Indian cotton blister mite, *Eriophyes puttarudriahi* (Channabasavanna). This mite causes hairy and feltlike outgrowths on its host. Damage is similar to that caused by the American mite on cotton. The host of the Indian mite is Levant cotton.

This Indian mite is distinguishable from *A. gossypii* by not only having the foretibial seta typical of the genus *Eriophyes*, but also by lacking longitudinal shield lines in the center. Instead, on *E. puttarudriahi* there is a strong transverse sinuate line across the shield just in front of the dorsal tubercles. As suggested above, this Indian mite may have different resistances to control chemicals than those possessed by the American mite.

Almond and plum bud gall mite, *Acalitus phloeocoptes* (Nal.). This species (fig. 124) ranges extensively throughout southern and central Europe, and occurs in Asia Minor. It has also appeared in the Eastern United States.

On almond the mite produces permanent irregular galls of various sizes around buds. It also deforms fruit spurs. The entire physiology of the attacked almond tree becomes upset: fruit buds fail to form, trees lose vigor, and damage appears to be progressive, and irreversible. This results in almond trees dying in 3 to 6 years. A physiological strain of almond grows in Lebanon which is immune to the attack of *A. phloeocoptes*.

On plum, small subspherical galls of between 1.3 and 1.8 mm in diameter grow around buds. These galls may be single, or grouped around the buds. The galls become woody, and deform fruit spurs. But in contrast to almond, plum trees often recover from mite attack, and show no lasting injury.

Adult *A. phloeocoptes* females overwinter in hardened galls. Early in spring, when temperatures become sufficiently mild and humidity drops, the galls crack and allow the mites to migrate to new buds. The mites crawl under bud scales, and as they feed, basal bud cells start proliferating. This starts in April and by the last of May consipcuous polylocular soft galls of different shapes and sizes develop around infested buds. Gall formation is the result of a coalescence of small chambers, or locules. These galls are watertight with highly convoluted surfaces. All galls are polylocular, having an average of 6 locules that are completely separate from each other by parenchymatous tissue.

In May, each locule contains 1 female mite. These females lay about 1 egg a day for 20 to 25 days. Eggs hatch about the first part of June, at which time parenchymatous cells lining the locules turn violet red. First generation adults are present by mid-June, and generations during July and August are completed in about 3 weeks. The locules become packed with as many as 4000 to 5000 mites and their eggs. Egg production decreases during late August, and ceases for the year in late October. Overwintering mites are more slender than summer forms and are slightly cuneate. Summer mites are more robust. These differences suggest the presence of structural deuterogyny in the species.

Dispersal of these mites to new host trees is accomplished by wind, and by insects that favor plum or almond. Birds may also carry this gall mite.

Cytologically, the tissue lining. *A. phloeocoptes* gall locules is parenchymatous, and usually consists of four rows of isodiametric nucleated cells with very small intercellular spaces, which are vacuolated with cellulose walls. Outside the parenchymatous cells are layers of typical parenchyma, with larger intercellular spaces. These cells vary from being typically isodiametric, to an elongate palisade form. Outside typical parenchyma cells there are several layers of thick cellulose-walled cells, which are collenchymatous in nature. These eventually become dead cork cells that constitute gall outside walls. Outermost gall layers consist of dead flattened cells, with cutin on the outside and suberized cells inside. In addition to the different groups of tissue layers described, one finds young vascular tissue masses running in the direction of the locules, but not reaching them. Between these masses and the cavity of a locule are loose and injured cells with spaces that might serve as diffusion centers to the locule. Bacteria and crystals are common in the region of these vascular tissue masses.

Predators of this plum gall mite, which play an important role in its control, include Chalcidoidea, Thysanoptera, and predacious mites of the families Phytoseidae and Anystidae (Philipp, 1957; Talhouk, 1963).

The female of *A. phloeocoptes* (Nal.) is whitish, wormlike, and somewhat robust (at least in the summer form). The mites vary from 135 to 150 μ long. The featherclaw is 5-rayed. There are no shield markings. Abdominal microtubercles are oval or rounded, and are smaller and more acuminate ventrally; all are set ahead of ring margins. Unlike many *Acalitus* species, this mite has a strong sternal line between the forecoxae, which line ends in a fork that has thick lines. The female genital coverflap has indistinct granules on its surface.

Blueberry bud mite, *Acalitus vaccinii* (K.). This mite (fig. 123, *h*) is native to eastern North America, especially coastal areas, and ranges from Canada to Florida and to Texas. It lives on certain ericaceous plants, mostly those belonging to the genus *Vaccinium*. These plants are known as blueberries and huckleberries. *Acalitus vaccinii* also lives on *Gaylussacia baccata* (Wangerin) another ericaceous plant.

This mite is of economic importance because it damages fruiting bodies on its hosts. Typical mite injury on wild plants appears as unnatural succulence and epidermal roughening or blistering of fruit bud scale bases, causing buds to hang in tight rosettes around the fruit stems. These rosettes are quite conspicuous and make visual survey for the mite relatively easy during the spring.

Bases of the infested fruit stems develop watery blisters and retain juvenile red coloring of growing tissue. Surface succulence and roughening of bud scales caused by the mite resembles to some extent the inside of certain leaf galls that other eriophyids make on other hosts. Perhaps this rosetting can be considered a pseudogall.

Injury at fruiting bud scale bases may not prevent the berry clusters from developing relatively normal berries, often on shortened stems. In more severe cases, blistering extends to calyx and petal parts, and the berry becomes deformed. Petals, when blistered, tend to fuse with the calyx. Severely attacked clusters on wild plants may remain through the summer as drying rosettes. Badly blistered and curled fruit stems make no fruit. Fruiting clusters on cultivated blueberry plants are more severely damaged by the mite than are those on wild plants, which may mean that the natural plants are in better locations for growth. Severe mite infestations may completely deform every berry cluster on a plant. Vegetative bush parts of infested plants show relatively little damage, except for some retardation of leaf growth.

Plants in deep shade are less damaged by bud mites, but those planted on dry sand flats or in open woods and at clearing margins, are most likely to have rosetting. As cultivated blueberry plants fruit better when grown in little or no shade, it may be that apparent differences in susceptibilities of different blueberry varieties are at least partially the result of location and of climate.

Eriophyids are not visible to the unaided eye, and these blueberry bud mites may be difficult to locate during a survey, unless berry cluster scale rosetting is present. These fruiting body rosettes are relatively loose, and do not offer the mites as much protection as would a regular gall. By early summer, if the clusters have dried and dropped, it may be impossible to find the mites except with a hand lens, or by laboratory examination. When the mites abandon the rosettes of scales they move to buds, usually terminal fruit buds, and those are the places to look for the mites.

Upon fruit scale cluster decline the mites start to move, with resulting mortality. As the mite populations decline in early summer, those fortunate enough to reach suitable buds gradually creep farther and farther into these buds, reaching the innermost parts by mid-September. In North Carolina the numbers of mites start diminishing in early summer with rosette drying, and the decline continues through the fall and winter, reaching its lowest point in February. With the spring flush of bush growth, the mites start actively breeding and their numbers reach a peak during the spring.

Migration by these mites is undoubtedly accomplished as with other eriophyids, that is, by wind and on insects that favor their particular hosts. One favorite location for mites wishing to travel seems to be within flowers where they can ride away on bees that are collecting pollen.

Eggs of the blueberry bud mite are present through most of the year, and their numbers are proportional to the mite population. The first nymph grows 4 or 5 days before molting. The second nymph takes about the same length of time to produce the adult. Maximum time periods from egg to adult are about 15 days (Edwards et al., 1935; Darrows, Wilcox, and Checkwith, 1944).

Practical control of blueberry bud mite is possible by applying post-harvest summer oil at 2% to 3% in water. Diazinon or chlorobenzilate are effective acaricides for the control of this mite. A fungus, *Hirsutella thompsonii,* is an effective pathogen on this blueberry bud mite when temperatures and humidity are high (Bailey and Bourne, 1946; Thomlinson, 1950).

A. vaccinii is a wormlike, light yellowish white mite, with females from 185 to 210 μ long. The shield has indistinct lines, with the admedians showing only as outwardly arched lines on the rear shield ½, between the dorsal tubercles. There is an ocellar spot at the lateral shield angles. Abdominal microtubercles are elliptical, and touch ring margins. Perhaps the most distinct features of this mite are, first, the strong sternal line between the forecoxae which is forked posteriorly, and second, the 8 to 10 longitudinal female coverflap ribs.

The Phyllocoptinae Nalepa

As indicated in the general key in appendix 3, the Phyllocoptinae are distinctive by having the combination of a prominent and broad-based anterior shield lobe, plus a thanosomal dorsoventral division of the rings into broader tergites and narrower sternites. Exceptions to these characters do occur. When, in response to habitat changes, there are alterations in the principal combination of structures, the anterior shield lobe nearly always remains in some form, but the tergites narrow to make dorsoventral relationship more equal. These mites are a diverse series of types that are basically eriophyoids that have left covered habitats for life on open green surfaces. As there is little or no overhead protection in open areas, the development of a rostrum-protecting anteior shield lobe and thick tergites is in response to the resulting exposure. But certain mites in buds, or that make erinea and galls, have morphological features that definitely ally them with the Phyllocoptinae. As discussed elsewhere in this text, the explanation for such species is that they represent mites that have reentered covered existances. Examples of this reentry are given under several genera below.

Acaphylla Keifer

This is a small genus, but species referable to it are widespread. Recognition anatomy for the genus is: divided featherclaws, a broad and rigid anterior shield

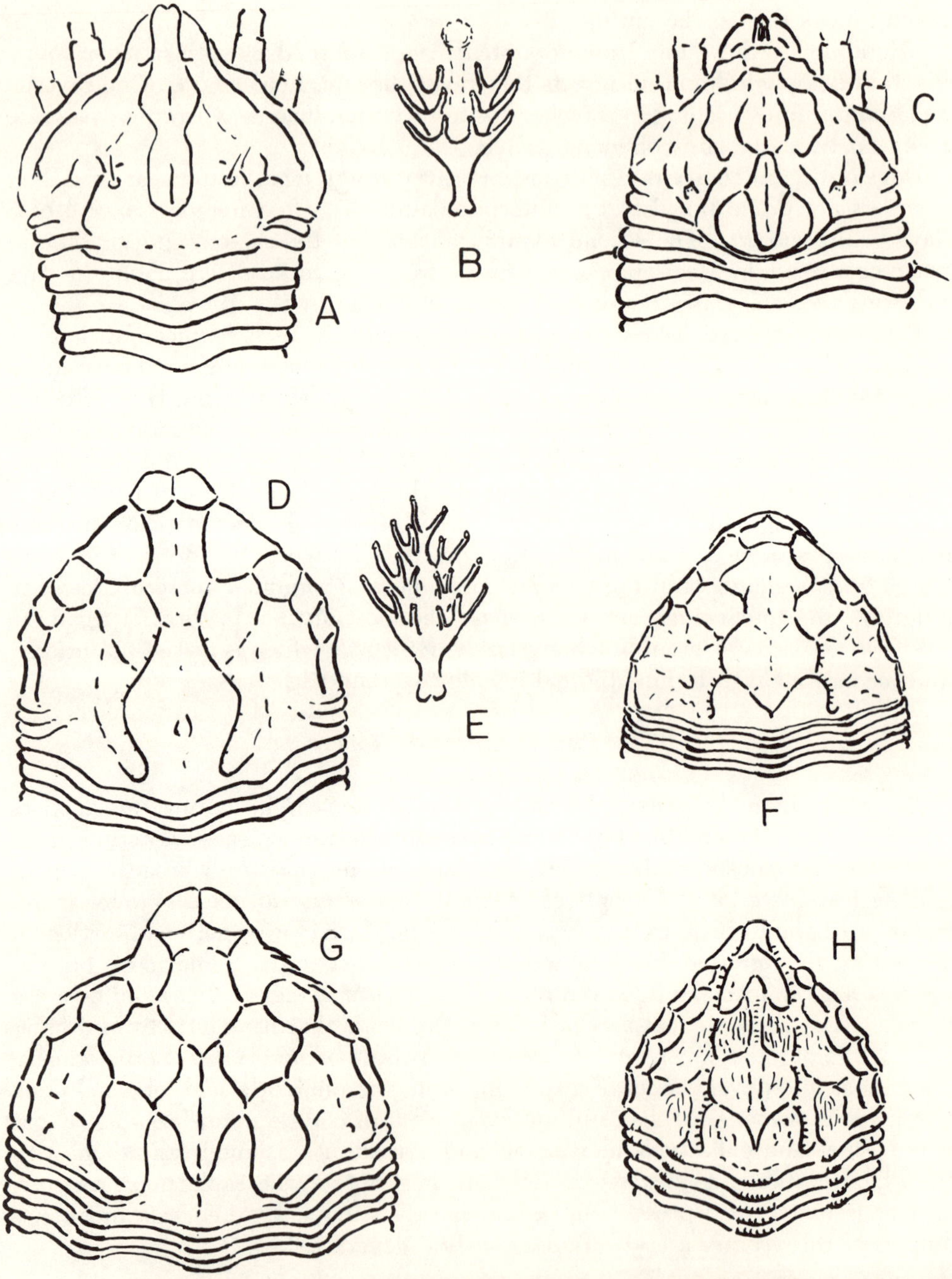

Fig. 125. Shields and featherclaws of *Acaphylla* and *Calacarus*: *a*, *Acaphylla steinwedeni* K., shield and anterior abdomen; *b*, *steinwedeni*, featherclaw; *c*, *Acaphylla indiae* K., shield *d*, *Calacarus brionesae* K., shield; *e*, *brionesae*, featherclaw; *f*, *Calacarus carinatus* (Green), shield; *g*, *Calacarus citrifolii* K., shield; *h*, *Calacarus coffeae* K., shield.

lobe, dorsal tubercles set well ahead of the rear shield margin, and a middorsal longitudinal thanosomal ridge that fades to the rear. While the principal pests in this genus are rust mites on tea and camellia, there are two alder rust mites. One of these on alder is *Acaphylla acromius* (Nal.), which occurs in Europe. The other alder mite is *Acaphylla distasa* K. that rusts alder leaves in eastern North America. *A. distasa* has the first forecoxal setiferous tubercle.

Pink tea rust mite, *Acaphylla theae* (Watt). The original description of this mite, in the 1903 issue of *The Pests and Blights of the Tea Plant*, by Watt and Man (1898), names the mite, gives its color, and describes the rusting it does. Nalepa, in his comprehensive host list (1929), considered this name to be nude. G. M. Das favors the use of this name since it is the one used for the mite through the years in India. Actually, *theae* is not a nude name.

A. theae ranges across southern Asia wherever tea bushes grow. It also extends into the East Indies. It feeds on both leaf surfaces, and may injure petioles and green stems. Attacked leaves become pale, often leathery, with veins and margins turning pinkish. Severely damaged leaves become crinkled, and the underside browning gives bushes a sickly appearance. Severe attacks retard bush growth. The mite favors Assam tea varieties over Chinese varieties.

Pink tea mites increase in numbers on leaves from March to June. Population decline occurs in dry weather and may not recover its former numbers that year even though there are subsequent wetter periods. Young tea plants suffer more severely than older plants.

Growth of the mite from egg to adult requires an average of 8.4 days under March conditions when the temperature has a mean of 24.3 C (77 F), and a relative humidity (RH) of 84 percent. In summer at 29 C (84 F) and 82 percent RH, the elapsed time is 6.2 days. Average adult life is 12 days. Females deposit 2 to 4 eggs a day, with a total egg production of from 30 to 40 in one life span. Eggs are laid over the leaf surface, with favored places being along the midrib and by lower leaf veins. Adults vary in color from orange to reddish. (Das and Sengupta, 1958).

Orange camellia rust mite, *Acaphylla steinwedeni* K. This species (fig. 125, *a*, *b*) is an orange mite that lives mostly on the underside of camellia leaves in California. It causes considerable rusting, especially in more protected places that have higher humidity. It must surely have originated in the general Asian area, but there is as yet no report of it being there.

A. steinwedeni when compared to *A. theae* discloses no features that can be used to separate the two populations. The hosts are closely related, but no experimental evidence exists as to whether or not the camellia mites can transfer to tea, or vice versa.

This camellia mite is a flattened, fusifrom species, with the anterior part of the thanosome being narrower than the shield. Females range in length from 175 to 190 μ. Each featherclaw bifurcation has 3 rays. Admedian shield lines extend back from the notched anterior lobe, have 3 curves, and run together centrally well ahead of the rear shield margin. The dorsal tubercles are small and well ahead of the rear margin. Each tubercle holds a short seta. Thanosomal tergites lack granules, and each covers about three sternites. The female coverflap has numerous fine basal longitudinal lines with numerous granules; the usual ribs are few and faint.

White tea rust mite, *Acaphylla indiae* K. This mite (fig. 125, *c*) lives on open leaf surfaces and makes some leaf discoloration. It was collected in Assam by G. M. Das. Females of this species are from 150 to 155 μ long, flattened, and whitish. The bifuracted featherclaws have about 8 rays on a side. The shield pattern is involved with admedian lines that branch 2 or 3 times, the central parts joining each other just in front of the rear margin. There are 2 particular differences between *Acaphylla indiae* and *A. theae*: the thanosome on *theae* is narrower than the shield, but bulges out wider on *indiae*; *indiae* has the first forecoxal setiferous tubercle, on *theae* it is absent. Basally the female coverflap on *indiae* has some faint converging lines, and there are about 14 longitudinal ribs.

Acaricalus Keifer

This genus has bifurcated featherclaws, as does *Acaphylla* (which is antedated by *Acaricalus*), but on *Acaricalus* spp. the middorsal thanosomal longitudinal ridge ends before the ending of the middorsal trough in which the ridge lies. In addition the subdorsal ridges on *Acaricalus* spp. that flank this line run a short distance diagonally outward from the rear shield edge from just behind the dorsal tubercle, and then turn caudad. (See Keifer, E.S., X, in reference list.) *Acaricalus* spp. are leaf vagrants.

One species, *Acaricalus paralobus* K., is an undersurface leaf rust mite on *Alnus rugosa* in eastern North America. *A. paralobus* differs from the type of the genus, *A. hederae* K., by lacking the dorsal thanosomal hump just behind the shield.

Acaricalus hederae K. This is an upper surface leaf vagrant on ivy, *Hedera helix* L. The only locality where this mite has been found so far is in San Francisco on the ocean front where humidity is highest. It does not come inland in California. It is amber or reddish and from 160 to 180 μ long. The divided featherclaw is apparently 3-rayed. The shield is acuminate anteriorly with a central low longitudinal ridge flanked by rather thick lines. Just behind the shield there is a pair of thanosomal humps, followed by a central longitudinal ridge extending back about ¾ of the thanosome length and ending in the dorsal trough. The female coverflap has thin longitudinal ribs.

Acaricalus hydrophylli K. This is an upper surface leaf vagrant found on holly, *Ilex aquifolium* L., in the California central valley. This mite is very similar to *A. hederae* but has one less featherclaw ray. It develops most strongly on new holly leaves. Ivy growing under infested holly never develops a population of *A. hydrophylli.*

Calacarus Keifer

This is one of the most distinctive of eriophyoid genera. The species referable to it are robust leaf vagrants, or rust mites. They have deep purplish bodies. They live in the most exposed places on leaves, often on the upper surfaces, and protect themselves in part by secreting white wax. On the shield this wax follows shield lines, and on the abdomen the wax stripes follow dorsal longitudinal ridges. Nearly all of the species so far known have 5 longitudinal ridges on the dorsal half of the thanosome. These mites lack prominent dorsal setae and usually lack the dorsal tubercles. When the tubercles are present they may have minute setae on them. *Calacarus* species usually have a characteristic lateral line of shield "cells" formed by longitudinal and diagonal lines. The genus ranges the length of Africa, across Eurasia, and onto the west coast of North America. No native species have so far come from the eastern part of North America, or from South America. Of the 3 native western North American species, two are somewhat structurally aberrant. One of these, *C. tejonis* K., has a reduced shield pattern; the other, *C. pulviferus* K., has but three thanosomal ridges.

Papaya leaf edgeroller, *Calacarus brionesae* K. This mite (fig. 125, *d, e*) is a minor pest of papaya in the Philippines. It rolls leaf edges and makes white spots along the rolls. It is a purplish robust mite, with 4-rayed featherclaws. The shield has a short anterior lobe. The median shield line is distinct only on the lobe. Admedian lines curve back from the sides of the anterior lobe, are first convex inwardly to a lateral cross line, then diagonally back to a central cross line. These admedians are broadly convex outwardly to their rear ⅔, and abruptly recurve out at rear shield margin. There are a few vertical short dashes along these lines. Laterally the shield has a series of enclosed "cells." The thanosome has the usual 5 ridges, with wax; the median ridge is not broadened anteriorly. The broad female genitalia have a basal pattern of short dashes, but apically the coverflap has but few longitudinal ribs.

Purple or ribbed tea mite, *Calacarus carinatus* (Green). In his 1929 host list Nalepa puts *carinatus* under *Eriophyes* (indicating how poorly it was described) and calls it a *nude name*. This rust mite is one of the more important tea pests in southern Asia. It does more tea leaf damage in Java than in India. During droughts, the mite causes some bush defoliation. Assam teas are more heavily damaged than Chinese varieties. It is more plentiful on mature leaves than on young leaves.

Badly infested leaves turn bronze, and both surfaces have a dusty appearance owing to cast skins and mite wax. Transparent, pale-colored spots, or blisters, appear on the leaves; most mites are on the upper surfaces, especially along margins. These mites stay on the leaves throughout the year, but rainy periods reduce them to negligible numbers. Low populations persist mostly on leaf undersurfaces. But by October injurious populations reappear, and serious leaf damage may occur before cold weather. There is a gradual increase in mite numbers from late February or early March, and this build-up continues until rainy weather.

Most eggs of *C. carinatus* are stated to be laid along veins. At 23 C (72 F) the eggs hatch in 6.5 to 8 days after oviposition. The first nymphal instar covers a period of from 1 to 3 days. The second nymphal instar grows for about 2 days before molting. Total development time from egg to adult averages 10 to 12 days, and the following preoviposition period is from 1 to 2 days. The life cycle of *C. carinatus*, on tea, is about 13.5 days long in January, 9 days in March and April, and 6 to 8 days in July and August.

Eggs of *C. carinatus* are minute, colorless, and almost transparent. They are circular, but flattened, with a diameter of about 75 μ and thickness of 11 μ. First nymphs are naked, shiny, and pale. The second nymph soon develops wax stripes. Adult females lay from 1 to 4 eggs a day, averaging about 21 eggs in a lifetime. Oviposition periods last from 10 to 13 days (King. 1937).

The use of acaricides is necessary to prevent serious damage to tea leaves. This purple mite is susceptible to sulfur, but application of sulfur during picking time taints the tea. Ethion and dicofol are effective, and have residue clearances (Das and Sengupta, 1962).

A purple mite, *Calacarus adornatus* (K.), was described from California; the original host was stated as *Viburnum.* Soon after this description it became clear that the mite had come to viburnum from nearby heavily infested camellia bushes. (Camellia is a close relative of tea.) There is as yet no record of such a mite on camellia in Asia, where it surely originated. Neither is there any information as to the transferability of such mites from camellia to tea, and vice versa.

In 1952, A. M. Boyce, of the University of California at Riverside, kindly secured ribbed tea mites from Ceylon. Examination of them revealed no structural difference between *C. adornatus* and *C. carinatus.* For that reason the recognition features of the camellia-tea complex of these mites, as recited below, are based on *C. adornatus* as it occurs in California. Figure 125, *f*, labeled *C. carinatus*, is from California camellia specimens.

Females of this purple, robust, fusiform mite, with white wax stripes, are from 180 to 225 μ long. The shield lobe is short and broadly rounded. The median line is practically absent. The admedian lines are strongly sinuate, from their origins at the lateral base of the anterior lobe. They curve in to meet a cross line at the anterior 1/5, meet lateral lines at 1/3 and 2/3, and fork at the rear, with the inner fork lines joining centrally at the rear shield margin where they form a large V. The outer bifurcation recurves slightly outward at the rear margin. The featherclaw is 5-rayed. Markings on the female coverflap are obscure, consisting of a few lines at the base, and slight subapical longitudinal ribs.

On camellia in California there is a reciprocal relationship between *Calacarus carinatus* (*adornatus*) and *Acaphylla theae* (*steinwedeni*) with regard to atmospheric humidity preferences and parts of leaves attacked. The *Calacarus* mite is most often found on the open leaf upper sides, and flourishes in warm weather. The *Acaphylla* species prefers more humid, partially protected places, and lives on the undersides of the leaves. This evidently describes the relationship on tea between the purple mite and the pink orange species.

Grey citrus mite or citrus blotch mite, *Calacarus citrifolii* K. This species (fig. 125, *g*) causes a condition on citrus leaves in South Africa known as *concentric ring blotch.* Various native euphorbias, including poinsetta, are listed as native hosts of this mite. Other hosts in literature for this mite are: *Dianthus*, peach, sumac, passionflower, arum lillies, papaya, buckthorn, *Pappea, Mimusops, Lippia*, and *Brunfelsia*. This plant series (all dicotyledons) is exceedingly broad, covering nine botancial orders, and eleven plant families. It suggests that a recheck, with careful consideration to shield patterns and featherclaw rays, might show that the purple mites on part of these hosts are not really *C. citrifolii.* An examination of specimens from euphorbia did indicate that the species on that host is *C. citrifolii.*

C. citrifolii makes leaf spots that may be circular, or irregular, slightly raised or sunken, and often are not sharply demarcated (pl. 67). If many spots develop on a mature leaf, the spots remain small, similar to stippling. But a single blotch may extend over the major portion of a leaf surface, being sometimes 35 mm in diameter or more. Blotches are typically marked with concentric rings, which vary from mere fine etching to broad circles, hence the name concentric ring blotch. Mite infestations along the midrib on very young leaves will result in blotches of an "oak leaf pattern." This oak leaf type usually covers more of the leaf surface than do other discrete blotches on the same leaf.

Mite feeding also often produces mottled chlorosis. This develops when infestation points or stippled blotches are so numerous on individual leaves as to become confluent before they can increase in size sufficiently to form discrete blotches. The mottled effect is either produced by innumerable chlorotic stipples on a green background, or by green stipples on a chlorotic field. Old leaves with stippled chlorosis from previous season show a partial recovery, and have a greyish green mottle. On such leaves the chlorosis is replaced by whitish specks or minute spots on an intensified green background.

In general there are 2 types of ring blotch on leaves. The first is typically chlorotic and concentrically ringed blotch, common on leaves of sweet orange. The second is a necrotic spot, with or without concentric markings, and surrounded by a broad halo of chlorotic tissue. The second type is prevalent on grapefruit, on tangelos, and occasionally on tangerines; it is present as often as not on sweet orange, both in nurseries and in old orchards. The plant host species seems to be the determining factor in the production of this type of leaf necrotic spot. The reason for the production of either chlorotic blotch or necrotic spot on sweet orange remains obscure. Exposure to intense sunlight often causes gummosis or resin formation on blotched tissue, resulting in dark brown centers and darkish rings of resinous tissue.

Twig and stem lesions. Pinpoint lesions on twigs develop into discrete or densely crowded, concentrically marked blotches. These often occur on vigorous and rapidly growing shoots and stems in nursery and orchard. Twig blotches eventually may assume an oval shape along the longitudinal axis and sometimes are 25 mm or more across, especially on stems and shoots of vigorous nursery stock. Blotches may appear somewhat water-soaked, raised, and bilstered, and may contain darkened resinous rings following exposure to strong sunlight. Young lesions are chlorotic, or of a lighter green color than surrounding tissue. Necrosis, longitudinal splitting, and copious exudation of gum, sometimes results from blotching on shoots and branches. Occasionally raised, scablike blisters, acquiring a rust-brown color, are seen in the necrotic areas of twig blotch. Entire girdling and the formation of a light brown or reddish brown scaly bark has likewise been encountered on two-year-old shoots of less than 15 mm in diameter. High mite populations may result in complete killing of shoots, or of branches above lesions.

Fruit blotch. Injury first shows up on green fruits less than 25 mm in diameter as whitish circular, pinpoint blotches, sometimes exhibiting depressed centers. The blotches increase in size as fruits enlarge, some attaining a diameter of 50 mm or more. The blotches are either irregular or circular, whitish-green, with the affected surface area depressed to a greater or lesser extent. Necrotic reddish brown spots may start to develop within blotches before the fruits reach full growth. On mature fruits the lesions vary immensely in color, size, and shape, and also in number. A list of these variations follows:

1. Depressed circular or irregular lesions, lemon or light yellow, with or without necrotic centers and scabs
2. Concentrically zoned blotches, with or without necrotic rings
3. Dark brown, circular, depressed spots, 2 to 12 mm in diameter
4. Scabbing of a "mud-cake" and nailhead rust type, showing circular cracks in the mud-cake surface
5. Lesions of the tear stain type
6. Bronzing, as in the rust mite injury, on the sun-exposed cheek of navel oranges
7. Deeply depressed, light-colored, furrowed, or circular lesions, with or without warts, as on tangerines.

Lesions, either on leaves, twigs, or on fruits, occur predominantly on sun-exposed surfaces.

The overall symptomatological effect of the feeding of *C. citrifolii* is summarized: (1) general chlorosis of the foliage; (2) partial or complete defoliation; (3) dieback of shoots and occasional killing of young nursery trees; (4) premature shedding of fruit that is approaching maturity.

Appearance and life history of *C. citrifolii.* The newly laid egg is flattish, transparent, and circular. It is about 70 μ in diameter. The young embryo shows in the egg, after about a day's incubation, as a whitish discoloration of the egg. After 2 days incubation the upper egg surface becomes spherical, and a yellow white embryo is visible. Eggs hatch in 5 to 6 days, the shell splitting dorsally. The first nymphs are initially smooth and pearly white, wedge-shaped, and 70 to 80 μ long. Waxy stripes develop dorsally, giving a roughened white appearance. The purplish grey color is visible in the newly emerged first nymph, but wax stripes tend to obscure body color. The second nymph is similar to the first, except that it grows larger, from the beginning 90 to 100 μ, to a length of 120 to 130 μ. Five days are required to complete each stage. The smooth gray color is striking in the newly emerged female. But the 5 double rows of waxy stripes, sometimes curved at the tips, soon become obvious. Shortly before death the mites assume a blue black body color.

At 20 C (68 F) the development from egg to adult requires about 10 days. A female may lay up to 243 eggs during her reproductive span, averaging about 1.4

eggs a day. While no males have been so far reported, they must surely be present, although undoubtedly in less numbers than the females. Since the climate of South Africa has mild winters, and since this mite lives on evergreen hosts, *C. citrifolii* has no overwintering stage. This means that activity continues throughout the year, and varies with temperature.

Control. Chemical control of the grey mite is necessary to prevent damage to citrus. As practically all fruit damage perpetrated by this mite arises from infestations occurring between petal fall and the half-grown stage of fruit, protective sprays should begin as soon after petal fall as possible. A second or third application may sometimes be necessary to protect fruit against late infestations, which cause bronzing and light blotch. Lime-sulfur, wettable sulfur sprays, and sulfur dusts, effectively control this mite. Dicofol and chlorobenzilate, and possibly zineb, are also effective, and are less likely to produce plant injury when applications are necessary during hot weather (Dippenaar, 1958; Merwe and Coates, 1965).

C. citrifolii is a grey to purplish, fusiform, robust mite, with longitudinal white stripes of wax on the abdomen in the fully developed adult. Females are 185 to 200 μ long, with 5-rayed featherclaws. The shield has a moderately projecting anterior lobe. The shield pattern is a network, but the admedian lines do not have the usual small vertical dashes along them, as do most species in the genus. The median line is present on the anterior lobe, and slightly indicated on the rear ½ of the shield. Admedian lines are strongly sinuate or undulating, arising from the sides of the anterior lobe, arching back to a lateral cross line at about ⅓, converging centrad to a slight cross line at about ½, then curving out to ¾ where they meet another lateral line and then fork. The inner fork lines of the admedians meet centrally a little ahead of the rear shield margin; the outer part of the bifurcation arches back and recurves outwardly at the rear margin. The submedian line, with branches, forms three large "cells" on the outside of the admedian. The forecoxae narrowly meet centrally, and are ornamented with numerous curved lines. The thanosome has the usual 5 longitudinal wax-bearing ridges. The female genital coverflap has numerous fine longitudinal lines in 2 ranks.

Coffee leaf rust mite, *Calacarus coffeae* K. This species (fig. 125, *h*) lives on underside of coffee bush leaves in Kivu area of Zaire. It is a robust purple mite that tapers strongly to the rear. Females are 185 to 200 μ long. The featherclaw is 5-rayed. The two distinctive features of *C. coffeae* are, the double median line in the shield center, and the large number of short vertical dashes appended to the admedian lines.

Scolocenus Keifer

Spiny coconut eriophyid, *Scolocenus spiniferus* K. This tiny mite (fig. 126, *f*) was present on coconut leaves submitted in connection with investigations for a possible vector of cadang-cadang disease. The principal adult characteristic of this mite is the large lateral spines projecting outward from the anterior thanosomal expansion. The nymphs are said to be sluggish and not at all like the adults

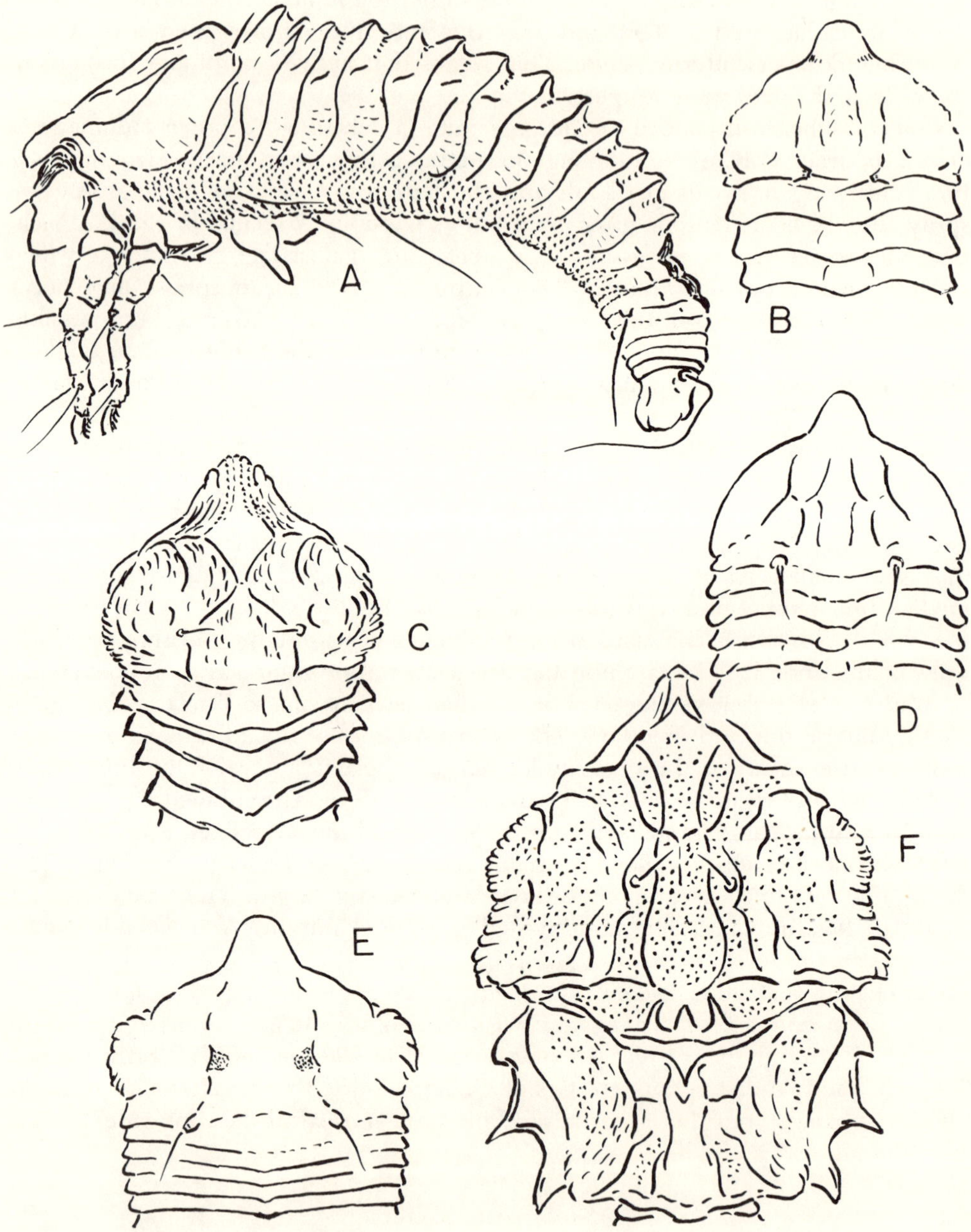

Fig. 126. Shields and anterior abdominal views of *Oxycenus*, *Tegonotus*, *Ditrymacus*, and *Scolocenus*: *a*, lateral view of *Oxycenus maxwelli* (K.); *b*, *Tegonotus aesculifoliae* (K.), shield and anterior abdomen; *c*, *Tegonotus mangiferae* (K.), shield and anterior abdomen; *d*, *T. scolocenus*; *e*, *Ditrymacus athiasellus* K., shield; *f*, *Scolocenus spiniferus* K., shield and anterior part of abdomen.

(Briones and Sill, 1963). Females of S. *spiniferus* are flat, 150 to 160 μ long. The featherclaws are 3-rayed on a side. The shield is broad, and is immediately followed by the broad expansion of the anterior part of the thanosome. This functions as a "pseudoshield." From the sides of this thanosomal expansion three spines project. Behind this anterior abdominal expansion, the rear part, which includes the caudal section of the thanosome and the telosome, narrows abruptly. The forecoxae lack the first setiferous tubercles. The female genital coverflap is completely covered with fine longitudinal lines.

Oxycenus Keifer

This genus has lateral tergal lobes, similar to *Tegonotus*, but differs principally by having a depression in the tergites, just ahead of the telosome. Scanning electron microscope pictures of this mite show longitudinal striations on the lateral tergal lobes.

Olive leaf and flower mite, *Oxycenus maxwelli* (K.). *O. maxwelli* (fig. 126, *a*) is normally found on upper surfaces of terminal olive leaves. Usually it causes no damage, but when present in large numbers on young leaves it silvers them and causes longitudinal curling. In the spring the mites collect on flower stalks, and, if in large numbers, cause flower drop. The native area of this mite is undoubtedly the Mediterranean. In Algeria, *O. maxwelli* and another olive leaf mite, *Ditrymacus athiasellus* K. (fig. 126, *e*), produce some leaf pitting, deformation, and discoloration. There is no evidence that *O. maxwelli* has a hibernating form. The lack of such a form is consistent with the evergreen nature of olive, and the relatively mild climates in which olives grow.

O. maxwelli is an orange mite, with spindleform body. Female length is from 140 to 160 μ. The shield is subtriangular, with a moderately long anterior lobe. While the median line is obscure, the shield admedians are complete, or angled, and flare at the rear margin. The featherclaw is 4-rayed (pl. 68, *a*, *b*). Broad thanosomal tergites have a middorsal ridge. Alternate tergites have laterally projecting rounded lobes. The female genital coverflap is principally characterized by having 18 to 20 longitudinal ribs. Males are 125 to 145 μ long.

Tegonotus Nalepa

Projecting lateral thanosomal tergal lobes are the feature defining this genus. Flattened rust mites referable to *Tegonotus* have a variety of these lateral extensions from their rather broad tergites. The more extreme forms display lobes that are long and pointed, like fairly large spines. But on most species the projections are relatively short, and only slightly pointed, or are rounded off. On some species, especially those with longer extensions, these lobes are alternately longer and shorter. Yet on each respective tergite the lobes are of the same length on each side. There is considerable diversity in the location of dorsal shield tubercles among these rust mites. On some kinds the tubercles are set well ahead of the rear shield margin, whereas on others the tubercles are on the rear margin. These lobe and tubercle variations indicate that the rust mites referable to *Tegonotus* arrived at the possession of lateral lobes from various sources. Unfortunately the name *Tegonotus* takes the place of the well-known *Oxypleurites* for this genus. As explained in appendix 3 key, Nalepa moved his original genotype, *heptacanthus*, from the homonynous *Acanthonotus* Nal., first to his new genus *Tegonotus*, and then to his next new genus for these mites, *Oxypleurites*. This clearly gives *Tegonotus* priority.

Sweet potato rust mite, *Tegonotus convolvuli* (Channabasavanna). This mite damages sweet potato in India. It is a yellowish species that feeds on upper leaf surfaces, and causes leaves to become bright brown. This damage eventually stunts the plants. The mite occurs on young and old plants during the cropping period.

Several species of morning-glory harbor *T. convolvuli* throughout the year. One of the more effective sources for this mite is a wild morning-glory that grows in both wet and dry seasons.

Mite infestations on sweet potato may be prevented by clearing morning-glory from sweet potato fields between plantings, also from areas around such fields (David, 1959).

T. convolvuli is a yellowish, flattened, spindleform species, with females ranging from 185 to 200 μ long. The shield of *T. convolvuli* is semicircular in anterior outline, with the rather small, and projecting, anterior lobe apically roughened. The shield has few design lines. Admedian lines are subparallel and close. Dorsal tubercles are close to each other and well ahead of the rear shield margin. The dorsal setae are 3 μ long and directed up. There are some curved lines on the coxae; the sternal line is moderately long, forking at the rear. Thanosomal tergites have a rather low, longitudinal middorsal ridge, and the lateral lobes end in small lateral teeth. Microtubercles are on the sternites only, and set slightly ahead of the ring margins. Female genitalia are shown as having numerous longitudinal ribs on the coverflap.

Horse chestnut rust mite, *Tegonotus carinatus* Nal. (fig. 126, *d*), browns the undersides of leaves of European horse chestnut. It probably attacks hybrids of this species, also. The mite is native to Europe, but has now appeared in Brooklyn, New York, and in the San Francisco Bay region of California. It causes some leaf browning on its host in these areas.

The female *T. carinatus* is from 180 to 200 μ long, and is flattened-fusifrom. The featherclaw is 4-rayed. The prominent anterior lobe that projects well over the rostrum is thick in side view. There is a faint shield pattern, with indications of admedian lines on the rear ⅔ of the shield. There are submedian lines indicated ahead of the dorsal tubercles. The dorsal tubercles are on the rear shield margin and direct the dorsal setae straight back. This character of dorsal setae projecting straight back is one of the principal distinctions between *T. carinatus* and the native California *T. aesculifoliae* (K.). The native California species, as described below, lives on California buckeye. *T. carinatus* has a middorsal thanosomal ridge, and moderately projecting lateral tergal lobes. The tergites have no granules. The coxae are ornamented with numerous curved lines. The female genital coverflap has faint basal marks, and an apical row of 14 to 16 short longitudinal ribs. Males of *T. carinatus* are similar to the protogynes, and are about 160 μ long.

T. carinatus deutogynes are about as long as the protogyne, but the deutogyne has much thinner anterior shield lobe, which curves down about half way over the rostrum. This deutogyne lacks both the middorsal longitudinal ridge, and lateral projectign lobes; both features are characteristic of the protogynes. The setation of all forms is the same. *Tegonotus ferruginator* (K.) is a synonym of *T. carinatus.*

The California buckeye rust mite, *Tegonotus aesculifoliae* (K.). The host tree of this mite is California buckeye, a large shrub or small tree. The mites live on both leaf surfaces and cause some rusting. The trees stay in leaf longer in intermediate coast range areas and those are the locations where this buckeye occurs in greatest abundance. In areas east of there the buckeye trees cast their leaves early in the season, and in the dry Sierra foothills leaf drop begins at early summer. Mite populations are lowest on, or absent from, these inland trees; and such rust mites as are present have to produce deutogynes early enough to avoid this early leaf drop.

T. aesculifoliae is not an economic pest, as the California buckeye has no economic value as a shade tree or shrub. The biology of this mite is interesting, however, as it illustrates the existence in eriophyids of structural deuterogyny.

The extreme differences betwen the protogyne (and males) and the deutogynes, both in *T. carinatus* and in *T. aesculifoliae,* originally made it inconceivable that the mites associated on these buckeye leaves were all of the same species.[6]

The *T. aesculifoliae* protogyne, with its middorsal thanosomal ridge and moderate lateral tergal lobes, runs to *Tegonotus* in the generic key in appendix 3. The deutogyne of this species, which lacks the dorsal ridge, and lateral lobes, runs to *Vasates* in the same key. As noted for *T. carinatus,* all forms of *aesculifoliae* have the same setation. If the setation had differences, that would mean at least an actual specific difference between these two associated types of rust mites.

The life history of *T. aesculifoliae* can begin with the hibernating forms in bark crevices during the fall season (fig. 97). At this time, after a period with no rain and with low atmospheric humidity, the aestivating or hibernating deutogynes are semidesiccated. After winter cold shock, rain, and early spring bud opening, the deutogynes reactivate. They crawl up to opening buds and begin to feed on embryo leaflets. They soon lay flat eggs. Protogynes hatch from these eggs, as shown by immediate continuation of reproduction in test leaf cells. New deutogynes appeared in about three weeks in the cells.

While it is easy enough during the spring for protogynes to reproduce in leaf cells, deutogynes formed at the same time do not reproduce. Deutogynes must aestivate or hiberinate, and go through the winter cold shock, before they will begin to lay eggs.

Full-fed deutogynes pick up sperm from spermatophores, and crawl down the twigs to repeat the life cycle.

Protogynes of *T. aesculifoliae* (fig. 126, *b*) are 170 to 190 μ long. The featherclaws are 3- to 4-rayed. The anterior shield lobe is rather thick, and turned up. The dorsal tubercles are on the rear shield margin, but the tubercle axes are nearly longitudinal, and direct the dorsal setae diagonally inward. This arrangement of the dorsal tubercles and setae is one of the principal distinctions between *T. aesculifoliae* and *T. carinatus*. The female coverflap has about 10 longitudinal ribs. Males of *T. aesculifoliae* (fig. 121, *b*) are similar to females, but smaller.

Deutogynes of *T. aesculifoliae* (fig. 121, *c*) are about 170 μ long. They lack the middorsal longitudinal ridge on the thanosome, and lateral tergal lobes. In contrast to the protogynes, they have no microtubercles on the rings. The anterior shield lobe is turned down.

Mango rust mite, *Tegonotus mangiferae* (K.). This mite (fig. 126, *c*) was named from examples received from Hilo, Hawaii. Females are 160 to 180 μ long. The featherclaw is 6-rayed. This mite has a subtriangular and rough shield. The dorsal tubercles are ahead of the rear shield margin. On the thanosome there is a middorsal longitudinal ridge, and the lateral tergal lobes are pointed, especially the anterior ones. The native home of this rust mite is unknown, and so far it has come only from the type locality. It rusts leaves on the underside, but is easily controlled with sulfur. (Note—Two other eriophyids that should not be confused with each other, or with this rust mite, are *Eriophyes mangiferae* (Sayed) and *Metaculus mangiferae* (Attiah), which see.)

Phyllocoptruta Keifer

The three species referred to this genus are tropical and subtropical rust mites. The principal characters defining the genus are as follows. The featherclaws are simple, that is, undivided. The anterior shield lobe projects well over the rostrum. The dorsal shield tubercles are set well ahead of the rear margin and have longitudinal axes; the dorsal setae are short and project up. The most notable feature of this genus is the broad dorsal longitudinal thanosomal trough, running almost its entire length. Undoubtedly more tropical species will be added to this genus.

Key to the Three Species of Phyllocoptruta

1. Shield surface covered with minute pits that have obliterated all lines; anterior coxae separated; (foretibia about 1½ times as long as tarsus; featherclaw 4-rayed); probable origin South America . . . . *sakimurae* K.
1. Shield with pattern of lines, no minute pits; anterior coxae touching . . . . 2.
2. Foretibia twice tarsal length, with small seta near base; featherclaw 4-rayed; coxae heavily lined; sternal line not clearly defined; known in Queenland . . . . *musae* K.
2. Foretibia and tarsus about of equal length, with foretibial seta at center; featherclaw 5-rayed; coxae only moderately lined, but with definite sternal line; probable origin southern Asia, but now widespread in tropics and elsewhere . . . . . . *oleivora* (Ashmead)

Banana rust mite, *Phyllocoptruta musae* K. This species (fig. 127, *b, d*) is reported to cause spotting on bananas in Queensland. *P. musae* is an elongate, flattened-fusiform mite, probably yellow. Females range from 200 to 230 μ long. The featherclaw is 4-rayed. The shield is subtriangular, with the anterior lobe projecting anteriorly ahead of the general frontal outline. This anterior lobe is thin in side view. The median shield line is present on the rear ⅔, running almost to the rear margin. Admedian lines are complete, rather close, somewhat sinuate, and flaring at rear margin. Cross lines meet the admedians at the anterior lobe base, at about ⅓, and another just within the rear shield margin. Submedian shield lines are unclear, but there are a row of "cells" along the lateral shield margins. The abdominal tergites are moderately wide and lack microtubercles, but the sternites are narrow and have microtubercles that project over ring margins. The female genital coverflap has a broad basal pattern of numerous short longitudinal dashes; there are 14 to 16 longitudinal ribs.

Citrus rust mite, *Phyllocoptruta oleivora* (Ashmead). This mite (fig. 127, *a*) causes considerable injury to citrus trees and fruits in many areas. From a worldwide standpoint this is one of the more serious citrus pests because it can reproduce rapidly, and its feeding lowers the quality of fruit.

Citrus rust mites live on both leaf surfaces and on fruit skin. Feeding by *P. oleivora*, probably owing to salivary injections, causes epidermal cells to collapse. Fruit skin turns silvery, reddish, or black. On grapefruit and lemons the feeding results in a silvered or "sharkskin" appearance (pl. 69). Rind of silvered fruit is thicker than on undamaged fruit, and silvered fruit loses water faster than undamaged fruit. One result of mite damage is small fruit, which not only appears substandard, but also deteriorates rapidly. Heavy populations of the rust mite cause bronzing of leaves and green twigs, and general loss of vitality by the whole tree.

These rust mites gather on sunny green surfaces during the day, but tend to scatter over all surfaces at night or on cloudy days. They prefer green fruit and the underside of leaves, but in heavy populations also move onto green twigs. Infestations are generally heavier on the outside of trees, and the mites leave damaged plant surfaces for fresher leaves and fruit.

Warm and humid conditions favor the development of this rust mite. In California, as already noted, these rust mites are limited or stopped by low inland atmoshperic humidity levels, and they ordinarily do not persist on citrus trees except on ocean frontages.

Mites do occur, however, in desert lemon orchards where they get protection from trees that are close together. The frequent irrigation in such lemon orchards may raise the atmospheric humidity levels to more than 20 percent higher than in surrounding areas immediately outside the orchard.

Citrus rust mites lay eggs in depressions on fruit and on leaf surfaces. At average maximum temperatures of 32 C (91 F) the mean incubation period is 3 days, and the duration of the first nymphal stage averages 1.8 days. The second nymph completes growth in 1.3 days. During the winter the incubation period, at an average of 22 C (72 F), is 5.5 days; the first nymphal stage is 4.3 days long, and the second stage lasts 6.4 days. A complete generation of these mites develops in 7 to 10 days in summer, and this lengthens to 14 days, or more, during winter. The maximum adult female life span is less than 20 days, during which time she can deposit as many as 20 eggs (Yothers and Mason, 1930).

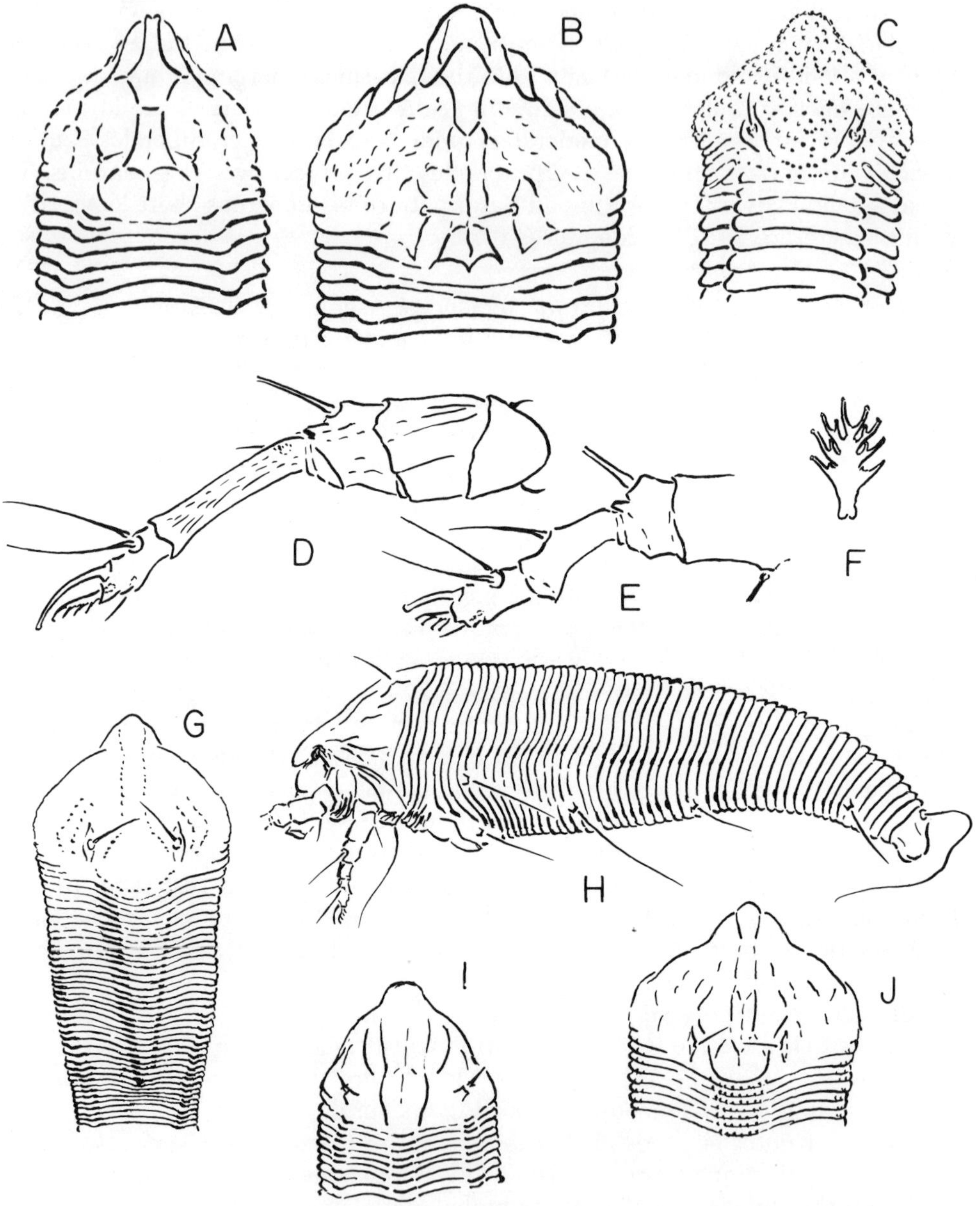

Fig. 127. Shields and anterior abdomen of *Phyllocoptruta* and *Calepitrimerus*: *a*, *Phyllocoptruta oleivora* (Ashmead) shield and anterior abdomen; *b*, *musae* K., shield; *c*, *sakimurae* K., shield and anterior abdomen; *d*, *musae* foreleg; *e*, *sakimurae* foreleg; *f*, *sakimurae* featherclaw; *g*, *Calepitrimerus baileyi* K., dorsal view of anterior section of protogyne; *h*, *C. vitis* (Nal.), side view of deutogyne; i, *C. muesebecki* K., shield; *j*, *C. vitis* (Nal.), protogyne.

The citrus rust mite has been controlled for many years by dust or spray applications of sulfur, or of lime-sulfur sprays. More recently, sprays with zineb and chlorobenzilate have been extensively used. There is some evidence that mite populations have become resistant to zineb in some Florida areas, in Israel, and in Lebanon. In Japan this mite seems to be developing a resistance to chlorobenzilate (Sternlicht, 1966).

P. oleivora is a yellow, flattened, fusiform mite, with females ranging from 150 to 165 μ long. The featherclaw is 5-rayed. The shield has a moderately produced anterior lobe, with a transverse furrow across the front edge. Admedian shield lines arise from each side of the anterior lobe, and curve out to the area just past the lobe base; from there they curve inward and meet a slight cross line at ⅓, then curve slightly out past the inner side of the dorsal tubercles at ⅔, and end in transverse arcs ahead of the rear shield margin. The abdomen has the broad longitudinal trough described for the genus, and the microtubercles are present only on sternites. Each tergite covers 2 or 3 sternites. The basal area of the female genital coverflap is granular, with a median longitudinal line; apically the coverflap has 14 to 16 longitudinal ribs. The male is about 135 μ long.

Pineapple fruit mite, *Phyllocoptruta sakimurae* K. This species (fig. 127, *c, e, f*) lives in shell grooves on pineapple in the Hawaiian Islands. It is a fusiform mite, probably yellowish. The female is 145 to 160 μ long. Shield characters are principally the general irroration of minute pits that have obliterated all possible lines. The mite undoubtedly originated in South America, and is the most distinct species in the genus at the present writing.

Calepitrimerus Keifer

Apple rust mite, or Bailey's rust mite, *Calepitrimerus baileyi* K. This mite (fig. 127, *g*) causes browning on the underside of apple leaves in interior regions of California. In California it stays well back of the coast range barrier ridges, and in that respect differs from the common apple rust mite, *Aculus schlechtendali* (Nal). Bailey's rust mite was previously known on apple only in California, but both of these mites are now known to occur in South Dakota. So the geographical range of *C. baileyi* requires further investigation.

C. baileyi is a yellow to amber spindleform mite. The female is 130 to 140 μ long, with subtriangular shield and moderately acute anterior lobe. The shield design is obscure. One of the more easily recognizable features of this mite is the elevated central abdominal ridge, or keel. The featherclaw is 4-rayed, and the female genital coverflap has 10 to 12 longitudinal ribs. The deutogyne, which was originally named *Phyllocoptes aphrastus* K. lacks the high abdominal keel and has no microtubercles. Its precise overwintering site is unknown, but the lack of microtubercles indicates that it can remain in dry crevices. This deutogyne has been noted in matted hairs around terminal buds.

Avocado rust mite, or leaf vagrant, *Calepitrimerus muesebecki* K. This species (fig. 127, *i*) occurs on leaves of its host both in Florida and in California. It is

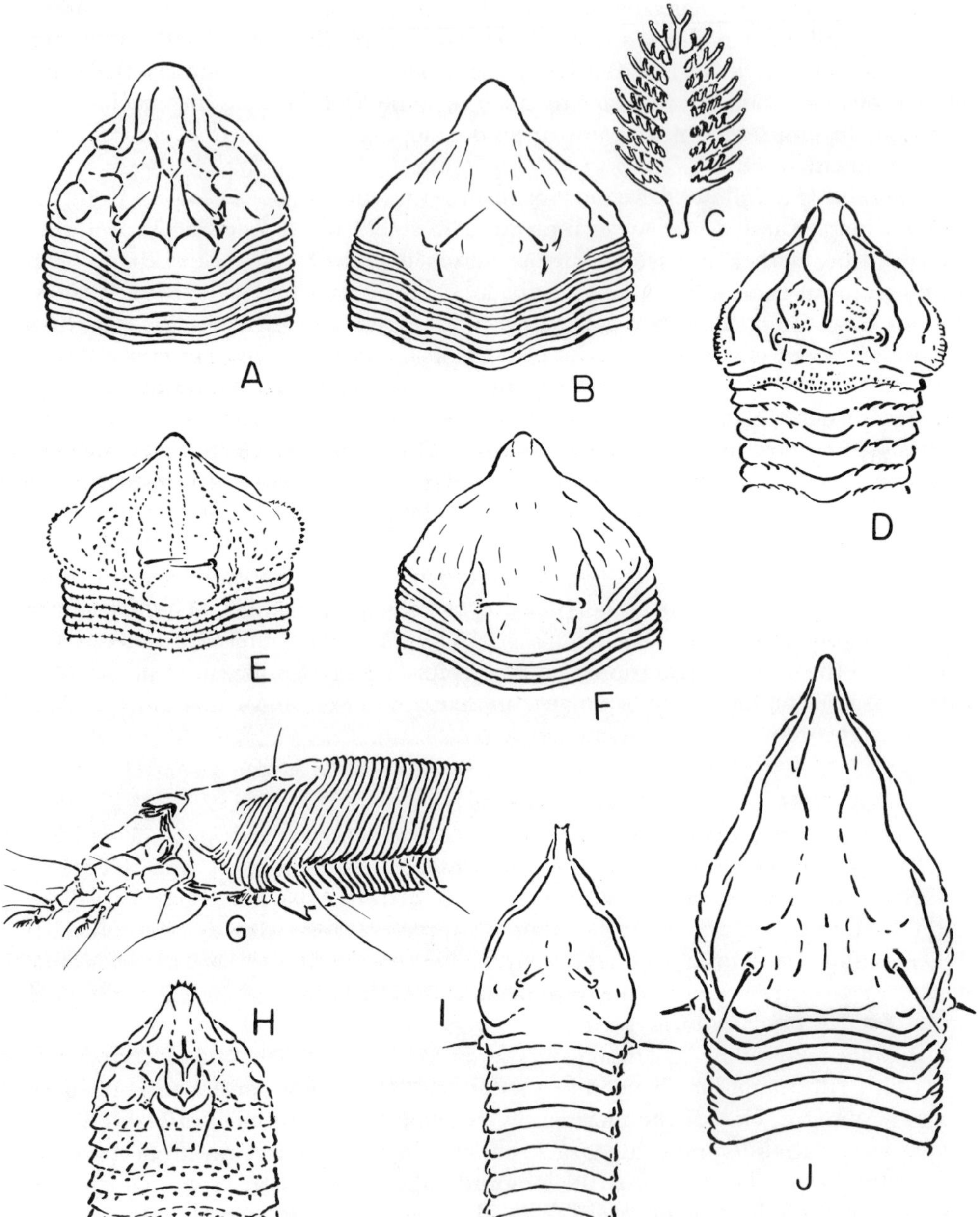

Fig. 128. Shield patterns: *a, Epitrimerus pseudotsugae* K.; *b, E. taxodii* K., shield; *c, E. taxodii* featherhclaw; *d, E. congoensis*, shield; *e, pyri* Nal., shield of protogyne; *f, pyri* shield of deutogyne; *g, Platyphytoptus sabinianae* K., lateral view of anterior section; *h, Tetraspinus capsicellus* (K.), dorsal view of anterior section; *i, Notostrix jamaciae*, shield and anterior abdomen; *j, Notostrix attenuata* K., shield, and anterior abdomen.

a rather slender fusiform species, yellowish, with extra short dorsal shield setae, and wax stripes on dorsal ridges. The female is 180 to 205 μ long.

Grape rust mite, *Calepitrimerus vitis* (Nal.). *C. vitis* (fig. 127, *h, j*) is a pest of grapevines. Heavy infestations of this species prevent vines from growing normally during the earlier parts of the season. Internodes are shortened, foliage becomes bunched, which interferes with proper pruning; grape production is reduced. Damage to grape clusters occurs either because flowers are injured, or because development is delayed. Deutogynes, often listed as *Phyllocoptes vitis* Nal., overwinter in cracks along twigs, or in hair masses around buds. Upon foliage development in spring, the mites promptly move to new leaves and as further shoot elongation takes place, these rust mites advance to keep up with the shoots. Treatments made when shoots are 5 to 10 cm long (2 to 3 inches) are most effective (Garcia, 1966).

The female of *C. vitis* is about 150 μ long, yellowish, and spindleform. The shield is subtriangular, with a moderately acute and rounded off anterior lobe. Admedian shield lines are complete from the sides of the anterior lobe, running back to and forking between the dorsal tubercles. The dorsal setae are moderately long and project diagonally up and centrad. The featherclaw of the protogyne is 4-rayed. The central abdominal ridge runs back to about 4/5 of the abdominal length, and has a moderate rear slope. The female genital coverflap has about 10 longitudinal or radial ribs.

The deutogyne of *C. vitis* lacks all microtuberculation, except on the ventral telosome surface. It has a 5-rayed featherclaw.

Epitrimerus Nalepa

Douglas fir needle vagrant, *Epitrimerus pseudotsugae* K. This mite (fig. 128, *a*) lives on Douglas fir needles along the Pacific Coast of North America. It is most numerous when new shoots are actively growing. The species has done no observable damage.

Bald cypress rust mite, *Epitrimerus taxodii* (K.). This species (fig. 128, *b, c*) occurs wherever bald cypress is native or grown in North America. It builds up in large numbers on needles and causes rusting. It is an orange yellow mite; the female is 180 to 210 μ long. The body is spindleform and usually covered with white wax. The anterior shield lobe is moderately acute; the shield design is obsolete, and the overall shape of the shield is subtriangular. This bald cypress mite has dorsal setae that project up 11 μ, and are set somewhat ahead of the rear margin. One notable feature of *taxodii* is the 10-rayed featherclaw. The abdominal convolutions show about four longitudinal furrows above the lateral ridge. The female coverflap has faint longitudinal ribs.

Pear rust mite, *Epitrimerus pyri* (Nal.). This mite (fig. 128, *e, f*) attacks both pear leaves and fruit in the spring. High infestations cause severe browning of leaves and russeting of fruit. Lower populations may injure only the calyx end of fruit. Clear-skinned fruit varieties show the most injury.

As treated in the section on deuterogyny, the primary forms of the rust mite occur on pear leaves in spring, but secondary females, or deutogynes that have no abdominal ring microtubercles, develop in early summer when the leaves harden, and these deutogynes soon migrate to resting niches—at the apex crevices—of the previous season's growth.

Deutogynes do not reactivate until the following spring, when they crawl to the developing leaves and lay eggs.

Timing of spring spray applications to prevent injury by *E. pyri* is critical and should be completed by the cluster-bud stage. Additional sprays may be required to prevent leaf and fruit injury. Acaricides currently suggested for cluster-bud application include flowable or lime-sulfur, dithiocarbamate fungicides, Karathane®, and Morestan®. Summer spray acaricides may include Thiodan®, dicofol, Morocide®, and sevin. Summer applications against this rust mite should take into account that the deutogynes, which are more resistant than the spring forms, are down in dry bark crevices.

E. pyri is a small yellowish flattened fusiform species. Protogynes are about 145 to 160 μ long, with males slightly smaller. The anterior lobe of the protogyne shield is rather small and acuminate, rounded apically. The shield design is not plain, being composed of fine lines or granules, with the median line absent. The admedian lines are nearly complete, close, and subparallel anteriorly, abruptly diverging between dorsal tubercles. Lateral shield lobes on protogynes and males are quite prominent, with finely serrate edges, and these lobes are a definitive feature of the mite. The dorsal setae are short and directed upward and inward. The featherclaw is 4-rayed. Flattened protogynes have rings with fine microtubercles, and the dorsum has a very shallow subdorsal longitudinal furrow on each side. The female genital coverflap has 2 or 3 basal cross lines, and about 12 ribs.

The pear leaf rust mite deutogyne is slightly smaller than the protogyne and, as mentionel above, lacks ring microtubercles. It has a larger anterior shield lobe than the protogyne. The name *Epitrimerus pirifoliae* K. was originally established because the pear rust mites collected in Sacramento, California, displayed the prominent, outwardly round and minutely serrate lateral shield lobes, but this did not match descriptions of *Epitrimerus pyri* in Europe. Since then specimens have been received from Europe that have these lateral lobes.

Coffee rust mite, *Epitrimerus congoensis* K. *E. congoensis* (fig. 128, *d*) lives on the underside of coffee leaves in the Kivue area of Zaire. It is an elongate-spindleform mite, and the shield is wider than the anterior part of the abdomen. The mite is probably yellow. The female is 165 to 175 μ long. The anterior shield lobe is prominent and acuminate, slightly rounded off. Admedian shield lines extend back from the sides of the anterior lobe, curving just behind the base of the lobe, then recurving to the center, the lines running close together and ending just before becoming even with the dorsal tubercles. Projecting lateral lobes are very prominent and have a pattern of curved lines; the outer lobe edges are finely serrate. The short dorsal setae project up and centrad. The featherclaw is 4-rayed. The abdomen has a middorsal longitudinal ridge which is moderately acute, more so caudally. Microtubercles are present on the tergites laterally, and are numerous and elongate; on the sternites the ring granules take the form of slightly elongate beads. The female genitalia have basal lines and granules and about 12 subapical ribs.

Platyphytoptus Keifer

Flat needle sheath pine mite, *Platyphytoptus sabinianae* K. This mite (fig. 128, *g*) lives on various pine species in North America, but the precise range is unknown. It was first found in California, but has since been observed in Kansas and further east. In California it is commonest on digger pine and on western yellow pine, where it lives in the same sheaths with *Trisetacus ehmanni* K. *P. sabinianae* occurs more often on the inner surface of the needles, whereas *T. ehmanni* usually lives next to the sheath. In the general Appalachian region of eastern North America *P. sabinianae* is found with *Setoptus* spp. It infests pitch pines, including Scotch pine and yellow pines, but is not an inhabitant of white pines. In eastern North America it, in company with *Setoptus jonesi* (K.), causes stunting and twisting of needles, and some yellowing.

P. sabinianae is an orange yellow species. Females are from 200 to 215 μ long. The body is flattened-fusiform, with a strong sublateral longitudinal groove on the thanosome. The shield has dorsal tubercles set well ahead of the rear margin; the anterior shield lobe is circular in anterior outline, but very thin. The shield surface is finely irrorated with granules. The forecoxae are finely granulate and separated. The thanosomal tergites and sternites are about equal in number. The tergites lack microtubercles, but sternites have these structures, touching margins. The female genitalia have a broad basal band of granules, but lack coverflap ribs. The featherclaw is 5-rayed.

There are three other species of *Platyphytoptus* on record. *Platyphytoptus monophyllae* K. occurs on needles of pinyon pine east of the Sierran crest in California. The host is one-leaf pinyon pine. The mite has a 5-rayed featherclaw, and lives on open needle surfaces. This is a more robust mite than *P. sabinianae*, and the shield design with granular lines is quite different from that species.

Two other names in *Platyphytoptus* refer to 4-rayed featherclaw species: *P. multisternatus* K., named originally from lodgepole pine, and *P. eldoradensis* K., which lives on ponderosa pine. The precise relationship of the populations bearing these latter two names is uncertain. *P. eldoradensis* is variable as to the pointed projections from the rear shield angles. Individuals having these rear shield angles are quite striking. But mites found on lodgepole pine in Yellowstone Park, that are otherwise similar to *P. multisternatus*, do have such rear shield angles. Collections of specimens from pitch pines in the eastern United States indicate that the mites from those pines also have similar rear shield angles.

Phyllocoptes Nalepa

Plum leaf vagrant, *Phyllocoptes abaenus* K. (fig. 129, *b*). This is a widely distributed species evidently originating in Europe. It lives on undersides of plum leaves, including ornamentals, and frequents basal hairs along midribs. Whether or not this mite stimulates growth of this basal hair is unknown.

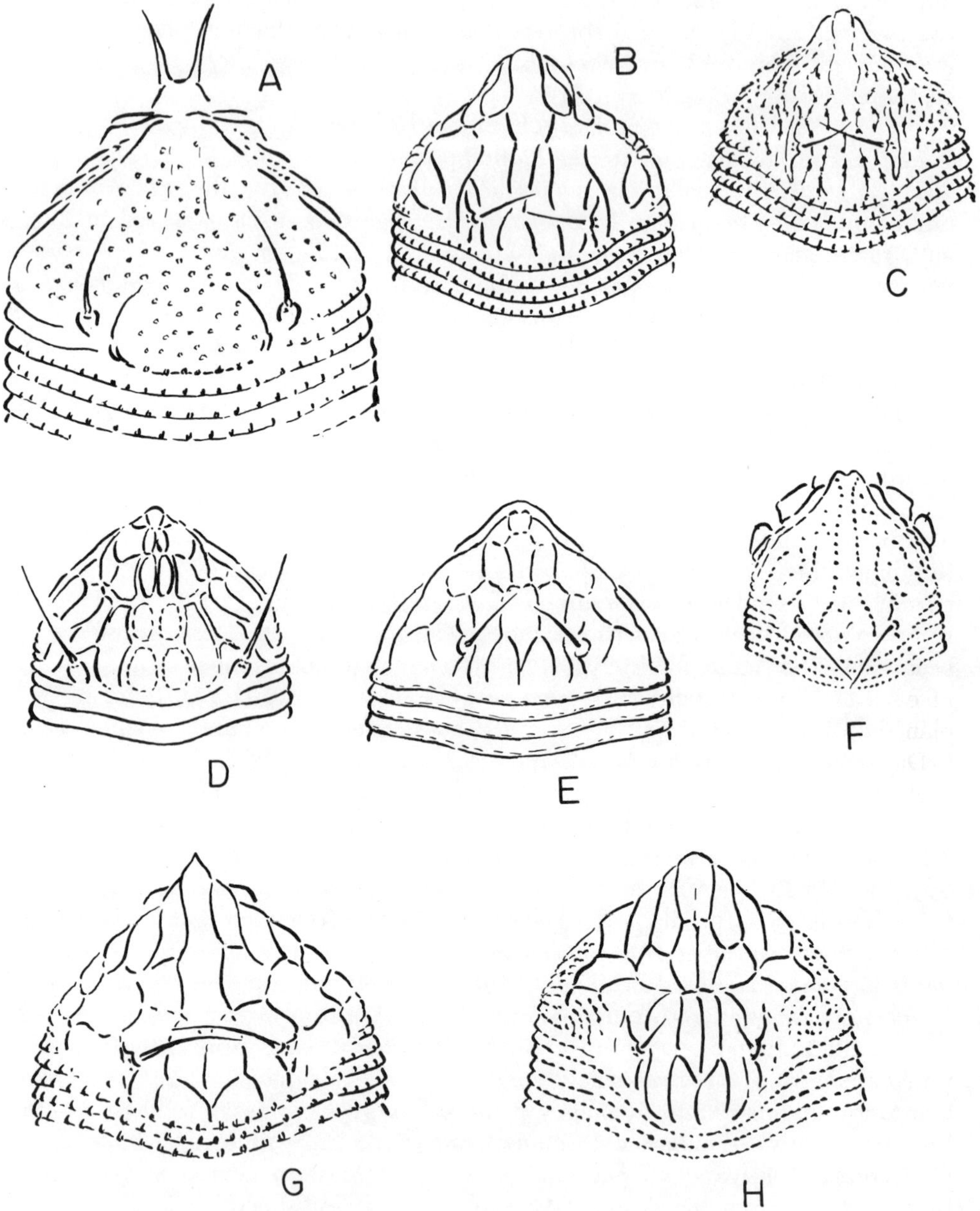

Fig. 129. Shield patterns of *Acadicrus* and *Phyllocoptes*: *a*, *Acadicrus mergiferus* K.; *b*, *Phyllocoptes abaenus* K. *c*, *Phyllocoptes malinus* (Nal.); *d*, *cribratus* K.; *e*, *P. graniti* K.; *f*, *P. gracilis* (Nal.); *g*, *P. fructiphilus* K.; *h*, *P. nageli* K.

A close relative of *P. abaenus* is *Phyllocoptes goniothorax* Nal. It is an erineum former on pomaceous Rosaceae, with hawthorne as the typical host. Nalepa has named four varieties of this erineum maker, including *P. malinus* on apple (fig. 129, *c*). This apple mite has granular shield lines and slight concentric rings on the coverflap. In California a *Sorbus* erineum mite, *Phyllocoptes calisorbi* K., is evidently in this *P. goniothorax* complex. *P. calisorbi* has solid shield lines, round microtubercles on the tergal and sternal margins, and has ribs on the coverflap that converge centrally to the rear. Nalepa characterizes typical *goniothorax*, on hawthorne, as cyclindrical, the female being 170 μ long. The shield is 5-sided, with three strong central lines and wrinkled lateral areas. He shows dorsal tubercles with elongate axes and setae directed outward. The featherclaw is 4-rayed. The sternal line is forked; the abdomen is entirely microtuberculate. There are no accessory setae. The female genital coverflap is smooth.

Wild rose bud and fruit mite, *Phyllocoptes fructiphilus* K. This mite (fig. 129, *g*) occurs on wild rose in the western half of North America. It is reported to be a vector of a disease of roses known as rose rosette. In contrast to normal buds, most of the auxiliary buds of infested plants break color, which results in viruslike symptoms. The symptoms in a new shoot usually appear first on auxiliary buds, then progress toward the shoot apex. Diseased shoots are thicker than normal, and usually have shortened internodes, particularly toward the end. Developing terminal leaves are small and misshapen, and short secondary shoots bearing very small leaves emerge from the auxiliary buds. This is brooming and gives the crowded nodes and leaves a rosetted appearance. Diseased young plants become dwarfed.

During the growing season all stages of *P. fructiphilus* live and breed in the protected areas between leaf petiole bases and lateral buds, particularly toward the end of the shoot. Mites overwinter in any protected place on the plant where green surfaces are appressed to each other. At times they move into open ends of fruit and live among the seeds. Surviving spring mites are few, and the populations of this mite do not become noticeably numerous until mid-June or early July. Populations increase until September or October, when the mites move to survival niches. Many mites are still present until December.

P. fructiphilus is light yellow, fusiform, with females from 140 to 170 μ long. The shield is subtriangular, with the anterior lobe acuminate and pointed. The shield design is a network, with the median line present only to the rear. Admedian lines are complete and sinuate. The dorsal setae are moderately long, pointing up and centrad. The featherclaws are 5-rayed. On this species there is little difference between tergites and sternites. The microtubercles on this mite vary in shape, depending on the locality. Perhaps these ring granule differences are specific in nature. On the typical *P. fructiphilus*, the microtubercles are rounded off and touch ring margins. The name *P. slinkardensis* refers to mites that are closely related to *fructiphilus* and occur within rose fruits on the eastern side of the Sierra Nevada, and which have pointed or spiny microtubercles. The female coverflap on these mites has about 6 to 8 longitudinal ribs.

North American persimmon rust mite, *Phyllocoptes cribratus* K. This species (fig. 129, *d*) probably infests native persimmon wherever it grows. These mites live on and rust—sometimes severely—the undersides of persimmon leaves. *P. cribratus* is a yellowish, fusiform species, with females from 165 to 180 μ long. The shield is subtriangular, with a small acuminate anterior lobe that is rounded off. The shield design is a network with thickened central lines that are somewhat obscured. The featherclaw is 4-rayed. The coxae have faint lines on them, and the sternal line between the forecoxae is fairly long. Microtubercles on tergites are restricted to the sides just above the sternites. Sternal microtubercles are elongate and rounded over margins. The female genitalia have some basal granules, but there are no coverflap ribs. The principal recognition features of this mite are the dorsal setae that project forward and divergently from the rear shield margin, the network shield design, and the ribless coverflap.

Douglas fir needle mite, *Phyllocoptes nageli* K. *P. nageli* (fig. 129, *h*) is a needle vagrant on *Pseudotsuga menziesii* (Mirbel) from central California, north to Oregon and possibly further. Its distribution is confined to coastal areas and in that respect it would seem to require fairly high atmospheric humidities. This needle vagrant has not as yet been associated with any damage.

P. nageli female is from 150 to 200 μ long, fusiform, and light yellow. The shield is subtriangular, with a somewhat acuminate but rather short, rounded off anterior lobe. The shield design is a network, with wide lateral granular areas. The median shield line is complete, but is crossed by lines from the admedians at ¼, ½, and ¾. Admedian lines are somewhat irregular owing to cross and lateral lines. Submedian shield lines form a complex series of "cells" with these lines turning centrad ahead of the dorsal tubercles. The featherclaw is 6-rayed. The coxae are generally granular, with a very short sternal line. There is not too much difference between tergites and sternites, perhaps reflecting the preference for a humid climate. The microtubercles are fine, somewhat elongate, pointed over margins. Female coverflap markings are weak, with some basal granules and short irregular ribs. Recognition characters for this mite are the rather elongate central part of the shield, with long median and admedian lines and their cross lines, the fine ring microtubercles, the 6-rayed featherclaws, and the first coxal tubercles pushed ahead of the anterior end of the short sternal line.

Dryberry mite, or blackberry leaf mite, *Phyllocoptes gracilis* (Nal.). This mite (fig. 129, *f*) is sometimes a serious pest of several *Rubus* species, including raspberry, blackberry, loganberry, Himalaya berry, and thimbleberry. It occurs in England and throughout Europe. In North America it is found in British Columbia, Washington, Oregon, and in California. In Germany *P. gracilis* lives on wild and cultivated raspberry.

Infestations of this mite cause the Lloyd George variety of red raspberry fruits to become hardened, somewhat misshapen, and fail to ripen normally. The term "sun scald" is descriptive of the effects of mite injury to the Washington variety of red raspberry. Microscopic examination of berries so injured discloses that each affected drupelet began dying and drying at the apex or style end, with the drying then spreading toward the periphery. Losses from this dry berry disease have been at times estimated at from 50 to 70 percent of the crop.

On loganberry the most notable feature of dryberry disease is the killing of early fruits soon after petal fall. Affected berries turn reddish, then brown, and then dry. Fruits that develop later are usually unaffected. Damaged mature berries may be withered, particularly at their distal ends. Boysenberries are often affected in a similar manner.

In Europe the feeding of this mite on red and yellow garden raspberries produces pale spots on fruit, because of "hair felt" destruction. On leaves the mites cause lower surface hair suppression.

Dryberry mites overwinter in semiexposed colonies near buds on canes of the current season's growth. The colonies do not hide beneath bud scales, which is in contrast to the habit of the redberry mite, *Acalitus essigi* (Hassan). In the spring as buds enlarge, *P. gracilis* mites crawl out from behind old buds and, as the leaflets separate, the mites begin to feed on them. There the eggs are laid and the spring and summer cycles begin. Since hibernation by *P. gracilis* is interruptible at any time the temperature rises above 11 C (52 F) the mite does not go into a true resting stage. Overwintering females that emerge in spring may lay 1 or 2 eggs in a 12-hour period. During the growing season these mites are present on both surfaces of the leaves, especially in hairy areas, and on green and ripe berries. They also occur on wild *Rubus* spp. in western North America (Breakey and Batchelor, 1957; Massee, 1932).

Timely applications of lime-sulfur or mineral oil sprays applied at dormant strength control the mite and prevent development of dryberry disease. Applications should start as cane buds enlarge in the spring to kill mites after they leave the winter habitat, and before they feed to any extent on expanding leaves. There should be a second application, this time at summer strengths, as soon as the flower buds form, but before any of them open. Time this treatemnt to kill the mites before they enter the blossoms and attack young fruits. Properly timed applications of chlorobenzilate are also effective.

Females of *P. gracilis* are from 115 to 130 μ long, the body is more wormlike than fusiform, and light yellow. Nalepa shows the shield as generally granular, and the design is mainly of three subparallel central lines; the median, flanked by the admedians. The shield is subtriangular, with an acuminate, apparently notched apex. The dorsal shield setae are moderately long and directed diagonally backward and centrad. The featherclaw is 5-rayed. The abdominal rings are approximately subequal dorsoventrally and they are entirely microtuberculate. The female genital coverflap has about ten longitudinal ribs.

Nalepa's statement concerning this mite is that it lives on the undersides of leaves. The type host is the red and yellow garden raspberry. Eriophyids of this type on the west coast of North America occur on various native *Rubus* spp., especially in the north. There is some variation throughout these populations in size and shield pattern. One kind, living in dense undersurface leaf hair on native thimbleberry in coastal central California, was named *Phyllocoptes parvi-*

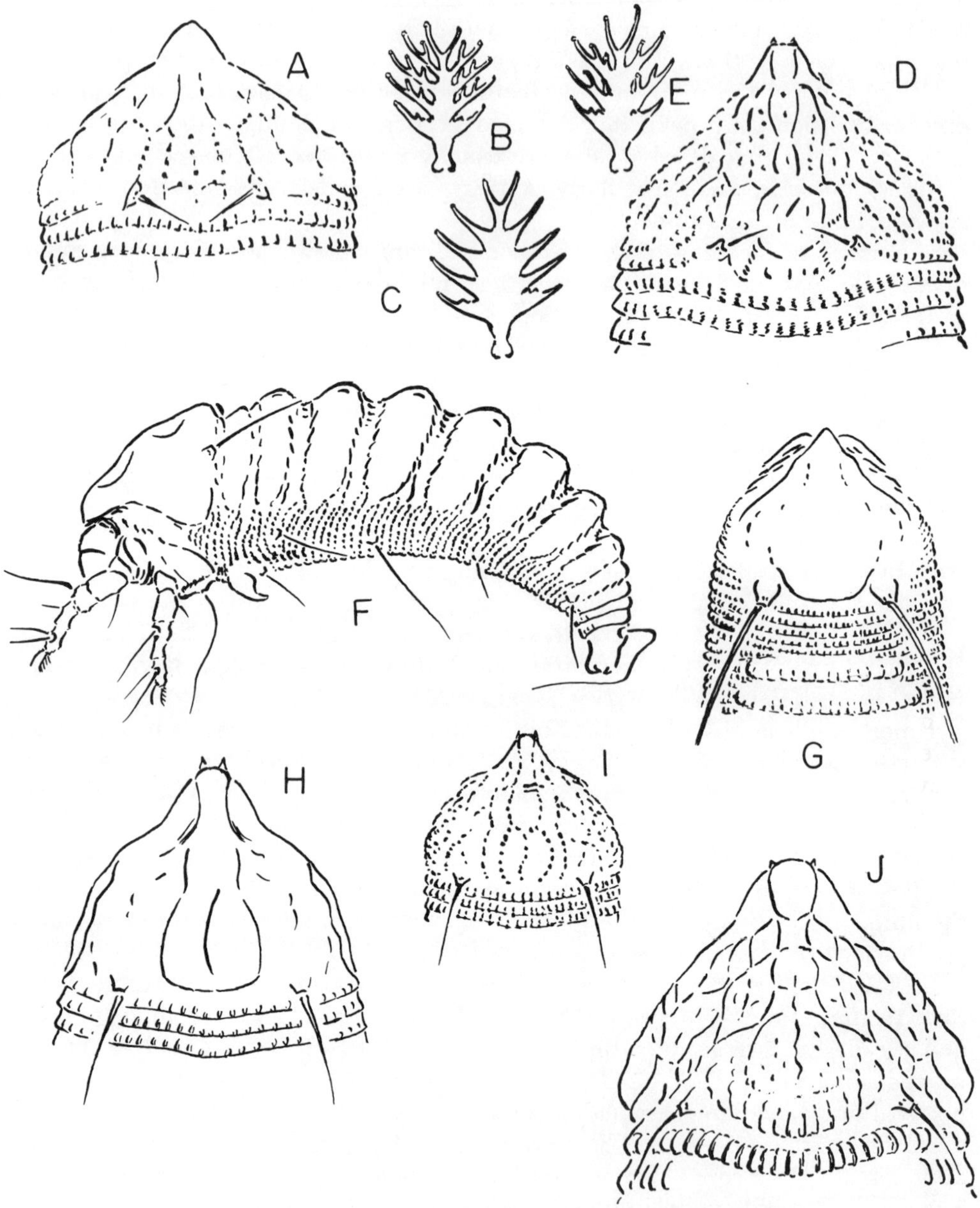

Fig. 130. Shields and featherclaws of *Vasates*, *Anthocoptes*, *Heterotergum*, and *Aculus*: *a*, *Vasates quadripedes* Shimer, shield; *b*, *Vasates quadripedes*, protogyne featherclaw; *c*, *Vasates quadripedes*, deutogyne featherclaw; *d*, *Vasates aceris-crumena* (Riley), shield; *e*, *aceris-crumena*, featherclaw; *f*, *Anthocoptes bakeri* K., female left side; *g*, *Heterotergum gossypii* K., shield and anterior abdominal dorsum; *h*, *Aculus atlantazaleae* (K.), shield; *i*, *Aculus schlechtendali* (Nal.), shield; *j*, *Aculus caryfoliae* K., shield.

florae K. The precise taxonomic status of these various types referable to the *gracilis* complex remains for investigation.

Almond erineum mite, *Phyllocoptes graniti* K. This mite (fig. 129, *e*) makes underside erineum patches on the leaves of almond in Italy. The erineum consists of capitate papillae. The form described was collected in August, and its structure suggests it may be a deutogyne, since it has suppressed dorsal microtubercles.

Vasates Shimer

Members of this small genus differ from the *Aculus-Aculops* type of mite by having dorsal tubercles with diagonal axes that direct the setae diagonally inward to the rear.

Maple bladder-gall mite, *Vasates quadripedes* Shimer . This species (fig. 130, *a-c*) is the common silver maple mite in eastern North America, where its globular leaf galls often cause some leaf deformation (pl. 70). Galls similar to this develop on red maple, and possibly on other maple species, but are unknown on sugar maple. These galls protrude from upper leaf surfaces, and have been likened in shape to a green pepper. The escape hole is on the leaf undersurface.

These globular galls start to grow on expanding spring leaves. They tend to turn dark brown or black by late July, at which time deutogynes of *V. quadripedes* start crawling out, looking for hibernation quarters, although this migration from galls can continue into early October. The deutogynes of this mite are often very numerous and may be found on many parts of the tree. They even crawl on surrounding shrubbery and it is possible to confuse them with other species, owing to their ubiquitousness in the latter part of the season. Their normal hibernation niches are in bark cracks or wounds, where they remain without food until the following spring. These *V. quadripedes* deutogynes lack abdominal microtubercles; thus they are adapted for dry resting locations.

Heavy infestations of this mite can badly disfigure silver maple trees, but injury is seldom sufficient to require chemical control. If treatments are necessary such acaricides as dicofol, chlorobenzilate, ethion, and chlorobenzide may be used to reduce populations that are not protected in galls (Herrick, 1935; Weidhaas and Reeves, 1963).

V. quadripedes is a fusiform species. Female protogynes are from 180 to 220 μ long, and yellowish. The protogyne shield has a moderately rounded off anterior lobe. The shield design is of granular lines with the median line indicated on the rear ½. Admedian lines extend back from the base of the anterior lobe to between the dorsal tubercles where they meet a cross line from the median. A submedian line forks from the admedian at about ½ and extends to the dorsal tubercle. The featherclaw of the protogyne is 4- or 5-rayed, and rays are not thickened basally. Thanosomal tergites are completely set with elongate microtubercles, and cover 2 or 3 sternites. Sternal microtubercles are more beadlike. The female genital coverflap has basal granules, and about 14 subapical radial ribs.

Deutogynes are smaller than protogynes and darker in color. They lack a shield design, lack all ring microtubercles, and have no ribs on the genital coverflap. The tergites uniformly cover but two sternites, and the featherclaw rays are thickened basally.

There is but one other gall-making species of *Vasates* known. It is the finger gall mite of sugar maple, *aceris-crumena* (Riley) (fig. 130, *d, e*). This mite differs from *V. quadripedes* in several ways, one of which is the presence of a pair of spinules projecting forward from the anterior shield lobe. The shield pattern is much more involved than on *V. quadripedes,* and the dorsal setae are directed diagonally ahead and inward. The featherclaws are uniformly 4-rayed and as on *V. quadripedes* the deutogyne featherclaw has thickened ray branches.

Heterotergum Keifer

The recognition characters of this genus include dorsal setae that project divergently back from the rear shield margin; the first few thanosomal rings are equally narrow entirely around the body. Wide tergites begin immediately after these first few narrow rings.

Cotton rust mite, *Heterotergum gossypii* K. This species (fig. 130, *g*) occurs on cotton in Brazil. Infestations cause bronzing of mature cotton leaves, and tip blighting of young leaves. The female of *H. gossypii* is yellow, rather robust-fusiform, and from 150 to 215 μ long. The shield is short, with an acuminate lobe over the rostrum base. There is no shield design. Featherclaws are 5-rayed. The abdomen dorsally has 5 narrow rings just behind the shield, followed by 13 or 14 broad tergites that have elongate microtubercles. The female coverflap has 12 to 14 longitudinal ribs.

Heterotergum gossypii, which is the genotype, is representative of a series of Western Hemisphere species. *Heterotergum olneyae* K., on *Olneya tesota* Gray, in Arizona, has an 8-rayed featherclaw and 4 postshield narrow rings. The admedian lines on *H. olneyae* extend full shield length. *Heterotergum schlingeri* Wilson and Oldfield, on *Franseria chenopodiella* in northern Mexico, has a 7-rayed featherclaw, 5 postshield rings, and prominent admedian shield lines. *Heterotergum tuttlei* K., on *Trixus californicus* Kelley, in Arizona, has a 7-rayed featherclaw, 4 post-shield narrow rings, and admedian lines only visible anteriorly on shield. *H. wilsoni* K., on *Simmondsia californica* Gray, has a 5-rayed featherclaw, 5 postshield rings, and no shield lines, although the shield has a broad anterior lobe. *Heterogum yumensis* K., on *Acacia greggi* Gray, has a 5-rayed featherclaw, 5 postshield narrow rings, and granular shield lines.

Anthocoptes Nalepa

The principal recognition features of this genus are the broad thanosomal tergites, followed by the sudden narrowing of the telosomal rings (see figs. 90, *b*; 92, *d*).

Honey locust rust mite, *Anthocoptes bakeri* K. This mite (fig. 130, *f*) infests honey locust throughout its natural range in eastern North America. In company with *Aculus gleditsiae* K., this mite feeds on the underleaf surfaces and often causes considerable leaf damage. *A. bakeri* is a dull yellowish, robust-fusiform mite, with females from 120 to 140 μ long. The featherclaw is 5-rayed. The shield is subcircular in anterior outline, with a broadly rounded off anterior lobe. The shield lacks lines except for the admedians, which start on each side of the anterior lobe, diverge to the shield center, then recurve slightly and arch back to the rear margin. The dorsal setae are moderately long, strongly diverging to the rear. There are about 13 very broad tergites, with faint and elongate microtubercles on the sides of the body. There are several sternites under each tergite, and the sternal microtubercles are elliptical, resting on ring margins. The female genital coverflap has 10 to 12 longitudinal ribs.

Metaculus Keifer

This genus is differentiated from the *Aculus-Aculops* series because it lacks the first setiferous coxal tubercle. There are three species now referable to *Metaculus*. The genotype is *Metaculus syzygii* K., which lives on *Syzygium cordatum* in central Africa. The *Syzygium* mite has thickened but obscure shield lines centrally, with the median line present on the rear half. The genital coverflap has ribs in two ranks, the basal rank being irregular, and the apical rank having 10 or 11 longitudinal ribs. A second species is *Metaculus rapistri* Carmona, 1968, living on *Rapistrum rugosum* (L.) in Portugal. *M. rapistri* lacks a median shield line and has a more pointed shield than the other two species. The female coverflap on *M. rapistri* has about 10 short ribs. The third species is *Metaculus mangiferae* (Attiah), as described below. It has a median line on the rear ½ of the shield. All three species have 4-rayed featherclaws.

Mango rust mite, *Metaculus mangiferae* (Attiah). This species (=*Vasates*) causes russeting of terminal leaves, buds, and inflorescences. It is an important pest of mango in Egypt, India, Palestine, and in Angola. Injury to young seedlings is called "bunchy top" because of brooming in the form of crowding of short shootlets. Buds and inflorescences stop growing and ultimately dry off. Severely infested trees have many of these stumpy and arrested terminals. Mango rust mites are active and quickly stunt and dry terminals and buds.

The mango bud mite, *Eriophyes mangiferae* (Sayed), which is described in the section with *Eriophyes* spp., is, according to Attiah, slower in its action than the *Metaculus*, and while this *Eriophyes* also stunts terminals, the terminals do grow out further than do those damaged by rust mite action.

Damage to inflorescences by the rust mite results in a compact mass of malformed flowers. The affected panicle continues to produce flower buds, even after fruit setting in a normal panicle has finished. The percentage of hermophrodite flowers in a malformed inflorescence is low, and the fruit that develops either drops or only reaches pea size.

Vegetative malformations may appear on seedling mangos when they are five months old. Mango trees four to eight years old sometimes have nearly 50 percent of the shoots injured. Late blooming varieties show less injury.

The rust mite lives more-or-less exposed and is fairly easy to control by such acaricides as wettable sulfur (Attiah, 1955; Singh, Singh, and Nirvan, 1961).

M. mangiferae females range from 170 to 240 μ long. They are fusiform and light yellowish. The shield is subtriangular in general outline, and the shield pattern is a network of lines. The median shield line is present on the rear ½. Admedian lines are irregular owing to cross lines and angles with lateral lines. There is a cross line on the shield at ¼, a second just behind the ½ point, and a third just ahead of the rear shield margin. The submedian lines produce network cells outside the admedian lines, and there is a row of lateral "cells" around the shield margins. The abdomen is entirely microtuberculate and the tergites are not much wider than the sternites. There is a shallow subdorsal furrow on the tergites for a short distance behind the shield. The female genital coverflap is represented as having granular cross lines basally, and with about 24 longitudinal ribs. (Note: The reader should not confuse this *Metaculus mangiferae* with *Eriophyes mangiferae* (Sayed) or with *Tegonotus mangiferae* (K.). These are three distinct species.)

Aculus Keifer

This is a fairly large genus of rust mites that often do considerable damage to their hosts. The recognition characters of the genus are the two or four anterior spinules projecting forward from just under the front edge of the broadly rounded anterior shield lobe, plus dorsal shield tubercles with transverse axes that rest on the rear margin and direct the setae divergently to the rear. The thanosome tends to have slight subdorsal furrows just behind the shield.

Azalea bud and rust mite, *Aculus atlantazaleae* (K.). This mite (fig. 130, *h*) lives at leaf bases around buds on native creeping azalea in eastern North America. The type host is Atlantic azalea. The mites can do some damage when numerous enough. It is a light amber mite, fusiform, with females from 160 to 175 μ long. The featherclaw is 4-rayed. The anterior shield lobe is moderately wide basally. The median shield line is slightly indicated on rear half of the shield. The admedian lines are complete, curving back and outward from the anterior lobe, then extending divergently to the rear shield area where they enclose an area between and anterior to the dorsal tubercles. The thansome is generally microtuberculate, with each tergite covering 2 or 3 sternites. The female genital coverflap has 12 or 14 longitudinal ribs.

Plum rust mite, *Aculus fockeui* (Nalepa and Trouessart). This is a pest of various plums and cherries in orchards and in nurseries. It thrives in more northern areas, such as Canada, and northern Europe. This mite produces a condition on plum leaves which has been called *asteroid spot* or *chlorotic fleck.* Spots occur in the spring and consist of more-or-less well-defined circular yellow areas ranging from pinpoint to 1 or 2 mm in diameter. Small flecks are often diffuse, but larger spots have well-defined margins, and may show concentric rings. There may be as many as 50 flecks to a leaf, under which condition the leaf is wavy or slightly twisted about its longitudinal axis. Heavy infestations of *A. fockeui* may rosette shoots and keep leaves from expanding to normal size. Green twig spots are more ovoid and generally larger than leaf spots (Hildebrand, 1945).

The yellow spots known as chlorotic fleck only occur when rust mites feed on emerging embryonic spring leaves. The flecks develop from 13 to 15 days after initial feeding. Fully expanded plum leaves, onto which rust mites were transferred, failed to develop the flecks. But on these same twigs, where the transfers were made to expanded leaves, the flecks showed up on leaves that were just emerging at the time of the transfers, and which were 4 or 5 leaves ahead of the transfer sites.

Deutogynes of the plum rust mite hibernate in niches on or near the current season's growth. They crawl into bark crevices, especially around injuries, but are also behind potentially active buds, or under available loose bud scales. As the buds expand on the spring the deutogynes crawl to, and feed upon, emerging embryonic leaves, and lay eggs. The next generation consists of males and protogynes. Populations increase rapidly under favorable conditions until midsummer, when tree growth wanes. New sucker growth will continue to have active rust mite populations for as long as it is fresh. New deutogynes appear as the result of summer leaf hardening, and when they are full-fed and inseminated they move to resting quarters (Gilmer and McEwen, 1958; Massee, 1932).

The life span of the plum rust mite in various times of the year is shown below. Temperatures are unstated, but are low or high according to the time of year. There is no indication as to which female is the subject of the study, but it is presumed that the figures refer to protogynes. A protogyne may lay up to 80 eggs during her life. These studies were made in Canada (Putman, 1940).

Month	*Incubation period (days)*	*Immature period (days)*	*Total life cycle (days)*
April	15.0		
May	5.0	12.8	18.7
June	4.3	5.0	11.4
July	3.7	3.3	8.3
August	3.3	3.3	9.0
September	5.2	5.5	14.2
October	7.0	8.0	—

The plum rust mite, *A. fockeui*, is essentially structurally inseparable from the peach silver mite, *Aculus cornutus* (Banks). But the life habits and host preferences of the 2 mites do not precisely coincide. The silver mite lives on both leaf surfaces, but the plum mite occupies only the undersurface of leaves. As already stated, the plum rust mite is more northern in its climatic preferences than the silver mite. In the Sacramento Valley, California, plum trees growing but a few feet from peaches with heavy silver mite populations did not become inhabited by this type of mite.

A. fockeui is a fusiform species with protogynes from 160 to 175 μ long. The featherclaw is 4-rayed. The shield is subtriangular in general, with a somewhat acute anterior shield lobe. The shield design is of lines, rather obscure centrally, except for the central line on the anterior lobe. (Both *A. fockeui* and *A. cornutus*

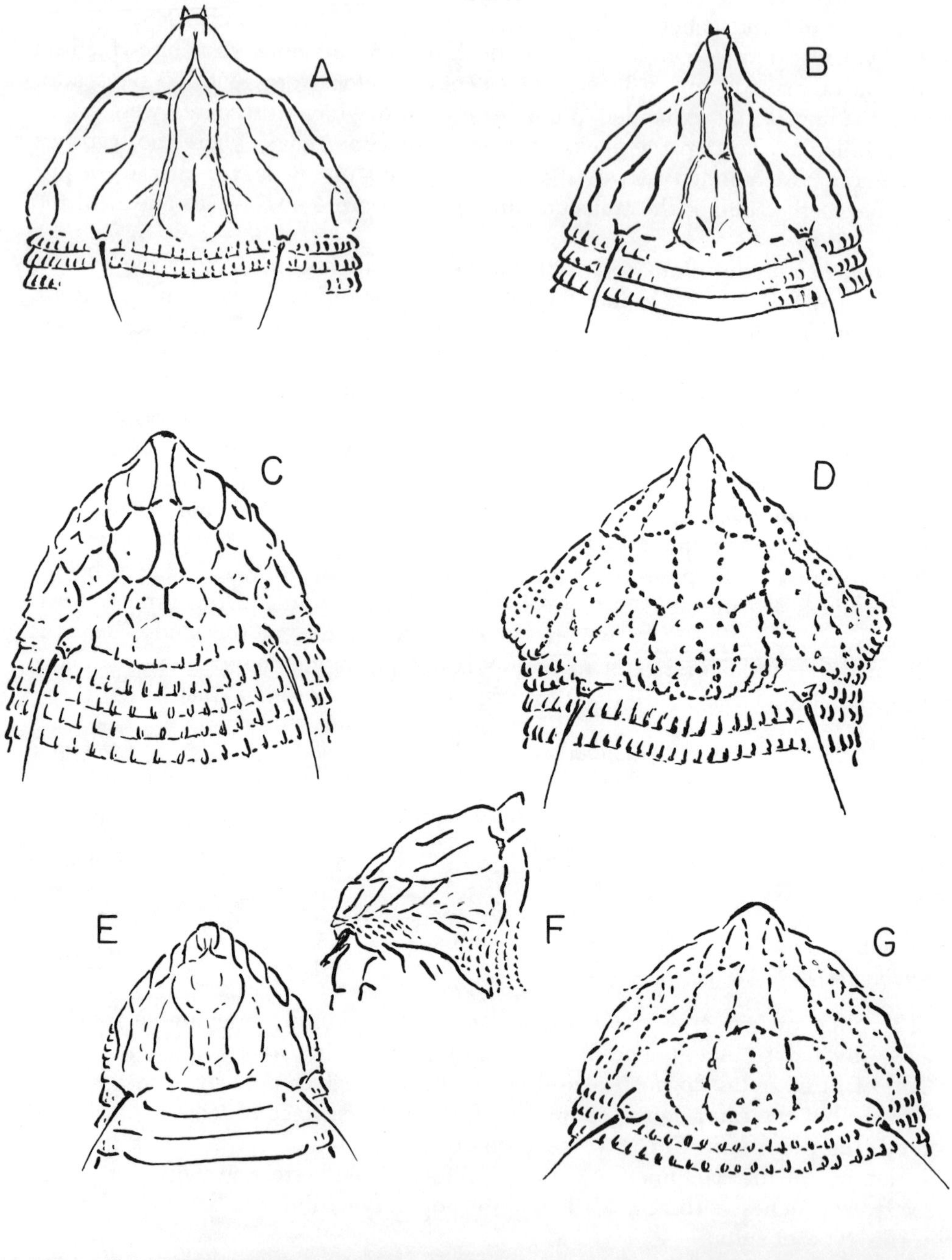

Fig. 131. Shield patterns of *Aculus* and *Aculops*: *a*, *Aculus cornutus* (Banks), shield; *b*, *Aculus ligustri* (K.), shield; *c*, *Aculops gleditsiae* (K.), shield; *d*, *Aculops lobuliferus* (K.), shield; *e*, *Aculops lycopersici* (Massee) shield from above; *f*, *Aculops lycopersici*, shield from side; *g*, *Aculops massalongoi* (Nal.), shield.

are characterized by this central longitudinal line over the anterior shield lobe.) The shield of *A. fockeui* has a fairly strong transverse line at ¼. The admedian lines, which are not very easily seen, start from the transverse line and run back divergently, being furthest apart when outside the median line. The abdomen has about 30 tergites and 50 sternites. The tergal microtubercles are elongate and rather faint. These ring granules on the sternites are elliptical and touch the margins. The female coverflap has curved transverse lines basally, and about 16 longitudinal subapical ribs.

The deutogyne is slightly smaller than the protogyne. It has narrower rings on the thanosome, especially dorsally. Dorsal microtubercles are obscured but shorter than on the protogyne. The deutogyne anterior shield lobe is thinner than on the protogyne and lacks the small pair of points.

The similarity of *A. fockeui* to *A. cornutus* causes some speculation as to the true relationship of these populations on plum and peach respectively. Examination of structure would tend to produce the conclusion that both kinds are the same species. Experiments are needed both with transplantations from host to host, and breeding tests exposing females form one host to spermatophores of males from the other host. Judging from examples received from the Mediterranean region, from plum and cherry, this type of mite has a considerable structural variation in the area. North American examples that have come from a relatively few importations show considerable uniformity in structure.

Peach silver mite, *Aculus cornutus* (Banks) (pl. 71; fig. 131, *a*), is injurious to peach, nectarine, and almond. It has a worldwide range on its hosts, the point of origin probably being the areas where peach originated. Some of the differences between this mite and *A. fockeui* are recited above. One of the characteristics of peach, of which the silver mite takes advantage, is the constant production of new leaves throughout the summer. This keeps the mite reproducing protogynes all of that time and so it can breed actively over a prolonged period.

The first type of damage that occurs to peach from the feeding of the silver mite is the yellow spotting of spring leaves. This is a viruslike damage. Accompanying this spotting is the upward longitudinal bending of the lateral edges of the leaves, producing a troughlike result. This troughing of the leaves is visible for some distance, and is usually much more general on trees that lack basal leaf glands. In that respect it is fairly easy in the spring to pick out seedlings without leaf glands, in an experimental planting, at a distance of fifty or more feet. Trees with active leaf glands, which can keep buds on embryonic leaves covered with sticky syrup, show much greater resistance to spring silver mite attack.

Peach leaf yellow spot was the subject of research by Wilson and Cochran (1952). They experimentally produced the spots, and found that silver mites could engender spotting in as short a period of feeding as 24 hours. The mites would only make these spots as the result of feeding on early-stage leaves; but when feeding on mature leaves the mites could not cause spots to develop. There was a delay of about 10 days between the mite feeding on embryonic leaves and spot appearance. Index grafting, and other manipulations, showed that no virus was involved.

Peach leaf silvering occurs on mature leaves and is the result of prolonged summer feeding by heavy mite populations. The general effect on peach trees owing to the feeding of this mite is to devitalize the trees, reduce fruit size, and cause premature fruit drop.

Peach silver mites overwinter as deutogynes principally around lateral buds along the terminal 6 to 8 inches of the twigs. The number of deutogynes below this terminal area will depend on the population. Deutogynes begin feeding on embryonic leaves in the spring, and the eggs they lay hatch into primary forms. New deutogynes do not appear on peach leaves until late September or October, after the tree has ceased developing fresh leaves.

A. cornutus is structurally about the same as the plum rust mite, as indicated above. Protogynes are from 190 to 205 μ long. Deutogynes of *A. cornutus* have narrower tergites and less conspicuous microtubercles than *A. fockeui*. The *A. cornutus* deutogynes have a suppressed shield pattern, they have a thinner anterior shield lobe, and this lobe lacks the pair of anterior points possessed by the protogyne.

Peach silver mites are susceptible to sulfur compounds. The mites are quite easily reached in summer with sprays or dusts because they live exposed on both leaf surfaces. Dormant or delayed dormant sprays, with lime-sulfur oil combinations, have generally controlled this mite for the season. Chlorobenzilate, diazinon, and other acaricides, have been used to control summer silver mite populations, when dormant or delayed dormant treatments were not applied, or were inadequate (Anthon, 1954; Webster, 1948).

Shagbark hickory rust mite, *Aculus caryfoliae* K. This mite (fig. 130, *j*) lives on the undersides of shag-bark hickory leaves on eastern North America, and can badly damage a tree. *A. caryfoliae* is a fusiform species, dull yellowish. Females are from 175 to 190 μ long. The featherclaw is 4-rayed. The shield is subtriangular, with a broad and rather blunt anterior lobe. This lobe has the anterior points, which are small and project under the edge. The median shield line is faint on the central part. Admedian lines are nearly complete, curving back from sides of the anterior lobe, meeting lateral lines at the lobe base, at ⅓, and at ½; the admedians fade to the rear on a wide central half disk that has rough rear edges at the rear shield margin. Submedian shield lines form "cells" beside the admedians. The abdomen has about 28 tergites and 60 sternites. Thanosomal tergites cover two sternites. Microtubercles on the tergites are elongate. Microtubercles on the sternites are elliptical. The female genital coverflap has basal transverse lines of granules, and about 12 longitudinal ribs subapically.

Apple rust mite, *Aculus schlechtendali* (Nal.). This species (=*Aculus malivagrans* K.; fig. 130, *i*) is a widespread pest of apples. In heavy infestations this mite damages terminal growth; the leaves curl lengthwise and become rusty brown, which gives the tree the appearance of being affected by drought. In the northwestern United States, apple damage is most common on young trees that do not receive a standard spray program.

While the structure of the deutogyne of this species has not been investigated, it is presumed that it is similar in this respect to the deutogynes of its congeners, which are *A. fockeui* and *A. cornutus.* The *schlechtendali* deutogynes find resting places under lateral buds not far below terminals, or in crevices on old wood. In the spring they move to opening buds, especially during the bloom period. Eggs are soon laid on the undersides of the leaves. The egg stage is variable in length, according to temperature, but eggs hatch in 10 days or less in the spring. There are several generations of apple rust mites during the year before deutogynes appear, and a protogyne may live as long as 30 days. Infestations increase in intensity until the hibernating females develop in July, when the mites diminish in numbers on the leaves.

Control of *A. schlechtendali* is best accomplished by prebloom sprays, followed by summer sprays. Effective prebloom sprays can have flowable sulfur, or lime-sulfur. Other useful acaricides for this mite are dithiocarbamate fungicides, Karathane, or Morestan. Summer spray applications have included thiodan, dicofol, Morocide®, and sevin (Westigard and Hoyt, 1965).

A. schlechtendali is 160 to 175 μ long in the protogyne stage. The body is fusiform and yellowish. The featherclaw is 4-rayed. The shield is subtriangular, with bulging sides. The anterior lobe is moderately acute. The shield on the protogyne has a pattern of granular lines. These granules show the median line is present only on the rear ⅓. The admedian lines are complete, sinuate, gradually diverging to the rear. Thanosomal tergites on the protogyne have rather faint but elongate microtubercles. Each tergite covers about 2 sternites. Microtubercles on the sternites are beadlike and on ring margins. The female genital coverflap has about 3 cross lines of granules, and 8 to 10 longitudinal subapical ribs.

The deutogyne is presumably smaller than the protogyne and should not only have a thinner anterior shield lobe lacking spinules, but an indistinct shield pattern. The tergites on the deutogyne should also be narrower than those on the protogyne.

The above protogyne characterization is from the description of *A. malivagrans.* This name is based on the granular lines on the protogyne shield, which were not indicated by Nalepa. It is now thought that *A. schlechtendali* actually has granular shield lines. Nalepa always listed this rust mite from both apple and pear. Examination of his original description indicates that he received his original collection of this mite from Schlechtendal, who reported he found these mites on both hosts. *Aculus schlechtendali* never appears on pear in California, so it is believed that the statement that the mite also occurs on pear is erroneous. Nalepa never reviewd Schlechtendal's original report to him to the effect that the mite lived on both hosts.

In checking the identity of *Aculus schlechtendali* on apple it is well to remember that there is another apple rust mite, *Calepitrimerus baileyi* K., which see. In California *A. schlechtendali* prefers higher coastal atmospheric humidities and does not come into the central valleys. On the other hand *C. baileyi* occupies the drier interiors. Investigation is necessary, however, to determine the actual North

American range of both of these mites. *C. baileyi* has dorsal tubercles on the shield which are located well ahead of the rear shield margin. *A. schlechtendali*, in contrast, has dorsal tubercles that are situated on the rear shield margin.

A leaf vagrant on apple reported originally as *Eriophyes malifoliae* Parrott was recorded from New York. For a discussion of the possible identity of this mite see *Cecidophyes malifoliae* (Parrott).

Hedge privet rust mite, *Aculus ligustri* (K.). This mite (fig. 131, *b*; pl. 72) is widespread on California hedge privet, but it does not attack tree privet. The mite causes spring browning and curling of leaves, and in severe infestations may produce leaf stunting. These mites live on both leaf surfaces. The life history of *A. ligustri* is recounted in the section on deuterogyny (chap. 12), including the habit of the species to retire to aestivation positions along lateral twig buds, in late spring, owing to temperature rise. This is the life cycle of the mite in central California, and is not necessarily indicative of its actions in another climate.

A. ligustri is a brownish-yellow, fusiform mite, with 4-rayed featherclaws. The shield is approximately triangular, with a moderately acute anterior lobe over the rostrum. Shield lines are distinct to a certain extent. The median line is present on the rear shield ½. Admedian lines are nearly complete, arising on the sides of the anterior lobe, describing gentle in-and-out curves, and reaching nearly to the rear margin. The admedians receive a cross line at about ¼, and a second at ½. The abdomen has about 36 tergites, with faint and elongate microtubercles, and about 66 sternites, with beadlike microtubercles. The female genitalia have 1 or 2 basal transverse curved lines, and about 9 longitudinal ribs. The protogyne is 185 to 200 μ long; the male is 130 to 150 μ long.

The *A. ligustri* deutogyne, as it occurs along the twigs in late winter, is about the same size as the protogyne, but browner. It has reduced microtubercles; none are present on the tergites. Females examined in late spring migrations to aestivation positions show various intergrades between typical protogynes, and deutogynes. Late winter survivors, however, found along twigs behind lateral buds, have rather uniformly reduced microtubercles and thin shield lobes.

The privet rust mite is similar to the peach silver mite (*A. cornutus*) in external structure. But the silver mite has united admedian lines on the anterior lobe, whereas on *A. ligustri* they are separate.

Aculops Keifer

This large "wastebasket" genus consists mostly of rust mites and leaf vagrants. It is chiefly characterized by having dorsal shield tubercles with transverse axes on the rear margin, which direct the setae divergently to the rear. It differs from *Aculus* by having a more acuminate anterior shield lobe that often comes to a single point. This lobe always lacks the small anterior-projecting spinules. Species referable to this genus have variable thanosomes, many species showing strong tergal-sternal differentiation. Many species have subequal tergal-sternal ring differentiation, resembling *Eriophyes* in that feature. *Aculus* spp. have more consistent tergal-sternal ring differentiation than *Aculops*. *Vasates* has the dorsal tubercles on the rear margin, but the tubercle axes converge diagonally ahead, and direct the dorsal setae diagonally toward each other to the rear.

Lilac rust mite, *Aculops massalongoi* (Nal.) (fig. 131, *g*), is a widespread rust mite occurring on the underside of lilac leaves. This mite is atypical in *Aculops* as it has a much reduced anterior lobe on the shield, and the abdomen has little tergal-sternal ring differentiation. The body is strongly fusiform. The female of *A. massalongoi* has a 4-rayed featherclaw and is 170 to 185 μ long. The shield design is a network of lines and granules, especially toward the rear. There is a prominent cross line of ⅔. The shield laterally has a granular band above the coxae. The dorsal setae are of moderate length. The tergal thanosomal microtubercles are longer than the sternal microtubercles, with the sternal ones slightly ahead of the ring margins. Rear abdominal microtubercles tend to be pointed over ring margins. The female genital coverflap has 3 or 4 transverse lines of granules basally, and 12 to 14 longitudinal ribs.

Tomato russet mite, *Aculops lycopersici* (Massee). This species (fig. 131, *e*, *f*) lives on leaves of many solanaceous plants. It is a particular pest of tomato and usually kills this host; but heavy infestations may cause injury to pepper, potato, tobacco, and petunia. This species was originally described as it occurs in Australia, but is now known to be nearly worldwide.

Tomato plant injury by *A. lycopersici* usually starts at the base of the plant and spreads upward over the stems and leaves, with the heaviest mite populations always ahead of the severest injury. Mite feeding on tomato produces bronzed or russeted aspects to the stems, which often crack longitudinally, or check. Leaves turn brown and paperlike but do not wilt down, as they do as a result of some tomato diseases. Some of the injured leaves may drop, even before turning color. Fruit usually does not russet, but often sunburns when leaves drop.

Potato does not show as much stem bronzing from mite feeding as does tomato, but potato leaves may dry, and the plants may die from the infestations. Eggplant supports heavy populations of *A. lycopersici*, but the plants remain alive. Eggplant damage is usually leaf distortion and crinkling, similar to 2, 4-D damage. Eggplants with an abundance of small glandular epidermal hairs do show russeting.

Black nightshade serves as a survival host in waste places in the vicinity of tomato fields. The mites are particularly able to carry over winters on nightshade growing close to abandoned buildings. Morning-glory or bindweed growing in tomato fields harbors the mite, and bindweed leaves appear dry, with silver sheen on both surfaces. The leaves may fall without russeting.

Tomato russet mites spread rapidly at picking time, being brushed off from pickers' clothing. Insects and wind also spread the mites. These mites live and feed on all green surfaces, and can exist in open sunlight. The species is a warm weather species and has no dormant stage. Winter cold will kill it except in protected places. Subfreezing temperatures kill all stages within a few hours or days. Survival by *A. lycopersici* is possible in the most southern areas of the

United States, and in greenhouses.

The most favorable conditions for development of the russet mite are near 26.5 C (80 F), and 30 percent relative humidity. As temperatures increase, the relative humidity must decrease to provide suitable developmental conditions. Light, or photoperiod, has little effect on the growth of this mite.

Under optimum conditions the incubation period for *A. lycopersici* is 2 days. The first instar nymph lasts 1 day; the second instar 2 days, and the total life cycle may be completed in 6 or 7 days. A female may lay up to 53 eggs during her lifetime. Females prefer to place eggs adjacent to some type of natural protection, such as leaf hairs, in small crevices, or next to a leaf vein. The first nymph moves little during its development, but activity increases until egg-laying begins. Laying females become more sedentary.

A dust containing 50 to 75 % sulfur, applied at the rate of 10 lbs actual sulfur per acre provides protection for 3 or 4 weeks. More than 1 application may be necessary when an infestation occurs early in the season (Anderson, 1954; Bailey and Keifer, 1943; Rice and Strong, 1962; Smith, 1955; Harpaz, 1955).

A. lycopersici structurally is unlike other species in the genus. The broad and short anterior shield lobe, which is precipitous anteriorly, is quite atypical for the group. This mite is an orange-yellow fusiform species, and rather robust. The featherclaw is 4-rayed. Females vary from 150 to 180 μ long. The blunt anterior shield lobe is topped by a transverse line that extends back along the sides to the rear margins. Laterally this line has several lower branches that form lateral "cells" above the coxae. The median shield line is faintly present on the rear ⅔. Admedian lines arch back from the anterior transverse line, curve inward at ½, extend toward the rear subparallel to each other on each side of the median line, and end by flaring out toward the dorsal tubercles. The dorsal setae are short. The abdomen has about 27 tergites and 60 sternites. Microtubercles are faint and elongate on tergites, but beadlike and pointed on the sternite margins. The female genitalia have faint basal granules and about 10 weak longitudinal ribs.

Olive yellow spot mite, *Aculops benakii* (Hatzinikolis). *A. benakii* lives on the underside of olive leaves where it inserts itself under the stellate hairs, causes these stellate structures to drop off, making yellow leaf spots. This mite is only known in Greece, but undoubtedly occurs elsewhere in the Mediterranean area. *A. benakii* is a fusiform, pinkish species, represented as having a network shield design and a long acuminate anterior shield lobe over the rostrum. The tergites are shown as broad, lacking microtubercles, and each covers about 4 or 5 sternites. Female length is given as from 165 to 175 μ. The featherclaw is 4-rayed. The shield design as drawn shows a median line on the rear ¼. There is a cross line at about ⅓, and a rear transverse line curving back from in front of the dorsal tubercles and crossing the shield just ahead of the rear margin. There are numerous lateral shield "cells." The dorsal setae are 21 μ long. The abdomen has about 20 tergites and 68 sternites. The microtubercles on the sternites are represented as beadlike on the ring margins. The telosomal rings are shown as completely microtuberculate, and thus contrast with the thanosome. Female genitalia have transverse basal lines, and about 14 longitudinal ribs on the coverflap.

Cottonwood rust mite, *Aculops lobuliferus* (K.) (fig. 131, *d*). This is one of the numerous species of leaf vagrants and rust mites which occur on various species of poplar and cottonwood. The precise host is the eastern cottonwood, and the mite was found damaging nursery grown young trees. This poplar can be called eastern cottonwood. The mite is a fusiform species with the female 180 to 205 μ long. The color of these mites is dull yellow, and the featherclaw is 4-rayed. The anterior shield lobe is acutely pointed over the rostrum. The general shape of the shield is triangular, with prominent lateral lobes projecting past the general body width. The shield design is a network, mostly of lines of granules. The microtubercles are strong on the tergites, being somewhat elongate. On the sternites the microtubercles are beadlike. The female genitalia have 3 basal transverse lines, and there are 12 to 14 longitudinal ribs on the coverflap.

The honey locust rust mite, *Aculops gleditsiae* (K.). This mite (fig. 131, *c*) causes severe rusting on young honey locust leaves in various areas of the eastern United States, and heavy populations of this rust mite have been associated with dying trees. The mite is a yellowish, fusiform species, with 5-rayed featherclaws. The female is from 175 to 190 μ long. The anterior shield lobe is moderately narrow and terminally rounded off. The shield design is a network of branching lines with several cross lines. The median line is present on the rear ⅓. The abdominal thanosome has strong differences in width between the tergites and sternites, laterally, with microtubercles elongate on the tergites, and beadlike on the sternite ring margins. The telosome is evenly ringed. Female genitalia have about 12 longitudinal ribs on the coverflaps.

Pink citrus rust mite, *Aculops pelekassi* (K.). This mite (fig. 132) is widespread, occurring in southern Asia (including Japan and Siam), in Italy and Sicily, in Florida, and in Brazil. It is potentially capable of causing more damage to its hosts than is the citrus rust mite (*Phyllocoptruta oleivora* Ashmead). Pink rust mites not only cause russeting of fruit and leaves, but also mild to severe distortion of new growth, brown lesions on lower surfaces and along midribs of immature leaves, and may produce mesophyll collapse, chlorosis, and leaf drop. The general appearance, biology and host injury is somewhat similar to that of the rust mite (*P. oleivora*). Damage by the pink rust mite may cause symptoms that can be confused with some virus diseases, such as infectious variegation.

Populations of the pink mite build up on the new growth of citrus. These mites seem to prefer "loose skin" oranges, mandarins, and clementines.

Eggs of *A. pelekassi* are a translucent white, and are laid indiscriminately over leaf and fruit surfaces. These eggs are large for the size of the mite, occupying about ⅕ the body of the female. The ratio of males to females is relatively low, being about 10 to 12 percent. Reproduction takes place throughout the year, the life cycle is short, and many generations develop annually.

This mite shows a definite response to light, especially to fluorescent light. This light stimulates more rapid movement, and causes the mites frequently to stand upright on their terminal lobes.

Distinguishing characteristics for separating *P. oleivora* from *A. pelekassi* (Burditt, Reed, and Chittenden, 1963) are several:

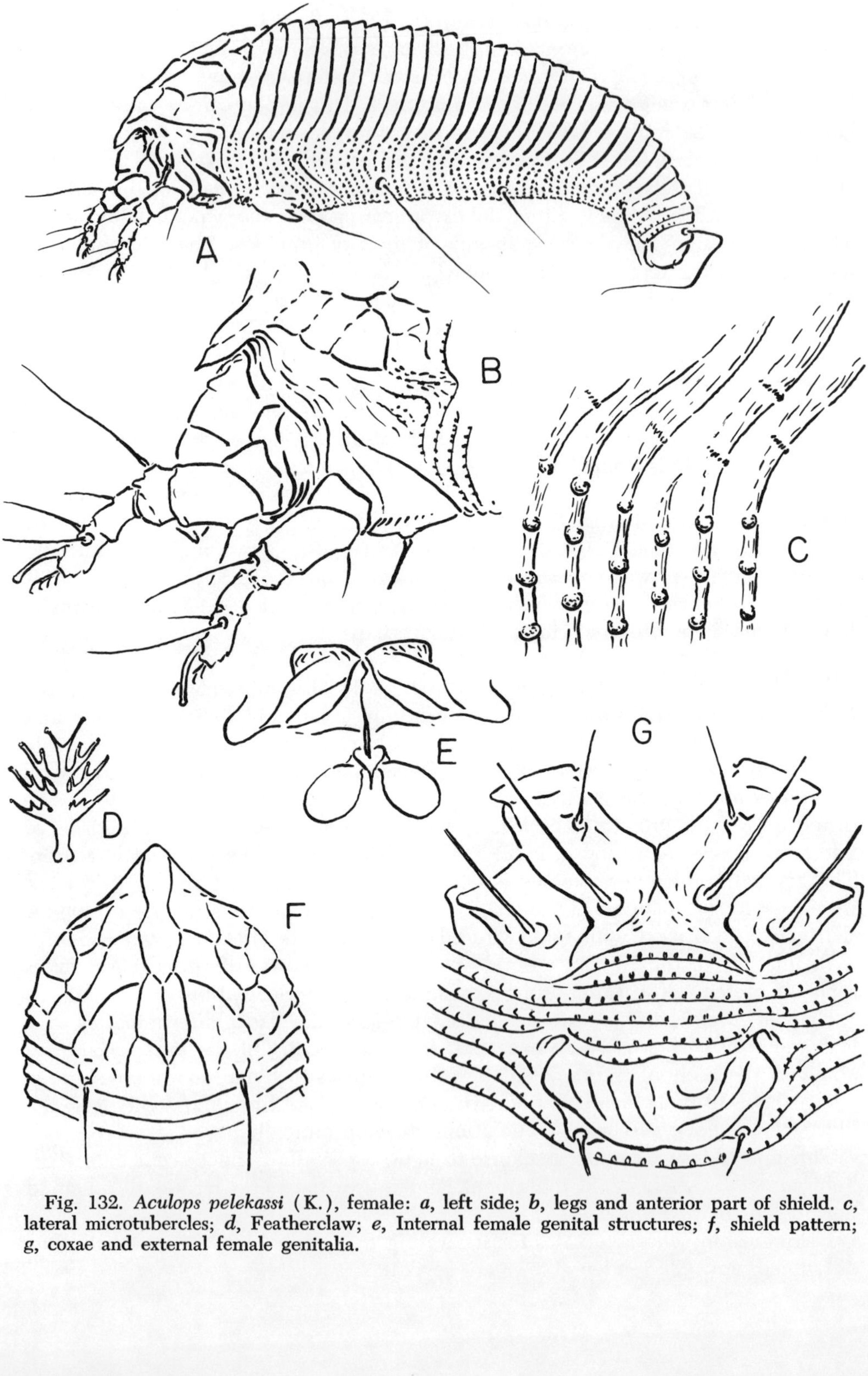

Fig. 132. *Aculops pelekassi* (K.), female: *a*, left side; *b*, legs and anterior part of shield. *c*, lateral microtubercles; *d*, Featherclaw; *e*, Internal female genital structures; *f*, shield pattern; *g*, coxae and external female genitalia.

As seen under 10 X to 30 X magnification

P. oleivora	*A. pelekassi*
Usually yellow in color	Usually pink in color
Eggs opaque	Eggs transulcent white
Eggs laid in hollows along midrib	Eggs scattered indiscriminately
Mites distribute themselves over leaf surfaces	Mites congregate at leaf edges so are easily distributed by wind
Only occasionally stand on terminal lobes	Frequently stand up on terminal pads
Mites favor mature leaves	Populations build up on new growth
Show no definite response to light	Light (esp. fluorescent) stimulates activity

As seen under phase contrast illumination at about 600 X magnification

P. oleivora	*A. pelekassi*
Dorsal shield setae arising from tubercles well ahead of rear shield margin and projecting up; these setae 9μ long.	Dorsal setae arising from tubercles on rear shield margin; setae projecting divergently to rear and 9μ long.
Abdominal dorsum with broad longitudinal trough, flanked on each side by longitudinal ridge.	Dorsum of abdomen evenly arched in cross section. No longitudinal trough or ridges present.

A. pelekassi is a fusiform mite, pinkish when alive, with a 4-rayed featherclaw. Females are from 140 to 155 μ long. The shield is subtriangular in anterior outline, with bulging sides, and with moderately acute but terminally rounded, anterior lobe over the rostrum. The shield design is a more-or-less obscure network of lines. The median line is present on the rear ½ or ⅔, and receives transverse lines at the ⅓ and ⅔ positions. The line at ⅓ extends laterally around to the rear shield margin. A subparallel line, below the transverse line at ⅓, runs from the side of the anterior lobe to the shield margin above the coxae. The admedian shield lines are complete, extending from the sides of the anterior lobe to the rear margin, somewhat sinuate, and furthest apart at the rear. These admedians also receive the cross lines at ½ and ⅔, which connect them with the median. The abdomen has about 36 tergites, and 50 sternites. Microtubercles on the tergites are faint, but are beadlike on the sternites, and rest on the ring margins. The female genitalia has 2 basal transverse lines with granules; the longitudinal ribs may take 2 forms, either about 10 longitudinal ribs, or transversely curved ribs.

Abacarus Keifer

The genus *Abacarus* contains rust mites that have a longitudinal dorsal thanosomal trough, formed by a subdorsal ridge on each side, and with a central longitudinal ridge that ends *before* the trough ends. This back structure is combined with dorsal shield tubercles that sit on the rear margin and direct the dorsal setae to the rear. The genus *Tegolophus* differs from *Abacarus* by having a middorsal ridge that does not end before the subdorsal ridges end. Various members of

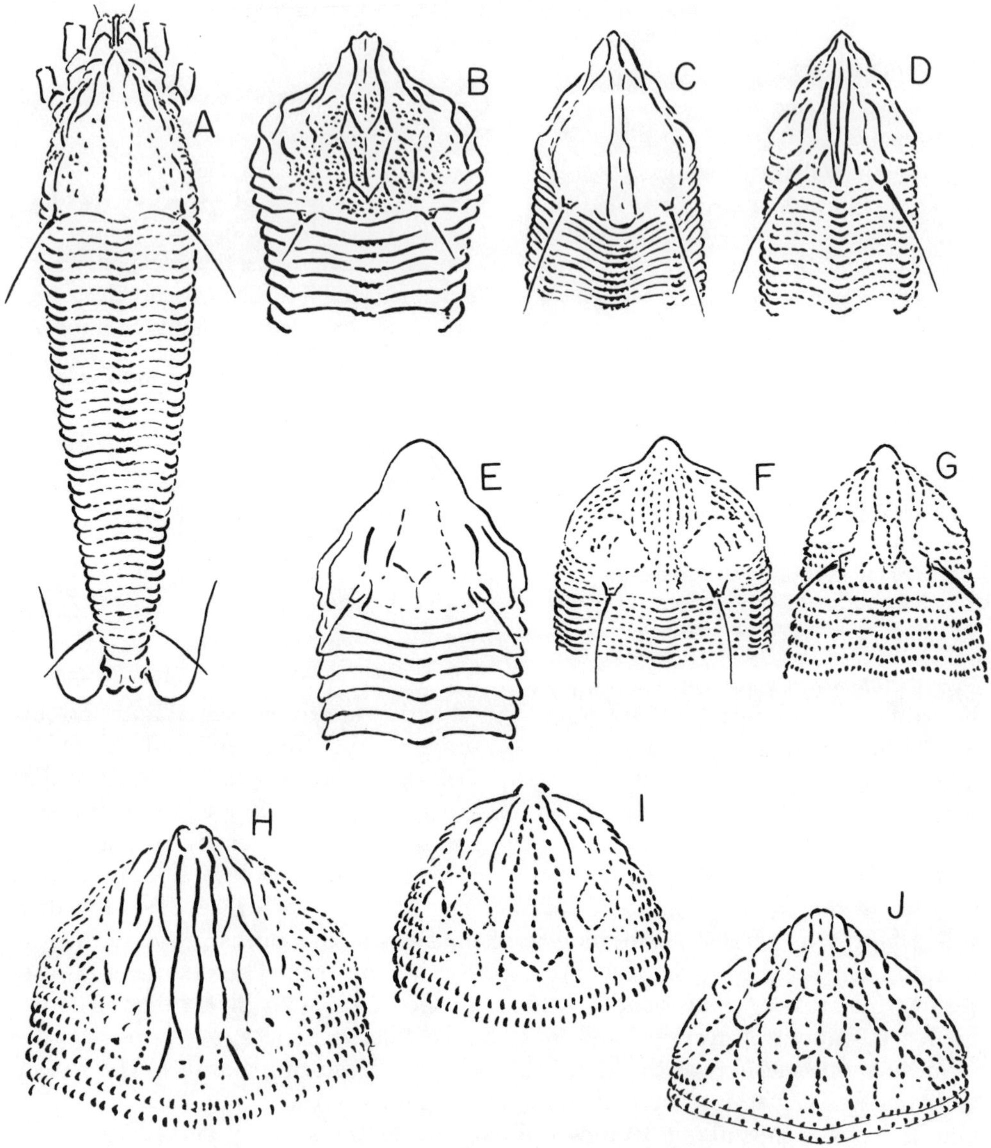

Fig. 133. Shield patterns of *Abacarus, Tegolophus, Cecidophyopsis,*and *Cecidophyes*: *a, Abacarus sacchari* Channabasavanna, dorsal view of mite; *b, Abacarus afer* K., shield; *c, Abacarus hystrix* (Nal); *d, Abacarus oryzae* K., dorsal view of anterior section; *e, Tegolophus hassani* (K.), dorsal view of anterior section; *f, Tegolophus myersi* (K.), shield; *g, Tegolophus perseaflorae* K., dorsal view of anterior section; *h, Cecidophyopsis selachodon* Van Eyndhoven, shield; *i, Cecidophyopsis psilaspis* (Nal.), shield; *j, Cecidophyes lyrata* K., shield.

these genera produce dorsal wax, usually along the ridges, but this wax is not the basis for the genera.

Cereal rust mite, *Abacarus hystrix* (Nal.). This species (pl. 73; fig. 133, *c*) causes injury to its grass hosts not only by its feeding, but also because it is able to transmit rye grass mosaic virus. This mite lives on perennial grasses for the most part and its hosts are English rye grass, Italian rye grass and cocksfoot, oats, barley, wheat, rice, meadow fescue, and timothy hay. There are additional grass hosts, and a short list of grass genera that have species that harbor this mite are: *Agropyron*, *Elymus*, *Festuca*, *Lolium*, *Phleum*, *Poa*, and *Triticum*.

Abacarus hystrix is a widespread species, being Holarctic in part. The virus disease it transmits occurs in England, elsewhere in Europe, and in North America.

Injury to timothy hay first appears as discolored areas, owing to mite feeding, which are at the tip of the plant where the mites congregate. On rye grass the mites and the virus disease reduce size, vigor, and yield. Severe infestations prevent blade tips from unfolding, which deforms the plants, more-or-less stops development, or even causes the new growth to die back. Virus-infected grasses generally show green-to-yellow mottling and streaking, but leaves of some rye grass strains develop brownish necrosis.

Adult *A. hystrix* females overwinter on fresh tissue at the tip of the plant. In the spring they attack tender timothy tips. At this time they congregate on the grooved upper blade surfaces; very few of the mites are on the smooth undersurfaces or on older leaves. Eggs are deposited in the grooves. Females may lay from 51 to 81 eggs during a 6-day period. Eggs hatch in 5 or 6 days at room temperature.

All active stages of this mite can acquire sufficient virus in two hours of feeding on diseased plants to enable them to infect healthy rye grass. However, feeding periods up to 12 hours result in a greater proportion of transmission. The mites lose their infectivity after two hours feeding on immune grass.

Cutting the first hay crop reduces the mite population. Exposure to hot weather following cutting causes the number of these mites to remain low until late August or early September, after which there is generally a rapid population increase. Burning stubble in September fails adequately to reduce mite population (Mulligan, 1960; Pepper, 1942).

A. hystrix is a slender mite, about 180 to 200 μ long. It typically lies in upper surface blade grooves, and is more common in warm months. In these grooves it is associated with *Aculodes mckenziei* (K.), but this latter mite is most numerous in winter, when the *A. hystrix* population is down. *A. hystrix* is recognizable by the dorsal longitudinal ridges that bear wax in the form of stripes. *Aculodes mckenziei*, although it has a powdery covering of fine wax, has no dorsal longitudinal ridges.

Coffee mite, *Abacarus afer* K. (fig. 133, *b*), is a rust mite occurring on coffee leaves in Zaire. Its effect is unknown except that it is an undersurface mite. It differs

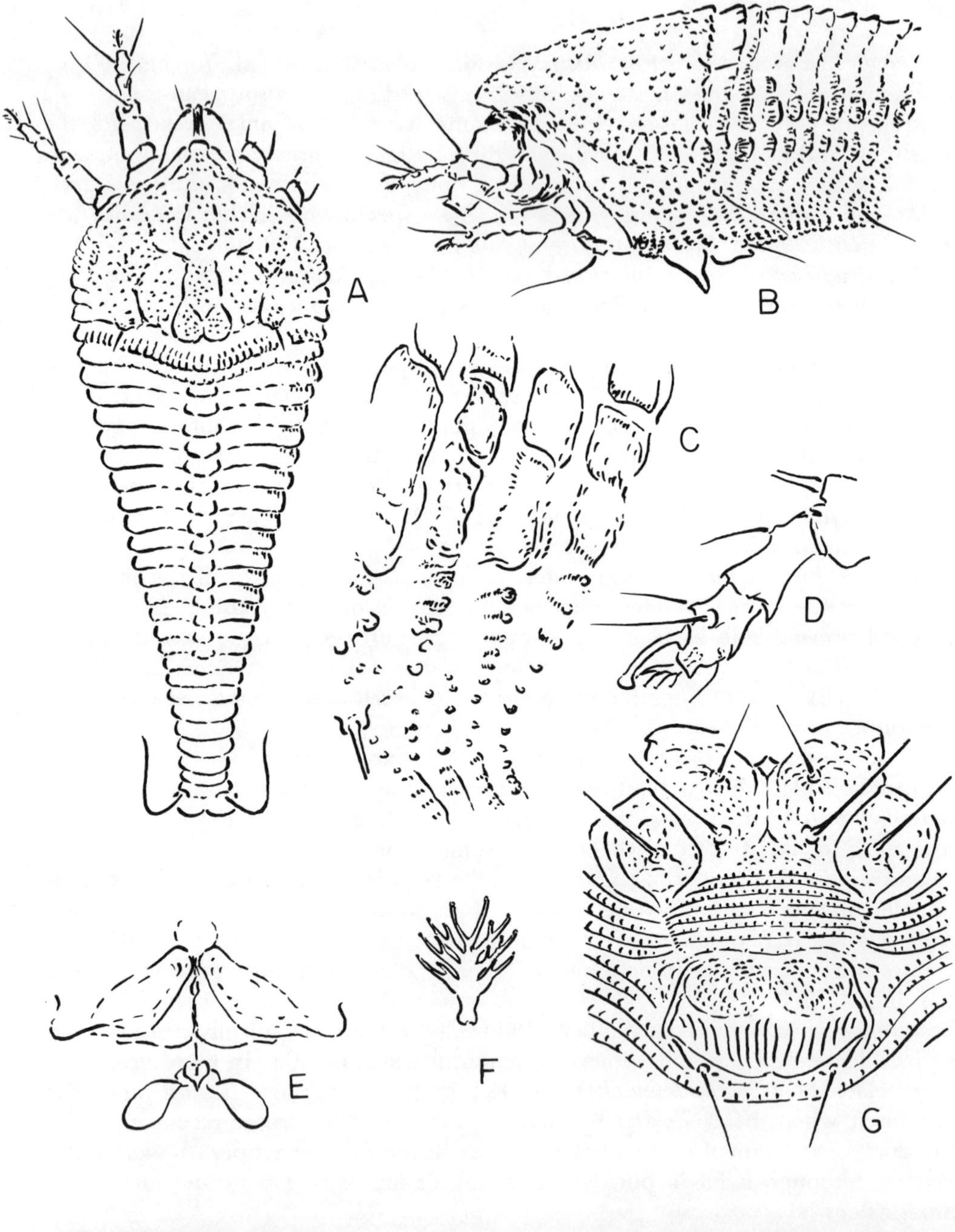

Fig. 134. *Tegolophus australis* K., female: *a*, dorsal view of mite; *b*, lateral view of anterior section; *c*, side skin; *d*, foreleg; *e*, internal female genitalia; *f*, featherclaw; *g*, coxae and external genitalia.

from *A. hystrix* in several details. The anterior shield lobe has two emarginations, forming three projections; the median line is strong and nearly complete; the featherclaw is 5-rayed (*A. hystrix* has an 8-rayed featherclaw); the shield surface is generally granular.

Rice rust mite, *Abacarus oryzae* K. This species (fig. 133,*d*) occurs in the Philippines. It is often associated with plants dwarfed by a disease known in the area as "tungo," but the precise relationship of the mite to the disease is unknown. *A. oryzae* is similar to *A. hystrix*, but *A. oryzae* has a 9-rayed featherclaw. The shield is pointed anteriorly, and the median shield line is nearly complete, with the admedians closely parallel to it.

Tegolophus Keifer

This genus contains species with dorsal tubercles situated on the rear shield margin, and directing the dorsal seta to the rear.

Olive rust mite, *Tegolophus hassani* (K.). Hassan's olive rust mite (fig. 133, *e*) lives on both surfaces of olive leaves in Egypt, and apparently causes russeting and some form of leaf deformation or defoliation. This is a fusiform rust mite, dull yellow, with no dorsal microtubercles on the tergites. The females range from 160 to 180 μ long. The shield is subtriangular, with a broad, rounded-off anterior lobe over the rostrum. The shield design is obscure, showing faint lines. The featherclaw is 4-rayed.

Brown citrus mite, *Tegolophus australis* K. This mite (fig. 134) occurs in coastal New South Wales and in Queensland, Australia. In New South Wales it is a minor pest of oranges and grapefruit, causing disfiguring blemishes on the fruit. The damaged spots are smooth and brown, and characteristically are located on the outer parts of the tree; where the blemishes are smooth, the skin appears as if polished with dark tan boot polish. In New South Wales, *Phyllocoptruta oleivora* (Ashmead), the citrus rust mite, prefers more sheltered inside fruit and inner fruit surfaces. Injury caused by *P. oleivora* is slightly rough and greyish brown, with edges of the area deeper brown (Hely, 1968).

T. australis is a rough fusiform mite, with brownish appearance. The female is from 180 to 190 μ long. The cephalothoracic shield is coarsely pitted, and the anterior lobe moderately acuminate over the rostrum. The rough surface of the shield has largely obliterated the basic design, although admedian lines are faintby discernible. The rear edge of the shield between the dorsal tubercles is partially furrowed. The dorsal tubercles are small and the setae very short, not reaching across the first tergite. The featherclaw is 4-rayed. The middorsal thanosomal ridge is narrow; the first tergite has longitudinal lines. There are about 22 tergites, which are broken laterally. The sternites number about 55, and the more ventral microtubercles are small and elongate. The basal part of the female genital coverflap is a pair of subcircular areas containing short dashes; there are about 16 longitudinal ribs on the coverflap.

Bud and flower avocado mite, *Tegolophus myersi* (K.). This mite (fig. 133, *f*) occurs in southern California, and undoubtedly in tropical America. It lives in buds and under fruit buttons. It blackens fruit to some extent and in buds discolors inner bud scale surfaces. *T. myersi* is from 155 to 170 μ long. It is an elongate-fusiform mite, with little differentiation between tergites and sternites. The shield has a small anterior lobe over the rostrum; the shield design is of lines of granules centrally, with broad granular areas laterally. The dorsal setae are long and directed divergently to the rear. The featherclaw is 5-rayed. The abdominal microtubercles are fine, elliptical, and touch ring margins. The accessory setae are present. The female genital coverflap has a few transverse lines of granules basally, and about 18 longitudinal ribs.

Tegolophus perseaflorae K. This species occurs in Brazil on avocado. It lives in flowers, where it reduces fruit production. It is similar morphologically to *T. myersi.* The shield design on *T. perseaflorae* has admedian lines (of granules) that meet near the rear margin and have a lateral branch from each line. In this respect it differs from *T. myersi.*

Tetraspinus Boczek

The characters of this genus associate it with *Aculus.* These features are the fine spinules projecting ahead from the anterior shield lobe, and backward-directed dorsal setae. *Tetraspinus* differs from *Aculus* by having a longitudinal dorsal trough.

Pepper rust mite, *Tetraspinus capsicellus* (K.). *Capsicellus* (fig. 128, *h*) is known only in Venezuela. It lives on all green surfaces of its host, but prefers the upper side of leaves. It damages leaves, stems, shoots, and flowers. *T. capsicellus* is a flattened, fusiform species, and is yellow. Females are from 195 to 210 μ long. The featherclaw is 5-rayed. The shield has a moderately acute anterior lobe, rounded off, and several small points projecting ahead from under the front edge. The shield design is a network of thickened central lines: the median line is present on the rear ⅔; the admedians are nearly complete, arising from the sides of the anterior lobe and turning together just ahead of the rear shield margin. There are lines lateral to the admedians, and the side of the shield has a row of "cells." The dorsal setae are directed somewhat diagonally inward to the rear. The thanosome has a moderately wide anterior trough with irregular microtuberculation. The sternites do not definitely begin until a little below the lateral ends of the tergites. Sternal microtubercles are beadlike and slightly ahead of the ring margins. The female genital coverflap design is weak, the ribs are irregular and about 10 in number.

THE RHYNCAPHYTOPTIDAE KEIFER

Rhyncaphytoptus Keifer

Members of this genus of big-beaked mites have simple featherclaws, and evenly arched thanosomes.

The fig leaf mite, *Rhyncaphytoptus ficifoliae* K. This (fig. 135, *a-g*) is a brownish mite that lives as a vagrant among the undersurface leaf hairs. It was first found in California but probably has a wide distribution; reported from India, it undoubtedly occurs widely in the Mediterranean region. Immature stages of this mite cover themselves with white flocculent wax, a characteristic of various

Fig leaf mite, *Rhyncaphytoptus ficifoliae* K. (fig. 135, *a-g*), is a brownish mite that lives as a vagrant among the undersurface leaf hairs. It was first found in California but probably has a wide distribution; reported from India, it undoubtedly occurs widely in the Mediterranean region. Immature stages of this mite cover themselves with white flocculent wax, a characteristic of various members of this family. It is of no known economic importance.

Rhyncaphytoptus ficifoliae is about 180 to 195 μ long. It is elongate-fusiform and the body is curved. Adults occur in various stages of light amber to brown. The featherclaws are 6-rayed. The shield is nearly without design the admedians are indicated only as opposing arcs toward the rear. Thanosomal tergites are much wider than the sternites. The tergal microtubercles are present only as fine lateral points on the margins. The narrower sternites have beadlike microtubercles, or ring granules. The female coverflap is unmarked.

Sycamore big-beaked leaf mite, *Rhyncaphytoptus platani* K. This mite (fig. 136, *c*) is a sycamore leaf vagrant that is occasionally abundant on the underside of plane tree leaves. The species is credited with some undersurface leaf browning in California. Hall (1967) reports that the species occurs in Kansas, as well as California. *R. platani* is a somewhat wormlike mite with females from 190 to 220 μ long. The shield has a blunt, or even somewhat concave anterior edge. Shield lines are well separated, with the median line complete. The admedian lines extend back to the vicinity of the dorsal tubercles. The featherclaw is 7-rayed. Abdominal tergites are entirely microtuberculate, the tergites slightly wider than the sternites. The sternites are also entirely microtuberculate. The female coverflap has no ribs.

Elm leaf vagrant, *Rhyncaphytoptus ulmivagrans* K. (fig. 136, *b, f*), is a widespread elm leaf mite that is generally distributed in Europe, and also lives on imported elms in California. It does not seem to do any damage to elm leaves. The life history of the species has already been discussed in the section on deuterogyny (chap. 12). The habits of the deutogyne limit the active summer life of the mite to June, July, and early August. Deutogynes retire to twig depressions for hibernation. The females of *R. ulmivagrans* are from 160 to 180 μ long. The body is robust, but tapers caudally. In life the mites are light amber, although the deutogyne may assume a reddish-brown aspect. The shield is partially acuminate anteriorly, but the anterior lobe over the rostrum is blunt. The median shield line is obsolete. The admedian lines are complete, widespread, and somewhat diverging. About 3 lateral lines branch from the outer side of the admedian. Dorsal setae are short, and point forward. The thanosomal tergites are considerably wider than the sternites. Microtubercles on the tergites are limited to being only points on the lateral angles. The sternites are completely set with

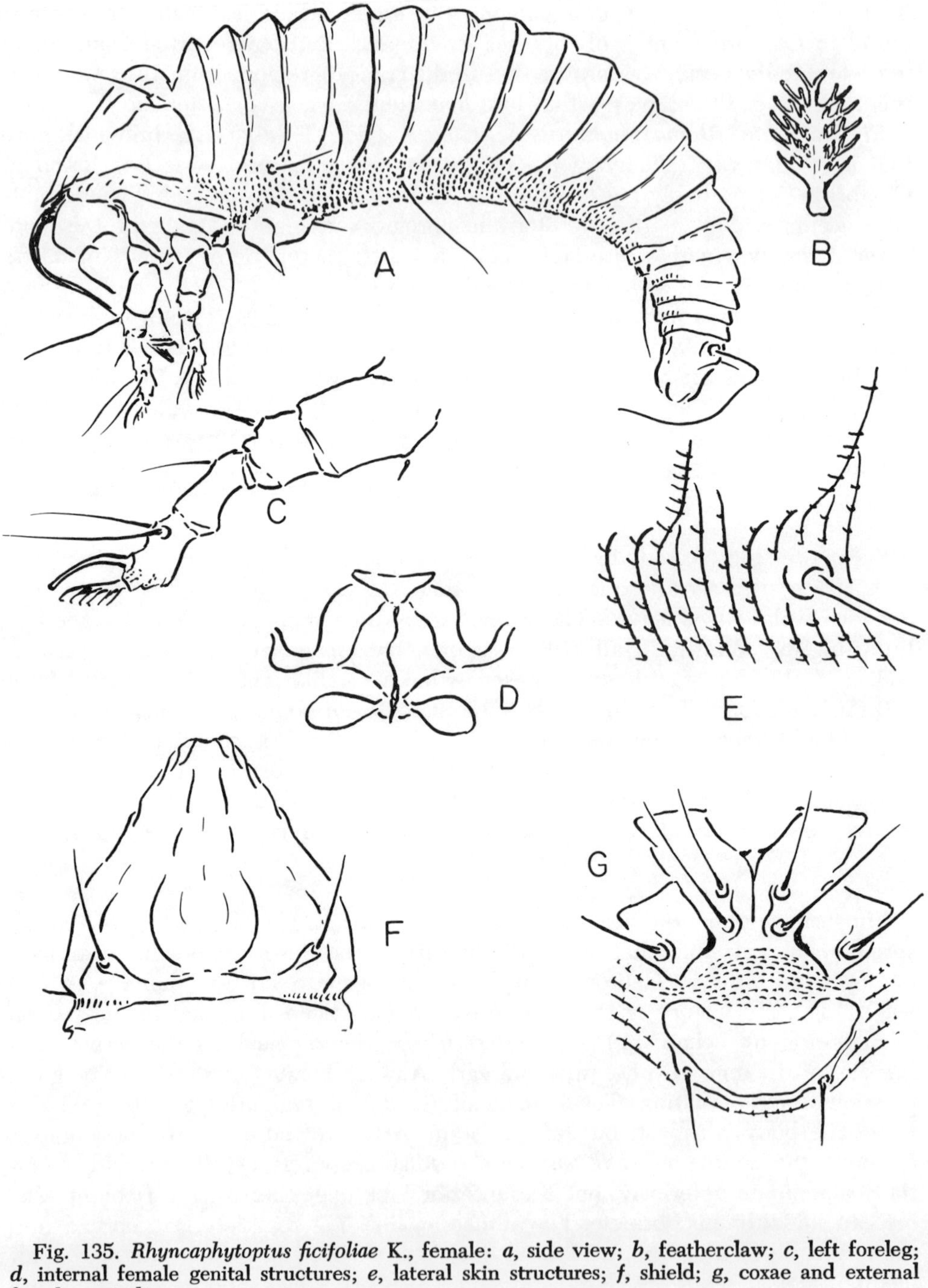

Fig. 135. ***Rhyncaphytoptus ficifoliae*** K., female: *a*, side view; *b*, featherclaw; *c*, left foreleg; *d*, internal female genital structures; *e*, lateral skin structures; *f*, shield; *g*, coxae and external female genitalia.

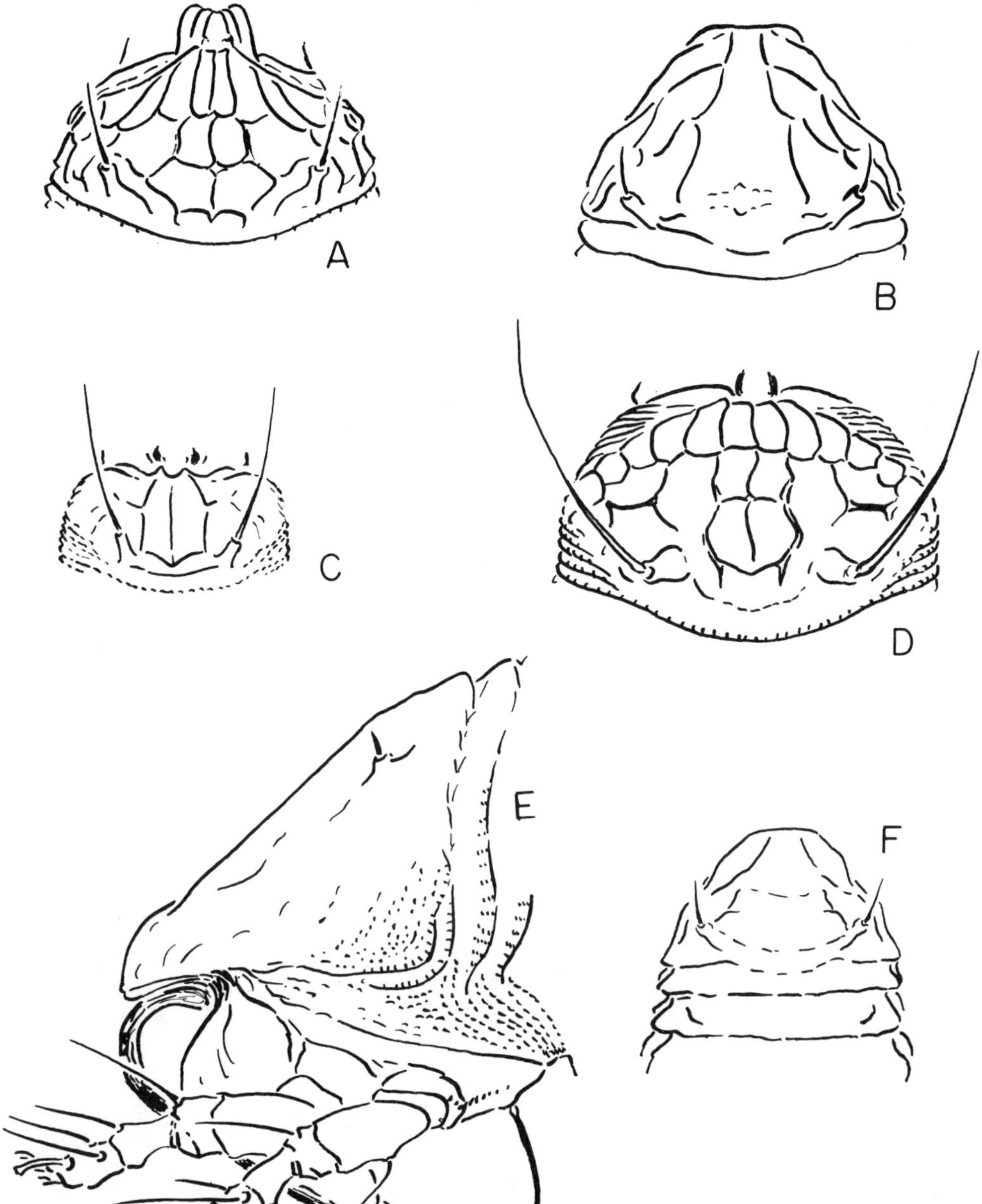

Fig. 136. Rhyncaphytoptid shields: *a, Rhyncaphytoptus atlanticus* K., shield; *b, ulmivagrans* K., shield of protogyne; *c, platani* K., shield; *d, amplus* K., shield; *e, Catarhinus tricholaenae* K. side of anterior section; *f, R. ulmivagrans* K., shield of deutogyne.

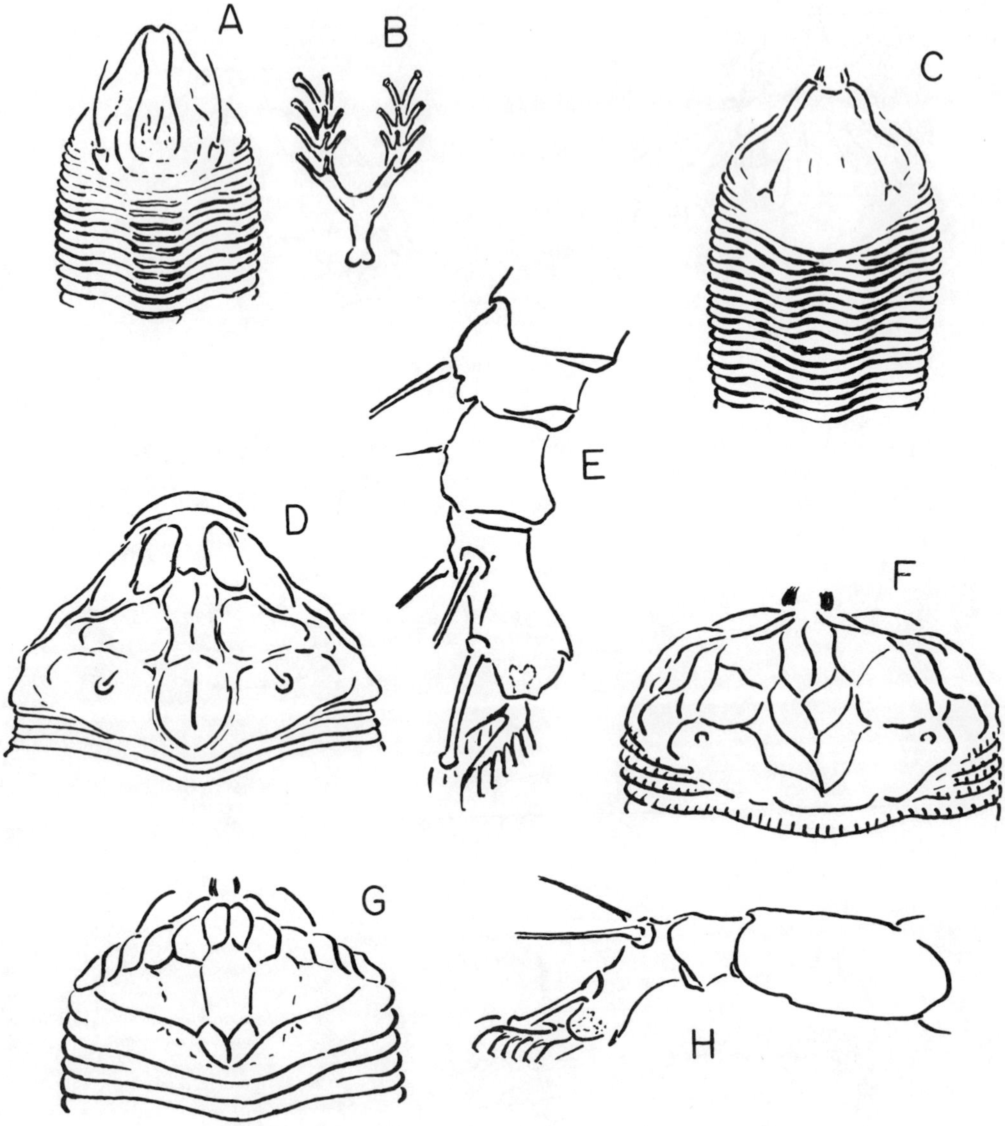

Fig. 137. Shield, featherclaws, body and leg characters: *a, Apodiptacus cordiformis* K., shield and anterior part of abdomen; *b, cordiformis*, featherclaw; *c, Trimeroptes aleyrodiformis* (K.), dorsal view of anterior part of body; *d, Diptacus swensoni* K., shield; *e, Rhyncus kraussii* K., foreleg; *f, krausii*, shield; *g, Diptilomiopus davisi* K. shield; *h, davisi*, foreleg.

elliptical microtubercles that touch ring margins. The female coverflap is unmarked. The featherclaw is 5-rayed.

The *R. ulmivagrans* deutogyne is about the same size as the protogyne, but lacks microtubercles. The deutogyne back has strongly ridged tergites that describe an overall flat dorsum. This is an unusual type of deutogyne that is only matched by the deutogyne of *Rhyncaphytoptus fagifoliae* K., which lives on eastern North American beech. *Rhyncaphytoptus atlanticus* K. lives on the American elm in eastern North America. This eastern species differs from *R. ulmivagrans* by having a much plainer network shield pattern and by having an 8-rayed featherclaw.

Catarhinus Keifer

Corn rust mite, *Catarhinus tricholaenae* K. This mite (fig. 136, *e*) occurs in Brazil and lives on various grasses, including corn and *Tricholaena.* It is said to rust corn leaves. The mite ranges in length from 180 to 205 μ. It is a robust-fusiform species, with a light yellow color. The shield design is absent and the dorsal setae are short and project up. The featherclaw is 6-rayed. The thanosomal tergites have a shallow subdorsal furrow on each side. The tergal microtubercles are most prominent as elongate dashes on the lateral curves. The sternal microtubercles are elliptical and touch ring margins. There are about 35 tergites and 80 sternites. The female genitalia are broad and shallow, with the coverflap having basal short dashes, but being apically smooth. The genus *Catarhinus* differs from *Rhyncaphytoptus* by lacking the forefemoral seta, and by having more noticeable subdorsal furrows. Both genera have simple featherclaws.

Diptilomiopus Nalepa

This genus has only been found in Africa, in southern Asia, and in the Indo-Malaysian region, extending to Australia. It is characterized by having large, deeply divided featherclaws. Members of *Diptilomiopus* lack dorsal tubercles and setae, and the patellae are missing from the legs, or are fused with the femora. The Western Hemisphere genus *Rhynacus* is similar to *Diptilomiopus,* but possesses the patellae.

Macadamia leaf vagrant, *Diptilomiopus davisi* K. (fig. 137, *g, h*), occurs on Macadamia nut as a leaf vagrant in Queenland. No information is available as to the significance of the species except that it lives on the undersides of the leaves.

D. davisi is an elongate-fusiform species that ranges from 190 to 215 μ long. The color is probably light yellow. The shield is broad and rather blunt anteriorly. The shield pattern consists of a strong row of "cells" across the whole anterior and lateral margins. The admedian lines are complete, but the median line is present only in front and at the rear. A curved lateral line extends across the shield, running just in front of the rudimentary dorsal tubercles, and arching back centrally. The featherclaw has about 6 rays on a side. The abdomen essentially lacks microtubercles dorsally, but the beadlike sternal granules are set just ahead of the margins. The markings on the female coverflap are transverse lines of granules, principally basal.

Diptilomiopus ficus Attiah. This is an Egyptian fig leaf vagrant. The female ranges from 143 to 172 μ long, is elongate-fusiform, and dark yellow. The shield design is basically the same as on *D. davisi.* The featherclaw is stated to be 8-rayed, and the abdominal rings are represented as being entirely microtuberculate.

Diptilomiopus assamica K. is a citrus leaf vagrant in India. It occurs associated with *Phyllocoptruta oleivora* (Ashmead). Females are from 190 to 210 μ long, are elongate-fusiform, and probably yellow in life. The shield design differs from *D. davisi* by being more of a network of lines. In this case the median line is present only at the rear. The rudimentary dorsal tubercles are present, but lack setae. The divided featherclaws have about 5 rays on a side. The abdominal rings are narrow, lack dorsal microtubercles, but ventrally these granules are small, round, and lie very close to the ring margins. The female coverflap lacks marks.

Coffee leaf vagrant, *Diptilomiopus jevremovici* K. This mite lives on the underside of coffee leaves in the Kivu area of Zaire. Its significance on this host is not now known. The females of this mite are from 190 to 210 μ long, the body is elongate-fusiform, and the living color is probably yellow. The shield is shallow subtriangular, with the anterior lobe short and slightly emarginate centrally. The shield design is a network, with the median line complete. The admedian lines are well separated, and all are cross-connected. Otherwise the design is as usual, with the standard row of marginal "cells." The dorsal tubercles are present as rudiments. The deeply divided featherclaws have 6 rays on a side. The anterior coxae are broadly connate centrally and pebbled. The thanosomal tergites are about 45 in number and with heavier lines than the sternites. Tergal microtubercles are only present laterally. There are about 70 thanosomal sternites, which are set with beadlike microtubercles. The female genital coverflap has a basal area with short longitudinal lines; apically there is a cross curved line on the coverflap.

Rhynacus Keifer

The only species so far known that belong to this genus are native to North America, Central America, and to South America. This genus differs from *Diptilomiopus* mainly by possessing the patellae on the legs. Otherwise it is very similar to *Diptilomiopus*, which inhabits the Old World tropics (fig. 137, *e*, *f*).

Cashew leaf vagrant, *Rhynacus globosus* K. This species lives on cashew leaves in Brazil. The females are 190 to 210 μ long, robust-fusiform, and probably purple in life. The shield is subsemicircular in anterior outline, with the anterior lobe hardly projecting. The design is of a few lines, but without a definite edging of "cells." The median line is faint, but the admedians are complete and centrally wide. There is a cross line at about ½, and the rudimentary dorsal tubercles are present. The featherclaw is 7-rayed on a side. The abdominal rings are nearly equal dorsoventrally, with beadlike microtubercles, although certain areas lack these granules. The female genital coverflap has some scattered basal granules, and apically has about 20 short longitudinal ribs.

Rhododendron leaf vagrant, *Rhynacus tampae* K. This mite lives on rhododendron leaves in Florida. Females of this species are 170 to 215 μ long. The mites are robust-fusiform, with the living color probably light yellow. The shield has a very short anterior lobe. The shield design is a network, with the median line present anteriorly and posteriorly, but absent centrally. The admedian lines are complete, but curving. There is a strong series of "cells" around the anterior and lateral margins, a strong line just ahead of the rear margin, with some short longitudinal lines after that line. The dorsal tubercles are not present. The abdomen has about 53 tergites, and nearly 80 sternites. Microtubercles are elongate, but do not extend above the lateral curves of the tergites. The featherclaw is 8-rayed. The female genital coverflap has longitudinal short dashes, and granules, basally, and 8 to 10 longitudinal ribs near rear margin.

Blackberry leaf mite, *Rhynacus abronius* (K.). This is a clear purple, elongate-fusiform species living on the undersurface of blackberry leaves along the Pacific Coast of North America. It requires fairly high atmospheric humidity, and is not found inland to any extent. It causes no known injury to its hosts. *R. abronius* is from 210 to 230 μ long. The shield is shortened longitudinally, and has an open network of lines. The median line is nearly complete. The admedian lines are complete, subparallel to the median, forking at about ½, the inner fork curving to the median, and the outer fork arching out and running together at the rear margin. The marginal row of "cells" is composed of large spaces. The rudimentary dorsal tubercles are present near the rear margin. The featherclaw is 6-rayed on a side. The abdomen is about the same dorsoventrally, and completely microtuberculate, each microtubercle consisting of a point over the ring margin. There are about 75 tergites and about 85 sternites. The female genital coverflap has some short dashes distributed basally, with slight lines representing longitudinal ribs over the subapical area.

Trimeroptes Keifer

This genus, with divided featherclaws, differs from *Rhyncaphytoptus* by having the middorsal longitudinal ridge bifurcate anteriorly just behind the shield.

Sweet gum leaf vagrant, *Trimeroptes aleyrodiformis* (K.) (fig. 137, *c*), is a yellowish, flattened-fusiform species that lives on the underside of sweet gum leaves. It has a wide range on its host in eastern North America, and is also found on planted ornamental sweet gums in the San Francisco Bay region. *T. aleyrodiformis* is particularly characterized by the wax pattern developed on its dorsum, which follows the ridges. The anterior shield lobe is emarginate centrally, anteriorly, but the shield has no pattern. The dorsal setae project up and diagonally inward. The tergites lack definite microtubercles, but the sternites have rounded granules that rest on the ring margins. The female genital coverflap has a basal section with irregular longitudinal dashes; apically the coverflap is unmarked. The females of *T. aleyrodiformis* are from 170 to 195 μ long.

Dialox Keifer

As well as having a deeply divided featherclaw, this genus has a broad, centrally emarginate anterior shield lobe, and 2 transverse creases across the rear of the shield. The dorsal setae are very small and point divergently anteriorly. The rostrum is long and thin, and the tibiae are unusually long and slender.

Coconut frond vagrant, *Dialox stellatus* K. This mite is found on the underside of coconut leaves in the Philippines, and can be recognized by the heavy covering of white wax that sends out various projections. The female is 240 to 260 μ long. The body is robust-fusiform, with tapering cauda. The body is probably light yellow. There is no particular shield design, but the broad anterior lobe has a section on each side of the emargination. The rear shield creases produce a transverse roll, at the ends of which are located the short setae. The abdomen has up to 100 dorsal rings, but about 80 ventral rings. The microtubercles are irregularly placed, scattered, and beadlike. The female genitalia are transverse, and the coverflap has numerous longitudinal short stripes.

Apodiptacus Keifer

The distinguishing features of this genus are the divided featherclaw, and the three dorsal ridges that are specialized for wax production. Representations of the genotype, *A. cordiformis*, should show these thickened ridges on the tergites.

Big-beaked hickory mite, *Apodiptacus cordiformis* K. (fig. 137, *a, b*), is a white waxy vagrant on the underside of hickory leaves in the eastern United States. It also lives on walnuts, and may be found on other hosts, probably in the vicinity of primary hosts. It evidently causes no damage, and can be recognized by the white wax that covers the body. Females of *A. cordiformis* are from 180 to 215 μ long. The body is elongate- fusiform. The color is light yellowish,, but the wax is present in stripes along the rdiges. The shield is triangular with a somewhat acuminate anterior lobe that has a small terminal notch. The median shield line is present only to the rear. Admedian lines are nearly complete, curving back from the sides of the anterior lobe notch and ending ahead of the rear margin by enclosing a subcircular central area. The dorsal setae are moderately long and project forward. The featherclaw has 4 rays on a side. The body has about 53 tergites and 3 dorsal longitudinal ridges. The middorsal ridge fades just before the end of the thanosome, whereas the lateral ones extend slightly further posteriorly. These ridges are topped by thickenings in the rings, which thickenings are specialized for wax production. There are no tergal microtubercles. There are about 80 to 85 sternites that have beadlike microtubercles resting on the margins. The coxae have coarse lines. The females genital coverflap has a transverse basal area with longitudinal lines; apically the coverflap has faint transverse lines.

Diptacus Keifer

This genus, which contains species with shallow subdorsal longitudinal furrows, has divided featherclaws. It is a genus of considerable size and range (fig. 137, *d*).

Dogwood white wax mite, *Diptacus flocculentus* K. This species lives on the underside of flowering dogwood leaves in the eastern United States. It is princi-

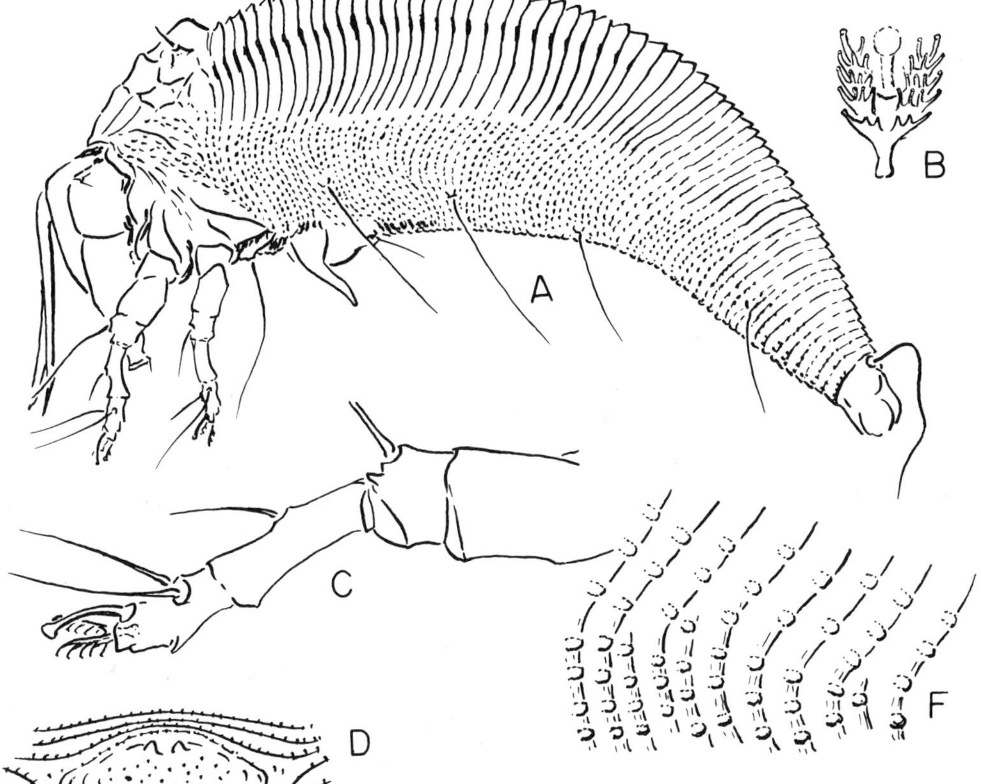

Fig. 138. ***Diptacus gigantorhynchus*** (Nal.), female: ***a***, left side; ***b***, featherclaw; ***c***, left foreleg; ***d***, male genitalia; ***e***, internal female genital structures; ***f***, lateral microtubercles and rings; ***g***, shield pattern; ***h***, coxae and external female genitalia.

pally characterized by the abundance of flocculent wax it produces around itself. It causes no known injury to its host. Females of this wax mite are from 225 to 240 μ long. The body is elongate-fusiform, and tapers caudally. The basic body color is light yellow. The anterior shield lobe is prominently acuminate, with an apical notch. The median shield line is absent. Admedian shield lines curve back from near the front end of the anterior lobe, form a semicircle at the base of this lobe, then run back to the rear ⅓ where they again curve out and encircle the whole area between the dorsal tubercles. The dorsal tubercles point forward and diagonally inward from a concave line that extends across the rear of the shield, nearly meeting the rear shield margin centrally. The dorsal setae are moderately long and project diagonally forward. The featherclaws have 7 rays on a side, but the central division narrows to the rear, the inner side of this division being convex inwardly. There are about 50 tergites. These tergites are not microtuberculate dorsally, but there is a longitudinal lateral band of coarse microtubercles on each side of the body for wax production. There are about 90 sternites, which are completely microtuberculate, with elliptical granules that rest on the margins. The coxae have coarse lines, and the female genital coverflap has slight, irregular lines in a transverse basal area.

Big-beaked plum mite, *Diptacus gigantorhynchus* (Nal.); *Epitrimerus gigantorhynchus* Nal.; *Diptilomiopus prunorum* K. (pl. 74, *a-d*; fig. 138). This sort of mite is a robust, purplish form, with 5-rayed featherclaws that are only partially divided. It has the large downturned rostrum characteristic of the Rhyncaphytoptidae. Mites of this kind have a very wide distribution in the Northern Hemishpere. They seem to favor rosaceous hosts, especially *Prunus* spp., but occur on other kinds of plants. Sometimes populations of the *D. gigantorhynchus* type are on nonrosaceous hosts because a favored host is nearby, but these mites living on *Nyssa*, in Virginia, remain for explanation. This sort of mite is evidently a complex, probably heading for a species proliferation.

Females of the *D. gigantorhynchus* mite as found on cultivated plum in central California range from 200 to 215 μ long. The shield anterior lobe is very short. The shield design is a network of lines with the median line complete, but broken by cross lines at ¼, ½, and ¾. The admedian lines more or less curve to accommodate the cross lines. Front and side edge "cells" are present in the lines. The rear shield margin is elevated above the adjacent part of the thanosome. Dorsal shield tubercles are ahead of the rear margin and the setae are very short. There is a ridge between the forecoxae. Microtubercles are weak on the tergites, but stronger on the sternites. There are about 60 rather narrow tergites, and 80 sternites. The female genital coverflap is unmarked.

A mite named *Diptacus fragarifoliae* K. is evidently a derivative of *D. gigantorhynchus*. It is much like the big-beaked plum mite, but has 6-rayed featherclaw and a less prominent ridge between the forecoxae. It occurs in Maryland, on wild strawberry.

Big-beaked holly mite, *Diptacus swensoni* K. This species (fig. 137, *d*) lives on commercial holly in the western United States, and is said to cause some foliage browning. Swenson, who has investigated the activities of this mite, found that in Oregon the populations increase during midsummer and into the fall. They do not decrease over winter unless unusual cold sets in. But decrease normally occurs in May. The mites tolerate a considerable range of temperature, being able to breed during winter, even when temperatures go below freezing. They also tolerate daytime temperatures that exceed 35 C (95 F). Evidence indicates that even when mite populations go above 600 individuals to the leaf, the leaf damage remains slight.

Diptacus swensoni females range from 230 to 250 μ long. The body is robust-fusiform and the color is purplish. The shield is somewhat acuminate, with rather straight sides, but the anterior lobe is broad and blunt over the rostrum. The shield design is a network of broad lines. The median line, which is slender, occurs on the rear ¾. The admedian lines receive a lateral line at ⅓ and another at ⅔. The dorsal tubercles are set ahead of the rear shield margin, and the setae are very short. The featherclaw is fully divided and has 5 rays on a side. There is no particular differentiation between tergites and sternites on the abdomen, the rings being entirely microtuberculate. The microtubercles are elongate dorsally and beadlike ventrally. The female genitalia have fine longitudinal lines on the coverflap, especially basally.

SELECTED BIBLIOGRAPHY

ALAN, M. Z., and M. A. WADUD. 1963. On the biology of the litchi mite, *Aceria litchii* (Eriophyidae, Acarina) in East Pakistan. Pakistan J. Sci. 15 (5) :231–240.

ALLINGTON, W. B., R. STAPLES, and G. VIEHMEYER. 1968. Transmission of rose rosette virus by the eriophyid mite, *Phyllocoptes fructiphilus* Keifer. J. Econ. Entomol. 61(5): 1137–1140.

ANDERSON, L. D. 1954. The tomato russet mite in the United States. J. Econ. Entomol. 47(6):1001–1005.

ANTHON, E. W. 1954. Peach silver mite control. J. Econ. Entomol. 47(5):866–868.

ATTIAH, H. H. 1955. A new eriphyid mite on Mango from Egypt. Bull. Roy. Entomol. Soc. Egypt 39:379–383.

BAILEY, J. S., and A. I. BOURNE. 1946. The control of the blueberry bud mite. J. Econ. Entomol. 39(1):89.

BAILEY, S. F., and H. H. KEIFER. 1943. The tomato russet mite, *Phyllocoptes destructor* Keifer: its present status. J. Econ Entomol. 36(5):702–712.

BAKER, E. W. 1939. The fig mite *Eriophyes ficus* Cotte. Bull. Dept. Agr. Calif. 28(4): 266–275 (*ficus* and *ficifoliae*).

BARNES, M. M. 1958. Relationships among pruning time response symptoms attributed to grape bud mite and temporary early season boron deficiency on grapes. Hilgardia 28(7):193–226.

BATCHELOR, G. S. 1952. Eriophyid mites of Washington. Wash. State Agr. Expt. Sta. Bull. 6.

BORDEN, A. D. 1932. The pear leaf blister mite as a cause of fruit-bud injury. Univ. Calif. Circ. 324.

Boyce, A. M., and R. B. Korsmeier. 1941. The citrus bud mite, *Eriophyes sheldoni* Ewing. J. Econ. Entomol. 34(6):745–756.

Boyce, A. M., R. B. Korsmeier, and C. O. Persing. 1942. The citrus bud mite and its control. Calif. Citrograph 27(5):124, 125, 136–138, 140, 141.

Breakey, E. P., and G. S. Batchelor. 1950. Phyllody of chrysanthemum and the eriophyid mite, *Paraphytoptus chrysanthemi* Keifer. Ann. Entomol. Soc. Am. 43(4):492–494.

———. 1957. Biology and control of the dryberry mite, *Phyllocoptes gracilis* (Nal.). Wash. State Agr. Expt. Sta. Bull. 574:1–17.

Briones, M. L., and W. H. Sill. 1963. Habitat, gross morphology, and geographical distribution of four new species of eriophyid species from coconuts in the Philippines. FAO Plant Prot. Bull. 11(2):25–30.

Browne, F. G. 1968. Animals as pests and diseases of forest plantation trees. p. 19. Oxford Uni. Press.

Burditt, A. K. Jr., D. K. Reed, and C. R. Chittenden. 1963. Observations on the mites, *Phyllocoptruta oleivora* Ashm. and *Aculus pelekassi* Keifer under laboratory conditions. Fla. Entomol. 46(1):1–5.

Burts, E. C. 1970. Biology of blister mites, *Eriophyes* spp. of pear and apple in the Pacific Northwest. Melanderia 4:41–53.

Buxton, G. 1969. Notice Calif. Dept. Agr. Bur. Entomol. No. 69D15–4, G–7, G18–42.

Callingwood, C. A. 1959. The control of black currant gall mite in garden and nursery. J. Hort. Soc. London 84(3):133–136.

Callingwood, C. A., and A. M. Brock. 1959. Ecology of the black currant gall mite (*Phytoptus ribis* Nal.) J. Hort. Sci. London 84(3):176–182.

Carmona, M. M. 1964. Contribuicao para o conhecimento dos acaros das plantas cultivadas em Portugal. III. Agronomia Lusitana 24(4):173–288. 1966. IV. Agronomia Lusitana 26(3):175–203.

Channabasavanna, G. P. 1966. Contribution to the knowledge of Indian eriophyid mites (Eriophyidae: Trombidiformes: Acarina) Univ. Agr. Sci. Hebbal, Bangalore. 153 pp.

Cooperative Econ. Insect Rept. 1969. Plant Prot. Div. Agr. Res. Ser., U. S. Dept. Agr. 19(29):550.

Costa, A. S. and P. D. Gonsales. 1950. Cinza on penugem branca do tomaterio. Bragantia 10(12):383–384. (From Biological Abstracts 32595, 1952.)

Darrow, G. M., R. B. Wilcox, and C. S. Ceckwith. 1944. Blueberry growing. U. S. Dept. Agr. Bull. 1951:36.

Das, G. M., and N. Sengupta. 1958. Observations on the pink mite, *Acaphylla theae* (Watt.) Keifer, of tea in North East India. J. Zool. Soc. India 10(1):39–48.

———. 1962. Biology and control of the purple mite, *Calacarus carinatus* (Green), a pest of tea in Northeast India. J. Zool. Soc. India 14(1, 2):64–72.

David, A. L. 1959. Host range of the sweet potato rust mite, *Epitrimerus* sp. Madras Agr. J. 46(8):313.

Dippenaar, B. J. 1958. Concentric ring blotch of citrus, its cause and control. S. African J. Agr. Sci. 1(1):83–106.

Domes, Von R. 1957. Zur biologie der gallmilbe, *Eriophyes gracilis* Nal. Z. Angew. Entomol. 41(4):411–424.

Edwards, E. T. 1935. A new virus disease of lucerne. J. Australian Inst. Agr. Sci. 1(1): 31–32.

Edwards, W. O., K. W. Gray, T. Wilcox, and D. C. Mote. 1935. The blackberry mite in Oregon. Oreg. Agr. Expt. Sta. Bull. 337:1–33.

Essig, E. O. 1925. The blackberry mite the cause of Redberry disease of the Himalaya blackberry and its control. Calif. Agr. Expt. Sta. Bull. 399:1–10.

Eyndhoven, G. L. van. 1967. The red currant gall mite, *Cecidophyopsis selachodon* n. sp. Entomol. Ber. 27(8):149–151.

FIFE, L. C. 1937. Damage to sea island cotton by West Indian blister mite, *Eriophyes gossipii* (Banks). J. Agr. Univ. Puerto Rico 21 (2) :169–177.

FLOCK, R. A., and J. M. WALLACE. 1955. Transmission of fig mosaic by the eriophyid mite, *Aceria ficus* (Cotte): Phytopathology 45(1):52–53.

FULTON, B. B. 1940. The Blueberry bud mite, a new pest. J. Econ. Entomol. 33(4):699.

GARCIA, E. A. 1966. A acariose ou no' curto parasitario da videira no Minho. Agricultura, Lisbon 30:20–23.

GIBSON, W. W. 1957. Biological and ecological studies of the wheat curl mite, *Aceria tulipae* (Keifer) on winter wheat in Kansas. Kans. State Coll. Manhattan Diss. Abstr. 17(1):697–698.

GILMER, R. M., and F. L. MCEWEN. 1958. Chlorotic fleck, an eriophyid mite injury of Myrobalan plum. J. Econ. Entomol. 5(3):335–341.

HAMILTON, ALICE. 1950. The blackberry mite (*Aceria essigi*). New Zealand J. Sci. and Tech. 31A(2):42–45.

HANSON, A. J. 1933. The blackberry mite and its control. Wash. State Agr. Expt. Sta. Bull. 279:1–20.

HARPAZ, I. 1955. Notes on the eriophyid mites of Israel. Bull. Res. Council Israel 5(1): 61–66.

HASSAN, A. S. 1928. Biology of the Eriophyidae with special reference to *Eriophyes tristriatus* (Nalepa). Univ. Calif. Publ. Entomol. 4:341–383.

HATZINIKOLIS, E. 1968. A new mite, *Aculus benakii* n. sp. from *Olea europea* L. Acarologia 10(4):450–652.

HELY, P. C. 1968. The entomology of citrus in New South Wales. Australian Entomol. Soc. Misc. Publ. 1:1–20.

HERIOT, A. D. 1935. Notes on the blister mite, *Eriophyes pyri*. Proc. Entomol. Soc. Britsh Columbia 31:41–42.

HERRICK, G. W. 1935. Insect enemies of shade trees. p. 139 Comstock Pub. Co., Ithaca, N. Y.

HILDEBRAND, E. M. 1945. Myrobalan mottle and asteriod spot. Phytopathology 35(1): 47–50.

HOBZA, R. F. 1971. Interrelationships of temperature and humidity on development of the citrus rust mite, *Phyllocoptruta oleivora* (Ashm.). M.S. Thesis, Univ. Calif., Riverside.

HODGKISS, H. E. 1930. The Eriophyidae of New York. II. The maple mites. N. Y. State Agr. Expt. Sta. Bull. 163.

HUGHES, T. E. 1959. Mites or the acari. Univ. London. The Athlone Press. Also Essential Books, Fair Lawn, New Jersey.

JEPPSON, L. R., M. J. JESSER, and J. O. COMPLIN. 1955. Control of mites on citrus with chlorobenzilate. J. Econ. Entomol. 48(4): 375–377.

KANERVO, V., and K. L. MAKINEN. 1967. Herukan äkämäpumkin torojunta. Maatalous Koetoim 21:148–158. (From Rev. Appl. Entomol. Ser. A. 56:232.)

KANTACH, E. J. and H. KNUTSON. 1958. Chemical control studies of the wheat curl mites. J. Econ. Entomol. 51(1):68–72.

KEIFER, H. H. 1946. A review of N. A. economic eriophyid mites. J. Econ. Entomol. 39(5):563–570.

———. 1952. Eriophyid mites of Calif. (Acarina: Eriophyidae). Bull. Calif. Ins. Survey 2(1):1–123.

———. 1955. Eriophyid mites—notes and new species (Acarina). Pan Pacific Entomol. 31(3):115.

———. 1959. Mouthparts of eriophyids. Eriophyid Studies XXVI. Bull. Calif. Dept. Agr. 47(4):278–281.

KEIFER, H. H., and J. L. SAUNDERS. 1972. *Trisetacus campnodus* n. sp., attacking *Pinus sylvestris*. Ann. Entomol. Soc. Am. 65(1):46–49.

Keifer, H. H., and N. S. Wilson. 1956. A new species of eriophyid responsible for vection of peach mosaic virus. Bull. Calif. Dept. Agr. 44:145.

Kido, H., and E. M. Stafford. 1948. The biology of the grape bud mite *Eriophyes vitis* (Pgst.). Hilgardia 24(6):119–142.

King, C. B. R. 1937. Notes on the life history of *Eriophyes carinatus* Green. Bull. Entomol. Res. 28:311–314.

Koenigsman, E. 1955. The Frondescence phenomena on caraway produced by gall mites. Pharmazie 10(12):749–752.

———. 1959. Fighting the gall mite, *A. carvi*. Pharmazie 14 (2) :99–104.

Kruel, W. 1963. Gallmilben an Keifer. Beitr. Entomol. 13(5–6):566–567.

Kuitert, L. C. 1962. Distortion of young cedar by an eriophyid mite, *Trisetacus cupressi* (Keifer). Proc. Fla. State Hort. Soc. 75:495–497.

Lange, W. H., Jr. 1953. *Aceria paratheni* Keifer. A new eriophyid mite injurious to guayule. J. Econ. Entomol. 46(1):162–163.

———. 1955. *Aceria tulipae* (K.) damaging garlic in California. J. Econ. Entomol. 48(5): 612–613.

Lange, W. H., Jr., and L. K. Man. 1960. Fumigation controls microscopic mite attacking garlic. Calif. Agr. 14(12):9–10.

Massee, A. M. 1928. Life history of the black currant gall mite, *Eriophyes ribis* (Westw. and Nal.). Bull. Entomol. Res. 18:297–309.

———. 1930. On some gall-mites (Eriophyidae) found on *Corylus avellana* L. Bull. Entomol. Res. 21:165–168.

———. 1932. Some injurious and beneficial mites on top and soft fruits. J. Pomology and Hort. Sci. 10:106–129.

Merwe, G. G. van, and T. J. Coates. 1965. Biological study of the grey mite, *Calacarus citrifolii* Keifer. S. African J. Agr. Sci. 8(3):817–824.

Meyer, M. K. P. 1968. The grass stunt mite, *Aceria neocynolonis* Keifer. S. Afr. J. Agr. Sci. 11(4):803–804.

Miles, P. M. 1964. Chrysanthemum semibud mite in Wales. Entomol. Mon. Mag. 100:97.

Morgan, C. V. G., and A. E. Hedlin. 1960. Notes on the juniper berry mite, *Trisetacus quadrisetus* (Thomas) in British Columbia. Can. Entomol. 42(8):608–610.

Morgan, C. V. G., P. T. Yee, and F. E. Brinton. 1962. Seven as a systemic miticide for the pear leaf blister mite, *Eriophyes pyri* (Pgst.). Acarina: Eriophyidae. Can. Entomol. 94(7):680–686.

Mulligan, T. E. 1960. The transmission by mites, host, range, and properties of rye grass mosaic virus. Ann. Appl. Biol. 48(3):575–579.

Nalepa, A. 1920. Zur kenntnis der Milbengallen einiger Ahornarten und ihrer Erzeuger. Marcellia 19:3–33.

Natcheff, P. D. 1966. Studies on eriophyid mites in Bulgaria II. Acarologia 8(3):415–420.

Nault, L. R. 1970. The wheat curl mite in Ohio wheat and corn. Ohio Reporter on Res. and Dev. 55(2):27–29.

Nault, L. R., M. L. Briones, L. E. Williams, and B. D. Barry. 1967. Relation of the wheat curl mite to kernel red streak of corn. Phytopathology 57(9):986–989.

Nault, L. R., and W. E. Styer. 1969. The dispersal of *Aceria tulipae* and three other grass-infesting eriophyid mites in Ohio. Ann. Entomol. Soc. Am. 62(6):1446–1455.

Nishida, T., and F. G. Holdoway. 1955. The erinose mites of lychee. Circ. Hawaiian Agr. Expt. Sta. 48. 10 pp.

Oldfield, G. N. 1969. The biology and morphology of *Eriophyes emarginatae*, a *Prunus* finger gall mite and notes on *E. prunidemissae*. Ann. Entomol. Soc. Am. 62(2):269–277.

———. 1970. Mite transmission of plant viruses. Ann. Rev. Entomol. 15:343–380. Ann. Reviews, Inc. 4139 El Camino Way, Palo Alto, Calif.

ORLOB, G. B. 1966. Feeding and transmission characteristics of *Aceria tulipae* (K.), as vector of wheat streak mosaic virus. Phytopathology 55:218–238.

ORTEGA, C. A., V. J. RODRIGUEZ, and V. C. GARIBAY. 1965. Investigaciones preliminares sobre el Eriófido del fruto del cocotero, *Aceria guerreronis* Keifer, en la Costa Grande de Guerrero. Agric. Técnica en México 2(5):222–226.

PARROTT, P. J., H. E. HODGKISS, and W. J. SCHOENE. 1906. The apple and pear mites. N. Y. Agr. Expt. Sta. Bull. 283:281–318.

PEPPER, B. B. 1942. Studies of the mite, *Epitrimerus hystrix*, on timothy. J. Econ. Entomol. 35(2):201–204.

PHILIPP, W. 1957. Intensified occurrence of the plum gall mite, *Aceria* (Eriophyes) *phloeocoptes* (Nal.) Z. Pflkrankh. 64(1):35–38.

PLANES, S., J. M. DEL RIVERO, F. MARTI, and J. M. CARRERO. 1965. Contribucion al estudio de la biologia de *Phytoptus avellanae* Nal. y de los medios de lucha contra la plaga. Boletin de Pathologia Vegetal y. Entomol. Agr. 28:27–55.

PRESCOTT, R. T. M. 1934. The pear leaf blister mite, *Eriophyes pyri* (Pgst.). J. Dept. Agr. Victoria 32(10):538–539.

PRITCHARD, A. E. 1957. New carnation pest found in California. Florist Exchange and Hort. Trade World 129(1):12.

PUTMAN, W. L. 1940. The plum nursery mite, (*Phyllocoptes fockeui* Nal. and Trt.). Rept. Entomol. Soc. Ont. 70(1939):33–40.

RICE, R. E., and F. E. STRONG. 1962. Bionomics of the tomato russet mite, *Vasates lycopersici* (Massee). J. Econ. Entomol. 55(4):431–435.

ROIVAINEN, H. E. 1947. Eriophyid News from Finland. Acta Entomol. Fen. 3:8.

ROSARIO, M. S. DEL, and H. S. WEBSTER, JR. 1964. Additional biological and ecological characteristics of *Aceria tulipae* (Acarina: Eriophyidae). J. Econ. Entomol. 57(6): 893–896.

SABA, V. F. 1964. Sonderabdruck anseiger fur Schadlingskunde 37(2):17–19.

SAYED, M. T. 1946. *Aceria mangiferae* nov. spec. (*Eriophyes* mangiferae Hassan M. S.). Bul. Fouad. ler Entomol. 30(7):7–10.

SCOTT, D. R. 1965. Carrot bud mite. Idaho current information. Ser. 10.

SHEVTCHENKO, V. G. 1957. The life-history of alder gall mite, *Eriophyes laevis* (Nalepa). [Summary in English] Revue d'entomol. de l'URSS 36(3):558–618.

———. 1961. Peculiarities of the postembryonic development of gall-mites and some notes on the classification of *Eriophyes laevis* (Nalepa). Zool. J. 60(8):1143–1158.

SHEVTCHENKO, V. G., and A. P. DEMILLO. 1968. The life cycle of *Triestacus kirghisorum* Schev. [Summary in English] Leningrad Universitet Vestink Seriaa Biolosii, 1(3): 60–67.

SINGH, L. B., S. M. SINGH, and R. S. NIRVAN. 1961. Studies on mango malformation. I. Review, symptoms, extent, intensity, and cause. Hort. Advance 5:197–217.

SLEPYAN, E. I., G. S. LANDSBERY, and N. I. MAL'CHENKOVA. 1969. The gall of the mite *Eriophyes vitis* Pgt. (Acarina, Eriophyidae) Entomol. Rev. 48(1):67–74.

SLYKHIUS, J. T. 1959. Current status of mite transmitted plant viruses. Proc. Entomol. Soc. Ont. 90:22–30.

———. 1967. Virus diseases of cereals. Rev. Appl. Mycol. 46(8):401–429.

SMITH, B. D. 1960. The behavior of the black currant gall mite, *Phytoptus ribis* Nal. during the free living phase of its life cycle. Ann. Rept. Agr. Hort. Res. Sta. Long Ashton (Bristol) 1959, 1960, pp. 130–136.

SMITH, L. M., and R. O. SCHUSTER. 1963. The nature and extent of *Eriophyes vitis* injury to *Vitis vinifera* L. Acarologia 5(4):530–539.

SMITH, L. M., and E. M. STAFFORD. 1948. The bud mite and the erineum mite of grapes. Hilgardia 18(7):317–334.

SMITH, R. H. 1944. Insects and mites injurious to sycamore trees (*Platanus* spp.) in western North America. Arborists News 9(2):9–15.

SMITH, W. A. 1955. The tomato mite. Queensland Agr. J. 81(2):90–91.

SOKOLOV, A. M. 1960. The pear gall mite (*Eriophyes pyri* Pgst.) and measures for its control in conditions of the Michurin raion of the Tambor ablast. [In Russian with English summary] Zool. Zhur. 39(4):521–526.

SOMSEN, H. W. 1966. Development of migratory form of wheat curl mite. J. Econ. Entomol. 59(5):1283–1285.

STERNLICHT, M. 1966. Trials in the control of the citrus bud mite, *Aceria sheldoni* (Ewing), in Israel. Israel J. Agr. Res. 16(3):115–124.

———. 1970. Contribution to the biology of the citrus bud mite, *Aceria sheldoni* (Ewing). Ann. Appl. Biol. 65(2):221–230.

STERNLICHT, M., and S. GOLDENBERG. 1971. Fertilization, sex ratio, and post embryonic stages of the citrus bud mite, *Aceria sheldoni* (Ewing). Bull. Entomol. Res. 60(3): 391–397.

STUBBINGS, W. A. K. 1947. Studies of the bionomics and control of the pear bud mite *Eriophyes pyri* Pagenstecher. Union S. Africa Dept. Agr. and Forest Sci. Bull. 270: 1–92.

STUBBS, L. L., and J. W. MEAGHER. 1965. A virosis-like proliferation (witches'-broom) of lucerne (*Medicago sative* L.) caused by an eriophyid mite, *Aceria medicaginis* (K.). Australian J. Agr. Res. 16(2):125–129.

SWIRSKI, E., and S. AMITAI. 1959. Contribution to the biology of the citrus rust mite, *Phyllocoptruta oleivora* (Ashm.) Katavim 9(3–4):281–285.

TALHOUK, A. S. 1963. *Aceria phloeocoptes* (Nal.), a serious eriophyid pest of almond in Lebanon and Syria. Univ. Beirut Publ. No. 22. (Original in Anz. f. schadlingskunde) 36(9):129–132.

THAKER, P. J., and M. H. DESA. 1929. Wooly-mites (*Eriophyes gossippi*) on cotton. Its status and economic importance in Gujart. Agr. J. India 24(3):175–182.

THOMLINSON, W. E. 1950. Summer oil sprays to control blueberry bud mite. J. Econ. Entomol. 43(5):727.

THRESH, J. M. 1963. Black currant reversion disease. Ann. Rept. East Malling Res. Sta. (1962):184–189.

———. 1964*a*. Association between black currant reversion virus and its gall mite vector (*Phytoptus ribis* Nal.). Nature 202:1085–1087.

———. 1964*b*. The chemical control of black currant reversion virus and its gall mite vector (*Phytoptus ribis* Nal.). Ann. Rept. East Malling Res. Sta. (1963):152–157.

———. 1964*c*. Increased susceptibility to mite vector (*Phytoptus ribis* Nal.) caused by infection with black currant reversion virus. Nature 202:1028.

TUTTLE, D. M., and G. D. BUTLER, JR. 1961. A new eriophyid mite infesting Bermuda grass. J. Econ. Entomol. 54(5):836–838.

VENABLES, E. P., and A. D. HERIOT. 1937. The blister mite of apple and pear. Publ. Dept. Agr. Can. 577. 3 pp.

VERNON, J. D. R. 1957. An eriophyid mite causing damage to chrysanthemums. Plant Pathol. 6(3):111.

VIDAL-BARRAQUER, R., et. al. 1966. *Phytoptus avellana* Nal. y otros Eriofides del avellano. Bol. Patol. Veg. Entomol. Agr. 29:133–256.

VRIE, M. VAN DE. 1967. De levenswije en de bestrijding van de rondknopmijt van zwarte bes *Cecidophyopsis ribis*. Neth. J. Plant Pathol. 73(5–6):170–180.

WALTHER, E. 1925. The pine needle mite, a new enemy of pines (*Eriophyes pini* Nal.). J. Econ. Entomol. 18(6):830–836.

WANG, C. S. 1964. A new blister mite on sugar cane in Taiwan, *Aceria saccharini* Wang. [In Japanese with English summary.] Rept. Taiwan Sugar Expt. Sta. 33:83–94.

WATT, G., and H. H. MANN. 1903. The pests and blights of the tea plant, 2d ed. Calcutta. Office of Supt. Govt. Printing. 416 pp.

WEBSTER, R. L. 1948. Mites of economic importance in the Pacific Northwest. J. Econ. Entomol. 41(5):677–683.

Weidhaas, J. A., and R. M. Reeves. 1963. The occurrence and importance of tetranychid and eriophyid mites on woody plants in New York. Advances in Acarology 1:25–29.

Westigard, P., and S. C. Hoyt. 1965. The rust mite apple and pear species, *Aculus schlechtendali* Nalepa and *Epitrimerus pyri* Nal. differ but look alike. Western Fruit Grower 19(4):31.

Wilson, C. E. 1923. Insect pests of cotton in St. Croix and means of combatting them. Virgin Island Expt. Sta. Bull. 3:1–20.

Wilson, N. S. 1959. Eight new species of eriophyid mites. Ann. Entomol. Soc. Am. 52(2):141–149.

———. 1965. A new species of blister forming mite on pear. Ann. Entomol. Soc. Am. 58(3):327–330.

———. 1970. Some eriophyid species from Mexico. Ann. Entomol. Soc. Am. 63(4): 1125–1136.

Wilson, N. S., and L. C. Cochran. 1952. Yellow spot, an eriophyid mite injury to peach. Phytopathology 42(8):443–447.

Wilson, N. S., and G. N. Oldfield. 1966. New species of eriophyid mites from western North America, with discussion of eriophyid mites on *Populus*. Ann. Entomol. Soc. Am. 59(3):585–599.

Yothers, W. W., and A. C. Mason. 1930. The citrus rust mite and its control. U. S. Dept. Agr. Tech. Bull. 176. 56 pp.

Zeck, E. H. 1955. Grape vine mites. Agr. Gaz. N. S. W. 66:155–157.

Publications on Eriophyoidea by H. H. Keifer

First series of Eriophyid Studies (ES), designated by roman numerals (except first installment), and published in the Bulletin of the California Department of Agriculture (BCDA).

E. S. no.	*Reference and date published*
(I)	Volume XXVII(2):181–206, June 22, 1938
II	Volume XXVII(3):301–323, September 7, 1938
III	Volume XXVIII(2):144–163, March 17, 1939
IV	Volume XXVIII(3):223–239, April 21, 1939
V	Volume XXVIII(5):328–345, July 7, 1939
VI	Volume XXVIII(6):416–426, August 10, 1939
VII	Volume XXVIII(7–8–9):484–505, November 17, 1939
VIII	Volume XXIX(1):21–46, March 1940
IX	Volume XXIX(2):112–117, June 21, 1940
X	Volume XXIX(3):160–179, October 23, 1940
XI	Volume XXX(2):196–216, May 26, 1941
XII	Volume XXXI(3):117–129, September 16, 1942
XIII	Volume XXXII(3):212–222, October 14, 1943
XIV	Volume XXXIII(1):18–38, April 13, 1944
XV	Volume XXIV(3):137–140, October 8, 1945
XVI	Volume XXV(1):39–48, April 16, 1946
XVII	Volume XL(3):93–103, September 26, 1951
XVIII	Volume XLI(1):31–41, March 18, 1952
XIX	Volume XLI(2):65–74, July 1, 1952
XX	Volume XLI(3):147–149, September 22, 1952
XXI	Volume XLII(2):65–79, June 30, 1953
XXII	Volume XLIII(3):121–131, September 1954
XXIII	Volume XLIV(3):126–130, October 17, 1955
XXIV	Volume XLIV(4):159–163, January 23, 1956
XXV	Volume XLVI(3):242–248, November 12, 1957
XXVI	Volume XLVII (4) :271–281, December 1959 (The last issue in the Bull.)

INSTALLMENTS IN OCCASIONAL PAPERS, BUREAU OF ENTOMOLOGY, CALIF. DEPT. AGR.

XXVII	Occ. Paper 1, 18 pp., May 8, 1959
XXVIII	Occ. Paper 2, 20 pp., December 30, 1959

INSTALLMENTS OF ERIOPHYID STUDIES
ISSUED AS THE "B" SERIES BY THE CALIF. DEPT. AGR.

Series no.	*Date Published*
B-1	20 pages, November 17, 1960

B-2	20 pages, May 1, 1961
B-3	20 pages, May 22, 1961
B-4	20 pages, December 4, 1961
B-5	20 pages, February 21, 1962
B-6	20 pages, March 7, 1962
B-7	20 pages, May 9, 1962
B-8	20 pages, July 9, 1962
B-9	20 pages, February 20, 1963
B-10	20 pages, August 12, 1963
B-11	20 pages, May 14, 1964
B-12	20 pages, June 11, 1964
B-13	20 pages, March 2, 1965
B-14	20 pages, May 17, 1965
B-15	20 pages, July 26, 1965
B-16	20 pages, August 27, 1965
B-17	20 pages, March 1, 1966
B-18	20 pages, April 22, 1966
B-19	20 pages, July 19, 1966
B-20	20 pages, September 21, 1966
B-21	24 pages, November 23, 1966

ERIOPHYID STUDIES INSTALLMENTS OF THE "C" SERIES, AGRICULTURAL RESEARCH SERVICE, U.S. DEPT. AGR.

Series no.	*Date Published*
C-1	20 pages, April 15, 1969
C-2	20 pages, May 20, 1969
C-3	24 pages, October 15, 1969
C-4	24 pages, October 30, 1970
C-5	24 pages, May 1, 1971 (part in collaboration with R. A. Newkirk)
C-6	24 pages, March 10, 1972
C-7	24 pages, December 18, 1972
C-8	24 pages, October 15, 1973
C-9	24 pages, April 19, 1974

ADDITIONAL PUBLICATIONS ON ERIOPHYOIDS BY H. H. KEIFER

BAILEY, S. F. and H. H. KEIFER, The Tomato Russet Mite, J. Econ. Entomol. 36 (5): 706–712, Oct. 1943 (*lycopersici* Massee).

KEIFER, H. H., The Eriophyid Mites of California, Bull. Cal. Insect Survey, 2(1):1–123, Dec. 12, 1952, U. C. Press.

KEIFER, H. H., Eriophyid Mites; Notes and New Species, Pan-Pacific Entomologist 31(3):109–116, July 1955.

KEIFER, H. H. and NORTON S. WILSON, Bull. Cal. Dept. Agr. 44(5): 2 pages, Jan. 23, 1956.

KEIFER, H. H., New Eriophyid Mites, Ann. Entomol. Soc. Am. 52 (6) :649–657, Nov. 1959.

Taxonomic References to Eriophyoid Species

The generic assignment we recognize for each species follows (in parenthesis) the name of the species, and of the author.

Nalepa's worldwide host list, in the cecidological journal *Marcellia* 25, (1929), has all species he then recognized and an accompanying reference for each. As this is a major reference, we indicate it as: "M25: page." Nalepa's Eriophyiden Gallenmilben in *Zoologica* 61 (1910), Stuttgart is entitled, "Die Zoocecidien Deutschlands und ihre Bewohner." It contains many of Nalepa's half-tone illustrations. We indicate it as Zoologica page."

G. P. Channabasavanna's important 154-page paper, "A Contribution to the Knowledge of the Indian Eriophyoid Mites," was published by the University of Agricultural Sciences, Hebbal, Banglalore, 1966. We refer to this below as "Bangalore page."

H. E. Hodgkiss treated maple mites in Technical Bulletin 163, New York State Agricultural Experiment Station, July, 1930. We call this publication "N. Y. Tech. Bull. 163: page."

Keifer's 56 Eriophyid Studies, with years of publication, are tabulated on preceding pages. Such applications as these studies have to species below are shown by roman numerals, followed by the page, or by B or C, followed by the installment number and page. Additional publications by Keifer have full references.

Some of the listed species appear in this book only in the generic key, appendix 3. This key is indicated "Key, app. 3." All of the genera in the key have their types named, with full references. Species named by authors after Nalepa published in 1929, all have full references.

Abbreviations: Chan. = Channabasavanna; K. = Keifer; Nal. = Nalepa.

Species and Author	*Genus*	*Reference*
abaenus K.	(Phyllocoptes)	VIII:30; XXVI:275; B13:15
abathus K.	(Mesalox)	C2:11; Key, app. 3
abietis Postner	(Trisetacus)	Z. Angew. Entomol. 62:106, 1968
abnormis (Garman)	(Phytocoptella)	M25:130
aboula K.	(Prothrix)	B13:2; Key, app. 3
abronius (K.)	(Rhynacus)	VII:492
acaciae Nal.	(Eriophyes)	M25:98
acalyptus (K.)	(Abacarus)	VII:490; XIV:28; Key, app. 3
acericola (Nal.)	(Aculops)	Zoologica p. 260
acericola (Garman)	(Vasates)	M25:122; see also general index
aceris (Nal.)	(Aculops)	Zoologica p. 260

Species and Author	*Genus*	*Reference*
aceris Hodgkiss	(Eriophyes)	N.Y. Tech. Bull. 163:19
aceris-crumena (Riley)	(Vasates)	M25:122; see also general index
acerivagrans (K.)	(Tegonotus)	XXI:66
acilius K.	(Rhyncaphytoptus)	VII:49; XXVIII:33
acromius (Nal.)	(Acaphylla)	M25:83, 84; see also general index
adoratus K.	(Acalitus)	C4:5
adornatus (K.)	(Calacarus)	VIII:32; Pan-Pac. Entomol. 31:115, 1955 syn. of *carinatus* Green
aesculifoliae (K.)	(Tegonotus)	II:307, 312 (as *carinatus* Nal.) III:152; XII:117; XXI:67
afer K.	(Abacarus)	B6:11; B11:11
africanus K.	(Colopodacus)	B1:16; Key, app. 3
ajoensis (K.)	(Achaetocoptes)	B2:16; Key, app. 3
alborum K.	(Trisetacus)	B10:5
aleyrodiformis (K.)	(Trimeroptes)	IX:168; XVII:98
alinae Liro	(Epitrimerus)	Ann. Zool. Soc. zool-bot. Fen, Vanamo 8(7):16, 1941
alnivagrans K.	(Sierraphytoptus)	III:152; Key, app. 3
aloinis K.	(Eriophyes)	XI:205
amandae (K.)	(Aculus)	C3:5
amiculus K.	(Eriophyes)	VIII:23, 25 (protogyne of brachytarsus)
anacardii K.	(Dicrothrix)	B18:17; Key, app. 3
antonimus (K.)	(Artacris)	B7:2; Key, app. 3
anysis (K.)	(Ectomerus)	C4:17; Key, app. 3
apalachi K.	(Phaulacus)	B3:1–3; Key, app. 3
aphrastus (K.)	(Calepitrimerus)	VIII:29 (deutogyne of baileyi)
arceuthobii K.	(Paraphytoptus)	XIX:68; B21:7
arctostaphyli (K.)	(Rhynacus)	II:305; XVII:98; Key, app. 3
arnaudi K.	(Paraphytoptella)	XXVII:18; Key, app. 3
asclepiadii K.	(Rhombacus)	C2:9
assamica K.	(Diptilomiopus)	XXVIII:14; B1:18
athiasellus K.	(Ditrymacus)	B1:14; Key, app. 3
atlantazaleae (K.)	(Aculus)	X:161
atlanticus K.	(Rhyncaphytoptus)	XXVII:17
attenuata K.	(Notostrix)	B9:7; Key, app. 3
australis K.	(Tegolophus)	B12:2
avellanae (Nal.)	(Phytocoptella)	M25:82; II:301; IX:112 (nymph); C5:3
baileyi K.	(Calepitrimerus)	II:310
bakeri K.	(Anthocoptes)	XXVIII:3
bakkeri (K.)	(Eriophyes)	C1:7
barbarae K.	(Asetacus)	XXVII:9
benakii (Hatzinikolis)	(Aculops)	Acarologia t. Xm fasc. 4:650, 1968
bifurcatus K.	(Acadicrus)	B14:7; Key, app. 3
biopsidia (K.)	(Eriophyes)	C2:5
bowlesia (Wilson)	(Eriophyes)	Ann. Entomol. Soc. Am. 52(2):144, 1955; orig. in Ceylon Independent 1840; genus *typhlodromus*
brachystegiae K.	(Criotacus)	B10:13; Key, app. 3
brachytarsus K.	(Eriophyes)	V:328; VIII:23
breechii (K.)	(Phytoptus)	VI:416 (poss. deutogyne of prunandersoni)

Species and Author	*Genus*	*Reference*
breitlowi Davis	(Rhynacus)	Fla. Entomol. 47:113, 1964
brevitarsus (Fockeu)	(Acalitus)	M25:85
brionesae K.	(Calacarus)	B9:11
cajani (Chan.)	(Eriophyes)	Bangalore p. 62
calaceris (K.)	(Eriophyes)	XVIII:33
caliberberis (K.)	(Eriophyes)	XIX:66
califraxini (K.)	(Tegolophus)	II:308; B2:4; Key, app. 3
calisorbi K.	(Phyllocoptes)	B13:15
calonyctionis K.	(Floracarus)	XXI:69; XXVIII:12; Key, app. 3
camelliae (K.)	(Cosetacus)	XV:137; B11:2; B20:7; Key, app. 3
camphorae K.	(Gammaphytoptus)	III:149; B20:17; Key, app. 3
capnodus K. and Saunders	(Trisetacus)	Ann. Entomol. Soc. Am. 65:46, 1972
capsicellus (K.)	(Tetraspinus)	C3:19
carinatus (Green)	(Calacarus)	M25:133; Pan-Pac. Entomol. 31:115, 1955; orig. in Ceylon Independent 1840; genus *Typhlodromas.*
carinatus Nal.	(Tegonotus)	M25:118
carpini Nal.	(Phyllocoptes)	M25:81; Key, app. 3
caryae K.	(Eriophyes)	VII:484; VIII:25; XXVIII:7
caryfoliae K.	(Aculus)	B2:14
caryvagrans K.	(Cecidophyes)	B11:15; B20:17, 19
caulis Cook	(Eriophyes)	M25:74
cedrelae K.	(Acerimina)	XXV:242; Key, app. 3
celtis Kendall	(Eriophyes)	Psyche 36 :300, 1929
cembrae (Tubeuf)	(Trisetacus)	M25:70, 71
centrolobii Flechtmann and K.	(Scolotosus)	Key, app. 3
cephaloneus (Nal.)	(Artacris)	M25:121; C4:9, 11 (in Artacris)
chinensis O'Gara	(Eriophyes)	M25:118; see litchii
chrondriphora (K.)	(Eriophyes)	B13:5
chrysanthemi K.	(Paraphytoptus)	VIII:27
chrysolepis K.	(Johnella)	Ann. Entomol. Soc. Am. 52:652, 1959
chrysophylli Cook	(Eriophyes)	M25:138
cinereae K.	(Eriophyes)	VIII:23, 25
citrifolii K.	(Calacarus)	XXIII:125
cocciformis K.	(Acarelliptus)	X:167; XVII:98; Key, app. 3
coconuciferae K.	(Acamina)	B6:5
coffeae K.	(Calacarus)	B1:14
collegiatus K.	(Cecidophyes)	B3:10, 11
comatus (Nal.)	(Aculus)	M25:82
concava (K.)	(Tetra)	VII:489; XIV:27; Key, app. 3
concinnus Liro	(Rhinophytoptus)	Key, app. 3
confusus Hodgkiss	(Vasates ?)	N.Y. Tech. Bull. 163:19 possibly protogyne of quadripedes
congoensis K.	(Epitrimerus)	B1:11
convolvuli (Chan.)	(Tegonotus)	Bangalore p. 94
cordai Oudemans	(Eriophyes)	Critico-Hist. Survey, Acarology III C, p. 1325, 1937 (see litchii)
cordiformis K.	(Apodiptacus)	B1:18
corniseminis (K.)	(Phytocoptella)	III:144; XXII:122; B9:19; Key, app. 3
cornutus (Banks)	(Aculus)	Proc. Entomol. Soc. Wash. 7:141, 1906
costae K.	(Aciota)	XXVII:9; Key, app. 3

Species and Author	*Genus*	*Reference*
cribratus K.	(Phyllocoptes)	B2:9
cupressi (K.)	(Trisetacus)	XIV:19
cuprifestor K.	(Cupacarus)	XIII:215; Key, app. 3
cynodoniensis (Sayed)	(Eriophyes)	Bull. Soc. Fouad Ier, Entomol. XXX: 149, 1946
cynodonis (Wilson)	(Eriophyes)	Ann. Entomol. Soc. Am. 52:142, 1959
davisi K.	(Diptilomiopus)	C1:17
deleoni (K.)	(Cosella)	XXIV:161; Key, app. 3
delhiensis Chan.	(Abacarus)	Bangalore p. 118
denmarki K.	(Acritonotus)	B8:6; Key, app. 3
dentatus (Hodgkiss)	(Tegonotus)	N.Y. Tech. Bull. 163:39
dentatus (Nal.)	(Tegoprionis)	M25:151; B2:4; Key, app. 3
dentilobus (Hodgkiss)	(Tegonotus)	N.Y. Tech. Bull. 163:28
depressus (Nal.)	(Tegonotus)	M25:82; VII:492; VIII:33
destructor (K.)	(Aculops)	X:160; Lamb, Bull. Entomol. Res. 44: 343, 1953 as syn. of lycopersici Massee
dianthi Lindroth	(Eriophyes)	M25:93
diospyri (K.)	(Eriophyes)	XVI:23
dipterochelus (Nal.)	(Diptacus)	M25:86; Key, app. 3; sacramentae may be synonym of this
distasa K.	(Acaphylla)	B4:6
dubius (Nal.)	(Aculodes)	M25:72; Key, app. 3
ecantyx (K.)	(Eriophyes)	C2:2
ehmanni K.	(Trisetacus)	B10:3; B12:19; not pini as originally identified
eldoradensis K.	(Platyphytoptus)	XXII:124
elongatus Hodgkiss	(Eriophyes)	N.Y. Tech. Bull. 163:25
emarginatae (K.)	(Phytoptus)	III:144; see also padi and pruni-crumena
emiliae Carmona	(Asetadiptacus)	Key, app. 3
epilis K.	(Dechela)	B13:3; B20:17; Key, app. 3
epidermi Donnadieu	(as Phytocoptes)	Ann. de las Soc. Linneene, 153–155, 1876; includes several species as larvae of tetranychids
erineus (Nal.)	(Eriophyes)	M25:74; I:184; VIII:24; 25; XX:148
eriobius Nal.	(Eriophyes)	M25:121, as E. macrochelus eriobius and vars. monspessulerinea, platanoidea, pseudoplatanea
erodii Gerber	(Trichostigma)	M25:117; Key, app. 3
essigi (Hassan)	(Acalitus)	XI:205; XVII:94; B16:15
eurynotus (Nal.)	(Aculus)	M25:136; XI:206
euzonus Nal.	(Cecidodectes)	M25:90; Key, app. 3
fagerinea (K.)	(Acalitus)	Ann. Entomol. Soc. Am. 52:650, 1959
fagifoliae K.	(Rhyncaphytoptus)	X:167
fagivagrans K.	(Caroloptes)	X:166; Key, app. 3
fastigatus (Nal.)	(Neotegonotus)	M25:119; C5:8; Key, app. 3
ferruginator (K.)	(Tegonotus)	XXIV:159; syn. of carinatus Nal.
ficifoliae K.	(Rhyncaphytoptus)	Key, app. 3
ficus Attiah	(Diptilomiopus)	Bull. Entomol. Soc. Egypt LI(1):1, 1967
ficus Cotte	(Eriophyes)	M25:92
filiformis (Nal.)	(Eriophyes)	M25:89

Species and Author	*Genus*	*Reference*
fleschneri (K.)	(Cosella)	XXVIII:12; Key, app. 3
flexilis K.	(Setoptus)	C7:23
flocculentus K.	(Diptacus)	XXVIII:19
floridanus K.	(Trisetacus)	B12:19
fockeui (Nal. and Trouessart)	(Aculus)	M25:109
fragarifoliae K.	(Diptacus)	Ann. Entomol. Soc. Am. 52:656, 1959
fraxinicolus Nal.	(Eriophyes)	M25:139
fraxinivorus Nal.	(Eriophyes)	M25:138
fremontiae K.	(Pareria)	XVIII:32; Key, app. 3
fructiphilus K.	(Phyllocoptes)	VIII:30; B21:21
gallarum Donnadieu	(as Phytocoptes)	Ann. de las Soc. Linneene 153–155, 1876, a composit of several species characterized as tetranychid larvae; Key, app. 3
gallicolus Nal.	(Phyllocoptyches)	M25:90; Key, app. 3
galii (Karpelles)	(Cecidophyes)	M25:150; Key, app. 3
gardeniella (K.)	(Colomerus)	B12:9; C5:6; Key, app. 3
garryana (K.)	(Phytocoptella)	XXII:122
gastrotrichus Nal.	(Eriophyes)	M25:149
gemmavitians Styer et al.	(Trisetacus)	Ann. Entomol. Soc. Am. 65(5):1089, 1972
georghioui (K.)	(Eriophyes)	XXVII:1
geratopus (K.)	(Phytoptus)	B14:9
gigantorhynchus (Nal.)	(Diptacus)	M25:109; III:149 (as prunorum); XXVI:279
glabri (K.)	(Aculops)	XVIII:33 (the protogyne, see paraglabri)
gleditsiae (K.)	(Aculops)	XXVI:275; XXVIII:3
globosus K.	(Rhynacus)	C1:15
glochidionis K.	(Colopodacus)	C2:19
goniothorax (Nal.)	(Phyllocoptes)	M25:104, 106; incl. vars.
gossypii (Banks)	(Acalitus)	J. N.Y. Entomol. Soc. 12:29, 1904; M25:127
gossypii K.	(Heterotergum)	XXIII:126; XXIV:161; Key, app. 3
gracilis (Nal.)	(Phyllocoptes)	M25:108 (see also parviflori)
granati (Canestrini and Massalongo)	(Eriophyes)	M25:134
graniti K.	(Phyllocoptes)	XXVI:275
grosmanni K.	(Trisetacus)	Ann. Entomol. Soc. Am. 52(6):649, 1959
guerreronis (K.)	(Eriophyes)	B14:7
gymnaspis (Nal.)	(Coptophylla)	M25:210
halourga K.	(Nalepella)	B16:3, 1965
harperi K.	(Acarolox)	B21:2; Key, app. 3
hassani (K.)	(Tegolophus)	XXVII:3
havrylenkonis K.	(Austracus)	XIV:20; Key, app. 3
hederae (K.)	(Acaricalus)	VII:490; X:164
hedericola (K.)	(Phytocoptella)	XIII:212
heptacanthus (Nal.)	(Tegonotus)	Zoologica p. 271; Key, app. 3
heteromeles (K.)	(Phytoptus)	IV:226; B8:9
heteronyx Nal.	(Eriophyes)	M25:88 (as subsp. of macrochelus)

Species and Author	*Genus*	*Reference*
heterophyllae K.	(Nalepella)	C1:19
hibisci Nal.	(Eriophyes)	M25:126
hibiscitellus Nal.	(Eriophyes ?)	M25:126
holmesi (K.)	(Diptilomiopus)	B7:18
holodisci (K.)	(Colomerus)	C4:13
hybridicola (K.)	(Phytoptus)	XXV:245
hydrophylli K.	(Acaricalus)	C7:13
hystrix (Nal.)	(Abacarus)	M25:72; XIV:28; B9:2; B11:11
ilicifolia K.	(Trimeroptes)	B11:17
ilicis (K.)	(Disella)	B13:9; Key, app. 3
ilicis (Canestrini)	(Eriophyes)	M25:87, 89; C2:2
inaequalis K.	(Paraphytoptus)	II:304
inaequalis (Wilson and Oldfield)	(Phytoptus)	Ann. Entomol. Soc. Am. 59(3):586, 1966
indiae K.	(Acaphylla)	XXII:126; B4:6
insidiosus (Keifer and Wilson)	(Phytoptus)	Bull. Cal. Dept. Agr. 44:145, 1955; XXV:245; B14:19
insolita K.	(Peralox)	B6:2; Key, app. 3
jamaciae K.	(Notostrix)	C4:21
jasmini (Chan.)	(Eriophyes)	Bangalore p. 75
jevremovici K.	(Diptilomiopus)	B1:18
johnstoni K.	(Anothopoda)	An:655; Key, app. 3
johnstoni K.	(Retracrus)	B16:7; Key, app. 3
jonesi (K.)	(Setoptus)	I:189; XIV:19; Key, app. 3
junipérici Boczek	(Keiferella)	Key, app. 3; see also general index
juniperinus (Nal.)	(Trisetacus)	M25:71, as subsp. of quadrisetus
kaweckii Boczek	(Bacculacus)	Key, app. 3; see also general index
kenyae K.	(Cisaberoptus)	B18:2
keiferi Southcott and Lange	(Aculops)	Rec. South Australian Mus., 16(7), June 1, 1971
kirghisorum Schevtchenko	(Trisetacus)	Leningrad Univ. Vestnik. Serian. Biolosu 3:60, 1967
kraftella (K.)	(Eriophyes)	XXVI:271
labiatiflorae Thomas	(Eriophyes)	Z. Gesam. Naturwis. n.s. 39:459, 1872
laevis Nal.	(Phytoptus)	M25:83, 86
lamimani (K.)	(Coptophylla)	VI:419; XIV:26; Key, app. 3
laniger Nal.	(Phyllocoptes ?)	M25:124
lantanae Cook	(Eriophyes)	M25:147; Rep. II Estac. Agron. Cuba 1909, p. 143
laricis Farkas	(Boczekella)	Key, app. 3; see also general index
laricis (Tubeuf)	(Trisetacus)	M25:71 as subsp. of pini
ledi K.	(Acalitus)	B14:2; Key, app. 3
lentus Boczek	(Tetraspinus)	Key, app. 3; see also general index
leucothonis (K.)	(Phytocoptella)	B9:19
ligustri (K.)	(Aculus)	I:190; XXVII:5; Key, app. 3
ligustri K.	(Eriophyes)	XIII:213
lineatus K.	(Anchiphytoptus)	XVIII:31; Key, app. 3
lionotus Nal.	(Phytoptus)	M25:83, 84 as subsp. of laevis
litchii K.	(Eriophyes)	XIII:212

Species and Author	*Genus*	*Reference*
lithocarpi K.	(Glyptacus)	XXI:67; Key, app. 3
lobatus K.	(Phantacrus)	B16:2; Key, app. 3
lobuliferus (K.)	(Aculops)	B2:13
loricatus (Nal.)	(Anthocoptes)	M25:82; Key, app. 3
löwi Nal.	(Phytoptus)	M25:140
lycopersici (Massee)	(Aculops)	Lamb, Bull. Entomol. Res. (BER) 44: 343–350, 1954. The name lycopersici Tryon 1917 considered nude. The name destructor K. considered a synonym.
lycopersici (Wolffenstein)	(Eriophyes)	M25:143; Lamb (BER) places calacladophora Rolfs as synonym
lyrata K.	(Cecidophyes)	Ann. Entomol. Soc. Am. 52:650, 1959
machaerii K.	(Catechela)	C2:15; Key, app. 3
mackiei K.	(Eriophyes)	II:302; XI:204; Wilson and Oldfield, Ann. Entomol. Soc. Am. 63(4):1119, 1970
macrochelus (Nal.)	(Eriophyes)	M25:119 includes form typicus Nal.
macrochelus eriobius Nal.	(Eriophyes)	M25:120–121, includes vars. monspessulerinea Nal., platanoidea Nal., pseudoplatanea Nal.
macrochelus var. pseudoplatani Corti	(Eriophyes)	M25:120
macrochelus psilomerus Nal.	(Eriophyes)	M25:119, and var. aceris campestris
macrorhynchus (Nal.)	(Artacris)	M25:119, 121; incl. vars. cephaloneus, var. aceris campestris; m. c. var. aceris monspessulani Nal.
macrorhynchus (Nal.)	(Artacris)	C4:9 as in Artacris
macrotrichus (Nal.)	(Eriophyes)	M25:81, 82
maculatus (Hodgkiss)	(Aculops)	N.Y. Tech. Bull. 163:25
madronae K.	(Asetacus)	XVIII:36; Key, app. 3
magnificus (Hodgkiss)	(Rhyncaphytoptus)	N.Y. Tech. Bull. 163:38
major Hodgkiss	(Eriophyes)	N.Y. Tech. Bull. 163:31
malifoliae (Parrott)	(Cecidophyes)	N.Y. Agr. Exp. Sta. Bull. 283:312, 1906
malivagrans (K.)	(Aculus)	XVI:41 (syn. of A. schlechtendali)
mangiferae (Sayed)	(Eriophyes)	Bull. Soc. Fouad Ier Entomol. XXX:7, 1947
mangiferae (Attiah)	(Metaculus)	Bull. Soc. Entomol. Egypt XXXIX:379, 1955; Bangalore p. 114
mangiferae Chan.	(Neocalacarus)	Bangalore p. 101; Key, app. 3
mangiferae (K.)	(Tegonotus)	XVI:43
marginemtorquens (Nal.)	(Phytoptus)	M25:105 (as subsp. of pyri)
marianae K.	(Baileyna)	XXII:124; Key, app. 3
massalongoi (Nal.)	(Aculops)	M25:140
masseei Nal.	(Phyllocoptes)	M25:112
maxwelli (K.)	(Oxycenus)	III:152; VIII:33; B2:7; Key, app. 3
mckenziei (K.)	(Aculodes)	XIV:26; B21:7; Key, app. 3
medicaginis K.	(Eriophyes)	XIV:26; B21:19
megarostris (K.)	(Rhyncaphytoptus)	II:305; XII:121
merwei Tucker	(Eriophyes)	M25:74
minutissimus Hodgkiss	(Eriophyes)	N.Y. Tech. Bull. 163:34 the deutogyne

Species and Author	*Genus*	*Reference*
modestus Hodgkiss	(Eriophyes)	N.Y. Tech. Bull. 163:29
monophylli K.	(Platyphytoptus)	XXI:70
montanus K.	(Phytocoptella)	XXII:121
morrisi K.	(Rhombacus)	B14:15
muesebecki K.	(Calepitrimerus)	IX:113
multisternatus K.	(Platyphytoptus)	III:146; XXII:124
murrayae Chan.	(Paratetra)	Bangalore p. 93
musae K.	(Phyllocoptruta)	XXIII:128
myersi (K.)	(Tegolophus)	III:150; B2:4
nageli K.	(Phyllocoptes)	B16:19
neglectus (Massee)	(Tegonotus)	M25:112
negundi Hodgkiss	(Eriophyes)	N.Y. Tech. Bull. 163:35
neobeevori (K.)	(Eriophyes)	XX:148; XXVI:272
neocynarae K.	(Eriophyes)	V:334
neocynodonis (K.)	(Eriophyes)	B1:2; B4:14; B8:13; syn. of cynodoniensis
neoessigi (K.)	(Eriophyes)	VIII:22
neolitsiae Chan.	(Keiferana)	Bangalore p. 103; Key, app. 3
neopiperis (Wilson)	(Colomerus)	Ann. Entomol. Soc. Am. 63:1125, 1970
nervorum Donnadieu	(as Phytocoptes)	Ann. de las Soc. Linneene, 153–155, 1876; includes composit of several species as tetranychid larvae
nolinae (K.)	(Acamina)	III:150; XIV:29; Key, app. 3
notus K.	(Dichopelmus)	XXVIII:10; Key, app. 3
nyssae K.	(Cenalox)	B3:7; Key, app. 3
occidentalis K.	(Acarelliptus)	XVII:98
oleae Nal.	(Eriophyes)	M25:140
oleivora (Ashmead)	(Phyllocoptruta)	Can. Entomol. 11:160, 1897; I:193; XIII:128; B18:3
olneyae K.	(Heterotergum)	B9:3
oregonus K.	(Acarelliptus)	C3:21
ornatus Hodgkiss	(Eriophyes)	N.Y. Tech. Bull. 163:32
orthiaspis Nal.	(Phytocoptochetus)	Key, app. 3
orthomera (K.)	(Acalitus)	XVII:94; B16:15
oryzae K.	(Abacarus)	B9:2; B11:11
padi Nal.	(Phytoptus)	M25:109
palomaris K.	(Stenacis)	C4:15; Key, app. 3
parabuxi (K.)	(Phytoptus)	B12:7
paradianthi (K.)	(Eriophyes)	XIX:65; XXVII:1
paradoxus Nal.	(Paraphytoptus)	M25:157; Key, app. 3
paraglabri (K.)	(Aculops)	XVIII:34 (deutogyne of glabri)
parallelus Hodgkiss	(Eriophyes)	N.Y. Tech. Bull. 163:40
paralobus K.	(Acaricalus)	B4:4; B11:13
paramackiei K.	(Eriophyes)	XI:204; Pan-Pac, Entomol. 31:112, 1955
parapopuli K.	(Eriophyes)	VIII:22; Wilson and Oldfield, Ann. Entomol. Soc. Am. 59(3):589, 1966
parthenii (K.)	(Eriophyes)	XX:147; B20:9; Lange, J. Entomol. 46(1):162, 1953
parviflori K.	(Phyllocoptes)	III:146; (syn. of gracilis)
pelekassi (K.)	(Aculops)	XXVII:6; B6:11
perseaflorae K.	(Tegolophus)	C1:9

Species and Author	*Genus*	*Reference*
peucedani (Canestrini)	(Eriophyes)	M25:136
peyerimhoffi Nal.	(Eriophyes)	M25:117
phloeocoptes (Nal.)	(Acalitus)	M25:108
phoenicis K.	(Mackiella)	III:147; XI:204; XIV:39; B11:19
phyllerites Nal.	(Eriophyes)	M25:120 (protogyne of pseudoplatani)
pickeringiae K.	(Anthocoptes)	XVIII:36
pini (Nal.)	(Trisetacus)	M25:70; B10:1
pipturella K.	(Brionesa)	Key, app. 3; B18:5
piri of authors—see pyri		
pirifoliae K.	(Epitrimerus)	II:309; XII:122 as syn. of pyri
pistaciae Nal.	(Eriophyes)	M25:125
platani K.	(Rhyncaphytoptus)	IV:230; Hall, Univ. of Kans. Sci. Bull. XLVII (9) :671, 1967
plectilis K.	(Acunda)	B16:19; Key, app. 3
plicator (Nal.)	(Eriophyes)	M25:101
podocarpi K.	(Paracalacarus)	B8:8; Key, app. 3
populi (Nal.)	(Eriophyes)	M25:80
populivagrans (K.)	(Aculops)	XXI:68; B21:9; Key, app. 3
prostantherae (K.)	(Eriophyes)	XXVII:17
prunandersoni (K.)	(Phytoptus)	VI:416; B14:9
pruni-crumena (Walsh)	(Phytoptus ?)	M25:110; the relation of this name to padi and emarginatae is uncertain
prunorum (K.)	(Diptacus)	III:149; a syn. of gigantorhynchus
pseudoinsidiosus (Wilson)	(Phytoptus)	Ann. Entomol. Soc. Am. 58:327, 1965
pseudoplatani (Corti)	(Eriophyes)	M25:120 as subsp. of macrochelus
pseudotsugae K.	(Epitrimerus)	XVI:42
pseudotsugae K.	(Trisetacus)	B16:5
psilaspis (Nal.)	(Cecidophyopsis)	M25:72
psilomerus Nal.	(Eriophyes)	M25:119 as subsp. of macrochelus
pulviferus K.	(Calacarus)	X:163; XVIII:35; Key, app. 3
pumilae K.	(Brachendus)	B11:2; Key, app. 3
purshivagrans K.	(Aculus)	C3:7 (protogyne of amandae)
pusilla K.	(Cecidophyes)	B5:2
puttarudriahi (Chan.)	(Eriophyes)	Bangalore p. 77
pyramidicus (K.)	(Tetraspinus)	B4:2; Key, app. 3
pyri (Nal.)	(Epitrimerus)	M25:105; B19:2; Key, app. 3
pyri Pagenstecher	(Phytoptus)	M25:104; I:183
pyri var. mali (Nal.)	(Phytoptus)	M25:100
pyri marginemtorquens (Nal.)	(Phytoptus)	see marginemtorquens
pyri marginemtorquens var. mali (Nal.)	(Phytoptus)	M25:106
pyri var. sorbi Canestrini	(Phytoptus)	M25:106
quadripedes Shimer	(Vasates)	N.Y. Tech. Bull. 163:16 represents the deutogyne; XIV:25; XXVII:3; B19:11; B21:23; Key, app. 3
quadrisetus (Thomas)	(Trisetacus)	M25:71; Morgan and Hedin, Can. Entomol. 92(8):608, 1960
querci (Garman)	(Eriophyes ?)	M25:89; The status of this species in relation to other oak erineum mites is uncertain at this writing

Species and Author	*Genus*	*Reference*
quercifolii Farkas	(Achaetocoptes)	Key, app. 3; see also general index
quercilobatae K.	(Caliphytoptus)	II:311; Key, app. 3
quinquilobus Hodgkiss	(Eriophyes)	N.Y. Tech. Bull. 163:33
ramosis (Hodgkiss)	(Trisetacus)	M25:71; considered a synonym of quadrisetus in this book
rapaneae K.	(Nothopoda)	XVII:96; XXI:69; Key, app. 3
rapistri Carmona	(Metaculus)	Agronomia lusit. 30:89, 1969
regulus Hodgkiss	(Eriophyes)	N.Y. Tech. Bull. 163:24
rhamni Pagenstecher	(Phytoptus ?)	M25:126 as annulatus Nal.?
rhamnicolus (K.)	(Thamnacus)	II:307; Key, app. 3
rhamnicroceae (K.)	(Pentamerus)	B17:2; Key, app. 3
rhamnivagrans (K.)	(Aculus)	VI:419
rheumella K.	(Rhombacus)	C4:19
ribis Massee	(Anthocoptes)	M25:112
ribis (Westwood)	(Cecidophyopsis)	M25:111
rossettonis (K.)	(Eriophyes)	C1:5
rostratae Roivainen	(Novophytoptus)	Key, app. 3; see also general index
rubigator K.	(Notalox)	B3:5; Key, app. 3
rudis (Canestrini)	(Acalitus)	M25:83, 84, 86
sabinianae K.	(Platyphytoptus)	I:188; II:304; XXI:70; Key, app. 3
sacchari (Chan.)	(Eriophyes)	Bangalore p. 55
saccharini K.	(Rhyncaphytoptus)	XXVII:15
sacramentae (K.)	(Diptacus)	IV:232; XIII:216; Key, app. 3
sakimurae K.	(Phyllocoptruta)	B18:3
salicicola Banks	(Cecidobia)	M25:80, Key, app. 3
samoae K.	(Aberoprus)	XVII:97; Key, app. 3
sawatchense (K.)	(Eriophyes)	B15:7
schlechtendali (Nal.)	(Aculus)	M25:105, 106
schlechtendali Nal.	(Callyntrotus)	M25:108; IV:234; Key, app. 3
schlingeri Wilson and Oldfield	(Heterotergum)	Ann. Entomol. Soc. Am. 59:589, 1966
secundus K.	(Dicrothrix)	B18:9
segundus K.	(Acaricalus)	X:164; XXVIII:11; Kéy, app. 3
selachodon vanEyndhoven	(Cecidophyopsis)	Entomol. Berich. 8:149, 1967
septemcarinatus (Liro)	(Pentamerus)	Key, app. 3
sequoiae K.	(Trisetacus)	C4:2
setiger (Nal.)	(Fragariocoptes)	M25:107; Key, app. 3
sheldoni Ewing	(Eriophyes)	Proc. Entomol. Soc. Wash. 29:193, 1937
similis Nal.	(Phytoptus)	M25:109
simondsi K.	(Cercodes)	B11:11; Key, app. 3
slinkardensis K.	(Phyllosoptes)	B21:21
slykhuisi (Hall.)	(Eriophyes)	J. Kans. Entomol. Soc. 31:233, 1958
solidaginis K.	(Leipothrix)	B19:2; Key, app. 3
sorbi Canestrini	(Phytoptus)	M25:106, as var. of pyri
sphaeralceae K.	(Acalitus)	B14:3
spiniferus K.	(Scolocenus)	B8:4; Key, app. 3
spinitibius (K.)	(Aculops)	B18:13
spiniventris K.	(Cymoptus)	XVI:40; Key, app. 3
stefanii Nal.	(Eriophyes)	M25:124
steinwedeni K.	(Acaphylla)	XIII:215; XXII:126; B4:6; Key, app. 3

Species and Author	*Genus*	*Reference*
stellatus K.	(Dialox)	B8:19; Key, app. 3
strobacus K.	(Setoptus)	B18:19
sulcatus (Nal.)	(Monochetus)	Zoologica p. 254; M25:87
swensoni K.	(Diptacus)	XXVII:13
syzygii K.	(Metaculus)	B6:13; Key, app. 3
tampae K.	(Rhynacus)	B10:19
taxodii (K.)	(Epitrimerus)	VII:486
tejonis K.	(Calacarus)	XVIII:35; XXIII:126
tenuis (Nal.)	(Eriophyes)	Zoologica p. 213; M25:72, 73
tetanothrix (Nal.)	(Aculops)	Zoologica p. 220; B20:13; C3:17
tetratrichus (Nal.)	(Phytocoptella)	Zoologica p. 226; M25:129, 131
theae (Watt)	(Acaphylla)	M25:133 (Considered a nude name); XXII:126; see also steinwedeni
theospyri (K.)	(Eriophyes)	B1:7
tiliae Pagenstecher	(Phytoptus)	Zoologica p. 225; M25:128, 31; Key, app. 3
tlaiae Trabut	(Eriophyes)	M25:133
tosichella (K.)	(Eriophyes)	C3:2 (see also tenuis)
trachycarpi K.	(Tumescoptes)	V:337; Key, app. 3
triceras (Börner)	(Nalepella)	Zoologica p. 225; M25:70 (as Phyllocoptes)
tricholaenae K.	(Catarhinus)	XXVII:12; Key, app. 3
trilobus (Hodgkiss)	(Tegonotus)	N.Y. Tech. Bull. 163:38; (deutogyne of T. dentatus)
triplacis (K.)	(Eriophyes)	B1:8; B17:13
tristriatus (Nal.)	(Eriophyes)	Zoologica p. 218; M25:74
tristriatus var. erineus (Nal.)	(Eriophyes)	M25:74 see erineus
tristichus Nal.	(Phytoptochetus)	Key, app. 3
tritici Schevtchenko	(Eriophyes)	? p. 232, 1970 (translation of serial title unavailable)
trymatus K.	(Acathrix)	Key, app. 3; B7:16
tsugifoliae K.	(Nalepella)	XXI:65; B16:3; also Marshall and Lindquist, Can. Entomol. 104:242, 1972
tucsonensis K.	(Vasates)	B19:11
tulipae K.	(Eriophyes)	1:185; XIV:22; Key, app. 3
tuttlei K.	(Heterotergum)	B2:7
tuttlei K.	(Mesalox)	B5:11; Key, app. 3
ulmivagrans K.	(Rhyncaphytoptus)	VI:420; VII:491 (deutogyne named as Abacoptes platynus); XIV:30
urticae K.	(Quadracus)	XIV:30 (syn. of urticarius)
urticarius (Canestrini and Massalongo)	(Quadracus)	M25:92 as Phyllocoptes
vaccinii (K.)	(Acalitus)	V:329
vaga (K.)	(Eriophyes)	B5:15
variabilis (Hodgkiss)	(Tegonotus)	N.Y. Tech. Bull. 163:26; deutogyne of T. dentilobus
verilicis (K.)	(Cecidophyopsis)	VI:418
vermicularis Nal.	(Eriophyes)	M25:118
vermiformis (Nal.)	(Cecidophyopsis)	M25:81, 83; XXVI:273; Key, app. 3
virginiana K.	(Johnella)	Ann. Entomol. Soc. Am. 52:652, 1959

Species and Author	*Genus*	*Reference*
vitis (Nal.)	(Calepitrimerus)	Zoologica p. 276 (as Epitrimerus); M25:126; XII:122
vitis (Pagenstecher)	(Colomerus)	Zoologica p. 230; M25:126 (Eriophyes); C5:6 (as Colomerus)
vitis (Nal.)	(Calepitrimerus)	M25:126 (as Phyllocoptes); XII:122 (as deutogyne of Calepitrimerus vitis)
wegorecki Boczek	(Latinotus)	Key, app. 3
wilsoni K.	(Heterotergum)	XXIV:161
yuccae (K.)	(Phytocoptella)	XXII:121
yumensis K.	(Heterotergum)	B18:11
zeasinis (K.)	(Eriophyes)	B18:13

Appendixes

Appendixes

APPENDIX 1

COMMON NAMES AND SCIENTIFIC NAMES OF PLANTS

Common Name	*Scientific Name*
alder,	*Alnus* Erhart
European	*incana* Moench
alfalfa	*Medicago sativa* L.
almond	*Prunus amygdalus* Batch
aloe	*Aloe humilus* Haworth
antelope bush	*Purshia tridentata* de Condolle
apple	*Malus sylvestris* Miller
apricot	*Prunus armeniaca* L.
arborvitae	*Thuja* Linnaeus
artichoke	*Cynara scolymus* L.
ash, California mountain	*Sorbus californica* Greene
European mountain	*aucuparia* L.
asparagus	*Asparagus officinalis altilis* L.
aspen, American or quaking	*Populus tremuloides* Micheaux
avocado	*Persea americana* Miller
azalea,	*Rhododendron* L.
dwarf or atlantic	*atlanticum* Rehder
bamboo,	*Bambasa* Schreber
sasa	*Phyllostachys reticulata* C. Koch
banana	*Musa* L.
barley,	*Hordeum vulgare* L.
foxtail barley	*jubatum* L.
barberry	*Berberis* L.
bean,	*Phaseolus vulgaris* L.
castor	*Ricinus communis* L.
lima	*Phaseolus limensis* MacFadyen
beach,	*Fagus* L.
American	*grandifolia* Erhart
begonia	*Begonia* L.
bindweed	*Convolvulus arvensis* L.
birch,	*Betula* L.
European white	*pendula* Roth (verrucosa Erhart)
white	*populifolia* Marshall
blackberry,	*Rubus* L.
Mammoth	*vitifolius* Chamisso and Schlechtendal
Himalaya	*thyrsanthus* Focke
evergreen or cut-leaved	*laciniatus* Willdenow

blackthorn (sloe)	*Prunus spinosa* L.
bloodweed	*Ambrosia aptera* (de Condolle)
boxelder	*Acer negundo* L.
boxwood,	*Buxus* L.
common	*sempervirens* L.
bramble	*Rubus* L.
buckeye, California	*Aesculus californicus* (Spach)
buckthorn	*Rhamnus* L.
butterfly bush	*Buddleja* L.
cabbage	*Brassica oleràcea capitata* L.
camellia	*Camellia* L.
camphor	*Cinnamomum camphora* Nees and Ebermaier
carnation	*Dianthus caryophyllus* L.
carob	*Ceratonia Siliqua* L.
carrot	*Daucus carota* var *sativa* de Condelle
cashew nut	*Anacardium* L.
cassava	*Manihot* Adans
castor bean	*Ricinus communis* L.
cedar, Atlas	*Cedrus atlantica* Manetti
japanese	*Cryptomeria japonica* David Don
red	*Juniperus virginiana* L.
incense	*Libocedrus* Endlicher
celery	*Apium graveolens* L.
cherry, sweet	*Prunus avium* L.
wild bitter	*emarginata* Douglas
wild black	*serotina* Ehrharts
chestnut	*Castanea* Miller
Asiatic or European	*sativa* Miller
chrysanthemum	*Chrysantheum* L.
cinquefoil (five-finger)	*Potentilla* L.
clover,	*Trifolium* L.
crimson	*incarnatum* L.
spotted bur	*Medicago arabica* Allioni
white	*Trifolium repens* L.
cocklebur	*Xanthium* (Tournefort) L.
cocksfoot	*Dactylis galmenata* L.
coconut	*Cocos nucifera* L.
coffee, Arabian or common	*Coffea arabica* L.
copper leaf	*Acalypha* L.
corn	*Zea mays* L.
cotoneaster	*Cotoneaster* Medicus
cotton,	*Gossypium* spp. L.
Levant	*herbaceum* L.
Sea-Island	*barbadense* L.
Upland	*hirsutum* L.
cottonwood, eastern or northern	*Populus deltoides* Bartram
cowpeas	*Vigna sinensis* Savi
crab apple	*Malus* spp. L.
cranberries	*Vaccinium* L.
croton	*Codlaeum variegatum* Blume
cucumber	*Cucumis sativus* L.
cucurbits	*Cucurbita* L.

currants,	*Ribes* L.
black	*americanum* Miller
northern red	*rubrum* Jancz
cyclamen	*Cyclamen* L.
cypress,	*Cupressus* L.
Monterey	*macrocarpa* Hartweg
mountain	*bakeri* Jepson
false	*Chamaecyparis* Span
dahlia	*Dahlia* Cavainiller
dates	*Phoenix dactylifera* L.
dewberry,	*Rubus* L.
garden	*geophilus* Blanch
Marvel	*mirus* Bailey
southern	*tribialis* Micheaux
wild	*macropetalus* Douglas
dill	*Anethum* L.
dogwood	*Cornus* L.
eggplant	*Solanum Melongena* L. var. *esculentum* Nees
Elaeagnus	*Elaeagnus* L.
elder or elderberry	*Sambucus* L.
elm,	*Ulmus* L.
American	*americana* L.
endive	*Cichorium endivia* L.
eucalyptus	*Eucalyptus* L'Heritier
Eugenia	*Eugenia* L.
fataque	*Panicum maximum* Jacquin
fescue, meadow or tall	*Festuca elatior* L.
sheeps	*ovina* L.
fig	*Ficus carica* L.
filbert, common or European	*Corylus avellana* L.
fir	*Abies* Miller
Douglas	*Pseudotsuga douglasii* Carriere (*P. taxifolia* Britton)
silver	*Abies alba* Miller (*A. pectinata* de Condolle)
frangipani	*Plumeria* L.
fuchsia	*Fuchsia* L.
gardenia	*Gardenia* Ellis
geranium	*Geranium* L.
gerbera	*Gerbera* Cassini
gladiolus	*Gladeolus* L.
gooseberry	*Ribes grossularia* L.
gourd	*Cucurbita* L.
grape	*Vitis* L.
grass, Bermuda	*Cynodon dactylon* Persoon
Beardless wild-rye	*Elymus triticoides* Buckley
brome grass	*Bromus* L.
buffalo	*Buchloe dactyloides* (Nuttall)
Canada blue	*Poa compressa* L.
Italian rye	*Lolium multiforum* Lamarck
Indian rice grass	*Oryzopsis hymenoides* (Roemer and Schult)

Johnson grass	*Sorghum halapense* L.
Kentucky blue	*Poa pratensis* L.
Natel grass	*Tricholaena* (*Agropyron*) *repens* Hitchcock
western wheat grass	*Agropyron smithii* Rydberg
perennial or English rye	*Lolium perenne* L.
Wheeler blue grass	*Poa nervósa* (Hook) Vasey
groundcherry	*Physalis* L.
guava	*Psidium* L.
hackberry	*Celtis* L.
hawthorne	*Crataegus* L.
hazelnut	*Corylus maxima* Miller
hemlock, common or Canadian	*Tsuga canadensis* Carr
western	*heterophylla* Sargent
hemp	*Cannabis sativa* L.
hibiscus	*Hibiscus micranthus* L.
hickory	*Carya* Nuttall
shagbark	*ovata* Koch
swamp	*cordiformis* Koch
Himalaya-berry	*Rubus procerus* Mueller
holly,	*Ilex* L.
American	*opaca* Aiton
English	*aquifolium* L.
hollyhock	*Althaea* L.
hops	*Humulus* L.
horehound	*Marrubium vulgare* L.
hornbeam,	*Carpinus* L
European	*betulus* L.
horse chestnut,	*Aesculus* L.
common or European	*hippocastanum* L.
horsemint	*Monarda* L.
hydrangea	*Hydrangea* L.
indigo	*Indigofera swaziensis* Bolus
incense cedar	*Libocedrus* Endlicher
iris	*Iris* L.
ivy	*Hedera helix* L.
jasmine	*Jasminum* L.
jimsonweed	*Datura* L.
juniper,	*Juniperus* L.
alligator	*pachyphloea* Martinez
California	*californica* Carriere
common	*communis* L.
Western or Sierra	*occidentalis* Hook
jute	*Corchorus capsularis* L.
Khadhawi trees (Khat, Cafta, or Arabian tea)	*Catha edulis* Forskal
knotweeds	*Polygonum* spp.
Kudsu vine	*Pueraria Thunbergiana* Bentham

larch	*Larix* Miller
larch, European	*decidua* Miller (*L. europa* de Condolle)
laurel	*Laurus* L.
lemon	*citrus limon* Burman
lettuce	*Lactuca sativa* L.
lilac	*Syringa vulgaris* L.
lily,	*Lilium* L.
arum	*Arum* L.
lily-of-the-valley vine	*Convallaria majalis* L.
lime, large-leafed	*Tilia platyphyllos* Scopoli
linden, basswood, lime	*Tilia* L.
American	*americana* L.
litchi	*Litchi chinensis* Sonnerat
locust, black	*Robinia pseudoacacia* L.
honey	*Gleditsia triacanthos* L.
loganberry	*Rubus ursinus* var. *loganobaccus* Bailey
loquat	*Eriobotrya japonica* Lindley
lucerne	*Medicago sativa* L.
magnolia tree	*Magnolia* L.
maize	*Zea mays* L.
mallow, common	*Malva* L.
globe	*Sphaeraleca orcutti* Rose
mango	*Mangifera indica* L.
maple,	*Acer* L.
hedge	*campestris* L.
mountain	*spicatum* L.
Norway	*platanoides* L.
red	*rubrum* L.
silver	*saccharium* L.
striped	*pennsylvanicum* L.
sugar	*saccharum* Marsh
sycamore	*pseudoplatanus* L.
western big leaf	*macrophyllum* Pursh
western mountain	*glabrum* Torrey
marigold	*Tagetes* L.
melon	*Cucumis* L.
milkweed	*Asclepias* L.
millet,	*Setaria* Beauvais
foxtail	*italica* Beauvais
mint	*Mentha* L.
morning-glory	*Ipomoea purpurea* Lamarck
mulberry	*Morus* L.
muskmelon (cantaloupe)	*Cucumis melo* L.
narcissus	*Narcissus* L.
nightshade, black	*Solanum modiflorum* L.
oak,	*Quercus* L.
black	*velutina* Lamark
California white	*lobata* Nee
coast live	*agrifolia* Nee

holly	*ilex* L.
pin	*palustris* Muenchhausen
red (northern red)	*borealis* Micheaux
scarlet	*coccifera* L.
white	*alba* L.
oats	*Avena sativa* L.
okra	*Hibiscus esculentus* L.
onion,	*Allium cepa* L.
wild	*fistulosum* L.
orange, citrus sweet	*Citrus sinensis* Osbeck
Osage	*Maclura pomifera* Schneider
orchid, dendrobium	*Vanda teres* Lindley
palm,	*Palmaceae*
date	*Phoenix daetylifera* L.
oil	*Elaeis guineensis* Jacquin
parlor	*Collinia elegans* Liebman
seaforthia	*Ptychosperma Labill*
papaya	*Carica papaya* L.
parsley	*Petroselinum crispum* Nyman
passionflower	*Passiflora* L.
Paulownia (Chinese tree)	*Paulownia tomentosa* (Thumberg) Baillon
pea,	*Pisum* L.
garden or winter	*sativum* L.
peach,	*Prunus persica* Batsch
desert	*andersoni* Grey
peanut	*Arachis* L.
pear	*Pyrus communis* L.
pecan	*Carya illinoensis* Koch
pepper, red	*Capsicum frutescens* L.
pepperidge	*Nyssa sylvatica* Marsh
persimmon	Diospyros virginiana L.
petunia	*Petunia* Jussieu
pigeon pea	*Cajanus cajan* Millspaugh
pikake	*Jasminum sambac* Aiton
pine,	*Pinus* L.
Aleppo	*halepensis* Miller
bishop	*muricata* James
canary	*canariensis* C. Smith
digger	*sabiniana* Douglas
limber	*flexilis* James
lodgepole	*cordata* var. *latifolia* Watson
Monterey	*radiata* Don
one-leaf pinyon	*monophylla* Torrey and Fremont
ponderosa	*ponderosa* Douglas
scots	*sylvestris* L.
sugar	*lambertiana* Douglas
swiss stone	*cembra* L.
western or mountain white	*monticola* Douglas
white	*strobus* L.
pineapple	*Ananas comosus* Merrill (A. sativus Schultes f.)
pistachio	*Pistacia vera* L.

plum,	*Prunus* L.
European	*domestica* L.
Pacific	*subcordata* Bentham
poinsettia	*Euphorbia pulcherrima* Willdenow
pomegranate	*Punica granatum* L.
poplar,	*Populus* (Tournefort) L.
black	*nigra* L.
poppy	*Papaver* L.
potato	*Solanum tuberosum* L.
primrose, evening	*Oenothera* L.
privet,	*Ligustrum* L.
california	*ovalifolium* Hasskarl
pulses	*Leguminosae*
pumpkins, squashes	*Cucurbita* L.
puncture vine	*Tribulus terrestris* L.
purshia	*Purshia* de Condolle
pussy-toes	*Antennaria* Gaerter
pyracantha	*Pyracantha* Roemer
pyrethrum	*Chrysanthemum coccineum* Willdenow
quince	*Cydonia oblonga* Miller
raspberry,	*Rubus* spp. L.
red and yellow garden	*idaeus* L.
redwood	*Sequoia* Endlicher
rice	*Oryza sativa* L.
rose	*Rosa* L.
Rowan	*Sorbus aucuparia* L.
rubber plant,	*Ficus elastica* Roxburgh
Para	*Hevea brasiliensis* Mueller of Aarau
rye,	*Secale certale* L.
Canada wild rye	*Elymus canadensis* L.
Virginia wild rye	*virginicus* L.
rye grass, Italian	*Lolium multiflorum* Lamarck
perennial	*perenne* L.
sage	*Salvia* L.
salmonberry	*Rubus spectabilis* Pursh
sasa bamboo	*Sasa* Makino and Shibata
serviceberry	*Amelanchier vulgaris* Möench
sloe	*Prunus spinosa* L.
snowberry	*Symphoricarpos albus* Blake
sorghum	*Sorghum vulgare* Persoon
sow-thistle	*Sonchus asper* Hill
soybean	*Glycine max* Merrill
speedwell	*Veronica* L.
spiderwort	*Tradescantia* L.
spinach	*Spinacia* L.
spirea	*Spiraea* L.
spruce,	*Picea* Dietrich
Sitka	*sitchensis* Carriere
squash	*Cucurbita* L.
stone leek	*Allium porrum* L.

storksbill	*Pelargonium* L'Heritier
strawberry	*Fragaria* spp. L.
sugarcane	*Saccharum officinarum* L.
sumac,	*Rhus* L.
staghorn	*typhina* L.
sunflower	*Helianthus* L.
sweet gum	*Liquidambar styraciflua* L.
sweet pea	*Lathyrus* odoratus L.
sweet potato	*Ipomoea batatas* Lamarck
sycamore	*Platanus racemosa* Nuttall
tamarisk	*Tamarix* L.
tapioca	*Manihot esculenta* Crantz
tea	*Thea sinensis* L.
thimbleberry	*Rubus odoratus* L.
tievine	*Ipomoea* L.
timothy hay (herds-grass)	*Phleum pratense* L.
tobacco	*Nicotiana tabacum* L.
tomato	*Lycopersicon* Miller
toyon	*Heteromeles arbutifolia* Lindlel
tuberose	*Polianthes tuberosa* L.
tulip	*Tulipa* L.
tulip tree	*Liriodendron tulipifera* L.
turnip	*Brassica rapa* L.
umbrella tree	*Magnolia tripetala* L.
velvet bean	*Stizolobium Deeringianum* Bort
verbena	*Verbena* L.
vetch,	*Vicia* L.
common	*sativa* L.
hairy	*villosa* Roth
violet, African	*Saintpaulia ionantha* Wendland
walnut,	*Juglans* L.
black	*nigra* L.
California black	*hindsii* Jepsen
Persian	*regia* L.
Southern California black	*californica* Watson
watercress	*Nasturtium officinale* R. Brown
water-hyacinth	*Eichhornia cassipes* Solms
wheat	*Triticum aestivum* L.
white beam tree	*Sorbus aria* Crantz
white birch	*Betula populifolia* Marsh
white clover	*Trifolium repens* L.
wild service tree	*Sorbus torminalis* Crantz
willow,	*Salix* (Tournefort) L.
pussy	*discolor* Muhlenberg
yew	*Taxus* L.
zinnia	*Zinnia* L.

APPENDIX 2

SCIENTIFIC NAMES OF PESTICIDES INCLUDED IN TEXT

Acarol®	isopropyl 4, 4′-dibromobenzilate
aldicarb, (Temick®, UC-21149)	2-methyl-2-(methylthio) propionaldehyde O-methylcarbamoyl) oxime
aldrin	not less than 95% of 1, 2, 3, 4, 10, 10-hexachloro-1, 4, 4a, 5, 8 8a-hexahydro-1, 4-endo-exo-5, 8-dimethanonaphthalene
Aramite®	2-(*p-tert*-butylphenoxy)-1-methylethyl 2-chloroethyl sulfite
azinphosmethyl (Guthion®, Gusathion®)	0, 0-dimethyl S(4-oxo-1, 2, 3-benzotriazin-3 (4H)-ylmethyl) phosphorodithioate
binapacryl	2-*sec*-butyl-4, 6-dinitrophenyl 3-methyl-2-butenoate
carbaryl	1-naphthyl methylcarbamate
carbophenothion (Trithion®)	S-[(*p*-chlorophenylthio)methyl]0, 0-diethyl phosphorodithioate
Cellosize	hydroxyethyl cellulose
chlorbenside	*p*-chlorobenzyl *p*-chlorophenyl sulfide
chlorobenzilate	ethyl 4, 4′-dichlorobenzilate
chlorofenson (see Ovex)	*p*-chlorophenyl *p*-chlorobenzenesulfonate
DCDM (Neotran®)	di(4-chlorophenoxy) methane
DCPC (See DMC®)	
DDT	1, 1, 1-trichloro-2, 2-bis(*p*-chlorophenyl) ethane
Delnav® (See dioxathion)	
demeton	0, 0-diethyl 0(and S)-2 (ethylthio) ethyl phosphorothioates
diazinon	0, 0-diethyl 0-(2-isopropyl-6-methyl-4-pyrimidyl) phosphorothioate
dicofol (Kelthane®)	1, 1-bis (*p*-chlorophenyl)2, 2, 2-trichloroethanol
dimethoate	0, 0-dimethyl S-(N-methylcarbamoylmethyl) phosphorodithioate
Dimite® (See DMC)	
dinitro compounds	(See below)
dinitrocresol (sodium salt) (DNOC, DNC, Sinox®, Elgetol 30®)	4, 6-dinitro-o-cresol, sodium salt
dinitrocyclohexylphenol (DNOCHP)	2-cyclohexyl-4, 6-dinitrophenol
dinobuton (Acrex®) (UC-19786)	2-*sec*-butyl-4, 6-dinitrophenyl isopropyl carbonate
dinocap (Karathane®, Arathane®)	2-(1-methylheptyl)-4, 6-dinitrophenyl crotonate

dinoseb (DNBP, DNOSBP, DNSBP, Elgetol 318®)	2-sec-butyl-4, 6-dinitrophenol, 2, 2′, 2″-nitrilotriethanol salt
dioxathion (Delnav®, Hercules AC-528®)	2, 3-*p*-dioxanedithiol S, S-bis(0, 0-diethyl phosphorodithioate)
DMC (Dimite®, DCPC)	1, 1-bis(*p*-chloropheny) ethanol
DN-111®	4, 6-dinitro-o-cyclohexylphenol, N, N-dicyclohexylamine salt
endosulfan	6, 7, 8, 9, 10, 10-hexachloro-1, 5, 5a, 6, 9, 9a-hexahydro-6, 9-methano-2, 4, 3-benzodioxathiepin 3-ocide
endrin	1, 2, 3, 4, 10, 10-hexachloro-6, 7-epoxy-1, 4, 4a, 5, 6, 7, 8, 8a-octahydro-1, 4-*endo-endo*-5, 8-dimethanonaphthalene
EPN	ethyl *p*-nitrophenyl benzene thiophosphonate
Eradex® (See thioquinox)	
ethion	0, 0, 0′, 0′-tetraethyl S, S′-methylene-bisphosphorodithioate
fenson (CPBS, PCPBS, GC-928®, TriFenson)	*p*-chlorophenyl benzenesulfonate
fentin hydroxide	tricyclohexyl tin hydroxide
ferbam	ferric dimethyl dithiocarbamate
formetanate, hydrochloride (Carzol SP®, EP-332®, Schering 36056®)	m- [(dimethylamino)methylene]amino phenyl methylcarbamate, hydrochloride
Fundal SP®	N′-4(4-chloro-o-tolyl)-N, N-dimethylformamidine hydrochloride
Galecron®	N′-(4-chloro-o-tolyl)-N, N-dimethylformamidine
Genite, (Genitol 923®)	2, 4-dichlorophenyl ester of benzenesulfonic acid
glyodin (34 percent)	2-heptadecyl-2-imidazoline acetate
Karathane® (See dinocap)	
Lovozal® (See Phenoflurazol)	
malathion	0, 0-dimethyl dithiophosphate of diethyl mercaptosuccinate
maneb	manganous ethylene bisdithiocarbamate
Metacide®	0, 0-dimethyl 0-*p*-nitrophenyl phosphorothioate, parathion mixture
Milbex® (Micasin®, CPAS®)	[(p-chlorophenyl)thio] (2, 4, 5-trichlorophenyl) diimide
Morestan®	6-methyl-2, 3-quinoxalinedithiol cyclic carbonate
Neotran® (DCDM)	di(4-chlorophenoxy)methane
Nissol®	2-fluoro-N-(1-naphthyl) acetamide

Omite®	2-(*p-tert*-butylphenoxy)-1-methylethyl 2-chloroethyl sulfite
OMPA (See schradan)	
ovex	p-chlorophenyl p-chlorobenzenesulfonate
Ovotran® (See ovex)	
parathion	0, 0-diethyl 0-*p*-nitrophenyl phosphorothioate
Pentac®	decachiorobi-2, 4-cyclopentadien-1-yl
phenoflurazol (Lovozal®)	(phenyl 5, 6-dichloro-2-trifluoromethyl)-1-benzimidazolecarboxylate
phosalone (Zolone®, RP-11974®)	S-[(6-chloro-2-oxo-3-benzoxazolinyl) methyl]0, 0-diethyl phosphorodithioate
phosphamidon	2-chloro-2-diethylcarbamoyl-1-methylvinyl dimethyl phosphate
Plictran®	tricyclohexyl tin hydroxide
rotenone	from plant species Derris and Lonchocarpus
Schradan® (OMPA)	octamethylpyrophosphoramide
Selocide	ammonium potassium selenosulfide
Sevin® (See carbaryl)	
Sulfotepp	tetraethyl dithiopyrophosphate
Sulphenone®	*p*-chlorophenyl phenyl sulfone and related sulfones
Systox® (See demeton)	
tedion (See tetradifon)	
TEPP	tetraethyl pyrophosphate
tetradifon	*p*-chlorophenyl 2, 4, 5-trichlorophenyl sulfone
tetrasul (Animert V-101®)	2, 4, 4′, 5-tetrachlorodiphenyl sulfide
Thiodan® (See endosulfan)	
thionazin	0, 0-diethyl 0-2-pyrazinyl phosphorothioate
thioquinox (See Eradex®)	2, 3-Quinoxalinedithiol cyclic trithiocarbonate
Torak® (Hercules 14503)	S-(2-chloro-1-phthalimidoethyl) 0, 0-diethyl phosphorodithioate
Trithion® (See carbophenothion)	
zineb	zinc, bis(dimethyl dithiocarbamate)
ziram	zinc dimethyl dithiocarbamate

APPENDIX 3

Eriophyoidea:

SYNOPTIC KEYS TO GROUPS AND GENERA

R. A. Newkirk and H. H. Keifer

In composing these synopses the writers have endeavored first to emphasize the respective type of each genus and to call attention to their more precise structures. A type name and pertinent reference accompanies each genus.

Another objective is to follow certain provisions in the *International Code of Zoological Nomenclature* as they apply to name availability and to type-setting. It has been necessary for us to designate new types for certain of Nalepa's genera, and these synopses bring the types into conformity with the code. (Note—The more important of these new designations and changes have been issued in Eriophyid Studies [USDA] C-5, 1971. Names by Newkirk and Keifer that are not indicated as new are in C-5).

While it has not in every case been possible (*Oxypleurites*, for example) the writers have endeavored to preserve names that Nalepa used. Preserving Nalepa's names is desirable and defensible and will cause a minimum of confusion, for workers in eriophyid taxonomy use his names. The Nalepa names are basic in the classification of the group. For a divergent opinion on the use of some of these higher category names, see E. E. Lindquist, *Canadian Entomologist*, 106 :209–212, Feb. 1974.

Under two code directives we have corrected types for *Eriophyes* von Siebold, Jan.–Mar. 1851, and for *Phytoptus* Dujardin, July 1851. While these genera retain their chronological priorities in relation to each other, their first bionomial uses came somewhat after the above dates. In the case of *Eriophyes* it is necessary to follow both Article 16 (a) (viii), and Article 69 (a) in the code. Article 16 provides that the description of the work of an animal, when accompanied by a binomial name for that animal, establishes a valid name for that animal. We view this directive as destructive to the quality of animal taxonomy because it allows careless workers to bypass the real information a taxonomist requires intelligently to place the animal in a classification system and relate it to its structural relatives.

Article 69 (a) states that the correct type of a genus can only come from the first binomial or binomials published with the genus name. When only one species name accompanies the first publication of a genus, binomially, that name automatically becomes the type of that genus by monotypy.

For *Eriophyes* the first binomial was by F. Thomas, 1872. He described the damage caused by an eriophyid to a mint plant, in general terms, and quoted the mite *Eriophyes labiatiflorae* Bremi, as the one causing the damage. While Thomas credits Bremi-Wolf with this name, he (Thomas) was the one actually publishing the name, accompanied by a description of the work of the mite. So the name is properly quoted as *Eriophyes labiatiflorae* F. Thomas.

While Thomas was equivocal about host identity (although Bremi had named it *Origanum vulgare* L.), we accept Bremi's host identification. Bremi had a considerable knowledge of plants. Also, since the short description of damage given by Thomas reads like Nalepa's characterization of eriophyid damage on this host by a mite he (Nalepa) later named *Phytoptus origani* Nal., we conclude that the mite species is the same as the one Nalepa named *origani*. We also follow Nalepa's first designation of *origani* as a separate species, and do not consider it a variety of *thomasi*, as Nalepa later placed it.

Eriophyes labiatiflorae (*origani*) has the dorsal shield tubercles on the rear shield margin, directing the dorsal setae divergently to the rear. This is one of the main features of *Aceria* Keifer, 1944, and makes this name a synonym of *Eriophyes*.

For *Phytoptus* Dujardin, article 69 (a) is the only consideration, and the first binomial use of this name was by Pagenstecher in 1857. Pagenstecher (Pgst.) quoted 4 species names under *Phytoptus*, one of which is the linden leaf gall maker, *tiliae* Pgst. We choose *tiliae* as the type of *Phytoptus*. This action then substitutes *Phytoptus* for the recent use of *Eriophyes*, and brings *Phytoptus* back into the Eriophyidae (in a restricted sense).

A few genera cannot be placed in the following keys because of lack of sufficient structural information on them. It has been the intention to list all genera that we are aware of, to put these insufficiently known ones as close to what seems their proper places as possible.

As will be evident, the following keys use female anatomy almost exclusively. The great majority of specimens collected are females, and males may at times be difficult to discover.

The principal problem in using the keys, aside from difficulties arising from faulty preparation of specimens, will arise from deuterogyny. Many species have no deutogynes, but those on deciduous plants often do. As explained in the section on life histories (chap. 12), deuterogyny is the existence of two or more different forms of females, including gradation, within a species. The terms *deuterogyny*, *deutogyne*, and *protogyne*, are defined in chapter 12. The female form resembling the male is the one the keys almost always use. It is the *protogyne*.

The keying out of deutogynes would complicate the keys more than do the protogynes, and probably could not be accomplished satisfactorily in the light of present knowledge. New facts about deutogynes continually develop. Deutogynes tend to have more generalized structures than males and protogynes, and in that respect show group relationships. Some deutogynes would bunch together in the best key that could be constructed for them.

If the worker, in attempting to use these keys, finds that he has only deutogynes and males, as sometimes happens in the latter part of the season, it may be possible to key out males correctly until protogyne anatomy stops progress, and then utilize deutogyne genitalia.

Deutogynes frequently tend to confuse eriophyoid classification. The worker should always bear in mind that what looks like the association of two different species, even representing different genera, on a deciduous host, may actually be a manifestation of deuterogyny. One rule is to search for males and to compare them with females for definition of protogynes. But those who fall into the trap of believing that these different forms are at least separate species are not the first ones to make this mistake, nor will they be the last.

Improper slide mounting of eriophyoids will also lead to faulty use of the keys. These keys are based on cleared examples that are three dimensional. Specimens must be clear to see the internal female anatomy, and then to see the oral stylets. Three dimensional specimens will have the dorsal ridges and furrows preserved, whereas flattened examples will have these features obscured or obliterated. To prevent flattening, use hair or thin fiber to hold up the coverslip. Also, orient the specimens by pushing the coverslip until the desired anatomy is visible. If the fiber props are not thicker than the mite the cover will rest lightly on the mite and it will remain steady.

Many of the references in these synoptic keys are abbreviated according to usual practice. Special abbreviations are: BCDA for Bulletin of the California Department of

Agriculture; CDA for California Department of Agriculture; E.S. for Eriophyid Studies; USDA for United States Department of Agriculture.

Data presentation formulas with each genus is in the following order as much as possible. (1) Genus name, author, and reference. (2) General distribution of genus. (3) Habit and hosts when possible to present a summary. (4) Type means the type species of the genus. (5) Monotype indicates that only one species in the genus is known to date. (6) The reference to the species will only appear if different from the genus reference. (7) Host of the type species and its relation to that host. (8) The term Note precedes particular statements of special discussion intended to clarify any given situation.

FAMILIES OF THE ERIOPHYOIDEA

Nalepellidae

Cephalothoracic shield with 3 or 4 shield setae, or rarely with dorsal seta pair minute or absent; rostrum of various sizes, often large, but evenly downcurved and containing short-form oral stylet. Legs with all standard setae, including foretibial seta; usually with lateral foretibial spur. All featherclaws (empodii) so far as known simple, not divided. Abdomen with all standard setae (lateral, first ventral, second ventral, telosomal or third ventral), and often with subdorsal anterior seta pair. Female genital coverflap not known to be ribbed; internal female genital anatomy with anterior apodeme always extending moderate forward distance; spermathecal tubes extending diagonally ahead from central rear attachment, then recurving caudad; this tube usually short but from 3 to 5 times or more longer than spermathecae in the Nalepellinae. Dimorphism of 3 types known, see *Trisetacus, Phytocoptella*, and *Sierraphytoptus*. Nalepellids make no leaf erineum but have otherwise moved into all of the plant habitats available to this type of Acarine. Many are on conifers. A few make galls on broadleaved plants.

Nalepellidae Newkirk and Keifer, E.S. (USDA) C-5:4, 1971.

Type genus: *Nalepella* Keifer, BCDA 33:21, 1944. Note—this family name became necessary owing to the removal of *Phytoptus* to the Eriophyidae.[1]

Eriophyidae

Shield with 2 rear (dorsal) setae, or none. Rostrum of various sizes, usually small in comparison to body size, but even when large containing short form oral stylet. Leg setae variable, most forms with standard setae (femoral, patellar, foretibial, 2 dorsal tarsal and 1 inner tarsal; the only variations are reductions); never with foretibial lateral spur. Featherclaws both simple, and divided. Abdomen with standard setae, never with anterior subdorsal pair. Female genital coverflap usually ribbed. Internal female anterior apodeme usually of moderate anterior length, but in Cecidophyinae folded up and appearing shortened in ventral view. Female spermathecal tubes always short, either projecting laterally from central rear attachment or diagonally to rear. Many species on deciduous plants are deuterogynous. Species of this family inhabit all places on plants useful to this type of mite. Some genera and subfamilies have moved independently into these habitats once the type became set, for example, some gall makers have rust mite structure, and Cecidophyines occupy various habitats. The great majority of gall makers are in this family, and erineum makers all belong to it. Species of the Aberoptinae have spatulate or shovel-shaped appendages for burrowing. Many species are serious plant pests and transmit virus diseases.

Eriophyidae Nalepa, Das Tierreich 4, 1898.

Tpye genus: Eriophyes von Siebold, Jahresber. Schles. Ges. 28:89, Jan.—Mar. 1851.

Rhyncaphytoptidae

Cephalothoracic shield with 2 or no setae. Rostrum large in comparison to body, abruptly bent down near base, attenuate, containing *long-form* oral stylet. Legs with usual segments, but some with reductions; standard number of setae usually present, unless reduced; never with foretibial spur. Featherclaws often large, either simple as in the Rhyncaphytoptinae, or deeply divided as in the Diptilomiopinae. Coxae usually with standard 3 pairs of setae. Abdomen usually with standard number of setae, unless reduced; never with subdorsal seta pair. Female coverflap usually smooth, less often ribbed. Anterior female genital apodeme of moderate anterior length, often acuminate. Internal spermathecal tubes short and either extending laterally or diagonally to rear. Deuterogyny similar to Eriophyidae present in many species. Mites of this family as so far known are leaf vagrants only, rarely causing much damage to their hosts.

Rhyncaphytoptidae Keifer, E.S. (CDA) B-4:20, 1961.

Type genus: *Rhyncaphytoptus* K., BCDA 28:149, 1939.

SUBFAMILIES OF THE NALEPELLIDAE

1\. Elongate wormlike mites with genitalia removed caudad from coxae by 12 to 16 unconstricted rings; abdominal rings subequal dorsoventrally. Four cephalothoracic shield setae present, rear pair pointing caudad; spermathecal tubes short.

Novophytoptinae Roivainen, Ann. Entomol. Fennica 19 (2) :86, 1953.

Only genus: *Novophytoptus* Roivainen 1947 (fig. 91, *b*).

Novophytoptus R., Acta Entomol. Fennica 3 :15–17 1947; Holarctic; most species on grasses.

Type: *rostratae* Roivainen; *Carex inflata* Sut., living in blade furrows; Finland.

Note—One unnamed species makes leaf blisters in *Vaccinium* in Colorado.

1\. Genitalia relatively close to coxae, separated by few more-or-less constricted rings. Abdomen variable, the mites being either wormlike, with rings subequal dorsoventrally, or with robust or fusiform bodies, often with strong tergal-sternal ring differentiation. One or two anterior shield setae . . . . 2.

2\. One central anterior shield seta, dorsal pair present or absent; internal spermathecal tubes 3 to 5 times longer than spermathecae. Bodies either wormlike and with subdorsal thanosomal seta pair present, or lacking these subdorsal setae and usually more robust.

Nalepellinae Roivanien, Ann. Entomol. Fennica 19 (2):86, 1953.

Type genus: *Nalepella* Keifer.

2\. Four shield setae, rear pair rarely minute; spermathecal tubes short. . . . . .3.

3\. Body wormlike, with abdominal rings subequal dorsoventrally; rear pair of shield setae pointing up if short, ahead if long; subdorsal abdominal seta pair present.

Phytocoptellinae Newkirk and Keifer, E.S. (USDA) C-5:5, 1971.

Type genus: *Phytocoptella* Newkirk and Keifer, 1971.

3\. Body more fusiform and often flattened; abdominal rings with lateral tergal-sternal differentiation; subdorsal anterior thanosomal seta pair present or absent.

Sierraphytoptinae Keifer, BCDA 33 (1):18, 1944.

Type genus: *Sierraphytoptus* K., 1939.

TRIBES OF THE NALEPELLINAE

Trisetacini Newkirk and Keifer, E.S. (USDA) C-5:5, 1971. Characterized by having anterior subdorsal thanosomal seta pair present.

Nalepellini Newkirk and Keifer, E.S. (USDA) C-5:5, 1971. Characterized by lacking subdorsal thanosomal pair of setae.

Genera of the Trisetacini

Dorsal pair of shield setae present; elongate wormlike species with abdominal rings subequal dorsoventrally.

Trisetacus Keifer (figs. 92, *b*; 103; 104; 105) BCDA 41:32, 1952. Holarctic; usually needle sheath mites on pines, but also bud mites on other conifers.

Type: *pini* Nal., SB. Akad. Wiss. Wien 96:133; 1887. Europe; *Pinus sylvestris* L., making twig galls or knots. Note—One species is a berry mite on juniper. *T. kirghisorum* Shev. is reported as having two forms of both males and females. This may be for hibernation purposes as males of other species hibernate with females; example *ehmanni* K. This dimorphism would not seem to be deuterogyny similar to the numerous species that have only two forms of females.

Dorsal or rear pair of shield setae absent; less wormlike species with some dorsoventral ring differentiation, according to figure.

Boczekella Farkas, Ann. Hist-nat. Mus. Nat. Hungary (Budapest) 57:467, 1965. Europe.

Monotype: *laricis* F. *Larix europa* L., the mites needle vagrants. Note—The information with this genus does not include data on the foretibial spur or spermathecal tubes.

Genera of the Nalepellini

1. Body more elongate and wormlike, thanosomal rings subequal dorsoventrally; either pine sheath mites or needle vagrants.

Setoptus Keifer (fig. 107). BCDA 33:19, 1944. North America; on pines.

Type: *jonesi* (K.), BCDA 27 (2) :189, 1938. California and elsewhere; a sheath mite on *Pinus torreyana* Parry. Note—The species *strobacus* K. stunts needles on eastern white pine.

1. Body more fusiform and robust; abdominal rings usually with more tergal-sternal differentiation .... 2.

2. Tergites with a pattern of large projecting lobes.

Phantacrus Keifer, E.S. (CDA) B-16:2, 1965. Western North America.

Monotype: *lobatus* K.; *Pseudotsuga menziesii* (Mirb.), the mites needle vagrants.

2. Tergites evenly curivng around abdomen, no projecting lobes; robust mites with variable dorsoventral ring differentiation.

Nalepella Keifer, (figs. 93, *a*; 96, *d*, *f*; 106) BCDA 33:21, 1944. Holarctic; on conifers.

Type: *triceras* (Börner), Arb. Bio. Anst. 5:149, 1906. Europe; needle vagrant on *Abies veitchi* Lindley. Note—The type of this genus has strong tergal-sternal differentiation along the lateral line, according to figure 106. Practically all of the North American species referred to this genus have much less differentiation, tending to resemble *Setoptus* except for more robustness.

Genera of the Phytocoptellinae

1 Dorsal setae minute; shield with rear central gland.

Acathrix Keifer, (figs. 92, *a*; 93, *b*) E.S. (CDA) B-7:16, 1962. Philippines (Guinobatan).

Monotype: *trymatus* K.; living in frond folds on *Cocos nucifera* L.

1. Dorsal setae short or long, but not minute .... 2.

2. Dorsal tubercles distributed at random on abdominal rings.

Phytocoptella Newkirk and Keifer, E.S. (USDA) C-5:3, 1971 (figs. 90, *a*; 94, *a–c*, *e*); Holarctic and possibly elsewhere; wide range of hosts including grasses.

Type: *corniseminis* K., BCDA 28:144, 1939. California; living in buds and among seeds on *Cornus nutalli* Aud. Note—The filbert big bud mite, *Phytocoptella avellanae* (Nal.) exhibits a surprising form change in the second nymph that lies on the underside of leaves during the summer. This nymph is flat and has sharp lateral projections, being shaped quite differently from the adult. The name *Phytcoptella* now takes the place of the recent use of *Phytoptus,* since *Phytoptus* goes into the Eriophyidae owing to a change in the type of the genus.

2. Dorsal microtubercles arranged in longitudinal rows on the thanosome.

Anchiphytoptus Keifer, BCDA 41 (1) :31, 1952. North America.

Type: *lineatus* K.; California; in buds on *Cercocarpus ledifolius* Nutt., and on *C. betuloides* Nutt. Note—On this latter host the mite may contribute to bud proliferation at times. A second species, *beeri* K., occurs on *Beaucarnea* in Mexico.

TRIBES OF THE SIERRAPHYTOPTINAE

Subdorsal anterior thanosomal seta pair present.

Sierraphytoptini Newkirk and Keifer, E.S. (USDA) C-5:5, 1971.

Subdorsal thanosomal setae absent.

Mackiellini Newkirk and Keifer, E.S. (USDA) C-5, 1971.

Note—This tribe takes the place of the subfamily designation of Mackiellinae by Keifer 1944, and Channabasavanna 1966.

Genera of the Sierraphytoptini

1. All four shield setae short and moved forward on shield.

Prothrix Keifer, E.S. (CDA) B-13:1–2, 1965. Possibly Philippines.

Monotype: *aboula* K., host unknown. Note—This species is a very unusual form.

1. Dorsal or second pair of shield setae at least halfway back on shield. ....2.

2. Body cylindrical and cyrtosomatic.

Austracus Keifer, BCDA 33 (1) :20, 1944; Argentina.

Monotype: *havrylenkonis* K.; the mites making fruit galls on *Nothofagus dombeyi* Blume.

2. Body fusiform or flattened.

Sierraphytoptus Keifer, BCDA 28:150, 1939. Possibly Holarctic.

Type: *alnivagrans* K.; western North America; rust mites on both surfaces of *Alnus tenuifolia* Nutt. Note—This species probably has several alder hosts. It is deuterogynous and closely resembles *Aculus* spp. of the Phyllocoptinae, not only in habit, but also in certain structures. The *alnivagrans* protogyne has small spines on the anterior shield lobe and the deutogyne lacks them. The deutogyne has reduced number of sternal microtubercles. These characters are precisely as on *Aculus* spp., even though the origins of the two types of mites are quite different. The deutogyne stage of *alnivagrans* hibernates behind lateral twig buds.

A possible synonym: *Fragariocoptes* Roivainen, Acta Entomol. Fennica. 8:51, 1951.

Type: *setiger* (Nal.), Nova Acta Leop. 55:368, 1894; Europe; the mites make leaf galls on *Fragaria collina* Ehrhart. Note—This gall-making habit is quite at variance with *alnivagrans*, and *Fragariocoptes* may actually be separable from *Sierraphytoptus*.

Genera of the Mackiellini

Shield tubercles not produced, the rear setae pointing forward.
Mackiella Keifer, (fig. 94, *f*) BCDA 28:147, 1939. Widespread on coconut palm.
Monotype: *phoenicis* K.; originally found in California, but undoubtedly occurring wherever the host grows; in folds on emerging fronds of *Phoenix dactylifera* L.
Shield tubercles bulbous and produced, rear or dorsal setae pointing caudad.
Retracrus Keifer, (fig. 94, *d*) E.S. (CDA) B-16:17, 1965. Central America.
Monotype: *johnstoni* K.; making black blotches on underside of *Chamaedorea* fronds. Note—The mites cover themselves with copious amounts of white wax.

SUBFAMILIES OF THE ERIOPHYIDAE

1. Tibiae reduced or completely fused with tarsi; foretibia never with seta..........2.
1. Tibiae always amply distinct from tarsi; foretibial seta nearly always present except in a few genera ....3.
2. Spatulate or shovel-shaped projections present on either rostral termen, or on tarsi; legs when without spatulate appendages very stout, segments shortened or combined; featherclaws large.
Aberoptinae Keifer, E.S. (CDA) B-18:2, 1966.
Type genus: *Aberoptus*.
2. Lacking spatulate projections for burrowing; legs of average thickness; forecoxae often fused across center line with sternal line faint or absent; first setiferous coxal tubercles usually absent; featherclaws relatively small.
Nothopodinae Keifer, BCDA 44 (4):163, 1956.
Type genus: *Nothopoda*.
3. Genitalia, especially female genitalia, noticeably projecting from thanosome venter, appressed to coxae, usually separating coxae more than normal; anterior female internal apodeme bent up and appearing shortened, usually present as heavy transverse line in ventral view; ribs of female coverflap typically in 2 uneven ranks: forecoxae usually narrowly connate at center line, sternal line shortened; coxae, especially forecoxae, with curving lines outlining produced coxal setiferous tubercles. Most genera lacking dorsal tubercles and setae. A generally distributed group that moved independently into various habitats on plants. The group includes rust mites, bud mites, erineum and gall mites.
Cecidophyinae Keifer, E.S. (CDA) B-20:15, 1966.
Type genus: *Cecidophyes* Nal.
3. Gentialia lying more on level with thanosome venter; genitalia not appressed to coxae and coxae not unusually spread apart; sternal line most often of moderate length; internal anterior female apodeme extending moderate distance forward; coxae often ornamented with granules, but curved lines when present, weak. Female coverflap variably embellished, ribs less often in 2 ranks; dorsal setae rarely absent from shield.4.
4. Body wormlike, thanosomal rings subequal dorsoventrally, at least on anterior ½ or ⅔; shield typically lacking anterior lobe, or with slight projection over rostrum base; if anterior shield projection present over rostrum, then this extension narrow, basally flexible, and combined with narrow thanosomal rings.
Eriophyinae Nalepa, Das Tierreich 4:5, 1898.
Type genus: *Eriophyes* von Siebold. 1851.
4. Body more fusiform and fitted for exterior living; shield usually with broad-based and rigid anterior lobe over rostrum; thanosome typically divided laterally into broad and heavy dorsal tergites, and narrow, soft sternites; if no anterior lobe present, or only a slight one, then some form of tergal-sternal differentiation discernible,

at least in larger dorsal microtubercles; if no dorsoventral contrast present, then broad shield lobe present.

Phyllocoptinae Nalepa, Das Tierreich 4:45, 1898.

Type genus: *Phyllocoptes* Nal.

Genera of the Aberoptinae

Spatulate or shovel-shaped projection for burrowing attached to foretarsi.

Aberoptus Keifer, BCDA 40 (3) :97, 1951. Samoa.

Monotype: *samoae* K.; on leaves of *Mangifera indica* L.

Spatulate burrowing projection on anterior end of rostrum.

Cisaberoptus Keifer (fig. 93, *d*; 96, *e*) E.S. (CDA) B-18:2, 1966. Probably holotropical.

Monotype: *kenyae* K.; on upper surface of leaves of *Mangifera indica* L.

Genera of the Nothopodinae

1. First setiferous coxal tubercle and seta present; forecoxae weakly divided; dorsal shield tubercles ahead of rear shield margin and directing setae up and centrad; foretibia completely fused with tarsi.

Colopodacus Keifer (fig. 108, *d, e*) E.S. (CDA) B-1:16, 1960. Africa and Australia.

Type: *africanus* K., rusting leaves of *Coffea arabica* L.; Australian species is *glochidionis* K., 1969, a leaf vagrant on *Glochidion supra-axillare* Benth., which differs from the type by having completely fused forecoxae.

1. First setiferous coxal tubercle and seta missing; forecoxae either separate or fused across midline .... 2.

2. Dorsal tubercles and setae absent from shield; foretibiae distinct, but short and without seta.

Anothopoda Keifer, Ann. Entomol. Soc. Am. 52:655, 1959. Australian.

Monotype: *johnstoni* K.; occurring in leaf erineum made by *Eriophyes prostantherae* (K.) on the host *Prostanthera lasianthus* Labill.

2. Dorsal setae present; foretibiae either fused with tarsi or but slightly distinct..... 3.

3. No appreciable projection over rostrum base; dorsal shield tubercles set somewhat near shield margin, but either tubercle axes longitudinal to body, or setae projecting up; foretibiae fused; forecoxae fused across midline.

Nothopoda Keifer, BCDA 40 (3) :96, 1951. Holotropical.

Type: *rapaneae* K.; Florida; making undersurface erineum on leaves of *Rapanea guinanensis* Aubl. Note—The type may very well range around the Caribbean. A second species, *footei* K., 1969, makes erinea on fern in the Philippines.

3. Dorsal shield with short projection over rostrum.......................... 4.

4. Dorsal tubercles on rear shield margin, slightly produced, projecting setae divergently to rear; forecoxae fused across center line; foretibiae fused with tarsi.

Floracarus Keifer, (figs. 91, *c*; 96, *h*) BCDA 41 (2) :69. 1953. Florida and possibly elsewhere in Caribbean area.

Monotype: *calonyctionis* K.; the mites blacken leaves of *Calonyction aculeatum* House.

4. Dorsal tubercles ahead of rear shield margin, usually projecting dorsal setae up; anterior shield lobe short but with broad base and with transverse furrows; foretibiae slightly discernible on leg underside .... 5.

5. Forecoxae separated by short or moderately long sternal line.

Disella Newkirk and Keifer, new genus.

Type: *ilicis* (K.), E.S. (CDA) B-13:19–20, 1965. Host: *Ilex glabra* L., the mites being leaf vagrants. Type found in Florida. A second species, *talisiae* (K.),

E.S. (USDA) C-2:17–18, 1969, lives as a rust mite on *Talisia olivaeformis* von Humbolt in Venezuela.

GENERIC DESCRIPTION: Nothopodine mites with fusiform, tapering body. Rostrum short, projecting down. Shield semicircular in anterior outline, slightly acuminate; shield patterns a rough network. Dorsal tubercles set well ahead of rear margin and directing setae up. Forecoxae separated by well-marked sternal line centrally; first setiferous coxal tubercles and seta absent. Forefemoral setae present; tibiae but slightly indicated on underside of legs. Thanosome with slight ventrad increase in ring number; microtubercles sparse dorsally. All standard thanosomal setae present. Telosome with seta and tapering from thanosome. Female genital coverflap with transverse curved line across center, and short ribs at right angles to edge on type species; internal anterior apodeme of moderate length. The genus name is *di*, meaning separate, and *sella*, referring to coxae.

5. Forecoxae united with each other and more or less united with suboral plate; coxal surface granular or with minute spinules.

Cosella Newkirk and Keifer, (figs. 108, *g*, *h*) new genus.

Type: *deleoni* (K), BCDA 44 (4):161, 1956. Host: *Pithecolobium guadaloupense* Chapm., the mites rust leaves. The type is in Florida; a second species, *fleschneri* (K.) is on citrus leaves in India.

GENERIC DESCRIPTION: Body fusiform, tapering. Rostrum short, projecting down. Shield semicircular in anterior outline; anterior lobe short, type with shield front declivitous. Dorsal tubercles with diagonal or longitudinal axes; setae short and pointing up. Forecoxae fused; first setiferous coxal tubercles absent. Tibiae slightly discernible. Thanosomal rings subequal dorsoventrally, all standard setae present. Microtubercles elongate, weaker or absent above. Telosome with seta and tapering evenly from thanosome. Female genital coverflap granular; interior apodeme projecting forward. The genus name is *co*, meaning together and *sella*, referring to coxae.

GENERA OF THE CECIDOPHYINAE

1. Dorsal setae absent.

Cecidophyini Newkirk and Keifer, new tribe.

Type genus: *Cecidophyes* Nalepa 1899 .... 2.

1. Dorsal setae present.

Colomerini Newkirk and Keifer, new tribe.

Type genus: *Colomerus* N. and K. 1971.

2. Wormlike bud mites with very small or no anterior shield projection over rostrum base.

Cecidophyopsis K., (figs. 109, *a-i*; 110, *a*, *b*, *c*) BCDA 47 (4) :273, 1959. Holarctic and possibly elsewhere.

Type: *vermiformis* (Nal.), SB. Akad. Wiss. Wien 98:18, 1889. Widespread on host; living as inquilines in big buds on *Corylus avellana* L., the big buds caused by *Phytocoptella avellanae* (Nal.).

2. More-or-less broad and rigid anterior shield lobe over rostrum.............. 3.
3. Foretarsal claw situated below featherclaw.

Dechela Keifer, E.S.(CDA) B-13:3, 1965. Philippines (Guinobatan).

Monotype: *epelis* K.; the mites make erineum on leaves of a plant that is possibly *Bixa* sp.

3. Foreclaw in normal position above featherclaw. .... 4.
4. Broad dorsal thanosomal trough; tergites almost as numerous as sternites.

Glyptacus Keifer, BCDA 41 (2) :67, 1953. California.

Monotype: *lithocarpi* K.; undersurface vagrants among leaf hairs on *Litho-*

carpus densiflora (Hook and Arnold).

4. Dorsum of thanosome evenly transversely arched.......................... 5.
5. Tergites more nearly as numerous as sternites and completely microtuberculate.
   Cecidophyes Nalepa, (figs. 133, *j*; 91, *a*; 92, *e*; 93, *e*) SB Akad. Wiss. Wien. 98: 141, 1889. Holarctic and possibly elsewhere; many species, the mites usually leaf vagrants, but also erineum makers and gall formers.
   Type: *galii* Karpelles, SB. Akad. Wiss. Wien 90:47, 1884. Europe; makes longitudinal leaf rolls on *Galium aparine* L.
5. Thanosomal tergites much broader than sternites and with no microtubercles... 6.
6. Tergites gradually narrowing to telosome.
   Coptophylla Keifer, BCDA 33 (1) :26, 1944. Widespread on host.
   Monotype: *lamimani* K., BCDA 28:419, 1939. The mites living as leaf vagrants on underside of leaves of *Corylus avellana* L. Note—Originally found in California, but European in origin.
6. Thanosomal tergites much broader than telosomal rings.................. 7.
7. First tergite separate from rear shield margin.
   Achaetocoptes Farkas, Acta Zool. Akad. Sci. Hungary 7 (1–2) 73, 1961. Europe and North America on oaks.
   Type: *quercifolii* F.; Europe; leaf vagrants on *Quercus cerris* L. Synonym: Pseudojohnella Keifer, E.S. (CDA) B-2:16, May 1, 1961. Arizona. Type: *ajoensis* K., Arizona; the mites undersurface leaf vagrants on *Quercus ajoensis* Mulsant. Note—The actual synonymy in this case will depend on the date of publication of *Achaetocoptes*.
7. First tergite fused with rear of shield.
   Johnella Keifer, Ann. Entomol. Soc. Am. 52:652, 1959. North America; on oaks.
   Type: *virginiana* K.; undersurface leaf vagrants on *Quercus virginiana* Mill. Note—A second species, *chrysolepis* K., is an undersurface leaf vagrant on *Lithocarpus densiflora* (Hook and Arnold) in California.
8. Body robust-fusiform and strongly tapering; shield with broad-based anterior basally flexible lobe; dorsal setae projecting to rear.
   Gammaphytoptus Keifer, BCDA 27 (2) :148, 1939. Probably occurs wherever camphor trees grow.
   Monotype: *camphorae* K.; only known in California to date; living in leaf glands at axils of main lateral veins near base of midrib, on *Cinnamomum camphora* N. and E.
8. Body wormlike; if anterior shield projection present, it is narrow and basally flexible; thanosomal rings subequal dorsoventrally.......................... 9.
9. Dorsal setae projecting to rear.
   Cosetacus Keifer, (fig. 108, *a*, *b*) B-20:7, 1966. Known on its host in California.
   Monotype: *camelliae* (K.), BCDA 34 (3):137, 1945. Living in buds, especially flower buds, on *Camellia japonica* L. Note—the mite probably contributes to flower bud drop at times. The host is Asiatic in origin but the mite is yet to be recorded as existing there. In this connection tea should also be investigated.
9. Dorsal setae projecting ahead or centrad, from tubercles situated slightly ahead of rear shield margin.. 10.
10. Shield with narrow basally flexible anterior projection over rostrum.
   Ectomerus Newkirk and Keifer, new genus.
   Type: *anysis* (K.), E.S. (USDA), C-4:17–18, 1970. Host: *Terminalia muelleri* Benth., the mites making leaf galls. Range of type: Australia. This type species originally described under the genus *Stenacis*.
   Generic Description: Wormlike mites with abdominal rings subequal dorsoventrally, completely microtuberculate. Rostrum with short form oral stylet. Shield subsemicircular in anterior outline. Narrow, somewhat acuminate anterior shield projection over rostrum base, about ¼ shield length. Dorsal tu-

bercles close to rear shield margin, but directing dorsal setae anteriorly. Forecoxae with moderate sternal line between them and with outlines partially around setiferous tubercles; all 3 standard setiferous tubercles on coxae. Legs with all standard setae and with foretibial seta. Abdomen with standard setae; thanosomal microtubercles either round or elliptical. Accessory seta minute. Female genitalia not strongly appressed to coxae; coverflap with longitudinal ribs, partly in 2 ranks. Internal female anterior apodeme transverse, not extending forward. The genus name is *ecto* meaning out of, referring to the anterior shield projection; *merus*, meaning part.

10. Shield lacking anterior projection over rostrum.

Colomerus Newkirk and Keifer. (fig. 108, *c*) E.S. (USDA) C-5:6, 1971. Contains a few widespread species.

Type: *gardeniella* (K.), E.S. (CDA) B-12:9, 1964. Occurs in Mexico and possibly elsewhere; mites live in petiole bases on *Gardenia jasminoides* Ellis. Two other named species are also referable to this genus: *holodisci* (K.), E.S. (USDA) C-4:13, 1970, making erineum on *Holodiscus microphyllus* Rydb., in California. The other species is *vitis* (Pgst.), the grape erineum mite, which is widespread.

Genera of the Eriophyinae

Dorsal tubercles more or less ahead of rear shield margin, and projecting dorsal setae ahead in some degree, or up; dorsal tubercles which are close to rear margin incline ahead, or have longitudinal axes.

Phytoptini Newkirk and Keifer, new tribe.

Type genus: *Phytoptus* Dujardin.

Dorsal tubercles on or very near shield margin, with transverse basal axes; directing setae to rear, usually divergently.

Eriophyini Newkirk and Keifer, new tribe.

Type genus: *Eriophyes* von Siebold.

Genera of the Phytoptini

1. Shield with narrow, basally flexible projection over rostrum, about ¼ shield length, or more.

Stenacis Keifer, E.S. (USDA) C-4:15, 1970. Widespread.

Type: *palomaris* K.; California; living in buds and Cecidomyid gall on *Salix laisiolepsis* Benth. Note—Another willow mite, *triradiatus* (Nal.), which is widespread, belongs here, as does *calisalicis* K.

1. Shield lacking narrow anterior projection of any length .... 2.

2. Thanosome with three dorsal longitudinal ridges.

Trimeracarus Farkas, Ann. hist.-nat. Mus. Hungary (Budapest) 55:509, 1963. Java

Monotype: *heptapleuri* F.; makes spherical undersurface leaf galls on *Heptapleurum pergamecum* Hassk.

2. Thanosomal dorsum evenly arched in cross section .... 3.

3. Internal female apodeme folded back and shortened in ventral view; female genitalia shortened but not appressed to coxae.

Brachendus Keifer, E.S. (CDA) B-11:1, 1964. Eastern North America.

Monotype: *pumila* K.; in buds on fruiting twigs on *Castanopsis pumila* L.

3. Female genitalia not shortened longitudinally and internal apodeme with normal anterior length .... 4.

4. First and second ventral thanosomal setae missing; abdomen wth broad rings that lack microtubercles.

Cecidodectes Nalepa, Anz. Akad. Wiss. Wien. 53–55:52, 1916–1918. Java.
Monotype: *euzonus* Nal.; making leaf galls on *Trema orientalis*. Note—The description indicates the shield is triangular, but there is probably no anterior projection over rostrum. Thanosomal rings reduced in number owing to broadness. No information available on precise leg setation, nor on female internal genital structures, which makes it necessary to place the genus here on an inferential basis.

4. Thanosome with all regular setae present .... 5.
5. Thanosome with dorsal microtubercles slanting anteriorly; rear dorsum of thanosome and adjacent telosome part lacking ring granules and with part of rings projecting unevenly above others.
Cercodes Keifer, E.S. (CDA) B-1:10; 1960. Western North America.
Monotype: *simondsi* K.; occurs on Nevada-California border; living around petiole bases, especially on fruit spurs, on *Prunus andersoni* Gray.
5. No caudal rings projecting above others, but one genus has rear tergal-sternal differentiation .... **6.**
6. Rings on thanosome and telosome subequal dorsoventrally for entire length.
Phytoptus Dujardin, (figs. 100, *a*; 111, 112) Ann. Sci. Nat. S.3 Zool. 15:166, 1851. A large genus with worldwide distribution.
Type: *tiliae* Pgst., Verh. Ver. Heidelberg 1:46, 1857. Europe and elsewhere; makes so-called upper surface leaf nail galls on *Tilia platyphyllos* Scop. See also Newkirk and Keifer, E.S. (USDA) C-5:2, 1971, for original selection of this genotype. Synonym: *Phytocoptes* Donnadieu, Ann. de la Soc. Linneenne 26: 153–155, 1876. Type: *gallarum* Don., the form making nail galls on *Tilia*. See also Newkirk and Keifer, E.S. (USDA) C-5:2, 1971. Note—The name *Phytocoptes* was first mentioned by F. Thomas, 1869, as a possible correction for *Phytoptus*. Donnadieu was first to publish the name binomially so he receives credit. Donnadieu assigned the name to the Tetranychidae on the theory that certain eight-legged mites had gall making larvae. The name therefore probably belongs in the Tetranychidae. As far as eriophyids are concerned the above synonymy should take care of it. Nalepa's entry into the use of this name came in 1891 when he declared that *Phytocoptes* Donnadieu did not exist and estabilshed *Phytocoptes dubuis* as a new genus and new species. But Nalepa soon abandoned *Phytocoptes*, referring *dubuis* to *Phyllocoptes* (see his 1898 Das Tierreich). We prefer to use the genus name *Aculodes* K., in connection with *dubuis*, but *dubuis* remains the type of Nalepa's version of the genus by original monotypy.
6. Rear fourth of thanosome with rings divided into wider tergites and narrow sternites.
Pareria Keifer, BCDA 41:32, 1952. California.
Monotype: *fremontiae* K.; the mites living in natural hairs around buds of *Fremontia californica* Torr.

Genera of the Eriophyini

1. Shield with narrow basally flexible anterior projection over rostrum, the projection from ⅕ to ⅓ shield length.
Artacris Keifer, (fig. 123, *a-e*) E.S. (USDA) C-4:9, 1970. Holarctic; on maples. Type: *antonimus* (K.), E.S. (CDA) B-7:1–2, 1962. California; the mites living in natural hairs around buds of *Acer negundo* L. Note—Two other species, both European, are referable to this genus. They are *macrorhynchus* (Nal.) which has a pointed anterior shield projection: cephalencus Nal. with an apically rounded anterior shield projection. The European members of the genus make minute leaf galls.

1. Either no anterior shield projection, or if one present then consisting of a very short lobe .... 2.
2. Thanosome with broad longitudinal dorsal trough.
   Acaralox Keifer, E.S. (CDA) B-21:1, 1966. California.
   Monotype: *harperi* K.; the mites living exposed and motionless on underside of leaves of *Cercocarpus betuloides* Nutt.
2. Thanosome with 3 close-set ridges, the middorsal ridge only present in central part of thanosome between subdorsal ridges.
   Baileyna Keifer, BCDA 43:124, 1954. Marianas Islands.
   Monotype: *marianae* K.; making upper surface leaf galls on *Premna* sp. Note—The galls have a narrow basal stem and moderately long internal fiilaments from walls. The mite has a triangular shield and rather broad rings on thanosome; there is some increase in ventral ring number. Absence of microtubercles dorsally, and numerous prominent pointed microtubercles ventrally give appearance of tergite-sternite relationship. Careful examination of these mites, however, discloses that the statement in the original description that there is an anterior shield projection is incorrect. While this genus might be put into either the Eriophyinae or Phyllocoptinae, we believe it fits better here. Second stage nymphs are completely set with sparse microtubercles.

A genus of uncertain position in these synopses.
Phytoptochetus Nalepa, Verh. Ges. Wien 68:73, 1918. Java.
Type: *tristichus* Nal.; Java; the mites make galls that project from leaf surface in both directions, on *Glochidion rubrum* Bl. Note—Nalepa states that the shield is triangular, but gives no information on possible anterior projection. Dorsal tubercles are described as being on rear shield margin and not close to each other. The abdomen has narrow rings that are completely microtuberculate, the ring granules described as small. The description further indicates the presence of 3 dorsal thanosomal ridges that are separated by wide furrows. Nalepa also referred an erineum maker from the same general area, *orthiaspis* Nal. on *Melochia umbellatum* Stapf., to this genus. The Nothopodine, *Colopodacus glochidionis* K. (C-2), which lives on *Glochidion supra-axillare* Benth. in Australia has the dorsal tubercles well ahead of shield margin and is a vagrant.

2. Thanosome evenly arched in cross section, or, in one case, with dorsal humps along midline .... 3.
3. Narrow elongate species with 6 spaced, middorsal humps or thickenings projecting up from dorsum, each covering several rings
   Acunda Keifer, E.S. (CDA) B-16:19, 1965. California.
   Monotype: *plectilis* K.; living in terminal rolled blades on *Distichlis spicata* (L.)
3. Dorsum of thanosome with even profile in side view, except for species in two genera with rear tergal-sternal differentiation .... 4.
4. First setiferous coxal tubercle and seta missing.
   Acerimina Keifer, BCDA 46 (3) :242, 1957; Australia.
   Monotype: *cedrelae* K.; Mites making clusters of irregular bead galls on leaves of *Cedrela australis* F. Muel.
4. First setiferous coxal tubercle and seta present .... 5.
5. Foretibial seta present .... 6.
5. Foretibial seta absent .... 8.
6. Rear part of thanosome with rings subequal dorsoventrally and not divided into terigtes and sternites.
   Eriophyes von Siebold, (figs. 113–120; 101, *a*, *b*, *c*) Jahresber. Schles. Ges. 28:89, *b*; 100, *b*, *c*; 101, *a*, *b*, *c*, 1851. One of the largest genera with species in all regions.
   Type: *labiatiflorae* F. Thomas, Z. Ges. Naturwiss. (N.S.5) 39:459, 1872. Europe; making tips into a hairy mass and causing flowers to remain green, on *Origanum vulgare* L. Synonym: *Aceria* Keifer, BCDA 33 (1) :22, 1944. Type:

tulipae K., BCDA 27:185, 1938; living in blade rolls on grasses and between layers in tulip and onion bulbs.

6. Rear 1/5 to 1/2 of thanosome with rings divided into broader tergites and narrower sternites .... 7.
7. Thanosome with all standard setae present.
   Paraphytoptus Nalepa, (fig. 121, *a*) Anz. Akad. Wiss. Wien 33:55, 1896. Contains many species, probably present in most regions; adapted to living natural hair masses.
   Type: *paradoxus* Nal.; Europe; browning leaves of *Artemisia absinthum* L. Note—This is evidently a polyphyletic genus as the anterior end of body exactly resembles *Eriophyes* spp. Diverse shield patterns and other dissimilar features suggest that the mites with this particular rear anatomy have developed it independently at various times. An example of a species with this morphology, but not living in natural hair, is *P. arceuthobii* K., on pine mistletoe.
7. Thanosome with second ventral seta missing.
   Paraphytoptella Keifer, Occ. Papers (Calif. Bur. Entomol.) 1:16–18, 1959. Western Mexico.
   Type: *arnaudi* K.; makes bead galls on leaves of *Cordia parviflora* DC. Note—A second but unnamed species also occurs in this general region.
8. Forefemoral seta present; forecoxae not fused and with distinct sternal line between; thanosomal dorsum of sinuate or undulating rings which lack microtubercles.
   Cymoptus Keifer, BCDA 35 (1) :40, 1946. Argentina.
   Monotype: *spiniventris* K.; making undersurface erineum on *Nothofagus dombeyi* Blume.
8. Both forefemoral and foretibial seta absent; forecoxae often fused across midline, with sternal line weak or absent; first coxal setiferous tubercle often situated well ahead of anterior coxal junction and close to coxal apex; second coxal tubercles also moved forward and near to a line between first and third tubercles.
   Acalitus Keifer, (figs. 101, *d*; 122; 123, *f-j*; 124; 96, *k*) E.S. (CDA) B14:1–2, 1965. A widespread genus; many are erineum makers, others bud mites.
   Type: *ledi* K. California; makes yellow erineum patches on leaf undersurfaces on *Ledum glandulosum* Nutt. Note—In addition to frequently fused coxae, the coxae are often granular, and these granules extend forward onto suboral plate, and back onto female coverflap. The female coverflap less often has longitudinal ribs, having either transverse lines, concave to front, or no lines. The setaless venter of forefemur often has a spine in place of the missing seta. Erineum made by these *Acalitus* spp., is often of the compound-capitate type.

Unplaced Genera

The following two genera cannot be placed in the synopses here because of lack of information as to the nature of the front of the shield, leg setation, abdominal microtuberculation, and internal female genital anatomy. Nalepa always associated these two genera with the Eriophyinae.

Monochetus Nalepa, Denk. Akad. Wiss. Wien 62:639, 1895. Europe.
Monotype: *sulcatus* Nal.; living either as vagrants on undersurface of leaves of *Fagus sylvaticus* L., or associated with erineum. Nalepa figures *sulcatus* as having some sort of a projection over the rostrum, which could associate it with *Artacris*. *Monochetus*, according to the figures, differs from *Artacris* by having a middorsal longitudinal thanosomal furrow. This furrow should be a main consideration in associating other species with *sulcatus* in *Monochetus*.

Trichostigma Gerber, Assoc. franc. L'Avanc. Sci. C. R. p. 254, 1901. Europe.
Monotype: *erodii* Gerber; probably a bud mite on *Erodium ciconium* Willd. Note—The main character quoted for this genus is that it is supposed to have microtubercles produced into hairlike structures.

Genera of the Phyllocoptinae

Key to Sections (These are for convenience, only part show relationship.)

1. Featherclaw divided .... Section I
1. Featherclaw simple, with central stem undivided .... 2.
2. No prominent dorsal shield setae; dorsal tubercles present or absent, if present with or without minute seta .... Section II
2. Dorsal tubercles and prominent shield setae present .... 3.
3. Tergites laterally extended into lobes or pointed projections, either from each tergite, or from part of them, or from lateral anterior abdominal expansion...... Section III
3. Terigtes rather evenly downcurved over abdominal sides and lacking lateral extenensions; thanosomal dorsum varying from evenly arched in cross section, to flattened, ridged, or furrowed .... 4.
4. Dorsal tubercles set more-or-less ahead of rear shield margin, and directing setae ahead, up, or diagonally centrad; if these tubercles are near rear margin, then they are subcylindrical and bent forward, or their axes longitudinal or diagonal to body .... Section IV
4. Dorsal tubercles on or very near rear shield margin, directing setae to rear, usually divergently; dorsal tubercles either subcyclindrical, or with axes at right angles to body .... Section V

Section I, Divided Featherclaws

1. Dorsal tubercles on rear shield margin, directing setae divergently to rear.
    Dichopelmus Keifer, Occ. Papers (Calif. Bur. Entomol.) 2:10, 1960 (XXVIII) Argentina.
    Monotype: *notus* K.; rusting leaves of *Ilex paraguarensis* St. Hil.
1. Dorsal tubercles ahead of rear shield margin .... 2.
2. First setiferous coxal tubercles and setae absent; featherclaw but partially divided; abdominal dorsum flattened.
    Brionesa Keifer, E.S. (CDA) B-18:5, 1966. Philippines (Guinobatan).
    Monotype: *pipturella* K.; the mites living in undersurface leaf hairs on *Pipturus arborescens* (Link.).
2. First setiferous coxal tubercles present; featherclaw more deeply divided; abdomen various .... 3.
3. Abdomen flattened and with broad expansion just behind shield.
    Tumescoptes Keifer, BCDA 28:336, 1939. Originally found in California, but probably native to the native area of host.
    Monotype: *trachycarpi* K.; in unfolded fronds on *Trachycarpus excelsa* Wendl.
3. Abdomen tapering regularly from rear shield margin .... 4.
4. Middorsal abdominal ridge fading gradually to rear; lateral ridge above sternites not turned centrad just behind shield.
    Acaphylla Keifer, (fig. 125, *a-c*) BCDA 32:214, 1943. California on camellia.
    Type: *steinwedeni* K.; the mites severe orange-colored rust mites on underside of leaves of *Camellia japonica* L. in California. Note—What is supposed to be this same eriophyid occurs on tea in Asia and goes under the name of *theae* Watt. A second species, *indiae* K., occurs on tea in India.
4. Middorsal thanosomal ridge ending in dorsal trough before lateral ridges end; lateral ridges originating on rear shield margin just lateral to dorsal tubercles, running diagonally out to sides of thanosome, and continuing caudad just above sternites; these lateral ridges tending to bulge on anterior diagonal part on some species.
    Acaricalus Keifer, BCDA 29 (3) :164, 1940. A widespread genus.
    Type: *segundus* K.; California; an undersurface leaf vagrant on *Quercus kelloggi* Newb.

Section II, No Dorsal Setae

1. Deuterogynous species; protogyne with tergites alternately extended into lateral lobes.
   Phaulacus Keifer, E.S. (CDA) B-3:1, 1961. Eastern North America.
   Monotype: *apalachi* K.; undersurface leaf vagrants on *Castanea dentata* Marsh. This genus may have affinities with the Cecidophyinae.
1. Lacking lateral thanosomal lobes; no deutogyne known .... 2.
2. Middorsal longitudinal thanosomal ridges fading with subdorsal ridges; femoral seta sometimes absent, and also second patellar seta; dorsal tubercles usually present and some species have minute dorsal setae.
   Calacarus Keifer, (figs. 125, *d-h*; 93, *c*; 95, *e*, *f*; 96, *g*) BCDA 29 (3) :163, 1940. Species native to Eurasia, Africa, and western North America.
   Type: *pulviferus* K.; California; upper surface leaf vagrants on *Quercus kelloggi* Newb. This is a very distinct genus of purple mites that have longitudinal thanosomal ridges, usually 5, and produce white wax stripes along the ridges. The genus contains the ribbed tea mite, *carinatus* (Green), of Asia. In California, camellia is always attacked by a mite that is structurally just like *carinatus*. The camellia mite was originally named *adornatus* K., but it has not been reported from the native lands of camellia. A species, *coffeae* K., rusts coffee leaves in Africa.
2. Middorsal ridge ending before lateral ridges; neither leg has a femoral seta.
   Paracalacarus Keifer, E.S. (CDA) B-9:8, 1962. Flordia, but probably native where the host is native.
   Monotype: *podocarpi* K.; the mites rusting needles of *Podocarpus* sp: This species is also purple with white stripes.

Section III, Lateral Spines or Lobes

1. Anterior part of thanosome with broad lateral expansion and bearing three stiff lateral spines; rear of thanosome abruptly narrower; femoral seta missing.
   Scolocenus Keifer, (fig. 126, *f*) E.S. (CDA) B-8:4, 1962. Philippines (Guinobattan).
   Monotype: *spiniferus* K.; underside vagrants on fronds of *Cocos nucifera* L.
1. Thanosome tapering gradually to telosome; most thanosomal tergites with lateral lobes, either of even length, or alternate length; lobes varying from rounded off to apically sharp; femoral seta present .... 2.
2. **Thanosome with dorsal depression just ahead of telosome; dorsal setae on rear shield margin and projecting to rear.**
   Oxycenus Keifer, (fig. 126, *a*) E.S. (CDA) B-2:7, 1961.
   Monotype: *maxwelli* (K.), BCDA 28 (2):152, 1939. Found originally in California but Mediterranean in origin, including North Africa; occurring on upper leaf surface on *Olea europa* L. Also causes spring flower drop.
2. Dorsum of thanosome running back evenly to telosome, lacking rear depression; dorsal setae varying from tubercles well ahead of rear shield margin to positions on margin.
   Tegonotus Nalepa, (figs. 97; 121, *b*, *c*; 126, *b-d*) Anz. Akad. Wiss. Wien 27:213, 1890. A widespread genus of rust mites.[2]
   Oxypleurites appears first in 1891 but without any species being named for it. Type: *heptacanthus* Nal., the first association of this species with *Oxypleurites* being in Zool. Jahr. Syst. 6:335, 1892, when Nalepa also included other species for the first time with *Oxypleurites*. As the code specifies that a replacement genus must take the type of the original genus, and since Nalepa had previously placed *heptacanthus* in *Tegonotus*, after moving it out of the preoccupied name *Acanthonotus*, *Tegonotus* takes priority over *Oxypleurites*. *Acanthonotus* Nalepa, 1889, was a homonym of its use in fishes, 1801.

Section IV

Dorsal tubercles set ahead of rear shield margin, or with longitudinal axes, or diagonal axes; dorsal setae directed ahead, centrad, or up.

1. Thanosome with subdorsal furrow extending diagonally down from rear shield margin, and fading at ¾; back somewhat flattened; dorsal tubercles near rear shield margin but with longitudinal axes and directing setae centrad to rear.

   Cenalox Keifer, E.S. (CDA) B-3:7, 1961. Eastern North America.

   Monotype: *nyssae* K.; undersurface leaf vagrants along midrib on *Nyssa sylvatica* Marsh.

1. Such subdorsal furrows as are present are longitudinal .... 2.
2. Antapical rostral setae long and bifurcated; anterior shield lobe large and transversely wrinkled; abdominal tergites just behind shield expanded into swollen humps on each side, the tergites coalescing; remainder of abdominal dorsum either ridged or furrowed.

   Dicrothrix Keifer, E.S. (CDA) B-18:7–10, 1966. Venezuela.

   Type: *anacardii* K.; rust mites on leaves of *Anacardium occidentali* Lam. A second form with *anacardii*, and named *secundus* K., should be investigated as a possible example of deuterogyny in the tropics.

2. Antapical rostral seta short and simple .... 3.
3. Anterior shield lobe with hinged flap underneath in front of rostrum; flattened species that remain motionless on underside of oak leaves for considerable periods of time .... 4.
3. Anterior shield lobe lacking underneath flap .... 5.
4. With a central dorsal longitudinal ridge, followed by a depression just ahead of telosome.

   Caliphytoptus Keifer, BCDA 27:311, 1938. California.

   Monotype: *quercilobatae* K.; mites lying along midrib on underside of leaves of *Quercus lobata* Nee.

4. Thanosome with central longitudinal furrow; subdorsal furrow either present or absent.

   Acarelliptus Keifer, BCDA 29 (3) :167, 1940. North America.

   Type: *cocciformis* K.; on underside of leaves of *Quercus prinus* L. in the eastern United States. Two additional species in the genus that occur in western North America are: *occidentalis* K. on *Quercus kelloggii* Newb., and *oregonus* K., on *Q. garryana* Dougl.

5. Dorsum of thanosome flat, or furrowed, or with longitudinal trough.......... 6.
5. Dorsum of thanosome with central ridge .... 10.
5. Dorsum of thanosome evenly arched, somewhat flattened in some genera..... 15.
6. Dorsum of thanosome flat; tergites somewhat broader than sternites and less numerous; anterior shield lobe short but rigid; dorsal tubercles near rear shield margin, but directing setae up and divergently ahead.

   Caroloptes Keifer, BCDA 29:166, 1940. Eastern North America.

   Monotype: *fagivagrans* K.; lying along midrib on leaf underside on *Fagus grandifolia* Ehrh.

6. Thanosome with middorsal furrow or trough .... 7.
7. Thanosome with narrow middorsal furrow, closely edged by subdorsal ridges; female coverflap with transverse curved lines.

   Cupacarus Keifer, BCDA 32 (3):215, 1943. California.

   Monotype: *cuprifestor* K.; lurking in twig tip depressions on *Cupressus macrocarpa* Hartw. Note—This species is limited to the hills on the immediate coast, requiring a constantly high humidity. Other cypresses infested are: *abramsiana* Wolf, *goveniana* Gord., and *Cupressus pygmaea* (Lem.)

7. Dorsal longitudinal thanosomal trough wide .... 8.
8. Patellar seta missing from second leg; elongate species with trough almost entire length of thanosome; dorsal tubercles set ahead of rear margin, but seta direction various.

Notostrix Keifer, (fig. 128, *i, j*) E.S. (CDA) B-9:7–8, (1963); apparently holotropical.

Type: *attenuata* K.; Philippines; mites living openly on fronds of *Cocos nucifera* L. Note—The species *attenuata* has dorsal setae directed to rear. A second species, *jamaicae* K., in the West Indies, but also on coconut, has dorsal setae directed up.

8. Patellar seta present on second legs, the mites not having a long attenuate body. . 9.

9. Trapezoidal species, widest across rear of shield, with central hump in start of dorsal longitudinal trough on thanosome; dorsal tubercles on rear shield margin but with longitudinal axes and directing setae centrad.

Notalox Keifer, E.S. (CDA) B-3:4–5, 1961. Eastern North America.

Monotype: *rubigator* K.; severely rusting undersurface of leaves on *Acer saccharinum* L.

9. Dorsal tubercles set well ahead of rear shield margin; no appreciable hump just behind shield; more elongate species.

Phyllocoptruta Keifer, (fig. 127, *a-f*) BCDA 27 (2) :193, 1938. Holotropical.

Type: *oleivora* (Asmead), Can. Entomol. 11:160, 1879 (oliiorus). Widespread; rusting leaves and fruit on *Citrus* spp. Note—The species *musae* K., browns banana leaves in Queensland; *sakimurae* K., occurs on pineapple in Hawaii.

10. Femoral seta absent . . . . 11.

10. Femoral seta present . . . . 12.

11. All standard abdominal setae present.

Leipothrix Keifer, E.S. (CDA) B-19:1–2, 1966. California.

Monotype: *solidaginis* K.; rusting underside of leaves of *Solidago californica* Nutt.

11. Either lateral and second ventral thanosomal seta pair missing, or lateral seta present and first and second ventrals missing.

Acamina Keifer, BCDA 33 (1):29, 1944. California and Florida.

Type: *nolinae* (K.), BCDA 28:151, 1939. Occurring at bases of leaf blades on *Nolina parryi* Wats. Note—The type occurs on *Yucca*. A second species, *coconuciferae* K., occurs on coconut in Florida.

12. Dorsal shield tubercles very small and on rear shield margin, far apart, directing setae anteriorly.

Latinotus Boczek, Jr. Kans. Entomol. Soc. 33 (1):9, 1960. Florida.

Monotype: *wegorecki* B; undersurface leaf vagrants on *Iva frutescens* L.

12. Dorsal tubercles well ahead of rear shield margin . . . . 13.

13. Thanosomal ridges low and topped with wax-bearing spines that have small bases; middorsal ridge divided centrally.

Callyntrotus Nalepa, Anz. Akad. Wiss. Wien 31:71, 1894. Europe and elsewhere.

Type: *schlechtendali* Nal., set as genotype by Keifer, BCDA 28 (3):234, 1939; on underside of leaves of *Rosa canina* L. Note—There are additional Eurasian species. The definitive character is the type of dorsal spine.

13. Middorsal ridge undivided centrally and bearing no spines; if waxbearing the ridge is thickened . . . . 14.

14. Middorsal ridge ending in broad trough before the end of the subdorsal ridges.

Caleptrimerus Keifer, (fig. 127, *g-j*) BCDA 27:310, 1938. A widespread genus.

Type: *cariniferus* K.; leaf vagrant, mostly on upper leaf surface on *Artemisia heterophylla* Nutt. Range of type: California. Note—The distinction between this genus and *Epitrimerus* is that on *Epitrimerus* the middorsal ridge does not terminate before the lateral ridges. Perhaps *Calepitrimerus* cannot be maintained indefinitely due to intergrades, but on some species this difference is striking.

14. Middorsal thanosomal ridge fading simultaneously with subdorsal or lateral ridges, the shallow subdorsal troughs making the ridge often weak.

Epitrimerus Nalepa, (fig. 128, *a-f*) Das Tierreich 4:61, 1898. A widespread genus.

Type: *pyri* (Nal.), set by Keifer, E.S. (CDA) B-19:2, 1966. A rust mite on fruit and leaves of *Pyrus communis* L. Note—The type was originally Eurasian but is now widespread. The genus is undoubtedly polyphyletic, the subdorsal longitudinal furrows likely developing many times independently. Some species with these ridges and furrows have thanosomal rings subequal dorsoventrally, others have wide tergites.

Keiferella Boczek, Bul. Acad. Aci. Poland Cl. V, 12 (9):393, 1964. Europe.

Type: *juniperici* B.; the mite occurring on *Juniperus communis* L. The genus is described as having a slight middorsal ridge and a prominent anterior shield lobe over rostrum, which is centrally emarginate. Some species referable to *Epitrimerus* have an emarginate anterior lobe. While it is necessary to learn more about this lobe emargination the most appropriate place for the genus seems next to *Epitrimerus*.

15. Anterior shield lobe small, narrow, ending in 2 spines over rostrum; dorsal tubercles very near rear margin but directing setae convergently anteriorly; tergal-sternal relationships on thanosome variable, wider tergites on some.

Acadicrus Keifer, E.S. (CDA) B-14:17, 1965. Australia.

Type: *bifurcatus* K.; brooming twigs on *Eucalyptus obliquus* L'Herit. Note—There are several species that broom twigs on various *Eucalyptus* spp.

15. Anterior shield lobe fairly broad, usually rigid, some species with minute spinules on anterior lobe; lobe not bifurcate; most species with strong differentiation between wider tergites and narrower sternites .... 16.

16. Body trapezoidal; flattened rust mites with low dorsal transverse arch; part of species with sinnuate second claw.

Rhombacus Keifer, E.S. (CDA) B-14:15, 1965. Australia.

Type: *morrisi* K.; taken from witches'-broom on *Eucalyptus viminalis* Labill. Note—The type came from Victoria. Another species, *asclepiadii* K., is on milkweed, perhaps accidentally. A third *rheumella* K., on rhubarb, lacks the peculiar sinuate second claw but has the body type to fit the genus otherwise.

16. Body more fusiform, not trapezoidal, never with recurved claw on second leg . . 17.

17. With sublateral thanosomal longitudinal furrow; body more-or-less flattened.

Platyphytoptus Keifer, (fig. 128, *g*) BCDA 27:188, 1939. North America and possibly elsewhere; apparently confined to conifers.

Type: *sabinianae* K.; wide range in North America; a partial needle vagrant, or needle sheath mite on *Pinus* spp. Note—The type was originally found in California on *Pinus sabiniana* Dougl. It is also common on *Pinus ponderosa* Laws. in the west. Other species in the genus are more free-living and tend to be less flattened.

17. No sublateral thanosomal furrow, the body less flattened.................. 18.

18. Tergites disorganized in center of dorsum.

Acritonotus Keifer, E.S. (CDA) B-8:6, 1962. Florida; on palms.

Monotype: *denmarki* K.; rusting fronds on *Roystonia elata* (Bartr.).

18. Tergites complete across dorsum .... 19.

19. Dorsal tubercles usually ahead of rear shield margin, directing setae ahead, or up and centrad; if dorsal tubercles close to rear margin then directing setae forward; tubercles on rear margin with longitudinal axes and directing setae either up, centrad or convergently to rear.

Phyllocoptes Nalepa, (fig. 98, *a*; 129, *b-h*) SB Akad. Wiss. Wien 98:116. 1889. A widespread and possibly polyphyletic genus; most kinds are leaf vagrants or rust mites.

Type: *carpini* Nal., SB. Akad. Wiss. Wien 96:127, 1887; set as genotype, Keifer, BCDA 27:191, 1938. The type lives on *Carpinus betulus* L. in Europe and is either a leaf vagrant or an inquiline in leaf creases. The leaf creases are the work of *Eriophyes macrotrichus* Nal., a mite with very long dorsal setae. *Phyllocoptes carpini* requires further investigation. The females available for study, from Poland and from Turkey, have the dorsal setae arising from tubercles set well ahead of the rear margin. Most of the specimens prove to be females with the deuterogynous features of suppressed and faint sternal microtubercles, and palmate featherclaws. Males, which are in the minority, also lack dorsal microtubercles, but the sternites have more definite elongate microtubercles. The male featherclaw, however, is also palmate, an unusual feature.

The vast group of species that find their way to this genus for want of a better assignment is difficult to visualize in helpful ways. Species vary from those with broad tergites and no microtubercles to those with hardly any difference between width or microtuberculation between the tergites and the sternites.

19. Dorsal tubercles on rear shield margin, the tubercles either having diagonal axes, or directing setae up .... 20.

20. Dorsal shield tubercles on rear margin but with axes converging anteriorly and directing setae convergently to rear; one species with small spinules on anterior shield lobe; thanosome with wide tergites and narrower sternites.

Vasates Shimer, (fig. 98, *b*, 130, *a-e*) Trans. Am. Entomol. Soc. 2:319, 1869. Eastern North America; mites with rust mite construction making galls on maple leaves.

Type by monotypy: *quadripedes* Shimer; making bladder galls on leaves of *Acer saccharinum* L. Note—A second species, *aceris-crumena* (Riley) makes finger galls on *Acer saccharum* L. and has a pair of anterior shield lobe spinules like *Aculus*. Another species, *tucsonensis* K., 1966, that occurs on *Larrea* in Arizona, has taxonomic features that as presently viewed prevent it from being separated generically from quadripedes. It, however, cannot actually be thus related, so requires further study.

20. Thanosomal rings broad and equal dorsoventrally; dorsal tubercles short-cylindrical, on rear margin, directing setae up.

Criotacus Keifer, E.S. (CDA) B-10:13, 1963. Africa; inquiline in erineum.

Monotype: *brachystegiae* K.; Rhodesia; on leaves of *Brachystegia spiciformis* Benth., the erineum made by *Eriophyes goldsmidii* (K.). The mite has long dorsal setae pointing forward and up and in a sense stands between the group with forward-directed setae and those with backward-pointing setae.

Section V

Dorsal tubercles on rear shield margin or extending over margin; tubercle axes transverse with body; dorsal setae directed to rear, usually divergently; featherclaw not divided.

1. Dorsum of thanosome flat .... 2.
1. Dorsum arched, furrowed, or ridged .... 3.
2. Foretibial seta and second patellar seta missing; lateral thanosomal ridge with tergites pointed at upper curve.

Aciota Keifer, Occ. Papers (Calif. Bur. Entomol.) 1:9, 1959. Brasil.

Monotype: *costae* K.; the mites living in natural hairs on underside of leaves on *Tibouchina mutabilis* Cogn.

2. Foretibial seta and second patellar seta present; thanosomal tergites on middorsum formed into hard transverse plates; small spine on foretibia.

Vittacus Keifer, E.S. (USDA) C-2:13–14, 1969. New Zealand; gall makers.

Monotype: *mansoni* K.; making wooly bead galls on leaves of *Urtica ferox* Forst.

3. Dorsum evenly transversely arched.................................... 4.
3. Thanosome with middorsal longitudinal furrow or trough................ 11.
3. Thanosome with central longitudinal ridge .... 13.
4. First setiferous coxal tubercles and seta missing.
   Metaculus Keifer, E.S. (CDA) B-6:12, 1962. Africa, Europe, Asia.
   Type: *syzygii* K.; Central Africa; rusting leaves on *Syzygium cordatum* Hochst. Note—*Metaculus mangiferae* (Attiah) Channabasavanna, damages terminal buds and inflorescences on *Mangifera indica* L. in India and Egypt. *Metaculus rapistri* Carmona, wrinkles and yellow leaves on *Rapistrum rugosum* (L.) in Portugal.
4. First setiferous coxal tubercle present .... 5.
5. First 4 to 6 narrow sternites continued as complete rings around thanosome just behind shield, abruptly followed by wide tergites.
   Heterotergum Keifer, (fig. 130, *g*) BCDA 44:126, 1955. Western Hemisphere; rust mites or leaf vagrants.
   Type: *gossypii* K.; Brasil; a rust mite on *Gossypium hirsutum* L.
5. Broad tergites begin at rear shield edge .... 6.
6. Foreclaw moved down below featherclaw; foreleg lacking femoral and tibial seta.
   Catachela Keifer, E.S. (USDA) C-2:15, 1969. Brasil.
   Monotype: *machaerii* K.; leaf vagrants on *Machaerium villosum* Vogel.
6. Foreclaw in normal position above featherclaw............................ 7.
7. Anterior shield lobe over rostrum flanked on each side by lateral forward extension of shield; female genitalia unusually elongate and coverflap acuminate.
   Keiferana Channabasavanna, Mysore Jr. Agr. Sci. I (4):264, 1967.
   Monotype: *neolitsiae* Chan., Univ. Agr. Sci., Hebbal, Bangalore, p. 103, 1966; India; as leaf vagrants on *Neolitsia zeylanica* Merrill.
7. No auxillary anterior shield projection beside anterior shield lobe; female genital coverflap bowl-shaped, not acuminate.................................... 8.
8. Thanosomal tergites unusually broad and abruptly contrasting with narrower telosomal rings.
   Anthocoptes Nalepa, (fig. 90, *b*; 92, *d*; 96, *a*; 130, *f*) Anz. Akad. Wiss. Wien 29 (5):16, 1892. A widespread type of Eriophyid; leaf vagrants or rust mites.
   Type: *loricatus* (Nal.), SB. Akad. Wiss. Wien 98:153, 1889. Europe; leaf vagrants on *Corylus avellana* L.
8. Thanosomal tergites narrowing to and blending with telosomal rings......... 9.
9. Anterior shield lobe rather broad and rounded off, with 2 to 4 small spines or spinules projecting forward from under front edge.
   Aculus Keifer, (fig. 130, *h-j*; 131, *a, b*) Occ. Papers (Calif. Bur. Entomol.) 1:5, 1959. A widespread genus of rust mites, many harmful to host.
   Type: *ligustri* K., BCDA 37 (2):190, 1938. A widespread species; rusting and curling leaves in spring on *Ligustrum ovalifolium* Haask.
9. Anterior shield lobe usually more acuminate, frequently ending in a sharp center point; never with spinules from under front edge .... 10.
10. Fusiform and often flattened mites living on various hosts; featherclaws of various sizes, usually in the 4– and 5–rayed group; thanosomal tergites various but usually wider than sternites; microtubercles tending to be elongate dorsally, more elliptical or beadlike ventrally, and in many species pulled ahead of ring margins.
   Aculops Keifer, (figs. 91, *d*; 93, *f*; 131, *c-f*; 132; 96, *j*). E.S. (CDA) B-21:9, 1966. Many species with wide range; habit varying from free-living species, to those that have returned to gall-making.
   Type: *populivagrans* K., BCDA 42 (2):68, 1953. California; a deuterogynous species rusting leaves of *Populus fremontii* Wats. Note—This is essentially an assemblage of species that are probably from various origins. A recent discovery of a 37 million-year-old fossil that fits into this genus indicates its antiquity. Mites of the *tetanothrix* (Nal.) complex make bead galls on numerous willows.

10. Wormlike eriophyids living on grasses; body elongate with rings subequal dorsoventrally; microtubercles present in the form of small round beads, on ring margins dorsally, slightly ahead of margins ventrally; featherclaws with 7 to 9 rays.

Aculodes Keifer, (fig. 108, *j, k*) E.S. (CDA), B-21:19, 1966. A widespread type of Eriophyid.

Type: *mckenziei* (K.) BCDA 33:26, 1944. Widespread; living in upper surface blade grooves on various grasses, the type grass being *Elymus triticoides* Buckley. Note—This type species breeds actively in winter, even in cold climates. Nalepa's version of *Phyllocoptes* fits here, along with its tpye species *dubius* (Nal.).

11. Middorsal thanosomal furrow terminated somewhat ahead of telosome by coalescence of subdorsal ridges.

Mesalox Keifer, E.S. (CDA) B-5:11, 1962. North and South America; rust mites or leaf vagrants.

Type: *tuttlei* K.; eastern North America; originally taken as a leaf vagrant on *Parthenocissus quinquefolia* (L.). Note—The type host is Virginia creeper, but grape is a host in Kansas (C.C. Hall, Jr.). Another species, *abathus* K., occurs on leaves of cashew in Brazil.

11. Middorsal trough wide, subdorsal ridges not joining ahead of telosome....... 12.

12. Anterior shield lobe broad and rounded, and with a pair of spinules projecting forward similar to *Aculus*; also with several smaller spinules on front of this lobe.

Tetraspinus Boczek, (fig. 128, *h*) Acarologia T. III, Fasc. 4, p. 565, 1961.

Type: *lentus* B.; rusting underside of leaves of *Syringa vulgaris* L. Note—The type is from Poland. Synonym: *Plataculus* Keifer, E.S. (CDA) B-4:2, 1961. California; upper surface leaf vagrants on *Sorbus californicus* Greene. Type of *Plataculus*: *pyramidicus* K. Note—Another species, *capsicellus* K., severely damages pepper in Venezuela, and the anterior shield lobe has a series of small points on its front.

12. Anterior shield lobe lacking spinules.

Tetra Keifer, BCDA 33:27, 1944. A widespread genus of leaf vagrants.

Type: *concava* (K.), BCDA 28 (7):489, 1939. Originally found in California but undoubtedly occurring in Europe; upper surface vagrants in vein depressions on *Ulmus procera* Salisbury. (*campestris* L.). Note—The generalized character of these mites would seem to indicate a polyphyletic origin of the group. The type of the genus has very long dorsal setae, and the third tergite projects up above the others.

Paratetra Channabasavanna, Univ. Agr. Sci., Hebbal, Bangalore, p. 92, 1966. India; rust mites.

Monotype: *murrayae* Chan.; browning tender shoots and leaves on *Murraya koenigii* Sprenger, Note—The author separates this genus from *Tetra* by dorsal setae direction. The figure shows dorsal tubercles projecting back straight over the rear shield margin, but the setae strongly curving outward.

13. Middorsal thanosomal ridge broken or discontinuous; first subdorsal ridges joining midridge toward rear; ridges topped with broad-based spines that secrete wax.

Pentamerus Roivainen, Acta Entomol. Fennica 8:52–53, 1951. Holarctic on *Rhamnus* spp.

Type: *septemcarinatus* Liro, 1941. Finland; leaf vagrants on *Rhamus frangula* L. Synonym: *Echinacris* Keifer, E.S. (CDA) B-17:1–2, 1966. California; Type: *rhamnicroceae* K.; underside vagrant on *Rhamnus crocea* Nuttal.

13. Ridges usually more prominent and not topped by broad-based spines; middorsal thanosomal ridge continuous until ending................................ 14.

14. Cephalothoracic shield with pair of central gland pits; thanosomal ridge low and most evident centrally.

Ditrymacus Keifer, (fig. 126, *e*) E.S. (CDA) B-1:12, 1960. North Africa.

Monotype: *athiasella* K.; upper surface leaf vagrant on *Olea europa* L.

14. Middorsal longitudinal ridge on thanosome usually more prominent and longer; no gland pits on shield .... 15.
15. Central thanosomal ridge extra broad; four anterior spinules on anterior shield lobe.
Thamnacus Keifer, BCDA 33 (1):27, 1944. California.
Monotype: *rhamnicolus* (K.), BCDA 27 (3), 1938. Causing constriction between two lateral leaf veins, the mites living on underside of leaf on *Rhamnus californica* Eschscholtz.
15. Central longitudinal ridge narrower .... 16.
16. Foretibial seta absent; 5 longitudinal thanosomal wax-bearing ridges on dorsum, separated by moderately broad furrows; dorsal tubercles on rear shield margin, and according to illustration unusually produced.
Neocalacarus Channabasavanna, Univ. Agr. Sci., Hebbal, Bangalore, p. 100, 1966. India.
Monotype: *mangiferae* Chan.; India; leaf vagrants on *Mangifera indica* L.
16. Foretibial seta present; thanosomal dorsum with at most three ridges .... 17.
17. Middorsal ridge shorter than subdorsal ridges and ending in dorsal trough.
Abacarus Keifer, (fig. 133, *a-d*) BCDA 33 (1) :28, 1944. A widespread type of mite.
Type: *acalyptus* (K.), BCDA 28:490, 1939. California; undersurface leaf vagrants on *Ceanothus cordulatus* Kellogg, and possibly other *Ceanothus* spp. Note—Eriophyids with this structure are widely distributed. Those living on broad-leaved plants having more blunt anterior lobes, and those on grasses with sharper anterior lobes. The best known species is *A. hystrix* (Nal.). This *hystrix* has a moderately broad central ridge, and all ridges are wax-bearing. While *hystrix* has been placed in *Callyntrotus*, and thus associated with mites that have narrow dorsal ridges that are topped with small spines, *hystrix* has broader ridges and no spines.
17. Middorsal ridge not ending before such lateral or subdorsal ridges as are present .... 18.
18. Tergites along middorsal ridge projecting up unevenly.
Tegoprionus Keifer, E.S. (CDA) B-2:4, 1961. Europe.
Monotype: *dentatus* (Nal.), Anz. Akad. Wiss. Wien 28:199, 1891. Causes leaf deformation on *Galium verum* L.
18. Most tergites, at least after first, presenting even profile along middorsal line of thanosome, in lateral view.. 19.
19. Tergites on thanosome of various widths according to species, but all reaching same level in lateral view.
Tegolophus Keifer, figs. 133, *e-g*; 134) E.S. (CDA) B-2:3, 1961. A widespread type.
Type: *califraxini* (K.), BCDA 27 (3):308, 1938. California; undersurface leaf vagrants on *Fraxinus dipetala* Hook and Arnold. Note—The type has narrow tergites, while *hassani* K. 1959, on olive, has wide tergites.
19. Deep cleft between shield and thanosome, with first tergite projecting up further than those following.
Neotegonotus Newkirk and Keifer, E.S. (USDA) C-5:8, 1971.
Type: *fastigatus* (Nal.), Zool. Jahrb. 6:332, 1892. Europe and North America; leaf vagrant on underside of leaves of *Acer* spp. Note—The type host is *Acer campestris* L., in Europe, but in eastern Northt America the mite occurs in natural hairs in vein angles on *Acer platanoides* L.

RHYNCAPHYTOPTIDAE: SUBFAMILIES AND GENERA

1. Featherclaw simple, that is, undivided.
Rhyncaphytoptinae Roivainen, Ann. Entomol. Fennica 19 (2):86, 1953.
Type genus: *Rhyncaphytoptus* Keifer .... 2.

1. Featherclaw divided, usually deeply so.
    Diptilomiopinae Newkirk and Keifer, E.S. (USDA) C-5:9, 1971.
        Type genus: *Diptilomiopus* Nalepa .... 7.
2. All leg setae present.. 3.
2. Femoral seta absent, at least on foreleg .... 6.
3. Tergites uneven along thanosome dorsum in lateral view.
    Quadracus Keifer, BCDA 33:30, 1944. Holarctic.
        Monotype: *urticarius* C. and M., Bul. Soc. Veneto-Trent. 5:152, 1893. Undersurface leaf vagrants among hairs on *Urtica* spp. Note—The European host is *U. dioica* L., and in California it is *U. holosericea* Nuttall. The genus reference has this species designated *urticae* n. sp. In Nalepa's Das Tierreich 4:50, 1898, this species is under *Phyllocoptes.*
3. Thanosomal tergites even dorsally in lateral view..........................4.
4. Tergites with sharp points at lateral tergal edges; deep indentation between shield and first tergite.
    Peralox Keifer, E.S. (CDA) B-6:1–2, 1962. Eastern North America.
        Monotype: *insolita* K.; undersurface leaf vagrants on *Ulmus americana* L.
4. Separation of shield from first tergite variable but not with a particularly deep indentation; no sharp points at lateral edges of tergites...................... 5.
5. Tergites and sternites of even width.
    Rhinophytoptus Liro. Ann. Entomol. Fennica 9 (3):39, 1943. Europe.
        Type: *concinnus* Liro; Finland; leaf vagrants on *Ulmus laevis* Pallas.
5. Tergites on thanosome wider than sternites.
    Rhyncaphytoptus Keifer, (fig. 135; 136, *a-d, f*) BCDA 28:149, 1939. A widespread genus of leaf vagrants.
        Type: *ficifoliae* K.; originally found in California but undoubtedly native in the area of origin of the host; among undersurface leaf hairs on *Ficus* spp.
6. Dorsal setae present.
(2)
    Catarhinus Keifer, (fig. 136, *e*) Occ. Papers (Calif. Bur. Entomol.) 1:11, 1959. South America.
        Monotype:*tricholaenae* K.; Brazil; leaf vagrants on *Tricholaena rosea* Nees. Note—This species is apparently a rust mite on corn.
6. Dorsal setae absent.
    Asetacus Keifer, BCDA 41:36, 1952. West Coast of North America.
        Type: *madronae* K.; undersurface leaf vagrants on *Arbutus menziesii* Pursh. Note—The type ranges from Washington to California. A second species, *barbarae* K., occurs in California on native *Prunus*, and on *Heteromeles.*
7. Dorsal setae missing.. 8.
7. Dorsal setae present.. 9.
8. Patella absent from legs.
    Diptilomiopus Nalepa, (figs. 92, *c*; 137, *g, h*; 96, *i*) Vehr. Ges. Wien 67:228, 1917. Range from Africa, across southern Asia, to Australia.
        Type: *javanicus* Nal.; Java; leaf vagrants on *Hemigraphis confinis* Cognniaux. A synonym: *Sectipes* Keifer, E.S. (CDA) B-7:18, 1962. Philippines. Type: *holmesi* K.; undersurface leaf vagrant on *Elephantopus mollis* von Humbolt.
8. Patella present but lacking patellar seta on both legs; femoral seta usually absent on both legs.
    Rhynacus Keifer, (fig. 137, *e, f*) BCDA 40:98, 1951. A large Western Hemisphere genus.
        Type: *arctostaphyli* (K.), BCDA 27 (3):305, 1938. California; leaf vagrants on coastal *Arctostaphylos* spp.
9. Extra long shield projection over rostrum; the mite characterized as having femoral setae.

Bacculacus Boczek, Acarologia t. 3 (4):560, 1961. Poland.
Monotype: *kaweckii* B.; leaf vagrant on *Quercus robur* L.

9. Forefemoral seta absent .... 10.
10. Middorsal thanosomal ridge divided for short distance just behind shield.
Trimeroptes Keifer, (fig. 137, *c*) BCDA 40:98, 1951. North America.
Type: *aleyrodiformis* K.; undersurface leaf vagrant on *Liquidambar styraciflua* L. Note—The type originated where *Liquidambar* is native, but has recently been found in California on planted trees. A second species, *ilicifolia* K.; occurs on *Ilex* in Virginia.
10. Middorsal ridge simple, that is, undivided .... 11.
11. Filament projecting from front of anterior shield lobe.
Acarhynchus Keifer, Occ. Papers (Calif. Bur. Entomol.) 1:12, 1959.
Monotype: *filamentus* K; eastern North America; blade vagrants on *Arundinaria gigantea* (Walter) Note—For comment concerning *Cecidobia* Banks, which was characterized as having an anterior shield fiilament, see the discussion at end of this key.
11. Anterior shield lobe lacking filament curving down in front of rostrum .... 12.
12. Two transverse grooves across on rear part of shield; dorsal setae very small.
Dialox Keifer, E.S. (CDA) B-8:19, 1962. Philippines.
Monotype: *stellatus* K.; frond vagrants on *Cocos nucifera* L.
12. If a transverse depression exists in connection with shield it is only between rear margin and first tergite .... 13.
13. Thanosomal tergal ridges considerably thickened for wax production.
Apodiptacus Keifer, (fig. 137, *a*, *b*) E.S. (CDA) B-1:18, 1960. Eastern North America.
Monotype: *cordiformis* K.; leaf vagrant white wax-covered mites, first found on underside of leaves of *Carya cordiformis* (Wangendeim) Note—Davis, 1964, lists this mite on several Juglandaceous hosts and on peach, sweet gum, mulberry, and oak.
13. Thanosomal ridges without noticeable extra thickening, but some species cover their bodies with flocculent wax.
Diptacus Keifer, (figs. 137, *d*; 138) BCDA 4:99, 1951. Widespread genus of many species.
Type: *sacramentae* K.; BCDA 28 (3):232, 1939. Originally found in California; waxy undersurface leaf vagrants on *Alnus rhombifolia* Nuttall. Note—type species may be a synonym of Nalepa's *dipterochelus*, and, if so, is Holarctic in range. It is probably deuterogynous. Some species of *Diptacus* have apically emarginate anterior shield lobes.

UNPLACED GENERA AND SPECIES

Phyllocoptyches gallicolus Nalepa was named in *Marcellia* 18:190, 1922. The host was stated to be *Ulmus pedunculata* Fougeroux also on *V. Campestris* L. The description further indicates it is a free-living mite in Austria. Nalepa's characterization of the rostrum would seem to place it in the Rhyncaphytoptidae. The peculiar character stressed by Nalepa for the mite is that the sternites are less numerous than the tergites. There are stated to be no microtubercles on the abdominal rings, suggesting the form named is a deutogyne, but no other rhyncaphytoptid is on record as living on this host, so no protogyne form is deducible at present.

Cecidobia salicicola Banks was named in the Entomological Society of Washington Proceedings 7:141–142, 1905. It was from specimens submitted by T.D.A. Cockerell from Ward, Colorado, from willow leaf galls. Banks' figure, which in some respects appears fanciful, shows a mite with long dorsal setae, a double longitudinal lateral abdominal stripe along the side, but what in part ties it in with the rhyncaphytoptid *Acarhynchus filamentus* K. is the indication that this *salicicola* has a filament projecting anteriorly from the anterior shield lobe. In the 1940's, while Cockerell was still alive, Keifer asked him for specimens from this Ward, Colorado, site, and he kindly sent them. The mites in the willow galls collected by Cockerell proved to be members of Nalepa's *Aculops tetanothrix* complex. The only resemblance between the mites from these galls and Banks' figure is the length of the dorsal setae. Perhaps it is best to refer this unusual description to some form of *tetanothrix*, where it may eventually have a place as a species in that complex. The *tetanothrix* mites have no anterior shield lobe filament or hair.

Notes

Chapter 1. Introduction to the Acari

1. This section was summarized from a review by Krantz (1970).

Chapter 3. History of Chemical Control

1. Chemical names are given in appendix 2.

Chapter 7. Tetranychidae Donnadieu

1. The student should compare the collection and slide mounting recommendations in this section, which treats with larger mites that have more rigid integuments, with the discussion on chemicals and handling methods for collection and preparation of the more delicate eriophyoids. There are similarities throughout these suggested procedures, but some chemicals useful for larger mites may harden the internal eriophyoid tissue too much, or be too astringent for the skeleton, to give best results.

Chapter 12. Eriophyoidea Nalepa

1. The eriophyid species referred to in this reference to fossils is *Aculops keiferi* Southcott and Lange. Reference: R. V. Southcott and R. T. Lang, Records of the South Australian Museum, vol. 16, no. 7, June 1, 1971. The type locality is the Maslin fossil beds, North Maslin Sands, South Australia. The age is the lower middle Eocene.

2. As Nalepa and Keifer have named most of the eriophyoid mites, their names are abbreviated in the text following the mite name, that is, Nal. = Nalepa; K. = Keifer.

Taxonomic references to economically important eriophyoid species are given in a special reference list at the end of chapter 13.

3. Further apparent correlation between lack of ring microtubercles and need for body water conservation is illustrated by eriophyids that make twig galls on leafless *Tamarix* sp. These tamarisk shrubs and trees grow on more-or-less salty soil in deserts surrounding the Mediterranean Sea. *Eriophyes tlaiae* Trabut (Bull. Soc. Afrique Nord 8:29, 1917) is one of the tamarisk-infesting gall mites. It lacks ring microtubercles on both sexes. This mite makes stem galls on *Tamarix articulata* Vahl in Morocco. Eriophyids living on desert tamarisks not only have to contend with frequent low atmospheric humidities, but, perhaps more important, the host sap evidently contains a relatively high level of salt. Considerable salt in a liquid makes extracting water from it more difficult than if the salt content were lower. Such tamarisk eriophyids as *E. tlaiae* have overcome the difficulties presented to them by low humidities and salt, as shown by twig galls on *T. articulata* numerous enough to be a practical source of commercial tannin.

4. In a report written for the Commonwealth Institute of Biological Control, Trinidad, Rachel Cruttwell recounts experiments with *Acalitus adoratus* K., which ordinarily makes small erineum patches on the underside of leaves of the composite plant, *Eupatorium adoratum* L. She states that erineum growth is started by female mites, or by second stage nymphs. New erineum patches develop on young leaves, but erineum cannot grow on leaves that have ceased to grow.

Rachel Cruttwell found that the best transfer method for moving mites from infested plants to uninfested ones was by contact. After contact erineum would develop on previously clean plants in 2 or 3 weeks. Using this contact method several other composites were tested, such as *Aster*, *Lactuca*; *Dahlia*, *Helianthus*, and two other species of *Euphatorium*. All of these latter tests were negative, showing the extreme host specificity of the mite.

Heavy infestations of *A. adoratus* on its host not only caused more extensive than normal erineum, but also resulted in stunted and deformed leaves. Predators prevented heavy infestations of *A. adoratus*. This eupatory mite is under consideration as an aid in the suppression of its host, which has become a pest weed in parts of Africa. This summary was taken from the original report by Rachel Cruttwell's permission, but is not her summary.

5. The term *nymph* is preferred here to *larva*, for the form of these mites which hatches from the egg. It essentially resembles the adult in all major features, except size and sexual maturity. If the first stage was noneriophyiform when hatching from the egg, then the term *larva* would be more nearly appropriate.

6. Those who make many slides of eriophyid mites will find that females with internal first stage nymphs, while almost rare, occur in a wide spectrum of species. Such internal nymphs can be found behind 1 fully formed egg in the oviduct, or behind 2 or 3 eggs. The existence of these internal nymphs leads to two suppositions: First, the eggs in which these internal nymphs have developed have likely not moved far enough ahead to have reached the spermathecal tubes, and therefore have not been fertilized. If so, the nymphs are haploid males. Second, the retention of the egg in the oviduct long enough to allow nymphal development, would seem to be a haphazard occurrence. A possible explanation for internal nymphs might be that this happens at the end of the life of a female, after she has become unable to extrude the egg. This second supposition is in opposition to the theory that internal nymphs represent the inception of ovovivipary in eriophyids. If it actually is the beginning of ovovivipary in this mite group, then it simultaneously but weakly appears in many unrelated species.

Chapter 13. Injurious Eriophyoid Mites

1. References to original descriptions of Eriophyoidea are given for each species at the end of chapter 13.

As Nalepa and Keifer have named most of the eriophyoid mites, their names are abbreviated in the text of this chapter. Nal. = Nalepa and K. = Keifer.

2. This grape bud and erineum mite is not the rust mite. The rust mite is *Calepitrimerus vitis* (Nal.), and is deuterogynous. The deutogyne of *Calepitrimerus vitis* usually goes under the name of *Phyllocoptes vitis* Nal. The rust mite has dorsal longitudinal ridges on the primary forms, and all adults of the rust mite have a prominent anterior shield lobe. In contrast, *Colomerus vitis* (Pagenstecher), has shown no deuterogynous changes and has no anterior shield lobe.

The coxal and genital characteristics on *Colomerus vitis* indicate it belongs to the Cecidophyinae, as already noted, and separate it from species now properly referred to *Phytoptus* (formerly *Eriophyes*). The genus *Colomerus* is a small but widespread taxon. *C. gardeniella* (K.) seems to be holotropical and is a gardenia bud mite. A third species, *C. holodisci* (K.), is an erineum mite on *Holodiscus* in California. The fourth species, *Colomerus neopiperis* (Wilson) (1970), makes bead or small finger galls on the upper surface of leaves of *Piper jaliscum* Watson, in northern Mexico. This *neopiperis* has a 6-rayed featherclaw, as opposed to the 5-rayed featherclaw on *Colomerus vitis*, and has heavier forecoxal lines than are on *C. vitis*. A new species on coconut now in the process of description belongs to this genus, and undoubtedly others already named will be found to associate themselves here when properly studied.

3. The lichi erineum mite has had three names. The first was *Eriophyes chinensis* O'Gara, but *chinensis* Trotter, on apricot, antedated it. Oudemans, in his 1937 *Critico-Historical Survey of Acarology* III:1325 gave it the name *cordai* nov. nom., associating it with *Erineum sixtaliae* Gorda (1840). There is no description of the mite accompanying the name *cordai*. The name *Eriophyes litchii* Keifer was published in *Eriophyid Studies* XIII:212, 1943. This 1943 publication contains a descripton of the mite and a drawing. However, since the name *cordai* might be considered to be associated with a "work description," perhaps the *Code* would give *cordai* a standing.

4. There are four eriophyids that live on mango that bear the name *mangiferae*. One is *Eriophyes mangiferae* (Sayed), as listed here. A second is *Metaculus mangiferae* (Attiah), a bud and rust mite. A third is *Tegonotus mangiferae* (K.), a leaf vagrant known only in Hawaii. The fourth is *Neocalacarus mangiferae* Channabasavanna, which is a stem and leaf vagrant in India.

5. *Eriophyes tulipae* K. (I, 1938), the females of which have 7-rayed featherclaws, and males

have 6-rayed featherclaws, is a member of a grass-infesting series of eriophyids the prototype of which is Nalepa's *tenuis* (1891). Nalepa characterized *tenuis* as having 5-rayed featherclaws, and stated he had found no males. Such grass mites as have been available for study from Europe in more recent years consist of females with 6-rayed featherclaws, and males with 5-rayed featherclaws. This suggests that 5-rayed featherclaw females are not common, or that Nalepa's statements need revising.

With more recent availability of samples of these *tenuis*-type grass mites from various parts of the Northern Hemisphere, it now appears that colonies of these mites not only exhibit geographical distinctions in featherclaw ray number, but that within colonies there is no variation in these ray numbers. This lack of variation in colonies suggests that either these different featherclaw ray numbers separate species (the view held here), or at most the feahterclaw ray numbers indicate subspecific status for the respective groups. Merely to call all of these mites *tenuis* would obscure the geographical and host distinctions these mites show, or would require careful explanations in each treatment as to the particular type of mite referred to.

In North America the only mites of this *tenuis* complex which occur on grasses and on onion and garlic, have females with 7-rayed featherclaws. All mites from garlic, whether collected in North America, or in Italy, or in Korea, have 7-rayed featherclaws.

Schevtchenko, 1970, has named a 7-rayed featherclaw mite collected on wheat in Russia, *tritici* n. sp. He differentiates this from typical *tulipae* on a statistical basis. He notes that mites on liliaceous plants (tulip, garlic, onion) average 82 dorsal abdominal rings, with the shield averaging 37 μ long, and the female genitalia averaging 24 μ wide. Mites of the *tritici* type from wheat and grasses, have an average of 74 dorsal abdominal rings, with the shield averaging 32 μ long, and the female genitalia averaging 19 μ wide. Schevtchenko states, however, that wheat mites can breed on onion.

A statistical check of North American mites from various hosts shows that those 7-rayed featherclaw types from grasses are usually smaller than those on the liliaceous hosts. So perhaps the name *tritici* is applicable to mites of this type found on grasses in North America.

A colony of these mites in Yugoslavia which has 8-rayed featherclaws on females, and 7-rayed featherclaws on males, bears the name *tosichella* K., 1969.

For an account of deutoerogyny that is apparently displayed by a European mite on *Festuca ovina* L., see the section on deuterogyny in chapter 12. The name of this mite is probably *tenuis* even though its female has 6-rayed featherclaws.

Undoubtedly all mites in the *tenuis* complex can carry viruses. As will be noted these mites are not narrowly host-specific.

6. The actual realization of the relationship of the two female forms of *Tegonotus aesculifoliae* (K.) came only after an October 1940 examination of a collection of peach silver mites that were moving into hibernation positions behind lateral twig buds. In the peach silver mite, *Aculus cornutus* (Banks), the protogyne and deutogyne are sufficiently similar to make it possible not only to recognize that both belonged to the same species, but also to discern the microtubercle and shield differences between the two female forms. The deutogyne of *aesculifoliae* is so different from the protogyne as to cause someone not knowing about deuterogyny to place the deutogyne in a genus not close to the genus to which the protogyne would be referred.

Appendix 3.

1. This citation of the family Nalepellidae and the authors of the original establishment of that family is for the purpose of avoiding confusion. According to Article 36 in the *International Code of Zoological Nomenclature* the correct citation should be Nalepellidae Roivainen 1953. The code states that what it calls "Categories-coordinate" requires that all names with the same genotype must be referred back to the first author who established such a name above the genus level. But in this case to search for the family designation in the 1953 reference would be futile. See Roivainen, Ann. Entomol. Fennica 19 (2):86, 1953.

2. The genus *Scolotosus* Flechtmann and Keifer (ref. Alguns Trombidiformes do Brasil e do Paraguai (Acari), a publication of the University of Sao Paulo, Piracicaba, p. 34, 1971), traces to *Tegonotus* in this key. The type of *Scolotosus* is *centrolobii* Flechtmann and Keifer, which species lives on *Centrolobium robustum* Mart. (Leguminosae). *Scolotosus* differs from *Tegonotus* by having the lateral tergal lobes set with spines. This genus also has dorsal tubercles on the rear shield margin, directing the setae diagonally outward and to the rear. There is a low middorsal thanosomal longitudinal ridge, and the ring margins are spiniferous.

Indexes

Host Plant Index

General Index

Plates

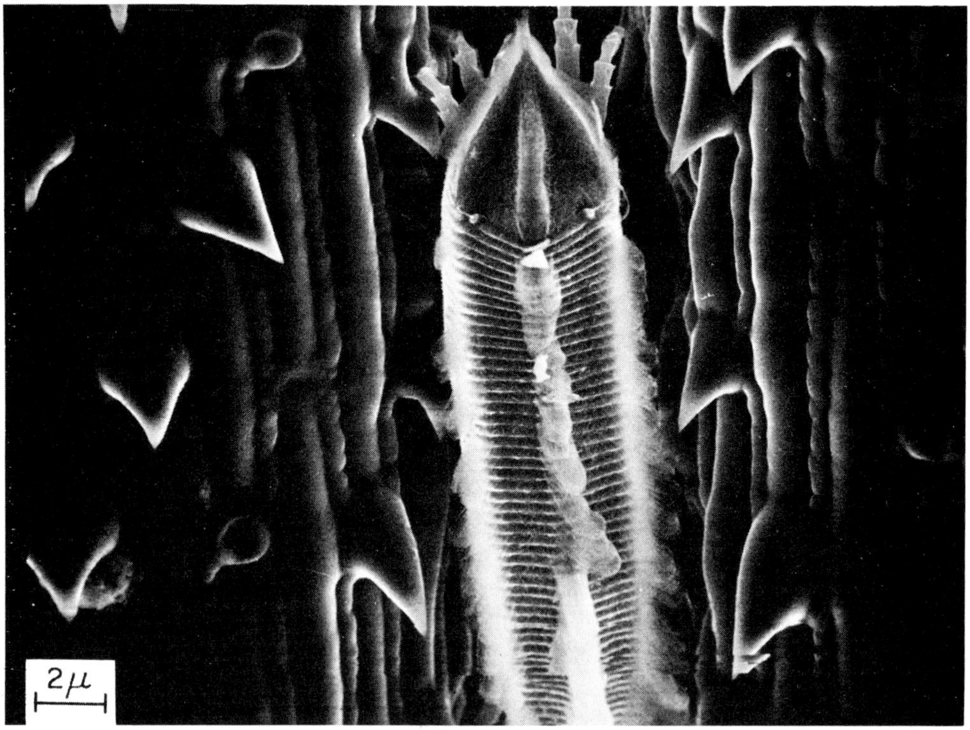

1. Grass rust mite, *Abacarus hystrix* (Nal.), in situ on grass leaf. The mites live in grooves of the grass blade, well protected by the spines of the grass leaf. (Scanning electron microscope photograph, original magnification; 510 X, 20 kv.).

2. Orange fruit russeted by feeding of citrus rust mite, *Phyllocoptruta oleivora* (Ashmead).

3. *a*, Mite-brushing machine. Original design by Henderson and McBurnie; *b*, Accessories for mite-brushing machine. Marked discs and slide holder. (Courtesy Leedom Engineering, Route 1, Box 325, Twain Harte, Calif.)

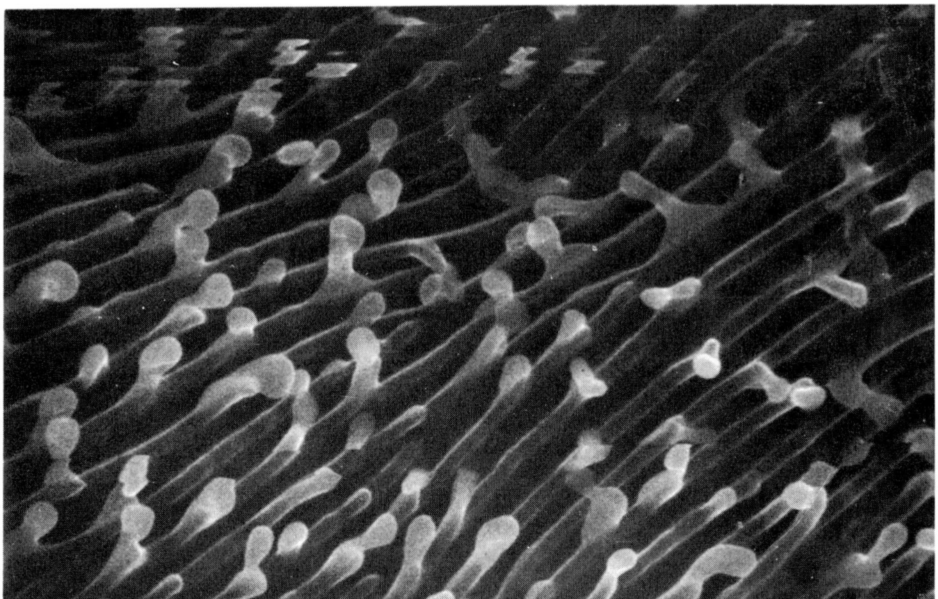

4. Striations and lobes on the dorsum of *Tetranychus* sp. Collected May, 1969, on violet at Berkeley, Calif. (Scanning electron microscope photograph, original magnification, about 9500 X).

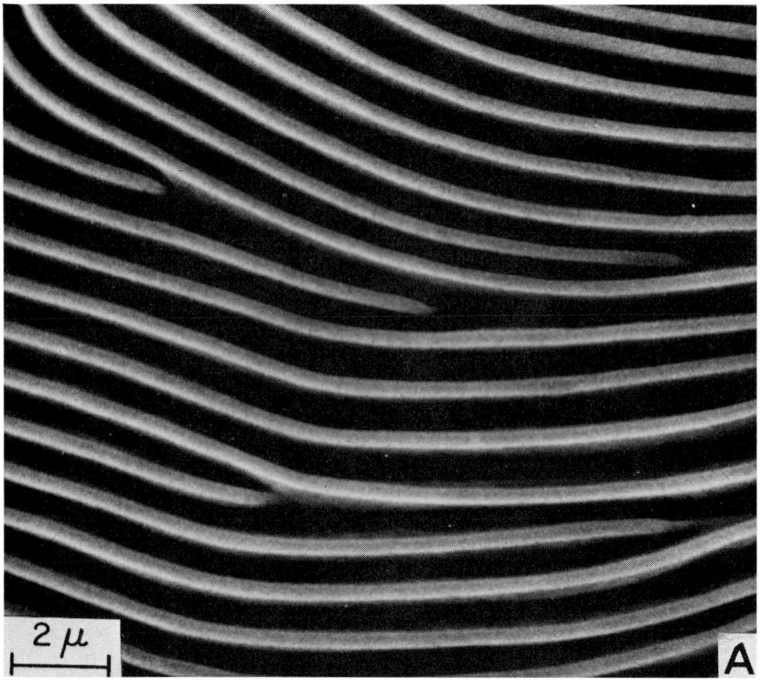

5. *a*, Striations on dorsum of hibernating two-spotted spider mite, *Tetranychus urticae* Koch. Collected from citrus in San Joaquin Valley, Calif., Dec. 1971; *b*, Striations and lobes on two-spotted spider mite, from first generation reared in laboratory from hibernating mites collected on citrus during Dec., 1971 (see 5 *a*).

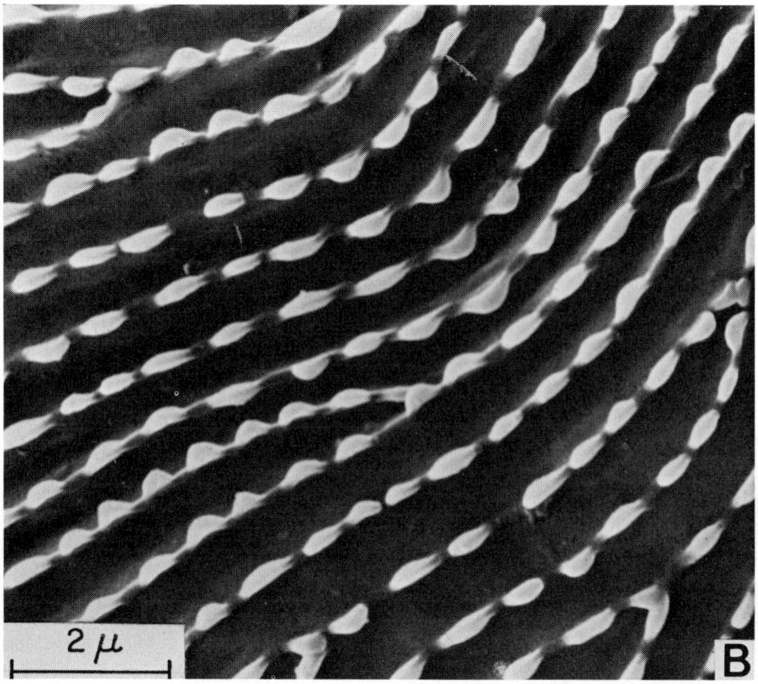

6. *Phytoseiulus persimilis* Athias-Henriot, a specialized acari predator of plant feeding mites which is dependent on mites for food. (Photograph supplied by J. A. McMurtry).

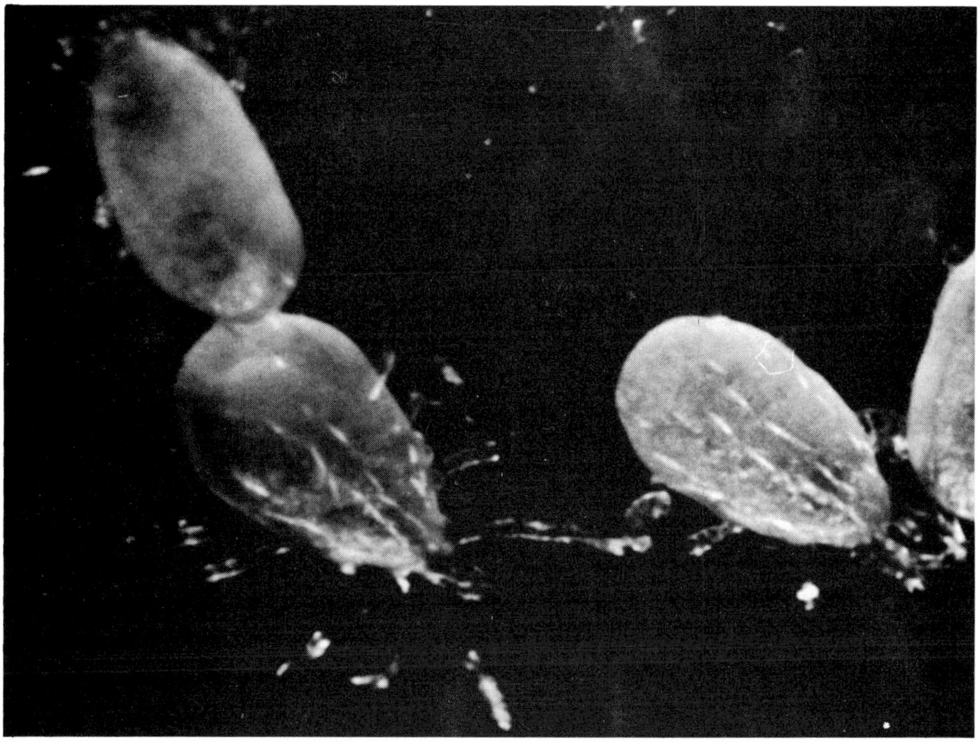

7. *Typhlodromus occidentalis* Nesbitt, a mite predator of the McDaniel mite, two-spotted spider mites, and other mites that web profusely. (Photograph supplied by M. M. Barnes and B. Croft.)

8. *Amblyseius limonicus* Garman and McGregor, a predator of European red mite. One, *P. ulmi*, shown in the photograph. (Photograph supplied by J. A. McMurtry).

9. *a*, *Stethorus picipes* Casey larva; *b*, *S. picipes* adult feeding on citrus red mite. (Photographs supplied by J. A. McMurtry).

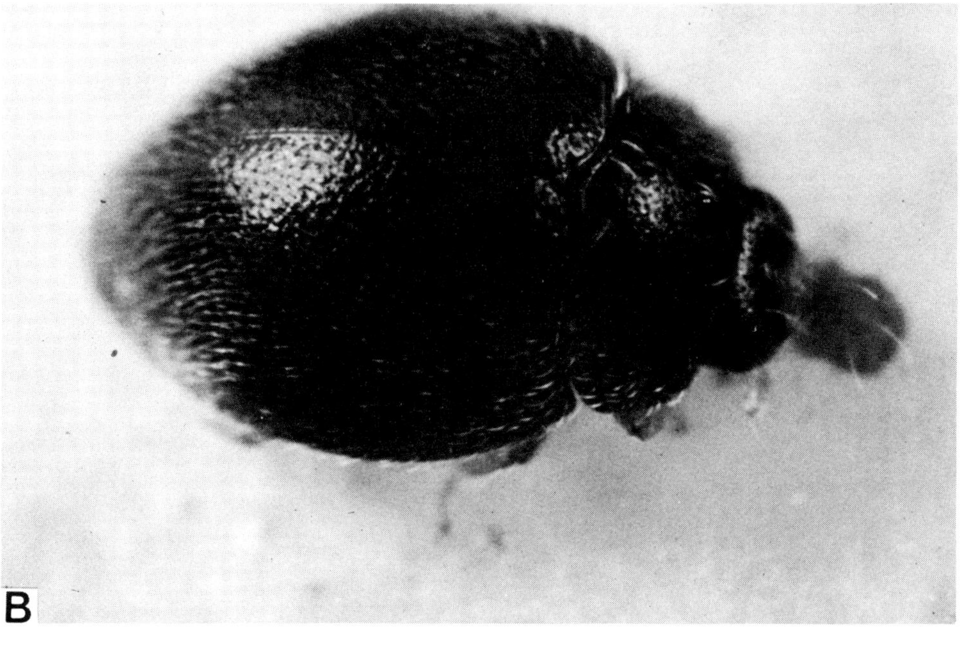

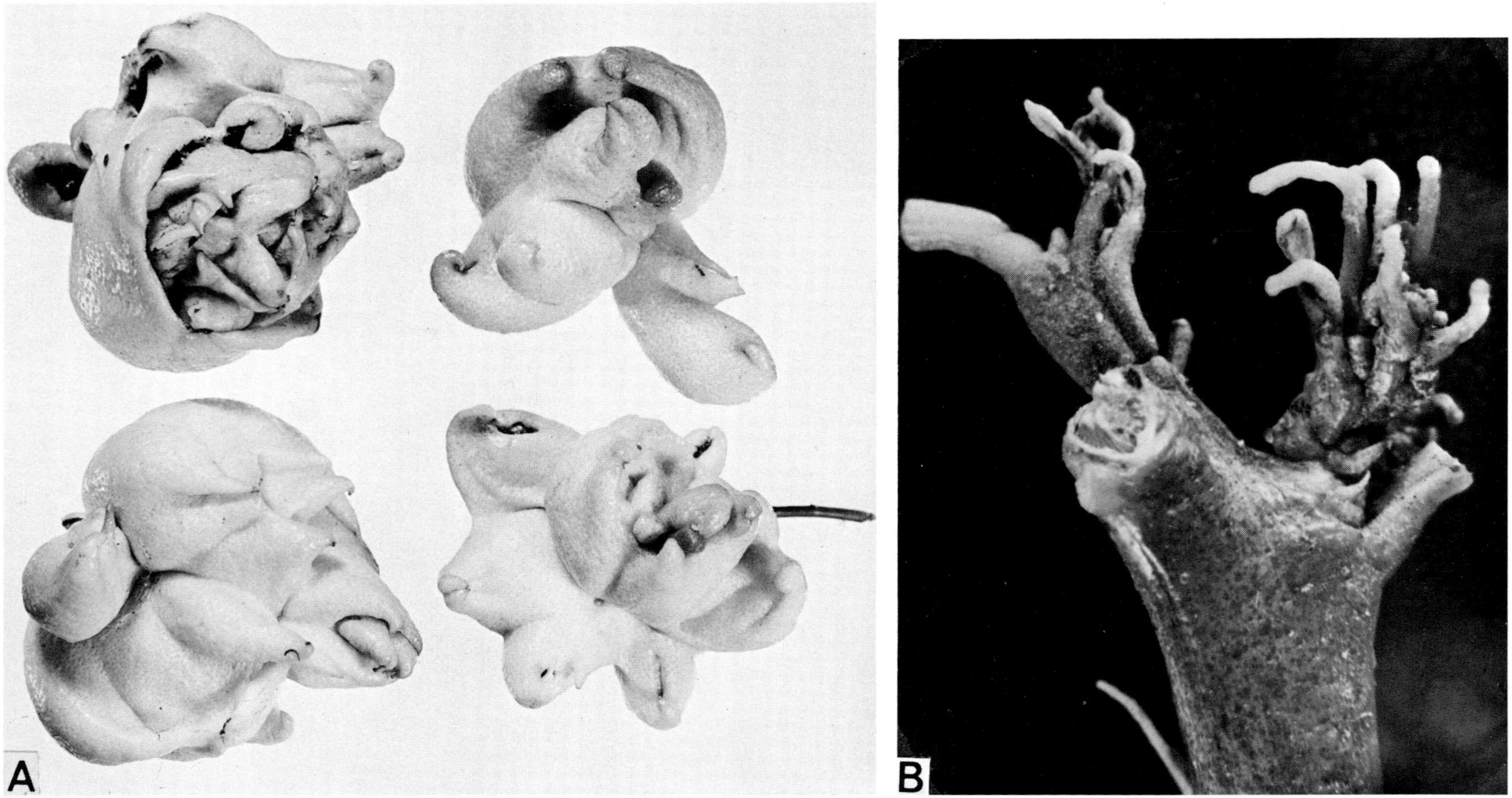

10. *a*, Deformed lemon fruit caused by feeding of citrus bud mite, *Eriophyes sheldoni* Ewing on the lemon fruit bud primordia; *b*, Multiple budding and deformed growth of lemon twig caused by citrus bud mite, *E. sheldoni*. (Photograph supplied by Norton Wilson.)

11. Leaf galls resulting from feeding of *Phytoptus emarginatae* Keifer on western chokecherry. (Photograph supplied by Norton Wilson.)

12. *a*, Concentric ring blotch (scattered infection) on navel orange leaves caused by *Calacarus citrifolii* Keifer; *b*, Concentric ring blotch on navel orange leaves, mostly vein infection. (Photographs supplied by E. C. Calavan.)

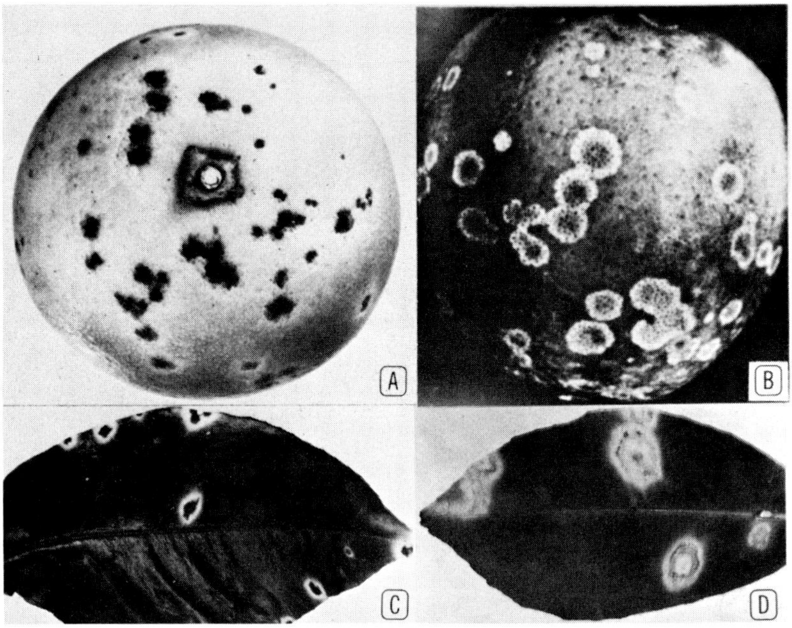

13. Leprosis: *a*, Fruit spotting: Florida; *b*, Fruit spotting: Venezuela; *c*, Leaf spotting: Florida; *d*, Leaf spotting: Venezuela; *e*, Twig lesions: Florida: (1) one-year-old lesions, (2) two-year-old lesions, (3) confluent lesions at advanced stage of twig scaling. (Photographs *a, c,* and *e*, supplied by L. C. Knorr; *b* and *d* by M. Hruskovec.)

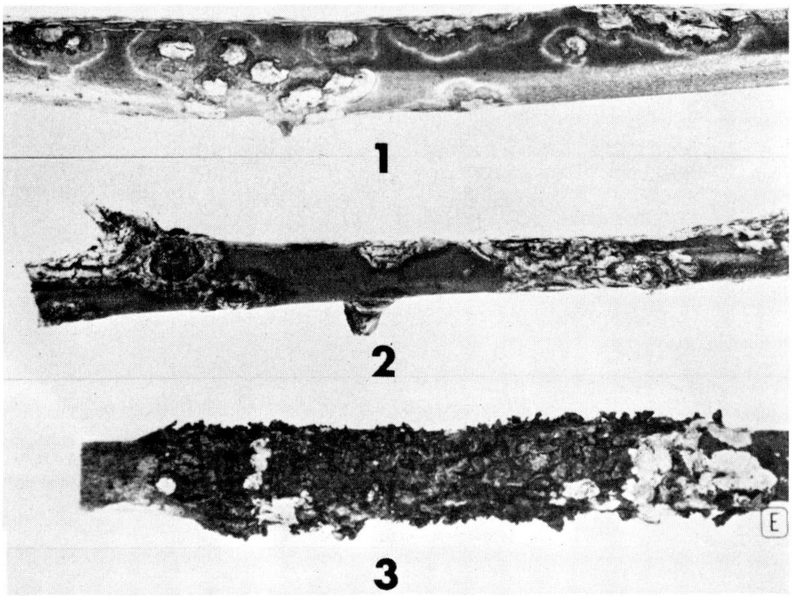

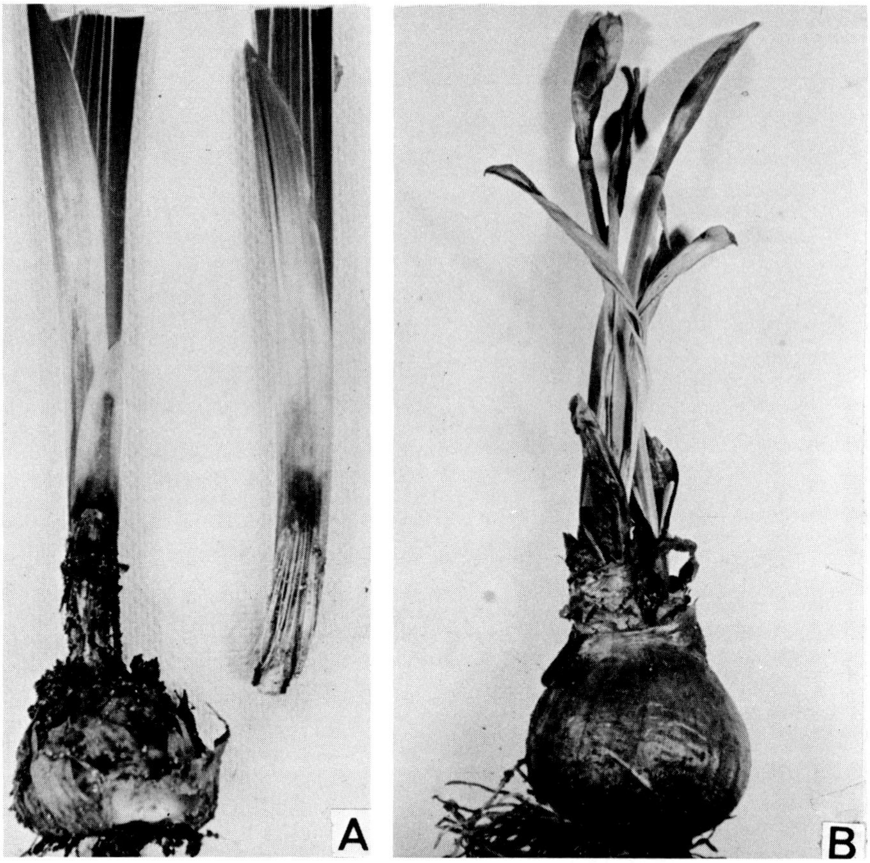

14. *a*, Injury to gladiolus bulb and roots by *Rhizoglyphus* mites; *b*, Bulb scale mite injury to narcissus; *c*, *Rhizoglyphus solani* Oudemans from gladiolus. (Photographs supplied by R. N. Jefferson.)

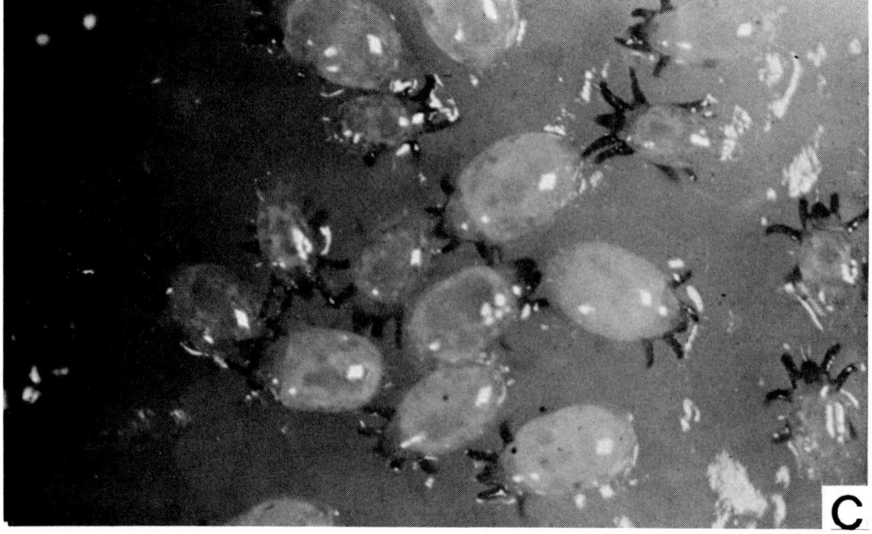

15. *a*, *Rhizoglyphus* sp.; *b*, Injury on gladiolus corm. (Photographs supplied by R. N. Jefferson.)

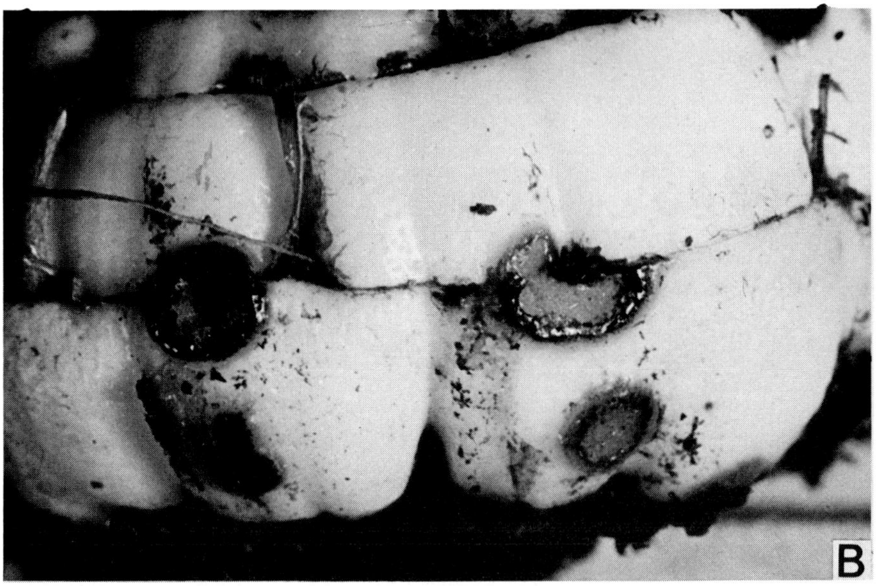

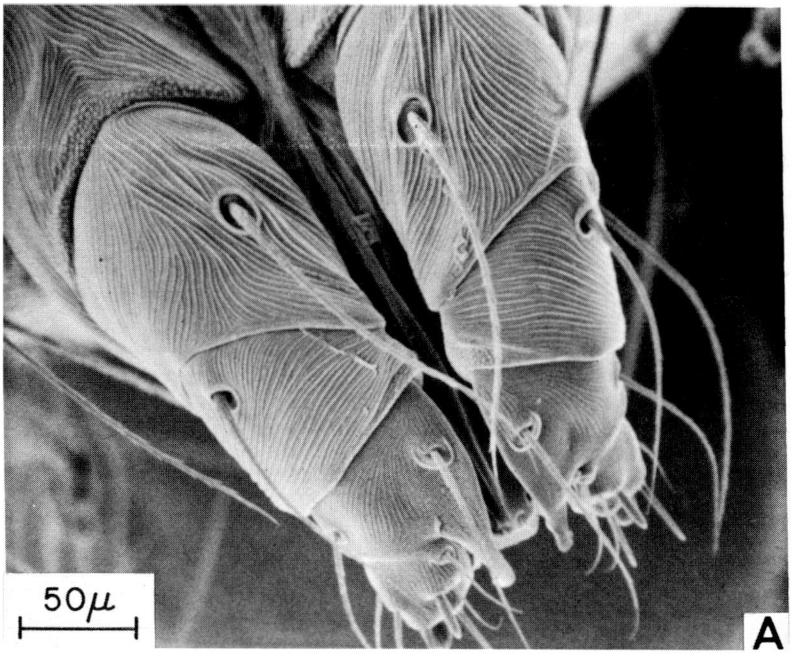

16. *a*, *Tetranychus pacificus* McGregor female, the pacific spider mite showing anterior-dorsal section with palps and stylets. (Scanning electron microscope photograph, original magnification, 300 X, 5 kv.); *b*, Same species showing base of tenent hairs and pulvillus. (Scanning electron microscope photographs; original magnification, 7,000 X, 15 kv.) (Note divided section near base of tenent hairs.)

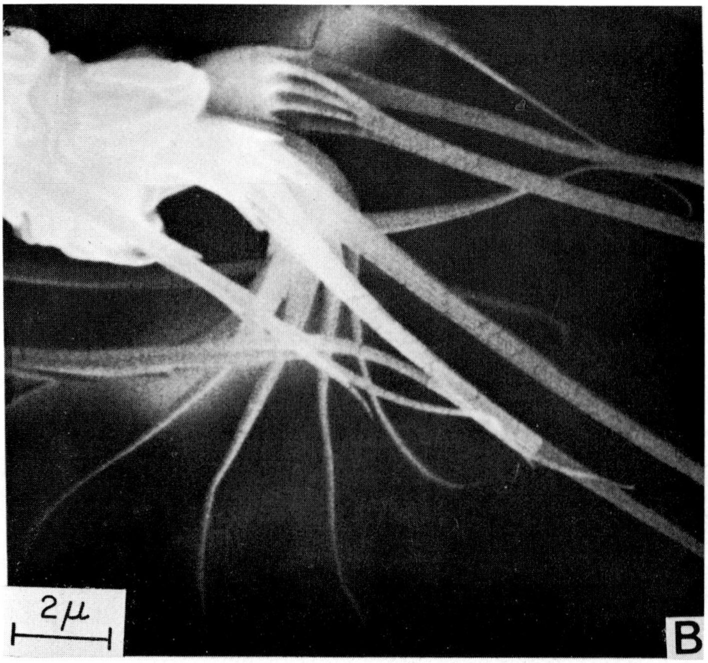

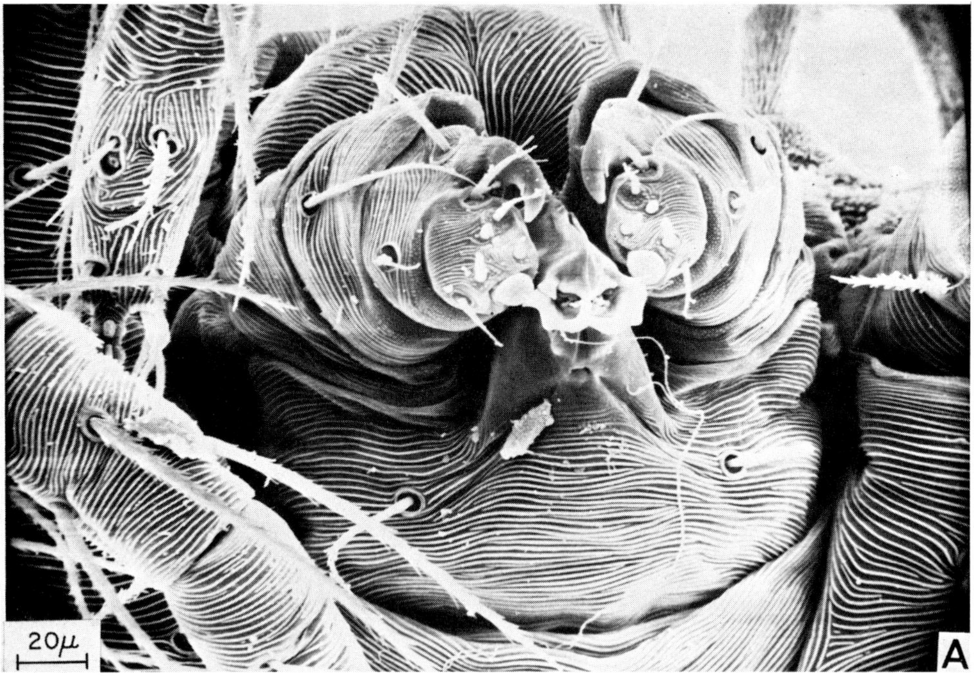

17. *a,* Citrus red mite, *Panonychus citri* McGregor, female rostrum from below (probing position). Mouth with flaps expanded (note hole below the mouth, function unknown). (Scanning electron microscope photograph, original magnification 630 x); *b,* Mouthparts of *P. citri,* showing rostrum and expanded "flaps" at its tip surrounding the openings through which stylets are extended; also the palpus with their silk-producing terminal euphathid with thread being extruded. (Scanning electron microscope photograph, original magnification, 4,160 X. Photographs supplied by Roberto H. Gonzalez.)

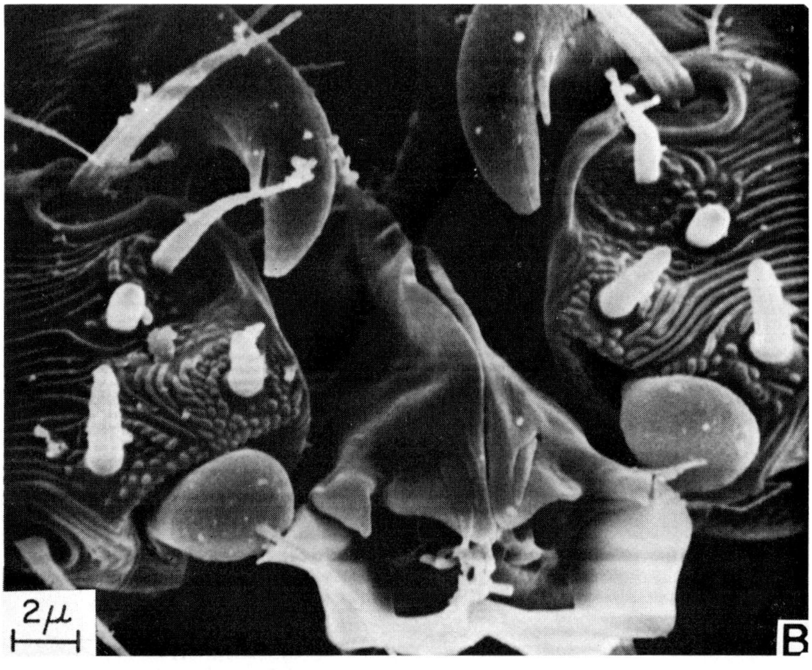

18. *Eotetranychus sexmaculatus* Riley female showing stylophore and peritreme groove at the hook end; also details of cuticle. (Scanning electron microscope photograph, original magnification 1,043 X. Photograph supplied by Roberto H. Gonzalez.)

19. *Tetranychus* sp. collected from violets at Berkeley, Calif. Genital area.

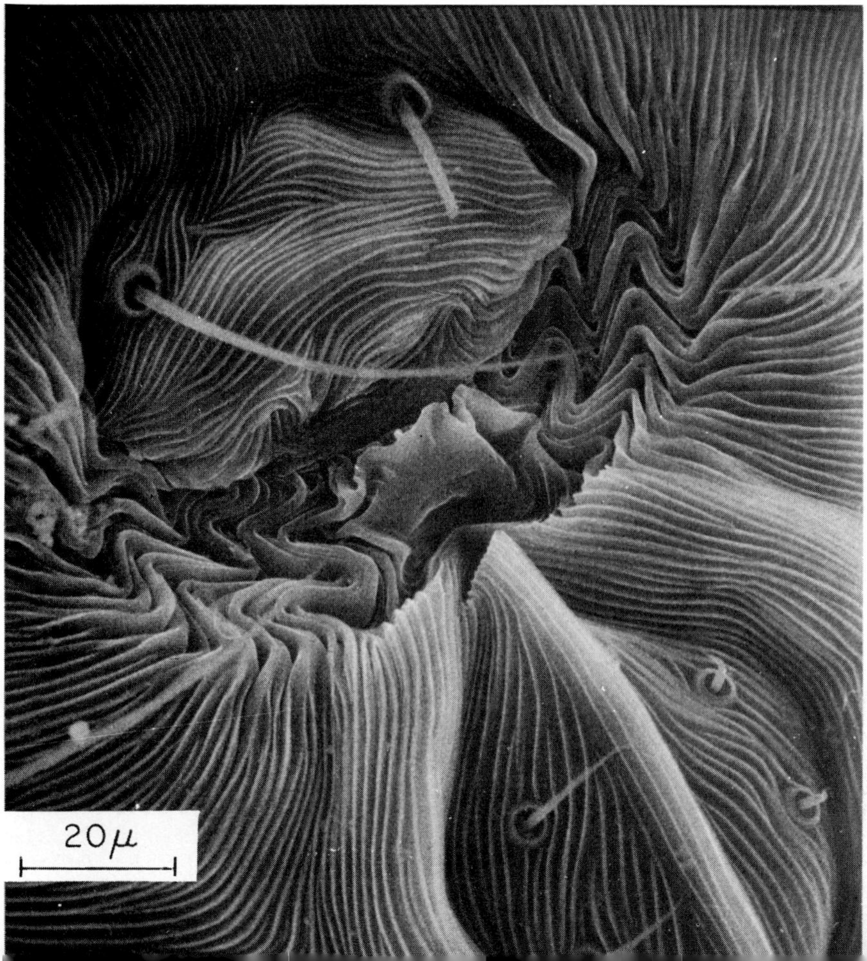

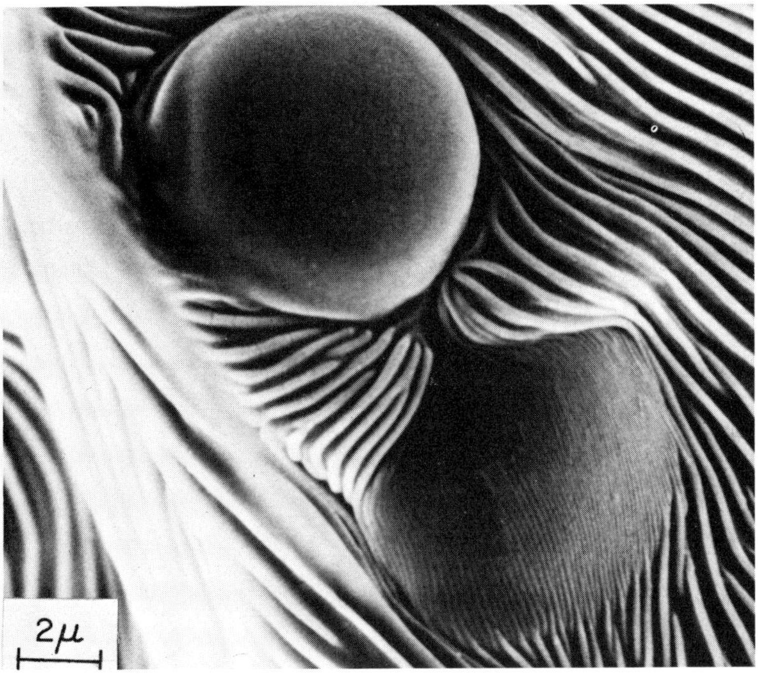

20. Eyes of the two-spotted spider mite, *T.* (*T.*) *urticae* Koch, hibernating form. (Scanning electron microscope photograph, original magnification, 4,000 X.)

21. McDaniel mite, *Tetranychus* (*Armenychus*) *mcdanieli* McGregor, injury to apple leaf at left; normal leaf at right.

22. "Blasting" of navel orange tree as a result of combined effect of citrus red mite, *Panonychus citri* (McGregor), feeding and drying winds. (Note twigs devoid of leaves, remaining leaves small and curled.)

23. The brown mite, *Bryobia rubrioculus* (Scheuten) adult. (Photograph supplied by M. M. Barnes.)

24. Brown mite, *Bryobia rubrioculus* (Scheuten), injury on apple leaf. (Photograph supplied by M. M. Barnes.)

25. *Petrobia* (*Tetranychina*) *harti* (Ewing) on oxalis. (Photograph supplied by F. F. Smith.)

26. *Eurytetranychus buxi* (Garman) injury to Buxus *semperivirens suffruticosa.* (Photograph supplied by F. F. Smith.)

27. Citrus red mite, *Panonychus citri* (McGregor) male, female, and egg.

28. Injury to lemon leaf (left) and normal leaf (right) by citrus red mite, *Panonychus citri* (McGregor).

29. Lemon tree (right) injured by feeding of citrus red mite. Uninjured tree at left.

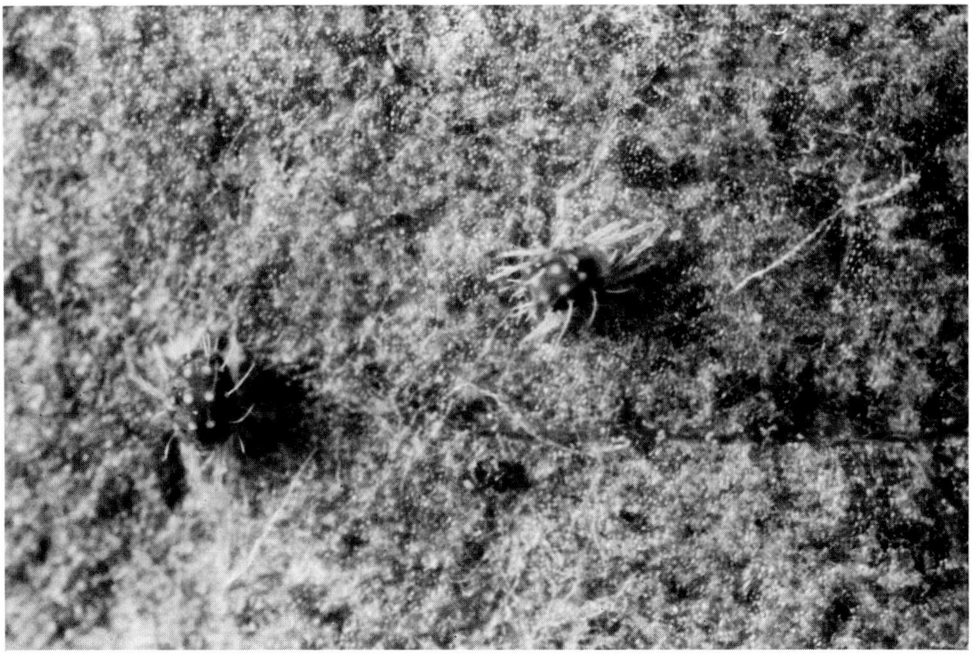

30. European red mite, *Panonychus ulmi* (Koch), adult females. (Photograph supplied by M. M. Barnes.)

31. The six-spotted spider mite, *Eotetranychus sexmaculatus* (Riley). Adult females and egg.

32. Injury to undersurface of lemon leaves by six-spotted spider mite.

33. *Oligonychus* (*O.*) *ilicis* (McGregor) injury to *Platanus racemosa* Nuttall (Plane tree). (Photographs supplied by L. R. Brown.)

34. *Oligonychus* (*O.*) *punicae* (Hirst). (Photograph supplied by L. R. Brown.)

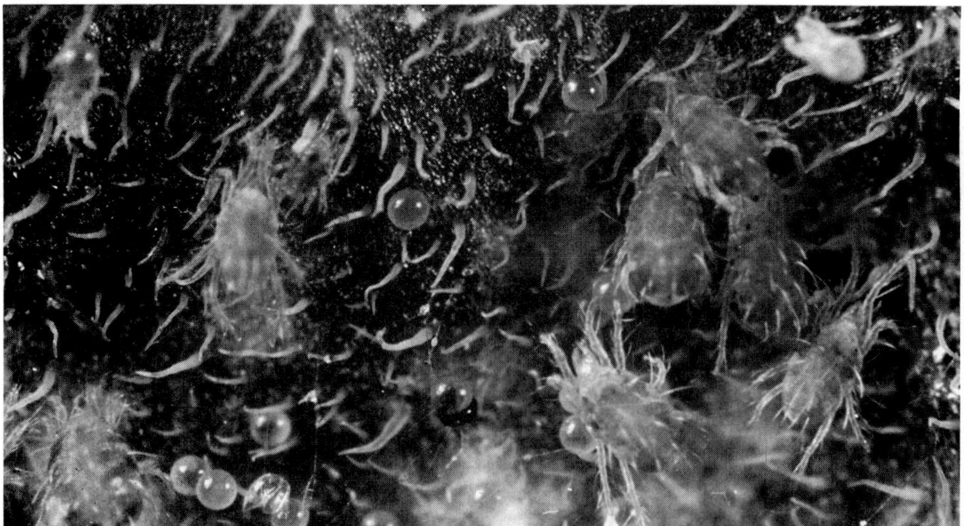

35. *Tetranychus* (*Tetranychus*) *evansi* Baker and Pritchard. (Photograph supplied by E. R. Oatman.)

36. Populations of *T.* (*T.*) *evansi* on lily-of-the-valley vine showing dense webbing and tendency of mites to congregate at top of the plant where they may be readily transferred by wind or flying insects or birds. (Photograph supplied by E. R. Oatman.)

37. Populations of *T.* (*T.*) *evansi* on lily-of-the-valley vine showing congregation of mites and webbing at the tip of the vine. (Photograph supplied by E. R. Oatman.)

38. The strawberry spider mite, *Tetranychus* (*Tetranychus*) *turkestani* (Ugarov and Nikolski), injury to Henderson bush, lima bean, and soybean. (Photograph supplied by F. F. Smith.)

39. The two-spotted spider mite, *Tetranychus* (*Tetranychus*) *urticae* Koch. Male, female, and egg. (Photograph supplied by M. M. Barnes.)

40. Two-spotted spider mite, *Tetranychus* (*Tetranychus*) *urticae* Koch, injury to lima bean leaves. (Photograph supplied by F. F. Smith.)

41. Carmine spider mite, *Tetranychus* (*T.*) *cinnabarinus* (Boisduval) injury to Henderson lima bean leaves. (Photograph supplied by F. F. Smith.)

42. The McDaniel mite, *Tetranychus* (*Armenychus*) *mcdanieli* McGregor. Adult females. (Photograph supplied by M. M. Barnes.)

43. Injury to apple leaf (left) by *Tetranychus* (*A.*) *mcdanieli*, and normal leaf (right). (Photograph supplied by M. M. Barnes.)

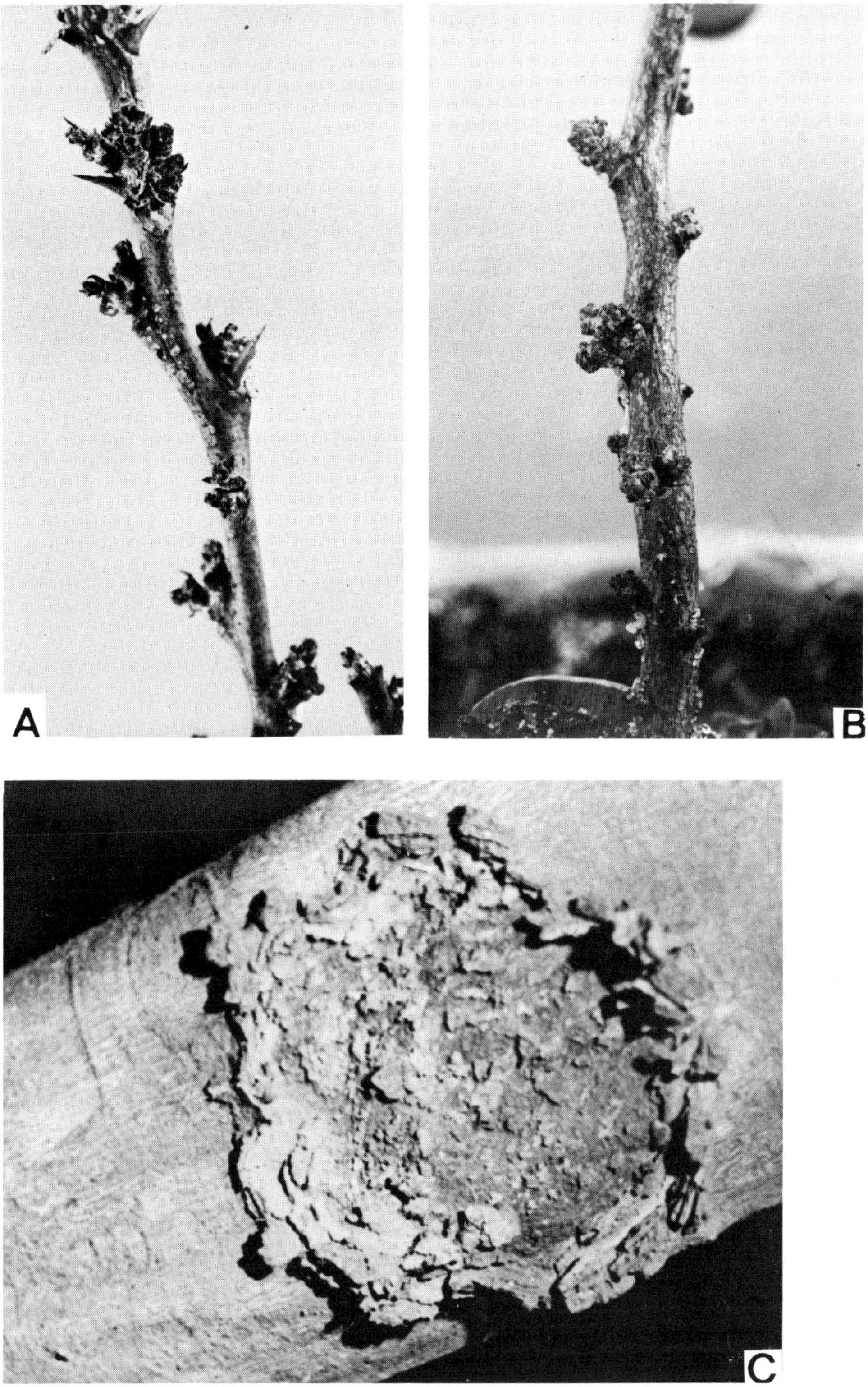

44. Injury to citrus by *Brevipalpus* mites (leprosis); *a*, Reproduction of *Brevipalpus* gall 120 days after infestation with *B. phoenicis* under controlled conditions; *b*, Natal galling on stem of sour orange seedling 90 days following initial infestation of seedling with *Brevipalpus phoenicis*; *c*, Old bark lesions with thick resinous scales on trunk of sweet orange, the result of an infestation of *B. californicus* some 70 years earlier. (Photographs supplied by L. C. Knorr.)

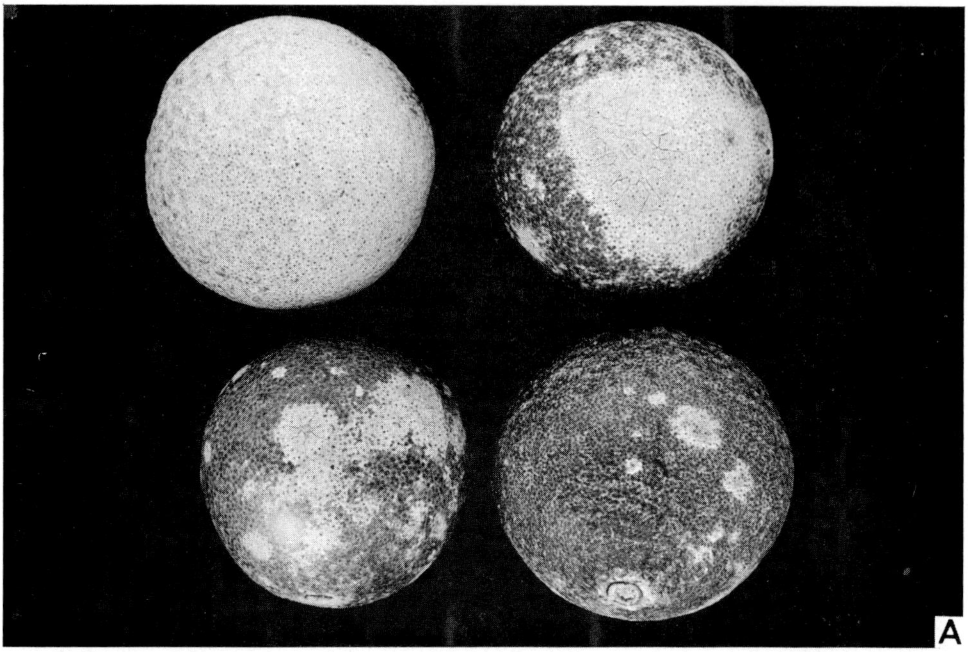

45. *a*, Citrus flat mite, *Brevipalpus lewisi* McGregor injury to oranges (whole fruit); *b*, Section of orange fruit showing injury by the citrus flat mite.

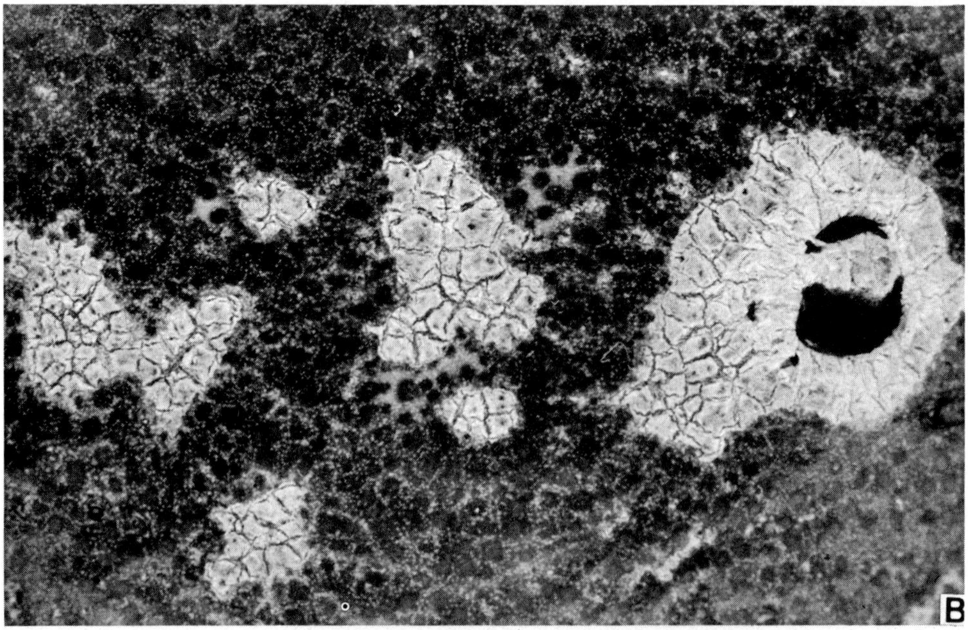

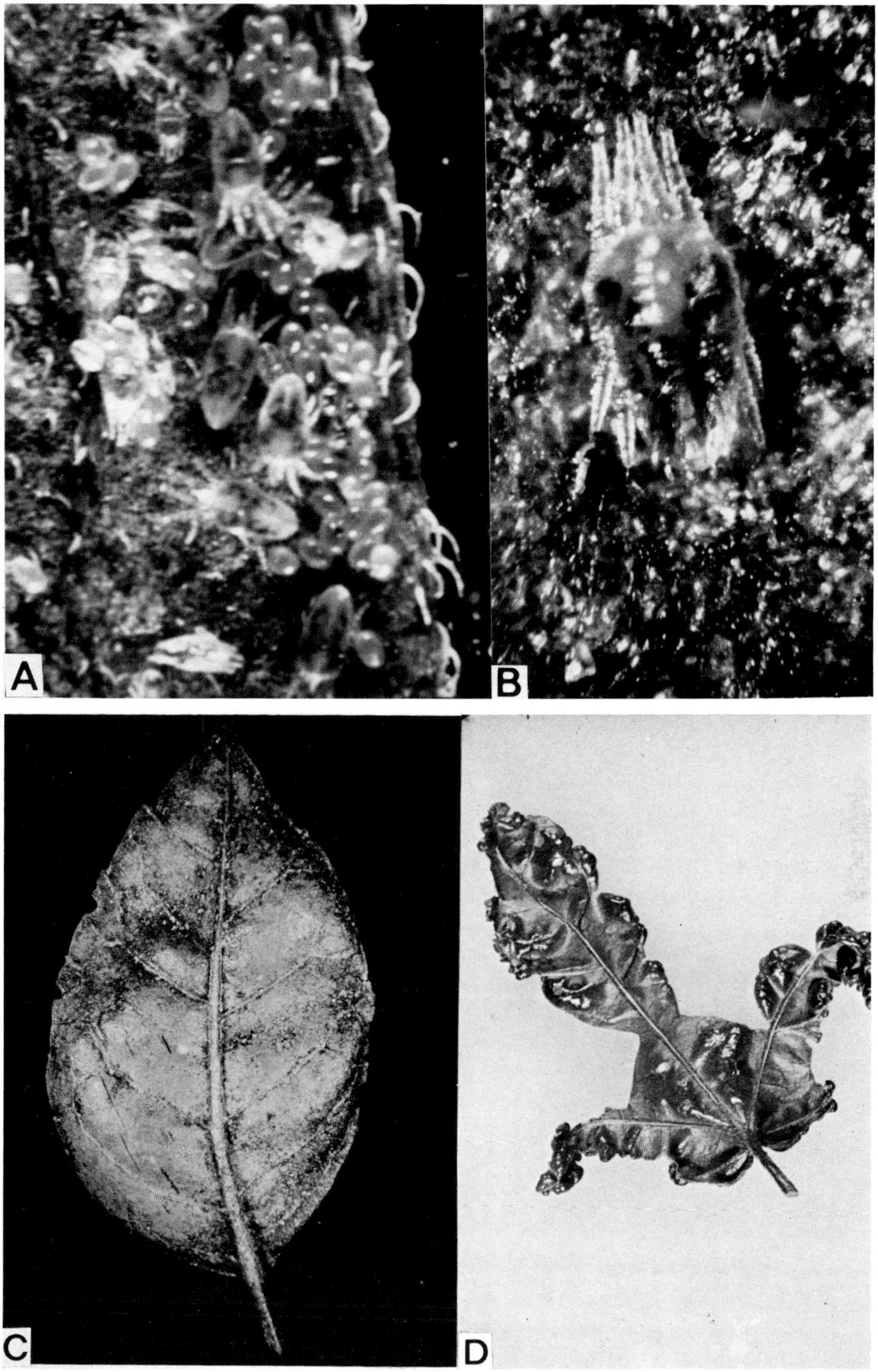

46. The privet mite, *Brevipalpus obovatus* Donnadieu; *a*, Adults, nymphs, and eggs; *b*, Immature; *c*, Injury to *Aucuba* leaf; *d*, Injury to *Trevesia.* (Photographs supplied by R. N. Jefferson.)

47. The privet mite, *Brevipalpus obovatus* Donnadieu injury on mums. (Photograph supplied by R. N. Jefferson.)

48. *Tenuipalpus pacificus* Baker injury to *Phalaenopsis*. Injured at left; normal at right (upper and lower surfaces). (Photograph supplied by R. N. Jefferson.)

49. Cyclamen mite, *Steneotarsonemus pallidus* (Banks) injury to: *a*, Cyclamen flowers; *b*, African violet leaves. (Photographs supplied by F. F. Smith.)

50. Cyclamen mite, *Steneotarsonemus pallidus* (Banks) injury to: *a,* Gloxinia; *b,* Azalea buds; *c,* Azalea flower at left; normal at right. (Photographs supplied by R. N. Jefferson.)

C

C

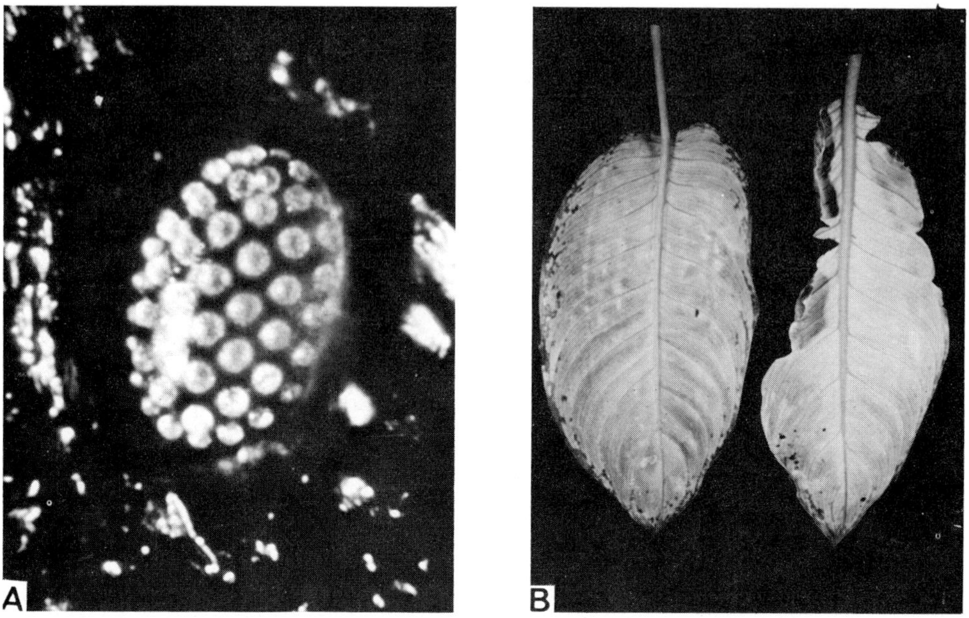

51. Broad mite, *Polyphagotarsonemus latus* (Banks); *a,* Eggs; *b,* Injury to *Dieffenbachia*; *c,* Left, injured Gerbera leaves; right, normal leaf. (Photographs supplied by R. N. Jefferson.)

52. *a, Eriophyes mackiei* K. galls on *Quercus agrifolia; b, Eriophyes mackiei* injury on *Quercus dumosa; c,* Plant galls produced by ***Phytoptus emarginatae*** (K.) on ***Prunus virginiana*** **L.** (choke-cherry). Photographs supplied by L. R. Brown.)

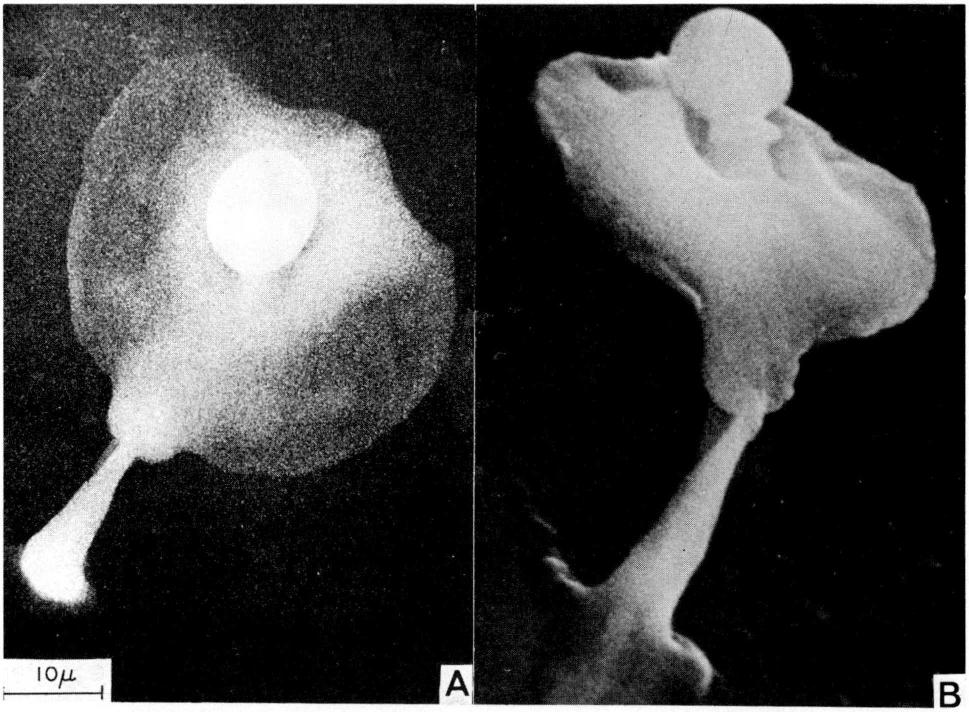

53. *a*, Spermatophore of *Phyllocoptruta oleivora* Ashmead.; *b*, Spermatophore of *Phyllocoptruta oleivora* Ashmead, partially collapsed showing more of sac. (Scanning microscope photographs supplied by George Oldfield, original magnification, 2,500 X, 20 kv.)

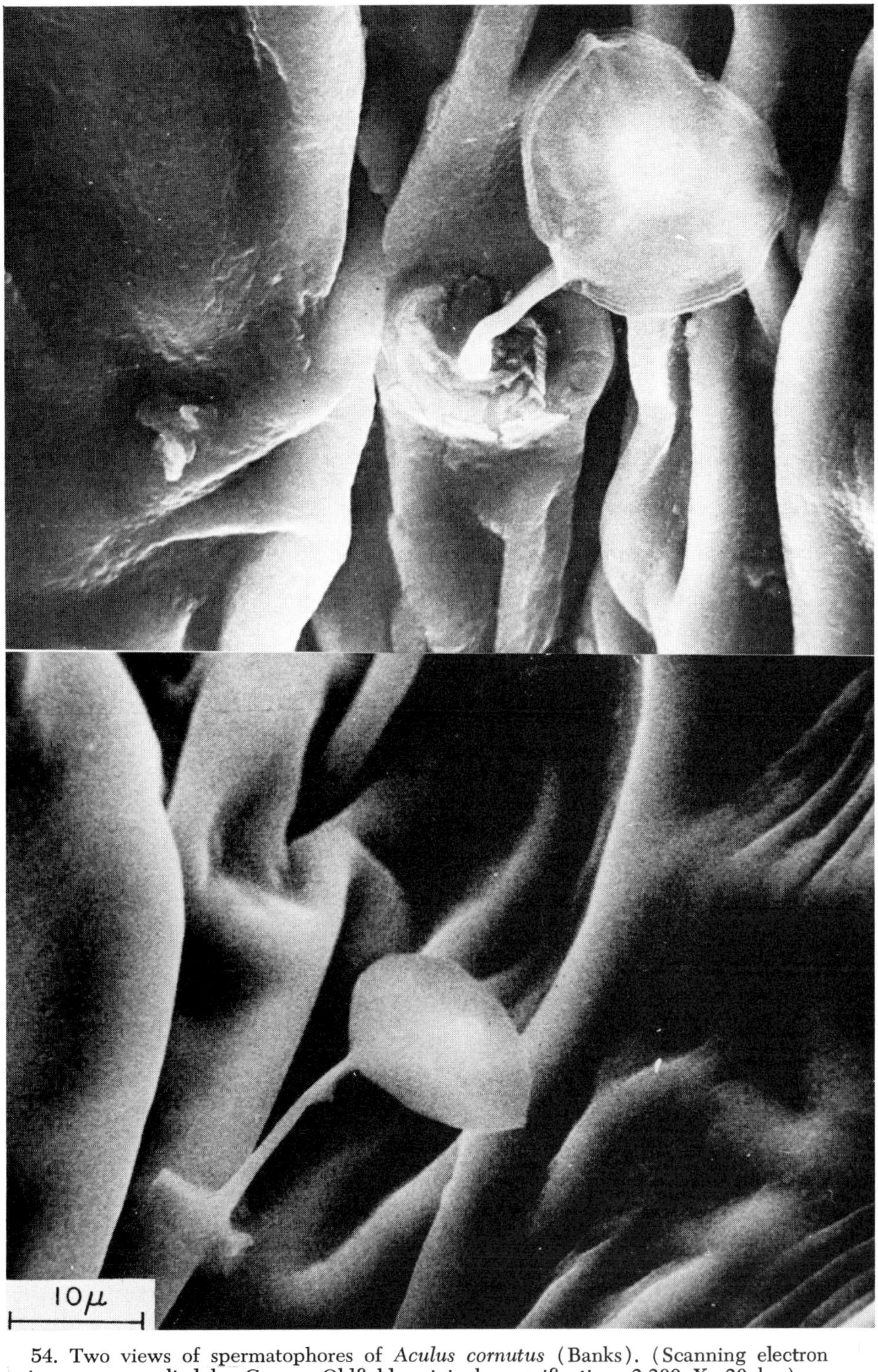

54. Two views of spermatophores of *Aculus cornutus* (Banks). (Scanning electron microscope supplied by George Oldfield, original magnification, 2,200 X, 20 kv.)

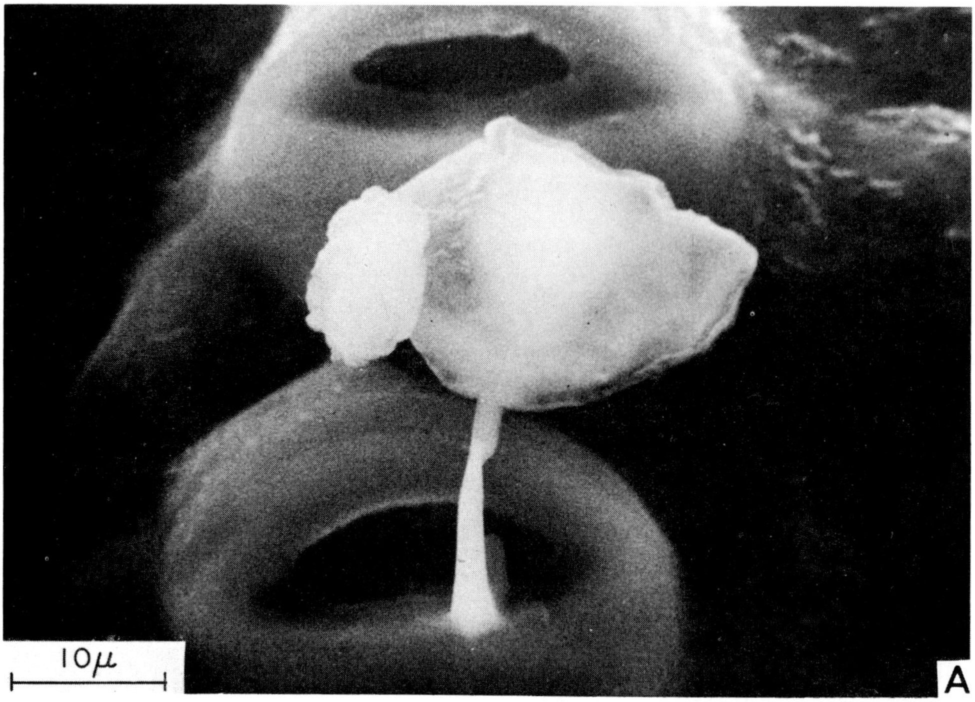

55. Spermatophores of *Diptacus gigantorhynchus* Nal.; *a,* Attached to rim of a leaf stoma. (Scanning electron microscope photograph, original magnification, 2,300 X, 20 kv.); *b,* On leaf amid stomata. (Scanning electron microscope photographs supplied by George Oldfield, original magnification, 570 X, 20 kv.)

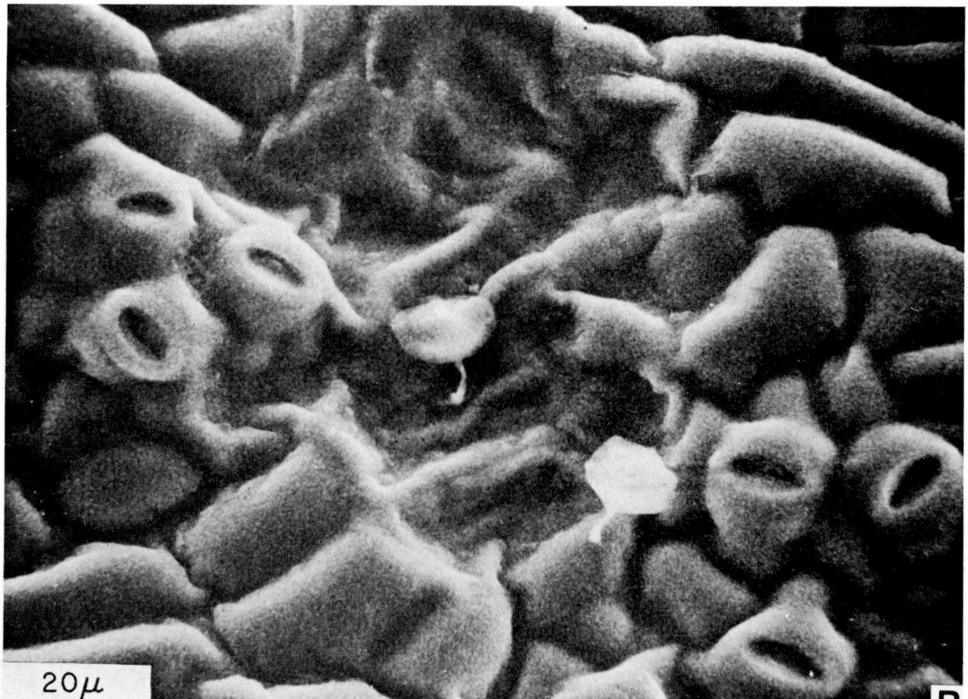

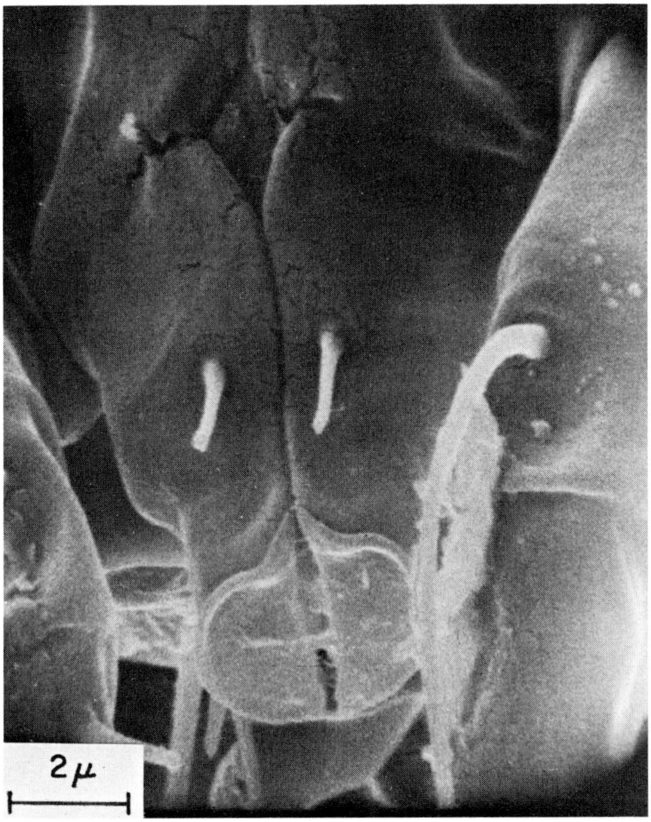

56. Tip of rostrum of *Aculops glabri* (K.) showing suction cups. (Scanning electron microscope photograph, original magnification, 7,450 X.)

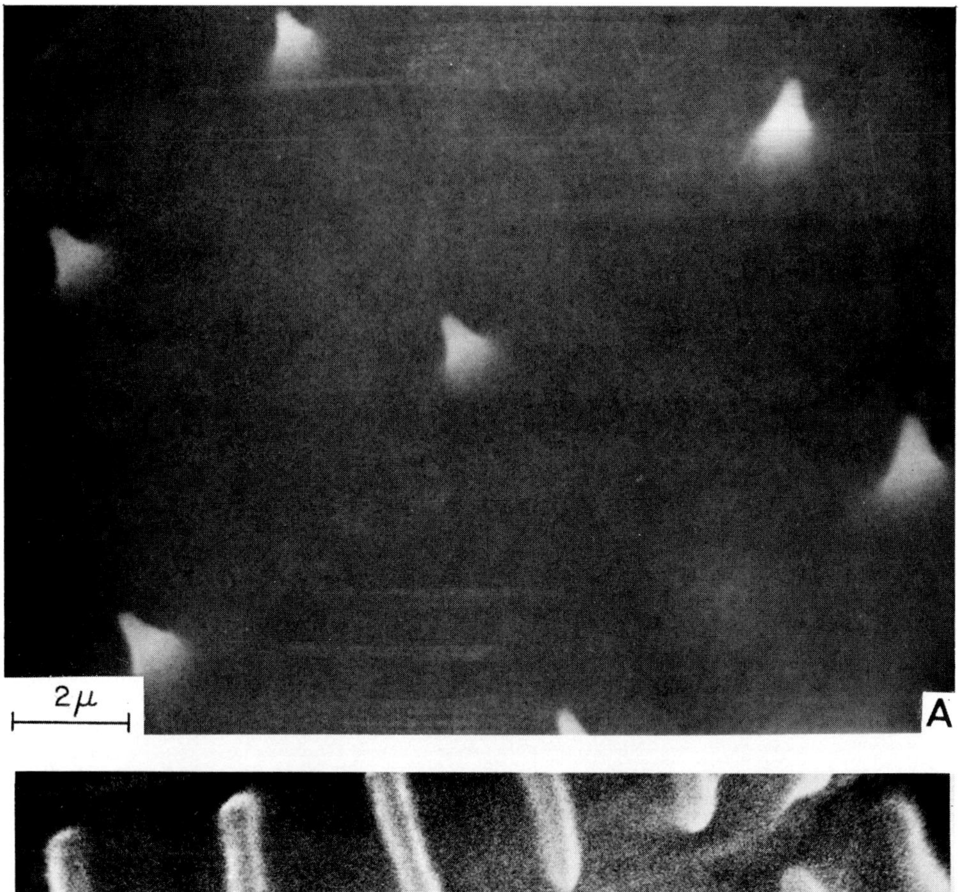
2μ
A
1μ
B

57. Three different types of dorsal microtubercles (Scanning electron microscope photographs.) *a*, Douglas fir bud mite, *Trisetacus pseudotsugae* K. (9,500 X); ***b***, Camellia bud mite, *Cosetacus camellia* (K.) (11,000 X); *c*, Olive leaf and flower mite, *Oxycenus maxwelli* (K.) (11,000 X.)

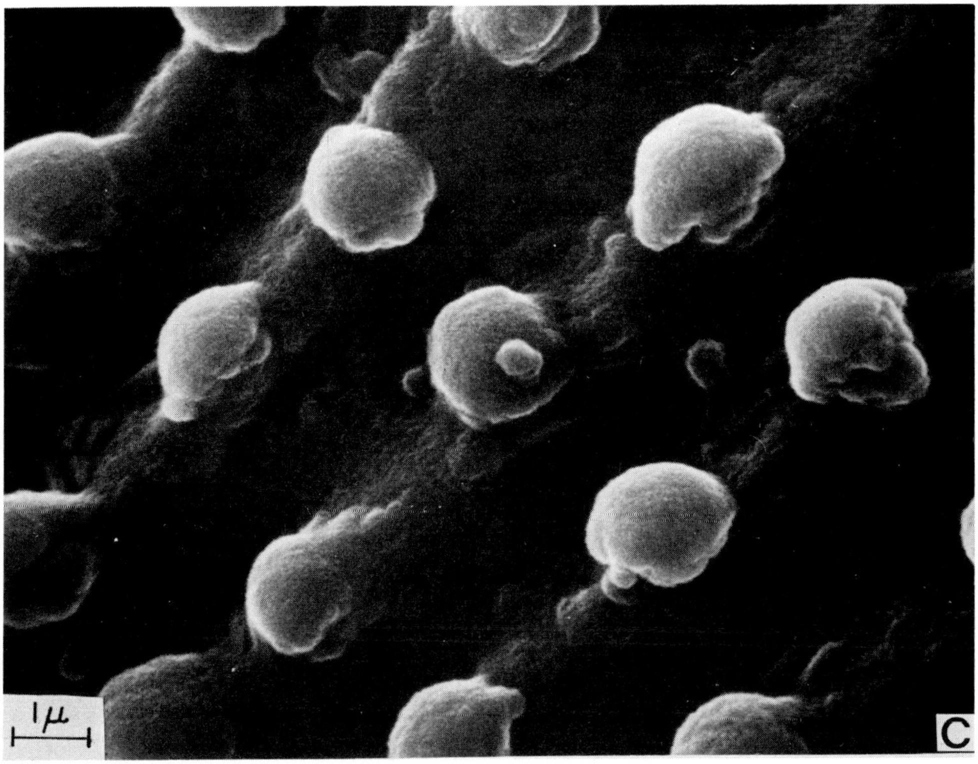

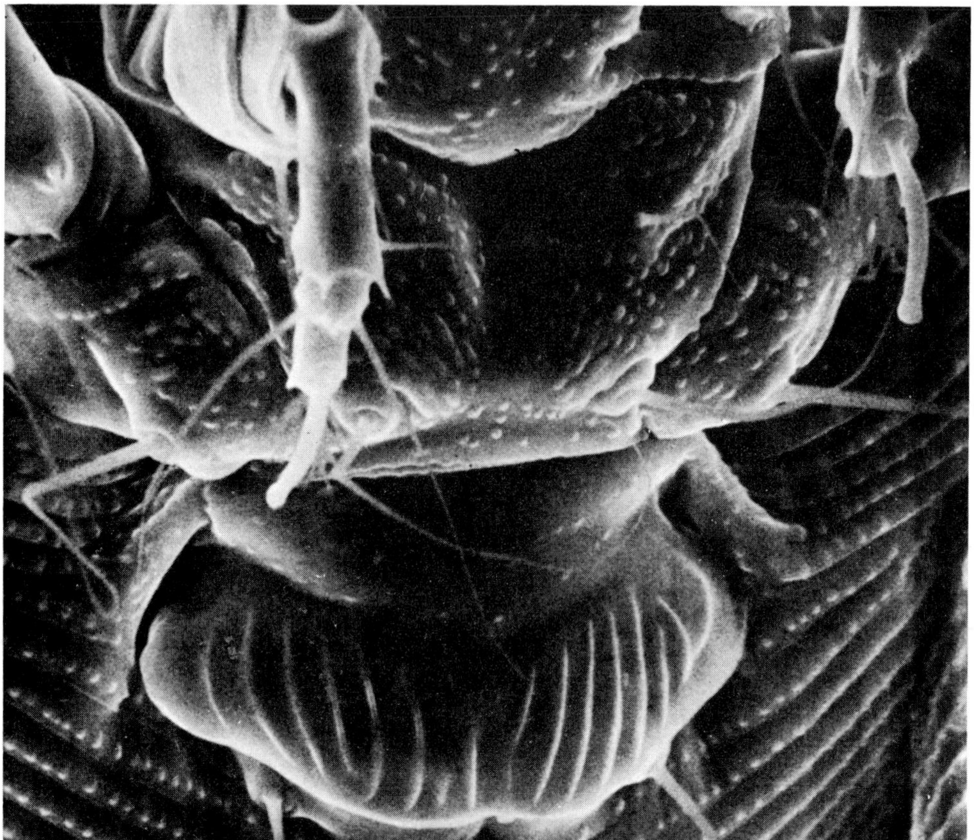
10μ
A

58. *a*, Olive leaf and flower mite, *Oxycenus maxwelli* K., female genital area (Scanning electron microscope photograph, original magnification, 2,200 X.) *b*, *Eriophyes brachytarsus* K., venter showing female genitalia and transverse aperture. (Scanning electron microscope photograph, original magnification 731 X, 1.9 kv.)

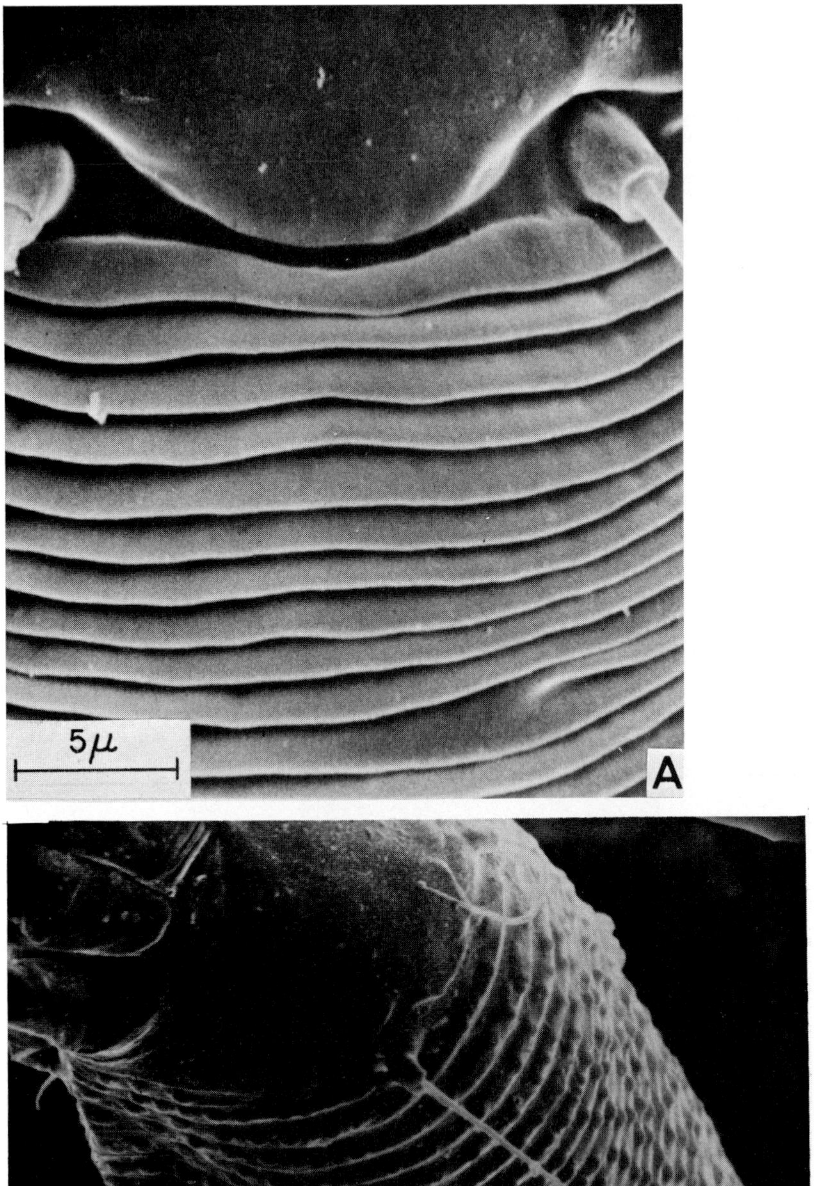
5μ
A
10μ
B

59. *a*, *Eriophyes calaceris* (K.) deutogyne, front dorsal view, from mountain maple red erineum. (Scanning electron microscope photograph, original magnification, 4,400 X.) *b*, Protogyne, front dorsal view. (Scanning electron microscope photograph, original magnification, 2,000 X.) *c*, *Eriophyes calaceris* (K.), side view of deutogyne showing doubling of dorsal rings. (Scanning electron microscope photograph, original magnification, 1,150 X.)

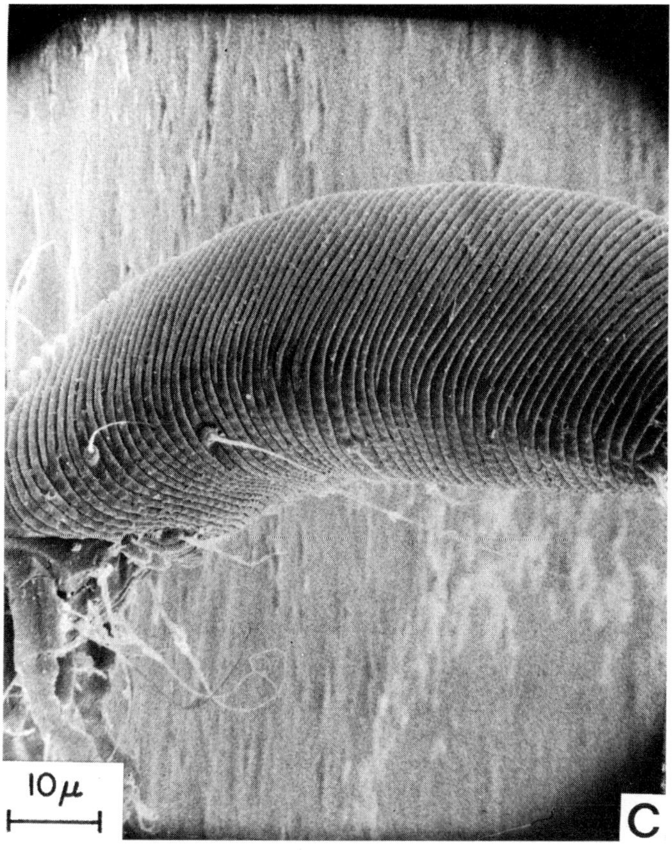

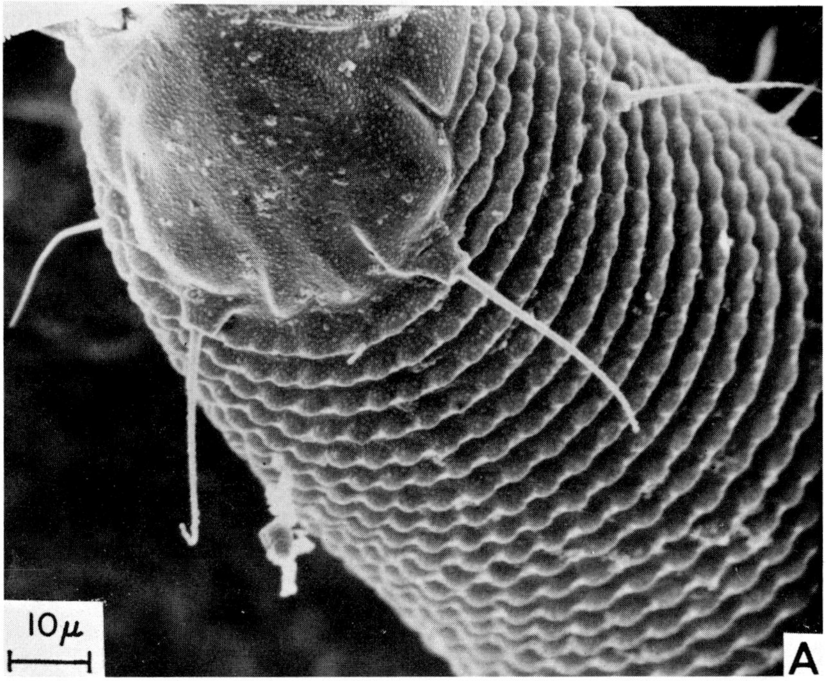

60. *a*, *Eriophyes brachytarsus* K., deutogyne, front dorsal view, from black walnut pouch galls. (Scanning electron microscope photograph, original magnification, 2,000 X.) *b*, *E. brachytarsus*, protogyne, showing shield and dorsal tubercles. (Scanning electron microscope photograph, original magnification, 5,500 X.) by Schuster.

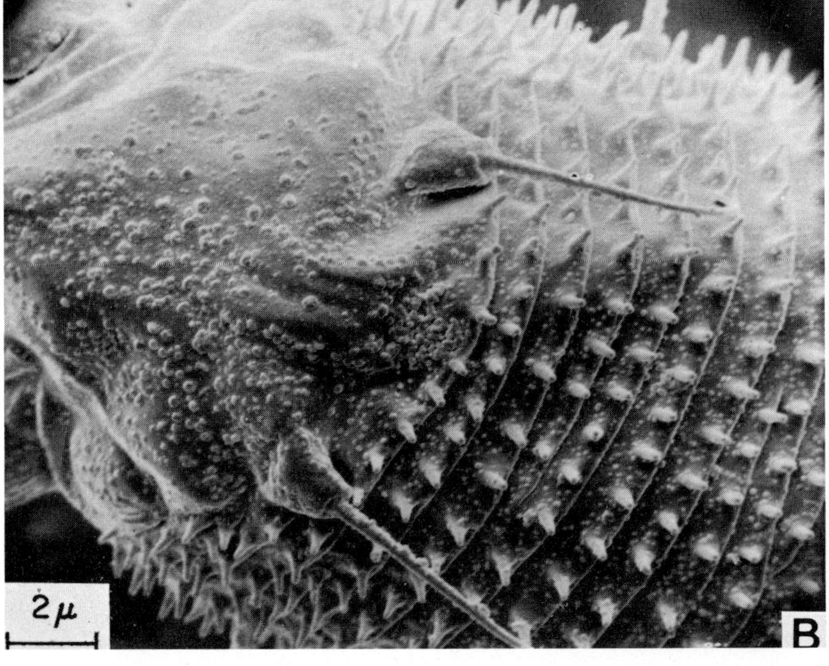

61. *Trisetacus sequoiae* K. gall on *Sequoia sempervirens* Endlicher (Redwood). (Photograph supplied by L. Brown.)

62. Featherclaws (transmission electron microscope photograph); *a*, First nymph of *Cecidophyopsis ribis* (Westwood); *b*, Second nymph of *C. ribis*; *c*, Adult of *C. ribis*; *d*, Adult of *C. selachodon* von Eyndhoven; *e*, *Erophyes tenuis* (Nal.); *f*, *Eriophyes ficus* Cotte. (Reproduced from Proeseler.)

63. The pear leaf blister mite, *Phytoptus pyri* Pagenstecher, injury to apples and leaves. (Photograph supplied by M. M. Barnes.)

64. Citrus bud mite *Eriophyes sheldoni* Ewing injury to leaf and fruit buds result in *a*, deformed leaves; *b*, deformed fruit.

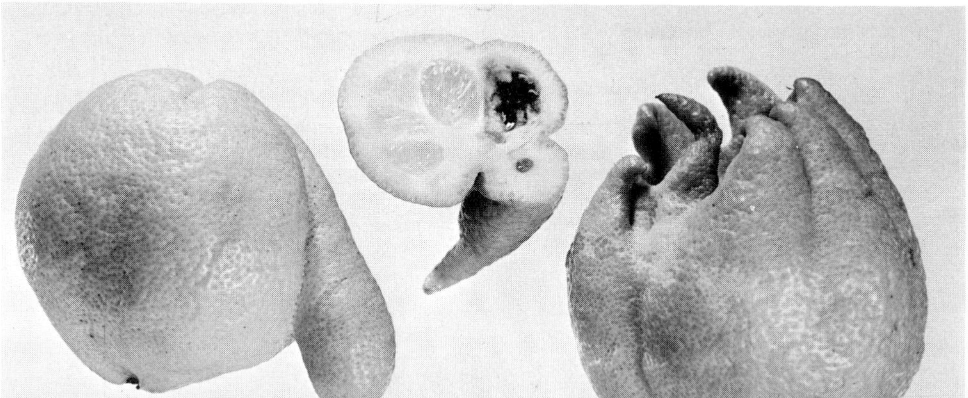

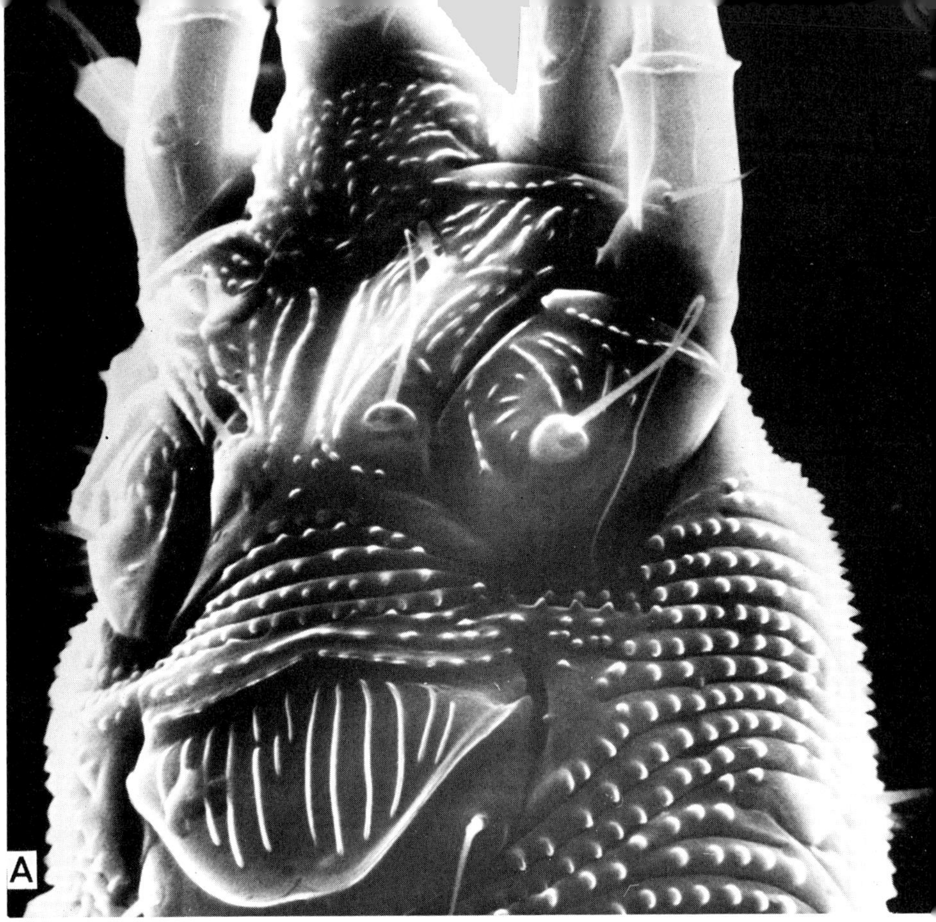

65. *a, Eriophyes tulipae* K. anterior-ventral view; *b, Eriophyes tulipae* (K.) anterior view. (Courtesy of G. W. Wharton.)

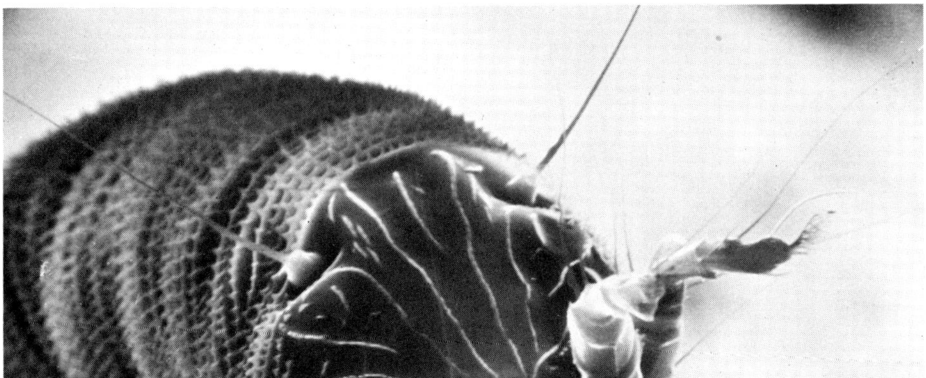

66. *Eriophyes cynodoniensis* (Sayed) injury to Bermuda grass. (Photograph supplied by R. N. Jefferson.)

67. Injury to citrus by the grey mite, *Calacarus citrifolii* K.; *a*, New leaves showing concentric ring blotch; *b*, Lesions on twigs. (Photographs supplied by B. J. Dippenaar, courtesy E. C. Calavan.)

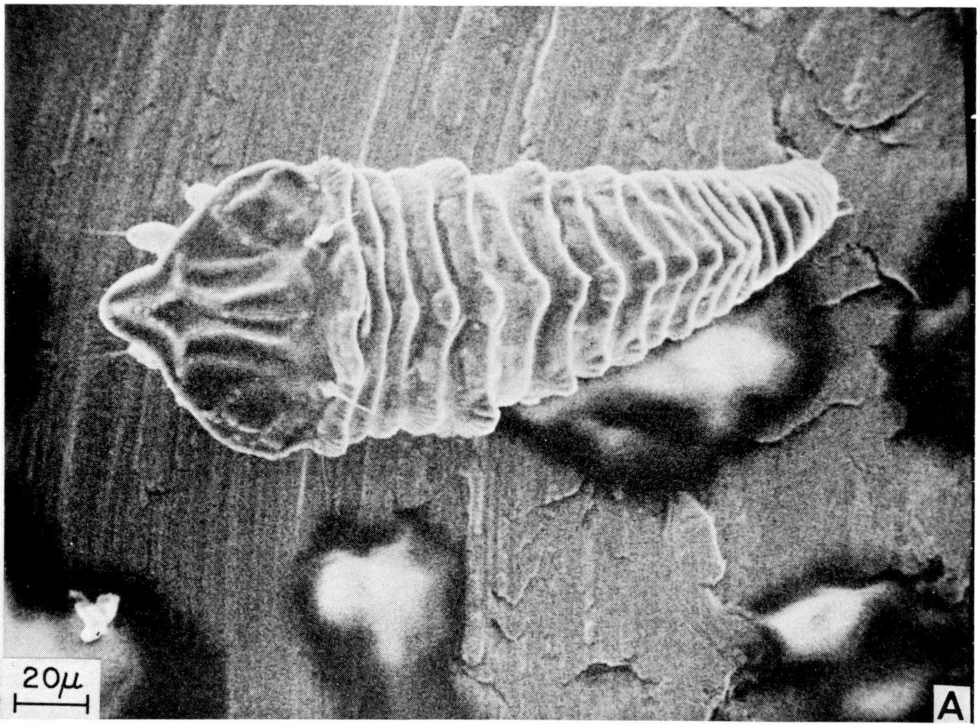

68. *a,* Dorsal view of the olive leaf and flower mite, *Oxycenus maxwelli* (K.) (Scanning electron microscope photograph, original magnification, 540 X, 20 kv.) *b,* Olive leaf and flower mite, *Oxycenus maxwelli* (K.) featherclaw. (Scanning electron microscope photograph, original magnification, 11,000 X, 20 kv.)

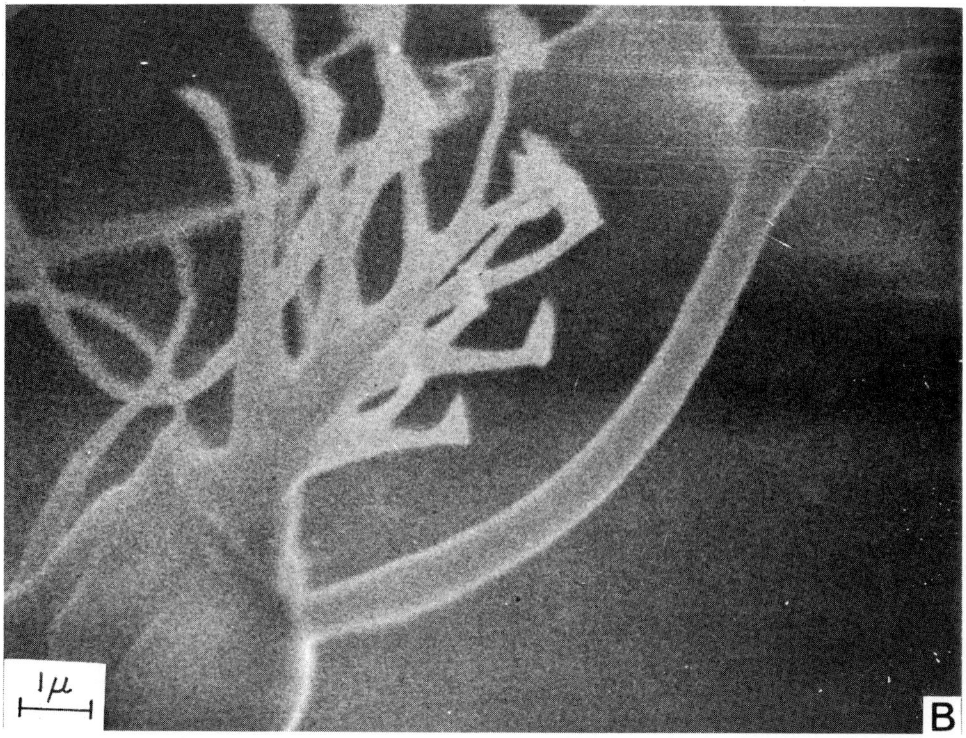

69. Lemon fruit silvered by feeding of citrus rust mite, *Phyllocoptruta oleivora* (Ashmead).

70. *Vasates quadripedes* Shimer, galls. (Photograph supplied by L. R. Brown.)

71. *a*, Peach silver mite, *Aculus cornutus* (Banks) dorsal view. (Scanning electron microscope photograph, original magnification, 570 X.) *b*, Peach silver mite, *Aculus cornutus* (Banks) featherclaw. (Scanning electron microscope photograph, original magnification, 11,000 X, 20 kv.)

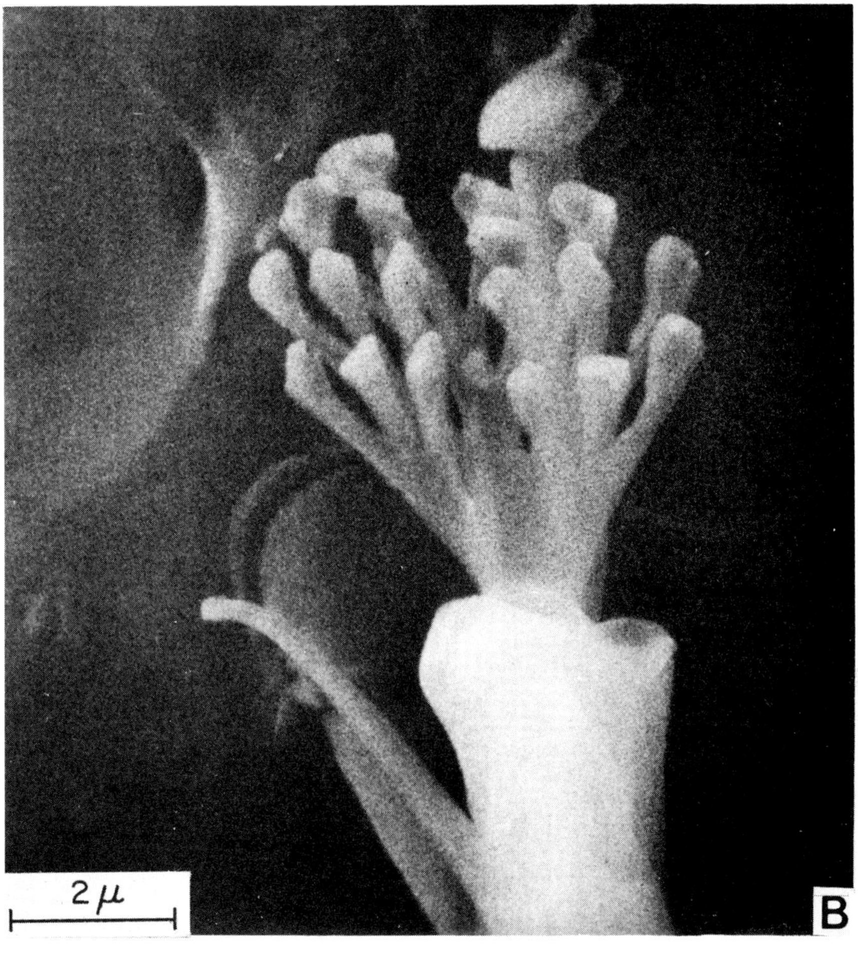
2 μ
B

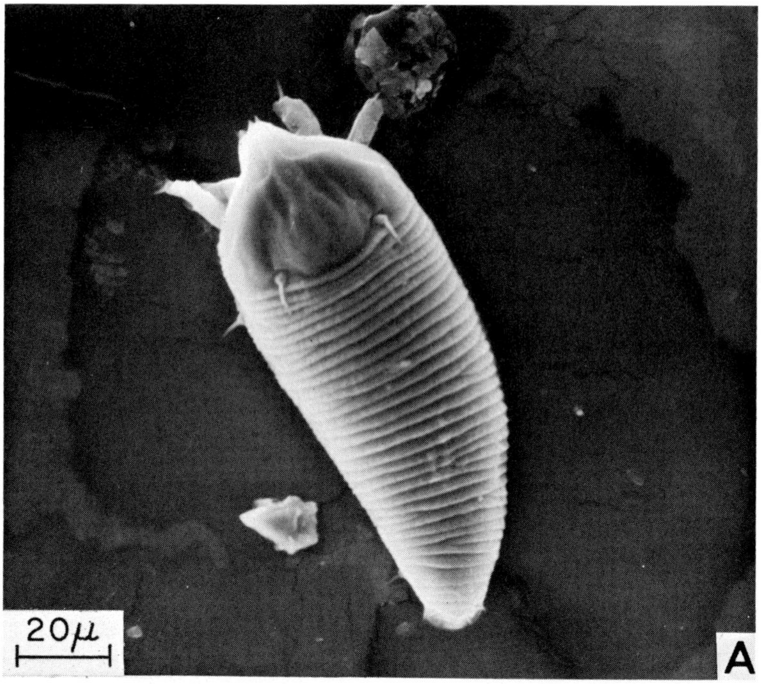

72. *a*, The privet rust mite, *Aculus ligustri* (K.) (Scanning electron microscope photograph, 600 X.) *b*, The privet rust mite, *Aculus ligustri* (K.), featherclaw (Scanning electron microscope original magnification, photograph, 10,000 X.)

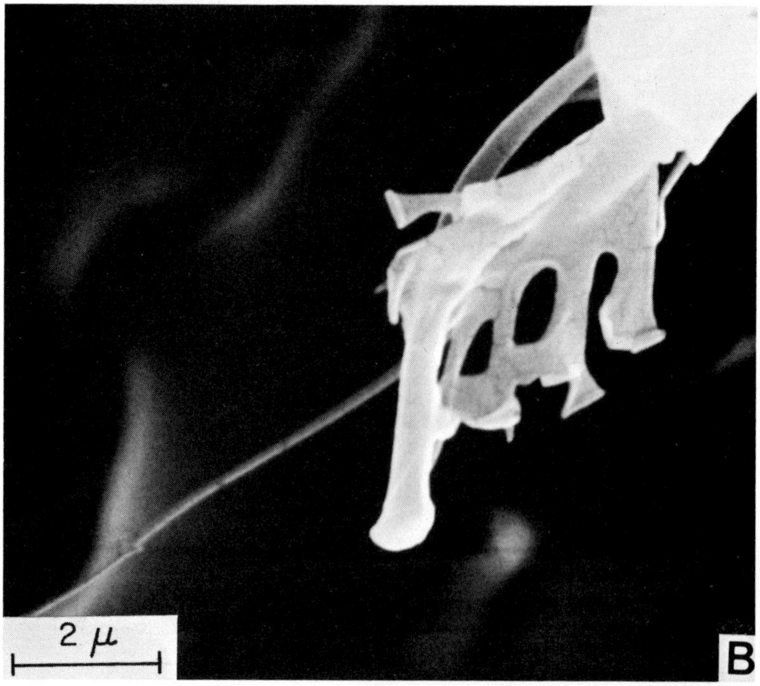

73. *a, Abacarus hystrix* (Nal.) dorsal view. (Scanning electron microscope photograph, courtesy of G. W. Wharton.) *b*, Featherclaw of second nymph of *Abacarus hystrix* Nal. *c*, Featherclaw of adult of *A. hystrix*. (Transmission electron microscope photograph, after Proeseler.)

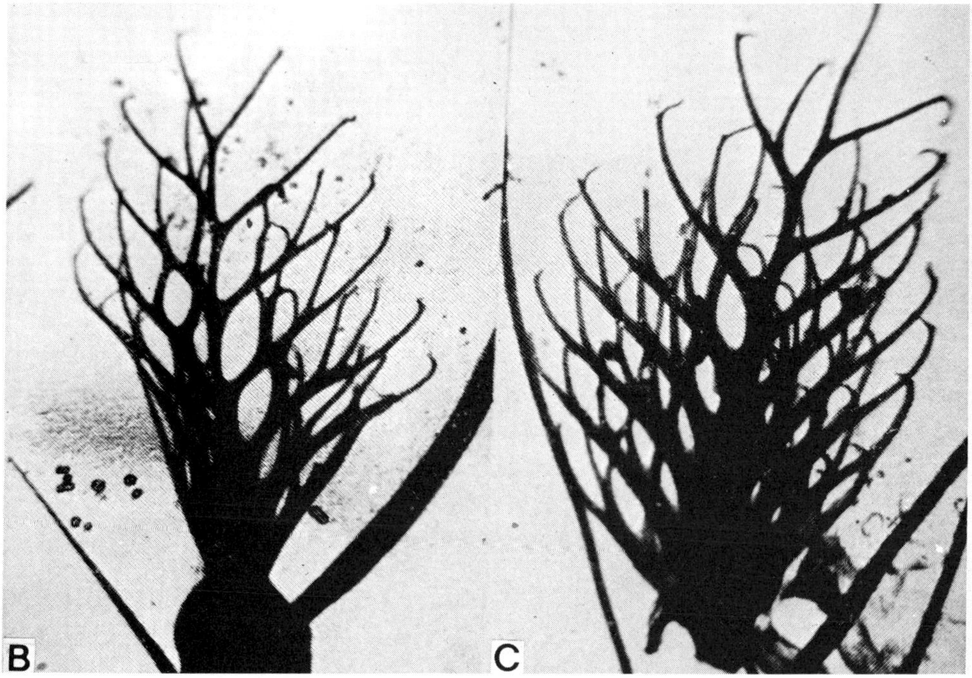

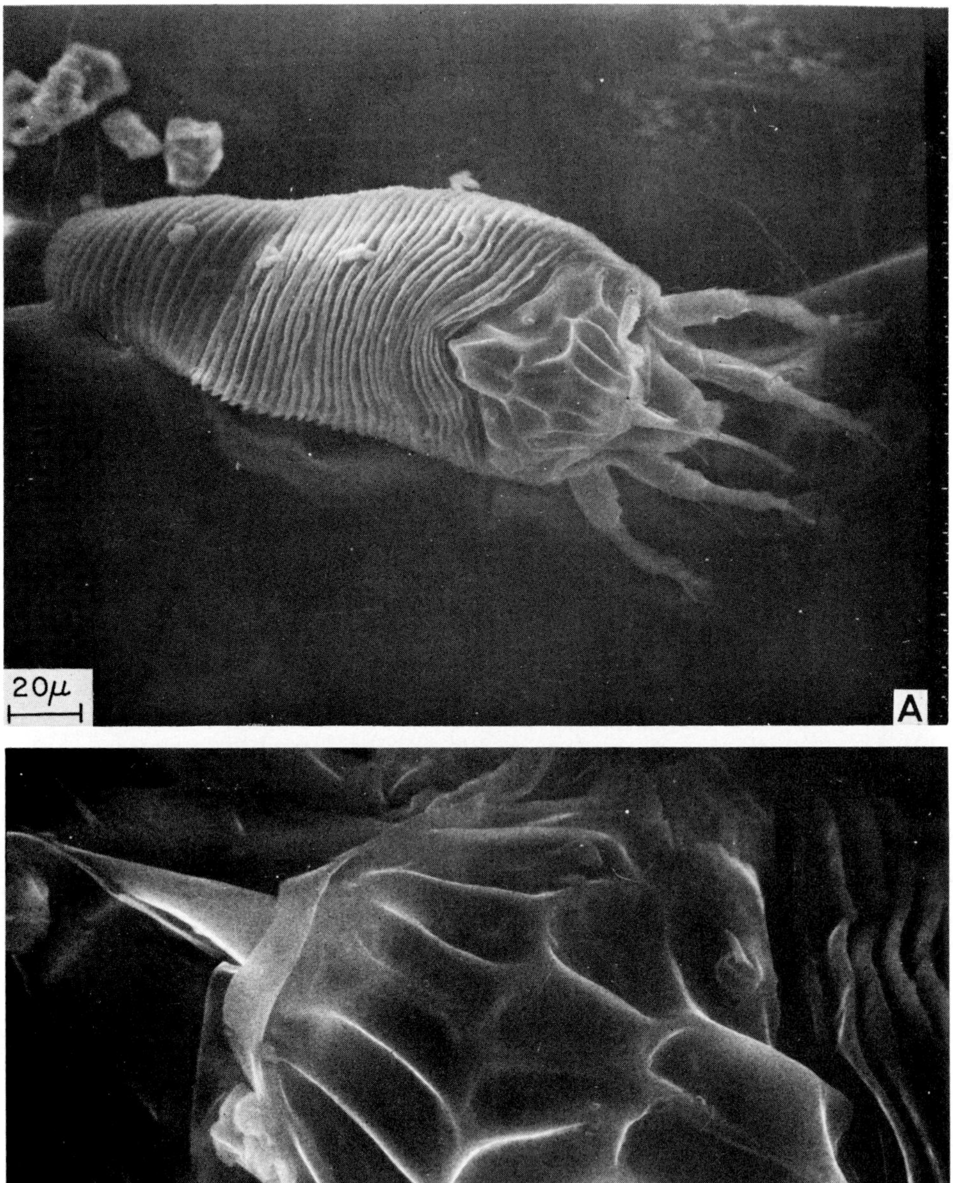
20μ
A
10μ
B

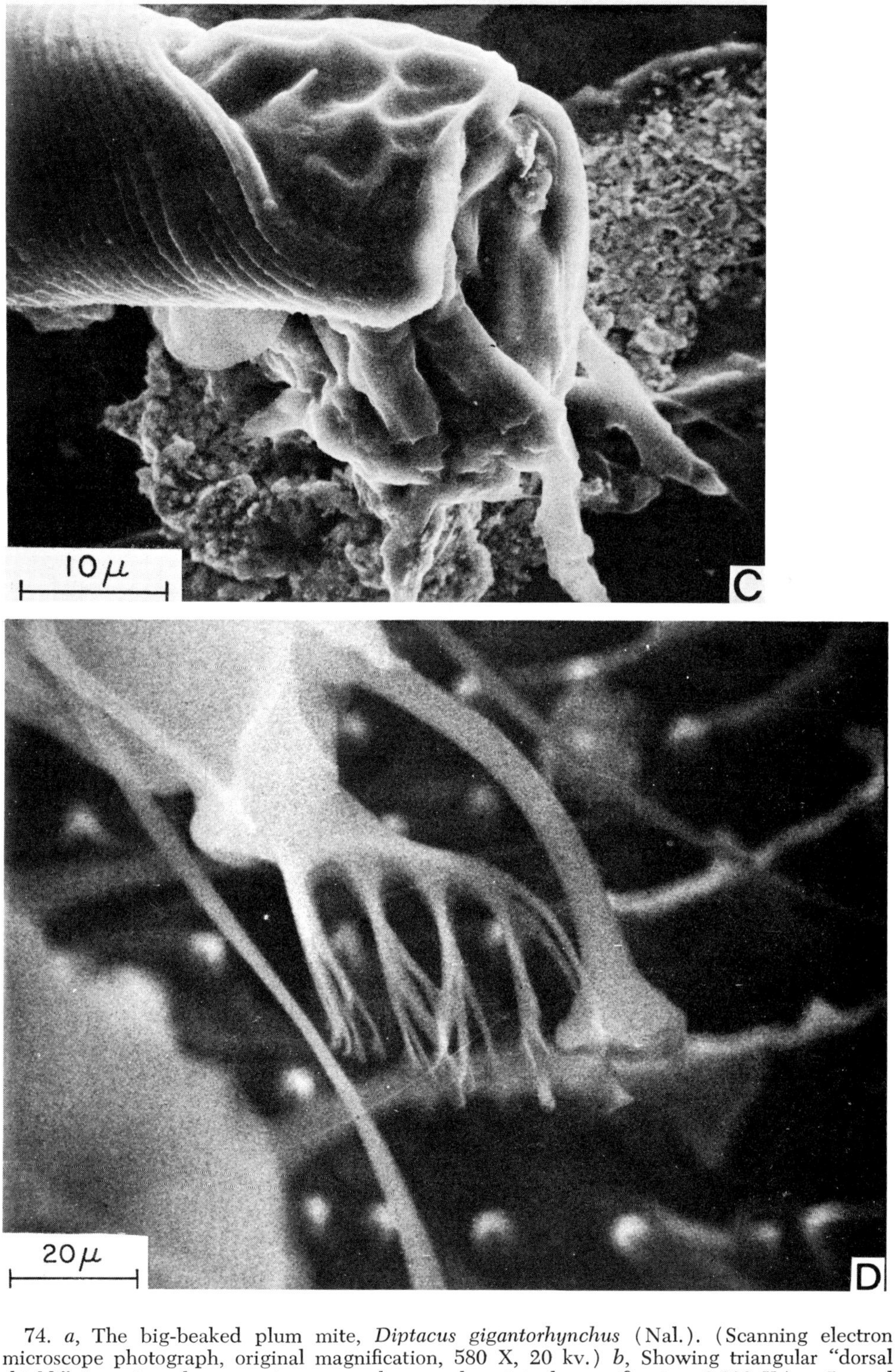

74. *a*, The big-beaked plum mite, *Diptacus gigantorhynchus* (Nal.). (Scanning electron microscope photograph, original magnification, 580 X, 20 kv.) *b*, Showing triangular "dorsal shield." Scanning electron microscope photograph, original magnification, 2,000 X.) *c*, Lateral view of anterior region including shield. (Scanning electron microscope original photograph, original magnification, 2,000 X.) *d*, Featherclaw. Scanning electron microscope photograph, original magnification, 10,000 X, 20 kv.)